Lecture Notes in Electrical Engineering

Volume 125

Zhihong Qian, Lei Cao, Weilian Su,
Tingkai Wang, and Huamin Yang (Eds.)

Recent Advances in Computer Science and Information Engineering

Volume 2

Editors
Zhihong Qian
Jilin University
China

Lei Cao
University of Mississippi
USA

Weilian Su
Naval Postgraduate School
USA

Tingkai Wang
London Metropolitan University
UK

Huamin Yang
Changchun University of Science
and Technology
China

ISSN 1876-1100
e-ISSN 1876-1119
ISBN 978-3-642-25788-9
e-ISBN 978-3-642-25789-6
DOI 10.1007/978-3-642-25789-6
Springer Heidelberg New York Dordrecht London

Library of Congress Control Number: 2011942930

Printed on acid-free paper

Springer is part of Springer Science+Business Media (www.springer.com)

Preface

On behalf of the organizing committee of the 2nd World Congress on Computer Science and Information Engineering (CSIE 2011), we would like to express our highest appreciation to all authors from all over the world.

CSIE 2011 is an international scientific Congress for distinguished scholars engaged in scientific, engineering and technological research, dedicated to build a platform for exploring and discussing the future of Computer Science and Information Engineering with existing and potential application scenarios. The professional interaction, afforded by this congress, will permit individuals with common interests the opportunity to share ideas and strategies. We believe that the congress will also develop a spirit of cooperation that leads to new friendship for addressing a wide variety of ongoing problems in this vibrant area of technology and fostering more collaboration in China and beyond.

The congress received 2483 full paper and abstract submissions from all over the world. Through a rigorous peer review process, all submissions were refereed based on their quality of content, level of innovation, significance, originality and legibility. We would like to apologize to those authors whose papers were declined due to the limited acceptance capacity. We are extremely grateful to each author, no matter whether his/her paper has been accepted or not.

We greatly appreciate all those who have contributed to the congress and express our grateful thanks to all supporters for their time and assistance. Thanks go to IEEE Harbin Section, Changchun University of Science and Technology, Jilin University, TPC members of the congress, for their support and hard work, without which we could not perform so efficiently and successfully. Thanks go to all the reviewers, speakers and participants for CSIE 2011.

Our day to day work in the CSIE 2011 field must always be sustained by a positive outlook and a real sense of joy from our awareness of the valuable work we do and the great contribution we make.

General Chair
Zhihong Qian
Jilin University
China

TPC Chair
Lei Cao
University of Mississippi
USA

TPC Chair
Weilian Su
Naval Postgraduate School
USA

TPC Chair
Tingkai Wang
London Metropolitan University
UK

Local Organizing Chair
Huamin Yang
Changchun University of Science and Technology, China

Conference Committee

General Chair

Zhihong Qian	Jilin University, China

Program Chairs

Lei Cao	The University of Mississippi, USA
Weilian Su	Naval Postgraduate School, USA
Tingkai Wang	London Metropolitan University, UK

Local Organizing Chair

Huamin Yang	Changchun University of Science and Technology, China

Publicity Chair

Ezendu Ariwa	London Metropolitan University, London Metropolitan Business School, UK
Jinghua Zhang	Winston-Salem State University, USA

Program Committee Members

Ibrahim Abualhaol	Khalifa University of Science, Technology & Research, UAE
Rajan Alex	West Texas A&M University, USA
Arnab Bhattacharya	Indian Institute of Technology, Kanpur, Indian
Mauro Birattari	Université Libre de Bruxelles, Belgium
Xianbin Cao	Beihang Univ., China
Wai-Kuen Cham	The Chinese University of Hong Kong, Hong Kong
Chung Yong Chan	University of Mississipi, USA
Nishchal Chaudhary	Atheros Communications, Inc., USA
Guotai Chen	Fujian Normal Univ., China
Huijun Chen	Epic systems Corp., USA

Minyou Chen	Chongqing University, China
Toly Chen	Feng Chia University, Taiwan
Weigang Chen	Tianjin University, China
Yixin Chen	The University of Mississippi, USA
Shi Cheng	Applied Micro Circuits Corporation, USA
Francisco Chiclana	De Montfort University, United Kingdom
Ryszard S. Choras	University of Technology & Life Sciences, Poland
Quan Dang	London Metropolitan University, United Kingdom
Fikret Ercal	Missouri University of Science & Technology, USA
Leonardo Garrido	Tecnológico de Monterrey, Campus Monterrey, México
Jihong Guan	Tongji University, China
Huiping Guo	California State University at LA, USA
Malka N. Halgamuge	University of Melbourne, Australia
Na Helian	University of Hertfordshire, United Kingdom
Liang Hong	Tennessee State University, USA
Yiguang Hong	Chinese Academy of Sciences, China
Edward Hung	The Hong Kong Polytechnic University, Hong Kong
Li Jin	University of Westminster, United Kingdom
Constantine Kotropoulos	Aristotle University of Thessaloniki, Greece
Vitus Sai Wa Lam	The University of Hong Kong, China
Cecilia Sik Lanyi	Pannon University, Hungary
Agapito Ledezma	Universidad Carlos III de Madrid, Spain
John Leis	University of Southern Queensland, Australia
Carson K Leung	The University of Manitoba, Canada
Robert Levinson	University of Minnesota, USA
Lin Li	Prairie View A&M University, USA
Ming Li	California State University, Fresno, USA
Tao Li	Florida International University, USA
Nianyu Li	Academy of Armed Force Engineering, China
Yen-Chun Lin	Chang Jung Christian University, Taiwan
Feng Liu	Beihang Univ., China
JiangBo Liu	Bradley University, USA
Ying Liu	University of Portsmouth, United Kingdom
Zhiqu Lu	University of Mississipi, USA
Wenjing Ma	Yahoo! Inc, USA
Valeri Mladenov	Technical University of Sofia, Bulgaria
Kalyan Mondal	Fairleigh Diskinson University, USA
Wasif Naeem	Queen's University Belfast, United Kingdom
Deok Hee Nam	Wilberforce University, USA
Fei Nan	Cisco Inc., USA
Daniel Neagu	University of Bradford, United Kingdom
Tang Hung Nguyen	California State University, Long Beach, USA
Philip Orlik	Mitsubishi Electric Research Laboratory, USA
George Pallis	University of Cyprus, Cyprus
Peiyuan Pan	London Metropolitan University, United Kingdom

Guangzhi Qu	Oakland University, USA
Mugizi Robert Rwebangira	Howard University, USA
Abdel-Badeeh Salem	Ain Shams University, Egypt
Alexei Sharpanskykh	Vrije Universiteit Amsterdam, The Netherlands
Tao Shi	Research in Motion, USA
Lingyang Song	Beijing University, China
Jonathan Sun	University of Southern Mississippi, USA
Weidong Sun	Tsinghua University, China
Yu Sun	University of Central Arkansas, USA
Jiacheng Tan	University of Portsmouth, United Kingdom
Shanyu Tang	London Metropolitan University, United Kingdom
Tong Boon Tang	The University of Edinburgh, United Kingdom
Eloisa Vargiu	University of Cagliari, Italy
Jørgen Villadsen	Technical University of Denmark, Denmark
Haixin Wang	Fort Valley State University, USA
Jing Wang	Bethune-Cookman University, USA
Lipo Wang	Nanyang Technological University, Singapore
Pan Wang	Wuhan University of Technology, China
Wenwu Wang	University of Surrey, United Kingdom
Changhua Wu	Kettering Univ., USA
Dan Wu	University of Windsor, Canada
Jingxian Wu	University of Arkansas, USA
Min Wu	Mako Surgical Corp., USA
Zhiqiang Wu	Wright State University, USA
Christos Xenakis	University of Piraeus, Greece
Weigang Xiang	University of Southern Queensland, Australia
Liang Xiao	Xiamen University, China
Jianxia Xue	University of Mississipi, USA
Li Yang	University of Tennessee at Chattanooga, USA
Yubin Yang	Nanjing University, China
Zhongpeng Yang	Putian University, China
Peng-Yeng Yin	National Chi Nan University, Taiwan
Jinghua Zhang	Winston-Salem State University, USA
Hong Zhao	Fairleigh Diskinson University, USA
Ying Zhao	Tsinghua University, China
Jiang-bin Zheng	Northwestern Polytechnic University, China
Shangming Zhu	East China Univ. of Science and Technology, China

Reviewers

Eiman Tamah Al-Shammari
Hassan Amin
Stefan Andrei
Guangwu Ao
Persis Urbana Ivy B
Yuhai Bao
Yuanlu Bao
Renata Baracho
Xu Bin
Ren Bo
Jianying Cao
Shuyan Cao
Min Cao
Luciana Cavalini
Guiran Chang

Tae Hyun Baek
Umesh Banodha
Wenxing Bao
Changqing Chen
Hongbin Chen
Ming Chen
Peng Chen
Min Chen
Li-Jia Chen
Pen-Yuan Chen
Shaoping Chen
Tinggui Chen
Xueli Chen
Zhibing Chen
Zhongwei Chen
Jau-Ming Chen
Min-Bin Chen
QiaoLi Chen
Ziwei Chen
Hong-Ren Chen
Jing Chen
Ken Chen
Li Chen
Minyou Chen
Stanley L. Chen
Ching Guo Chen
Yaowen Chen
Ze Cheng
Hongmei Chi
Chih-Ping Chu
Rong-yi Cui
Guangcai Cui
Dandan Dang
Delvin Defoe
Jayanthi Devaraj
Adolfo Di Mare
Jianbo Ding
Jianbo Ding
Hui Dong
Xiaomei Dong
Guang Dong
Sihui Dong
Xiaomei Dong
Carlos A. Dos Reis Filho
Huijing Dou
Jiawei Dou
Jorgen Boegh
Andrea Bottino
Hongping Cao
Fouzia Elbahhar
Honghai Fan
Jyh perng Fang
Junxiao Feng
Tao Feng
Benedito Renê Fischer
Dajie Fu
Guangquan Fu
Weihong Fu
Xiao-ning Fu
We Fu
Yong Gan
Zhiyin Gan
Chongzhi Gao
Feng Gao
Tiegang Gao
Zhilei Ge
Ruhai Ge
Jie Gong
Miaosen Gong
Yue Gong
Tao Gu
Junzhong Gu
Tian Guan
You-qing Guan
Dahai Guo
Jia Guo
Yinjing Guo
Xingming Guo
Kai Hao
Xinhong Hao
Fuyun He
Teruo Hikita
Sachio Hirokawa
Zih-Ping Ho
Liqiang Hou
Huei-Chen Hsu
Shih-Tsung Hsu
Shaolin Hu
Hongping Hu
Xiaochun Hu
Min Hu
Hanbing Hua
Horng Jinh Chang
Chung-Hsing Chao
Bo Chen
Zhiping Huang
Leijun Huang
Yong-Ren Huang
Xuewen Huang
Wang Hui
Ruo-Wei Hung
Qiuyan Huo
Jiuming Ji
Bei Jia
Xiaojing Jia
Feng Jiang
Dongmei Jiang
Rui Jiang
Yannan Jiang
Ruili Jiao
Ding Jue
Dae-Seong Kang
Hian Chye Koh
Yefu Kou
Chun-Hsiung Lan
Pranav Lapsiwala
Byoung-Dai Lee
Jung-Ju Lee
Guobin Li
Bin Li
Zengyuan Li
Xin Li
Gongquan Li
Xu Li
Yunqing Li
Hong Li
Li Li
Mingshun Li
Qi Li
Yanmei Li
Yun Li
Liu Li
Qingguo Li
Su Li
Ming Li
Hong-Yi Li(Lee)
Wei Liming
Hongbo Lin

Keynote Speakers

Ivan Stojmenovic

Title: Contribution of applied algorithms to applied computing

Abstract: There are many attempts to bring together computer scientists, applied mathematician and engineers to discuss advanced computing for scientific, engineering, and practical problems. This talk is about the role and contribution of applied algorithms within applied computing. It will discuss some specific areas where design and analysis of algorithms is believed to be the key ingredient in solving problems, which are often large and complex and cope with tight timing schedules. The talk is based on recent Handbook of Applied Algorithms (Wiley, March 2008), co-edited by the speaker. The featured application areas for algorithms and discrete mathematics include computational biology, computational chemistry, wireless networks, Internet data streams, computer vision, and emergent systems. Techniques identified as important include graph theory, game theory, data mining, evolutionary, combinatorial and cryptographic, routing and localized algorithms.

Biography: Ivan Stojmenovic received his Ph.D. degree in mathematics. He held regular and visiting positions in Serbia, Japan, USA, Canada, France, Mexico, Spain, UK (as Chair in Applied Computing at the University of Birmingham), Hong Kong, Brazil, Taiwan, and China, and is Full Professor at the University of Ottawa, Canada and Adjunct Professor at the University of Novi Sad, Serbia. He published over 250 different papers, and edited seven books on wireless, ad hoc, sensor and actuator networks and applied algorithms with Wiley. He is editor of over dozen journals, editor-in-chief of IEEE Transactions on Parallel and Distributed Systems (from January 2010), and founder and editor-in-chief of three journals (MVLSC, IJPEDS and AHSWN). Stojmenovic is one of about 260 computer science researchers with h-index at least 40 and has >10000 citations. He received three best paper awards and the Fast Breaking Paper for October 2003, by Thomson ISI ESI. He is recipient of the Royal Society Research Merit Award, UK. He is elected to IEEE Fellow status (Communications Society, class 2008), and is IEEE CS Distinguished Visitor 2010-12. He received Excellence in Research Award of the University of Ottawa 2009. Stojmenovic chaired and/or organized >60 workshops and conferences, and served in >200 program committees. He was program co-chair at IEEE PIMRC 2008, IEEE AINA-07, IEEE MASS-04&07, EUC-05&08-10, AdHocNow08, IFIP WSAN08, WONS-05, MSN-05&06, ISPA-05&07, founded workshop series at IEEE MASS, ICDCS, DCOSS, WoWMoM, ACM Mobihoc, IEEE/ACM CPSCom, FCST, MSN, and is/was Workshop Chair at IEEE INFOCOM 2011, IEEE MASS-09, ACM Mobihoc-07&08.

Andreas F. Molisch

Title: Wireless propagation and its impact on wireless system design

Abstract: Wireless propagation channels determine the fundamental performance limits of communications over the air. Furthermore, the propagation channels also determine the practical system performance of actual, deployable, systems. It is thus vital to establish models that are "as complicated as required to reproduce all RELEVANT effects, but no more complicated than that". As new systems and applications have emerged, what is "relevant" has changed significantly. Thus, the wireless propagation models we need today have to be suitable for wireless systems with large bandwidth, multiple antenna elements, and possibly operating in highly mobile environments. The talk will give an outline of the basic modeling principles for channel models that are suitable for modern systems, and will also show a few case studies that demonstrate the importance of realistic modeling.

A short discussion of standardization of channel models and application in system testing will conclude the talk.

Biography: Andy Molisch received the Dr. techn., and habilitation degrees from the Technical University Vienna (Austria) in 1994, and 1999, respectively. After working at AT&T (Bell) Laboratories, he joined Mitsubishi Electric Research Labs, Cambridge, MA, USA, where he rose to Distinguished Member of Technical Staff and Chief Wireless Standards Architect. Concurrently he was also Professor and Chairholder for radio systems at Lund University, Sweden. Since 2009, he is Professor of Electrical Engineering at the University of Southern California, Los Angeles, CA, USA. Dr. Molisch's current research interests are measurement and modeling of mobile radio channels, UWB, cooperative communications, and MIMO systems. He has authored, co-authored or edited four books (among them the textbook "Wireless Communications"), fourteen book chapters, more than 130 journal papers, and numerous conference contributions, as well as more than 70 patents and 60 standards contributions.

Dr. Molisch has been an editor of a number of journals and special issues, General Chair, TPC Chair, or Symposium Chair of multiple international conferences, and chairman of various international standardization groups. He is a Fellow of the IEEE, a Fellow of the IET, an IEEE Distinguished Lecturer, and recipient of several awards, most recently the IEEE's Donald Fink Award.

Arun Somani

Title: Aggressive and Reliable High-Performance Architectures

Abstract: As the transistor count on a chip goes up, the system becomes extremely sensitive to any voltage, temperature or process variations. One approach to immunize the system from the adverse effects of these variations is to add sufficient safety margins to the operating clock frequency. Timing Speculation (TS) provides a silver lining by providing better-than-worst-case systems. We introduce an aggressive yet reliable framework for energy efficient thermal control. We bring out the inter-relationship between power, temperature and reliability of aggressively clocked systems. We provide solutions to improve the existing power management in chip multiprocessors to dynamically maximize system utilization and satisfy the power constraints within safe thermal limits. We observe that up to 75% Energy-Delay squared product savings relative to base architecture is possible.

Biography: Arun K. Somani is currently Anson Marston Distinguished Professor of Electrical and Computer Engineering at Iowa State University. Prior to that, he was a Professor in the Department of Electrical Engineering and Department of Computer Science and Engineering at the University of Washington, Seattle, WA and Scientific Officer for Govt. of India, New Delhi from. He earned his MSEE and PhD degrees in electrical engineering from the McGill University, Montreal, Canada, in 1983 and 1985, respectively.

Professor Somani's research interests are in the area of computer system design and architecture, fault tolerant computing, computer interconnection networks, WDM-based optical networking, and reconfigurable and parallel computer systems. He has published more than 250 technical papers, several book chapters, and has supervised more than 100 graduate students (35 PhD students). He is the chief architects of an anti-submarine warfare system for Indian navy, Meshkin fault-tolerant computer system architecture for the Boeing Company, Proteus multi-computer cluster-based system for US Coastal Navy, and HIMAP design tool for the Boeing Commercial Company.

He has served on several program committees of various conferences in his research areas, served as IEEE distinguished visitor and IEEE distinguished tutorial speaker, and delivered several key note speeches, tutorials and distinguished and invited talks all over the world. He received commonwealth fellowship for his post-graduate work from Canada during 1982-85, awarded Distinguished Engineer member of ACM, and elected a Fellow of IEEE for his contributions to "theory and applications of computer networks."

Nei Kato

Title: Robust and Efficient Stream Delivery for Application Layer Multicasting in Heterogeneous Networks

Abstract: Application Layer Multicast (ALM) is highly expected to replace IP multicasting as the new technological choice for content delivery. Depending on the streaming application, ALM nodes will construct a multicast tree and deliver the stream through this tree. However, if a node resides in the tree leaves, it cannot deliver the stream to its descendant nodes. In this case, Quality of Service (QoS) will be compromised dramatically. To overcome this problem, Topology-aware Hierarchical Arrangement Graph (THAG) was proposed. By employing Multiple Description Coding (MDC), THAG first splits the stream into a number of descriptions, and then uses Arrangement Graph (AG) to construct node-disjoint multicast trees for each description. However, using a constant AG size in THAG creates difficulty in delivering descriptions appropriately across a heterogeneous network. In this talk, a new method, referred to as Network-aware Hierarchical Arrangement Graph (NHAG), to change the AG size dynamically to enhance THAG performance, even in heterogeneous networks, will be introduced. By comparing this new method to THAG and Split-Stream, the new method can be considered with better performance in terms of throughput and QoS. Meanwhile, some other related topics such as how to detect streaming content in high speed networks will also be touched upon.

Biography: Nei Kato received his M.S. and Ph.D. Degrees in information engineering from Tohoku University, Japan, in 1988 and 1991, respectively. He joined Computer Center of Tohoku University at 1991, and has been a full professor at the Graduate School of Information Sciences since 2003. He has been engaged in research on computer networking, wireless mobile communications, image processing and neural networks. He has published more than 200 papers in journals and peer-reviewed conference proceedings.

Nei Kato currently serves as the chair of IEEE Satellite and Space Communications TC, the secretary of IEEE Ad Hoc & Sensor Networks TC, the chair of IEICE Satellite Communications TC, a technical editor of IEEE Wireless Communications(2006~), an editor of IEEE Transactions on Wireless Communications(2008~), an associate editor of IEEE Transactions on Vehicular Technology(2009~). He has served as co-guest-editor for many IEEE journals and magazines, symposium co-chair for GLOBECOM'07, ICC'10, ICC'11, ChinaCom'08, ChinaCom'09, and WCNC2010-2011 TPC Vice Chair.

His awards include Minoru Ishida Foundation Research Encouragement Prize(2003), Distinguished Contributions to Satellite Communications Award from the IEEE Communications Society, Satellite and Space Communications Technical Committee(2005), the FUNAI information Science Award(2007), the TELCOM System Technology Award from Foundation for Electrical Communications Diffusion(2008), the IEICE Network System Research Award(2009), and best paper awards from many prestigious international conferences such as IEEE GLOBECOM, IWCMC, etc.

Besides his academic activities, he also serves as a member on the expert committee of Telecommunications Council, the special commissioner of Telecommunications Business Dispute Settlement Commission, Ministry of Internal Affairs and Communications, Japan, and as the chairperson of ITU-R SG4 and SG7, Japan. Nei Kato is a member of the Institute of Electronics, Information and Communication Engineers (IEICE) and a senior member of IEEE.

Yasushi Yamao

Title: An Intelligent WDN for Future Ubiquitous Society

Abstract: Intelligence is an essential feature of advanced systems. The most important ability given by intelligence is adaptation, which keeps system performance high under the change of its environment. One of the interesting areas to apply intelligence is Wireless Distributed Network (WDN), which is an important technology of future ubiquitous society. Under the time-varying wireless environments that severely suffer from fading, quality control of multihop communication is a critical issue. This speech discusses how multi-hop communication quality in WDN can be maintained by the intelligence of distributed nodes that always watch surrounding node's behavior and take cognitive action. Cross-layer cooperation at each node enables real-time local path optimization including creation of bypass and shortcut paths. Packet communication quality improvements in terms of delivery ratio and delay are shown in some examples.

Biography: Dr. Yasushi Yamao received his B.S., M.S., and Ph.D. degrees in electronics engineering from Kyoto University, Kyoto, Japan, in 1977, 1979, and 1998, respectively.

He started his research career of mobile communications from the measurement and analysis of urban radio propagation as his M.S. thesis. In 1979, he joined the Nippon Telegraph and Telephone Corporation (NTT) Laboratories, Japan, where his major activities included leading research on GMSK modulator /demodulator and GaAs RF ICs for digital mobile communications, and development of PDC digital cellular handheld phones. In 1993, he moved to NTT DoCoMo Inc. and directed standardization of high-speed paging system (FLEX-TD) and development of 3G radio network system. He also joined European IST research programs for IP-based 4th generation mobile communication.

In 2005, he moved to the University of Electro-Communications as a professor of the Advanced Wireless Communication Research Center (AWCC). His current interests focus on wireless ubiquitous communication networks and protocols, as well as high-efficiency and reconfigurable wireless circuit technologies both in RF and Digital Signal Processing. He is a Fellow of IEICE and member of IEEE. He served as Vice President of IEICE Communications Society (2003-2004), Chairman of the

IEICE Technical Group on Radio Communication Systems (2006-2008) and Chief Editor of IEICE Communication Magazine (2008-2010). He is currently Vice Chairman of IEEE VTS Japan Chapter.

Michael Small

Title: Complex Networks – Chaotic Dynamics

Abstract: In the last decade, physicists and then biological scientists have found evidence of complex networks in a stunning range of physical and biological systems. In this talk, I will focus on a more basic, and possibly more interesting question: what can complex networks and the methods of complexity theory actually tell us about the dynamics underlying observed time series data?

A variety of methods have been introduced to transform time series data into complex networks. The complex network representation of the time series can then be used to gain new insight (information not readily available from other methods) about the underlying dynamics. We show that the structure of the complex network, and more specifically, the motif frequency distribution, depends on the nature of the underlying dynamics. In particular, low dimensional chaotic dynamics are associated with one particular class of complex network; and hyper-chaotic, periodic and stochastic motion are each associated with others. This complex network approach can then be used to identify the nature of the dynamics underlying a particular time series. Application of these methods will be demonstrated with several experimental systems: from musical composition, to sound production, and population dynamics.

Biography: Michael Small got his PhD in applied mathematics from the University of Western Australia, and then did post docs at UWA, Heroit-Watt University (Edinburgh) and Hong Kong Polytechnic University. Michael Small is now an Associate Professor in the department of Electronic and Information Engineering at the Hong Kong Polytechnic University. His research interests focus on complex systems and nonlinear time series analysis. His work emphasises the application of these methods in a diverse range of fields: disease propagation, neurophysiology, cardiac dynamics and many others. Workshop Chair at IEEE INFOCOM 2011, IEEE MASS-09, ACM Mobihoc-07&08.

Contents

A Biologically-Inspired Network for Generic Object Recognition Using CUDA

Yuekai Wang, Xiaofeng Wu, Xiaoying Song, Wenqiang Zhang, and Juyang Weng*

Abstract. Generic object recognition is one of the most important fields in the artificial intelligence. Some cortex-like networks for generic object recognition are proposed these years. But most of them concentrated on the discussion about the recognition performance (such as recognition rate, number of objects to be recognized), not the practicability, i.e., implementation with ubiquitous devices and application in real time. This paper reports a try on implementation of a biologically-inspired where-what network (WWN), which integrates object recognition and attention in a single network, via parallelizing the various stages of the network training with CUDA on GPU to shorten the training time. The experiment on HAIBAO Robot exhibited in 2010 Shanghai Expo shows that this optimization can achieve a speedup of almost 16 times compared to the C-based program on an Intel Core 2 DUO 3.00 GHZ CPU in real environments.

1 Introduction

Up to now, many researchers have tried to realize the generic object recognition in cluttered backgrounds based on the methods in computer vision field, but it is

Yuekai Wang · Xiaofeng Wu
State Key Lab. of ASIC & System, Fudan University, Shanghai, 200433, China

Yuekai Wang · Xiaofeng Wu
Department of Electronic Engineering, Fudan University, Shanghai, 200433, China

Xiaoying Song · Wenqiang Zhang · Juyang Weng
School of Computer Science, Fudan University, Shanghai, 200433, China

Juyang Weng
Dept. of Computer Sci. and Eng., Michigan State University, East Lansing, MI 48824, USA

* This work was supported by the Fund of State Key Lab. of ASIC & System (11MS008) and the Fundamental Research Funds for the Central Universities to XW, and a Changjiang Visiting Scholar Fund of the Ministry of Education to JW.

Z. Qian et al. (Eds.): Recent Advances in CSIE 2011, LNEE 125, pp. 1–6.
springerlink.com

still a challenging task. The appearance-based feature descriptors are quite selective for a target shape but limited in tolerance to the object transformations, while the histogram-based descriptors, for an example, the SIFT features, show great tolerance to the object transformations but seemed no good enough to perform well on generic object recognition in some experiments [1]. On another hand, human vision systems can accomplish such tasks easily and quickly. Thus to create a proper network by simulating the human vision systems is thought as one possible approach to break the neck-bottle of generic object recognition.

In recent decades, with the advances of the studies on object recognition in visual cortex [2] in physiology and neuroscience, several biologically-inspired network models are proposed. One famous model is HMAX, introduced by Riesenhuber and Poggio [3], is proved to be robust in generic object recognition by its invariance to the object sizes and positions, which is based on hierarchical feedforward architecture similar to the organization of visual cortex. It analyzes the input image via Gabor function and builds an increasingly complex and invariant feature representation by maximum pooling operation [4]. The where-what network (WWN) introduced by Juyang Weng [5] and his coworkers is a biologically plausible developmental model which is designed to integrate the object recognition and attention (i.e., what and where information in the ventral stream and dorsal stream respectively) interactively in a single network for some unspecific tasks by using both feedforward (bottom-up) and feedback (top-down) connections.

Except the feasibility, the practicability is another important issue for these biologically-inspired models. With general CPU-based implementation, it may spend hours or even days to train a network if the size of the input image meets the demands of real application. Considering visual processing in cortex is collective parallel, parallel processing of above models might be the best way for real application. At present, multi-core GPUs become more and more popular which provide a hardware solution to implement parallel computation.

In our research, we try to realize object recognition and attention in real environments on HAIBAO robot. Considering the individual features of HMAX and WWN (integrating object recognition and attention in a single network), the later one is chosen although the examples of the WWN are still given under ideal conditions.

2 Architecture Modification of WWN For Real Application

Before description of network optimization using parallel processing technique in details, firstly, the architecture of the original WWN is simply reviewed. The architecture of original WWN is illustrated in Fig. 1 (a), in which there are three cortical areas, V2, IT/PP and motor (the stream from V2, through PP, and to PM corresponds to the dorsal pathway and the stream from V2, through IT, and to TM corresponds to the ventral pathway in human vision systems).

V2 neurons have the local receptive fields from the retina (i.e., input image). Each neuron in V2 perceives $a \times a$ area of the input image and the distance of the two adjacent receptive field centers in horizontal or vertical directions is 1 pixel.

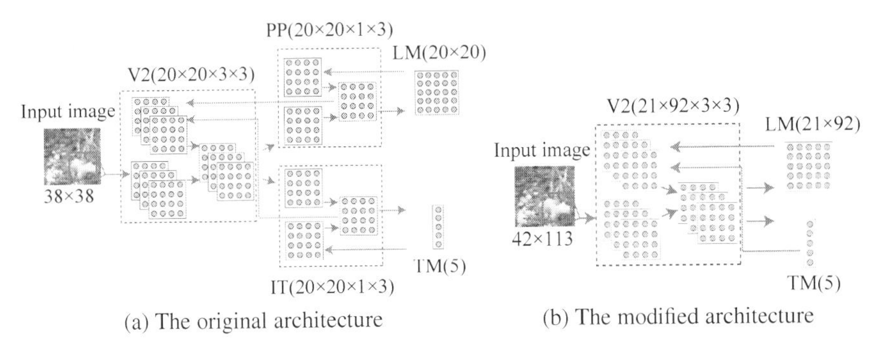

(a) The original architecture (b) The modified architecture

Fig. 1 Illustration diagrams of the WWN architecture.

Suppose the size of the input image is $w \times h$, and the depth of V2 is d, therefore, the number of neurons in V2 is $(w-a+1) \times (h-a+1) \times d$ totally. Cortical area IT/PP turns the local features learned in V2 into global features and/or combines the individual feature to the multi-feature set. Since only single features exist not shared features in our experiment, IT/PP can be omitted which is proved to achieve a better performance [6]. The motor area (TM and LM) represents two concepts: "what is the object" and "where is the object".

In the example of original WWN, a is 21, w and h are 40 and d is 3 [7]. It is meaningless for real applications with such small input images. Usually, the size of images captured by cameras is 320×240 or 640×480 in general application. The approximate memory needed by WWN can be estimated by the following formula (suppose all data in computation using float type):

$$\text{memory (byte)} \approx d \times [(h-a+1) \times (w-a+1)]^2 \times 4.$$

In the case of 320×240, h is 320, w is 240, d is 3 and a is 19, the necessary memory is about 54GB. Considering that in general, the memory on a GPU is no more than 2GB and the situation of the demonstration in 2010 World Expo (objects only can be learnt in limited area), we finally set the size of input images 42×113. Thus the size of V2 area is $21 \times 92 \times 3$ which is about 5.7 times larger than that of the original network. After expanding the size and omitting IT/PP, the modified network is shown as Fig. 1(b) in which V2 connects to the motor area directly.

3 Parallelization of Network Training

GPU is specialized for compute-intensive and highly parallel computation what graphics rendering requires, which provides an inexpensive, highly parallel system for general developers. In fact, parallel computations on a GPU are realized through the high ratio of arithmetic operations to memory operations which means the arithmetic operations instead of big data caches hiding the memory access latency.

The CUDA (Compute Unified Device Architecture) introduced by NVIDIA is designed for the GPU programming and becomes more and more popular recently, which comes with a software environment that allows developers to use C as a high-level programming language[1].

In WWN, the in-place learning algorithm, i.e., each neuron is responsible for the learning of its own signal processing characteristics, is well-suited to the parallel processing. In the whole process of object recognition and localization, the most time consuming part is network training mainly including three parts: the computations of pre-response, top-k competition and Hebbian learning, which is optimized using CUDA described in detail in the following part.

3.1 Parallelization of Computation of Pre-responses

The pre-response of each V2 neuron (e.g., neuron (i,j)), $z^a_{i,j}(t)$, is calculated as:

$$z^a_{i,j}(t) = \frac{\mathbf{w}^a_{i,j}(t) \cdot \mathbf{x}_{i,j}(t)}{\|\mathbf{w}^a_{i,j}(t)\| \|\mathbf{x}_{i,j}(t)\|}$$

where $\mathbf{w}^a_{i,j}(t)$ is the ascending weight of the neuron (i,j) and $\mathbf{x}_{i,j}(t)$ is the ascending local input (i.e., the $a \times a$ area perceived by neuron (i,j)).

Specifically, we assign the above calculation of each V2 neuron (i.e., inner product of $\mathbf{w}^a(t)$ and $\mathbf{x}(t)$) to one thread, i.e., $2280 \times 3 = 6840$ neurons correspond to 6840 threads, so that the computation of pre-responses is executed concurrently.

3.2 Parallelization of Top-k Competition

Essentially, the top-k competition is a sorting problem. We apply bitonic sorting network, a sorting algorithm designed specially for parallel machines[2] to realize the top-k competition.

3.3 Parallelization of Hebbian Learning

The weights of the top-k neurons in V2 and motor area will be updated by Hebbian learning as follows:

$$\mathbf{w}_{i,j}(t+1) = w_1(t)\mathbf{w}_{i,j}(t) + w_2(t)z_{i,j}(t)\mathbf{p}_{i,j}(t)$$

where $\mathbf{w}_{i,j}(t)$, $\mathbf{p}_{i,j}(t)$ and $z_{i,j}(t)$ is the weight vector, input vector and response of the neuron (i,j). $w_1(t)$ and $w_2(t)$ are two parameters.

---

[1] Available from `http://www.wenku.baidu.com/view/e576bb4733687e21af45a93c.html`

[2] `http://www.tools-of-computing.com/tc/CS/Sorts/bitonic sort.htm`

Table 1 Performances evaluation of each method using the tool Compute Visual Profile.

Method	GPU (usec)	Glob mem read throughput (GB/s)	Glob mem write throughput (GB/s)
DataCUDA	15315	10.6	8.9
Pre-response computation	2072	16.7	13.5
Top-k competition	37	0.3	0.2
Hebbian learning	23	14.4	5.1
MemcopyDtoH	18626	—	—
MemcopyHtoD	2118	—	—

Similar to the processing in parallelization of computation of pre-responses, we assign the above operation to the threads. The only difference is that in this part every thread corresponds to one or two elements of a vector instead of all elements of a vector due to the small amount of data (only top-k neurons need Hebbian learning).

4 Experiments and Results

In order to evaluate the accleration performance and recognition/attention performance after parallelization, we compared the CUDA-based (running on a GeForce GT 240 with 1GB memory which possesses 12 Stream Multiprocessors composed of 96 Stream Processors) and the C-based (running on a machine equipped with an Intel Core 2 Duo 3.00GHz CPU with 3GB memory) programs by the training time of one training cycle, and evaluated the network performances in real environments on HAIBAO robot exhibited in 2010 Shanghai EXPO.

On speed performance, it spends 1423.2 seconds for the C-based version and 91.7 seconds for the GPU-based version achieving a speedup of almost 16 times. Table 1 lists the time each part costs and the global memory read and write throughput of each part. In the table, Pre-responses computation, Top-k competition and Hebbian learning are parallelized using CUDA described in section 3. The part DataCUDA refers to the data preparations before the network training. The most time-consuming operation is data transfer, including transfer from GPU to CPU (MemoryDtoH) and from CPU to GPU (MemoryHtoD). As the bandwidth of GeForce GT 240 is 50.4GB/s, the throughput of three optimized parts (three stages in network training) has not reached the limit.

On recognition performance, shown as Fig.2 in real environments (the foreground images and background images are got through a camera and the arbitrary foreground object contours are used), the recognition performance is satisfied and the network is tolerant to little changes in scale, illumination and viewpoint. But when the changes in scale, illumination and viewpoint are obvious, the recognition performance of the network is poor.

Fig. 2 Examples of network performance test using free viewing mode which was performed in real environments of 2010 Shanghai EXPO.

5 Conclusion and Future Work

To prompt the real application of WWN, there are still a lot of works to be done, such as to improve the generic object recognition capability by adding scale-invariance, viewpoint invariance and illumination tolerance, and more important functionality - the autonomous developmental of the network (incremental learning and adaptive adjustment of the network).

References

1. Lowe, D.: Object recognition from local scale-invariant features. In: Proc. International Conference on Computer Vision, Kerkyra, September 20-27, vol. 2, pp. 1150–1157 (1999)
2. Roelfsema, P.R.: Cortical algorigthms for perceptual grouping. Annual Review of Neuroscience 29, 203–227 (2006)
3. Riesenhuber, M., Poggio, T.: Hierachical models of object recognition in cortex. Nature Neuroscience 2(11), 1019–1025 (1999)
4. Serre, T., Wolf, L., Bileschi, S., Riesenhuber, M., Poggio, T.: Robust object recognition with cortex-like mechanisms. IEEE Trans. Pattern Analysis and Machine Intelligence 29(3), 411–426 (2007)
5. Ji, Z., Weng, J., Prokhorov, D.: Where-what network 1: "Where" and "What" assist each other through top-down connections. In: Proc. IEEE International Conference on Development and Learning, Monterey, CA, August 9-12, pp. 61–66 (2008)
6. Luciw, M., Weng, J.: Where-what network-4: The effect of multiple internal areas. In: Proc. IEEE International Joint Conference on Neural Networks, Ann Arbor, MI, August 18-21, pp. 311–316 (2010)
7. Ji, Z., Weng, J.: WWN-2: A biologically inspired neural network for concurrent visual attention and recognition. In: Proc. IEEE International Joint Conference on Neural Networks, Barcelona, Spain, July 18-23, pp. 1–8 (2010)

On Lagrangians of Hypergraphs and Cliques

Yuejian Peng and Cheng Zhao

Abstract. Motzkin and Straus establishes a remarkable connection between the maximum clique problem and the Lagrangian of a graph in [4]. It is useful in practice if similar results hold for hypergraphs. In this paper, we provide evidence that the Lagrangian of a 3-uniform hypergraph is related to the size of its maximum cliques under some conditions.

1 Introduction

In 1965, a classical paper by Motzkin and Straus [4] provided a new proof of Turán's theorem [11] based on a continuous characterization of the clique number of a graph using the Lagrangian of a graph. This new proof aroused interests in the study of Lagrangians of hypergraphs. Furthermore, the Motzkin-Straus result and its extension were successfully employed in optimization to provide heuristics for the maximum clique problem, and the Motzkin-Straus theorem has been also generalized to vertex-weighted graphs [2] and edge-weighted graphs with applications to pattern recognition in image analysis (see [2], [5], [8]). It is useful in practice if similar results hold for hypergraphs. In this paper, we provide evidence that the Lagrangian of a 3-uniform hypergraph is related to the size of maximum cliques under some conditions. We first state a few definitions.

For a set V and a positive integer r we denote by $V^{(r)}$ the family of all r-subsets of V. An r-uniform graph or r-graph G consists of a set $V(G)$ of vertices and a set $E(G) \subseteq V(G)^{(r)}$ of edges. An edge $e = \{a_1, a_2, \ldots, a_r\}$ will be simply denoted by $a_1 a_2 \ldots a_r$. Let N be the set of all positive integers. For any integer $n \in N$, we denote the set $\{1, 2, 3, \ldots, n\}$ by $[n]$. Let $K_t^{(r)}$ denote the complete r-graph on t vertices, that is the r-graph on t vertices containing all possible edges. A complete r-graph

Yuejian Peng · Cheng Zhao
Indiana State University, Terre Haute, IN 47809 USA. Also,Hunan University, Changsha, China
e-mail: {yuejian.peng,cheng.zhao}@indstate.edu

Z. Qian et al. (Eds.): Recent Advances in CSIE 2011, LNEE 125, pp. 7–12.
springerlink.com

on t vertices is also called a clique with order t. We also let $[n]^{(r)}$ represent the complete r-uniform graph on the vertex set $[n]$. When $r = 2$, an r-uniform graph is a simple graph. When $r \geq 3$, an r-graph is often called a hypergraph.We now give the definition of the Lagrangian of an r-uniform graph below. More studies of Lagrangians can be found in [1], [4], and [10].

Definition 1.1. *For an r-uniform graph G with vertex set $\{1,2,\ldots,n\}$, edge set $E(G)$ and a vector $\mathbf{x} = (x_1,\ldots,x_n) \in R^n$, define*

$$\lambda(G,\mathbf{x}) = \sum_{i_1 i_2 \cdots i_r \in E(G)} x_{i_1} x_{i_2} \ldots x_{i_r}.$$

Let $S = \{\mathbf{x} = (x_1,x_2,\ldots,x_n) : \sum_{i=1}^{n} x_i = 1, x_i \geq 0$ for $i = 1,2,\ldots,n\}$. The Lagrangian of G, denote by $\lambda(G)$, is defined as

$$\lambda(G) = \max\{\lambda(G,\mathbf{x}) : \mathbf{x} \in S\}.$$

A vector $\mathbf{y} \in S$ is called an *optimal vector* for G if $\lambda(G,\mathbf{y}) = \lambda(G)$.

In [4], Motzkin and Straus provided the following simple expression for the Lagrangian of a 2-graph.

Theorem 1.1. *(Motzkin and Straus [4]) If G is a 2-graph in which a largest clique has order l then $\lambda(G) = \lambda(K_l^{(2)}) = \frac{1}{2}(1 - \frac{1}{l})$.*

An attempt to generalize the Motzkin-Straus theorem to hypergraphs is due to Sós and Straus [9]. Recently, in [7] Rota Buló and Pelillo generalized the Motzkin and Straus' result to r-graphs in some way using a continuous characterization of maximal cliques with applications in image analysis. Due to the difficulty of Turán's problem [11] for $r \geq 3$, determining the Lagrangian of a general r-graph is non-trivial when $r \geq 3$ (see [10]). As it turns out, $\lambda(G)$ cannot be directly used to extend the Motzkin-Straus theorem to r-graphs. Indeed the obvious generalization of Motzkin and Straus' result is false because there are many examples of r-graphs that do not achieve their Lagrangian on any proper subhypergraph.

For distinct $A,B \in N^{(r)}$ we say that A is less than B in the *colex ordering* if $max(A \triangle B) \in B$, where $A \triangle B = (A \setminus B) \cup (B \setminus A)$. For example we have $246 < 156$ in $N^{(3)}$ since $max(\{2,4,6\} \triangle \{1,5,6\}) \in \{1,5,6\}$. Let $C_{r,m}$ denote the r-graph with m edges formed by taking the first m elements in the colex ordering of $N^{(r)}$. Talbot in [10] also gave the following.

Lemma 1. *(Talbot [10]) For any integers m, l, and r satisfying $\binom{l-1}{r} \leq m \leq \binom{l-1}{r} + \binom{l-2}{r-1}$, we have $\lambda(C_{r,m}) = \lambda([l-1]^{(r)})$.*

Question 1. *Let l and m be positive integers satisfying $\binom{l-1}{r} \leq m \leq \binom{l-1}{r} + \binom{l-2}{r-1}$. Let G be an r-graph with m edges. Is it true that $\lambda(G) \leq \lambda([l-1]^{(r)})$?*

In [6], we give a partial answer to the above question.

Theorem 1.2. *(see [6]) Let m and l be positive integers satisfying $\binom{l-1}{3} \le m \le \binom{l-1}{3} + \binom{l-2}{2}$. Let G be a 3-graph with m edges and G contain a clique of order $l-1$. Then $\lambda(G) = \lambda([l-1]^{(3)})$.*

This is a generalization of Theorem 1.1 to 3-graphs when the number of edges is in this range. Indeed, the upper bound $\binom{l-1}{3} + \binom{l-2}{2}$ in this theorem is the best possible since if $m > \binom{l-1}{3} + \binom{l-2}{2}$ then $\lambda(C_{3,m}) > \lambda([l-1]^3)$.

Question 2. *Let G be a 3-graph with m edges and G contains no clique of size $l-1$, where $\binom{l-1}{3} \le m \le \binom{l-1}{3} + \binom{l-2}{2}$. Is it true that $\lambda(G) < \lambda([l-1]^{(3)})$?*

In this paper, we provide some evidence for the above question. Let us state some preliminary results in the following section.

2 Preliminary Results

For an r-graph $G = (V,E)$ we denote the $(r-1)$-neighborhood of a vertex $i \in V$ by $E_i = \{A \in V^{(r-1)} : A \cup \{i\} \in E\}$. Similarly, we will denote the $(r-2)$-neighborhood of a pair of vertices $i, j \in V$ by $E_{ij} = \{B \in V^{(r-2)} : B \cup \{i,j\} \in E\}$. We denote the complement of E_i by $E_i^c = \{A \in V^{(r-1)} : A \cup \{i\} \in V^{(r)} \backslash E\}$. Also, we will denote the complement of E_{ij} by $E_{ij}^c = \{B \in V^{(r-2)} : B \cup \{i,j\} \in V^{(r)} \backslash E\}$. Denote $E_{i \backslash j} = (E_i \cap E_j^c) \setminus \{\{j\} \cup B,\ B \in E_{ij}\}$. We will impose one additional condition on any optimal weighting $\mathbf{x} = (x_1, x_2, \ldots, x_n)$ for an r-graph G:

$$\begin{aligned} &|\{i : x_i > 0\}| \text{ is minimal, i.e. if } \mathbf{y} \text{ is a legal weighting} \\ &\mathit{for}\ G \text{ satisfying } |\{i : y_i > 0\}| < |\{i : x_i > 0\}|, \\ &\text{then } \lambda(G, \mathbf{y}) < \lambda(G). \end{aligned} \tag{1}$$

The following lemma gives some necessary condition of an optimal vector of $\lambda(G)$.

Lemma 2. *(Frankl and Rödl [1]) Let $G = (V,E)$ be an r-graph and $\mathbf{x} = (x_1, x_2, \ldots, x_n)$ be an optimal legal weighting for G with $k\ (\le n)$ non-zero weights satisfying condition (1). Then for every $\{i,j\} \in [k]^{(2)}$, (a) $\lambda(E_i, \mathbf{x}) = \lambda(E_j, \mathbf{x}) = r\lambda(G)$, (b) there is an edge in E containing both i and j.*

We say that an r-graph $G = (V,E)$ on vertex set $\{1, 2, \cdots, n\}$ is *left compressed* if $E_{j \backslash i} = \emptyset$ for any $1 \le i < j \le n$. In other words, for any $i < j$, if $k_1 k_2 \ldots k_{r-1} \in E_j$, where $k_1, k_2, \ldots, k_{r-1} \ne i$, then $k_1 k_2 \ldots k_{r-1} \in E_i$.

Remark 1. (a) In Lemma 2, part(a) implies that $x_j \lambda(E_{ij}, \mathbf{x}) + \lambda(E_{i \backslash j}, \mathbf{x}) = x_i \lambda(E_{ij}, \mathbf{x}) + \lambda(E_{j \backslash i}, \mathbf{x})$. In particular, if G is left compressed, then

$$(x_i - x_j)\lambda(E_{ij}, \mathbf{x}) = \lambda(E_{i \backslash j}, \mathbf{x}) \tag{2}$$

for any i, j satisfying $1 \le i < j \le k$ since $E_{j \backslash i} = \emptyset$.

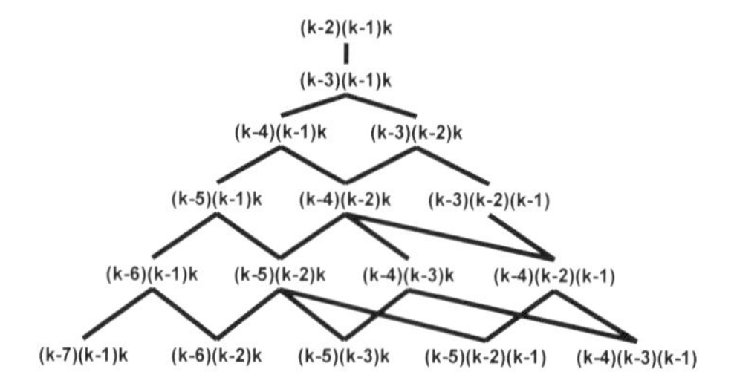

Fig. 1

(b) If G is left-compressed, then an optimal legal weighting $\mathbf{x} = (x_1, x_2, \ldots, x_n)$ for G must satisfy

$$x_1 \geq x_2 \geq \ldots \geq x_n \geq 0 \tag{3}$$

by (2).

3 Evidence for Question 2

Proposition 3.1. *Let G be a 3-graph with m edges and G contains no clique of size $l-1$, where $\binom{l-1}{3} \leq m \leq \binom{l-1}{3} + \binom{l-2}{2}$. Then $\lambda(G) < \lambda([l-1]^{(3)})$ for $l = 6, 7, 8$.*

Denote $\lambda^{r-}_{(m,l)} = \max\{\lambda(G) : G$ is an $r-$graph with m edges and G does not contain a clique of size $l\}$. We need the following lemma.

Lemma 3. *Let m and $l \geq 6$ be positive integers satisfying $\binom{l-1}{3} \leq m \leq \binom{l-1}{3} + \binom{l-2}{2}$. Then there exists a left compressed r-graph G with m edges and on vertex set $[l]$ such that $\lambda(G) = \lambda^{r-}_{(m,l-1)}$ and there exists an optimal weight $\mathbf{x} = (x_1, x_2, \ldots, x_n)$ of G satisfying $x_i \geq x_j$ when $i < j$.*

A triple $i_1 j_1 k_1$ is called an *ancestor* of a triple $i_2 j_2 k_2$ if $i_1 \geq i_2$, $j_1 \geq j_2$, $k_1 \geq k_2$, and $i_1 + j_1 + k_1 > i_2 + j_2 + k_2$. In this case, triple $i_2 j_2 k_2$ is called a *descendant* of $i_1 j_1 k_1$. We say that $i_1 j_1 k_1$ has higher hierarchy than $i_2 j_2 k_2$ if $i_1 j_1 k_1$ is an ancestor of $i_2 j_2 k_2$. Note that hierarchy is a partial ordering. Figure 1 shows hierarchy relationship of triples in $[k]^{(3)}$. Note that an 3-graph G is left-compressed if and only if for any edge in G, all its descendants should be in G as well.

Proof of Lemma 3. Let G' be a 3-graph with m edges without containing a clique of size $l-1$ such that $\lambda(G') = \lambda^{r-}_{(m,l-1)}$. We call such a G' an extremal r-graph for m and $l-1$. Let $\mathbf{x} = (x_1, x_2, \ldots, x_n)$ be an optimal weight of G'. We can assume that $x_i \geq x_j$ when $i < j$ since otherwise we can just relabel the vertices of G' and obtain another extremal 3-graph for m and $l-1$ with an optimal weight $\mathbf{x} = (x_1, x_2, \ldots, x_n)$ satisfying $x_i \geq x_j$ when $i < j$. Replace each triple in $[l]^{(3)} \setminus E(G')$ by its current available ancestor with the highest hierarchy and make sure that the 4th triple taken is in $[l-1]^{(3)} \setminus E(G')$ so that this triple is replaced by $(l-3)(l-2)(l-1)$. This is possible since there are at least $l-2 \geq 4$ triples in $[l]^{(3)} \setminus E(G')$ and at least one of them is in $[l-1]^{(3)} \setminus E(G')$. Let the new resulting 3-graph be G''. Notice that $\lambda(G'') \geq \lambda(G')$, the induced subgraph of G'' is left-compressed and it does not contain the edge $(l-3)(l-2)(l-1)$, therefore it does not contain a clique of size $l-1$.

If s edges of G'' contain a vertex not in $[l]$, then replace these edges by s triples in $[l]^{(3)} \setminus E(G'')$ with hierarchy as low as possible (in other words, if a triple is selected, then all its descendants should be selected) excluding $(l-3)(l-2)(l-1)$. This is possible since $[l]^{(3)} \setminus E(G'')$ has $l-2+s \geq 4+s$ edges. The resulting new 3-graph G is left-compressed on vertex set $[l]$ and it does not contain a clique of size $l-1$, and $\lambda(G) \geq \lambda(G'')$. This proves Lemma 3. □

Proof of Proposition 3.1. We just verify the case when $l=8$. The argument for $l=6,7$ can be done similarly but slightly simpler.

The verification in this case goes as follows. Note that $\binom{8-1}{3} \leq m \leq 50 = \binom{8-1}{3} + \binom{8-2}{2}$. Since $\lambda^{3-}_{(m,7)}$ increases as m increases, it is sufficient to show that $\lambda^{3-}_{(m,7)} < \lambda([8-1]^{(3)}) = \binom{7}{3}/7^3 = \frac{5}{49}$ for $m=50$.

Let $G=(V,E)$ be a 3-graph satisfying $\lambda(G) = \lambda^{3-}_{(50,7)}$ for $|E|=50$. By Lemma 3, we can assume that G is left-compressed and on vertex set $[8] = \{1,2,3,4,5,6,7,8\}$. Let $\mathbf{x} = (x_1,x_2,x_3,x_4,x_5,x_6,x_7,x_8)$ be an optimal legal weighting for G satisfying condition (1) on Page 9 and $x_i \geq x_j$ for $i<j$. Since G contains no clique of size 7, so $x_8 \neq 0$. By Lemma 2, the edge $178 \in E$. Note that E is formed by removing 6 edges from $[8]^{(3)}$. Since G is left compressed, $178 \in E$ and G does not contain a clique of size 7 , then the set $\{678,578,478,378,568,567\}$ or the set $\{678,578,478,568,567,468\}$ is removed from $[8]^{(3)}$. Note that $[8]^{(3)} = [7]^{(3)} \cup \{8ij, \text{where } ij \in [6]^{(2)}\} \cup \{178,278,378,478,578,678\}$.

Case 1. The set $\{678,578,478,378,568,567\}$ is removed from $[8]^{(3)}$. In this case,

$$E = ([7]^{(3)} \setminus \{567\}) \cup \{8ij, ij \in [6]^{(2)} \setminus \{56\}\} \cup \{178,278\}.$$

Observe that $E_1 \cap E_2^c = \emptyset$. By (2), $x_1 = x_2 \stackrel{\text{def}}{=} a$. Similarly, $E_3 \cap E_4^c = \emptyset$, $E_5 \cap E_6^c = \emptyset$, and $E_7 \cap E_8^c = \emptyset$ imply that $x_3 = x_4 \stackrel{\text{def}}{=} b$, $x_5 = x_6 \stackrel{\text{def}}{=} c$, $x_7 = x_8 \stackrel{\text{def}}{=} d$. Note that $2a+2b+2c+2d=1$. In viewing of edges in E,

$$\begin{aligned}
\lambda(G) &= a^2(2b+2c+d) + 2a(b^2+4bc+c^2+2bd+2cd) \\
&+(2bc^2+2b^2c+b^2d+4bcd) + d(a^2+4ab+4ac+b^2+4bc) + 2ad^2 \\
&\stackrel{\text{def}}{=} X(a,b,c,d).
\end{aligned}$$

Using the software Matlab, testing result shows that the maximum value of $X(a,b,c,d)$ under the constraint $2a+2b+2c+2d=1, a \geq b \geq c \geq d \geq 0$ is $\leq 0.1011297 < \frac{5}{49}$.

Case 2. The set $\{678,578,478,568,567,468\}$ is removed from $[8]^{(3)}$. In this case, $E = ([7]^{(3)} \setminus \{567\}) \cup \{8ij, ij \in [6]^{(2)} \setminus \{56,46\}\} \cup \{178,278,378\}$. Observe that

$E_{i \setminus j} = \emptyset$ for $1 \le i < j \le 3$. By (2), $x_1 = x_2 = x_3 \stackrel{\text{def}}{=} a$. Let $x_4 \stackrel{\text{def}}{=} b$, $x_5 \stackrel{\text{def}}{=} c$, $x_6 \stackrel{\text{def}}{=} d$, $x_7 \stackrel{\text{def}}{=} e$, $x_8 \stackrel{\text{def}}{=} f$. Note that $3a+b+c+d+e+f = 1$. In viewing of edges in E,

$$\begin{aligned}\lambda(G) &= a^3 + 3a^2(b+c+d+e) + 3a(bc+bd+be+cd+ce+de) \\ &\quad +bcd+bce+bde+f[3a^2+3a(b+c+d)+bc]+3aef \\ &\stackrel{\text{def}}{=} Y(a,b,c,d,e,f).\end{aligned}$$

Using the Software Matlab, testing result shows that the maximum value of $Y(a,b,c,d,e,f)$ under the constraint $3a+b+c+d+e+f=1, a \ge b \ge c \ge d \ge e \ge f \ge 0$ is $\le 0.1010085 < \frac{5}{49}$. This completes the verification. □

4 Further Study

We will study Proposition 3.1 for more values of l. Since as l gets bigger, the analysis will be more tedious although it is possible. Also, we believe Question 2 should hold for general r-hypergraphs as well.

References

1. Frankl, P., Rödl, V.: Hypergraphs do not jump. Combinatorica 4, 149–159 (1984)
2. Gibbons, L.E., Hearn, D.W., Pardalos, P.M., Ramana, M.V.: Continuous characterizations of the maximum clique problem. Math. Oper. Res. 22, 754–768 (1997)
3. He, G., Peng, Y., Zhao, C.: On finding Lagrangians of 3-uniform hypergraphs, Ars Combinatoria (accepted)
4. Motzkin, T.S., Straus, E.G.: Maxima for graphs and a new proof of a theorem of Turán. Canad. J. Math. 17, 533–540 (1965)
5. Pavan, M., Pelillo, M.: Generalizing the Motzkin-straus Theorem to Edge-Weighted Graphs, with Applications to Image Segmentation. In: Rangarajan, A., Figueiredo, M.A.T., Zerubia, J. (eds.) EMMCVPR 2003. LNCS, vol. 2683, pp. 485–500. Springer, Heidelberg (2003)
6. Peng, Y., Zhao, C.: On Motzkin-Straus Type Results for 3-uniform Hypergraphs (submitted)
7. Rota Buló, S., Pelillo, M.: A generalization of the Motzkin-Straus theorem to hypergraphs. Optim. Letters 3(2), 287–295 (2009)
8. Rota Buló, S., Torsello, A., Pelillo, M.: A Continuous-Based Approach for Partial Clique Enumeration. In: Escolano, F., Vento, M. (eds.) GbRPR. LNCS, vol. 4538, pp. 61–70. Springer, Heidelberg (2007)
9. Sós, V., Straus, E.G.: Extremal of functions on graphs with applications to graphs and hypergraphs. J. Combin. Theory Series B 63, 189–207 (1982)
10. Talbot, J.: Lagrangians of hypergraphs. Combinatorics, Probability & Computing 11, 199–216 (2002)
11. Turán, P.: On an extremal problem in graph theory. Mat. Fiz. Lapok 48, 436–452 (1941) (in Hungarian)

Practice of Semi-automatic Formal Description of Terms in Knowledge Organization Systems

Zhang Yunliang, Zhu Lijun, Qiao Xiaodong, Zhang Quan, and Miao Jianming

Abstract. The interoperation problem between knowledge organization systems become more critical because the complication and reuse of KOS. Enlightened by dictionary complication and the characteristics of Chinese language, we use formal description instead of relationships between items in KOS. We use the conceptual primitives and related instance base of Hierarchical Network of Concepts (HNC) theory and develop both semi-automatic and full-automatic methods for different applications. It is an exploration and should be realized and revised in the future.

1 Introduction

In a typical knowledge organization system for library of information, such as thesaurus, there are a lot of items and relation connections between items. When the KOS is the only one used in an application system, it usually works perfectly. But more and more applications need 2 or more knowledge organization systems that already constructed by different researchers. In these cases, the relations sometimes are different in types and granularity, researchers must integrate or merge or map the systems with interoperation techniques [1-2]. How to deal with this problem? Recall what experts do in dictionary compilation [3], they select a limited amount of words or characters (in Chinese), may be 1000~2000, which are commonly understood by the public. All the words are interpreted by these selected words or characters. But there is a big problem of the dictionary compilation model, namely the interpretation is very difficult for computer to understand. To cope with this problem, there are at least two approaches. The first is to improve the natural language understanding techniques and make the computer has

Zhang Yunliang · Zhu Lijun · Qiao Xiaodong
Institute of Scientific & Technical Information of China, Beijing 100038, P.R. China

Zhang Quan · Miao Jianming
The Institute of Acoustics, CAS, Beijing 100080, P.R. China

Z. Qian et al. (Eds.): Recent Advances in CSI E 2011, LNEE 125, pp. 13–18.
springerlink.com

the common sense of a human being with basic education experience, which is not very practical now. The other approach is to design a limited formal primitive set, and the primitives are unambiguous and linked by some joint marks that provide the exact meaning of a word [4].

There are at least three theories for formal description of Chinese words for computer applications. They are conceptual Graph theory developed by professor Lu [5-6], HowNet theory developed by professor Dong[7-8], HNC (Hierarchical Network of Concepts) theory of Professor Huang[9-10]. The first theory is still in variation, the other two theories are relatively mature and the primitive sets are basically fixed. Now the HowNet and HNC both have several tens of thousands instances for general information processing constructed by manual work. In this paper, we adapt HNC theory to do some experiment on automatic method of formal description of some items in our research knowledge organization system.

2 The Characteristics of Chinese Language

Chinese is very different from western languages. Most Chinese characters are not only form units and phonetic units of information, but also semantic units. In traditional Chinese linguistics, Chinese language has the composition character of: 1) Assemble characters into word, assemble words into clauses, assemble clauses into chapter; 2) From the characters understand the word, from the words understand the clauses, from the clause understand the chapter [9]. So the traditional Chinese linguistics pays more attention to characters, and there is a discipline named Chinese Exegetics. Especially on characters and words, the meanings of characters are usually primitives of the meanings of words composed with these characters. For example the Chinese word 政府(government)has two Chinese characters 政and 府, 政means politics related and 府means office or residence. Now the amount of the common used Chinese characters is less than 4000. Considering the polysemy and synonym phenomena, there are several thousand primitives of meanings. So some scientists, such as Lu, Dong and Huang, propose some expression methods of the semantic primitives. The semantic primitives of course can be expressed with Chinese characters, and can also be expressed with a new symbol system, which is composed with characters and figures.

3 Formal Description Method

In this paper, we use HNC theory to do the formal description work. HNC is a comprehensive theory for Chinese language processing with consideration of traditional linguistics and modern computer techniques [11], which includes theories of words expression with HNC symbols, sentence category analysis, and sentence group analysis and so on. HNC symbol is the conceptual meaning expression of the natural language words (including words with only one character or a lot of characters). A HNC symbol of a specific meaning of a word is a character string of letters, figures and special signs such as #, $ and so on. The conceptual primitives of HNC are organized in a tree structure. Though in fact it is a forest with 18 categories, 101groups and 456 trees, of course, it can be considered as a tree with some imaginary nodes. In a primitive tree, the depth is 2.The amount of the genuine primitives in all levels is 6580. All the 6580 nodes in the tree, whether they are leaf nodes or not, can be used to construct the HNC symbol. The formal description process of HNC theory is analyze the connotation of a specific meaning of a word, then select proper HNC conceptual primitives and at last linked the conceptual primitives with proper operator(expressed as special signs)(Huang, 1998).

The merit that with rich operators is very important for formal description. For example, in table 1, the word拥戴 has two primitives: "v943e61" and "v71101", means "support" and "respect" respectively, but "support" is not equal to "v943e61", it is only an approximate natural language expression. In this HNC symbol, the two conceptual primitives are both important for description of word 拥戴 and without any of them, the connotation cannot be integral. So we use a operator "and" two express the equivalent importance. Then about another example of 巡警. At first 巡警 is "pa41", that is a human being who belong to a militarized organization. In some scenarios, this connotation is enough for application. But in another scenario, maybe we also to know "va11", namely that what patrols do is a professional activity serve for government and to avoid sins. Without the "va11", the basic connotation is integral roughly, so a refinement operator"+" is used to link the two primitives. It is a little similar to series expansion in mathematics, the former is important than the latter, and with more expansion, the connotation is more accurate.

4 Semi-automatic Method

There are some constraints when the words expressed by formal description. The words should be traditional Chinese words or loan words from free translation. If from transliteration, it doesn't work.

The basic of the automatic formal description is material or instances of the word-HNC symbol pairs. Now the pairs are constructed by manual work, now the instances divided into two parts. The first part has about 3,000 pairs of monosyllabic words and HNC symbols. The second part has about 40, 000 pairs of polysyllabic words and HNC symbols. All the words are collected as a dictionary.

The ideal processing procedure is as follows:

- **Step1.** Resolute the input word with the dictionary derived from pair instances. In this step a forward maximum matching method is used and the word is separated into words in dictionaries, or words and separated Chinese characters. We name this kind of words as word-components, and characters as character-components.
- **Step2.** List the HNC symbol of word-components with the nature sequence and link them with "and" operator.
- **Step3.** If there are separated Chinese characters. Select the words which contain these characters and count all the semantic primitives, give the most 3

frequent conceptual primitives as candidate and insert them into the sequence of step 2 in proper position. If there are no such characters in the instance base, use "null" as the result.

- **Step4.** Now Check the candidate sequences for knowledge engineers and they decide which sequence is better, and revise the operator. Perhaps some revision on primitives is also necessary.

For example, 情报业(Information industry) is a word we want to describe the connotation, and get a word-component 情报 and a character-component 业. The corresponding HNC symbol of情报 is jw03|(ga4;ga2;ga1;ga6) . In this expression, jw03 means 基本信息物（a basic object type of what human communicate). The vertical bar means that what in the brackets is the concrete content of jw03 The symbol a means 专业活动（professional activities） and a4,a2,a1,a6 means军事(military field),经济(economy field),政治(politics field),科技(science & technology field). The symbol g means a 静态表示(statistic description of a concept, in English we can say describe is a dynamic description and description a statistic description). And there is not a word 业, so we do a statistics, find 39 words that has this character, and find the most frequent semantic primitives are "ga219", "g661"and "l630a8". And "a219" means建造（produce and build）, "661"means 生命体基本劳作（labor of livings）, "l630a8"means 表示过去的时态说明符(past tense indicator).So the candidate series are:

- jw03|(ga4;ga2;ga1;ga6),ga219
- jw03|(ga4;ga2;ga1;ga6),g661
- jw03|(ga4;ga2;ga1;ga6),l630a8
-

And knowledge engineer will choose a) from the understanding to "情报业", usually the HNC symbol as a) is enough, but it is not perfect. So knowledge engineer may use another operator "|" to replace "," and replace "a219" with a more specific primitive "ga219\25". The final HNC symbol of 情报业 is "ga219\25|(jw03|(ga4;ga2;ga1))" . Though the formal description needs human interaction with the computer, the result of the full-automatic method is acceptable in common applications.

Though there is human interaction with the computer in formal description, it is difficult to decide the operators or 5-tuple symbols, so in practice, all operators and 5-tuple symbols are removed and primitives are separated by space, so the concept description of 情报业in practice is "a219\25 jw03 a4 a2 a1 a6".

And we have developed a tool for formal description, the tool can split the input term into word segments and look for them in the word-HNC symbol pairs base to find the exact or similar pairs for reference. With the method, we have developed a word-HNC symbol pairs base of 10260 terms of the region of new energy vehicles and which provided in the service platform that can be accessed by the site of www.vocgrid.org .

5 Applications and Problem Analysis

The formal description of terms is very useful and can be used in a lot of applications. One of the application is computing of term colony, that is, if you give some terms, use the formal description, it is easy to return all terms has same primitives with the given ones. An example is showed in Fig.2. In a KOS of new energy vehicles, if input terms 电动汽车（electric vehicles）,新能源汽车（new energy vehicles）,燃料电池车（fuel cell vehicles ,FCV）， there will be return a list of 10 terms composed into a term colony.

There are also some problems in the semi-automatic method of formal description with HNC primitives. The words they can processed is limited to traditional Chinese words or loan words from free translation, but it is a difficult problem for computer to recognize the proper words from the improper ones. If the input is not suit for the method, the output will be meaningless.

The statistics usually give good output, but there are cases which should adopt the low-frequency conceptual primitives. We count the primitives in a HNC symbol, but only some primitives are mapping of the specific Chinese characters, there are primitives from other Chinese characters in these words. The appearance position is also need to be considered. In the example of 情报业，primitive“l630a8” only for Chinese character 业 in the beginning location of the words, and it should be excluded in the statistics.

This method may be also effective for English and other western languages. Some English etyma and words also have the compositionality, just like Chinese characters and words. Though HNC conceptual primitives derived from Chinese ， the kernel is suit for all Languages. But to do this for English, the concept primitive forest should be customized for English, that is, some primitives should be expanded and some should be compressed. Of course an English formal description instance base should be manually built or mapped from Chinese base at first.

6 Conclusion

In knowledge organization system, we use relationships to reveal the relevance between items. If we want to make the KOS more useful, more and more thinning relationships should be designed and added into items. And some knowledge organization systems will be used together. But the inconsistency will also increase. Enlightened by the dictionary compilation approach and the characteristics of Chinese language, we try to give the formal description of a new item in a knowledge organization system from the component of Chinese characters and words. In most applications, the result of semi-automatic formal description method is acceptable. We developed a base of 10260 word-HNC symbol pairs with our semi-automatic tool. But there are still a lot of problems, which should be resolved or explored in future study.

Acknowledgments. This work is supported by the "11th Five-Year" national science and technology support project 2006BAH03B03, 2006BAH03B06; the National Fundamental Research Project (973 Project) 2004CB318104 and ISTIC key project 2009KP01-3-2, ZD-2010-3-2,ZD-2011-3-2.

References

1. Li, S.: Interoperability and Its Implementation Among Knowledge Organization Systems. New Technology of Library and Information Service 3, 29–34 (2007)
2. Liu, C., Zhao, P., Liu, H.: Semantic Interoperability Models of Digital Library Based on Ontology. Modern Information 29(8), 66–69, 74 (2009)
3. Wang, D., Wang, L.: Study of Definitions in a Learners' Dictionary of Terminology for Educational Purpose. Lexicographical Studies (2), 111–123 (2010)
4. Zhang, Y., Zhu, L., Xue, C., Qiao, X.: Problem Analy-sis in Construction of Chinese S&T Vocabulary System –Term Resources, Knowledge Structure and Conceptual Description. Research Report of Institute of Scientific and Technical Information of China (2009)
5. Lu, R., Jin, G.: A New Perspective on the Study of Modern Chinese. Applied Linguistics (5), 94–98 (2004)
6. Wu, B., Hu, Y., Lu, R.: Research on Recursive Conceptual Graph Based Text Retrieval Model. Journal of the China Society for Scientific and Technical Information 27(6), 825–831 (2008)
7. Dong, Z., Dong, Q.: How Net [EB/OL], `http://www.keenage.com/zhiwang/e_zhiwang.html`
8. Dong, Z., Dong, Q.: Theoretical Findings of How Net. Journal of Chinese Information Processing 21(4), 3–9 (2007)
9. Huang, Z.: HNC Theory. Tsinghua University Press, Beijing (1998)
10. Huang, Z.: A Summary of HNC Theory. Journal of Chinese Informa-tion Processing 11(4), 11–20 (1997)
11. Jin, Y.: Language processing technologies and applications based HNC theory. Sciences Publication Inc., Beijing (2006)

Specific Dimension Stagnation Optimal Mutation of Particle Swarm Optimization Algorithm

Daqing Zhang and Ling Wang

Abstract. A new specific dimension stagnation optimal mutation of PSO algorithm (SMPSO) is proposed to overcome the shortcomings of the standard PSO algorithm such as premature convergence, easily trapping into a local optimum and lowly searching precision. The presented algorithm can prevent the iterative process from falling into the local extremum. Five test functions are employed to test performance of the algorithm. Comparisons with some existing algorithms show that the new algorithm has a good performance.

Keywords: particle swarm, optimization, mutation, premature convergence.

1 Introduction

Particle swarm optimization (PSO) algorithm [2, 5], which is a group of intelligent technology based on the research of the behavior of birds and fish, was put forward by Kennedy and Eberhart in 1995. Because of the PSO algorithm has many pretty properties, such as coding in real number, easy to be realized, and having a small set of adjustable parameters, it has been widely used in different areas of engineering optimization problems.

However, in practice, people find that in solving complex optimization question, the PSO algorithm is easy to fall into a local extremum, and appear "premature" convergence [3]. And the population diversity may down too fast with iterative times increase, which leads the algorithm can not converge to the global optimum point.

In the past decades, there has been much literature introduces the strategy of mutation to overcome the shortcomings of the PSO algorithm. In [4] the gaussian mutation operator was applied to select the individual mutation at a predetermined

Daqing Zhang · Ling Wang
Institute of Applied Mathematics, Liaoning University of Science and Technology, Anshan, Liaoning Province, China
e-mail: d.q.zhang@ustl.edu.cn, wangling45121@163.com

Z. Qian et al. (Eds.): Recent Advances in CSIE 2011, LNEE 125, pp. 19–24.
springerlink.com

probability. [7] introduced the concept of average distance of the population. If the average distance of the population is less the given value, or the global optimal has no change for a long time, the mutation will be carried out in part of the population. In [8], all the parameters of each particle are adjusted dynamically according to the overall performance evaluation of the whole population. The optimal solution of particles mutate according to dynamic probability to ensure the diversity of the particles and to prevent the local extremum. In [9] a mutation operator was introduced into the algorithm to control the mutation of the optimal particles according to the given probability, to prevent the premature convergence. However, the existing mutation strategies are with blindness and randomness in general.

In this paper, the PSO algorithm is improved by the following idea. Firstly, some dimensions of the optimal particles are focused on, if they have no change after K (mutation time) times iterations. Then mutations are made only on these specific dimensions of the optimal particles. Note that the mutations just happen in the specific dimensions of the particles, not all of the dimensions of the particle. This is the main difference from the existing results. Because the mutation may happens before the fitness value being immovable, the trend of the global optimal value falling into the local extremum can be terminated earlier, and the diversity of the population can be maintained.

2 Specific Dimension Stagnation Optimal Variable PSO Algorithm

2.1 Standard PSO Algorithm

Suppose that there are m particles moving in the D dimensional space. The following notation are adopted. The i-th particle's position is denoted as $x_i = (x_{i1}, x_{i2}, \cdots, x_{iD})$; velocity as $v_i = (v_{i1}, v_{i2}, \cdots, v_{id}, \cdots, v_{iD})$; the best location in the history as $p_i = (p_{i1}, p_{i2}, \cdots, p_{id}, \cdots, p_{iD})$; and the best position of the particle of the population is denoted as $p_g = (p_{g1}, p_{g2}, \cdots, p_{gd}, \cdots, p_{gD})$.

In standard PSO algorithm, the values of position and velocity of the particles are adjusted according to the following iterative algorithm.

$$v_{id}(t+1) = \omega v_{id}(t) + c_1 rand_1 (p_{id} - x_{id}(t)) + c_2 rand_2 (p_{gd} - x_{id}(t)) \tag{1}$$

$$x_{id}(t+1) = x_{id}(t) + v_{id}(t+1) \tag{2}$$

where, ω is the inertia weight; c_1 and c_2 are accelerate constants; $rand_1$ and $rand_2$ are random values changing in the interval of $(0,1)$. The maximum speed v_{max} is predetermined before the algorithm start. During the iterative process, if the speed of one dimension is greater than the v_{max}, then the velocity of the dimension will be set to v_{max}.

2.2 The Improved PSO Algorithm

Suppose that the j-th dimension of the global optimal particle makes search in the interval (l_{dj}, l_{uj}). The average value of the j-th dimension is $l_{middle} = \frac{l_{dj}+l_{uj}}{2}$. Here, we introduce a counter $SG(j)$ to every dimension of the particles. The initial value of the counter is set to zero. If the p_{gj} changes at one step of the iterative process, the corresponding counter holds its value, else, the counter increases one. If the counter arrives at K, which is a prior given value to set the timing of mutation, then the p_{gj} takes a new value in the interval $(l_{middle} - |p_{gj}|, l_{middle} + |p_{gi}|)$, as

$$p_{gj} = rand\left(l_{middle} - \left|p_{gi}\right|, l_{middle} + \left|p_{gi}\right|\right) \tag{3}$$

Algorithm (Specific dimension stagnation optimal variable PSO algorithm (SMPSO)):

Step 1. Initial the swarm (the population size is m). Let the particles take position values and velocity values randomly within the acceptable bounds. Initial the value of p_i of each particles. Let p_g take the best value of p_i. Set the counter SG to zero.

Step 2. Compute the value of the target function over every particle to get the fitness value of the each particle.

Step 3. Compare the fitness value of particles with their historical optimum. If the current fitness is better than p_i, then p_i takes the value of the current fitness value, else, holds the p_i with no change.

Step 4. Compare the current p_i of each particle with the global optimum p_g. If the current p_i is better than p_g, then let the p_g take the value, else, holds the global optimum p_g with no change.

Step 5. Recompute the velocity and position of each particle by using equations (1) and (2).

Step 6. Check the counters $SG(j)$ of the dimensions of the global optimum. If one of the counters has arrived at K, the given value, then reset the p_{gj} by equation (3), and reset the counter $SG(j)$ to zero.

Step 7. Check the conditions for termination, such as the iterative steps has come to its given maximum, the fitness value has been good enough, or the global optimal solution has no change anymore. If one of the termination conditions is satisfied, then stop the iteration, else, goto Step 2.

Remark 1. Due to the counters of the dimension of the optimal particles, the stagnation dimension can be found before the global optimum falls into the local optimum. Furthermore, the mutations happen on the specific dimension, which has pertinence. The diversity of the population can be maintained too.

3 Simulation and Comparison

Consider the following five well known test functions.

$$f_1(x) = \sum_{i=1}^{n} (x_i^2 - 10\cos(2\pi x_i) + 10) \tag{4}$$

$$f_2(x) = \frac{1}{4000}\sum_{i=1}^{n} x_i^2 - \prod_{i=1}^{n}\cos(\frac{x_i}{\sqrt{i}}) + 1 \tag{5}$$

$$f_3(x) = -20\exp\left(-0.2\sqrt{\frac{1}{30}\sum_{i=1}^{n} x_i^2}\right) - \exp\left(\frac{1}{30}\sum_{i=1}^{n}\cos 2\pi x_i\right) + 20 + e \tag{6}$$

$$f_4(x) = 0.5 + \frac{\left(\sin\sqrt{x_1^2 + x_2^2}\right)^2 - 0.5}{\left(1 + 0.001(x_1^2 + x_2^2)\right)^2} \tag{7}$$

$$f_5(x) = \sum_{i=1}^{n}\left(100(x_{i+1} - x_i^2)^2 + (x_i - 1)^2\right) \tag{8}$$

In the following, linear decrease strategy [1] of the inertia weights are adopted, that is

$$\omega = \omega_{\max} - \omega_{\min}\frac{t}{T} \tag{9}$$

where, $\omega_{max} = 0.9$ denotes the maximum of the inertia weights, $\omega_{min} = 0.4$ the minimum of the inertia weights, t the times of current iteration, and T the maximum of the iterative steps. The size of the population m is set to 30. The accelerating factor $c_1 = c_2 = 2$. The mutation timing $K = 30$. Test functions (4),(5),(6),(8) take values in 10-dimensional space, and (7) takes value in 2-dimensional space. The maximum of the iterative steps T is set to 2000.

To illustrate the performance of the algorithm presented here, we compare the algorithm with the existing methods in [4, 5, 6, 8, 9].

Run the algorithms 20 times independently. The mean values, the standard deviations, the optimal values and the worst values are recorded in Table 2–6 for the five test functions.

From Table 2–6, it can be seen that GPSO performs well for functions f_2 and f_4, but for f_1 and f_5, it falls into premature convergence. APSOw, MOPSO and AMPSO behave well only when the test functions are simple. The SMPSO presented here shows strong advantage in the overall performance.

Table 1 The initial value of v_{max} and the bounds of x_{min} and x_{max}.

Functions	x_{min}	x_{max}	Bounds	v_{max}
f_1	-5	5	$(-2.56, 5.12)$	5
f_2	-600	600	$(-300, 600)$	600
f_3	-32	32	$(-5, 10)$	10
f_4	-10	10	$(-5, 10)$	10
f_5	-100	100	$(-50, 100)$	100

Table 2 Performance comparison with f_1.

Methods	Mean value	Optimal value	Worst value	Standard deviation
PSO[5]	7.1454	0.8578	33.2863	6.7406
LPSO[6]	3.2834	0	5.9698	1.9164
GPSO[4]	32.1876	25.1890	39.2623	4.3666
APSOw[7]	4.8753	0.9950	16.9143	3.5303
MOPSO[8]	39.8220	13.7938	77.6689	18.9360
AMPSO[9]	20.1897	12.9481	38.1970	6.5238
SMPSO	7.0344×10^{-14}	0	7.5318×10^{-13}	1.7510×10^{-13}

Table 3 Performance comparison with f_2.

Methods	Mean value	Optimal value	Worst value	Standard deviation
PSO[5]	0.4306	0.2185	0.6650	0.1302
LPSO[6]	0.0811	0.0222	0.1427	0.0322
GPSO[4]	0	0	0	0
APSOw[7]	0.0672	0.0320	0.1378	0.0262
MOPSO[8]	2.7549	0.3012	13.1561	3.8844
AMPSO[9]	2.1701	1.5331	2.9741	0.3308
SMPSO	0.0029	0	0.0221	0.0063

Table 4 Performance comparison with f_3.

Methods	Mean value	Optimal value	Worst value	Standard deviation
PSO[5]	0.0060	5.5956×10^{-4}	0.0162	0.0051
LPSO[6]	4.6185^{-15}	4.4409×10^{-15}	7.9936×10^{-15}	7.7430×10^{-16}
GPSO[4]	8.8818×10^{-16}	8.8818×10^{-16}	8.8818×10^{-16}	0
APSOw[7]	4.7962×10^{-15}	4.4409×10^{-15}	7.9936×10^{-15}	1.0658×10^{-15}
MOPSO[8]	0.4482	1.2879×10^{-13}	5.3656	1.3594
AMPSO[9]	5.2975	3.1803	9.3348	1.6698
SMPSO	1.0658×10^{-15}	8.8818×10^{-16}	4.4409×10^{-15}	7.7430×10^{-16}

Table 5 Performance comparison with f_4.

Methods	Mean value	Optimal value	Worst value	Standard deviation
PSO[5]	8.3267×10^{-18}	0	1.6653×10^{-16}	3.6295×10^{-17}
LPSO[6]	0	0	0	0
GPSO[4]	0	0	0	0
APSOw[7]	0.0024	0	0.0097	0.0042
MOPSO[8]	0.0058	0	0.0097	0.0048
AMPSO[9]	0	0	0	0
SMPSO	0	0	0	0

Table 6 Performance comparison with f_5.

Methods	Mean value	Optimal value	Worst value	Standard deviation
PSO[5]	1.3131×10^4	4.7413	2.5001×10^5	5.4387×10^4
LPSO[6]	2.6197×10^4	1.0136	2.500×10^5	7.6170×10^4
GPSO[4]	8.9714	8.7746	9	0.0686
APSOw[7]	1.5541×10^4	0.0060	2.5000×10^5	5.3983×10^4
MOPSO[8]	2.9965×10^6	5.6792×10^5	8.4744×10^6	1.6989×10^6
AMPSO[9]	1.6986×10^5	36.6095	1.3058×10^6	3.1708×10^5
SMPSO	3.2529	0.0549	5.5657	1.0011

4 Conclusion

A new mutation strategy for the PSO algorithm was presented. Due to the specific stagnation dimension in the iterative process can be focused on, the mutation becomes more accuracy. Simulation shows that the the algorithm presented a good performance in solving the function optimization problems.

References

1. Chen, J.Y., Yang, D.Y., Lu, J.: Self-adaptive crossover particle swarm optimization based on penalty mechanism. Computer Science 37, 249–254 (2010) (in Chinese)
2. Eberhart, R., Kennedy, J.: A new optimizer using particle swarm theory. In: Proc. of the Sixth International Symposium on Micro Machine and Human Science, Nagoya, Japan, pp. 39–43 (1995)
3. Eberhart, R., Shi, Y.: Particle swarm optimization: developments, applications and resources. In: Proc. of Congress on Evolutionary Computation, Seoul, South Korea, pp. 81–86 (2001)
4. Higashi, N., Iba, H.: Particle swarm optimization with gaussian mutation. In: Proc. of Swarm Intelligence Symposium (2003)
5. Kennedy, J., Eberhart, R.: Particle swarm optimization. In: Proc. of IEEE International Conference on Neural Networks, Perth, WA, Australia, pp. 1942–1948 (1995)
6. Shi, Y., Eberhart, R.: Empirical study of particle swarm optimization. In: Proc. of Congress on Evolutionary Computation, Washington, DC, USA, pp. 1945–1950 (1999)
7. Yang, C.H., Gu, L.S., Gui, L.S.: Particle swarm optimization algorithm with adaptive mutation. Computer Engineering 16, 188–190 (2008) (in Chinese)
8. Zhao, Q.Y., Pan, B.C., Zheng, S.L.: Mutational particle swarm optimization algorithm based on warm evaluation. Computer Engineering and Applications 45, 57–59 (2009) (in Chinese)
9. Zhu, Y.P.: Particle swarm optimization algorithm witn optimal mutation. Computer Era 11, 37–39 (2008) (in Chinese)

The Study of Natural Language Processing Based on Artificial Intelligence

Xia Yunye, Zhu Meizheng, and Li Xin

Abstract. In this paper, we study the Natural Language Processing in the communication management system and utilize the principle of Finite Automata Machine (FAM) in the AI theory to propose a simple and effective algorithm for Natural Language Processing. Furthermore, we present the structure model of the new algorithm and parameters configuration. The experimental results of our testing show that if we use this new algorithm neatly, it will bring us very good performance.

Keywords: Natural Language Processing (NLP), Artificial Intelligence, Finite Automata Machine, State Transaction Function, Grammar Rule Matching; Production Interpreter.

1 Introduction

Along with the development of information technology, interaction between human being and computer systems are becoming more and more frequent. During the interaction, people would like to adopt the natural language as the major way to communicate with computer system, which makes the natural-language-processing -based dialogue system become a hot issue of current dialogue management system research, so that make Natural language processing technique an important part of the dialogue management system. From the view of computer processing, we find the main task of the technique is to establish a model, so that the computer can extract key information, which decide the machine comprehension, from natural language information.

Applying the grammar rules matching in artificial intelligence theory We conduct the natural language processing and through Grammar Production, we

Xia Yunye · Zhu Meizheng · Li Xin
North Computation Institute (NCI)
Beijing 100083, China
e-mail: xyyhyf@gmail.com

Z. Qian et al. (Eds.): Recent Advances in CSIE 2011, LNEE 125, pp. 25–31.
springerlink.com

abstract lots of sentences available for human-computer interaction to Gram-mar Rules and tag the key information, so that the system can directly transfer language information to parameter sequence through finite state automata generated from rule set, and lead it to corresponding function of information processing. It not only improves the identifying efficiency of natural language information, but also enhances the expansibility of the rule set.

2 System Modeling

The main task of the grammar-rule-matching-based natural language processing system is to resolve the natural language information to parameter information that is comprehensible for computer, it mainly composed by three modules, Segmentation, Parameter Tagging, and Grammar Rules matching.

3 The Pipeline of System Processing

3.1 Pre-processing

The main task of pre-processing is to tag and resolve the source text to improve the segmentation efficiency and ac-curacy.

1. Pre-segmentation: In the source text, there are some mixed messages that cannot be easily divided by the segmentation algorithm in correct way, such as floating-point number, IP address, email address, time, date and so on. However, these information might be key parameters effecting computer comprehension and should be processed in advance to avoid segmentation algorithm divides them abnormally. We use regular expressions to match the source text in our system, and tags the eligible word or clauses as parameters, so that the segmentation algorithm don‘t need to divide the tagged text any more.
2. Sentence division: In Chinese text, no symbols are included in Chinese words. When we divide sentences, take the punctuations as basis to dividing sentences is a good way to reduce the amount of processed information each time and improve segmentation speed. In consideration of the fact that some special parameters (e.g. IP address) contains sentence-division-related punctuations, so that the pre-segmentation should be processed prior to sentence division, and the words or clauses tagged with parameters should also serves as the basis of sentence division.

3.2 Segmentation

Current segmentation methods can be divided into three categories: character-string-matching based segmentation method, comprehension based segmentation method, and statistics based segmentation method.

Among them, the character-string-matching based segmentation method is also known as mechanical segmentation method; it matches the character strings to be analyzed and vocabulary entries from a "sufficiently large" machine dictionary according to certain policies. If a string is found in the dictionary, then the match succeeds (to identify a word.) According to the scanning direction, matching segmentation methods can be divided into forward match and reverse match; and if the match priority is according to length, the methods can be divided into the largest (longest) match and the smallest (minimum) match; and according to whether combined with the part-of-speech tagging process, they can be divided into simple segmentation method and integrated method combining tag and word segmentation.

Our system adopts character string matching segmentation method of double-dictionary-structure; besides a segmentation dictionary with a large number of entries, it also include a parameter dictionary consisted of parameter lists as an additional dictionary. The parameter dictionary de-fines the storage format of parameter lists, and which parameter type, parameter tag and parameter vocabulary each parameter list should include.

3.3 *Post-processing*

The main task of post-processing is to optimize the segmentation results, and to improve the recognition ability of computer.

1. Stop-Word processing: Stop-Words are the words which appear in text frequently, but are actually of little significance, and Stop-Words mainly refers to adverbs, function words, modals and so on. Our system uses the Stop-Words list to traversing the word sequences that have been processed by seg-mentation algorithm, and remove the Stop -Words the sequences contain.
2. Standardization: The complexity of Chinese gram-mar, extensive vocabulary and the non-standardization of common-use words usually results in the expression of the same meaning in several different ways. With synonym-set generating standardized mapping table in our system, all non-standard words in the segmentation results can be replaced by standard word, in order to facilitate machine recognition.

3.4 *Parameter Tagging*

For different types of questions in different fields, our system needs to extract key information for querying answers, and the set of key information is the system pre-defined set of parameters. For instance, when query about weather, one may ask "how's the weather like in Beijing to-day?" This sentence contains two types of parameters; one is time parameter "today" and the other is region parameter "Beijing" respectively. Since for machine, parameter plays an important auxiliary role in understanding natural language information, parameter tagging becomes an important preparation before using word segmentation results to match Grammar Rules.

Parameter tagging is similar to part of speech tagging, however, the target of tagging is not the part of speech, but the parameter type the word contains. Take the word "Tiananmen" for example, when parameter tagging processes, two parameters will be tagged: "location" and "scenic spot", corresponding to information query in two fields respectively: "inquiry the way" and" tourism". A word may not have any parameter type, or it may have a variety of parameter types, which is determined by the system's specific application area.

3.5 Grammar Rules Matching

We define Grammar G = ({E, W, L},{ (,), [,], {,}, <, >, $, num, word}, P, S),

While P ={
S→E,
E→EE,
E→ (W),
W→E$E|W$E,
E→ []| [L],
E→ {}| {L},
L→num,
E→<E>,
E→word }

We consider L (G) = {w|w∈T*, S=> w} as the language produced by Grammar G. For any w∈L (G), we consider was a sentence produced by Grammar G. The sentences which we utilize to classify sentences in Grammar G, we call them Rules.

In the process of NLP, we can consider the understanding of language information as an implementation of a FAM. The parameter sequences achieved from Segmentation and Parameter Tagging will be used as Rules to match the input sequence of FAM. Well, content and parameter type in eve-ry word of the sequence are the transaction conditions of the FAM which lead it implemented. When the segmentation sequence input over, the state of the FAM turns to be the terminal node, then it identify a sentence successfully. In this way, we can utilize Grammar Production to produce specific Rules to identify very language information and consider Rules as credentials to create state nodes and transaction conditions of the FAM. Its efficiency can be enhanced by upgrading the set of Grammar Rules to adjust the structure and states of the FAM.

3.6 Production Interpreter

Above Grammar G contains Production P which can be an abstract of natural language rules set. For those sentences, such as A1, A2 … An, complied with Grammar Rules, there are formats as below:

1. Connection: “A1A2” means two rules A1 and A2 have been connected into a new rule A1A2;
2. Alternative: “(A1$A2$...$An)”, as n≥2, means sentence only has to comply with one rules in them;
3. Eliminable: “<A1>” means sentence don‘t has to comply with A1;
4. Parameter Tag: “[L]” means a word whose Parameter Tag is L;
5. Any Parameter Tag: “[]” means a word whose Parameter Tag can be any;
6. Group Tag: “{L}” means a group of words whose Parameter Tags are L;
7. Any Group Tag: “{}” means a group of words whose Parameter Tags can be any;

According to the formats of Grammar Production, we can define corresponding Grammar Rules Expression. E.g. < How is > the weather < of > ([101][102]$[102][101]). As [101] represents data parameter; [102] represents city parameter; (E1$E2) means just choose one rule in E1 or E2 to match; <> means the content inside it can be eliminate. Through Production Interpreter we can utilize a small quantity of Grammar Rule Expressions to identify a large set of sentences.

3.7 *State Transaction*

In order to ensure while information in a sentence has been identified successfully, the state transaction path which this sentence come through is only corresponding to one Grammar Rule. Therefore, the structure of FAM from initial node S to terminal node should be a tree structure and its root node has to be node S, so that the FAM will also be a Nondeterministic Finite Automaton (NFA). As for a word matching on a node, the FAM will follow a fixed sequence to choose State Transaction Function. If a mismatch occurs in a State Transaction Function, it will be traced back to the previous node and choose a new State Transaction Function to attempt. It won‘t stop until match successfully or all State Transaction Functions on the nodes fail to matching.

3.8 *Achieve Matching*

When the execution of FAM has been over, if current node is the terminal node, it will read a rule flag. Through that rule flag, the FAM can extract parameters information of the deserved rule from the nodes through traversing the state transaction path, and it will be saved as a Parameters Sequence. Thus, FAM can achieve Segmentation Sequence and Parameters Sequence successively and accomplish the task of Grammar Rules Matching and Parameters Tagging.

4 Basic Tests and Evaluation

As a beginning test, we can use XML Creation Function convert predefined Grammar Rules, Processing Algorithm and Parameter Tagging information needed by Finite Automata Machine (FAM) to XML format label.

For instance, the expression "< How is > the weather < of > [101] [102] $[102] [101])" will be converted to labels as below:

```
<rule><e><w> How is </w></e><w> the weather
</w><e><w> of </w></e><g><o><t para ="Date">101</t><t
para ="City">102</t></o> <o><t para ="City">102</t><t
para ="Date">101</t></o></g><goto>

Weather|WeathcrSearch</goto></rule> begin
```

So that when we need to identify the sentence "How is whether of Shanghai today", the NLP system will extract in-formation in those 2 ways:

1. Use filters WeathcrSearch to collect information from Weather domain;
2. Use the parameters have been obtained, such as: City | Shanghai, Date | today.

As a result, the NLP system implemented in this way can be very easily upgraded by extending the set of Dictionary and Grammar Rules if need to enhance the ability of natural language understanding, so that the NLP system will have a good expansibility. On other hands, for those sentences which have not been identified well, we just need to follow the specific Grammar structure to abstract new Grammar Rules and add them into the NLP system. Indeed, we can get a considerable enhance on identifying of this kind of sentences, simple and efficient.

Through review the implantation of this NLP algorithm, we can also find that resting upon FAM, the Grammar Rules Expression has becomes more logical, and design of this kind of system has been made more guidable than those NLP processing mechanisms based on Bayesian and Boolean model.

5 Conclusions

While AI theory has been investigated more universally and deeply nowadays, it brings us a better opportunity to study NLP system design than ever before. Such algorithm and modeling research of NLP system we present in this paper is just a small piece of an iceberg. In order to optimize our NLP algorithm to higher performance and flexibility, we still have lots of work to do in the machine learning and decision-making. On the other hand, there are still difficulties in the Segmentation of multi-pronunciation words need to be solved.

References

1. Jurafsky, D., Martin, J.H., Kehler, A., Vander Linden, K.: Speech and language processing: An introduction to natural language processing, computational linguistics, and speech recognition. MIT Press (2000)
2. Lapata, M.: Image and Natural Language Processing for Multimedia Information Retrieval. In: Gurrin, C., He, Y., Kazai, G., Kruschwitz, U., Little, S., Roelleke, T., Rüger, S., van Rijsbergen, K. (eds.) ECIR 2010. LNCS, vol. 5993, pp. 12–12. Springer, Heidelberg (2010)
3. Suen, C.Y.: N-gram statistics for natural language understanding and text processing. IEEE Pattern Analysis and Machine Intelligence (2009)
4. Litman, D., Moore, J., Dzikovska, M.O.: Using Natural Language Processing to Analyze Tutorial Dialogue Corpora Across Domains Modalities. In: Proceeding of the 2009 Conference on Artificial Intelligence in Education (2009)
5. Snow, R., O'Connor, B., Jurafsky, D.: Spam and the ongoing battle for the inbox. In: EMNLP 2008 Proceedings of the Conference on Empirical Methods in Natural Language Processing (2008)
6. Gabrilovich, E., Markovitch, S.: Wikipedia-based semantic interpretation for natural language processing. Journal of Artificial Intelligence Research 34, 443–498 (2009)
7. Plath, W.J.: REQUEST: a natural language question-answering system. In: Network and Parallel Computing Workshops (2010); IBM Journal of Research and Development
8. Knight, K., May, J.: Applications of weighted automata in natural language processing. In: Handbook of Weighted Automata. Springer, Heidelberg (2009)

Using Spatial Reasoning in CPM Based Image Retrieval

Shengsheng Wang, Chuo Dong*, and Dayou Liu

Abstract. Traditional image retrieval methods are based on the spatial relationship regarding objects as centroids, and take the advantage of point based spatial relational models to measure the similarity of images. They can not distinguish the directional or topological relationships, so it will return irrelative results. We give a new image retrieval algorithm RACPM by using spatial reasoning and common pattern method. The RACPM compares the rectangle algebra expression of the objects both in the query image and the database image on the x axis and the y axis, and these two pairs are "similar" if and only if all the expressions of them are identical respectively. Compared with the original CPM, the experimental results show that the RACPM outperforms CPM, not only on the time consuming, but also provides more precise retrieval results.

1 Introduction

Content-Based Image Retrieval (CBIR) is one of the trends in Image Retrieval. Nowadays, some algorithms have been proposed for the similarity retrieval in CBIR. In 2002, Eugenio D.S et al. proposed a structured approach which transforms an image into a tree according to the basic shapes of objects, positions of shapes and composite shape descriptions, the image retrieval is viewed as retrieval in the tree [1]. In 2003,Ying-Hong Wang proposed 2D BE string and calculated LCS between two strings [2]. In 2004, Po-Whei et al. proposed spatial relational representation based on 9D-SPA[3]. In 2006, P.Punitha and D.S.Guru et al. proposed a method for representing symbolic images[4]. In 2008, they developed TSR model to represent symbolic images and constructed B-Tree for retrieval [5]. In 2008, Shu-Ming Hsieh et al. proposed CPM [6]. In this paper, we propose RACPM based on the CPM and take use of rectangle algebra which is one of the qualitative spatial reasoning methods.

Shengsheng Wang · Chuo Dong · Dayou Liu
College of computer science and Technology, Jilin University
e-mail: {wss,liudy}@jlu.edu.cn, dongchuo_2009@yahoo.cn

* Correspoding author.

Z. Qian et al. (Eds.): Recent Advances in CSIE 2011, LNEE 125, pp. 33–38.
springerlink.com

2 RACPM Algorithm

Constructs a directed acyclic graph for image retrieval, every edge in the graph represents a topological relationship between two objects. Before constructing the graph, we range these objects on the basis of their MBR coordinates in ascending order rather than type-i rules. We only check the spatial relationships between the current object and the objects ordered ahead of it, so we can guarantee that every edge starts from a vertex with a larger number and ends with a vertex with a smaller number. Within a non-directed graph, there may exist several maximum common subimages, therefore the directional relationships may be nubilous.

After the above steps, a non-directed graph has been converted to a directed acyclic graph, at the meantime, all edges in nondirected graph are assigned a path ID. Now, we can search the longest path in the directed acyclic graph.

Based on the analysis above, we propose RACPM based on rectangle algebra and CPM. First, a data structure will be defined and it's named rectangle algebra similarity graph(RASG);after that we will introduce the algorithm RACPM, it searches the longest path in the RASG, the vertices along the longest path form a set consists of common objects which have identical spatial relationships in both query image and database image. The number of objects in the set is more, the corresponding images are more similar. RACPM is as follows:

```
Algorithm RACPM(a1a2...an1, b1b2...bn2)
/*RACPM constructs RASG for each object-pair which have the same symbol, in the
RASG there is a longest path. Ultimately, RACPM returns the maximum value of all the
lengths, and we also find the maximum similar set.  */
  S1.[Constructs RASG for each object-pair]
 FOR i=1 TO n1 DO
   FOR j=1 TO n2 DO
     IF M[i][j]=1 THEN          /*M is match table*/
       Construct_RASG(i, j)/*constructs RASG(i, j)*/
     EndIF
   End FOR
  EndFOR
  S2.[Returns the maximum value of all the lengths]
  RETURN maximal value from all values
```

Match table and Construct_RASG(i, j) are defined as follows. Let $\alpha=a_1a_2...a_{n1}$ be the sorted objects in query image f_1, let $\beta=b_1b_2...b_{n2}$ be the sorted objects in a database image f_2. Let match table $M_{n1\times n2}$ shows that for every object-pair in the both images, whether or not these symbols are identical. When the symbol of a_i and the symbol of b_j are the same, $M[i][j]=1$, otherwise $M[i][j]=0$.

For calculating the similarity of spatial relationships, a directed rectangle algebra similarity graph (RASG) is in need. Let $RASG=\{V, E\}$,$V=\{(i, j) / M[i][j]=1\}$,$E=\{<v_k, v_h>\}$,there is an edge from v_k to v_h if and only if the spatial relationships between v_k and v_h in both images are identical and $k<h$.

As every *RASG* beings with the vertex (i, j), we can label the graph $RASG(i, j)$. All paths in the $RASG(i, j)$ represent that the object-pair $<v_i, v_j>$ in both images has the same spatial relationships with the other vertices along a path. Now, the algorithm Construct_RASG(i, j) is introduced as follows:

Algorithm Construct_RASG(i, j)

C1.[RASG(i, j) begins with the vertex (i, j)]

VertexSet of $RASG(i, j)=\{(i, j)\}$

C2.[Check the graph wether or not should be expanded]

IF i=1 or j=1 THEN

RETURN *value*=1

FOR $k=i$-1 TO 1 DO

FOR $h=j$-1 TO 0 DO

IF $M[k][h]$=1 AND RA(i, j, k, h) THEN

/ RA(i, j, k, h) is a function which returns bool type, it returns Ture if and only if the vertex (i, j) and (k, h) has the same spatial relationships in both images, otherwise it returns False.*/*

Construct_condense_RASG(k, h, i, j)

Merge_condense_RASG(k, h, i, j)

EndIF

EndFOR

EndFOR

C3.[Returns the length of the longest path]

RETURN *value*=the length of the *longest path*

In the algorithm Construct_RASG(i, j) there are two other functions called: Construct_condense_RASG(k, h, i, j) constructs a condense $RASG(k, h)$, Merge_condense_RASG(k, h, i, j) merges the condense graph into $RASG(i, j)$. The two algorithms are as follows:

Algorithm Construct_condense_RASG(k, h, i, j)

*/*condense_RASG(k, h, i, j) is a condense representation of RASG(k, h), in the condense graph, every vertices* $<v_p,v_q>$ *has the same spatial relationships in both images with* $<v_i,v_j>$*.*/*

C1.[Constructs the vertex set of *condense_RASG*(k, h, i, j)]

FOR each vertex in $RASG(k, h)$ DO

```
    IF RA(i, j, k, h) THEN
          mark (k, h) is qualified
          add the vertex (k, h) into VertexSet of
          condense_RASG(k, h, i, j)
   EndIF
 EndFOR
  C2.[Constructs the path of condense_RASG(k, h, i, j)]
  FOR each path p_id in RASG(k, h) DO
    id'=find a brand new id
    add the path p_id' into condense_RASG(k, h, i, j) along
      the qualified vertexs which are originally on p_id
   EndFOR
  Algorithm Merge_condense_RASG(k, h, i, j)
  M1.[Gets the vertex set of RASG(i, j)]
  VertexSet of RASG(i, j) =VertexSet of condense_RASG(k, h,
    i, j) ∪ VertexSet of RASG(i, j)
  M2.[Expands the path of RASG(i, j)]
  IF |VertexSet of condense_RASG(k, h, i, j) |=1 THEN
      id'=find a brand new id
      add the edge with p_id' into RASG(i, j) from the vertex
        (i, j) to vertex (k, h)
   ELSE
     FOR each path p_id in condense_RASG(k,h,i,j) DO
         add the edge p_id into RASG(i, j) from
           the vertex (i, j) to vertex (k, h)
         copy all the edges in p_id from
           condense_RASG(k, h, i, j) to RASG(i, j)
     EndFOR
   EndIF
```

3 Experiments and Results

In these experiments we use a synthetic symbolic image database for identify efficiency and effectiveness. The database is divided into 6 groups: in the first three groups there are 1000 images in every group and each image contains 5 objects, 10 objects, 15 objects respectively; in the last three groups there are 2000 images in every group and each image contains 5 objects, 10 objects, 15 objects respectively. In these experiments, the visual features are not considered in that the visual features are irrelevant in calculating the maximum common subimage.

These experiments were implemented on Visual Studio 2008 and were run on Intel (R)Core(TM)2 Duo CPU E7500 2.93GHz, with 2G RAM running Windows XP. The average comparison time of RACPM and CPM are in Fig.1.

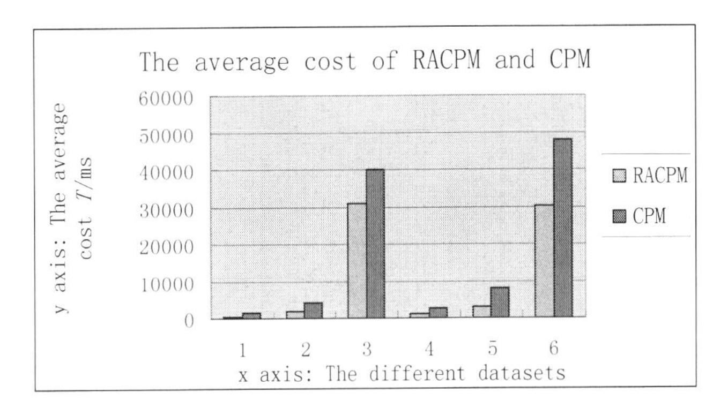

Fig. 1 The average comparison time of RACPM AND CPM

In Fig. 1, the comparison time of RACPM is much less than the comparison time of CPM in all 6 groups. Due to the spatial relationships of rectangle algebra need more time to identify than type-*i* rules, the procedure of RACPM is much more complex than CPM, but just because spatial relationships of rectangle algebra is stricter than type-*i* rules, the procedure of construing graph can be cut down by reducing the number of objects which are candidate for expanding the RASG, so the total search time is shorter than CPM.

Table 1 represents the number of objects in maximum common subimages calculating in CPM and RACPM, and the experiment is running on the third group. We show the top 10 results among 1000 images.

Table 1 The comparison of maximum common subimage

Test Set	1	2	3	4	5	6	7	8	9	10
CPM	15	6	6	6	6	6	6	6	6	6
RACPM	15	4	4	3	4	4	4	4	4	4

As it shown in Table 1, the number of objects in maximum common subimage calculating in RACPM is more than CPM, these similar images searched by RACPM are more precise, and they are more in line with human's recognition.

4 Conclusion

This thesis proposed an improved algorithm RACPM based on rectangle algebra and CPM. RACPM can identify more precise spatial relationships in calculating the maximum common subimages, and rectangle algebra defines 269 topological relationships which represent all the relationships in 2D. RACPM and CPM run on the same database and the six results demonstrate that RACPM use some stricter rules to identify relationships and provide users more precise images, also dramatically shorten the search time.

Acknowledgments. Chuo Dong is the corresponding author. This work is supported by National Natural Science Foundation of China (No. 61133011), Jilin University Research Project (200903178), Seed Foundation of Jilin University and Open Project of Key laboratory of symbolic computing and knowledge engineering of ministry of education in China.

References

1. Eugenio, D.S., Francesco, M.D., Marina, M.: Structured Knowledge Representation for Image Retrieval. Journal of Artificial Intelligence Research 16, 209–257 (2002)
2. Ying, H.W.: Image Indexing and Similarity Retrieval based on Spatial Relationship Model. Information Sciences 154, 39–58 (2003)
3. Huang, P.-W., Lee, C.-H.: Image Database Design Based on 9D-SPA Representation for Spatial Relations. IEEE Transactions on Knowledge and Data Engineering 16(12), 1486–1496 (2004)
4. Punitha, P., Guru, D.S.: An Effective and Efficient Exact Match Retrieval Scheme for Symbolic Image Database Systems. IEEE Transactions on Knowledge and Data Engineering 18(10) (October 2006)
5. Punitha, P., Guru, D.S.: Symbolic Image Indexing and Retrieval by Spatial Similarity: An Approach based on B-tree. Pattern Recognition 41, 2068–2085 (2008)
6. Shu, M.H., Chiun, C.H.: Retrieval of Images by Spatial and Object Similarities. Information Processing and Management 44, 1214–1233 (2008)

A WIA-PA Network Oriented Routing Algorithm Based on VCR

Xiushuang Yi, Peijun Jiang, Xingwei Wang, and Weixing Wu

Abstract. Based on the IEEE-802.15.4 standard and WIA-PA networks with VCR in it, the minimum hop routing algorithm and energy-efficient routing algorithm in wireless networks are introduced. The merits and weaknesses of the two algorithms are analyzed. A routing algorithm based on VCR is designed and simulated. The results show that the novel routing algorithm is efficient. By comparing the WIA-PA VCR application with different algorithms, achieved lower delay management data packet forwarding and energy consumption of data packets transmitting, and be more suitable for the WIA-PA network applications.

Keywords: Industrial wireless network, WIA-PA, Routing, VCR.

1 Introduction

WIA-PA[1] which is based on IEEE-802.15.4 standard[2-3] and compatible with Wireless HART standards is researched and developed independently by China. WIA-PA provides high reliability and real-time wireless communications services.As illustrated in Fig.1[1], Mesh structure and star structure are combined to a two-layer network topology. The first level network is mesh topology in which routing devices and gateway devices are deployed. The second level network is star topology[1] in which field devices are deployed.

In WIA-PA network, on account of different user applications, the communication resources are divided into communication resources used by management services and data services. It uses a VCR (Virtual Communication Relationships) to distinguish the communication resources.

Xiushuang Yi
Network Center, Northeastern University
Shenyang, P.R.China
e-mail: xsyi@mail.neu.edu.cn

Peijun Jiang
Computer Technology, Northeastern University
Shenyang, P.R.China.
e-mail: Jiangpeijun2009@126.com

Z. Qian et al. (Eds.): Recent Advances in CSIE 2011, LNEE 125, pp. 39–44.
springerlink.com

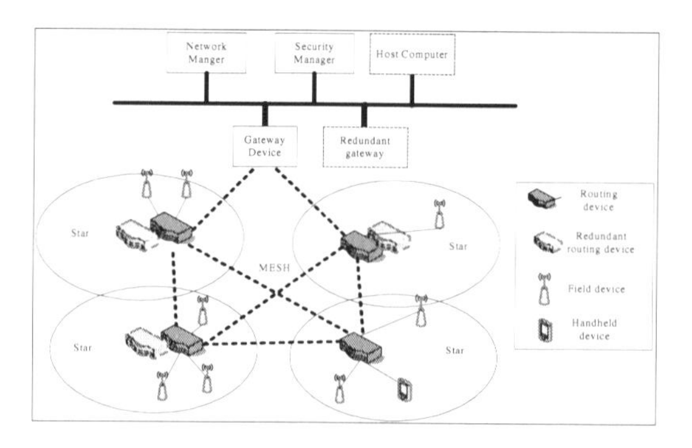

Fig. 1 WIA-PA Network Structure.

According to VCR in the WIA-PA, this paper analyzes the merits and weaknesses of both algorithms in the WSNs. Meanwhile, it combines the merits of both with the characteristics of VCR in the WIA-PA, and puts forward a new VCR-based routing algorithm which can select the appropriate route according to the different VCRs to meet the actual needs of industrial wireless applications.

2 Two Important Routing Algorithms

There are many routing algorithms[4] being used in the WIA-PA networks. The minimum hop routing algorithm and energy-efficient routing algorithm are the more commonly used in the WIA-PA network.

The minimum hop routing algorithm mainly focuses on the real-time requirements. It implements simply and can automatically accomplish the shortest path and the smallest delay. It doesn't need to maintain the global network topology. It is so suitable for management message that the delay requirement of network is higher. It can't take full use of some advantages of the node wired power supply in the industrial environment, and the node in the key position will make the network appear the energy black hole because it may forward much data.

Energy-efficient routing algorithm weights the path on the basis of the percentage of the node's residual energy. It can reduce the possibility that energy black hole appears. Thus prolongs the lifetime of the whole network. It is suitable for the vast majority of data packets which require less for real-time nature to be forwarded in the wireless network. Because of focusing more on energy efficiency, the path which energy-efficient routing algorithm selects out of is not the path with the least delay. So using this method to manage packet routing is not optimal.

3 The Routing Based on VCR

As above analysis, although the minimum hop routing algorithm and energy-efficient routing algorithm are important in the WIA-PA network, they both have some limitations and shortcomings. This paper combines the merits of the two routing algorithms and proposes a new routing algorithm based on VCR in the

WIA-PA network. We can assign different VCRs according to different user applications and select the minimum delay routing for the management message and the energy-efficient path for data packet, when the message forwarded.

3.1 The Indication of Power Information

In general, the power information of node is usually expressed as a percentage of the remaining power. As time goes by, the node's energy consumption gradually decreases and the value also reduces. When a node runs out of its energy, the value goes down to 0 finally. For the active node, the remaining capacity has been 100%, so the value has always been 1.0. Therefore, in the WIA-PA network, the value $r_i(t)$ of the node i 's power information can be expressed as:

$$r_i(t) = \begin{cases} 1 & i \in R \\ \dfrac{(E_i - \alpha_i(t))}{E_i} & i \notin R \end{cases} \tag{1}$$

In this expression, R is the set of active nodes. E_i as the initial energy of the node i . $\alpha_i(t) \in [0, E_i]$ is the energy that the node i consumes in the t time. The(1) is the power information of different nodes in the WIA-PA network.

3.2 Routing Update Messages

This paper improves the traditional routing update message.

Table 1 Route update message format

Parameter name	Date type	Range
ControlInfo	Unsigned8	0～255
DstAddr	Unsigned16	0～65535
SrcAddr	Unsigned16	0～65535
PreHop	Unsigned16	0～65535
NextHop	Unsigned16	0～65535
JobID	Unsigned32	0～4294967295
TTL	Unsigned8	0～255
BatteryState	Float	0～1.0
Hops	Unsigned8	0～255

In the WIA-PA network, the node power information is a very important performance indicator. It often determines the network connectivity and survival time. We add a power information field(BatteryState) and a hop information field(Hops) to the traditional routing update message. It is shown in table 1.

3.3 Routing Table Format

We also improve the routing table format of the routing node. We add the power information and the number of hop information to the routing table entry. The improved routing table format is shown in table 2.

Table 2 Example of routing

Destination addr	The next hop	Message ID	Power Information	Hop count information
N1	N3	4	0.5	2
GW	N2	6	0.7	2
......	......	......	......	......

3.4 Routing Algorithm

Gateway nodes send the routing update messages to the sensor network in the flooding way. The implementation process of the gateway node is described in Algorithm 1:

1. hop count information Hops=1;
2. the power information BatteryState=1;
3. while the message uniquely identity JobID=0 do;
4. send the routing update messages to other adjacent nodes;
5. JobID=JobID +1;
6. flooding after a period of sleep time;
7. end while.

After the routing node receives the routing update messages sent by the gateway node, the node routing table will be updated. The routing node implementation process is described in algorithm 2:

1. The routing node receives the update message from the gateway node;
2. Search the routing table entry in the current routing table;
3. If the message identity of the routing entry is smaller than the one of the routing update message;
4. Delete the routing table entry; End if;
5. If the routing table entry is empty or the power value in the routing update message is larger than the one in the routing table entry or the hop count value in the routing update message is less than the one in the routing table entry;
6. Call the function to insert a new routing table entry;
7. Mark the insert action; End if;

8. If the mark has an insert action;
9. Power value=power value in the routing update message * the current message * the current power value;
10. Add 1 to the hop count value in the routing update message;
11. Forward the modified routing update message to the neighbor node;
12. End if;
13. Function to insert a new routing table entry;
14. Set the destination addr. to the source address of the routing update message;
15. Set the next hop to the previous address of the routing update message;
16. Set the message ID to the unique identity in the routing update message;
17. Set the power information to the one in the routing update message;
18. Set the hop count to the one in the routing update message;
19. End Function.

When searching the routing in the management VCR, in accordance with the same destination address, it can achieve the low latency by searching the minimum hop of the routing table entry as the final search result. When searching the routing in the data VCR, in accordance with the same destination address, it can achieve the high energy efficiency by find the largest power value of the routing table entry as the final search result.

4 Analysis of Simulation Results

In this paper, we use NS2 simulation platform to compare their survival time of nodes and the average delay of the network among the minimum hop routing algorithm, energy-efficient routing algorithm and the routing algorithm based on VCR. We simulate two rectangular machines side by side in the industrial site.

There are 8 routing nodes being distributed into each machine. We set a gateway node in the middle of the two machines. The topology is shown in Fig.2.The active nodes are Host[0,2,3,6,7], and the others use the battery.

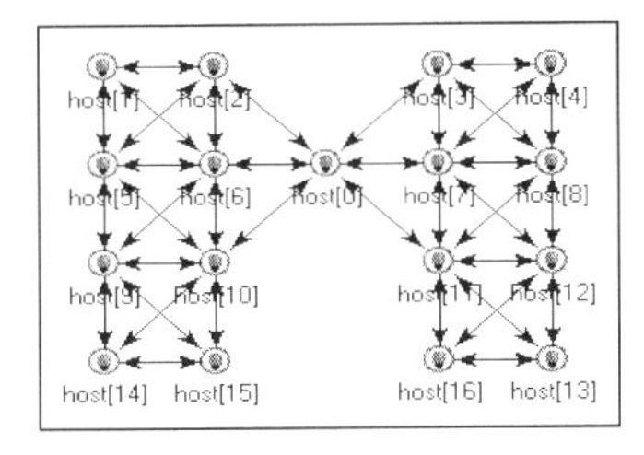

Fig. 2 Simulation topology

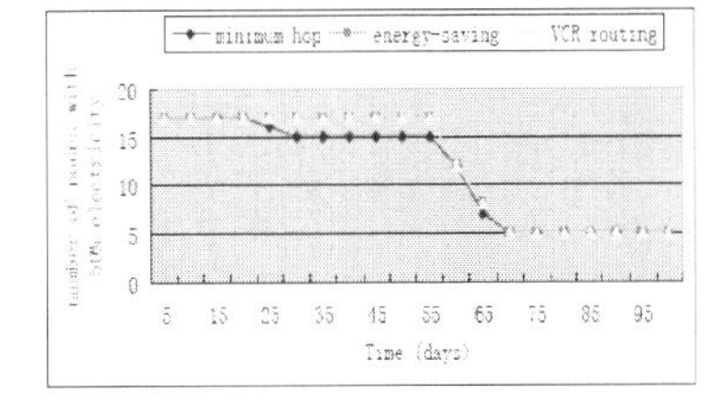

Fig. 3 The comparison of the number of nodes with 50% electricity

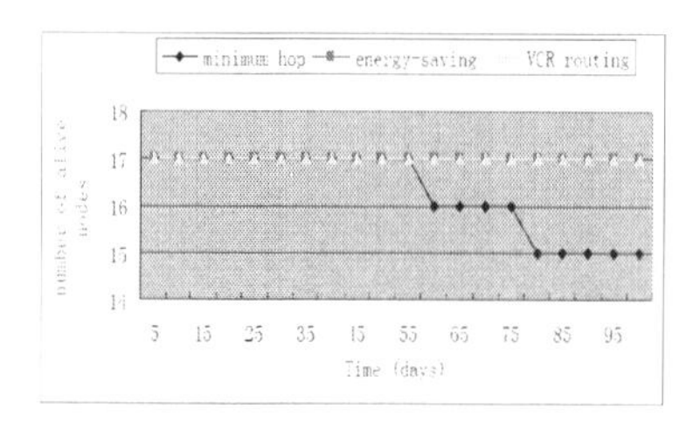

Fig. 4 The comparison of the number of alive nodes

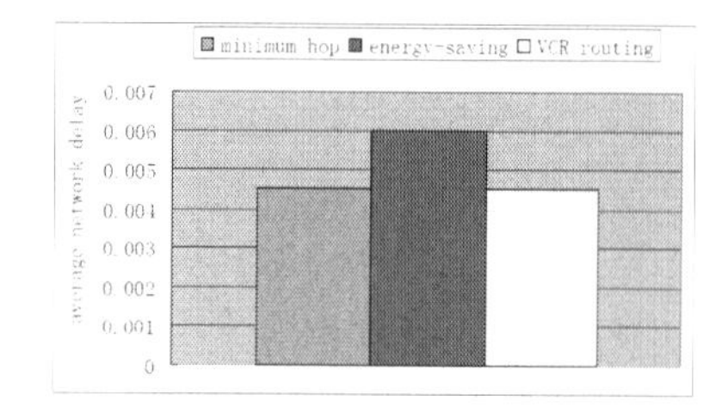

Fig. 5 The comparison of the average network delay

In the same simulation environment, the power consumption of each node based on VCR routing algorithm is basically consistent with that based on the energy-efficient routing algorithm. The average network latency based on VCR is basically consistent with that based on the minimum hop routing algorithm.

5 Conclusion

Distinguishing among communication resources used by different user application objects is conducive to aim to the characteristics on the communication resource requirements of different user application objects. It can arrange for communication resources in a targeted manner. Thus it can meet the requirements of real-time for industrial wireless application, and also take account of the requirements for saving energy.

The simulation results show that the routing algorithm based on the VCR combines the merit of the minimum hop routing algorithm with the energy-saving routing algorithm. On account of the different routing algorithms based on VCR in the WIA-PA network, it can achieve the low latency of management packet forwarded and the low energy consumption of data packets forwarded. The further work will focus on the problem that evaluates the convergence rate of the algorithm in order to accommodate large-scale industrial wireless network applications.

References

1. CIWA (Chinese Industrial Wireless Alliance), `http://www.industrialwireless.cn/`
2. Knlik, J., Heinzelman, W.R.: Negotiation-based protocols for disse. minating information in wireless sensor networks. Wireless Networks 8(223), 169–185 (2002)
3. Intanagonwiwat, C., Govindan, R., Estrin, D., et al.: Directed diffusion for wireless sensor networking. IEEE/ACM Transactions on Networking 11(1), 2–16 (2003)
4. Braginsky, D., Estrin, D.: Rumor routing algonithm for sensor networks. In: Proceedings of the 1st Workshop on Sensor Networks and Applications, pp. 22–31 (2002)

Design and Implementation of Image-Based Monitoring and Tracking System

Jen-Chao Tai and Hsin-Ming Lo

Abstract. Image tracking has become an important technique for monitoring and surveillance applications. This paper presents a novel design of image-based monitoring and tracking system. The system uses eight infrared sensors to monitor the intruder. When an intruder enters, the system detects the locality of the burglar according to the sensory information. A pan-tilt-zoom (PTZ) camera is immediately rotated to aims the intruder according to the detection results of locality. Because the rotation range of PTZ camera is less than 360 degree, a rotation mechanism is designed to rotate the PTZ camera. Image processing technologies are then utilized to find a rectangle region that covers the image of the intruder for tracking initialization. Color feature analysis is executed for tracking. The mean shift method measures the image motion of moving object in the image frame. Fuzzy set rules is designed to rotate the PTZ camera to track the intruder according to the position of moving image. Practical experimental studies are conducted to evaluate the performance of the proposed method and results obtained verify its detection and tracking ability.

1 Introduction

It is widely recognized that image-based monitoring systems are flexible and versatile for monitoring and surveillance applications. The system can assist security officer to ensure the safety of the monitored area. Especially the system can track the intruder by image tracking techniques. The study of image tracking for monitoring and surveillance applications has gained increasingly attention in recent years [1, 2, 3, 4].

In this paper, we propose an image-based monitoring and tracking system that automatically detects and tracks moving man in any portion of a house. The system monitors the environment with infrared sensors and a PTZ camera. Eight

Jen-Chao Tai · Hsin-Ming Lo
Inst. of Precision Mechatronic Engineering, Minghsin University of Science and Technology, No.1, Xinxing Rd., Xinfeng Hsinchu 304, Taiwan, R.O.C.
e-mail: tjc@must.edu.tw, kevin750126@hotmail.com

Z. Qian et al. (Eds.): Recent Advances in CSIE 2011, LNEE 125, pp. 45–54.
springerlink.com

infrared sensors, installed in an octagon platform, are used to detect the intruder. Mean shift method is employed to track the moving target. An initialization method is proposed to detect the intruder for tracking. Fuzzy rule controllers are designed to rotate the PTZ camera to ensure the image of moving target will locate in the image. Fig. 1 shows the system architecture of the proposed contour initialization and tracking system.

The rest of this paper is organized as follows. Sect. 2 presents the image tracking system and the proposed method. Experimental results of traffic parameter estimation of the proposed method will be presented in Sect. 3. Sect. 4 summarizes the contribution of this work and discusses some of its future research directions.

2 The Proposed Method

This image tracking system consists of four parts: the intruder detection module, the initialization module, the motion detection module and Fuzzy tracking module. The intruder detection module detects the intruder with eight infrared sensors. The detection module uses a server motor to rotate the PTZ camera to aim at the intruder according to the detection result of sensory information. The initialization module uses a difference method to segment the image of moving object for tracking operation. Once initialized, the motion detection module will determine the motion of moving object by using the mean shift method. The mean shift method measures the image motion of moving object in the image frame. The fuzzy tracking module rotates the pan and tilt angle of PTZ camera to ensure that the image of moving object locates in the image frame.

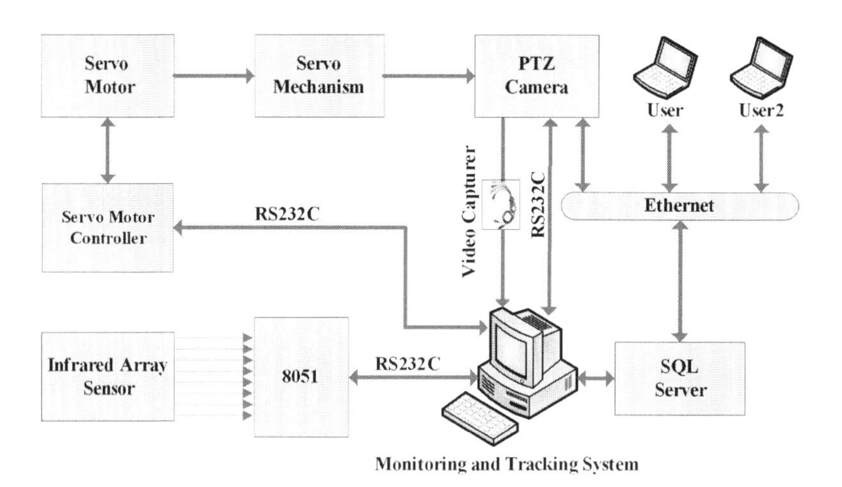

Fig. 1 Block diagram of the system architecture

2.1 Intruder Detection Module

The intruder detection module uses eight infrared sensors to detect the burglar. The sensors are installed in a fixed octagon platform, as shown in Fig. 2(a). A PTZ camera is also installed in the platform. Because the rotation range of PTZ camera is less than 360°, a rotational mechanism is designed to rotate the PTZ camera. The mechanism is directly driven by a servo motor. The detection angle of the sensor is 120°and the detection distance is 5m. When a burglar enters the detection region, the change of thermal radiation will activate some sensors. According to the response of these sensors, we can determine 16 regions: POS1, POS2, POS3, POS4, POS5, POS6, POS7, POS8, POS9, POS10, POS11, POS12, POS13, POS14, POS15 and POS16, as shown in Fig. 2(b). The developed system detects the locality of the burglar according to the sensory information. A pan-tilt-zoom (PTZ) camera is immediately rotated to aims the burglar according to the detection results of locality. The hardware of the developed system is shown in Fig. 3.

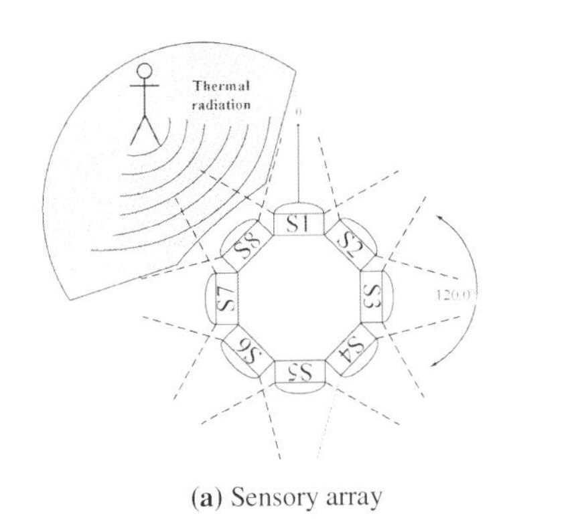

(a) Sensory array

(b) 16 detection regions

Fig. 2 Infrared sensors array

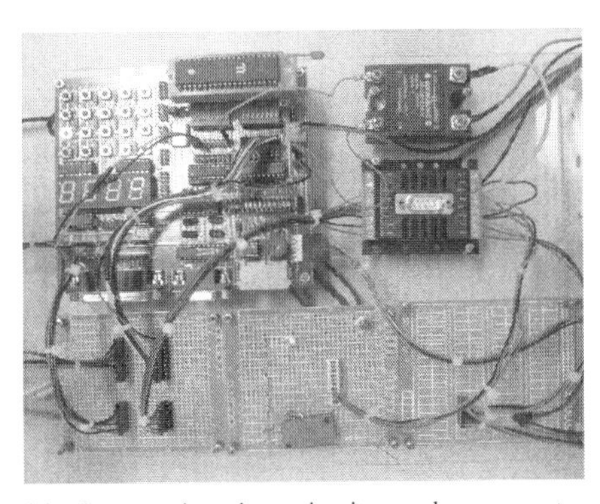

(a) Sensor detection circuits and servomotor controller

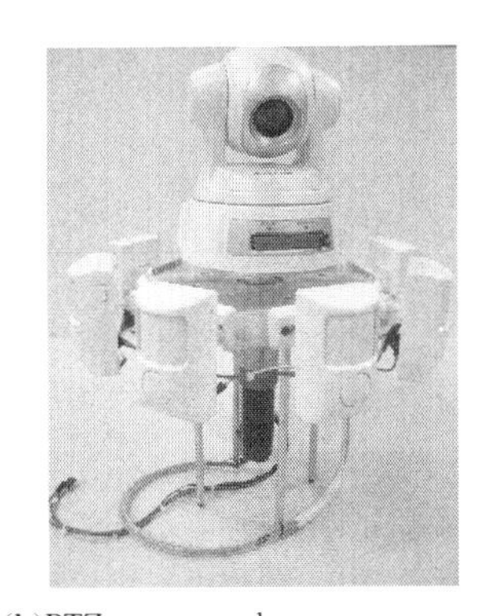

(b)PTZ camera and sensory array

Fig. 3 Hardware of the developed system

2.2 Initialization Module

After the PTZ camera aims at the intruder, three successive frames are employed to determine the image region of the moving object. The absolute difference between successive frames can be used to divide an image frame into changed and unchanged regions. The changed region is the image of moving object. We use the horizontal and vertical projection to measure the occupancy of the image of moving object. Accordingly, a rectangle region is determined to fit the image of moving object. The PTZ camera will adjust the zoom setting of the lens to make the image of moving object completely locate within the image frame. Finally the difference method detects the moving rectangle region again. The center region of the rectangle region is used as a template region for tracking initialization. The dimension of the template region is 30x30 pixels. The color features of the center region are employed to track with mean shift method and Fuzzy rule.

2.3 Motion Detection Module

By mean shift method [4, 5], the location of the intruder can be estimated by the similarity measure between the template region and a candidate region. The candidate region is predicted by mean shift method. The motion detection procedure over one time-step is summarized as follows:

1. *Selection of template region:* The dimension of template region cannot be chosen as a large scale for the tracking efficiency. In our design, the region is selected as described in Sect. 2.2.
2. *Compute the probability of feature in the template region:* To decrease the computation load, the 24-bit RGB representation of the truecolor has been adjusted to 12-bit RGB representation. Each R/G/B color only uses 4 leftmost bits for its color model. Edge information is also employed to define the **C** feature of each pixel:

$$\mathbf{C} = 4096\mathbf{E} + 256\mathbf{R} + 16\mathbf{G} + \mathbf{B} \tag{1}$$

   where **E** is edge value, **R**,**G**, or **B** is the value in red, green and blue color model. Histogram has been made for **C** features of all pixels in the template region.
3. *Compute the probability of color feature in candidate region in the current frame:* Probability of Computation is the same as *step 2*.
4. *Mean shift calculation:* A mean shift can be found by maximizing the similarity function (Bhattacharyya coefficient) between the probabilities computed from the template region and the candidate region [6].
5. *The prediction of candidate region:* The first candidate region is the template region. The candidate region is predicted by mean shift method. The mean shift is the center of new candidate region.
6. Go to *step 3* for next iteration.

2.4 *Fuzzy Tracking Module*

During the tracking operation, the image of moving object moves randomly. The moving region sometimes will leave out the image frame. The camera must rotate its view to follow the moving path of the moving region. In our design, we propose a method to maintain the center position of the tracked region in the center of the image frame. Fuzzy tracking method is developed to rotate the PTZ camera. The inputs of fuzzy controller are the difference between the center position of the candidate region and the center position of the image frame. The outputs are the rotation speeds of PTZ camera. The positions and the rotation speeds both have two degree of freedom. To simplify the design, we break them into two groups. Two individual controllers are developed accordingly. One is designed to adjust the pan angle and the other is developed to adjust the tilt angle.

2.4.1 Fuzzification

There are two fuzzy controllers in our design. One is designed to control the tilt rotation and the other is used to control the pan rotation. Horizontal position and vertical position, respectively, is the input of the controller. The membership functions for horizontal position and vertical position are shown in Fig. 4(a) and Fig. 4(b). The membership functions for pan rotation speed and tilt rotation speed are shown in Fig. 5(a) and Fig. 5(b).

2.4.2 Fuzzy Rules

Ten fuzzy rules are designed to control the rotation. The rules used are of the following type:

If the horizontal position is very left then the pan rotation speed is very fast (right). If the horizontal position is left then the pan rotation speed is fast (right).

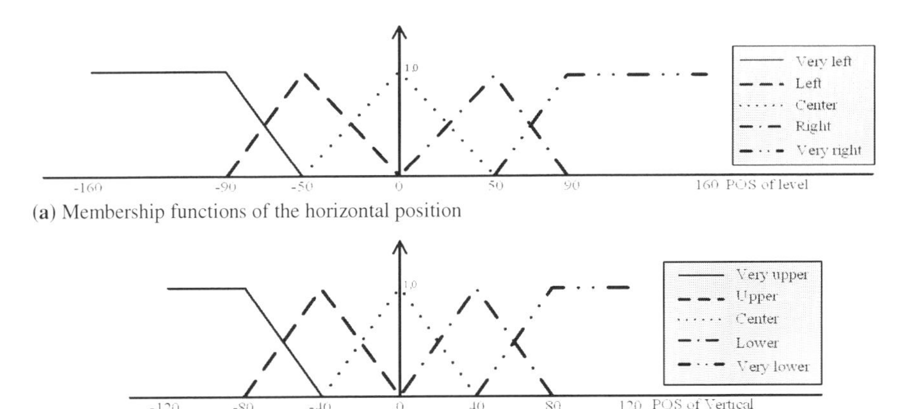

(**a**) Membership functions of the horizontal position

(**b**) Membership functions of the vertical position

Fig. 4 The Fuzzy sets of the horizontal position and the vertical position

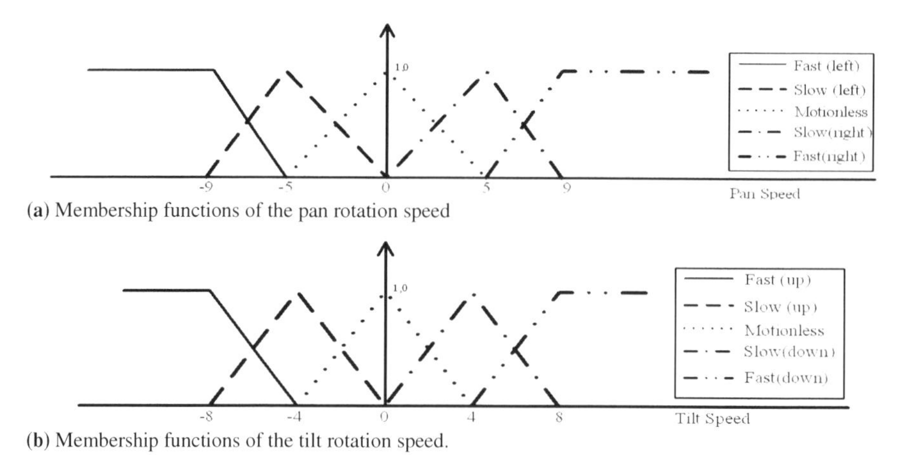

(**a**) Membership functions of the pan rotation speed

(**b**) Membership functions of the tilt rotation speed.

Fig. 5 The Fuzzy set of pan rotation speed and tilt rotation speed

2.4.3 Fuzzy Inference Engine and Decision-Lookup Table

In this design, *max-min* composition is used for the inference engine to obtain the fuzzy set describing the fuzzy value of the control output [7, 8]. The center of gravity defuzzification method is used to obtain the control output. To meet the real-time requirement, Fuzzy control rules, *max-min* composition and defuzzification method can be built into a decision-lookup table off-line. In real-time mode, the controller uses decision-lookup tables to decide the rotation control according to the input quantization level of the center position.

2.5 System Overview

The developed system consists of an octagon platform with eight infrared sensors, a PTZ camera installed in a servomechanism, a USB 2.0 image grabber for acquiring the image, a special-designed sensor circuit and a personal computer. The special-designed circuit is developed with 8051 microcontroller. The circuit detects infrared sensors. The circuit sends the sensory information to computer by RS232C if it detects some intruder.

In the remote site, the user can send commands to control the PTZ camera by Internet. The local computer makes the desired response after it receives the command. The communication between these two computers is not peer-to-peer. Database communications are used to achieve remote control operation. Apache HTTP Server is installed in our computer. The database management system is MySQL [9]. For the communication, we create a specified field in a table. The control code is the data of the field. The user can modify the data of the field from a remote site. The local computer sends commands to PTZ camera according to the field data. The PTZ camera will rotate its lens to aim at the specified location.

3 Experimental Results

Three practical experiments have been conducted to evaluate the detection and tracking performance of the developed system. One experiment is designed to test the detection and the others are used to test the tracking. Fig. 6(a) depicts the display of image-based tracking system. The user can set manual or automatic mode by clicking on the option buttons in the box of upper right corner. He can start or stop the tracking operation by clicking on the command buttons near the upper right corner. The upper center region shows the captured image and the upper left region shows the tracking result. The lower left corner shows the dimension of the tracked target. The command buttons of PTZ camera control are placed in the lower center box. The command buttons of servo motor control are placed in the lower right box. The experiments are performed in our laboratory, as shown in Fig. 6(b). The proposed system is installed in the table of lower left region. The circle depicts the detection range of infrared sensor array.

First experiment is designed to test the performance of intruder detection. In the beginning, PTZ camera aims to the left, the view of PTZ camera is shown in Fig. 7(a). When a man enters into the detection range of sensors, the system detects him and the servomechanism rotates the PTZ camera to aim at him, as shown in Fig. 7(b). Once the system aims at him, the image of moving object is segmented by difference method for tracking initialization, as shown in Fig. 7(c).

Second experiment is designed to verify the performance of tracking algorithm. A man walks from right to left. The PTZ camera has the ability to adjust pan or tilt angle to ensure that the image of the man is within in the frame. The experimental results are shown in Fig. 8(a)-(d). The PTZ camera adjusts the pan (or tilt) angle to track the moving man successfully.

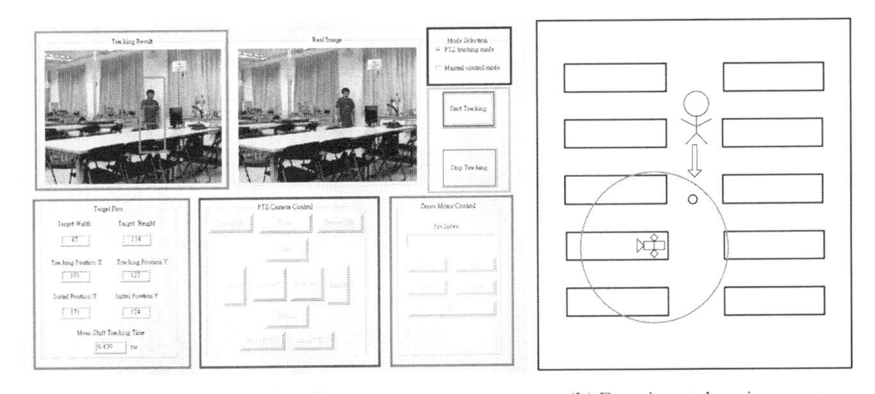

(a) The display of image-based tracking system (b) Experimental environment

Fig. 6 Experimental setup

Fig. 7 Results of tracking initialization. (a) The view captured during the system has not yet detected the intruder. (b) The view captured after the system detects the intruder and the system rotates the PTZ to aim the intruder. (c) Moving image has been segmented for tracking initialization.

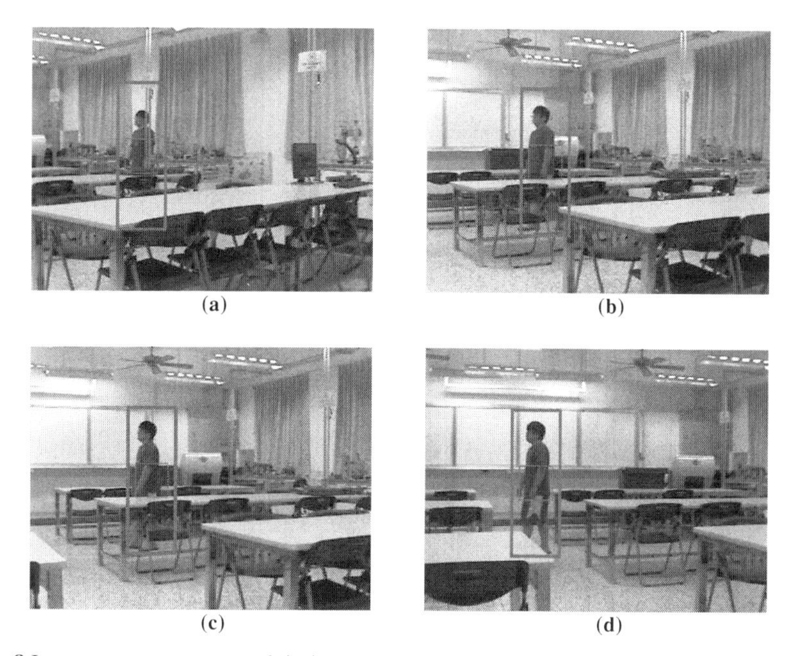

Fig. 8 Images sequence captured during a man walks from right to left.

Final experiment is carried out to test the performance of tracking algorithm. A man moves from left to right. The PTZ camera can rotate its lens to follow the path of the moving man. The rotation can make that the image of the man is within in the frame. The experimental results are shown in Fig. 9(a)-(d). The PTZ camera successfully adjusts the pan or tilt angle to track the moving man.

A video clip of image processing sequence for automatically tracking can be found at http://me.must.edu.tw/tracking/NSC98_MUST1.avi. A video clip of manual operation can be found at http://me.must.edu.tw/tracking/NSC98_MUST2.wmv.

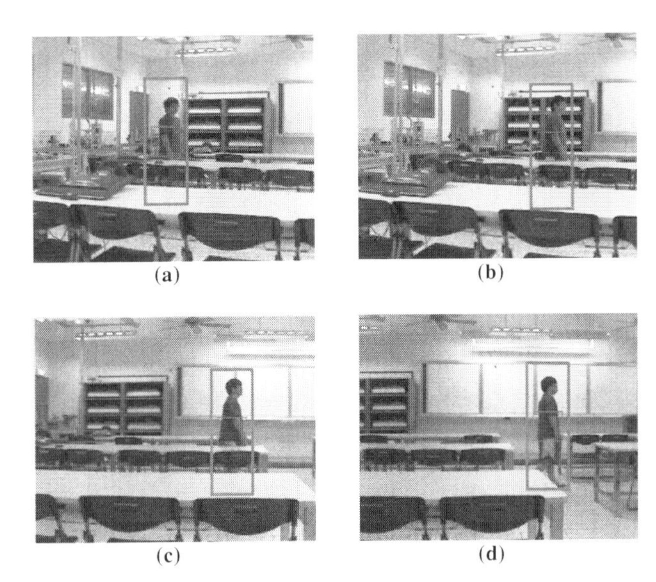

Fig. 9 Images sequence captured during a man walks from left to right

4 Conclusions and Future Works

A novel design of image-based monitoring and tracking system has been developed. An initialization method has been designed for monitoring and tracking application. The system uses eight infrared sensors to detect the intruder. Servomechanism rotates the PTZ camera according to the sensory information. Mean shift method and Fuzzy rules are employed to track the intruder. Experimental results reveal that the proposed method successfully detects and tracks the intruder.

The system only has one camera to track the person, so it tracks only one person. If more than one intruder exist, the one that has been detected or scanned first by the system will be tracked. The system scans the intruder from POS1 to POS16. If the intruder has been detected then the system will not check the rest region. Color feature and image tracking will be initiated by the detected region.

Several directions deserve further study in the future. On one hand, color feature influences the accuracy of tracking operation. Color feature is sometimes affected by the light source. Hue-Saturation-Value color model or other color models are invariant to a change in the illumination conditions. The proper selection of color model can increase the accuracy. On the other hand, in order to increase the performance of image tracking, Pan and tilt rotation control of the camera can be directly driven by a special-designed servomechanism. In this design, the speed control has some limitation because PTZ camera is controlled by a command sent from computer from a RS232C serial communication.

Acknowledgments. This work was supported by NSC of Taiwan under Grant No. NSC 98-2211-E-159-012.

References

1. Everts, I., Sebe, N., Jones, G.A.: Cooperative Object Tracking with Multiple PTZ Cameras. In: IEEE 14th International Conference on Image Analysis and Processing (2007)
2. Hu, F.W., Tan, T., Wang, L., Maybank, S.: A survey on visual surveillance of object motion and behaviors. IEEE Trans. Syst., Man, Cybern. C, Appl. Rev. 34(3), 334–352 (2004)
3. Song, K.T., Tai, J.C.: Image-Based Traffic Monitoring with Shadow Suppression. Proceedings of the IEEE 95(2), 413–426 (2007)
4. Comaniciu, D., Meer, P.: Mean shift: A robust approach toward feature space analysis. IEEE Trans. Pattern Anal. Mach. Intell. 24(5), 603–619 (2002)
5. Comaniciu, D., Ramesh, V., Meer, P.: Kernel-based object tracking. IEEE Trans. Pattern Anal. Mach. Intell. 25(5), 564–575 (2003)
6. Bhattacharyya, A.: On a measure of divergence between two statistical populations defined by their probability distributions. Bull. Calcutta Math. Soc. 35(1), 99–110 (1943)
7. Song, K.T., Tai, J.C.: Fuzzy Navigation of a Mobile Robot. In: Marks II, R.J. (ed.) Fuzzy Logic Technology and Application. IEEE Press (1994)
8. Lin, C.T., Lee, C.S.G.: Neural Fuzzy Systems: A Neuro-Fuzzy Synergism to Intelligent Systems. Prentice-Hall, New Jersey (1996)
9. Gilmore, W.J.: Beginning PHP and MySQL From Novice to Professional, 4th edn. Apress, New York (2010)

Face Recognition Based on Grain-Shape Features

Weijun Dong, Mingquan Zhou, and Guohua Geng

Abstract. Traditional face recognition method was mainly dependent on single vision features such as color, texture, shape and so on. So the recognition result was not satisfactory.To cure this problem, propose a new face recognition method to get the texture features and shape features of face image based on wavelet transform. The corresponding shape and texture features are then processed by linear discriminant analysis. The PIE face database was used to test the proposed method. The experiment result shows that the proposed method has better recognizing effect and is not sensitive to the pose and expression of human faces. Experimental result also shows that the method is superior to the PCA and DCT method.

Keywords: Face Recognition, Texture Features, Shape Feature, Feature Abstraction.

1 Introduction

Compared to other biological recognition technologies, face recognition is advanced in easiness, intuitiveness, covertness and no special requirements on collection equipment. Face recognition has become one of the hottest research field of image analysis and understanding after 50 years' development[1],[2]. Though face recognition has made great progress, it still faces tremendous challenges by taking advantage of computer for some problems such as non-rigidness, changing appearance with the ages and being covered by other objects.

Face features have great influence on the choice of core recognition ways. Shape feature is one of the key information recognized by human vision system[3]. As a stable feature, it will not change with the change of the image color. The retrieval by using shape features can be more correct and efficient. Texture feature is an overall feature, which describes the apparent characteristics of the image or the object in the zone of the image [4]. Texture feature is characterized by being unchanging

Weijun Dong · Mingquan Zhou · Guohua Geng
College of Information Science and Technology, Northwest University, 710127
Xi an, China

Z. Qian et al. (Eds.): Recent Advances in CSIE 2011, LNEE 125, pp. 55–61.
springerlink.com

while whirling and anti-noise. It is an effective retrieval way to search texture images different in thickness and density. Applying both the shape and texture information of human face to human face image can be an effective choice to improve the recognition ratio.

Wavelet has become an instrument with powerful signal thanks to favorable time-frequency[5]. It has been widely applied into the field of image process and model recognition. In view of the limitations of traditional face recognition algorithm, a new method of face recognition based on wavelet change has been proposed, namely, to acquire texture-shape characteristics by breaking down wavelet of face image and linear discriminant analysis to achieve classified recognition of human face. Recognition algorithm can recognize human face of high efficiency and overcome the negative effect on face recognition brought by the robustness of change of gesture, light, facial expression, accessories and background.

2 Human Face Detection

To fix the position of facial features is the basis of human face recognition. In fact, the position of eyes and the relative distance is constant for the most of the people and the basic criterion of further fixing the position of other face features. In the face image analysis, we can find that the eyes area is darker than other areas, in other word, the grey scale of the former is lower than that of the latter. This feature can be used to fix the position of eyes[6]. To be specific, based on eight neighboring threshold value, we find all the suspect pixels in the eyes area in face image; then, reject sparse pixels and connect suspect neighboring pixels; at last, we screen all the suspect eyes areas according to the size of the eyes[7].To set grey scale image with the size of:

$N_1 \times N_2$: $p(x,y), x \in [0, N_1 - 1], y \in [0, N_2 - 1], p(x,y) \in [0,1]$. Let us suppose that the length and width of the eyes are len and wid respectively and compare each pixel with the average grey scale of its neighboring area. If the pixel value is lower than the average grey scale, the pixel is thought to be darker than its neighboring area. thus, it is called suspect eyes pixel, which is in accordance with the pixel of formula:

$$p(x,y) < \alpha \cdot average(P, x_i, y_i, len_i, wid_i) \quad (1)$$

In the above formula, α is comparability factor. Its value is about 0.9 acquired by experiment. $i = 1,2,3,\cdots 8$, (x_i, y_i) is in the upper-left position in the neighboring area i. len_i is the length of the neighboring area i while wid_i is its width and $average(P, x_i, y_i, len_i, wid_i)$ is its average grey scale. The formula is as follows:

$$average(P, x_i, y_i, len_i, wid_i) = \frac{\sum_{m=x_i}^{x_i+len_i-1} \sum_{n=y_i}^{y_i+wid_i-1} P(m,n)}{len_i \times wid_i} \quad (2)$$

By revolving and zooming the image according to suspect eyes area and face model, we can get normalized suspect face area. We discriminate suspect face image with model match technology. Standardized correlation coefficients represent the two similar images[8]. T is set to be an image model, while I_T is an image as big as T. The formula of Standardized correlation coefficients of I_T and T is as follows:

$$C_N(I_T) = \frac{< I_T T > - < I_T >< T >}{\sigma(I_T)\sigma(T)} \tag{3}$$

In formula, <> represents the average, $I_T T$ represents the multiplication of the corresponding pixel value of I_T and T, and σ represents the standard deviation of pixels value in the image area. If $C_N(I_T)$ is greater than a certain threshold value, the image is regarded as a human face image.

3 Abstract the Features

3.1 Texture Features

To conduct three-level wavelet decomposition on the original image by Gabor wavelet. The definition of a two-dimensional Gabor wavelet is as follows:

$$g(x, y) = \frac{1}{2\pi\sigma_x\sigma_y}\exp[2\pi jwx]\exp\left[-\frac{1}{2}\left(\frac{x^2}{\sigma_x^2} + \frac{y^2}{\sigma_y^2}\right)\right] \tag{4}$$

μ and σ represent average value and standard variance of energy distribution of wavelet transform coefficients. The scale number is M ; the orientation is N. The texture features of the image can be represented by the average value and standard variance of energy distribution in the frequency channel on each decomposition level. Therefore, the horizontal, vertical and diagonal distribution of the image based on different resolution ratio is shown. The description of the texture features is as follows:

$$g = \lfloor \mu_{00}\sigma_{00}\mu_{01}\sigma_{01}\cdots\mu_{(M-1)(N-1)}\sigma_{(M-1)(N-1)} \rfloor \tag{5}$$

3.2 Shape Features

If the integral value of $\theta(x, y)$ on the surface is 1, and it is convergent to 0 at infinity, then, $\theta(x, y)$ is defined as two-dimensional smooth function.

According to gradient definition, if the gradient mold reaches the maximum value on (x, y), then, $f * \theta_{2^j}(x, y)$ has the maximum directional derivative.

So, at this point, there is a mutation happened on $f(x, y)$, and the direction of the mutation is along that of the gradient. Thus, (x, y) can be regarded as the edge point of the image $f(x, y)$. In other words, the spots with dramatic changes on $f * \theta_{2^j}(x, y)$ is those spots with maximum value of mode $M_{2^j} f(x, y)$ along the gradient direction $A_{2^j} f(x, y)$.We can only record the position of those spots with maximum mode value, corresponding mode $M_{2^j} f(x, y)$ and the angle $A_{2^j} f(x, y)$.

We conduct adaptive threshold processing on coefficients in the wavelet transform domain to obtain multi-scale exterior joints. Adaptive threshold processing retains the feature information of the original image and has good effect when dealing with low-quality images as well. Generally, moment is the moment of an area. But the moment here refers to that of curve. The formula of area moment cannot be used to calculate curve moment. Therefore, we revise the calculation formula of invariant moment to meet the requirement of boundary moment.

For the area $f(x, y)$, if the scale is $x' = ax$, $y' = ay$, its moment should be multiplied by a^p, a^q, a^2. Factor a^2 is the result of target change caused by scale change. The central moment of $f(x', y')$ becomes $u'_{pq} = u_{pq} * a^{p+q+2}$.

The definition of normalized moment is:

$$\eta_{pq} = \frac{u_{pq}}{(u_{00})^r} \tag{6}$$

$\eta'_{pq} = \eta_{pq}$ is the condition for scale invariant of normalized moment. From $\frac{u_{pq} * a^{p+q+2}}{(u_{00} * a^2)^r} = \frac{u_{pq}}{(u_{00})^r}$, we can obtain $r = \frac{p+q+2}{2}$. However, regarding curve, a is the change factor leading to circumference change caused by scale change. Under this circumstance, the central moment $u'_{pq} = u_{pq} * a^{p+q+1}$. To ensure the scale invariant of normalized moment, we should get the formula $\eta'_{pq} = \eta_{pq}$, namely $\frac{u_{pq} * a^{p+q+1}}{(u_{00} * a^2)^r} = \frac{u_{pq}}{(u_{00})^r}$, from which, we obtain:

$$r = p + q + 1 \tag{7}$$

4 Experiment

PIE face database, created by Carnegie Mellon University in the US, is an important test set in human face recognition field. It consists of more than 41368 facial images with various pose, lighting and expressions of 68 volunteers. The images with variations of pose and lighting were collected in strictly controlled environment. We carry out the experiment on PIE face database to verify the algorithm. Each person select ten images as the training sample on random basis and other images are used for test. Each set of experiment is to be repeated for ten times. Then, we take the mean value as the recognition result. Test hardware platform is standard PC of P42.08G/512MB.

Table 1 The Recognition rate comparison between different Decomposition series

Decomposition series	Recognition rate	Mean time(ms)
1	91.5	82
2	93.2	95
3	98.6	107
4	98.7	143
5	98.7	179

When the face image is decomposed by wavelet, different decomposition series will affect the recognition rate and speed. The fewer the decomposition series are, the quicker the decomposition is but the lower the recognition rate is; on the contrary, the more the series are, the more information included on different scale, the higher recognition rate is but the slower the decompostion is. Table 1 is the recognition and time-consumption comparison between different decompostion series by the new method. We can find that the recognition rate has been improved and time-consumption has been increased with growing decompostion series. Taking time and performance into consideration, we regard there level decomposition as the suitable choice.

Table 2 The Recognition rate comparison between different methods

methods	Recognition rate(%)	Mean time(ms)
DCT (36 coefficient)	94.9	89
DCT+LDA (36 coefficient)	96.1	113
PCA (32dimension)	90.2	105
New method (three level decomposition)	98.6	107

Tradition face recognition method based on DCT chooses low frequency DCT coefficient as the recognition feature. According to the method, the low frequency coefficient in the top left corner of the square in DCT coefficient matrix is directly applied to the classification[9]. With regard to the method of combing DCT and

LDA, first, we reduce the dimension of the image; then conduct discrete cosine transform on the image; at last, we select the low frequency coefficient in the top left corner of the square in DCT coefficient matrix to make linear discriminant analysis and extract recognition features[10]. On the basis of wavelet analysis, the new method is applied to achieve face classification by linear discriminant analysis in the area of wavelet transform. Table 2 is the comparison between the new method and other recognition methods. We can find that compared to traditional methods such as DCT, the combination of DCT and LDA, and PCA, the new method is advanced in high recognition rate, low time-consumption and better overall performance.

5 Conclusion

Human face recognition is one of the important research subject in image analysis and understanding field. It is difficult to recognized human face automaticly due to some problems such as non-rigidness, changing appearance with the ages and being covered by other objects. Since sigle visual feature such as color, texture, shape and space relation includes only partial information of the iamge, the effect is not ideal.

On the contrary, method of extracting different features can provide more information.Various features of face images are complementary. Taking the above, consideration, put forward a new face recognition method based on wavelet transform. To be specific, we try to acquire texture-shape characteristics by breaking down wavelet of face image and linear discriminant analysis to achieve recognition of human face. The method overcomes the limitations brought by the single visual features and fully utilizes the overall visual features of face iamge. It is in accordance with human eyes recognition mode. Simulation experiment was carried out on the new method to verify its great recognition effect, fast arithmetic speed , high recognition rate and nice robustness.

Acknowledgments. The research is supported by the National Natural Science Foundation Project (60873094), National Post Doctoral Foundation (2008043124), Natural Science Foundation Project of Shaanxi Province (2009JM8004-01) and Natural Science Foundation Project of Education Department of Shaanxi Provincial Government (09JK739).

References

1. Deng, W., Hu, J., Guo, J., et al.: Emulating biological strategies for uncontrolled face recognition. Pattern Recognition 43(6), 2210–2223 (2010)
2. Liu, Z.M., Liu, C.J.: A hybrid color and Frequency Features method for face recognition. IEEE Transactions on Image Processing 17(10), 1975–1980 (2008)
3. Kennedy, G.J., Orbach, H.S., Loffler, G.: Global shape versus local feature: An angle illusion. Vision Research 48(11), 1281–1289 (2008)
4. Fazekas, S., Amiaz, T., Chetverikov, D., et al.: Dynamic texture detection based on motion analysis. International Journal of Computer Vision 82(1), 48–63 (2009)

5. Cheng, Z.: Wavelet anlysis and applications, pp. 156–223. Xi'an Jiaotong University Press, Xi'an (1998) (in Chinese)
6. Feng, G.C., Yuen, P.C.: Multi cues eye detection on gray intensity images. Pattern Recognition 34(5), 1033–1046 (2001)
7. Liu, X.-D., Chen, Z.-Q.: Research on Several Key Problems in Face Recognition. Journal of Computer Research and Development 41(7), 1075–1080 (2004)
8. Bruneli, R., Poggio, T.: Face recognition: Features versus templates. IEEE Trans. on Pattern Analysis and Machine Intelligence 15(10), 1042–1052 (1993)
9. Hafed, Z.M., Levine, M.D.: Face recognition using the discrete cosine transform. International Journal of Computer Vision 43(3), 167–188 (2001)
10. Zhang, Y.K., Liu, C.Q.: A Novel Face Recognition Method Based on Linear Discriminant Analysis. Journal Infrared Millimeter and Waves 22(5), 327–330 (2003)

Fault Identification for Industrial Process Based on KPCA-SSVM

Yinghua Yang, Qingchao Yu, and Shukai Qin

Abstract. Industrial fault identification is significant for finding fault reason and remedying the potential safety problems. As kernel principal components analysis (KPCA) has excellent performance in nonlinear data processing, a kind of fault identification method is proposed based on KPCA and simple support vector machine (SSVM). KPCA was applied to choose the nonlinear principal component of the model input data space, and SSVM was applied to establish fault identification modeling, which could not only enhance the efficiency of calculation, but also could improve the fault identification ability. The proposed KPCA-SSVM was applied to the Tennessee Eastman Process (TEP). Simulation indicates that this method features high learning speed and good identification ability compared with SVM, PCA-SSVM and KPCA-SVM, and is proved to be an efficient fault identification modeling method.

Keywords: KPCA, SSVM, fault identification, modeling.

1 Introduction

In the Chemical Industry, the informal conditions make the products unacceptable, even lead to major accidents. So it is very important to enhance the production benefit and eliminate the safety risks with faults detection and diagnosis technology. The technique of data driving can extract the worthwhile information from great quantities of data, and improve the security and reliability of process. PCA is a data driving technology of widely used, such as the fault detection of industrial boiler system based on PCA [1]. When the data is nonlinear, PCA does not work well. But KPCA is suitable method for nonlinear data processing. In reference [2], the KPCA was used to detect faults in the sewage treatment plant and gained a good result.

Yinghua Yang · Qingchao Yu · Shukai Qin
College of Information Science and Engineering, Northeastern University, Shenyang, China
{Yhyang,qinshukai}@mail.neu.edu.cn, yuqingchao333@126.com

Z. Qian et al. (Eds.): Recent Advances in CSIE 2011, LNEE 125, pp. 63–69.
springerlink.com

Once faults are detected, we further want to determine the fault reasons and remove faults. Data classification methods are often used to identify faults. At present linear classification methods include of KNN, PCA, FDA and LSDA algorithm, neural network and SVM belongs to nonlinear methods. SVM is a new type of statistical learning method [3], it has good generalization ability and can overcomes the dimension disasters and local extreme problem, so it has been widely used in data classification. However, for SSVM adopts iterative working set support vector algorithm, so it is more efficient and fast [4]. This method can not only improve the accuracy of the data processing, and also improve the data processing speed.

In this paper, a KPCA-SSVM algorithm is proposed to identify faults. Firstly, in order to eliminate the correlations of datasets, the KPCA is used to extract the features. These new features as the inputs of SSVM are trained by SSVM to establishing the fault recognition model.

2 The Theory of KPCA and SSVM

2.1 *The Principle of KPCA*

The idea of KPCA is firstly map the original inputs into a high dimensional feature space using a nonlinear kernel function, and then calculate the kernel matrix's eigenvalues and eigenvectors in the high dimensional feature space. However the direction of eigenvectors projection of the inputs is the linear combination of the kernel matrix. For this reason, the computation is greatly simplified. We can establish the principal components model of arbitrary precision, due to enough principal components in the feature space [5, 6].

Assuming the sample data is $X = \{x_1, x_2, \cdots x_l\} \in R^N$, φ is the nonlinear mapping function and F is the high dimensional feature space. The data satisfies $\sum_{j=1}^{l} \varphi(x_j) = 0$. In the F space, the covariance operator is

$$C = \frac{1}{l} \sum_{j=1}^{l} \varphi(x_j) \varphi(x_j)^T \tag{1}$$

The principal components are computed by solving the non-zero eigenvalue and the corresponding eigenvectors of covariance matrix C:

$$\lambda V = CV = [\frac{1}{l} \sum_{j=1}^{l} \varphi(x_j) \varphi(x_j)^T]V = \frac{1}{l} \sum_{j=1}^{l} \langle \varphi(x_j), V \rangle \varphi(x_j) \tag{2}$$

where $\langle x, y \rangle$ denotes inner product. Multiplying both sides of Eq.2 by $\varphi(x_k)$, we obtain the following equation:

$$\lambda(\varphi(x_k), V) = \langle \varphi(x_k), CV \rangle, \quad k = 1, 2..., l \tag{3}$$

Therefore, the V vector can be linearly expended by

$$V = \sum_{i=1}^{l} \alpha_i \varphi(x_i) \quad (4)$$

Substituting (4) into (3), thus leads to the following equation:

$$l\lambda\alpha = K\alpha \quad (5)$$

where $\alpha = [\alpha_1, \alpha_2 \cdots \alpha_i]^T$, λ_k is the eigenvalue and α^k is the eigenvector, K is the l dimensional kernel matrix. The value is $K_{i,j} = \langle \varphi(x_i), \varphi(x_j) \rangle$.

Normalizing the eigenvector V^k, we can get the kth principal component t_k :

$$t_k = \langle V^k, \varphi(x) \rangle = \sum_{j=1}^{l} \alpha_i^k K(x_i, x), k = 1, 2..., l \quad (6)$$

However, the sample data in real do not always satisfy $\sum_{j=1}^{l} \varphi(x_j) = 0$, and then the K in Eq.5 is replaced by:

$$\bar{K} = K - L_{l\times l}K - Kl_{l\times l} + L_{l\times l}KL_{l\times l} \quad (7)$$

where the matrix $L_{l\times l} = (\frac{1}{l})_{l\times l}$.

2.2 *Support Vector Machine*

Support Vector Machine (SVM) has gained prominence in the field of machine learning and pattern classification. Given a training sample $\{x_i, y_i\}_{i=1}^{n}$, where $x \in R^n$, $y \in \{+1, -1\}$, and n is the sample number. It is possible to determine the hyperplane $(w \cdot x) + b = 0$ that separates the given data when two classes are linearly separable. The optimal hyperplane can be obtained by the following optimization problem:

$$R(w, \xi) = \frac{1}{2}\|w\|^2 + C\sum_{i=1}^{l} \xi_i \quad s.t. \quad y_i(w \cdot x + b) \geq 1 - \xi_i \quad \xi_i \geq 0 \; i = 1, 2, \cdots, l \quad (8)$$

where C(C>0) is the penalty factor and ξ_i is the slack variable. With the Lagrange multiplier method, we can gain the dual form of Eq.8, which has the following form:

$$\max W(\alpha) = \sum_{i=1}^{l} \alpha_i - \frac{1}{2}\sum_{i,j=1}^{l} \alpha_i \alpha_j y_i y_j (x_i \cdot x_j) \quad s.t. \sum_{i=1}^{n} \alpha_i y_i = 0, \; 0 \leq \alpha_i \leq C \; i = 1, 2, ..., n \quad (9)$$

In these optimal solutions, some solutions are not zero. The corresponding training samples of theirs are support vectors.

For the non-linear problem, we solve it by extending the simple data in a high dimensional feature space with the map Φ , and then construct a hyperplane in that high dimensional feature space. Substituting $K(x_i, x) = (\Phi(x_i) \cdot \Phi(x))$ into (9), we can get the nonlinear support vector machine's decision function:

$$f(x) = \text{sgn}(\sum_{i=1}^{l} a_i y_i K(x_i, x) + b) \tag{10}$$

2.3 Simple SVM

The idea of Simple SVM algorithm is that some processes are mixed into every separated point, through adding data points into the solution. This algorithm reduces the computational time greatly than the traditional SVM algorithm. The new constraint condition of (9) is shown as follows:

$$\begin{aligned} &\sum_{i=1}^{n} \alpha_i y_i = 0 \\ &s.t. \quad 0 < \alpha_i < C \quad i \in I_s; \alpha_i = 0 \; i \in I_o; \alpha_i = C \quad i \in I_c \end{aligned} \tag{11}$$

The Simple SVM algorithm decomposes the database into three groups (the working set I_s , the inactive set I_o and the bounded set I_c). Only the working set is needed to solve the optimization problem, if all sets are known. Meanwhile, this algorithm detects the correlation of all sets. If they are not related, all sets should be updated. The particular iterative steps are given in [7].

3 Fault Diagnosis Procedure Based KPCA and SSVM

In this paper, the KPCA-SSVM algorithm is used to recognize and classify the faults, and the polynomial basis function $K(x, y) = (\theta + x \cdot y)^d$ is chosen as the kernel function of SSVM. d is the polynomial order variable.

The particular fault diagnosis procedure can be summarized as follows:

1) Decompose the fault data into two groups (the training set and the test set), and establish one-to-one relationship between fault vector and fault type.
2) Normalize $\overline{X} = [X - (1 \ 1 \cdots 1)^T M] \text{diag}(1/s_1, 1/s_2, \cdots, 1/s_m)$; $M = [m_1 \ m_2 \cdots m_m]$ and $s = [s_1 \ s_2 \cdots s_m]$ are the mean and standard deviation of each variable.
3) Calculate the kernel principal components using Eq.1~Eq.7.
4) Select k principal components as the input samples of SSVM and t group samples as the training set.
5) Determine the normalized parameter set and kernel parameter set.
6) Combine some parameters from two parameter sets to train SSVM.
7) Select the best parameters combination to build the SSVM model, according to the lowest fault classification rate.
8) Predict the faults using the built model.

4 Application of TEP

The Tennessee Eastman process (TEP) is an open benchmark chemical process for researchers. In order to verify the feasibility and efficiency of the proposed algorithm, three faults including faults 4, 9 and 11 of TEP are chose as simulation data.

For each fault, the datasets are divided into the training and testing datasets. In the lab experiments, the training and testing datasets are generated and contain 480 and 800 observations respectively. Each observation contains 52 process variables. However, only variable 9 and variable 51 are important in partitioning the three faults in engineering perspective.

Fig. 1 shows that fault 4 and fault 9 are separated briefly using variable 9 and variable 51, however, these two faults and fault 11 are partially overlapped. So it is a certain difficulty to recognize these three faults.

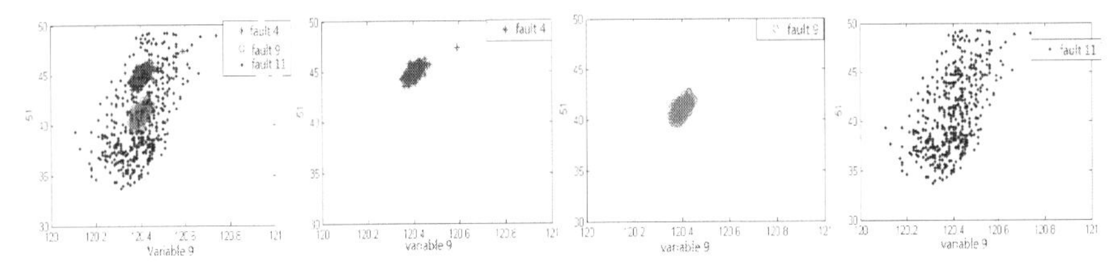

Fig. 1 Bivariate plot of variables 9 and 51 for classes in the training dataset

Not only did we compare the KPCA-SVM algorithm with SVM algorithm, but also we compared the similar SVM algorithm proposed in reference [8]. The σ of RBF kernel, d and θ of poly kernel and regularization parameter C are set as 0.3, 7, 0.1 and 920000 in this experiment via 1-Vs-all respectively. The results are shown in Table 1.

Table 1 Comparison of the classification results using SVM, SSVM, KPCA-SVM, and KPCA-SSVM

	SVM	SSVM	KPCA-SVM	KPCA-SSVM
Percentage of support vector	7,40,53	15,66,73	56,85,90	17,10,19
Regularization parameter (C)	920000	1000	60	1000
Step	-	50	-	690
Misclassification percent	44	40	39	32

From Table 1, the overall misclassification rate of KPCA-SSVM algorithm is only 32%, significantly lower than other methods. But the overall misclassification rate of all methods ranges from 32% to 44%. However, the overlaps rate is 6%, that is to say this model does not obtain the data features completely. In order to simplify the problem, key variables were selected using the contribution charts.

The contribution charts of fault 4 and fault 11 are shown in Fig.2, it indicates that variable 51 is the only important variable for fault 4, while variables9 and 51 are both important variables for fault 11. There is no larger contribution variable in contribution charts of fault 9.

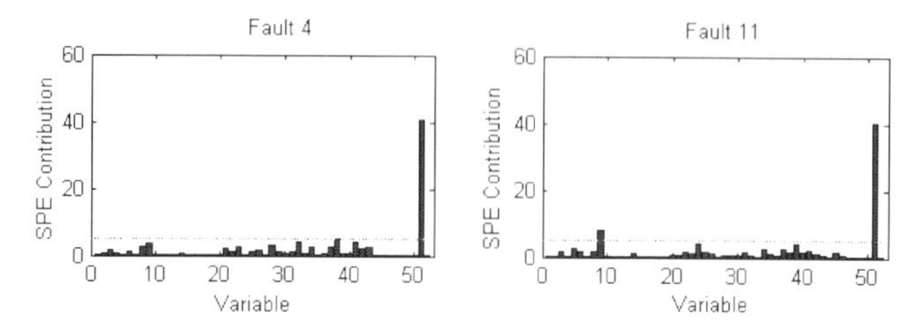

Fig. 2 The SPE contribution charts of fault 4 and fault 11

So we extract variable 9 and variable 51 as the key variables to build the new models of SVM, KPCA-SVM and KPCA-SSVM. The simulation results are shown in Table 2.

Table 2 Comparison of the classification results with selected variables

	SVM	KPCA-SVM	PCA-SSVM	KPCA-SSVM
Percentage of support vector	0.4,25,25	30,21,8	15,56,21	9,9,10.8
Regularization parameter (C)	920000	2	38	1000
Step	-	-	-	690
Misclassification percent	6.5	6.0	15	6.0

In the simulations, the parameters are similar to the simulations of table 1. From table 2, three faults are divided well via the new model. The proposed algorithm improves the accuracy and the computational speed.

5 Conclusions

In this paper, a new fault diagnosis method (KPCA-SSVM) is proposed, and the commendable results are obtained with applications in TEP process. The KPCA algorithm is a powerful approach of extracting nonlinear features. The SSVM algorithm can put the low dimensional nonlinear inputs mapping to the high dimension linear outputs, in addition, this method increases the fault recognition accuracy through the iterating working sets. The experimental results show that the accuracy of the proposed algorithm is significantly better than other methods.

Acknowledgments. This research is sponsored by the Fundamental Research Funds for the Central Universities (N100404018).

References

1. Misra, M., Yue, H.H., Qin, S.J., Ling, C.: Multivariate process monitoring and fault diagnosis by multi-scale PCA. Computers and Chemical Engineering 26, 1281–1293 (2002)
2. Lee, J.M., Yoo, C.K., Choi, S.W.: Nonlinear process monitoring using kernel principal component analysis. Chemical Engineering Science 59, 223–234 (2004)
3. Vapnik, V.N.: The Nature of statistical learning Theory. Springer, New York (1995)
4. Vishwanathan, S.V.N., Smola, A.J., Murty, M.N.: SimpleSVM. In: Proceedings of the Twentieth International Conference on Machine Learning, Washington, DC (2003)
5. Huang, Y.W., Peng, T.G.: Fault diagnosis for nonlinear dynamic system using kernel principal component analysis. Journal of System Simulation 17(9), 2291–2294 (2005) (in Chinese)
6. Scholkopf, B., Smola, A., Muller, K.R.: Kernel principal component analysis. In: Advances in Kernel Method-Support Vector Learning, pp. 327–352. MIT Press, Cambrighe (1999)
7. Loosli, G., Canu, S., Vishwanathan, S.V.N., Smola, A.J.: Invariance in Classification: an efficient SVM implementation. In: International Symposium on Applied Stochastic Models and Data Analysis (2005)
8. Chiang, L.H., Kotanchek, M.E., Kordon, A.K.: Fault diagnosis based on Fisher discriminates analysis and support vector machines. Computers and Chemical Engineering 28, 1389–1401 (2004)

High Accuracy Temperature Control Research on Charge Stable Colloidal Crystals

Shangqi Gao, Hao Yang, Zhibin Sun, Yuanda Jiang, Guangjie Zhai, and Ming Li

Abstract. Colloidal crystal phase dynamics is one of the hotspots on Condensed Matter Physics. Kossel-line diffraction method is an important way to measure the inner structure of colloidal crystal. The research on the model of colloidal crystal phase dynamics changing as the temperature will provide scientific basis to the preparation of colloidal crystal materials, especially in the condition of microgravity. Achieving high accuracy temperature control on the suspension of colloid crystal is a key work on the research of colloidal crystal phase dynamics. A proportional-integral-derivative (PID) control method based on pulse-width modulation (PWM) is proposed. Then, system identification on the heating system of the sample solution is used to get transfer function, and Hooke-Jeeves pattern search method is used to get optimal parameters of the PID controller. We obtain best temperature control accuracy (±0.1°C) as well as simulation results, after actual temperature control system uses the optimal parameters.

Keywords: Colloidal Crystals, Kossel-line Diffraction, PWM, PID, System Identification, Temperature Control.

1 Introdution

Colloidal crystal phase dynamics is one of hotpots on the field of condensed matter physics [1-5], especially, the research under microgravity condition. Diffraction method is useful to observe the inner structure of colloidal crystal which size is submicro.1935, Kossel observed a diffraction cone from a copper single crystal using fluorescent x-ray[6]. Clark[7] and Ackerson[8] used Kossel-line diffraction method on colloidal crystals. Since the diameter of colloidal crystal is proportion to visible wavelength, we use 473nm laser beams to illuminate the sample vertically. When the Bragg condition is established, Kossel-line will form on the imaging screen.

Shangqi Gao · Hao Yang · Zhibin Sun · Yuanda Jiang · Guangjie Zhai
Center for Space Science and Applied Research, Chinese Academy of Sciences, Beijing, China

Ming Li
Institute of Physics, Chinese Academy of Sciences, Beijing, China

Z. Qian et al. (Eds.): Recent Advances in CSIE 2011, LNEE 125, pp. 71–77.
springerlink.com

In order to research the charge stable colloidal crystal change with temperature under microgravity condition, high accuracy temperature control is necessary. A PID controller based on PWM is a very efficient way. Firstly, system identification based on correlation coefficient and recursive extended least squares is used to describe mathematical tools and algorithms of heated module that build dynamical models from measured data. Secondly, Hooke-Jeeves method is used to determine the optimal parameter of PID controller. Finally, the real temperature control system uses the optimal parameter to obtain best temperature control accuracy (±0.1°C) as well as simulation result.

2 Theory and Method

2.1 Principles of Optics

1984, Carlson[9] introduced the theory of Kossel line diffraction method on colloidal crystals. According to Bragg diffraction equation $n\lambda = 2d\sin\theta$, where λ is the wavelength of light which illuminates the sample. d is the lattice spacing. θ is the Bragg angle. When the angle of scattered light is equal to Bragg angle θ, the light can be diffracted and cannot propagate through the lattice. Therefore dark Kossel rings are projected onto a target screen while the bright ones are broadened relatively due to multiple Bragg diffraction (see Fig1).

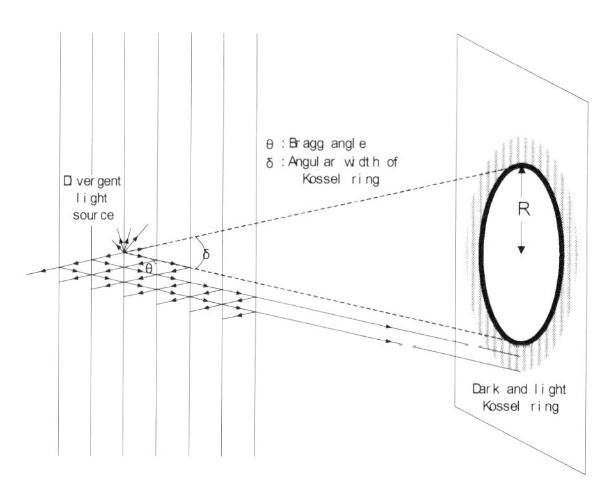

Fig. 1 Schematic diagram of Kossel-line diffraction principle

We can describ the Bragg angle as $\tan\theta = \tan\delta/2 = R/L$, where R the radius of the Kossel rings. L is the distance between the sample and imaging plate. Therefore the lattice spacing can be obtained from Bragg equation.

We use 473nm wavelength laser (Nichia Corp., NDHA210APAE1) to illuminate the samples. The colloidal crystals are prepared in a flat capillary cell (0.5mm×15mm×25mm). We observe the Kossel ring by optical path (see Fig2).

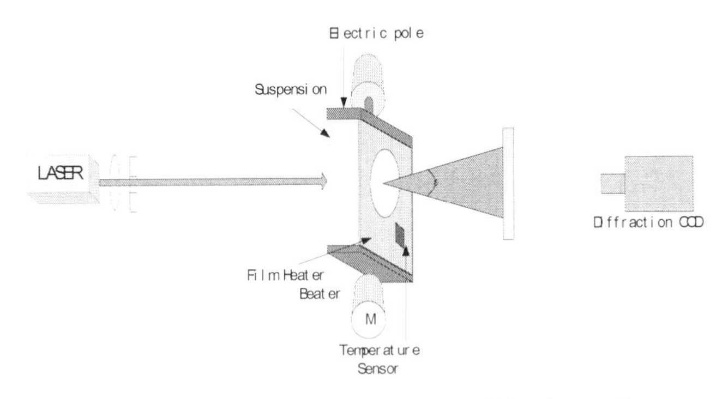

Fig. 2 Schematic diagram of optical path of Kossel-line diffraction station

2.2 System Identification

1962, L.A.Zadeh[10] defined system identification: It is a process of finding a model equaled to the measurement system with input and output data from a given model class. Correlation Coefficient and Recursive Extended Least Squares are uesed in our work.

2.2.1 Recursive Extended Least Squares

Model description:

$$A(q^{-1})\cdot y(k) = B(q^{-1})\cdot u(k-d) + C(q^{-1})\cdot e(k) \tag{1}$$

In order to obtained output temperature data, step response method is used in our work. The algorithm of recursive extended least squares as follow:

$$\hat{\theta}_k = \hat{\theta}_{k-1} + \frac{P_{k-1}\hat{\phi}_k}{\hat{\phi}_k^T P_{k-1}\hat{\phi}_k + 1}[y(n+k) - \hat{\phi}_k^T \hat{\theta}_{k-1}] \tag{2}$$

$$P_k = P_{k-1} - \frac{P_{k-1}\hat{\phi}_k \hat{\phi}_k^T P_{k-1}}{\hat{\phi}_k^T P_{k-1}\hat{\phi}_k + 1} \tag{3}$$

Where, $\hat{\theta}^T = [a_1,\cdots,a_{n_a},b_1,\cdots,b_{n_b},c_1,\cdots,c_{n_c}]$

$$\hat{\phi}_k^T = [-y(k-1),\cdots,-y(k-n_a),u(k-d-1),$$
$$\cdots,u(k-d-n_b),e(k-1),\cdots,e(k-n_c)]$$
$$\hat{\theta}^T = zeros(n_a+n_b+n_c),$$
$$P = 10^6 \cdot eye(n_a+n_b+n_c)$$

2.2.2 Correlation Coefficient

The principle of correlation coefficient is: we measure actual sampling value Y and model output value Y_m for N times, the result data are $y(i)$ and $y_m(i)$ respectively, where $i=1,2,\ldots,N$. The sample correlation coefficient is written.

$$\gamma = \frac{\sum_{i=1}^{n}(y(i)-\overline{y})(y_m(i)-\overline{y}_m)}{\sqrt{\sum_{i=1}^{n}(y(i)-\overline{y})^2 \cdot \sum_{i=1}^{n}(y_m(i)-\overline{y}_m)^2}} \tag{4}$$

where, $\overline{y} = \frac{1}{N}\sum_{i=1}^{N} y(i)$, $\overline{y}_m = \frac{1}{N}\sum_{i=1}^{N} y_m(i)$

2.3 *Principles and Implementation of PID Algorithm*

Classic control theory had been developed a lot of modern advanced control technology, but it is still widely used in various control field with the advantage of simple and adaptability [11].

The ideal PID controller is descriped as follow:

$$u(t) = P[e(t) + \frac{1}{I}\int e(t)dt + D\frac{de(t)}{dt}] \tag{5}$$

Where, P is the proportional coefficient, I is the integral coefficient, and D is derivative codfficient.

Commonly, the digital incremental PID control algorithm is used in the computer control system.

$$\begin{aligned}\Delta u(k) = {} & K_P[e(k)-e(k-1)] + K_I e(k) \\ & + K_D[e(k)-2e(k-1)+e(k-2)]\end{aligned} \tag{6}$$

3 Exiperiment and Results

According to system identification of Correlation Coefficient and Recursive Extended Least Squares, transfer function of the system is obtained. The compare between actual temperature control and model tempreature control is given (see Fig3).

$$G(s)=\frac{0.2193s^2+0.3581s+0.2414}{s^2+0.1954s+0.001101}\cdot e^{-s} \tag{7}$$

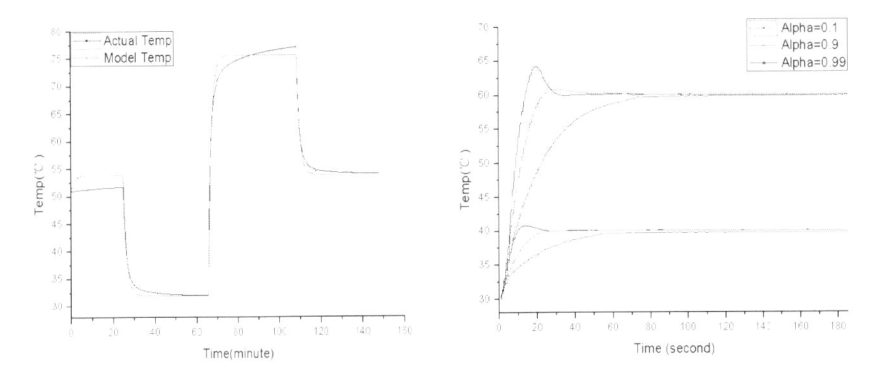

Fig. 3 Actual temperature and model temperature

Fig. 4 Response of the system in different α (different PID parameters)

Hooke-Jeeves search method [12] is used to determine the optimal parameter of PID controller (see Tabel1). We get the optimal parameter by Hooke-Jeeves search method: P=10, I=0.5, D=0.1, and the response of the system si perfect (see Fig4).

Table 1 PID parameters

α	P	I	D	超调 (40°C/60°C)	调节时间 (40°C/60°C)
0.1	10	0.07	0.005	0.8125°C/4.2500°C	25s/31s
0.9	10	0.05	0.004	0.1825°C/0.8125°C	25s/68s
0.99	10	0.04	0.0004	无	>100s

The result of control temperature from 35°C to 65°C (see Fig5).

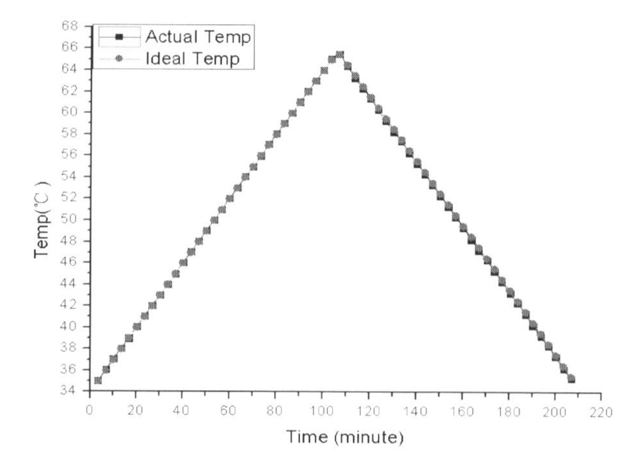

Fig. 5 The temperature control from 35°C to 65°C

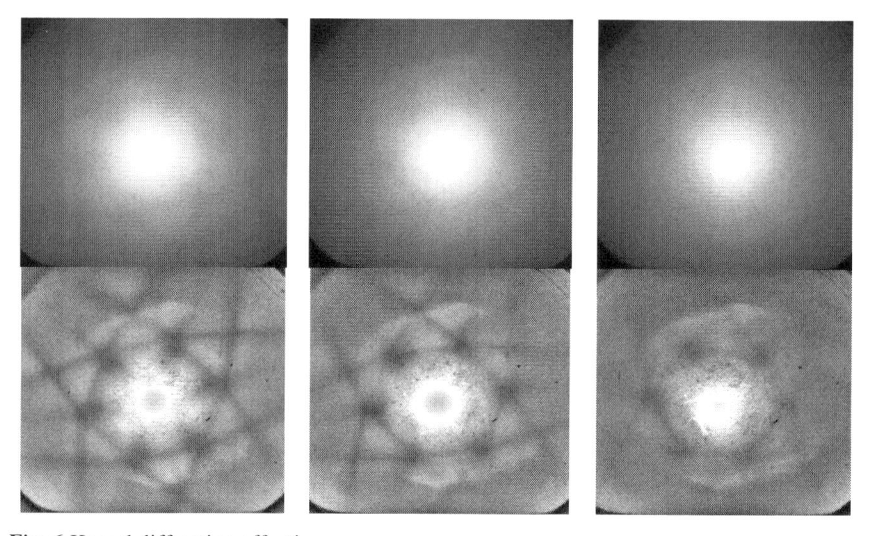

Fig. 6 Kossel diffraction effection

Phase transition of colloidal crystals is observed by Kossel diffraction, after control the temperature at 30°C、50°C and 55°C. Kossel diffraction effection is perfect (see Fig6).

4 Conclution

Temperature control accuracy is critical during studying colloidal crystals in a specific crystallization temperature. The final control accuracy is ±0.1°C, the overshoot within1°C, and the response time within 100s after adopting the system identification and designing parameters of PID controller.

Acknowledgments. This work was supported by National Science Foundation of China (40804032).

References

[1] Zhu, J., Li, M., Rogers, R., Meyer, W., Ottewill, R.H., Russel, W.B., Chaikin, P.M.: Crystallization of Hard-Sphere Colloids in Microgravity. Nature 387, 883–885 (1997)
[2] Cheng, Z., Zhu, J., Russel, W.B., Meyer, W.V., Chaikin, P.M.: Colloidal Hard-Sphere Crystallization Kinetics in Microgravity and Normal Gravity. Applied Optics 40, 4146–4151 (2001)
[3] Vailati, A., Cerbino, R., Mazzoni, S., Giglio, M., Nikolaenko, G., Takacs, C.J., Cannell, D.S., Meyer, W.V., Smart, A.E.: Gradient-Driven Fluctuations Experiment Fluid Fluctuations in Microgravity. Applied Opitics 45, 2155–2165 (2006)
[4] Cheng, Z., Chaikin, P.M., Zhu, J., Russel, W.B., Meyer, W.V.: Crystallization Kinetics of Hard Spheres in Microgravity in the Coexistence Regime. Physical Review Letters 88, 015501-1–015501-4 (2002)
[5] Yoshiyama, T., Sogami, I., Ise, N.: Kossel Line Analysis On Colloidal Crystals in Semidilute Aqueous Solutions. Physical Review Letters 53, 2153–2156 (1984)
[6] Kossel, W., Vogues, H.: X-Rav Interferences in a Monocrystal Cathode. Ann. Physik. 23 (1935)
[7] Clark, N.A., Hurd, A.J., Ackerson, B.J.: Single Colloidal Crystals. Nature 281, 57–60 (1979)
[8] Ackerson, B.J., Clark, N.A.: Shear-Induced Melting. Physical Review Letters 46, 123–126 (1981)
[9] Carlson, R.J., Ashier, S.A.: Characterization of Optical Diffraction and Crystal Structure in Monodisperse Polystyrene Colloids. Applied Spectroscopy 38, 297–304 (1984)
[10] Zadeh, L.A.: From Circuit Theory to System Theory. Proc. IRE 50(5), 856–865 (1962)
[11] Astrom, K.J., Hagglund, T.: The future of PID control. Control Engineering Practice 9, 1163–1175 (2001)
[12] Moser, I., Chiong, R.: A Hooke-Jeeves Based Memetic Algorithm for Solving Dynamic Optimisation Problems. In: Corchado, E., Wu, X., Oja, E., Herrero, Á., Baruque, B. (eds.) HAIS 2009. LNCS, vol. 5572, pp. 301–309. Springer, Heidelberg (2009)

Optimum Design of PID Controller Parameters by Improved Particle Swarm Optimization Algorithm

Xiaodong Chen and Yumin Zhang

Abstract. To avoid the slow convergence and the premature problem of basic particle swarm optimization algorithm (basic PSO), an improved particle swarm optimization algorithm (IPSO) was presented to used for optimizing PID controller parameters. On the basic of the basic PSO with contraction factor, the IPSO with mutation probability was proposed to get a good population diversity and to avoid basic PSO getting into local best result. In order to gain satisfied transient process dynamic characteristics, Integral Time Absolute Error (ITAE) is adopted as the fitness function for parameter selection. The IPSO applied to optimizing PID controller parameters are much better than those of basic PSO.

Keywords: Particle swarm optimization algorithm, Mutation probability, PID; Parameter optimization.

1 Introduction

With advantages of stable operation, simple structure, high reliability, and good robustness, PID (Proportional-Integral-Derivative) controller is widely applied to all kinds of industrial process control. However, traditional PID controller is not good enough in parameters tuning, and its adaptability towards operation condition is bad. In this thesis, a kind of improved particle swarm algorithm is put forward, which is used to optimize PID controller parameters, so as to achieve the aim for improving the performance of control system.

On the basis of particle swarm algorithm with contraction factor, in this thesis, a kind of particle swarm algorithm with mutation probability is proposed, applying to optimize PID controller parameters. The specific particle swarm optimization algorithm is provided, and simulation results indicate the effectiveness and rationality of this algorithm in the optimal design of PID controller parameters.

Xiaodong Chen · Yumin Zhang
College of Chemistry Jilin University, Qianjin Street 2699, 130012 Changchun, China
e-mail:{cxd,zhang_ym}@jlu.edu.cn

Z. Qian et al. (Eds.): Recent Advances in CSIE 2011, LNEE 125, pp. 79–84.
springerlink.com

2 Particle Swarm Optimization Algorithm

Particle swarm optimization (PSO) algorithm is a kind of optimization algorithm [1,2] based on swarm intelligence theory, which was officially put forward by Kennedy and Eberhart in IEEE international neural network academic meeting in 1995. This algorithm guides optimal search through the swarm intelligence produced by cooperation and competition among particles in the group, it keeps the advantage for using swarm intelligence of traditional optimization algorithm, and at the same time it also possesses features, such as simple operational model and easy to be understood and achieved, so it is applied successfully in many optimization problems [3, 4, 5, 6].

Inertia weight ω [6] is introduced in the item of speed in PSO algorithm by Shi and Eberthart, and this improved algorithm is generally known as basic PSO algorithm. In D- dimensional target search space, a particle swarm is made up of m particles, among them, the particle is shown as $X_i=(x_{i1}, x_{i2}, x_{i3},\ldots, x_{id})$, $i=1, 2, 3,\ldots, m$, and the best position experienced on the wing with the speed of V_i is the optimal position of particle i, marked as P_{id} (local optimal position). In addition, the optimal position passed by the whole group is the best position found by the whole group at present, marked as P_{gd} (global optimal position). Each particle renews their own position through the above two extreme values, consequently, the group of new generation is produced. In actual operation, fitness function decided by optimization problems is used to evaluate the good or bad of particle position. The speed of particle i is shown with $V_i= (v_{i1}, v_{i2}, v_{i3},\ldots, v_{id})$, $i=1, 2, 3,\ldots, m$. As for each generation, its d-dimension ($1\leq d\leq D$) is evolved gradually according to the following formulas:

$$V_{id}(t+1)=\omega\times V_{id}(t)+c_1\times r_1\times(P_{id}(t)-X_{id}(t))+c_1\times r_1\times(P_{gd}(t)-X_{id}(t)) \quad (1)$$

$$_{id}(t+1)=X_{id}(t)+V_{id}(t+1) \quad (2)$$

In the formula, c_1 and c_2 are constants, named as positive accelerated factors; r_1 and r_2 are random numbers distributed evenly in [0, 1]; ω is inertia weight, using to confirm the reservation condition of this particle's previous speed; P_{id} is the best position experienced by particle i; P_{gd} is the best position among positions passed by all particles. What's more, search is limited among the maximum and the minimum position and the maximum and the minimum speed.

In order to improve the astringency of algorithm, a kind of PSO algorithm with contraction factor is raised in literature [7]. The position of particle i is evolved according to formula (2), and iterative formula of speed is:

$$V_{id}(t+1)=\chi\times(\omega\times V_{id}(t)+c_1\times r_1\times(P_{id}(t)-X_{id}(t))+c_2\times r_2\times(P_{gd}(t)-X_{id}(t))) \quad (3)$$

$$\chi=\frac{2}{|2-\phi-\sqrt{\phi^2-4\phi}|} \qquad \phi=c_1+c_2\ , \quad \phi>4$$

In the formula, X is a contraction factor,Φ is no need to limit search between the maximum and the minimum speed. It indicates that PSO algorithm with contraction factor has better performance when comparing with PSO algorithm without contraction factor [8].

3 The PSO with Contraction Factor

Through the research towards formula (1), we have found that with the progress of evolution, if the local optimal position P_{id} and the global optimal position P_{gd} keep invariant for a long time, particle speed will change along the direction of$\omega \times V_{id}(t)$, and particle has strong homoplasy, so algorithm may be low efficiency and local optimum. In order to solve this problem, the speed and position of the half particles with low fitness shall not be renewed according to formula (1) and (2), while they shall initialized randomly again, so as to improve particles' diversity, expand search range, and avoid that the algorithm will be involved in local optimum, which are put forward in literature [9]. However, we can not make sure that optimal solution can be obtained in each calculation because particles' speed and position are initialized randomly.

In order to solve this problem, mutation operator method of genetic algorithm is used for reference. In this thesis, on the basis of using PSO algorithm with contraction factor, we have designed Improved Particle Swarm Optimization (hereinafter referred to as IPSO) varying the speed and position of particles with low fitness according to the mutation probability of gradual decrease in space of real numbers. Specific practices are that the speed and position of the half particles with low fitness shall not be renewed according to formula (1) and (2), but varied according to mutation probability P_m, $P_m=1/(1+1.5\times\log_e(gen))$. In the formula, gen is the present iteration number, at the beginning of evolution, the scope for mutation is relatively big, with the evolution of population, the scope for mutation becomes smaller and smaller, and it is this regulating action that accelerates the convergence rate of this algorithm. Figure 1 shows the relational graph between mutation probability and iteration number.

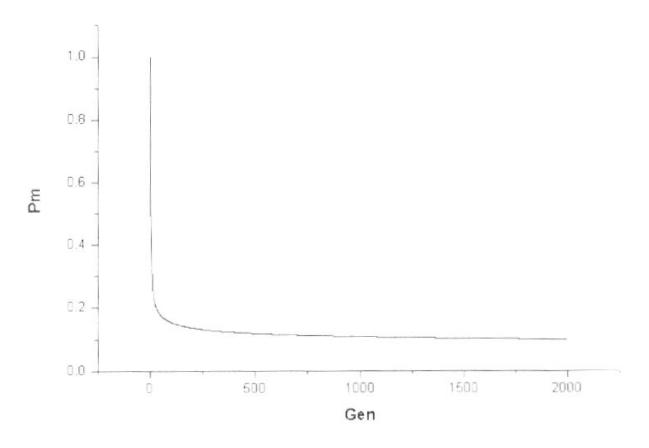

Fig. 1 The relational graph between mutation probability and iteration number.

4 PID Parameter Optimization

Suppose that the controlled object is the second-order transfer function: $G(s)=400/(s^2+50s)$. Sampling time is 1 ms, and input instruction is a first-step signal. The so-called optimizing PID control parameter optimization is searching a group of controls parameters kp, ki, and kd, so as to achieve dynamic performance indexes, such as high system response speed, small overshoot, and short adjusting time and so on.

In order to gain satisfied transient process dynamic characteristics, Integral Time Absolute Error (ITAE) is adopted as the fitness function (optimal index) for parameter selection. For the sake of preventing oversized control energy, the square of control input is added in fitness function. Punishment function is adopted to avoid overshoot, namely once overshoot emerges, the overshoot will be regarded as one item of fitness function, and the fitness function of parameter selection is as follows:

$$\begin{cases} J=\int_0^\infty (w_1 \cdot t \cdot |e(t)|+w_2 u^2(t))\,dt+w_3 \cdot t_u & \text{without overshoot} \\ J=\int_0^\infty (w_1 \cdot t \cdot |e(t)|+w_2 u^2(t)+w_4|ey(t)|)\,dt+w_3 \cdot t_u & \text{if } ey(t)<0 \end{cases} \tag{4}$$

In the formula, $e(t)$ is system error, $u(t)$ is controller output, t_u is rise time, and w_1, w_2, w_3 and w_4 are weights. When $w_4 >> w_1$, $ey(t)=y(t)-y(t-1)$, $y(t)$ is controlled object output.

Optimizing PID controls parameter particle swarm algorithm processes are as follows:

Step 1. Initializing the population. Confirming the population size as 30, the numeric area of parameter *kp* as [0, 20], and the numeric area of parameters *ki* and *kd* as [0, 1], adopting real numbers to encode, providing the initial position and speed of each particle randomly. What's more, the number of iterations is 100.

Step 2. Evaluation. The fitness function value of each particle shall be calculated according to formula (4), comparing with the best fitness value. And then, the best fitness value of each particle and the population shall be saved.

Step 3. Finding out the best position P_{id} of each particle in the search process by now.

Step 4. Finding out the whole best position P_{gd} of all particles in the search process by now.

Step 5. Renewing the position and speed of each particle according to formula (2) and formula (3). In this algorithm, c_1=2.8, c_2=1.3.

Step 6. Sequencing the whole population according to particle optimal fitness, and renewing the position and speed of the half particles with low fitness in accordance with mutation probability P_m rather than formula (2) and formula (3).

Step 7. Returning to step 2 to execute continually, till a satisfactory result is obtained or end condition is satisfied.

On the basis of IPSO algorithm in this thesis, the process for optimizing PID controls parameter is gained with the adoption of Matlab 7.0 language. w_1=0.01, w_2=0.01, w_3=0.2, and w_4=100, iterating for 100 times, the tuning results of PID are *kp*=19.1975, *ki*=0.2275, and *kd*=0.0019, and the optimal fitness function value is J = 11.8943. The comparison between the optimization process for fitness function of IPSO algorithm and basic PSO algorithm is shown in Figure 2, and PID step response after optimized by IPSO algorithm and basic PSO algorithm is shown in Figure 3.

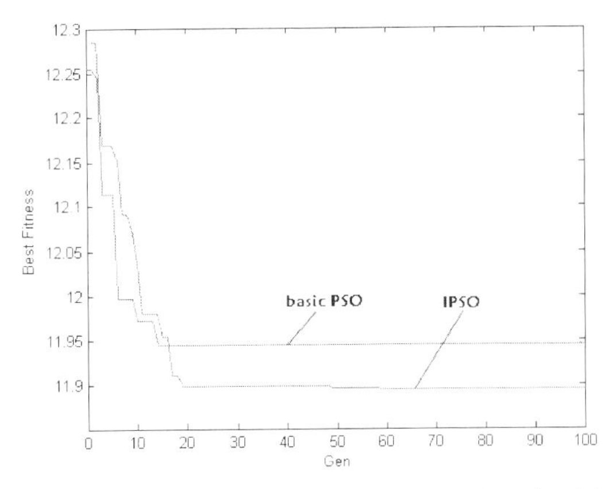

Fig. 2 The optimization process for fitness function of IPSO algorithm and basic PSO algorithm.

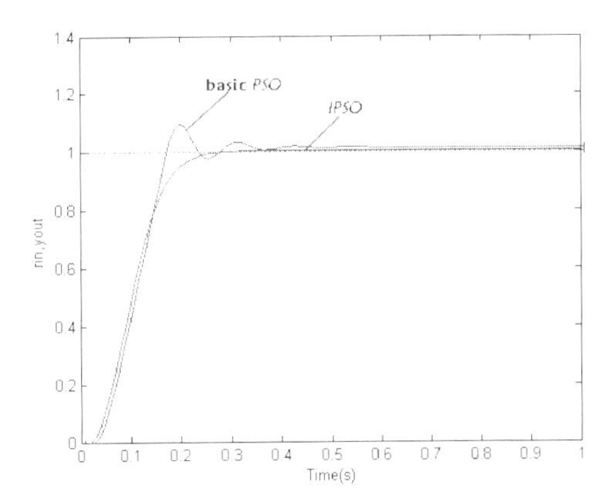

Fig. 3 PID step response after optimized by IPSO algorithm and basic PSO algorithm.

It can be seen from Figure 2 and Figure 3 that there are some advantages when optimizing PID controller parameter with the use of IPSO algorithm rather than basic PSO algorithm. Firstly, no overshoot, high response speed, and short adjusting time when using IPSO algorithm to optimize; on the contrary, with overshoot and easy to be early-maturing when using basic PSO algorithm. Secondly, IPSO algorithm optimization is relatively stable, which can converge optimal solutions faster.

5 Conclusion

In this thesis, a kind of particle swarm algorithm with contraction factor and mutation probability is put forward to optimize PID controller parameter. Renewing the position and speed of the half particles with low fitness in accordance with mutation probability P_m rather than formula (2) and formula (3), so as to improve particles' diversity, expand search range, and avoid that the algorithm will be involved in local optimum. In this thesis, specific particle swarm algorithm is raised, and simulation results prove that in the aspect of avoiding early-mature and accelerating convergence IPSO algorithm is superior to basic PSO algorithm.

References

1. Kennedy, J., Eberhart, R.: Particle swarm optimization. In: Proceeding of IEEE International Conference on Neural Networks, pp. 1942–1948. IEEE Service Center, Piscataway (1995)
2. Eberhart, R., Kennedy, J.: A new optimizer using particle swarm theory. In: Proceedings of the Sixth International Symposium on Micro Machine and Human Science, Nagoya, Japan, pp. 39–43 (1995)
3. Parsopoulos, K.E., Vrahatis, M.N.: Particle swarm optimization method for constrained optimization problems. In: Intelligent Technologies Theory and Applications: New Trends in Intelligent Technologies, pp. 214–220 (2002)
4. Eberhart, R., Hu, X.: Human tremor analysis using particle swarm optimization. In: Proceedings of the IEEE Congress on Evolutionary Computation (CEC 1999), Washington, DC, pp. 1927–1930 (1999)
5. Yoshida, H., Kawata, K., Fukuyama, Y., et al.: A particle swarm optimization for reactive power and voltage control considering voltage security assessment. IEEE Transactions on Power Systems, 1232–1239 (November 2000)
6. Shi, Y., Eberhart, R.: A modified particle swarm optimizer. In: IEEE Int'l Conf. on Evolutionary Computation, Anchorage, Alaska, pp. 69–73 (1998)
7. Clerc, M.: The swarm and the queen: towards a deterministic and adaptive particle swarm optimization. In: Proceedings of the 1999 Congress on Evolutionary Computation, Piscataway, NJ, USA, pp. 1927–1930 (1999)
8. Eberhart, R., Shi, Y.: Comparing inertia weights and constriction factors in particle swarm optimization. In: Proceedings of the 2000 Congress on Evolutionary Computation, Piscataway, NJ, USA, pp. 84–88 (2000)
9. Zhang, L.B., Zhou, C.G., Liu, X.H.: Application of particle swarm optimization for solving optimization problems. Journal of Jilin University (Information Science Edition) 23, 385–389 (2005)

Research on Fuzzy-Neural Networks Controller in Thermostatic and Humidistatic Aircondition System

Xing Li, Dingguo Shao, and Wei Lv

Abstract. The thermostatic and humidistatic aircondition control system is a typical nonlinear, time-variant system with great lag and strong interference. It is difficult to control complicated process to get nice control precision by classical control technology or modern control theory, due to rigorous mathematics model.Non optimized control method reduces the efficiency of aircondition, causes enormous loss of energy. A new method of the perfect combining the fuzzy control and the neural Networks control is proposed in this paper. The fuzzy control could generate control laws for structural fuzzy controller based on the expert experience, without the need of the accurate mathematical model of the object. The neural Networks control can adjust the fuzzy rule base online and fine tune parameters in real time to achieve satisfactory precision and good energy saving. The result of the simulation shows, in the condition of the system parameter's change or the external disturbances, these methods of improvement show high precision and obviously lower energy expenses.

Keywords: Thermostatic and humidistatic aircondition, Fuzzy-neural networks, Energy saving.

1 Introduction

The thermostatic and humidistatic aircondition which is usually taken as an intact system to design has been widely used in the places where the environment and air quality is strictly demanded and aircondition needs to work for long time continuously, such as electrical manufacturing, food productions, high-end printing, pharmaceutical production, hospital surgery room, computer rooms, science laboratory, etc. With more control processes, strong coupling, strong interference, scattered control object, the thermostatic and humidistatic aircondition system is different from ordinary aircondition. Studies of this control system are just in the elementary stages.

Xing Li · Dingguo Shao · Wei Lv
Mechanical and Electronic Engineering and Automation, Shanghai University
Shanghai 200072, China

Z. Qian et al. (Eds.): Recent Advances in CSIE 2011, LNEE 125, pp. 85–91.
springerlink.com

The existing classical control can't satisfy high accuracy of the temperature and moisture which is needed to be considered in the high-end printing room or science laboratory. In practical projects, the general method is that the temperature is preferentially adjusted, the moisture is adjusted when the temperature basically reached the expected temperature requirements. As the intractable problem of the dewing and the leaking, the moisture control is an additional function and is hard to meet the accuracy demanded.

The parameters of aircondition and controller are designed for the most harsh circumstances of the thermal and moisture load, but in the most running time, the aircondition usually loaded partly, so the control parameters often are mismatching. Furthermore, when the systems run for a while as a number of causes, such as the changing of the self parameters of the system or the new violent disturbance disturbance's joining, the control performance deteriorates progressively and energy consumption is obviously raised. As this kind of aircondition is usually working for long time continuously, there is great scope for energy savings.

It has been an important research field to optimize the aircondition control system to meet the expectation of High accuracy, to make the aircondition control system to self-adjust online effectively and save energy as far as possible.

2 The Mathematical Model of Thermostatic and Humidistatic Aircondition System

2.1 The Establishment of the Room's Mathematical Model

A Novel Approach to the thermo and hydrograph load of rooms modeling considers simultaneously the temperature and relative humidity in this paper. It considers the total enthalpy and the total water vapour of the room, views all the disturbances and the aircondition controlled quantity as the change of the total enthalpy and the total water vapour of the room, then in terms of thermodynamics relation of the total enthalpy and the total water vapour, figures out the temperature and relative humidity of the rooms. Expressions of mathematic relation of the rooms are shown as below:

The total enthalpy in the room=The original total enthalpy in the room+The enthalpy of the supply air of aircondition-The enthalpy of the return air of aircondition+The heat/cold sources in the room-The heat transfeering from the wall-The heat of air leaking

The total water in the room=The original total water in the room+The water of the supply air of aircondition-The water of the return air of aircondition+The water sources in the room-The water of air leaking.

2.2 *The Establishment of the Aircondition's Mathematical Model*

Through the methods of working mechanism and the measuring result of experiments, the mathematical model of every part of the aircondition is established. Fig. 1 shows the structure of the sample aircondition.

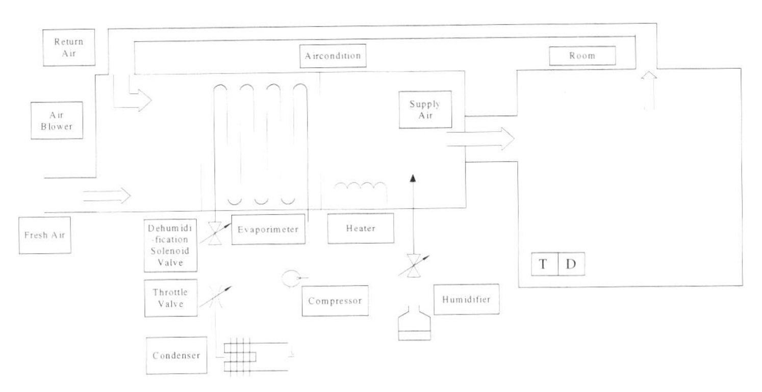

Fig. 1 The structure of the aircondition

The typical modules of thermostatic and humidistatic aircondition includes:the forced draught fans, supply air valve, the fresh air valve, return air valve, surface cooling coils, compressor, condenser, evaporator, electrical heater, steam humidifier valve, electrical expansion valves, dehumidification solenoid valve, etc.

The control apparatus contains three air valves, three flow valves, three ventilator, etc. Most of them can both influence the temperature and moisture of the flowing air. Therefore, the system possesses the characteristic of having complicated control processes and strong coupling. The inner of the aircondition would waste energy, just like 'the offset of heat and cold', if the processes is not handled properly. Also, if the proportions and treatments of the fresh air is improper, the efficiency of aircondition would be reduced. Only when the contradictions between heating and Cooling, dehumidifying and humidifying, proportions and treatments of the fresh air have been solved appropriately, the efficiency of aircondition can be improved.

3 The Study of Intelligent Control for Thermostatic Aircondition System

3.1 *The Fuzzy Self-adaptive Control for Thermostatic and Humidistatic Aircondition System*

When fuzzy rules are designed, the disturbance of strategic point should be monitored, for instance, the monitor of the district of calorification and damping. When the immediate interference influences the system, the monitor can tell the controller beforehand, to reach the aim of predictive control.

The surface cooling coils are divided into two sections: One section is placed in the entrance of the fresh air, and is used to process the fresh air collectively beforehand, the other section placed in the common position used to process the mixed air of the fresh air and the return air. The fresh air is processed by the fuzzy rule and then is mixed with the return air. By the new method, the fuzzy rule is also used to adjust the processing of the fresh air to reduce the energy waste of heating and humidification after cooling and dehumidification, and reach the aim of energy conservation.

3.2 The Fuzzy-Neural Networks Controll for Thermostatic and Humidistatic Aircondition System

However, the fuzzy control can't reach high precision, and when the systems run for a while, as the a number of causes, such as the changing of the self parameters of the system or the new violent disturbance's joining, if using the fuzzy control yet, the control performance deteriorates progressively and energy consumption is obviously raised.

The neural Networks control is brought to this system for two aspects of effection: One is adjusting the PID parameter continually and in real time, giving the secondary adjustment to the parameter to achieve the control precision. The other is , according to the monitor and the records to the temperature and humidity indoors and the control quantities, adjusting the fuzzy rule of fuzzy control to improve the mismatching of parameter caused by the system variation.

Aiming at the multiple inputs and multiple outputs of the thermostatic and humidistatic aircondition system, we express it as a full network structure based on the associating mechanism. The structure of the multi-layered feed forward neural network is shown as Fig.2.

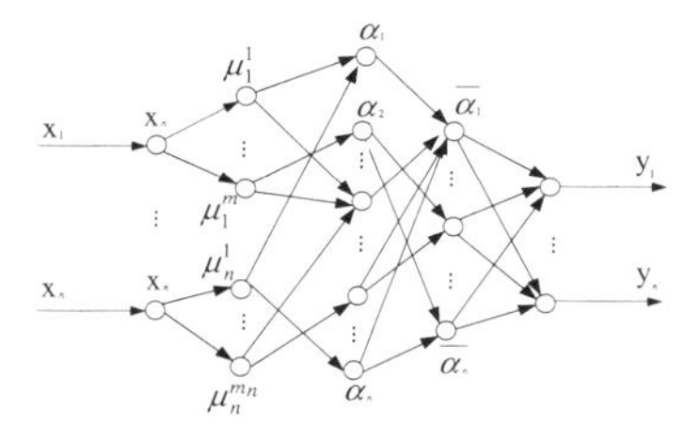

Fig. 2 The full network structure of neural network

As the Fig. 2 shows, this network consists of five layers. The implication of them is as follows:

The first layer: The Input layer. The output of this layer connects with the input vector x_i directly. It serves as transferring the input value $x=[x_1,x_2]^T$ to the next layer.

The second layer: The fuzzification layer. Every node of this layer stands for a linguistic variables, such as NB(very close), NS(a little close), ZO(hold still), PS(a little far), PB(very far), etc.It serves as calculating the membership degree u_{ij} between the inputs and the linguistic variables.

Where c_{ij} and σ_{ij} denote, respectively, the center and the width membership degree of the input i and the linguistic variables j. They are the parameters waiting for learning.

The third layer: The fuzzy inference layer. Every node of this layer stands for a fuzzy rule. It serves as the antecedent of matching the fuzzy rules, calculating the suitability of every rule. The count of nodes is $N_3 = m$, which means there are m fuzzy rules.

The fourth layer: The normalized transfer layer. The outputs of the nodes are the usability of every rule.

The fifth layer: The output layer. Nodes in this layer stand for the output variables. Each node severs as a defuzzifier and computing the output value.

It is shown in the network structure above that the every layer of the neural network corresponds to the calculating of every step of fuzzy logic controller. The network structure expresses and describes the whole process of fuzzy reasoning. The optimization of the neural network for the fuzzy control is just to adjust the center c_{ij} and the width σ_{ij} of the membership degree and the connection weights ω_{ij}.

4 The Simulation of Fuzzy-Neural Networks Control

The graph of simulation is shown as Fig.3.This simulation consists of the control module, the aircondition module, the power monitor module, the neuro-predictive module, fuzzy-neuro network module, the interference module, etc.

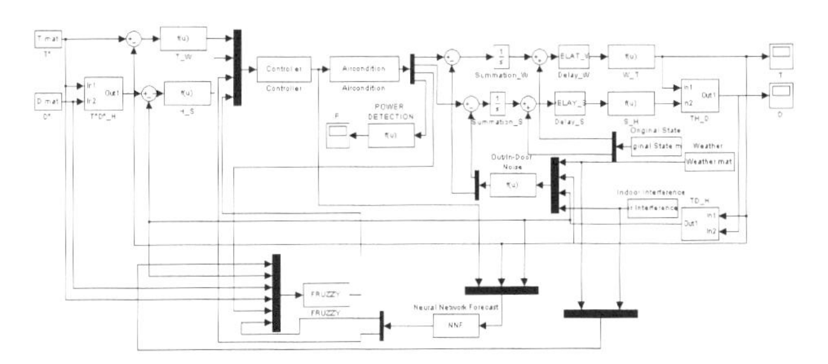

Fig. 3 The graph of simulation based on the Matlab/Simulink software

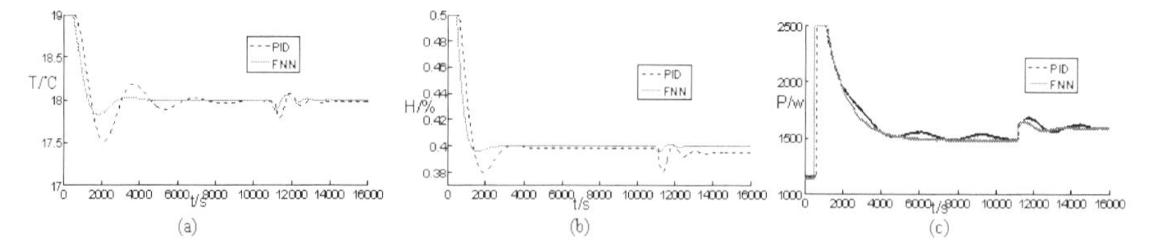

Fig. 4 The contrasting effects of the two control methods

Fig. 4 is the compare of the system response, during the course of the disturbance of the temperature and humidity. At the 500th second, the expecting value of the temperature and the relative humidity is respectively changed from 19 °C to 18°C（Fig.a）,from 50% to 40%（Fig.b）. The graph shows the advantages of quick response, small over modulation and high precision. At the 1100th second, the disturbance of the temperature and humidity is added into the system, the former shows better robustness. As the Fig. c shows, their differences in energy consumption are not significant.

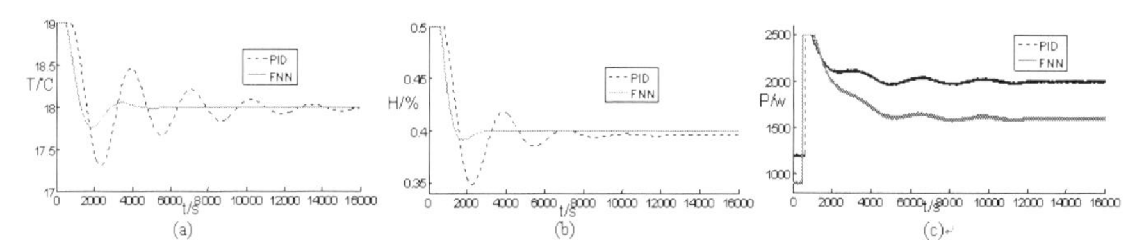

Fig. 5 The contrasting effect when the system parameters are changed

Fig. 5 is the compare of the system response when the system parameters are changed. The expecting value of the temperature and the relative humidity is respectively changed formerly. It shows the advantages of quick response, small over modulation and high precision more obviously. At the 1100th second, the disturbance of the temperature and humidity is added into the system, the former shows better robustness. As the Fig. c shows, their differences in energy consumption are so significant that 20% of energy is saved.

5 Conclusion

Aiming at the thermostatic and humidistatic aircondition system, system model is established first of all, then fuzzy control and the neural Networks control are combined to increase efficiency and raise applicability in the paper. The simulink results show the Fuzzy-neural Networks controller possessing high performance and ideal energy conservation.

References

[1] Yang, Y.: Direct robust adaptive fuzzy control(DRAFC) for uncertain nonlinear systems using small gain theorem. Fuzzy Sets and Systems 151, 79–97 (2005)
[2] Liu, H.-J., Li, K., Liu, J.-Y.: Application of Fuzzy PID Control Based Flux Control of Coal Gas BP Neural Network on Furnace. Hydrometallurgy of China 27(2), 1–4 (2008)
[3] Juuso, E.K.: Integration of intelligent systems in development of smart adaptive systems. International Journal of Approximate Reasoning 35, 307–337 (2004)
[4] Suna, Q., Li, R., Zhang, P.: Stable and optimal adaptive fuzzy control of complex systems using fuzzy dynamic model. Fuzzy Sets and Systems 133, 1–17 (2003)
[5] Wu, Z., Mizumoto, M.: PID type Fuzzy Controller and Parameters Adaptive Method. Fuzzy Set and Systems 78(1), 23–26 (1996)
[6] Zong, X.-P., Feng, H.-P.: Neural Network-Based Predictive Control for Time-delay Systems. Control Theory and Application 24(12), 1–3 (2005)

A String Matching Algorithm Based on Real Scaling

Zhang Ying, Chang Guiran, and Jia Jie

Abstract. A good intrusion detection algorithm is important not only in traditional networks, but also in Ad Hoc networks. This paper proposes a novel intrusion detection algorithm based on the real scale of the string matching. Based on properties of Ad Hoc networks, we design a real indexing tree which gives a complete implementation of the algorithm based on the real scale of the string matching. The algorithm can find out all the possible text scaling, and identify all the possible strings with the text scaling. In order to evaluate efficiencies of our algorithms, we use an evaluation data set, widely used in intrusion detection, to test our algorithms. The experimental results show that our algorithm is superior over other sate of art algorithms, and Ad Hoc network security is improved.

Keywords: Intrusion Detection System, real scaling, string matching.

1 Introduction

Ad Hoc Network is a dynamic network topology of wireless mobile networks, in order to protect the Ad Hoc network security, intrusion detection system is needed. But its processing speed and detection rate can not keep up with network data's transfer speed, the detection system will lost some data packets to ensure the normal operation of IDS, resulting in omission affect the system's accuracy and effectiveness. So how to improve the detection speed and detection accuracy of IDS have become a hot research content [1-3].Detection speed and detection accuracy are mainly reflected in the search speed for attack information, that is string matching speed. Thus selecting the appropriate string matching algorithms has become an important part of IDS research [4-6]. Intrusion detection system establishs a set of characteristic patterns of intrusion for a variety of intrusion

---

Zhang Ying · Chang Guiran · Jia Jie
School of Information Science and Engineering
Northeastern University, Shenyang, China
e-mail: engleyy@163.com, chang@neu.edu.cn

Z. Qian et al. (Eds.): Recent Advances in CSIE 2011, LNEE 125, pp. 93–98.
springerlink.com

characteristics while adding a variety of behavioral measures in intrusion detection system which should be taken to constitute the system's security rules. Therefore, an intrusion detection system's main job is listening to the data and matching with safe rules, then the invasion to take appropriate measures according to the situation [3]. In the intrusion detection system, the exact string matching techniques has not only been widely used, but also a decisive factor affecting system performance. String matching algorithms have a variety of categories, according to pattern matching, pattern matching algorithms can be divided into single and multi-pattern string matching algorithm. More used in IDS is multi-pattern string matching algorithm. Classic multi-pattern matching algorithms include Aho-Corasick algorithm [7] and Wu-Manber algorithm [8] and FS algorithm [9] and so on.

This paper designs and implements an algorithm based on the real scale of the string matching and the algorithm is integrated into the Snort system. In or-der to test the efficiency of our algorithms, we comparing with classical algo-rithm, the results show that real scaling based on string matching algorithms has improved the efficiency of the detection system Snort.

2 Algorithm Description

2.1 *Idea of the Real Scaling*

In this paper, we use real scaling ideas for string matching, so we describe the matching algorithm through an example before introduce the real scaling ideas. Let the text is aaaabbbcc;compressed text is a4b3c2;match string is abbbc; compressed string is a1b3c1。 Scaling factor of each letter are a:1,2,3,4, $\frac{4}{3}$,b:1,2,3, $\frac{3}{2}$,c:1,2 , After the intersection is taken:1,2,3,4, $\frac{4}{3}$, $\frac{3}{2}$, Scaled text is:T1=a4b3c2, T2= a2b1c1, T3=a1b1, T4= a1, $T_{\frac{3}{2}}$ = a2b2c1, $T_{\frac{4}{3}}$ = a3b2c1. Beacase the matching string is a1b3c1, and the factor of the middle letter b is 3. In addition, the scaled text has the same T1 a4b3c2. We compare the coefficients a and c and the results are greater than the coefficients of the matching characters a and c. Finally we have a successful matching.

2.2 *The Construction of Real Scale Tree*

First, we indenfiy the scaling string by building a real constructing tree to determine the real text aabbbbaa , compressed to be a2b4a2.

In the algorithm of real scaling, we first list the range of all real numbers: we find the largest index of article n, then specify the scaling range T1'~Tn'。 T: a2b4a2 ; T1' : $ a2b4a2$; T2' : $ a1b2a1$; T3' : $ b1$; T4' : $ b1$. We set an example show the method of real scaling, T=aabbbbaa(a2b4a2).

(1)Find the longest characters in front of a2b4a2. K=1, a2b4a2 ; k=2,ab2a; Construct { a2b4a2, ab2a }tree, as shown in Fig. 1. (2)Find the longest in front of ab4a2.k=1, ab4a2 ; k=2, none ; Construct { ab4a2} tree, as shown in Fig. 2. (3)Find the longest in front of b4a2.K=1, b4a2 ; k=2, b2a ; k=3, b3 ; k=4, b; Construct {b4a2, b2a, b3, b } tree, as shown in Fig. 3. (4)Find the longest in front of b3a2. K=1, b3a2 ; k=2, b2 ; k=3, b ; Construct { b3a2, b2, b } tree, as shown in Fig. 4. (5)Find the longest in front of b2a2. K=1, b2a2 ; k=2, ba; Construct { b2a2, ba} tree, as shown in Fig. 5. (6)Find the longest in front of ba2.K=1, ba2 ; k=2, none ; Construct { ba2 } tree, as shown in Fig.6. (7)Find the longest in front of a2. K=1, a2 ; k=2,a ; Construct {a2,a} tree, as shown in Fig. 7. (8)Find the longest in front of a.K=1, a;Construct {a} tree,as shown in Fig.8

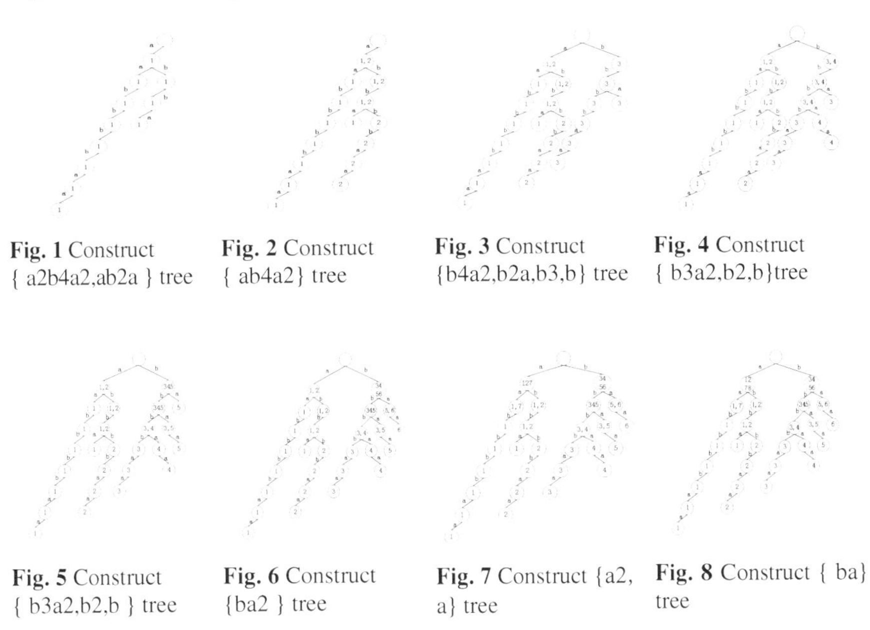

Fig. 1 Construct { a2b4a2,ab2a } tree

Fig. 2 Construct { ab4a2} tree

Fig. 3 Construct {b4a2,b2a,b3,b} tree

Fig. 4 Construct { b3a2,b2,b}tree

Fig. 5 Construct { b3a2,b2,b } tree

Fig. 6 Construct {ba2 } tree

Fig. 7 Construct {a2, a} tree

Fig. 8 Construct { ba} tree

2.3 Matching Algorithm

We obtain T # (= t1, ..., tn) from the index of the symbol string T^{Σ}. Similarly, we obtain P # (= p1, ..., pn) from the index of the symbol string P^{Σ}. Find all the P in the relative position of T, meanwhile satisfy the following conditions A and B:

Condition A : P^{Σ} has a right position i corresponding to T^{Σ} 。

Condition B1 : $t_i \geq kp_1$

Condition B2 : $t_{i+1} \geq kp_2, \ldots\ldots, \quad t_{i+m-2} \geq kp_{m-1}$

Condition B3 : $t_{i+m-1} \geq kp_m$

(1)Assume $m \geq 3$, obtain quotient string

$T' = t_2 / t_1, t_3 / t_2, \ldots, t_n / t_{n-1}$ from $T^{\#}$; Similarly, obtain quotient string $P' = p_2 / p_1, p_3 / p_2, \ldots, p_n / p_{n-1}$ from P#。

Observation 2 : From observation 1, the condition can be drawn as to meet B.2 string position in the string condition, if and only if, $k = t_2 / p_2$.

(2)Identify all the strings P^{Σ} in the string T^{Σ}, and find all the strings P' in string T'. This can be obtained in any linear time by the string matching algorithm. T^{Σ} occurs in each location i, P' occurs in the position i+1 of T'. We extend conditions B.1 and B.3 to determine whether T' has the i location and position $i + m - 1$. It takes O (1) in each location.

From above, we know the condition of $m < 3$,

(1)obtain quotient string

$T' = t_2 / t_1, t_3 / t_2, \ldots, t_n / t_{n-1}$ from $T^{\#}$; Similarly, obtain quotient string $P' = p_2 / p_1, p_3 / p_2, \ldots, p_n / p_{n-1}$ from P#.

Observation 2 : From observation 1, the condition can be drawn as to meet B.2 string position in the string condition, if and only if $k = t_2 / p_2$.

(2)Identify all the strings P^{Σ} in the string T^{Σ}, and find all the strings P' in string T'. This can be obtained in any linear time by the string matching algorithm.

The time complexity and scope has the same as the above time complexity.

In summary, when $m \geq 3$ and $m < 3$, we can get the result using the same process.

3 Experiment

This paper uses the intrusion detection system evaluation data set in Massachusetts Institute of Technology Lincoln Laboratory. We compare with the string matching algorithms based on real scale and other typical string matching algorithms (Aho-Corasick (AC) algorithm, Modified Wu-Manber (MWM) algorithm).

3.1 Accept the Same Packet Test

Test uses a packet size of 325.2MB, a total of 1,532,453 packets, which contains 43 328 warnings, where packet transmission protocol TCP accounted for 91.961%, UDP accounted for 6.999%, ARP accounted for 0.368%, ETHLOOP accounted for 0.516%, other accounting 0.086%. In order to obtain accurate test values, test uses 10 times the average of the experimental data obtained by the detection. All the time it takes to send the packets is shown in Table 1.

Table 1 All the time it takes to send the packets

Rule	AC	real scaling	MWM
10	11.99 ±0.66	11.07 ±0.86	11.29 ±1.19
15	12.38 ±1.18	11.23 ±0.34	12.34 ±0.57
20	12.57 ±0.71	11.48 ±0.95	12.94 ±0.49
25	13.51 ±0.66	12.98 ±0.37	14.05 ±1.15
30	17.47 ±1.09	17.35 ±0.92	18.16 ±1.70

The size of test packets per second of data is shown in Figure 9, and the test includes 1,532,453 packets. From Fig. 9, we know that with increasing of the number of rules, the efficiency of each algorithm has reduced, but the proposed algorithm is superior to other real scaling algorithms. The time of each packet inspection is shown in Figure 10. With the increase of rules, each algorithm has increased the detection time, but the real time scaling algorithm is less than the AC algorithm.

The number of packets per second test is shown in Figure 11. With the increase in the number of rules, the number of packets per second for each algorithm has reduced, but the real scaling algorithm reduces slower than the other two.

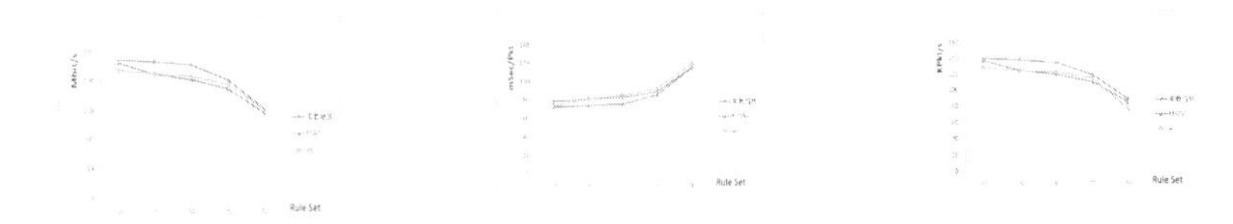

Fig. 9 Size of packets per second

Fig. 10 Inspecting time of per packet

Fig. 11 Number of inspected packets per second

In summary, to accept the same packet in the test, the real scaling algorithm have certain advantages than the other two algorithms.

3.2 Ping Text

The tests take the PING program in the attack side, and use the fastest speed to send the size of 60kbyte packets for 30 times, where packet transmission protocol ICMP is 100%. In order to obtain accurate test values, we use the average of testing 10 times to obtain experimental data, the times are shown in Table 2.

Table 2 All the time it takes to send the packets

Rule	AC	real scaling	MWM
10	11.99 ±0.66	11.07 ±0.86	11.29 ±1.19
15	12.38 ±1.18	11.23 ±0.34	12.34 ±0.57
20	12.57 ±0.71	11.48 ±0.95	12.94 ±0.49
25	13.51 ±0.66	12.98 ±0.37	14.05 ±1.15
30	17.47 ±1.09	17.35 ±0.92	18.16 ±1.70

The size of test packets per second has been shown in Figure 12. With the number of rules increases, the efficiency of each algorithm has reduced, but the real scaling algorithm performs better than the other two. The time of each packet inspected has been shown in Figure 13. With the rule number increase, the inspecting time of each algorithm increases, but real scaling algorithm uses less time than the other two.The time of each packet inspected has been shown in Figure 14.When the rule number is from 10 to 30, the efficiency of each algorithm has decreased.

We conclude the results as follows. Kbit/s : That is bits transmitted in per second, whether Lincoln Laboratory test data or PING test data, the efficiency of real scale is higher than those of AC and MWM algorithms.mSec/Pkt : That is the time of getting a packet. When using the Lincoln Laboratory data, the real scaling algorithm use more time than MWM algorithm. But in the PING test, the real scaling algorithm is better than AC and MWM algorithm. Pkt/Sec : That is how many second is enough to get a packet. In the two test, the real scaling algorithm is better than AC and MWM algorithm.

4 Conclusion

We study the string matching algorithm in IDS. We proposed the real scaled string matching algorithm. The algorithm gives the real number tree construction methods, and the entire matching process. The algorithm is integrated into the Snort system, and is compared with the AC and MWM algorithms in performance evaluations. Experimental results show that our proposed algorithms can more effectively improve the matching efficiency and the Ad Hoc network security.

References

1. Duan, X., Jia, C., Liu, C.: Intrusion detection method based on the hierarchical hidden Marco model and variable-length semantic pattern. Journal of Communication 31(3), 109–114 (2010)
2. Lee, W., Xiang, D.: Information theoretic measures for anomaly detection. In: Proceedings of the 2001 IEEE Symposium on Research Security and Privacy, Oakland, California, pp. 130–134 (2001)
3. Murakami, M., Honda, N.: A study on the modeling ability of the IDS method: A soft computing technique using pattern-based information processing 45(3), 470–487 (2007)
4. Zhang, Y., Xu, J., Chang, G., Jia, J.: Two-way fast string matching algorithm in Ad Hoc Network. Computer Science 37(10), 42–47 (2010)
5. Alicherry, M., Muthuprasanna, M., Kumar, V.: High Speed Pattern Matching for Network IDS/IPS. In: Proceedings the 2006 IEEE International Conference on Network Protocols, NW Washington, DC, USA, pp. 187–196 (2006)
6. Sheu, T.-F., Huang, N.-F., Leeb, H.-P.: Hierarchical multi-pattern matching algorithm for network content inspection. Information Sciences 178(14), 2880–2898 (2010)
7. Charras, C., Lecroq, T.: Handbook of exact string—matching algorithms [DB/OL] (May 2010), `http://www.igm.univ-mlv.Fr/qecroq/tring/`
8. Sunday, D.M.: A very fast substring search algorithm. Communications of the ACM 33(8), 132–142 (1990)

An Approach for System Model Identification

Xiaoping Xu, Fucai Qian, and Feng Wang

Abstract. A method is investigated for system model identification in this paper. The idea of the scheme employs a system model composed with classical models so as to transform the system structure identification into a combinational problem. The bacterial foraging optimization technique is then applied to implement the identification on the structure and parameters. Finally, simulation results indicate the rationality of the proposed method.

1 Introduction

Recently, system identification has been widely applied to the design and analysis of control system, and has been gone deep into many fields of science and technology. Consequently, system identification becomes one of the current very active subjects [3, 6].In identification, one often dispose a series of sample data to obtain the model of a system, and apply it to practical task. However, most existing identification methods can not completely solve the system identification problem which the structure and parameter are completely unknown [1, 5].

In this paper, a new system model identification method is proposed. First of all, the idea of the method employs a model composed with classical models so as to transform the system structure identification into a combinational problem. Then, a bacterial foraging optimization (BFO) algorithm was adopted to implement the identification on the system structure and parameters. Finally, Simulation results showed the rationality and effectiveness of the presented method.

Xiaoping Xu · Fucai Qian
School of Sciences, Xi'an University of Technology, Xi'an, China
e-mail: xuxp@xaut.edu.cn

Feng Wang
State Key Laboratory for Manufacturing Systems Engineering and Systems Engineering Institute, Xi'an Jiaotong University, Xi'an, China
e-mail: wangf@mail.xjtu.edu.cn

Z. Qian et al. (Eds.): Recent Advances in CSIE 2011, LNEE 125, pp. 99–104.
springerlink.com

2 Problem Description

Let y is an observable system output variable, the input variables $x_1, x_2, \cdots, x_m$ are likely to impact the output, thereby, the n groups of samples data obtained from the system can be described as follows:

$$(y_i, x_{1i}, x_{2i}, \cdots, x_{ji}, \cdots, x_{mi}) \tag{1}$$

where x_{ji} denotes the jth sample data of the ith group of samples, y_i expresses the output values of the ith group of samples, j=1,2,$\cdots$,m; i=1,2,$\cdots$,n.

Definition 1: Let x_i be a single variable, it takes $f(x_i)$ as form to influence the output of the system, then, the $f(x_i)$ is called a single-variable meta-model.

Assuming the number of single-variable meta-model is N_1, and considering each input variables may be influence system output via all model form, therefore, the sample data model can be expressed as:

$$y = p_0 + \sum_{k=1}^{N_1} \sum_{i=1}^{m} f_k(x_i) \tag{2}$$

where, p_0 is the constant term.

Definition 2: There are two variables x_i and x_j, they take $f(x_i, x_j)$ as form to influence the output of the system, and the $f(x_i, x_j)$ can not be decomposed into the form $f(x_i)+f(x_j)$, then, the $f(x_i, x_j)$ is called a two-variable meta-model.

Assuming the number of two-variable meta-model is N_2, and taking into account each input variables may be combine by various possible form, then, the sample data model can be described as:

$$y = p_0 + \sum_{k=1}^{N_1} \sum_{i=1}^{m} f_k(x_i) + \sum_{k=1}^{N_2} \sum_{i=1}^{m} \sum_{\substack{j=1 \\ i \neq j}}^{m} f_k(x_i, x_j) \tag{3}$$

In summary, the combinational term number of the number of the sub-model of the sample data model is increased exponentially along with the increase of the number of the variable, meta-model and meta-model's independent variables number. The general form of the sample data model can be described as follows:

$$y = p_0 + \sum_{k=1}^{N} M_k(x, p_{k1}, p_{k2}, \cdots, p_{km_k}) \tag{4}$$

where, p_0 is constant term; N is the number of sub-model; $M_k(x,p_{k1},p_{k2},\cdots,p_{k,mk})$ is the kth sub-model constituted by meta-model and its variables, where x can denote both single-variable and multi-variable, k=1,2,$\cdots$,N.

All in all, the sample data model essentially describes that the various variable may influence the system by various sub-model. For a real system, if we can select the appropriate meta-model, the model of the actual system will be included in the sample data model, or it will be close to the model of the actual system.

Generally, the purpose of identification is that the system outputs $y(t)$ can be b-to approach the known system output $y_0(t)$. Thereby, we may minimize the following cost function.

$$P = \sum_t [y(t) - y_0(t)]^2 \tag{5}$$

Because the minimizing (5) is an optimization problem, and the minimal value of (5) can be obtained by the following BFO algorithm. Accordingly, we achieve the identification of the structure and parameters of the system model.

3 Solving Model via BFO Algorithm

The given system identification method of this paper can be divided into the selection of the sub-model problem and model parameter identification problem. The selection of the sub-model from multitudinous sub-model, belongs to a class of combination optimization problem and parameter identification problem also belongs to an optimization problem.

This paper considers the foraging behavior of E. coli, which is a common type of bacterial as in reference [4]. Currently, BFO algorithm has been successfully applied to engineering problems [2]. Of course, BFO algorithm also provides an important way for system identification problem.

The main goal based on BFO algorithm is to apply in order to find the minimum of $P(\varphi)$, $\varphi \in R^n$, not in the gradient $\Diamond P(\varphi)$. Here, when φ is the position of a bacterium, and $J(\varphi)$ is an attractant-repellant profile. A neutral medium, and the presence of noxious substances, respectively can be showed by

$$H(j,k,l) = \{\varphi^i(j,k,l) \mid i = 1,2,\cdots,N\} \tag{6}$$

Eq. (6) represents the positions of each member in the population of the N bacteria at jth chemotactic step, kth reproduction step, and lth elimination-dispersal event. $P(i,j,k,l)$ denote the cost at the location of ith bacterium $\varphi^i(j,k,l) \in R^n$. Let

$$\varphi^i(j+1,k,l) = \varphi^i(j,k,l) + C(i)\varphi(j) \tag{7}$$

so that $C(i)>0$ is the size of the step taken in the random direction specified by the tumble. If at $\varphi^i(j+1,k,l)$ the cost $J(i,j+1,k,l)$ is better (lower) than at $\varphi^i(j,k,l)$, then another chemotatic step of size $C(i)$ in this same direction will be taken and repeated up to a maximum number of steps N_S. Functions $P_c^i(\varphi)$, i=1,2,···, S, to model the cell-to-cell signaling via an attractant and a repellant is represented by

$$P_c(\varphi) = \sum_{i=1}^{N} P_{cc}^i = \sum_{i=1}^{N}[-L_{\text{attract}} \exp(-\delta_{\text{attract}} \sum_{j=1}^{n}(\varphi_j - \varphi_j^i)^2] + \sum_{i=1}^{N}[-K_{\text{repellant}} \exp(-\delta_{\text{repellant}} \sum_{j=1}^{n}(\varphi_j - \varphi_j^i)^2] \tag{8}$$

where, L_{attract} is the depth of the attractant released by the cell, δ_{attrac} is a measure of the width of the attractant signal, $K_{\text{reppllant}} = L_{\text{attract}}$ is the height of the repellant effect magnitude, and $\delta_{\text{repellant}}$ is a measure of the width of the repellant. This paper describes method in the form of an algorithm to search optimal value of system parameter.The steps of BFO algorithm can be briefly summarized as follows.

Step 1: Initialize parameters n, N, N_C, N_S, N_{re}, N_{ed}, P_{ed}, $C(i)$, φ^i, and random values of system model parameter.

Step 2: Elimination-dispersal loop: $l=l+1$

Step 3: Reproduction loop: $k=k+1$

Step 4: Chemotaxis loop: $j=j+1$

Step 5: If $j<N_C$, go to step 3.

Step 6: Reproduction: The least healthy bacteria eventually die while each of the healthier bacteria split into two bacterial which are placed in the same location.

Step 7: If $k<N_{re}$, go to Step 3. In this case, we have not reached the number of specified reproduction steps, so we start next generation in the chemotatic loop.

Step 8: Elimination-dispersal: For $i=1,2,\cdots,N$, with probability P_{ed}, eliminate and disperse each bacterium. If $l<N_{ed}$, then go to Step2; otherwise, end. We obtain the estimate values of system model parameter.

4 Meta-model Selection

The sample model is composed by various sub-model, and each sub-model are constituted by the combination of the meta-model and independent variables. So the selection of the meta-model for system identification plays a decisive role. The selection of the meta-model should follow the following principles: 1) Common character; 2) Typical character; 3) Cover character.

In the simulation study of this paper, we selected the following typical meta-model: 1)Linear model: $y=bx$; 2)Exponential function model: $y=ae^{bx}$; 3)Negative exponential function model: $y=ae^{bx}$; 4)Power function model: $y=ax^b$; 5)Logarithm function model: $y=a\ln(b+x)$; 6)Periodic function model: $y=a\sin(bx)$.

5 Simulation Examples

Example 1. Set single-variable system model:

$$y=1+1.2\sin(0.3x)-0.2\ln(0.6+x) \tag{9}$$

According to (9), 30 groups of sample data are produced, assuming no prior knowledge of model structure, we select all meta-model of this paper to implement identification of the system model. In identification, the parameters of BFO algorithm are set: $N=10$, $N_C=100$, $N_S=4$, $N_{re}=4$, $N_{ed}=2$, $P_{ed}=0.25$, and $C(i)=0.025$. Let system model parameters' searching ranges are the same, i.e., [-1.2,1.2].

After multiple simulations are carried out, a following better result is chosen.

$$y=0.9977+1.1892\sin(0.2985x)-0.1962\ln(0.6011+x) \tag{10}$$

The maximum deviation is 0.0320, mean square error is 0.0227, and the fitting result is shown in Fig. 1.

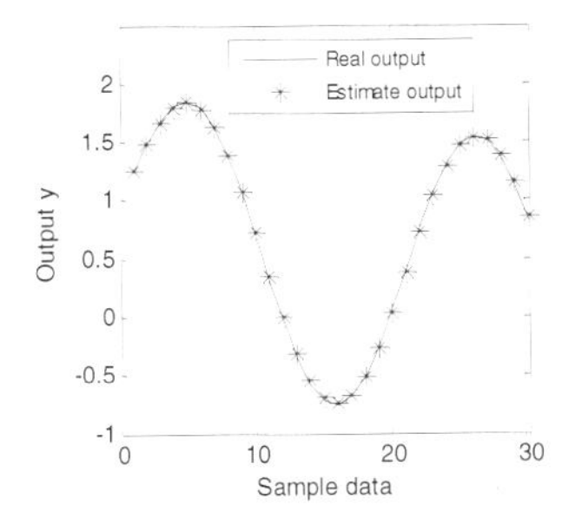

Fig. 1 Identification results of Example 1.

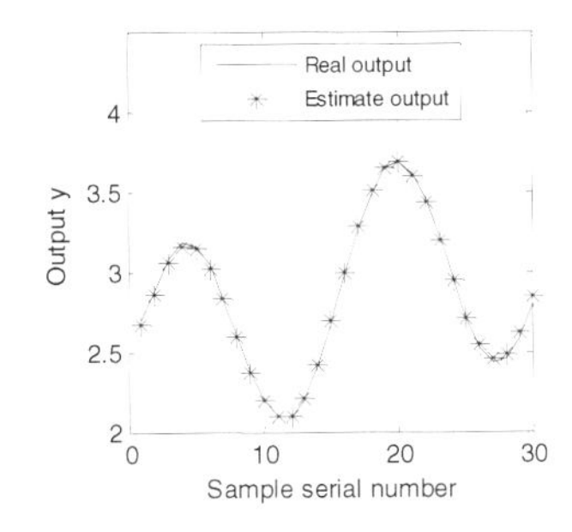

Fig. 2 Identification results of Example 2.

Example 2. Set multiple-variable system model:

$$y = 1 + 0.6x_1^{0.3} + 0.5e^{0.5x_2^{-1}} + 0.7\sin(0.4x_3) \tag{11}$$

Using (11), 30 sets of sample data is produced, we still do not the structure of system model, and select all meta-model of this paper to implement identification of system. In simulation, the parameters of BFO algorithm are same as those of Example 1. After multiple simulations are carried out, we choose a better result.

$$y = 0.9695 + 0.5903x_1^{0.3104} + 0.5172e^{0.4894x_2^{-1}} + 0.6898\sin(0.4011x_3) \tag{12}$$

The maximum deviation is 0.0437, mean square error is 0.0190, and the fitting result is also shown in Fig. 2.

From the simulation results of Examples 1 and 2, we can see that the error of the proposed identification algorithm is relatively minor, the stability of the proposed identification algorithm is good.

6 Conclusion

In this paper, we presented an identification method for a static MISO system. The idea is to employ a system model composed with classical models so as to change the system structure identification problem into a combinational problem. And then, the bacterial foraging optimization technique is applied to implement the identification of the system model structure and parameters. However, the system identification problems of other cases need to further explore and study.

Acknowledgments. This work is supported by Scientific Research Program Funded by Shaanxi Provincial Education Department (Program No. 2010JK708), the National Natural Science Foundation of China (Grant No. 60874033, 60971127), the Ph. D. Scientific Research Start-up Funds of Teachers of Xi'an University of Technology of China (Grant No. 108-211006, 108-210905), the Project Item of Technological Innovation of Xi'an University of Technology of China, and the Campus Foundation of Xi'an Jiaotong University of China.

References

1. Liu, S.A., Tang, F.: Study on system identification based on genetic algorithm method. System Engineering Theory and Practice 23(1), 134–137 (2007)
2. Liu, Y., Passino, K.M.: Biomimicry of social foraging bacterial for distributed optimization: models, principles, and emergent behaviors. Journal of Optimization Theory and Applications 115(4), 603–628 (2002)
3. Liu, X.G., Lu, J.: Least squares based iterative identification for a class of multirate systems. Automatical 46(3), 549–554 (2010)
4. Passino, K.M.: Biomimicry of bacterial foraging for distributed optimization and control. IEEE Control Systems Magazine 22, 52–67 (2002)
5. Pillonetto, G., De Nicolao, G.: A new kernel-based approach for linear system identification. Automatica 46(1), 81–93 (2010)
6. Wang, L.Y., Yin, G.G.: Quantized identification with dependent noise and Fisher information ratio of communication channels. IEEE Transaction on Automatic Control 55(3), 674–690 (2010)

Extracting Characteristics from Corrosion Surface of Carbon Steel Based on WPT and SVD

Li Guo-bin and Li Ting-ju

Abstract. In order to study the early corrosion behavior of marine carbon steel in electrolyzed ballast water, a novel approach for extracting feature of the corrosion morphology images was proposed using waveletpacket transform (WPT) and singularity value decomposition (SVD). The waveletpacket transform and the singularity value decomposition arithmetic were introduced. The characteristic parameters $p_1$and p_2 were defined based on the singular value characteristic vector. The intrinsic relations between the characteristic parameters and the corrosion behavior of marine carbon steel were discussed. It has been shown that the characteristic parameters can reveal the early corrosion behavior of marine carbon steel. The parameter p_1 reflects the corrosion rate of carbon steel, which is coincident with the evolution of corrosion weight loss, the larger characteristic parameter p_1, the larger corrosion weight loss and the higher corrosion rate. The parameter p_2 reflects the size and quantity of pitting corrosion, the larger characteristic parameter p_2, the larger and few pitting corrosion. Therefore the corrosion morphology can be identified by the characteristic parameters $p_1$and p_2 .

Keywords: Carbon steel, Corrosion morphology, WPT, SVD, Characteristic parameter.

1 Introduction

Surface corrosion morphology contains large amount of corrosion information that is one of the important characteristics to be used to evaluate the material corrosion

Li Guo-bin · Li Ting-ju
Postdoctoral Station of School of Materials Science and Engineering, Dalian University of Technology, Dalian 116024, China
e-mail: guobinli88@yahoo.com.cn

Li Guo-bin
Marine Engineering College, Dalian Maritime University, Dalian 116026, China

Z. Qian et al. (Eds.): Recent Advances in CSIE 2011, LNEE 125, pp. 105–112.
springerlink.com

performance. Through the characteristic collecting of the corrosion image, we can obtain useful corrosion data. But different corrosion image has different texture characteristic. Texture diversity and complexity make the characteristic collecting of the corrosion image have difficulty. In recent years, the WPT obtains abundant achievement in the characteristic collecting of the corrosion images, that cause many corrosion science workers' attention[1-4]. Song Shizhe etal. used the continuous WPT to disassemble and collect energy value as the characteristic information, studied the relationship between the image characteristic and sample corrosion weightlessness data[1], Livens etal[3]. brought in the WPT into corrosion morphology image texture characteristic analysis and original corrosion image disassembling, at the same time, treated the corrosion subsidiary image energy value as the image characteristic value to classify two kinds of corrosion. Fnarztiksonis etal[4]. explained the way of obtaining corrosion image under more scale conditions in detail. While the study shows the WPT using for obtaining the image texture characteristic will lose part of high frequency details information[5-6], after the texture image is disassembled by waveletpacket, we get different disassembling quotient under different scales, and these quotients compose high dimension characteristic matrix reflect image texture characteristic[7], but how to obtain the characteristic parameter reflecting the texture characteristic from high dimension characteristic matrix should be further study.

Waveletpacket disassembling has the ability to analysis texture image high frequency, the SVD in the matrix information dealing again has unique superiority, high dimension matrix that is decomposed by singularity value can get one sole vector singularity value to reflect matrix essence characteristic. Therefore, this article discussed the application of WPT and SVD in the collecting corrosion image characteristic, and come up with arithmetic. Through the definition characteristic parameters, we studied the corrosion behavior of marine carbon steel in the electrolyzed seawater.

2 WPT and SVD for Image

2.1 WPT and Characteristic Matrix

The WPT is a meticulous signal analysis method coming from the wavelet transform. The wavelet analysis can make the two dimension grey image decompose low frequency rough part and high frequency detail part, and then continue decomposing low frequency information, then get low frequency information and high frequency information by parity of reasoning, not doing anything for the high frequency information, while the waveletpacket decomposition not only deals with the low frequency information but also the high frequency information. We use the waveletpacket decomposition to collect corrosion morphology image characteristic. The wavelet packet decomposition tree is shown in Fig. 1.

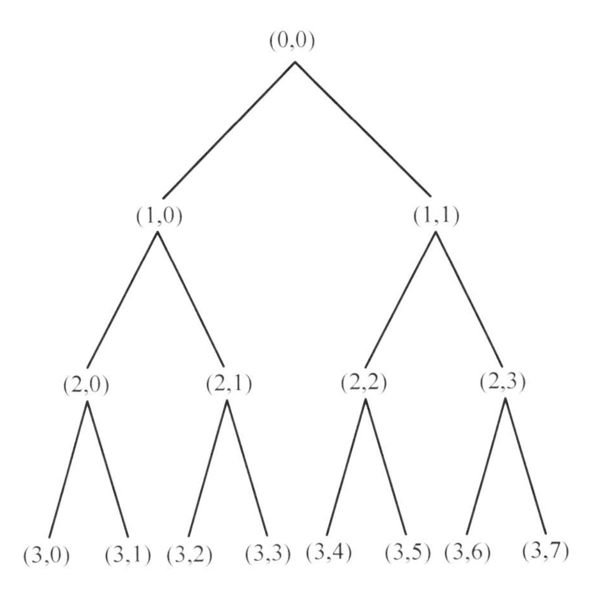

Fig. 1 Tree structure of waveletpacket decomposes

As shown in Fig.1, the wavelet packet decomposition has the relationship: $S_{0,0}=S_{3,0}+S_{3,1}+S_{3,2}+S_{3,3}+S_{3,4}+S_{3,5}+S_{3,6}+S_{3,7}$. Therefore waveletpacket decomposition divides every frequency area, makes up for the lack of high-frequency information in the wavelet analysis and can describe the image characteristic information comprehensively.

In the WPT, the 256 gray image is indicated by the wavelet packet coefficients. The increasing of decomposition level makes the image high frequency and low frequency components reached a very fine degree. Therefore, The original grayscale image is decomposed into j layer by the WPT, the wavelet packet coefficients can constitute a pixel gray matrix A as follows:

$$A = \begin{bmatrix} a_{11} & a_{12} & \cdots & a_{1n} \\ a_{21} & a_{22} & \cdots & a_{2n} \\ \vdots & \vdots & & \vdots \\ a_{m1} & a_{m2} & \cdots & a_{mn} \end{bmatrix} \quad (1)$$

In above formula, $n=N/2^j$, $m=2^j$, U, N is total points of the image pixels. According to the characteristics of machined surface image, the decomposition was taken up to the level j = 5 and the db20 wavelet was chose.

2.2 SVD and Characteristic Parameters

Taking the image matrix A as m×n's non-negative matrix, rank (A)=r, $r \leq m$, then resolving this to Matrix A's singular value as follows:

$$A = UDV^T = [u_1, u_2, \cdots u_n] \begin{pmatrix} \lambda_1 & & 0 \\ & \ddots & \\ 0 & & \lambda_n \end{pmatrix} [v_1, v_2, \cdots v_n]^T = \sum_{i=1}^{r} \lambda_i u_i v_i^T \tag{2}$$

In above formula, U, V are respectively the m×m 's matrix and the n×n 's matrix, their column vectors are respectively u_i, v_i, D is m×n 's diagonal matrix and its singular value λ_i satisfies the equation:

$$\lambda_1 \geq \lambda_2 \geq \ldots \geq \lambda_r \geq \lambda_{r+1} = \ldots = \lambda_n = 0 \tag{3}$$

Non-zero singular values are equal to matrix's rank. If these non-zero singular values form a characteristic vector *X*, and it can be obtained as follows:

$$X = (\lambda_1, \lambda_2, \ldots, \lambda_r) \tag{4}$$

Matrix A's singular value is the average non-negative real number and is also unique. The singular value has relative stability towards disturbance and unchangeableness towards matrix transformation. In linear algebra, the matrix feature value shows the matrix feature, while the matrix singular value is better than its feature value in manifesting its feature. So the image matrix singular value reflects the image's "energy feature" while its corresponding singular vector reflects the image's "geometrical feature".

Texture characteristic parametersp1and p2 is defined according to the characteristic vector *X*, as follows:

$$p_1 = \bar{x} \tag{5}$$

$$p_2 = \frac{(x_1 - \bar{x})^2 + (x_2 - \bar{x})^2 + \ldots + (x_n - \bar{x})^2}{n} \tag{6}$$

p_1 is the mean value of the characteristic vector x, reflect the size of the pixel gray values, and closely related to the development of surface corrosion. The bigger P_1 is, the larger the pixel gray value is, the greater the overall brightness of the image is, and the greater the corrosion of the surface is. P_2 is the variance value of the characteristic vector x, can reflect the degree of fluctuation of pixel gray value, and closely related to the size and number of corrosion surface pit. The bigger P_2 is, the greater the fluctuations in gray value is, the greater the corrosion of the surface pit is, the fewer the number is.

3 Feature Extractions and Analysis for Surface Corrosion

3.1 Test Method

The experiment metal was the low carbon steel from the ship's ballast tanks plate, and its chemical compositions (wt %) were: 0.21 C, 0.5 Si, 0.6 Mn, 0.035 P, 0.035 S and balance Fe. The samples with 70 mm × 40 mm × 37 mm were prepared by electrospark-machined from the experiment metal. Surfaces of the samples were mechanically polished with the waterproof silicon carbide paper down to 2000 grade. The chlorinated ballast water was selected as the corrosive media, and the chlorine concentrations were 0 mg/L, 7.5 mg/L, 17 mg/L and 28 mg/L respectively. The test temperature was 24℃ and the corrosion time was 7 days. Mettler Toledo-Al204-type electronic balance with an accuracy of 1 mg was employed to measure the mass of the samples before and after the test. The corrosion morphology of the specimen surface was collected by Philips-XL-30 scanning electron microscope.

3.2 Consequence and Discussion

The surface topography of the specimen in different corrosive media after experiment is shown in Fig. 2.

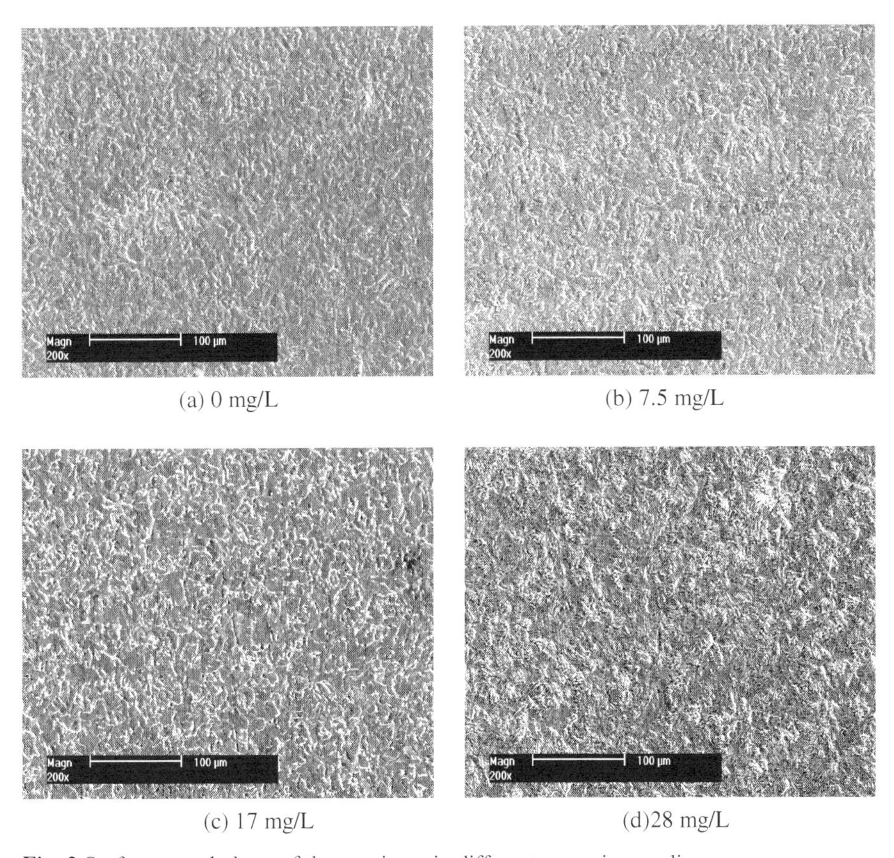

(a) 0 mg/L (b) 7.5 mg/L

(c) 17 mg/L (d)28 mg/L

Fig. 2 Surface morphology of the specimen in different corrosive media

It can be seen that the samples in different corrosive media are corrosive homogeneous. Corrosive topography appear the sheet features, the entire corrosion region is divided into some small unit regions with the white boundary. With the chlorine concentration increasing, the unit region reduces, the white boundary increases, the refinement lamellar morphology of corrosion appears. Form the perspective of microstructure, the corrosive topography is coincided with actual corrosive process. The microstructures of sample are produced by ferrite and pearlite. Owing to the inhomogeneity of tissues, cementite becomes micro- negative pole compared with ferrite, therefore, ferrite will be corrosived, resulting in corrosion topography with lamellar, as shown in Fig. 2. With the chlorine concentration increasing, the corrosion increases. When the chlorine concentration is 28mg/L, the corrosion becomes severe and appears refinement lamellar morphology, as shown in Fig. 2 (d).

The character parameters of corrosion surface are calculated by applying the method put forward in the paper. Table 1 shows the parameter p_1 and p_2. In order to analyze and compare, the corrosion weight loss also is recorded in Table 1. According to the result of Table 1, the corrosion analysis for the marine carbon steel is detailed as the following.

Table 1 Parameters and corrosion weight loss

chlorine concentration (mg/L)	weight loss (mg/)	P_1 ($\times 10^3$)	p_2 ($\times 10^8$)
0	1.92	8.3271	12.3960
7.5	2.43	9.0025	8.9280
17	2.81	9.8710	7.5891
28	3.32	10.373	7.5501

We can see from Table 1 that the characteristic parameters p_1 and corrosion weight loss show the same variation. The corrosion weight loss and the characteristic parameters p_1 increase with the increasing chlorine concentration. According to the definition of character parameter p_1, it reflects the image gray values and reveals the corrosion of the surface, the larger character parameter p_1, the larger image gray values, the more severe corrosion. When the chlorine concentration in the chlorinated ballast water is 0 mg/L, the image gray value is minimum, and the corrosion is slight; When the chlorine concentration in the chlorinated ballast water is 7.5 mg/L and 17 mg/L, the image gray value is gradually increasing, and the corrosion is gradually increasing too; When the chlorine concentration in the chlorinated ballast water is 28 mg/L, the image gray value is maximum, and the corrosion is Severe. The analysis of character parameters p_1 shows that the corrosion of the surface can be increased by increasing chlorine concentration in the chlorinated ballast water. The corrosion weight loss further verifies this result.

Seen from Table 1, the characteristic parameters p_2 decreases with the increasing chlorine concentration. According to the definition of character parameter p_2, it reflects the variation of image gray values and reveals the size and number of corrosion pit in the surface, the larger character parameter p_2, the larger variation of image gray values, and the smaller and the more corrosion pit. When the chlorine concentration in the chlorinated ballast water is 0 mg/L, the variation of image gray values is maximum, the size of corrosion pit is maximum too, but the number of corrosion pit is minimum; When the chlorine concentration in the chlorinated ballast water is 7.5 mg/L and 17 mg/L, the variation of image gray values is gradually steady, the size of corrosion pit is gradually decreasing, but the number of corrosion pit is gradually increasing; When the chlorine concentration in the chlorinated ballast water is 28 mg/L, the variation of image gray values is minimum, the size of corrosion pit is minimum, but the number of corrosion pit is maximum. The analysis results show that the size and number of corrosion pit in the surface can be reflected by the character parameters p_2.

4 Conclusion

The characteristic parameter p_1 and p_2 of the corrosion image is put forward by applying the WPT and the SVD. The characteristic parameter p_1 and p_2 can be used to analyze the corrosion process.

The marine carbon steel corrosion analysis show that the characteristic parameter p_1 reveals the corrosion of the surface, the larger character parameter p_1, the more severe corrosion, and is consistent with the changes of corrosion weight loss; The character parameter p_2 reveals the size and number of corrosion pit in the surface, the larger character parameter p_2, the smaller and the more corrosion pit.

References

1. Song, S.Z., Wang, S.Y., Gao, Z.M., et al.: Atmospheric Forepart Corrosion Behaviors of Nonferrous Metal Based on Image Recognition. Acta Metall. Sin. 38(8), 893–896 (2002)
2. Wang, S.Y., Gao, Z.M., Song, S.Z.: Feature Extraction and Analysis of Corrosion Image of Materials in Seawater. Corrosion Science and Technology Protection 13(Z), 461–463 (2002)
3. Livens, S., Seheunders, R., de Wbuwer, G.V., et al.: A Texture Analysis Approach to Corrosion Image Classification. Microscopy, Microanalysis 7(2), 1–10 (1996)
4. Frantziskonis, G.N., Simon, L.B., Woo, J., et al.: Multiscale Characterization of Pitting Corrosion and Cpplication to An Aluminum alloy. European Journal of Mechanics-A/Solids 19(2), 309–318 (2000)
5. Correia, S.E.N., Sabourin, R.: On the Performance of Wavelets for Handwritten Numerals Recognition. In: International Conference on Pattern Recognition, pp. 127–130 (2002)

6. Jian, M.W., Dong, J.Y., Zhang, Y.: Image Fusion Based on Wavelet Transform. In: Proceedings of the 8th ACIS International Conference on Software Engineering, Artificial Intelligence, Networking, and Parallel/Distributed Computing, pp. 713–718. IEEE, Chicago (2007)
7. Wang, S.H., Du, D., Zeng, K., et al.: Weld Recognition Based on Texture Feature. Transactions of The China Welding Institution 29(11), 5–8 (2008)

Study Actuality of Immune Optimization Algoriithm and It's Future Development

Shi Weili, Zhang Wenbo, Bai Baoxing, Xu Honghua, and Miao Yu

Abstract. In this paper, an optimization algorithm base immune principle is expatiated, explain its basic theory and process. And discuss immune algorithm's advantage than other heuristic algorithms, such as: genetic algorithm and evolution strategy. And introduce several better algorithms base immune algorithm, present application in optimization problems. At last we propose immune algorithm's further development in optimization problems' application.

Keywords: immune algorithm, genetic algorithm, evolution strategy, optimization problem.

1 Introduction

The nature immune system is one of the most intricate bodily systems. It is a very important effect in make sure living healthy. In so complexity nature, immune system protects the body from a large variety of bacteria, viruses, and other pathogenic organisms.

A novel immune algorithm was proposed by imitating the defending process and mechanism of a nature immune system, and it takes objective functions and constraints of a problem as antigens, feasible solution of the problem as an antibody. Base on the nature immune theory, the nature immune system produce antibody corresponding antigen invading bodily via that cellular divide and polarization effect by self-motion, and the process called immune response. In the immune response processing, some of the antibodies are saved, when the same kind of the

Shi Weili
Changchun University of Science and Technology, School of Computer Science and Technology. Weixing Road 7089, Changchun, China
e-mail: shiwl@cust.edu.cn

Zhang Wenbo
CNR Chuangchun Railway Vehicles Facilitycs Co., Ltd Information Department. Qingyin Road 435, Changchun, China

Z. Qian et al. (Eds.): Recent Advances in CSIE 2011, LNEE 125, pp. 113–118.
springerlink.com

antigens invade bodily again, the memory cell will be started and produce larger antibodies rapidly than last times, make the immune response more stronger than the last response, incarnate the memory function in immune system. After combine antibody and antigen, via a series of reactions to destroy antigen. At the same time, between antibody and antigen promote and suppression each other, to keeping antibody diversity and immune balance, the balance is keeping base on the consistency mechanism, namely the more antibodies consistency high, the more it is suppression; The more lower, the more be promote. Namely the immune body consistency goes past higher, then more receives suppresses; thickly goes past lower, then more receives promotes, incarnate immune system have self-adjust function.

2 Composite Optimization Algorithm

2.1 Immune Evolution Algorithm

(1) Introduce Algorithm

Article [21] design an immune evolution algorithm, first introduce adjacent field concept, via definite expand radius and mutation radius, construct small adjacent field and large adjacent field. Utilizing the two adjacent fields, via expand operation and mutation operation, process local search and overall search at the same time, achieve two levels adjacent field search mechanism.

(2) Immune Evolution Algorithm Approaches

① Selective operation, produce n real-coding units randomly, and called initial population A, and calculate every unit evaluate value. Selective m units that evaluate values are most high, compose group B (m<n).

② Simulation immune system clone expand process, article [21] construct a small adjacent field, every unit of group B produce several new unit randomly in corresponding adjacent field, m units produce n new units and compose group C.

③ Mutation operation, construct a larger adjacent field, select n-m lowest evaluate units in group C, mutation and become some units in the larger adjacent field. Mutation produced new units and not mutation units in group C compose group D.

④ Replace operation, produce l units randomly replace l lowest evaluate units in group D, and produce new group called group E. Replace operation simulate marrow produce new B cell process to increasing group diversity.

(3) Algorithm Merit and Application

Solves three models multi-peaks function problem (De Jong's F5 function, Schaffer's F6 function and Schaffer's F7 function) with immune evolution algorithm[21].This algorithm had guaranteed the overall situation seeks superior ability and high precision, also has the high search efficiency. This algorithm has provided an effective way for the complex continual parameter optimization question.

2.2 *Immune Genetic Algorithm*

(1) Immune Genetic Algorithm Base Approach

Article [8] [9] combine immune algorithm and genetic algorithm design a optimization modal base immune genetic theory.

For optimization problem, base on article [9] introduce immune genetic algorithm, the base flow chart as show as fig. 5, calculate process as follow:

① Initialization: Setting immune genetic calculate parameter;

② Antigen identity: Antigen identity correspond data input of optimization problem, objective function of optimization design to take the immune algorithm antigen;

③ Initial antibody produce: Parameter of optimize design to take the immune algorithm antibody. Base on the constraint of optimization problem, algorithm produce n groups of initial design vectors randomly, as the immune system initial antibody population N.

④ Affinity calculate: Calculate the affinity between antibody v and antigen, and the affinity between antibody v and w, and arrangement base the affinity.

⑤ Antibody concentration calculate: Calculate antibody concentration, and arrangement base concentration.

⑥ Antibody encourage and suppress: Base on the calculate result of the antibody and antigen affinity, selective antibodies of the affinity value higher than valve value T enter the next round iteration, and eliminate the partial antibodies that have higher concentration.

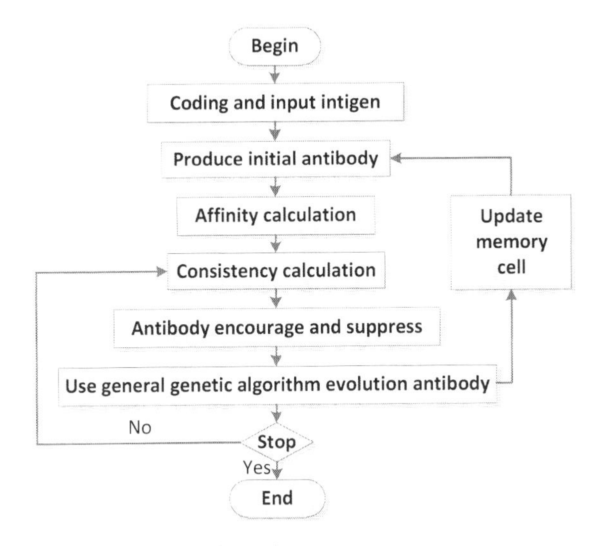

Fig. 1 Immune genetic algorithm base flow chart

⑦ New antibodies produce: Use normal genetic algorithm produce new antibodies.

⑧ Terminate condition judgment: Differentiate terminate condition, if satisfy, the calculation terminate, else calculation continue.

⑨ Get the optimization solution: From the finally generate antibody group, get the biggest affinity antibody. The antibody is the optimization solution of the problem.

(2) Algorithm Advantage and Application

Unify immune algorithm and genetic algorithm, construct an immune genetic algorithm modal, avoid genetic algorithm disadvantage, for example, early mature, search efficiency low, can't keep unit diversity etc.

Immune genetic algorithms have a very big superiority and good application prospect in optimized design. Based on the immune genetic optimized algorithm supports the distribution coordination optimization, It may through the living system inherent distributed computing mechanism, realizes the design modeling, controls variable, intelligence adjustment, the coordination seeks superiorly, intelligent diagnosis, process analysis and result evaluation and so on. Further research based on immunity genetic distributed computing function, through different antigen and antibody synthesis coordination, produce the multi-objectives design question optimal solution.

3 Conclusion and Prospect Forecast

3.1 Conclusion

Article [22] via experiment prove base immune algorithm lack convergence characteristic. Analyze the reason, the antibody evaluate manner and utilize memory library are main problem. Immune algorithm base vaccine inoculation and information entropy improve on the antibody evaluate manner, encourage high matching degree antibody, at the same time, suppress high concentration antibody, make sure the diversity of group, improve the algorithm efficiency; Base on immune recognition and immune evolution, the algorithm improve on the memory library (gene library), make memory library keep and update optimization all the time, algorithm can find more suitable antigen, at the same time spend less time.

The immunity evolution algorithm improved the basic immunity algorithm to have the strong overall importance thick search, lacked carries on the high accuracy search ability in the partial region. It carries on overall large adjacent field search, finds evaluate value high area, and then carries on the local accuracy search in this area. Through from thick to accuracy the two adjacent field search, had achieved guaranteed its overall search optimize solution and local seek accuracy solution. Immune genetic algorithm has inherited the immune algorithm and genetic algorithm merit, can more effective find the overall situation optimal solution. Article [23] design a improvement immune genetic algorithm, introduced the pattern memory library, causes a memory cell more effective renewal, has solved the problem which the general immunity genetic algorithms is short of memory function.

3.2 Development and Prospect Forecast

According to immune algorithm research experience, external violation examination and internal study mechanism optimized become two big difficulties [24].At present uses the external violation examination algorithm extremely simple, has the very big disparity with organism function, can't satisfy need of the actual optimized question application. Present immune algorithm general establishment in precise mathematical model or in formula foundation. The mathematical model no doubt simple, is easy to realize, but the function is not strong, the result often distorts, intelligent is not high, also is not advantageous for the improvement.

Although the immune algorithm development has many difficulties, but the immune algorithm prospects for development are still extremely optimistic, the application immune algorithm carries on optimized the example to be more and more in the real life. Along with the immune algorithm mature, gradually becomes hot spot[20] [19] [25] [18] of the artificial immunization system the application.

The next immune optimization algorithm research will approximately concentrate with emphasis in: discriminate self and non-self capability; immune optimization algorithm intellectualization and several other aspects. Along with the immune optimization algorithm development and the consummate, will open the new development space for the stochastic optimized method.

References

[1] Mo, H., Jin, H.: Immune algorithm and application. Aeronautical Computer Technique 32(4), 49–51 (2002)

[2] Fu, H., Wang, F.: Intrusion detection system model based on mathematic statistic and immunlogical principle. Computer Engineering and Design 29(12), 3037–3039 (2008)

[3] Chun, J.-S., Jung, H.-K., Hahn, S.-Y.: A Study on Comparison of Optimization Performances between Immune Algorithm and other euristie Algorithms. IEEE Transactions On Magnetics 34, 2972–2975 (1998)

[4] Jiao, L., Du, H.: Development and Prospect of the Artificial Immune System. Acta Electronica Sinica 32(10), 1540–1548 (2003)

[5] Tan, G.-X., Mao, Z.-Y.: Study on Pareto Front of Muliti-Objective Optimization using Immune Algorithm. In: Proceedings of the Fourth International Conference on Machine Learning and Cybernetics, pp. 2923–2928 (2005)

[6] Chun, J.-S., Kim, M.-K., Jung, H.-K.: Shape Optimization of electromagnetic devices using Immune Algorithm. IEEE Transactions on Magnetics 33, 1876–1879 (1997)

[7] Yang, K., Wang, X.: Research and im plement of adaptive multimodal im mune evolution algorithm. Control and Decision 20(6), 717–720 (2005)

[8] Liao, G.-C., Tsao, T.-P.: Application embedded chaos search immune genetic algorithm for short-term unit commitment. Electric Power Systems Research 71, 135–144 (2004)
[9] Yang, J., Li, B., Yu, L.: Based on immune genetic algorithm in Optimization Design. Journal of Machine Design 19(9), 14–17 (2002)
[10] Branke, J., Kaubler, T., Schmeck, H.: Guidance in evolutionary multiobjective optimization. Advances in Engineering Software 32, 499–507 (2001)
[11] Bosman, P.A., Thierens, D.: Multi-objective optimization with diversity preserving mixture-based iterated density estimation evolutionary algorithms. International Journal of Approximate Reasoning 31, 259–289 (2002)
[12] Summanwar, V., Jayaraman, V., Kulkarni, B., Kusumakar, H., Gupta, K., Rajesh, J.: Solution of constrained optimization problems by multiobjective genetic algorithm. Computers and Chemical Engineering 26, 1492 (2002)
[13] Deb, K., Gulati, S.: Design of truss-structures for minimum weight using genetic algorithms. Finite Elements in Analysis and Design 37, 447–465 (2001)
[14] Wang, L., Pan, J., Jiao, L.: The Immune Algorithm. Acta Electronica Sinica 28(7), 74–78 (2000)
[15] Su, C., Zhu, X.: An Immune Optimal Algorithm and Its Application. Journal of Southwest Jiaotong University 37(6), 677–680 (2002)
[16] Zhang, S., Cao, X., Wang, X.: Immune Algorithm Based on Immune Recognition. Acta Electronica Sinica 30(12), 1840–1844 (2002)
[17] Shao, X., Yu, Z., Sun, L.: Immune algorithms in analytical chemistry. Trends in Analytical Chemistry 22, 59–69 (2003)
[18] Castro, L., Timmis, J.I.: Artificial immune systems as a novel soft computing paradigm. Soft Computing 7, 526–544 (2003)
[19] Castro, L., Zuben, F.J.: Artificial Immune Systems:Part I – Basic Theory and Applications (1999)
[20] Castro, L., Zuben, F.J.: Artificial Immune Systems:Part II – A Survey of Applications (2000)
[21] Zuo, X., Mo, H.: Survey on immune scheduling algorithms. Control and Decision 24(12), 1761–1768 (2009)
[22] Yi, Z.: An Improved Immune Algorithm. Journal of East China Jiaotong University 24(1), 123–128 (2007)
[23] Wu, X., Zhu, Z.: Schema Control for Immune Genetic Algorithm. Computer Simulation 22(8), 98–101 (2005)
[24] Cai, Z., Gong, T.: Advance in research on immune algorithms. Control and Decision 19(8), 841–846 (2004)
[25] Cayzer, S.: Artificial Immune Systems (2005)

Fuzzy Rough Set Based on Dominance Relations

Zhang Xiaoyan and Xu Weihua

Abstract This model for fuzzy rough sets is one of the most important parts in rough set theory. Moreover, it is based on an equivalence relation (indiscernibility relation). However, many systems are not only concerned with fuzzy sets, but also based on a dominance relation because of various factors in practice. To acquire knowledge from the systems, construction of model for fuzzy rough sets based on dominance relations is very necessary. The main aim to this paper is to study this issue. Concepts of the lower and the upper approximations of fuzzy rough sets based on dominance relations are proposed. Furthermore, model for fuzzy rough sets based on dominance relations is constructed, and some properties are discussed.

1 Introduction

The rough set theory [10,11], proposed by Pawlak in the early 1980s, is an extension of set theory for the study of intelligent systems. It can serve as a new mathematical tool to soft computing, and deal with inexact, uncertain or vague information. Moreover, this theory has been applied successfully in discovering hidden patterns in data, recognizing partial or total dependencies in systems, removing redundant knowledge, and many others [7,12,13,15]. Since its introduction, the theory has received wide attention on the research areas in both of the real-life applications and the theory itself.

Theory of fuzzy sets initiated by Zedeh [9] also provides useful ways of describing and modeling vagueness in ill-defined environment. Naturally, Doubois and Prade [8] combined fuzzy sets and rough sets. Attempts to combine these two

Zhang Xiaoyan
School of Mathematics and Statistics, Chongqing University of Technology, Chongqing 400054, P.R. China
e-mail: zhangxyms@gmail.com

Xu Weihua
School of Management, Xi'an Jiaotong University, P.R. China
e-mail: chxuwh@gmail.com

Z. Qian et al. (Eds.): Recent Advances in CSIE 2011, LNEE 125, pp. 119–125.
springerlink.com

theories lead to some new notions [1,5,7], and some progresses were made [2,3,4,5,6,14]. The combination involves many types of approximations and the construction of fuzzy rough sets give a good model for solving this problem [5]. However, most of systems are not only concerned with fuzzy data, but also based on a dominance relation because of various factors. In order to obtain the succinct knowledge from the systems, construction of model for fuzzy rough sets based on dominance relations is needed.

The main aim of the paper is to discuss the issue. In present paper, a dominance relation is introduced and instead of the equivalence relation (discernibility relation) in the standard fuzzy rough set theory. The lower and the upper approximation of a fuzzy rough set based on dominance relations are proposed. Thus a model for fuzzy rough sets based on dominance relations is constructed, and some properties are studied. Finally, we conclude the paper and look ahead the further research.

2 Preliminaries

This section recalls necessary some concepts used in the paper. Detailed description can be found in [15].

Definition 2.1. Let U be a set called universe and let R be an equivalence relation (indiscernibility) on U. The pair $S=(U,R)$ is called a Pawlak approximation space. Then for any non-empty subset X of U, the sets $\underline{R}(X)=\{x\in U: [x]_R\subseteq X\}$ and $\overline{R}(X)=\{x\in U: [x]_R\cap X\neq\varnothing\}$ are respectively, called the lower and the upper approximations of X in S, where $[x]_R$ denotes the equivalence class of the relation R containing the element x.

Then X is said to be definable set, if $\underline{R}(X)=\overline{R}(X)$. Otherwise X is said to be rough set. Pawlak approximation space $S=(U,R)$ derives mainly from an equivalence relation. However, there exist a great of systems based on dominance relations in practice.

Definition 2.2. If we denote $R_B^{\leq}=\{(x_i,x_j)\in U\times U: f_l(x_i)\leq f_l(x_j), \forall a_l\in B\}$ where B is a subset of attributes set, and $f_l(x)$ is the value of attribute a_l, then $R_B^{\leq}$ is referred to as dominance relation of information system S. Moreover, we denote approximation space based on dominance relations by $S^{\leq}=(U,R^{\leq})$.

For any non-empty subset X of U, denote $\underline{R^{\leq}}(X)=\{x\in U: [x]_{R^{\leq}}\subseteq X\}$ and $\overline{R^{\leq}}(X)=\{x\in U: [x]_{R^{\leq}}\cap X\neq\varnothing\}$ $\underline{R^{\leq}}$ and $\overline{R^{\leq}}$ are respectively said to be the lower and the upper approximations of X with respect to a dominance relation $R^{\leq}$.

If we denote $[x_i]_{B^{\leq}}=\{x_j\in U:(x_i,x_j)\in R_B^{\leq}\}=\{x_j\in U: f_l(x_i)\leq f_l(x_j), \forall a_l\in B\}$ then the following properties of a dominance relation are trivial.

Proposition 2.1. Let $R_B^{\leq}$ be a dominance relation.

(1) $R_B^{\leq}$ is reflexive and transitive, but not symmetric, so it isn't an equivalence relation generally.

(2) If $B_1 \subseteq B_2 \subseteq A$, then $R_A^{\leq} \subseteq R_{B_2}^{\leq} \subseteq R_{B_1}^{\leq}$.

(3) If $B_1 \subseteq B_2 \subseteq A$, then $[x_i]_A^{\leq} \subseteq [x_i]_{B_2}^{\leq} \subseteq [x_i]_{B_1}^{\leq}$.

(4) If $x_j \in [x_i]_B^{\leq}$, then $[x_j]_B^{\leq} \subseteq [x_i]_B^{\leq}$.

Next, we will review some notions of fuzzy sets. The notion of fuzzy sets provides a convenient tool for representing vague concepts by allowing partial membership. In fuzzy systems, a fuzzy set can be defined using standard set operators.

Definition 2.3. Let U be a finite and non-empty set called universe. A fuzzy set A of U is defined by a membership function $A: U \to [0,1]$ .The membership value may be interpreted in term of the membership degree. In generally, let $F(U)$ denote the set of all fuzzy sets, i.e., the set of all functions from U to [0, 1].

From above description, we can find that a crisp set can be regarded as a generated fuzzy set in which the membership is restricted to the extreme points $\{0,1\}$ of $[0,1]$.

Definition 2.4. Let $A, B \in F(U)$. For any $x \in U$, if $A(x) \leq B(X)$ is true, then we say that B contain A or A is contained by B, which denoted by $A \subseteq B$.

If both $A \subseteq B$ and $B \subseteq A$ are all true, then we say A is equal to B, denoted by $A = B$. Empty set $\varnothing$ denotes the fuzzy set whose membership function is 0, and set U denotes the fuzzy set whose membership function is 1.

Let denote intersection, union of A and B by $A \cap B$, $A \cup B$ respectively. Moreover, denote complement of A by $\sim A$ or A^C. Membership functions of these fuzzy sets are defined as

$$A \cap B = A(x) \wedge B(x) \quad = \min\{A(x), B(x)\}\ ;$$
$$A \cup B = A(x) \vee B(x) \quad = \max\{A(x), B(x)\}\ ;$$
$$\sim A = A^C = 1 - A(x)\ .$$

There many properties of these operators in fuzzy sets which similar with crisp sets. Detailed description can be found easily.

Definition 2.5. The λ-level set or λ-cut, denoted by A_λ, A_λ of a fuzzy set A in U comprise all elements of U whose degree membership in A are all greater than or equal to λ, where $0 < \lambda \leq 1$. In other words, $A_\lambda = \{x \in U : A(x) \geq \lambda\}$ is a non-fuzzy set, and be called the λ-level set or λ-cut. Moreover, the set $\{x \in U : A(x) > 0\}$ is defined the supports of fuzzy set A, and denoted by supp A.

3 Fuzzy Rough Sets Based on Dominance Relations

Model for fuzzy rough sets is generalized by the standard Pawlak approximation space, and it is concerned with fuzzy sets on universe U. But the model is still depended on an equivalence relation. In practice, most of systems are not only related to fuzzy sets, but also based on dominance relations. In order to deal with this problem, the model of fuzzy rough sets based on dominance relations is proposed.

Definition 3.1. Let $S^{\leq}=(U,R^{\leq})$ be an approximation space based on dominance relation $R^{\leq}$. For a fuzzy set A of U, the lower and the upper approximation of A denoted by $\underline{R^{\leq}}(A)$ and $\overline{R^{\leq}}(A)$, are defined respectively, by two fuzzy sets, whose membership functions are $\underline{R^{\leq}}(A)(x)=\min\{A(y): y\in[x]_{R^{\leq}}\}, x\in U$ and $\overline{R^{\leq}}(A)(x)= \max\{A(y): y\in[x]_{R^{\leq}}\}, x\in U$ where $[x]_{R^{\leq}}$ denotes the dominance class of the relation $R^{\leq}$.

Fuzzy set A is called fuzzy definable set, if $\underline{R^{\leq}}=\overline{R^{\leq}}$. Otherwise, A is called fuzzy rough set. $\underline{R^{\leq}}$ is called positive field of A in $S^{\leq}=(U,R^{\leq})$, and $\sim\overline{R^{\leq}}$ is called negative field of A in $S^{\leq}=(U,R^{\leq})$. In addition, $\overline{R^{\leq}}(A)\cap(\sim\underline{R^{\leq}}(A))$ is called boundary of A in $S^{\leq}=(U,R^{\leq})$.

It can be easily verified that $\underline{R^{\leq}}(A)$ and $\overline{R^{\leq}}(A)$ will become the lower and the upper approximation of standard approximation space based on dominance relation, when A is a crisp set.

Theorem 3.1. Let $A\in F(U)$, $\underline{R^{\leq}}(A)$ and $\overline{R^{\leq}}(A)$ be the lower and the upper approximation of A respectively. The following always hold.

(1) $\underline{R^{\leq}}(A)\subseteq A\subseteq\overline{R^{\leq}}(A)$.

(2) $\overline{R^{\leq}}(A\cup B)=\overline{R^{\leq}}(A)\cup\overline{R^{\leq}}(B)$; $\underline{R^{\leq}}(A\cap B)=\underline{R^{\leq}}(A)\cap\underline{R^{\leq}}(B)$.

(3) $\underline{R^{\leq}}(A)\cup\underline{R^{\leq}}(B)\subseteq\underline{R^{\leq}}(A\cup B)$; $\overline{R^{\leq}}(A\cap B)\subseteq\overline{R^{\leq}}(A)\cap\overline{R^{\leq}}(B)$.

(4) $\underline{R^{\leq}}(\sim A)=\sim\overline{R^{\leq}}(A)$; $\overline{R^{\leq}}(\sim A)=\sim\underline{R^{\leq}}(A)$;

(5) $\underline{R^{\leq}}(U)=U$; $\overline{R^{\leq}}(\emptyset)=\emptyset$.

(6) $\underline{R^{\leq}}(A)\subseteq\underline{R^{\leq}}(\underline{R^{\leq}}(A))$; $\overline{R^{\leq}}(\overline{R^{\leq}}(A))\subseteq\overline{R^{\leq}}(A)$.

(7) If $A\subseteq B$, then $\underline{R^{\leq}}(A)\subseteq\underline{R^{\leq}}(B)$ and $\overline{R^{\leq}}(A)\subseteq\overline{R^{\leq}}(B)$.

Proof. These items are obvious from definitions.

Definition 3.2. Let $S^{\leq}=(U,R^{\leq})$ be an approximation space based on dominance relation $R^{\leq}$. For $A\in F(U)$, the lower and the upper approximation with respect to parameters $\alpha,\beta(0<\alpha\leq\beta\leq 1)$, denoted by $\underline{R^{\leq}}(A)_{\alpha}$ and $\overline{R^{\leq}}(A)_{\beta}$, are defined by $\underline{R^{\leq}}(A)_{\alpha}=\{x\in U:\underline{R^{\leq}}(A)(x)\geq\alpha\}$; $\overline{R^{\leq}}(A)_{\beta}=\{x\in U:\overline{R^{\leq}}(A)(x)\geq\beta\}$.

From above definition, we can know that $\underline{R^{\le}}(A)_{\alpha}$ is the crisp set of some elements of U whose degree of membership in A certainly are not less than α, $\overline{R^{\le}}(A)_{\beta}$ is the crisp set of some elements of U whose degree of membership in A possibly are not less than β.

Remark. For $\forall\alpha,\beta\in[0,1]$,above definition $\underline{R^{\le}}(A)_{\alpha}$ and $\overline{R^{\le}}(A)_{\beta}$ will become $\underline{R^{\le}}(A)$ and $\overline{R^{\le}}(A)$ respectively, when A is crisp set.

In fact $\forall\alpha,\beta\in[0,1]$, since A is a crisp set, thus $A(x)\in[0,1]$. So we have

$$\begin{aligned}\underline{R^{\le}}(A)_{\alpha}&=\{x\in U:\underline{R^{\le}}(A)(x)\ge\alpha\}\\&=\{x\in U:\underline{R^{\le}}(A)(x)=1\}\\&=\{x\in U:\underline{R^{\le}}(A)(y)=1,\forall y\in[x]_{R^{\le}}\}=\underline{R^{\le}}(A)\end{aligned}$$

Similarly, we can show that $\overline{R^{\le}}(A)_{\beta}=\overline{R^{\le}}(A)$, when A is crisp set.

Theorem 3.2. Let $S^{\le}=(U,R^{\le})$ be an approximation space based on dominance relation $R^{\le}$, and $A,B\in F(U)$. For $\alpha,\beta(0<\alpha\le\beta\le1)$, we have

(1) $\overline{R^{\le}}(A\cup B)_{\beta}=\overline{R^{\le}}(A)_{\beta}\cup\overline{R^{\le}}(B)_{\beta}$, $\underline{R^{\le}}(A\cap B)_{\alpha}=\underline{R^{\le}}(A)_{\alpha}\cap\underline{R^{\le}}(B)_{\alpha}$.

(2) $\underline{R^{\le}}(A)_{\alpha}\cup\underline{R^{\le}}(B)_{\alpha}\subseteq\underline{R^{\le}}(A\cup B)_{\alpha}$, $\overline{R^{\le}}(A\cap B)_{\beta}\subseteq\overline{R^{\le}}(A)_{\beta}\cap\overline{R^{\le}}(B)_{\beta}$.

(3) If $A\subseteq B$, then $\overline{R^{\le}}(A)_{\beta}\subseteq\overline{R^{\le}}(B)_{\beta}$, $\underline{R^{\le}}(A)_{\alpha}\subseteq\underline{R^{\le}}(B)_{\alpha}$.

(4) $\underline{R^{\le}}(A)_{\alpha}\subseteq\overline{R^{\le}}(B)_{\beta}$.

(5) $\overline{R^{\le}}(\sim A)_{\beta}=\sim\underline{R^{\le}}(A)_{1-\beta}$, $\underline{R^{\le}}(\sim A)_{\alpha}=\sim\overline{R^{\le}}(A)_{1-\alpha}$.

Proof. It can be achieved obviously by definitions and Theorem 3.1.

Definition 3.3. Let $S^{\le}=(U,R^{\le})$ be an approximation space based dominance relation $R^{\le}$, and $A\in F(U)$. Roughness measure of A in $S^{\le}$, denoted by $\rho_{R^{\le}}(A)$, is be defined as

$$\rho_{R^{\le}}(A)=1-\frac{|\underline{R^{\le}}(A)|}{|\overline{R^{\le}}(A)|}$$

when $|\overline{R^{\le}}(A)|=0$, $\rho_{R^{\le}}(A)=0$ is ordered.

Clearly, there is $0\le\rho_{R^{\le}}(A)\le1$. Moreover, $\rho_{R^{\le}}(A)=0$, if A is definable fuzzy set.

Definition 3.4. Let $S^{\le}=(U,R^{\le})$ be an approximation space based dominance relation $R^{\le}$, and $A\in F(U)$. Roughness measure with respect to parameters $\alpha,\beta(0<\alpha\le\beta\le1)$ of A denoted by $\rho_{R^{\le}}^{\alpha,\beta}(A)$, is be defined as

$$\rho_{R^{\leq}}{}^{\alpha,\beta}(A)=1-\frac{|\underline{R^{\leq}}(A)_{\alpha}|}{|\overline{R^{\leq}}(A)_{\beta}|}$$

when $|\overline{R^{\leq}}(A)_{\beta}|=0$, $\rho_{R^{\leq}}{}^{\alpha,\beta}(A)=0$ is ordered.

From the definition, we have easily following properties.

Theorem 3.3. (1) $0\leq\rho_{R^{\leq}}{}^{\alpha,\beta}(A)\leq1$.

(2) If β is fixed, then we have that $|\underline{R^{\leq}}(A)_{\alpha}|$ increases. Thus, $\rho_{R^{\leq}}{}^{\alpha,\beta}(A)$ is to increase, when α increases.

(3) If α is fixed, then we have that $|\overline{R^{\leq}}(A)_{\beta}|$ is to decrease, when β increases. Thus, $\rho_{R^{\leq}}{}^{\alpha,\beta}(A)$ is to increase, when β increases.

Theorem 3.4. If membership function of fuzzy set A is a constant, i.e., there exists $\delta>0$ such that $A(x)=\delta$ for any $x\in U$, then we have $\rho_{R^{\leq}}{}^{\alpha,\beta}(A)=1$ when α,β $(0<\alpha\leq\beta\leq1)$. Otherwise, $\rho_{R^{\leq}}{}^{\alpha,\beta}(A)=0$.

Proof. When $\alpha,\beta(0<\alpha\leq\beta\leq1)$, it is clear that $|\underline{R^{\leq}}(A)_{\alpha}|=\varnothing$, $|\overline{R^{\leq}}(A)_{\beta}|=1$. So $\rho_{R^{\leq}}{}^{\alpha,\beta}(A)=1$.

Otherwise, there exist two cases.

Case 1. When $\delta<\beta\leq\alpha$, we can know $|\underline{R^{\leq}}(A)_{\alpha}|=|\overline{R^{\leq}}(A)_{\beta}|=\varnothing$. So $\rho_{R^{\leq}}{}^{\alpha,\beta}(A)=0$ is true.

Case 2. When $\beta\leq\alpha\leq\delta$, we can know $|\underline{R^{\leq}}(A)_{\alpha}|=|\overline{R^{\leq}}(A)_{\beta}|=U$. So $\rho_{R^{\leq}}{}^{\alpha,\beta}(A)=0$ is true from the definition.

4 Conclusions

It is well know that there exist most of systems, which are not only concerned with fuzzy sets but also based on dominance relations in practice. Therefore, it is meaningful to study the fuzzy rough set based on dominance relations. In this paper, we discussed this problem mainly. We introduced concepts of the lower and the upper approximations of fuzzy rough sets based on dominance relations, and constructed the model for fuzzy rough sets based on dominance relation additionally. Furthermore some properties are obtained. In next work, we will consider the information systems which are based on dominance relations and with fuzzy decisions.

Acknowledgments. This work is supported by the Postdoctoral Science Foundation of China (No. 20100481331) and the National Natural Science Foundation of China (No. 61105041, 71071124 and 11001227).

References

1. Cornelis, C., Cock, M.D., Kerre, E.E.: Intuitionistic Fuzzy Rough Sets: At the Crossroads of Imperfect Knowledge. Expert Systems 20(5), 260–270 (2003)
2. Banerjee, M., Miltra, S., Pal, S.K.: Rough fuzzy MLP: knowledge encoding and classification. IEEE Trans. Neural Network (9), 1203–1216 (1998)
3. Chakrabarty, K., Biswas, R., Nanda, S.: Fuzziness in Rough Sets. Fuzzy Set and Systems 110(2), 247–251 (2000)
4. Mordeson, J.N.: Rough Sets Theory Applied to (Fuzzy) Ideal Theory. Fuzzy Sets and Systems 121(2), 315–324 (2001)
5. Radzikowska, A.M., Keere, E.E.: A Comparative Study Rough Sets. Fuzzy Sets and Systems 126(6), 137–156 (2002)
6. Sarkar, M.: Rough-fuzzy Functions in Classification. Fuzzy Sets and Systems 132(3), 353–369 (2002)
7. Yao, Y.Y.: Combination of Rough and Fuzzy Sets Based on -Level Sets. In: Lin, T.Y., Cercone, N. (eds.) Rough Sets and Data Mining: Analysis for Imprecise Data, pp. 301–321. Kluwer Academic, Boston (1997)
8. Dubois, D., Prade, H.: Rough Fuzzy Sets and fuzzy Rough Sets. International Journal of General Systems 17(1), 191–208 (1990)
9. Zadeh, L.A.: Fuzzy Sets. Information and Control 8, 338–353 (1965)
10. Pawlak, Z.: Rough Sets. Int. J. Computer Information Sciences 11, 145–172 (1982)
11. Pawlak, Z.: Rough Sets-Theoretical Aspects to Reasoning about Data. Kluwer Academic Publisher, Boston (1991)
12. Yao, Y.Y.: Two Views of The Theory of Rough Sets in Finite Universes. International Journal of Approximation Reasoning 15, 291–317 (1996)
13. Yao, Y.Y., Lin, T.Y.: Generalization of Rough Sets Using Modal Logic. Intelligent Automation and Soft Computing, An International Journal 2, 103–120 (1996)
14. Banerjee, M., Pal, S.K.: Roughness of A Fuzzy Set. Information Sciences 93(1), 235–264 (1996)
15. Zhang, W.X., Wu, W.Z., Liang, J.Y., Li, D.Y.: Theory and Method of rough sets. Science Press, Beijing (2001)

The Study on the Internal Diameter Measuring System for the Small Bores

Runzhong Miao, Zhanfang Chen, Rui Li, Zhiwen Yang, and Shufang Wu

Abstract. For measuring internal diameter of a bore for Φ25～Φ37mm, the measuring method was proposed on the internal diameter measurement for a small bore by the grating sensor with data processing technology. The rating sensor was specially improved because of the small internal diameter and the reliability for automatically detection, the construction and measuring principle is described. More, the measuring accuracy of the system is verified by experiments. The results indicate that the measuring accuracy of inspecting system is superior to ±0.005 mm.

Keywords: technology of instrument and meter, bore inner diameter, grating sensor, data processing, optic-electronic inspecting.

1 Introduction

The quality of the bores directly related to the reliability and service life, in order to ensure the reliability and the accuracy of the shooting, high precision detection for inner diameter of the bores could be on achieved. The work conditions of the bores is very bad , and the bores should not only bear the role of high temperature and pressure of the gunpowder gas. The quality of the bores directly influence on the using safety, shooting effect and running life of the artillery. Thus bores diameter inspection was very important for its whole manufacturing, acceptance and maintenance. the developed inner diameter detection system for small bores were carried out by using the reflex grating sensor design which aimed to the measurement of the device of small size and the reliability of the measurement accuracy.

Runzhong Miao · Zhanfang Chen · Rui Li · Zhiwen Yang · Shufang Wu
Changchun University of Science and Technology, Changchun, 130022, China
e-mail: miao1921@vip.sina.com

Z. Qian et al. (Eds.): Recent Advances in CSIE 2011, LNEE 125, pp. 127–132.
springerlink.com

2 Detection System Composition and Structure

2.1 Bores Diameter Basic Detecting Methods [1-3]

The traditional bores diameter basic detecting methods were used by mechanical method, optical testing and magnetic leakage testing etc. Mechanical detection had been widely used as its simple principle, the easy operation and low cost, thus the measurement. The mistake examining would be done by the wearing between inspection heads and the inner wall of the bores, temperature change, measuring force caused by the influence of such factors as the diameter.

Optical detection method can realize the contact detection, in the process of detecting, by no wear and no deformation. When the bores barrel surface ablation, have a color change cannot be detected when, at present the photoelectric detection devices are used on-line size detection for bores tube of on the bores production.

In recent years it had some other detection methods, such as ultrasonic method, magnetic flux leakage test, etc[4-5]. Methods of magnetic were heavy effected by the surface roughness of the bores, and the measurement precision is not higher, measurement range is limited, and is not suitable for bores detection.

The grating detecting methods as modern developed, grating sensor compared with other sensors, have the following characteristics. The grating sensor can be done in the large range measuring length or linear displacement of precision. It could realize dynamic measurement, easy for data processing. Its detection system has strong anti-jamming ability to the environmental conditions' and did not like laser sensor that were requested strictly. It would make the displacement (linear displacement or angular displacement) direct converted into digital output, etc.

The bores inner diameter detection, the main performance had particularity in high precision measurement requirements and radial dimensions would be adopted to small line displacement sensor. The bores by long length, need to be longer depth detection. The light and the data transmission system would be mini-size designed for detecting and these devices were set in the inner of the bores.

2.2 Inner Diameter Detections Composition and Structure

Grating sensor type diameter detector were composed with the pilot, centering device, detecting head, single-chip microcomputer, grating sensor, depth signal generator, bracket and sending devices, digital display meter, printer, computers and other more parts. The diameter caliper system structure as shown in figure 1 below.

The Bragg grating measuring diameter system applied the measuring method for relative measurement. Caliper gauge before work, the detecting head into the standard rings which named calibration rings were the same standard size as the bores with inner diameter, and let the heads were contacted onto the inner wall of the grating sensor calibration rings, the data of the standard diameter were set to standard size, by counting system and record store. Then caliper gauge measurement head would be conducted into testing bores, travelled and measured by driving devices push head go-back in the tubes with mean speed. The heads controlled by computer were stopped when a certain distance measurement travel, by this time, the changes data between the Bragg grating measuring sensor relative to the inner diameter of the standard size would be transferred into an electrical signal, the amount again through the data acquisition card were converted numerical signal and were send into a computer processing system, the test data could be directly display and storage, or draw the wear curve and print built finally, etc.

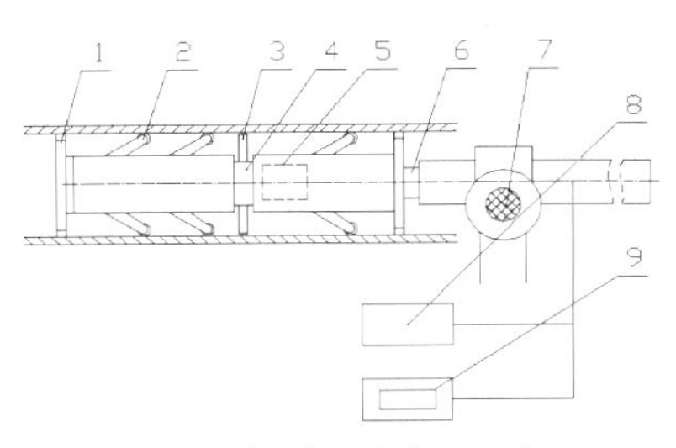

1—pilot, 2—centering, 3—detecting head, 4—grating sensor, 5—microcomputer, 6—sending, 7—bracket and feeding, 8—digital display meter, 9—data process

Fig. 1 Overall structure of the grating sensor system

2.3 *Reflex Grating Sensor Design*

Reflex grating sensor is general by light source, the condenser, the mirror, the field lens, the indicating grating, ruler grating, optical lens and photosensitive components. Its principle was as shown in figure 2.

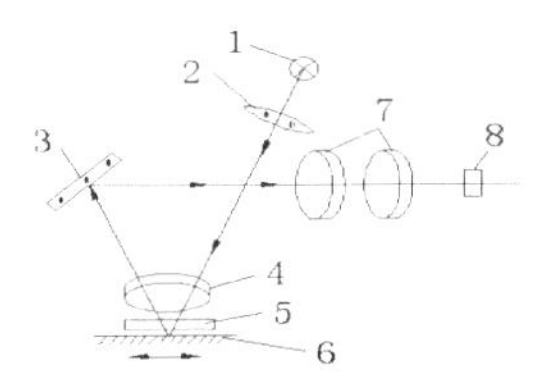

1—light source, 2-condenser, 3—mirror, 4—field lens, 5—indicating grating, 6—ruler grating, 7—optical lens, 8—photosensitive components

Fig. 2 Reflecting principle of the grating sensor

The parallel light beams, which formed by the condenser and were from reflex grating sensor source of light emitting diode, could be sending onto the crack phase grating with a certain angle. The moiré fringe was formed by rod grating reflected light and formations. The photoelectric device received moiré fringe of this light. The small size of the design and process of the high reliability reflex grating sensor real figure were as shown in figure 3 below.

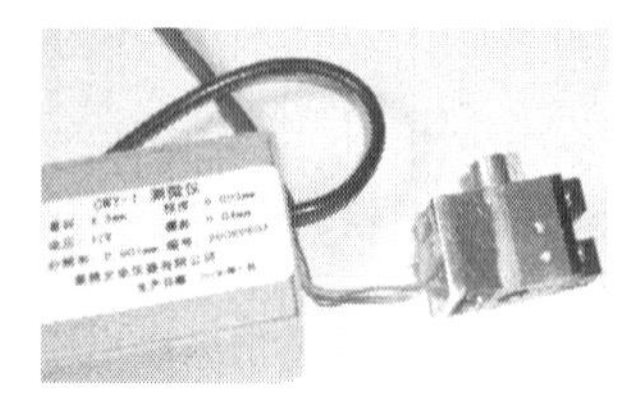

Fig. 3 Photo of the Reflex grating sensor

3 Signal Acquisition and Real-Time Data Processing System Design

3.1 *Control Circuit Design*

The detecting heads of the sensors could slide along the surface of the bores, the displacement changes could be transferred by sensor into the electrical signals, then by the cable, into digital display table and computer display and processing. Electronic principle was as shown in figure 4.

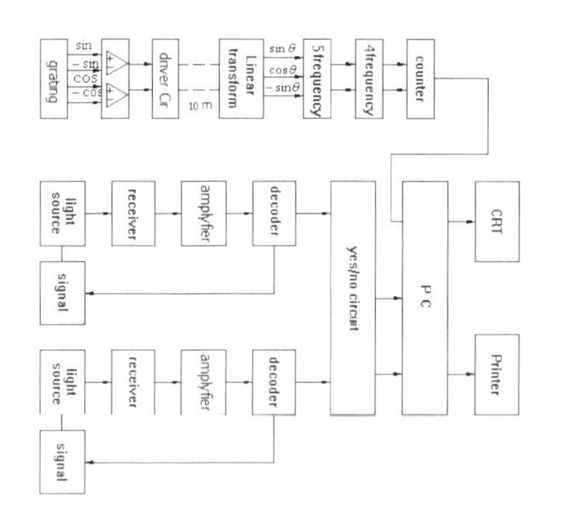

Fig. 4 Design diagram of the electronic system

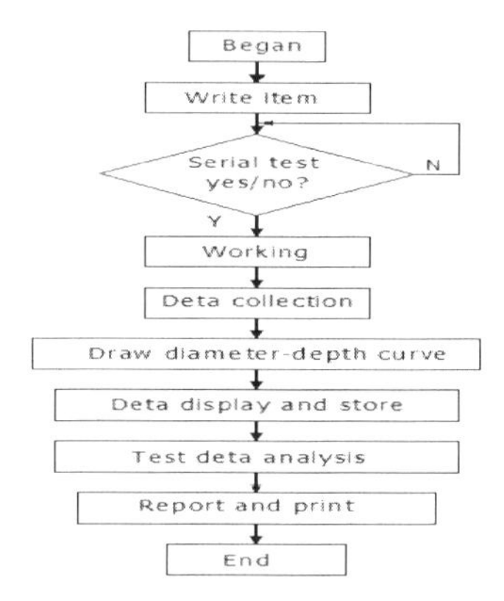

Fig. 5 Block diagram of data processing

3.2 Data Processing System Design

According to the concept of software engineering, and the principles of module partition, the software system mainly would be divided into four modules: real-time measuring display module, measurement data processing module, measurement data storage module, curve data printing module, the whole system flow chart as shown in figure 5 shows. In the process of data collection, it need to save all kinds of real-time data acquisition, data type more and request for data playback, These data need to be stored in the data source files, so as to be accessed the programming interface.

The system is based on ODBC (open database connection) database application program, through the using of driver software, it could provide the independence of the databases, and for all types of database management system it could provide the unified used SQL language programming interface. This software system in Microsoft Visual Studio integrated development environment by using Visual C++ 6.0 and NI Company Studio Measurement as development tools.

4 The Results and Conclusions

The practical measurement were carried out with this detecting system for the bores inner diameters as the size ofΦ25 mm, Φ30 mm, Φ35 mm, Φ37 mm, and repetitive measures had been done for seven times at the same depths of the any size of the diameters. The results and the measurement errors measured atΦ25 mm inner diameter were shown as the table 1 and table 2.

Table 1 Inner diameter detecting error data to Φ25 mm bores (step by 100 mm)

Head Length	diameter change between detected and standard(7 times), μm							Error σ,μm
	1	2	3	4	5	6	7	
210	0	0	0	0	0	0	0	0
310	0	1	1	1	0	3	0	1.1
410	10	10	11	11	10	10	10	0.5
510	13	13	13	10	12	12	12	1.1
610	11	11	10	12	10	10	10	0.8
710	13	12	12	13	12	13	12	0.5
810	13	13	13	13	13	13	13	0
910	12	13	13	13	13	13	13	0.4
1010	13	12	12	12	12	12	12	0.4
1110	17	15	14	15	14	14	14	1.1
1210	16	15	14	15	15	26	15	4.2
1310	20	20	17	17	19	20	18	1.4

Note: technique requirement σ=8 μm.

The measurement results showed on table 1 were detected by selecting the point-distance long 100 mm after the initial position, and the measurement results on table 2 were detected by change the step length. The measured results showed that the system is very good repeatability, testing error is not more than + 0.005 mm, could satisfy ф 25 to 37 mm the inner diameter precision requirements of the bores, the measurement system were versatile, it could also be applied to other small parts aperture detection.

Table 2 Inner diameter detecting error data to ф 25 mm bores (step by length random)

Step Length	ϕ_1	ϕ_2	ϕ_3	ϕ_4	ϕ_5	ϕ_6	ϕ_7	$\bar{\phi}$	σ
406	0.000	0.000	-0.009	0.000	-0.002	-0.002	-0.002	-0.0021	0.0032
446	-0.004	-0.007	-0.014	-0.002	-0.004	-0.007	-0.005	-0.0061	0.0039
486	-0.010	-0.010	-0.009	-0.010	-0.012	-0.012	-0.012	-0.0107	0.0013
566	0.000	-0.003	0.003	-0.002	-0.002	-0.002	-0.002	-0.0011	0.0020
746	0.003	0.002	0.003	-0.001	0.001	0.001	0.001	0.0014	0.0014
864	0.002	0.003	0.004	0.001	0.000	0.001	0.001	0.0017	0.0014
986	0.006	0.005	0.004	0.003	0.001	0.002	0.003	0.0034	0.0017
1046	0.007	0.005	0.004	0.004	0.002	0.004	0.005	0.0044	0.0015

References

1. Chen, Y., Yu, P.: Optical techinal for auto-NDT to inner surface of the tubes. Measurement Technology 6, 7–10 (2000)
2. Ma, Y., Fang, R.: Auto detecting system for beelines anto the artillery bores. Instrumentation Technology 1, 11–12 (2002) (in Chinese)
3. Yoffe, G.W., Krug, P.A., OueUette, F.: Passive temperature compensating package for optical fiber gratings. Appl. Optics 34(30), 6859–6861 (1995)
4. Kleuver, W.: New triangulation sensor for inside inspection of small drillings and hollows with integrated image processing. Paul-Bonatz-Str., 9–11 (1999)
5. Finkelstrtein, L.: Measurement and Instrumention Science-An Analytical Review Measurement, pp. 38–50 (1994)

Moving Object Detection and Tracking in Mobile Robot System Based on Omni-Vision and Laser Rangefinder

Jin-xiang Wang and Xiao-feng Jin*

Abstract. To meet the requirements of real-time and accurate information processing in robot tracking system, a real-time object detection and tracking method based on Multi-Sensor Information Fusion technique is proposed. Firstly, the real-time environmental data was obtained by using omni-vision and laser rangefinder, so that the moving object was detected; Secondly, the average color of color-similar point set within the neighborhood of a sampling point on the moving target was extracted as color feature, and the color feature was enhanced by using the real-time data obtained from the laser in the direction of object. Finally, the orientation and distance information were calculated between robot and the moving object, and the moving object tracking was achieved by using the robot tracking tactic. Experiments were carried out on the AS-R mobile robot, and the experimental results show that the proposed method can meet the requirements of real time and robust in the robot tracking system.

Keyword: Mobile Robot, Omni-vision, laser ranging, MSIF, Moving object tracking.

1 Introduction

The real-time detection and tracking of moving object is a hotspot issue in the field of robotics research. There are many researches on this issue at home and abroad. References [1-4] emphasize on the research based on vision sensor; References [5-7] emphasize on the research based on laser ranging sensor; References [8-10] emphasizes on the research based on vision and laser ranging sensors. The problems of accuracy and real-time property of the detection and tracking on the moving object are difficult problems that need to be solved.

Jin-xiang Wang · Xiao-feng Jin
Intelligent Information Processing Lab. Yanji, Jilin, China
e-mail: {wangjinxiang,xfjin}@ybu.edu.cn

* Corresponding author.

Z. Qian et al. (Eds.): Recent Advances in CSIE 2011, LNEE 125, pp. 133–139.
springerlink.com

The laser ranging sensor and vision sensor is complementary. These two kinds of sensors are adopted to enhance the perception ability on three dimensional en vironment of the robot in this paper, and the information acquired from the two sensors is integrated effectively to improve the detecting accuracy and real-time of tracking. The method has been validated by using AS-R robot to detect and track human body. The research is based on the following premises in this paper:

- The color of the legs in moving human is single or similar and can be distinguished from other objects.
- The laser at the placed height can successfully scan the leg of moving man.
- The space between the robot and human is not obstructed by any barrier.

2 Detection of Moving Object

The picture shot by omni-vision camera is expanded into rectangular image quickly according to the method of nearest neighbor sampling[11], the area of moving object is detected accurately by applying frame subtraction method and technology of Multi-Sensor Information Fusion (MSIF), specifically in Fig. 1.

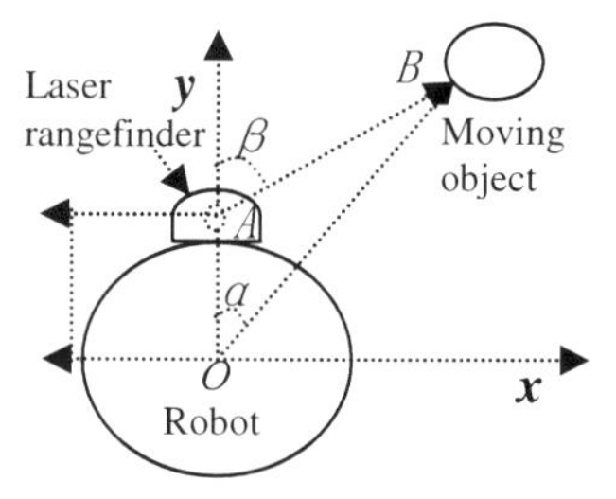

Fig. 1 The positions of robot, sensor and moving object.In Figure 1, the central point O of robot which is 83cm away from the ground, is the projection point of omni-vision to the vertical direction of xoy plane. Point A is the laser emitting point of laser rangefinder which is 41cm away from the ground. Point B is the central point of moving object, AO=48cm, $\beta\in[-\pi/2, \pi/2]$.

Assume that the angle interval of moving object gained from vision and laser sensors is $[\theta_{v1},\theta_{v2}]$ and $[\theta_{l1},\theta_{l2}]$, respectively, which θ_{v1}, θ_{l1}, θ_{v2} and θ_{l2} are angles of moving object's left and right borders and the opposite direction of x-axis.

Assume that the laser detection range is correct only if $[\theta_{l1},\theta_{l2}]\subseteq[\theta_{tl1},\theta_{tl2}]$. The span of knee is $W\in[W_1,W_2]$ during the normal or rapid walking of human. θ_{tl} is a angle which is determined by the following formula(1):

$$\theta_{tl} = \lceil 2\times \arctan(W/(2\times AB)) \rceil \tag{1}$$

Obtained,$\theta_{tl1}=(\theta_{l1}+\theta_{l2})/2-\theta_{tl}/2$, $\theta_{tl2}=(\theta_{l1}+\theta_{l2})/2+\theta_{tl}/2$, W_1=0.15m, W_2=0.6m.

According to the correspondence relationship between AB, angle β and Fig. 1, using cosine and arccosine, the values of OB and α can be calculated. Then, replace AB in formula(1) with the calculated OB, so that the angle range is $[\theta_{tlv1},\theta_{tlv2}]$, $\theta_{tlv1}=\alpha-\theta_{tl}/2$, $\theta_{tlv2}=\alpha+\theta_{tl}/2$.According to $[\theta_{l1},\theta_{l2}]$ and value AB, the actual width of moving object O_w can be calculated by using formula (2).

$$O_w = \frac{2 \times AB \times \sin((\theta_{l2} - \theta_{l1})/2)}{\sin(\pi/2 - (\theta_{l2} - \theta_{l1})/2)} \quad (2)$$

So the actual target detection area $[\theta_{lv1},\theta_{lv2}]$ be calculated by O_w in the video image, $\theta_{lv1}=\alpha$ -arctan(O_w/(2×OB)), $\theta_{lv2}=\alpha$ + arctan(O_w/(2×OB)).

The detection method is as follows. If $[\theta_{v1},\theta_{v2}]\subseteq[\theta_{tlv1},\theta_{tlv2}]$, $[\theta_{l1},\theta_{l2}]\subseteq[\theta_{tl1},\theta_{tl2}]$, $|\theta_{v1}-\theta_{lv1}|<\pi/18$ and $|\theta_{v2}-\theta_{lv2}|<\pi/18$, then the detection of the vision is correct, record $[\theta_{v1},\theta_{v2}]$; If $[\theta_{v1},\theta_{v2}]\subseteq[\theta_{tlv1},\theta_{tlv2}]$,$[\theta_{l1},\theta_{l2}]\not\subset[\theta_{tl1},\theta_{tl2}]$, then record $[\theta_{v1},\theta_{v2}]$;If $[\theta_{v1},\theta_{v2}]\not\subset[\theta_{tlv1},\theta_{tlv2}]$, $[\theta_{l1},\theta_{l2}]\subseteq[\theta_{tl1},\theta_{tl2}]$, then record $[\theta_{lv1},\theta_{lv2}]$; If $[\theta_{v1},\theta_{v2}]\not\subset[\theta_{tlv1},\theta_{tlv2}]$, $[\theta_{l1},\theta_{l2}]\not\subset[\theta_{tl1},\theta_{tl2}]$, then give up records.

3 Extract and Enhance the Color Feature

Determine the average color of the color-similar point set within the neighborhood of sampling point as the seed color which is used to fill the real-time image, and enhance the color feature by using the real-time data from the laser.

Extract the color feature of object. The sampling point is located by angle and distance information. The collect window is obtained by the sampling point. Collect the seed color when the following conditions are satisfied. $AB'\in[AB_1,AB_3]$, $\alpha<\pi/6$; Calculate the average value of all the colors similar to sampling point within the collect window ; Condition 1 and 2 are continuously satisfied for three times, the average values of the three collected colors are similar.

After collecting the seed color, filling the color to tracking window is needed. The window should include the moving object and the position that the moving object may get to in the next frame. Then the width P_w of tracking window be calculated by people's walking speed.

Assume that P_w corresponds to column O_{x1} and O_{x2} , $P_w=|O_{x2}-O_{x1}|$. If the color is similar to seed color in tracking window, then it is filled with red, otherwise filled with blue. Next, setting a two-dimensional array S, recording the position S_x of each pixel column within $[O_{x1},O_{x2}]$ and the amount S_{sum} of the red pixels in this column.

Enhance the color feature of object. With the increase of S_{sum} and In the middle of P_w the more obvious characteristics of the object, The difference are divided into 90 equally, which are expressed as value of $[0,\pi/2]$ by 1° assigned to $w1$,$w2$. The w can be calculated by formula(3).

$$w = \sin(w1) + \sin(w2) \quad (3)$$

S_{sum} is assigned with the product of S_{sum} and w within P_w.

4 Tracking Method of Moving Object

The object is located by following algorithm, the robot alters own state by tracking strategy and tracks object.

Step1: Initialize counter *Count* and accumulator *Sum*;

Step2: Scanning from O_{x1} to O_{x2} in turn, finding out the maximum value of S_{sum} in S, and record it as $S_{sum\text{-}max}$;

Step3: Scanning again from O_{x1} to O_{x2} in turn, if $S_{sum\text{-}max}\text{-}S_{sum} \leq 15$, then $Sum=Sum+S_x$, $Count=Count+1$;

Step4: Divide *Sum* value by *Count*, figure out the average of characteristics series of current moving object, it is $S_{x\text{-}sav}$;

Step5: Figure out the angle θ by $S_{x\text{-}sav}$, and calculate AB by laser and θ.

After ascertaining the current position relationship between moving object and robot, then control the robot to move. First, divide the area according to the length of AB. Assume that it is near area when $AB<AB_1$, it is middle area when $AB\in[AB_1,AB_3]$, it is far area when $AB>AB_3$.

When the object in middle area, object is locked by controlling the angular velocity of robot. The ideal state of robot is $|\theta|=0$. The selection of angular velocity of robot is mainly determined by two factors, the θ_Δ is the difference of $|\theta|$ between the current and previous frame. The rotary acceleration is fixed at $\pm a_\omega$.

The rotary direction and angle are judged by the plus-minus of θ and θ_Δ. Given $\theta_1>\theta_2>\theta_3>0$, dividing the angle θ into four intervals, which are $\theta\in[0, \theta_3)$, $\theta\in[\theta_3, \theta_2)$ and $\theta\in[\theta_2, \theta_1]$, $\theta>\theta_1$. Divide angular velocity into four intervals, which are the slowest rotary speed ω_1, the quickest rotary speed ω_3, the middle rotary speed ω_2 and 0 (stillness). Divide angle θ_Δ and θ respectively into two intervals according to their plus-minus. Corresponding rotary rules are given as follow.

Rule 1: If $\theta>0$, then ω will clockwise rotate. If $\theta<0$, then ω will counter clockwise rotate, which are adjusted by $\pm a_\omega$;

Rule 2: If $\theta_\Delta \geq 0$, then adjusting $\pm a_\omega$ to satisfy when $\theta>\theta_1$, $\theta\in[\theta_2,\theta_1]$,$\theta\in[\theta_3,\theta_2)$,$\theta\in[0, \theta_3)$, the velocity is ω_3, ω_2, ω_1, 0;

Rule 3: If $\theta_\Delta <0$, then adjusting $\pm a_\omega$to satisfy when $\theta>\theta_1$, $\theta\in[\theta_2,\theta_1]$,$\theta\in[0, \theta_2)$, the velocity is ω_2, ω_1, 0. In experiment, $\theta_1=\pi/6$, $\theta_2=\pi/12$, $\theta_3=\pi/36$, $a_\omega=1.5\text{rad/s}^2$, $\omega_3=2\text{rad/s}$,$\omega_2=1.5\text{rad/s}$,$\omega_1=1\text{rad/s}$,$v_b=0.5\text{m/s}$,$v_m=0.75\text{m/s}$,$v_h=1\text{m/s}$,$\alpha_v=\pm1.5\text{m/s}^2$.

When the object in far area, object is tracked by controlling robot rectilinear or curvilinear motion. Assume that AB is at $[AB_1, AB_3]$ and make the object in front of the robot. Suppose $AB\in(AB_3,AB_4]$ is low-speed area, $AB\in(AB_4,AB_5]$ is middle-speed area and $AB_5<AB$ is high-speed area. The distance difference of current and previous frame is Δ_{AB}. Assume that the low-speed tracking speed of robot is v_b, middle-speed speed is v_m, high-speed speed is v_h, the linear acceleration is $\pm a_v$. Assume angle $\theta=\theta_k$ is rectilinear tracking, otherwise curvilinear tracking. The tracking rules are given as follows.

Rule 4: If $\Delta_{AB}\geq 0$, then controlling left and right wheel speed, when $AB\in(AB_3,AB_4]$, $AB\in(AB_4,AB_5]$, $AB>AB_5$, $AB\leq AB_3$, the speed is v_b, v_m, v_h, 0;

Rule 5: If $\Delta_{AB}<0$, then controlling wheel speed, when $AB\leq AB_4$, $AB\in(AB_4, AB_5]$, $AB>AB_5$, the speed is 0, v_b, v_m ;

Rule 6: When $\theta_k\in(-\theta_2,\theta_2)$, the robot satisfy rule 4 and 5. When $\theta_k\leq-\theta_2$, according to the value of $|\theta_k|$, the robot turns left by α_v, otherwise turning right.

When object in near area, robot is backward until object in the middle area.

When object lost, robot looks for the object by using Omni-Vision camera. If robot finds the object, then robot turns and continues to track the object, if robot does not find the object after a period of time, then ending. In experiment, $AB_1=0.6\text{m}$, $AB_2=3.5\text{m}$, $AB_3=2\text{m}$, $AB_4=2.5\text{m}$, $AB_5=3\text{m}$.

5 Experimental Results and Analysis

To validate the performance of the algorithm, many experiments have been done on wheeled ASR mobile robot. The front two wheels of robot are differential driving wheels, and the back wheel is a universal driven supporting wheel. The configuration of car-mounted computer is P4 3.0G, 256M memory.

Fuse the data got from the laser to the video image, ascertain the color sampling point and sampling area, which is shown in Fig. 2:

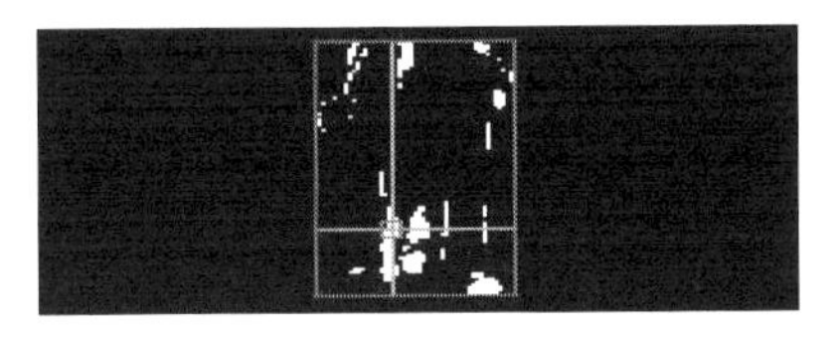

Fig. 2 Fusion image of when human is 0.9m from the robot. The red area is the moving object area gained from vision sensor, the intersection of green lines is the distance and direction information.

The following images are images of color filling, horizontal projecting on color filling area and enhancing the projection area. Fig. 3 is shown.

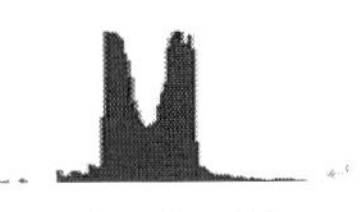

(a) Filling object area (b) Color projecting on object area (c) Enhancing the object area

Fig. 3 Color features extraction and enhancement of moving object

The moving object, sampling point and sampling area can be detected accurately by analyzing the consistent data gained from the two sensors for the same moving object, so the color feature is enhanced and the robustness of robot is improved. Fig. 4 is shown.

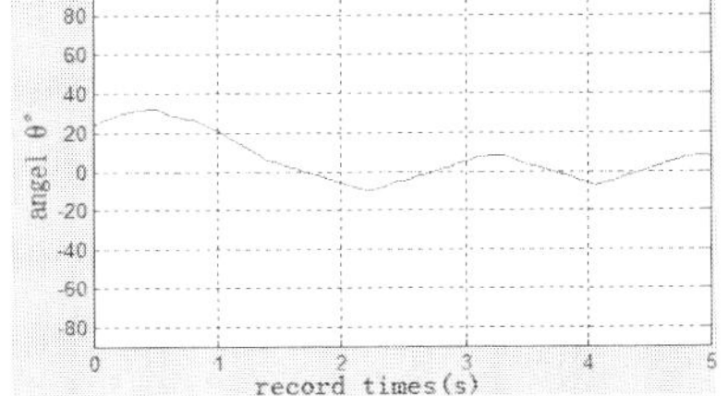

Fig. 4 Change curve of angle θ during the tracking course. It is known that angle θ is changed dynamically with the object motion from fig. 5. Angle θ decreases and maintained [-π/36,π/36], which can keep the moving object directly ahead of the robot.

On the real-time property, both the omni-vision camera and laser rangefinder adopt frame rate of 20fps. Compared with refer[8], the frame rate has been improved for about three times. Obviously, the real-time property of the method used in this paper has been greatly improved. The decision time of tracking method is recorded ten times in Table 1 and the average time is 2.77ms.

Table 1 The real-time property of tracking method

Number	1	2	3	4	5	6	7	8	9	10	average
Time(ms)	3.3	2.6	2.1	2.4	3.0	2.8	3.0	3.3	2.8	2.4	2.77

Concluding the experimental results above, the advantages of this tracking method are as follows.

1. The accuracy of detecting and recognizing object is improved for robot by using information fusion.
2. The real-time of tracking object is improved for robot by applying the tracking strategy.

6 Conclusion

The robot uses two kinds of sensors to perceive the external environment. By applying the MSIF technology to the detection of moving object, the moving object in three-dimensional space can be detected effectively. Moreover, the color feature of object is extracted and enhanced. The robot locks the position of object by locking the object features. According to the relative position and distance information to the object, the robot adopts tracking tactic to adjust its motion state and track the moving object effectively in real time.

References

1. Kuang, J.M., Liu, M.: Stereo Vision based Moving Target Tracking for Omnidirectional Mobile Robots in 2D Space. In: Proceedings of the 2007 International Conference on Information Acquisition, Jeju, Korea, pp. 602–606 (2007)
2. Maki, K.: Fiber-Grating-Based Vision System for Real-Time Tracking, Monitoring, and Obstacle Detection. IEEE Sensors Journal 7(1), 105–120 (2007)
3. Yang, G., Liu, H.: Survey of visual tracking algorithms. CAAI Transactions on Intelligent Systems 5(2), 95–105 (2010)
4. Menegatti, E., Pretto, A., Scarpa, A., et al.: Omnidirectional Vision Scan Matching for Robot Localization in Dynamic Environments. IEEE Transactions on Robotics 22(3), 523–534 (2006)
5. Rigling, B.D.: Amplitude space–time adaptive processing for ground-moving target indication laser radar. IEEE Proc.-Radar Sonar Navig. 153(4), 361–364 (2006)
6. Robert, A.: Tracking of Moving Objects from a Moving Vehicle Using a Scanning Laser Rangefinder. In: IEEE Intelligent Transportation Systems Conference, Toronto, Canada, pp. 301–306 (2006)
7. Amarasinghe, D., Mann, G.K.I., Gosine, R.G.: Moving Object Detection in Indoor Environments Using Laser Range Data. In: International Conference on Intelligent Robots and Systems, Beijing, China, pp. 802–807 (2006)

8. Jung, B., Sukhatme, G.S.: Real-time Motion Tracking from a Mobile Robot. Int. J. Soc. Robot. 2, 63–78 (2010)
9. Dai, W., Cuhadar, A., Liu, X.: Robot Tracking Using Vision and Laser Sensors. In: 4th IEEE Conference on Automation Science and Engineering, Washington DC, USA, pp. 169–174 (2008)
10. Liu, H., Dong, N., Zha, H.: Omni-directional Vision based Human Motion Detection for Autonomous Mobile Robots. In: IEEE SMC 2005, Hawaii, USA, pp. 2236–2241 (2005)
11. Chen, Y., Yang, D.Y., Shen, Z.W.: Generation of panoramic and perspective views from omnidirectional images. Computer Engineering 33(2), 183–185 (2007)

Robot Path Planning Based on Random Expansion of Ant Colony Optimization

Jinke Bai, Lijia Chen, He Jin, Ruixia Chen, and Haitao Mao

Abstract. Aimed at the shortcomings of the ant colony algorithm in robot path planning, which need much time and easy to fall in premature stagnation. This paper proposes a random expansion ant colony optimization algorithm through giving a possible way in the initial pheromone distribution to narrow the searching range of algorithm and raise the searching speed. At the same time random expansion factor is introduced to ant colony optimization algorithm, improved the diversity of routes and global optimization properties. The simulations result shows that the algorithm has excellent global optimization property and fast convergence speed.

Keywords: path planning, ant colony optimization, random expansion.

1 Introduction

Robot route planning is a kind of multi-constrains minimum problems. Robot route planning algorithms consist of global planning with known environment information and local planning with unknown environment information. In global planning, environment information known before planning is used to find out the best route from start point to end point [6]. The popular algorithms includes grid method, visibility graph, artificial potential field method and other algorithms [5,7,8]. Grid method can find the best route in some given conditions, but the selection of grid sizes affects the quality of solutions and more store space is needed when seaching space is big. Visibility graph method searches the route through reconstructing visibility graphs continuously, that decreases the efficiency of the algorithm. On the other hand, the route found in grid method and visibility graph is not smooth, which increases the difficulty of movement of the robot. Artificial potential field method has simple construction and easy method to complete it. Furthermore the route gotten in the method is smooth and safe. But some problems is

Jinke Bai · Lijia Chen · He Jin · Ruixia Chen · Haitao Mao
School of Physics and Electronics, Henan University, Kaifeng, China
e-mail: chenlijia_just@163.com

Z. Qian et al. (Eds.): Recent Advances in CSIE 2011, LNEE 125, pp. 141–146.
springerlink.com

difficult to avoid in it, such as local minimum, unreachable goal, and vibration around obstacles. Recently, genetic algorithm, ant colony algorithm and neural network have been proposed to plan the route for robots.

ACO (Ant Colony Optimization) is a simulating optimization algorithm [1,2] which can be used in many fields. However, when ACO is used to robot route planning, pheromones are lack in the initial stage which results in slow convergence speed. On the other hand, ACO often gets into premature stagnation in the middle stage. All of them affect the performance of ACO in robot route planning. Some literatures have made improvements to ACO. In [9], pheromones are updated by using an inter updating system. In [3], three local optimization operators are added to ACO to exchange route information. This paper will improve ACO in route searching and pheromone construction.

A random expansion of ACO is proposed in this paper. First, a route is found rapidly using a detour method. Pheromones are initialized based on the route. Thus, the problem of the lack of pheromones is solved and fastens the convergence speed of the algorithm. Second, a random factor is introduced into the ACO to avoid the premature phenomena. The algorithm shortens the planning time and expands searching space based on initial pheromone distribution. It gets a route with the biggest pheromone density. Simulation results show that the route found approach the performance of the best route.

2 The Improved ACO (IACO)

Assuming that some static obstacles distribut in the 2D working area C and robots know the information. The goal of planning is to find the point set, along which the robot can go across the C from start point to end point with avoidance of obstacles. We model the environment as follows. Robots is consider as points, and the obstacles are considered as convex polygon. Denote the obstacles in C by C_{obs} and free space by C_{free}. Both of them has the same boundary.

2.1 Construction of Pheromones

A detour method is proposed to find a route rapidly, along which pheromones are configured in the initial stage of ACO. The detour method originates from [4], and we make some improvements.

The detour method is a heuristic algorithm from a simulation of finding paths of human. Robots divide the goal into some subgoals when they meet obstacles. Subgoals will conduct them to detour these obstacles one by one and finnaly reach the goal. The movement direction of the robot is determined as follows:

1. The robot is moving to the goal when the movement direction is $\overrightarrow{P_k} = (P - G) / |P - G|$
2. The robot is moving along the boundary of an obstacle with current position point S_k. Assuming some points in the obstacle are $M_1 M_2 \cdots M_n$ which are

smaller than a given value δ. Let $\overrightarrow{S_kM_k} = \overrightarrow{S_kM_1} + \overrightarrow{S_kM_2} \cdots \overrightarrow{S_kM_n}$, and compute the unit vector $\overrightarrow{P_k}$ which is vertical to $\overrightarrow{S_kM_k}$ and has a sharp or right angle with $\overrightarrow{S_{k-1}S_k}$ The movement direction is $\overrightarrow{P_k}$ or opposite direction.

The detour around obstacles has steps as follows:

1. When the robot encounters an obstacle, the robot is divided into two robots. The two robots move to the direction determined by steps (2) described above but with opposite directions along the boundary of the obstacle.
2. When the two robots moves along the boundary of the obstacle, consider the line connecting the point where the robots are with the goal. If the line doesn't intersects with the obstacle, it shows the robots have detoured the obstacle. If only one robot detours the obstacle, abort another robot and continue the process on untill the goal.

A route will be found by following the above steps. The route changes the pheromones distribution with an addtional value $\Delta\tau_{i,j}$. The initial pheromones are

$$\tau_{i,j} = \tau_{i,,j}(0) + \Delta\tau_{i,j} \tag{1}$$

where $\tau_{i,j}(t)$ presents the residual amount of pheromones along the route <*i*,*j*> at time *t*, *i* and *j* are arbitrary two points in C_{free}. $\tau_{i,j}(0)$ presents initial pheromones computed from environment information.

2.2 Selection of the Next Point

ACO is easy to fall in prematurity when it plans a route. We introduce a random expand factor into the process of finding routes, which avoids the prematurity and produces bigger route space than ACO.

At first, some symbols are introduced. *W* is the number of ants in the ant colony, d_{ij} presents the distance between point *i* and point *j* in *C*, $b_i(t)$ presents the number of ants at point *i* at time *t*. Ant *k*, *k*=1,2...*W*, moves toward the direction where the amount of pheromones is maximum. At time *t*, ant *k* moves from point *i* to point *j* which has to be adjacent to *i* and never be visited by ant *k*. We introduce a random number between (0,1) into ACO, which is used to change the decision of ant *k*. When $random(0,1) \le \lambda$, ant *k* desides as described above in ACO; otherwise, ant *k* randomly moves to a free point around point *i*. The process is

$$\delta_{i,j}^k(t) = \begin{cases} \dfrac{\tau_{i,j}^{\alpha}(t)\eta_{i,j}^{\beta}(t)}{\sum_{r\in s_i^k}(\tau_{i,j}^{\alpha}\eta_{i,j}^{\beta}(t))}, Random(0,1) \le \lambda \\ Random(allowed), Random(0,1) > \lambda \end{cases} \tag{2}$$

where $\eta_{i,j}(t) = Q / d_{i,j}(t)$, Q presents the constant of pheromone strength, α presents the relative importance degree of pheromones, β presents the relative importance degree of distance. *Random*(*allowed*) randomly return a point in point set *allowed*, every point of which point *i* can reach , λ presents the probability of random expansion with value field [0,1]. λ is dominated by

$$\lambda = \lambda_0 (1 - R_T) \sin(\pi T / T_{\max})$$
$$R_T = \begin{cases} Random(0, \Delta), & Random(0,1) \geq \lambda \\ T / T_{\max}, & Random(0,1) < \lambda \end{cases} \tag{3}$$

where T_{max} presents the biggest number of iterations, T presents the current number of iterations, *Random*(0, Δ) produces a number between (0, Δ), $0<\Delta<1$. R_T is configured to manage random expansion with finite times in every iteration. When the number of iterations increases, λ stable in early stage increases. The random expansion affects ACO more in the middle stage than in the early stage. In the early stage, the random expansion will break down the convergence of pheromones, so it must be weakened in configuration. In the middle stage of ACO, the random expansion is strengthened to avoid the prematurity which appears in the stage. Finally, the random expansion fades in end stage of ACO which will converge at the stage.

2.3 Updation of Pheromones

With time goes, pheromones will decrease. Denote the residual pheromones by ρ . The updation of pheromones is formulated as

$$\tau_{i,j}(t+h) = \rho\tau_{i,j}(t) + \sum_{k-1}^{W} \Delta\tau_{i,j}^{k} \tag{4}$$

where $\Delta\tau_{i,j}^{k}$ presents the residual amount of pheromones along the route <*i,j*> by ant *k*. $\Delta\tau_{i,j}^{k} = Q / L_k$, where L_k presents the sum of all route lengths which ant *k* moves along in current iteration.

The steps are as follows:

1. Get a route using the detour method. Set the number of ants and maximum of iteration number. All ants are initialized at start point, and pheromones are also initialized by the relation (1).
2. Start ant colony, and every ant moves toward its next point determined by relation (2).
3. Repeat step (2) untill all ants reach the goal. Compute the length of routes which every ant finds, and record the best route. Update pheromones along those routes by relation (4).
4. When ant colony converges to one route or the maximum of iteration number is reached, abort the algorithm; otherwise, turn to step (2).
5. Output the best route found by above steps.

3 Simulation and Results

Simulation is done by MATLAB. Ant colony is set with these parameters: W=30, T_{max} =50, α =1, β =2, Q=100, ρ =0.9. Figure 1 shows the best routes found using IACO. In Fig. 1, the route is smooth across the environment I along which the robot is easy to move. The algorithm converges with longer time in environment II because the environment becomes more complicated. Other simulation results show the algorithm always converges in complicated environments in which at least a route exists.

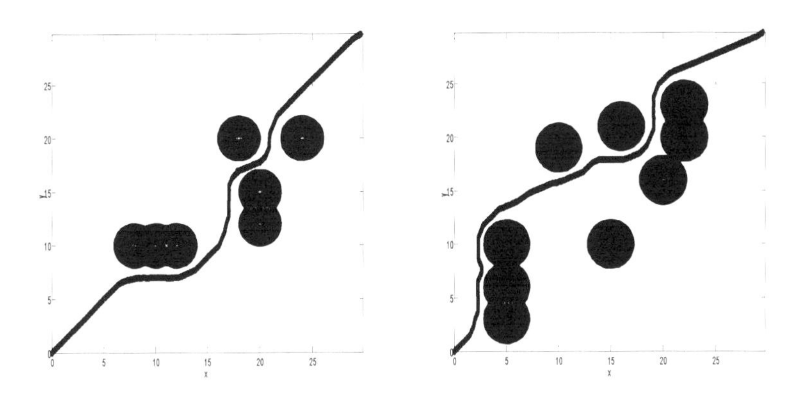

Fig. 1 The route found by IACO in environment I (left) and environment II (right)

In Table 1, the IACO and ACO are simulated 50 times respectively in MATLAB. IACO finds shorter routes with less time expended averagely as the table shows. In fact, IACO converges to the best route with high probability and less time than ACO.

Table 1 The comparison of ACO and IACO

Environment	Average length of the routes of ACO	Average time of ACO (s)	Average length of the routes of IACO	Average time of IACO (s)
I	79.1	46.3	65.8	32.4
II	83.4	73.5	69.6	53.1

4 Conclusion

ACO has the defects of slow convergence and prematurity in robot route planning. An improve ACO (IACO) is proposed in this paper, which changes the initial pheromone distribution by the detour method and introduces the random expansion factor. The simulation results show that IACO converges faster and the route found is shorter than ACO.

Acknowledgments. This paper is supported by the Natural Science Foundation of He'nan Educational Committee Grant No. 2010A510001 and The Science Foundation of He'nan University Grant No. 2008 YBZR028.

References

[1] Bonabeau, E., Dorigo, M., Theraulaz, G.: Swarm Intelligence: From Natural to Artificial Systems. Oxford University Press, New York (1999)

[2] Dorigo, M., Caro, G.D.: The Modified Swarm Optimization Metaheuristic. In: Come, D., Mdorigo, Glover, F. (eds.) New Ideas in Optimization, pp. 11–32. Graw Hill, Mc London (1999)

[3] Gong, B.C., Li, L.Y., Jiang, Y.T.: Ant colony algorithm based on local optimization for TSP. Application Research of Computers 25(7), 1974–1976 (2008)

[4] Hua, J.N., Zhao, Y.W., Wang, Y.C.: New Global Path Planning Algorithm for Mobile Robot. Robot 28(6), 548–597 (2006)

[5] Keron, Y., Borenstein, J.: Potential field methods and their inherent limitations for mobile robot navigation. In: Proceedings of the IEEE International Conference on Robotics and Automation, pp. 1398–1404. IEEE, Piscata way (1991)

[6] Li, L., Ye, T., Tan, M.: Present state and future development of mobile robot technology research. Robot 24(5), 475–480 (2002)

[7] Ma, Z.Q., Yuan, Z.R.: Real time navigation and obstacle avoidance based on grids method for fast mobile robot. Robot 18(6), 344–348 (1996)

[8] Oommen, B., Iyengar, S., Rao, N., Kashyap, R.: Robot navigation in unknown terrains using learned visibility graphs. Part I: The disjoint convex obstacle case. Robotics and Automation 3(6), 672–681 (1987)

[9] Pintea, C.M., Dumitrescu, D.: Improving ant systems using a local updating rule. In: Proc. of the 7th International Symposium on Symbolic and Numeric Algorithms for Scientific Computing, pp. 295–298 (2005)

Moving Human Head Detection for Automatic Passenger Counting System

Xiaowei Liu, Shasha Tian, Jiafu Jiang, and Jing Shen

Abstract. Moving human head detection is the basis of automatic passenger counting system and the speed and the detection accuracy of existing algorithms need to be improved.This paper puts forward a fast and effective algorithm for detecting moving human head.At first,the background is updated by block symmetric difference,then,the moving objects are extracted by culminating symmetric difference and background subtraction,finally,the human head contours are detected using the random Hough transform based on gradient.Experimental results show that this algorithm can detect moving human heads against illumination level changes and extraneous motion efficiently in automatic passenger counting system.

1 Introduction

Most APC technology based on image processing firstly detects and extracts moving objects,then, identifies and tracks objects,finally counts them.Therefore, accurate detecting and extracting moving objects is particularly important and is the premise and foundation of subsequent processings,such as object identification and object tracking.The method through the edge detection[1] can effectively detect moving human objects,but the detection efficiency reduces and the amount of calculation is large when objects is many.The adaptive background updating algorithm [2] may cause error detection of non-human objects such as baggage.The adaptive moving object detection algorithm[3] will lost heads like balds,heads with caps or heads whose color are similar to clothes.

Therefore,a fast and effective human head detection method for automatic passenger counting technology is proposed in this article.First,the background is updated using the block symmetry difference method to ensure the authenticity and reliability,then,moving objects are extracted by combining the symmetric difference method with the background subtraction method to narrow the search area,finally, head contours are detected using the random Hough transform based on gradient because moving human heads are arc-shaped.

Xiaowei Liu · Shasha Tian · Jiafu Jiang · Jing Shen
Changsha University of Science & Technology, Changsha, China

Z. Qian et al. (Eds.): Recent Advances in CSIE 2011, LNEE 125, pp. 147–152.
springerlink.com

2 Moving Human Head Detection Algorithm

2.1 Obtain and Update the Background

The camera which is used to obtain video images should be installed directly above the door in order to avoid passengers' cover when they get on and off.In this scene,the primary task is to extract the real and reliable background.Methods people usually use are Gaussian mixture model method [4],symmetric difference method [5] and pixel gray classification method [6]. The analysis reveals the change of the background on bus mainly dues to the external condition like the illumination and the weather.The GMM method has large computation amount and doesn't apply to the bus automatic passenger counting system.So,the background is updated using the block symmetric difference method [7],which not only reduces the system resource consumption and the complex computation of the symmetric difference method,and also improves the intelligence of the block method.The steps are as follows:

(1) Initialize the background model $B_k(x, y)$.

(2)Calculate the symmetric difference.Three images whose interval time is t are labeled respectively.The symmetric difference is calculated according to the formula (1).

$$D_k(x,y)=\begin{cases} 1, \left|W\times I_{k+1}(x,y)-W\times I_k(x,y)\right| \geq T_1 \\ 0, \left|W\times I_{k+1}(x,y)-W\times I_k(x,y)\right| < T_1 \end{cases} \quad (1)$$

(3)Calculate the mean difference of each block. D_k Will be divided into M blocks whose size is $p\times q$ (p is the number of rows, q is the number of columns) .The set of points of the block m is $b_m[s][t]$ and the average difference of each block is calculated according to the formula (2).

$$d_m^k = \frac{\sum\sum_{(i,j)\in b_m[s][t]} D_k[i][j]}{p\times q} \quad (2)$$

(4)Update the background.A counter $N=0$ is set and is used to count the mutative blocks.If $\left|d_m^k - d_m^{k-1}\right| > T_d$,the counter will be added by 1.Then the rate of mutations is calculated.If p is less than 70%,the background will not be updated;otherwise,the background will be updated according to the formula (3).

$$B_{k+1}(x,y)=\alpha I_k(x,y)+(1-\alpha)B_k(x,y) \quad (3)$$

2.2 Extract the Moving Objects

A more reliable background has been gotten after the background updating.But the noise still exists.In order to narrow the search area and improve the detection efficiency,the symmetric difference method and the background subtraction method are combined to determine the movement area and extract the moving objects[8].The detailed steps are as follows:

(1)Calculated the symmetric differential binary image and determine the moving area.The symmetric difference image are got by taking the logical AND operation of D_{k-1} and D_k according to the formula (4),as shown in Figure 1(a).According to the symmetric difference image,the maximum horizontal x_{max} ,the minimum horizontal x_{min} ,the maximum ordinate y_{max} and the vertical axis minimum y_{min} are got.The moving area $J_k(x,y)$,whose vertexs are $(x_{min,} y_{min})$, $(x_{min,} y_{max})$, $(x_{max,} y_{min})$ and $(x_{max,} y_{max})$,is established as shown in Figure 1 (b).

$$D_k^s(x,y)=\begin{cases}1,D_{k-1}(x,y)\cap D_k(x,y)=1\\0,D_{k-1}(x,y)\cap D_k(x,y)=0\end{cases} \tag{4}$$

(2)Calculate the background subtraction image.Based on the background in 2.1,the background subtraction image is got using the background subtraction method as shown in Figure 1 (c) .The formula is as follow:

$$D_k^b(x,y)=\begin{cases}1,\left|\mathrm{W}\times\mathrm{I_k(x,y)}\text{-}\mathrm{W}\times\mathrm{B_k(x,y)}\right|\geq \mathrm{T_3}\\0,\ \left|\mathrm{W}\times\mathrm{I_k(x,y)}\text{-}\mathrm{W}\times B_k\mathrm{(x,y)}\right|<\mathrm{T_3}\end{cases} \tag{5}$$

(3)Extract the moving objects.The moving object binary image $F_k(x,y)$ is got by taking the logical AND operation of the symmetrical difference image $D_k^s(x,y)$ and the background difference image $D_k^b(x,y)$ in the moving area.The formula is as follow:

$$F_k(x,y)=\begin{cases}1,D_k^s(x,y)\bigcup D_k^b(x,y)=1\\0,D_k^s(x,y)\bigcup D_k^b(x,y)=0\end{cases} \tag{6}$$

2.3 Detect the Moving Human Head Objects

The moving objects are obtained through the above method.But those objects also contain non-human heads,such as baggage or other objects.Therefore,it also needs to be dealed with to detect real moving human heads.It is not difficult to find that the human head edge is arc-shaped.Using this feature,the moving human head detection can change into the head contour arc-shaped detection.The current circle detection is mainly based on the Hough transform and its improvement.In this paper,we use the random Hough transform based on gradient [9].The algorithm is as follows:

(1)Detect the edge of the image $F_k(x, y)$ and get the edge image $E_k(x, y)$,as shown in Figure 1 (e);

(2)Initialize the unit set of the circle parameters C is empty;

(3)Search for continuous moving curves in the edge image, mark the location and length of each curve, save the curves which meet the conditions, and add them to the curve set $M = (L_1, L_2, ..., L_n)$;

(4)Select the curve L_i and detect circles.

(5)When $i = n$, the circle detection in the moving area is over. The locations of the circles in the unit set is the locations of the moving human heads,as shown in Figure 1 (f).

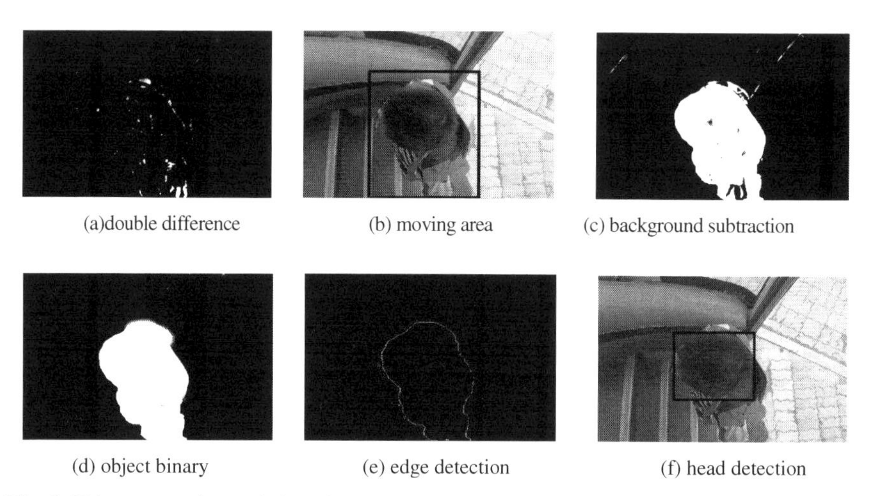

(a)double difference (b) moving area (c) background subtraction

(d) object binary (e) edge detection (f) head detection

Fig. 1 Object extraction and detection

2.4 *Moving Human Head Detection Algorithm*

In this moving human head detection algorithm, firstly,the background is updated using the block symmetric difference method, and then, moving objects are extracted by combining the symmetric difference method with background subtraction method,finally,the human head contours are detected using the random Hough transform based on gradient.The algorithm flow is shown in Figure 2. The specific steps are as follows:

STEP 1. Get the background image and judge whether it needs to be updated using the block symmetry method.If it needs, turn to STEP 2;otherwise turn to STEP 3.

STEP 2. Update the background.

STEP 3. Extract moving objects using the symmetric difference and the background subtraction method.

STEP 4. Detect moving human heads using the random Hough transform based on gradient.

3 Simulations and Analysis

To analyze the detection rate of this method, one to six people are respectively experimented on thirty times.Illumination changes strong ten times, non-human heads appear ten times and human heads with caps appear tem times.The experiment result is shown in Table 1.From the result in Table 1,we can see this method can effectively detect moving human heads.

Table 1 The detection of this method

persons	times	illumination change	non-human heads	complex heads	detection rate
1	30	10	10	10	99.8%
2	30	10	10	10	99.7%
3	30	10	10	10	99.4%
4	30	10	10	10	98.9%
5	30	10	10	10	98.8%
6	30	10	10	10	98.6%

4 Conclusion

This moving human head detection algorithm can quickly and accurately detect moving human heads in the bus video sequences. Experiments show that this algorithm solves the problem of strong illumination change,non-human heads and complex human heads, ensures the accuracy and effectiveness of moving human head detection,improves the value of the symmetric difference,and has a good robustness.

Acknowledgments. This paper is supported by Hunan Provincial Natural Science Foundation of China (10jj2050).

References

[1] Yu, S.-S., Chen, X.-P., Sun, W.-P., Xie, D.-P.: A robust method for detecting and counting people. In: Proceedings of International Conference on Audio, Language and Image Processing (2008)

[2] Damien, L., Alaya, C.F., Yngve, H.J., Pierre, G., Romain, P.-C.: Real-time people counting system using a single video camera. The International Society for Optical Engineering (2008)

[3] Sun, D.-H., Zhao, M.: Automatic Passenger Counting Based on Multi_objects Recognition Using Dynamic Images. The International Society for Optical Engineering (2005)
[4] Humayun, S., Kamran, K., Ahmed, Q.W.: Using modified mixture of gaussians for background modeling in video surveillance. In: Proceedings of 2nd International Conference on Advances in Space Technologies - Space in the Service of Mankind (2008)
[5] Tang, Z., Miao, Z., Wan, Y.: Background subtraction using running Gaussian average and frame difference. Lecture Notes in Computer Science (including subseries Lecture Notes in Artificial Intelligence and Lecture Notes in Bioinformatics) (2007)
[6] Hou, Z.-Q., Zhang, Q.: Effective background reconstruction algorithm in image sequence analysis. Yi Qi Yi Biao Xue Bao/Chinese Journal of Scientific Instrument (2008)
[7] Deng, X., Bu, J., Yang, Z., Chen, C., Liu, Y.: A block-based background model for video surveillance. In: Proceedings of IEEE International Conference on Acoustics, Speech and Signal Processing (2008)
[8] Zhao, Y., Wang, Z.: Detecting moving objects by background difference and frame-difference. The International Society for Optical Engineering (2007)
[9] Lu, H., Zhang, J., Zhang, M.: Head Detecting and Tracking Based on Hough Transform. Journal of System Simulation (2008)

Study on the Longitudinal and Lateral Coupled Controlling Method of an Intelligent Vehicle Platoon

Cui Shengmin, Zhang Kun, and Wang JiMeng

Abstract. Intelligent Vehicle(IV) is an important component of Intelligent Transport System(ITS) which includes automatic navigation system, high efficient traffic manage system, vehicle platoon system and so on. An efficient longitudinal and lateral coupled controller of an intelligent vehicle platoon is designed in this paper, through which vehicles in the platoon can realize the path-following task on both a straight lane and a curvilinear one. The model of this controller is designed in Matlab/Simulink environment and the simulation result is analyzed at the end of this paper.

Keywords: ITS, Intelligent Vehicle, Platoon, longitudinal and lateral coupled control.

1 Introduction

Study of the intelligent vehicle platoon method has always been a hotspot of the recent research about ITS. Much effort has been spent on various control laws of intelligent vehicle platoons. In 1999, Zhang Y. proposed a cruise intelligent control law which uses speed and space information from both the vehicle in the front and the one behind. The controller guarantees the vehicle stability as well as the platoon stability [1]. Zhang J. and Ioannou P. A. build a PID controller to realize the longitudinal control of a vehicle platoon which can ensure the string stability of the platoon [2]. [3] uses the fuzzy control method, [4] uses the SMC(Sliding Mode Control) method and so on. All the above study focused on the longitudinal control without taking longitudinal and lateral coupled into consideration. In fact the longitudinal and lateral movement of the vehicle is coupled together, they affects each other specially when the vehicle is travelling along a curvilinear line. This paper indicated to find an easier and more precise way for the build of the longitudinal and lateral coupled controller of an intelligent vehicle platoon.

Cui Shengmin · Zhang Kun · Wang JiMeng
School of Automobile engineering, Harbin Institute of Technology, Weihai, China

Z. Qian et al. (Eds.): Recent Advances in CSIE 2011, LNEE 125, pp. 153–158.
springerlink.com

2 Vehicle Modeling

In this work, the vehicles are considered to perform the path following task on both a straight lane and a curvilinear one. The vehicle used in this paper is called bicycle model as shown in figure 1, which is the simplified mathematical model for a four-wheel vehicle and the roll and pitch are neglected in the model [5]. The dynamic and kinematic differential equation of this model is shown as equation (1).

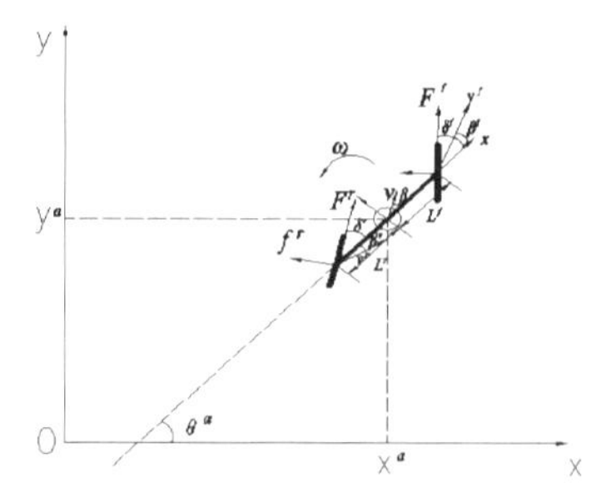

Fig. 1 Vehicle model

$$\begin{cases} \dot{x}_i = v_i \cos(\theta_i + \beta_i) \\ \dot{y}_i = v_i \cos(\theta_i + \beta_i) \\ \dot{\theta} = \omega_i \\ m\dot{v}_i = f_i^x \cos\beta_i + f_i^y \sin\beta_i \\ \dot{\beta}_i = -\frac{1}{mv_i} f_i^x \sin\beta_i + \frac{1}{mv_i} f_i^y \cos\beta_i - \omega_i \\ \dot{w}_i = \frac{1}{J}\tau_i^z \end{cases} \tag{1}$$

3 Platoon Modeling

The platoon is modeled as figure 2.

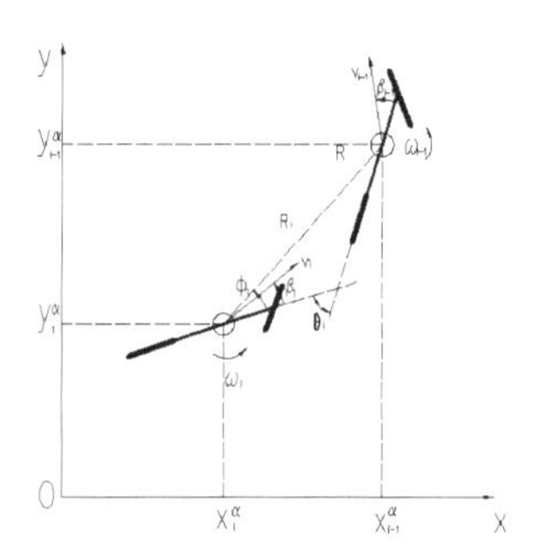

Fig. 2 Platoon model

The kinematic differential equation of the platoon model is shown as equation (2), and the kinematic differential equations are the same as the vehicle model as shown in equation (1).

$$\begin{cases} \dot{R}_i = -v_i \cos(\phi_i - \beta_i) + v_{i-1} \cos(\theta_i - \phi_i + \beta_{i-1}) \\ \dot{\phi}_i = \frac{1}{R_i} v_i \sin(\phi_i - \beta_i) - \omega_i + \frac{1}{R_i} v_{i-1} \sin(\theta_i - \phi_i + \beta_{i-1}) \\ \dot{\theta}_i = -\omega_i + \omega_{i-1} \end{cases} \tag{2}$$

4 Design of Controller

The objective is to control the inter-platoon space within a desired one Rref. The controller to be used for the platoon is assumed to be able to realize the stability of the platoon and the vehicle itself, at the same time it should be as simple as possible so as to be used through hardware. A simple PID controller is designed here to accomplish the above requirements. A platoon without wireless communication system a intelligent vehicle can only acquire limited information from the one above through sensors such as CCD and laser radar. The error function of the PID controller [6] is designed as equation (3).

$$\begin{cases} e_i^{\phi} = \phi_i - \phi^{ref} \\ e_i^{R} = R_i - R^{ref} \\ E_i^{\phi} = e_i^{\phi} \cos\phi^{ref} + \omega_{12} e_i^{R} \sin\phi^{ref} \\ E_i^{R} = \omega_{21} e_i^{\phi} \sin\phi^{ref} + e_i^{R} \cos\phi^{ref} \end{cases} \tag{3}$$

For a vehicle with rear-wheel-traction and front–wheel-steer, the PID controller can be designed as equation (4).

$$\begin{cases} F_i^{r} = (K_{P,i}^{R} + \frac{K_{I,i}^{R}}{s} + K_{D,i}^{R} s) E_i^{R} \\ \delta^{f} = (K_{P,i}^{\phi} + \frac{K_{I,i}^{\phi}}{s} + K_{D,i}^{\phi} s) E_i^{\phi} \end{cases} \tag{4}$$

The parameters Kp, Ki, Kd will be calculated later in this paper.

5 Simulation Result

5.1 Path Following Task on a Straight Lane

The simulation parameters are chosen as follows: Lf=0.8m; Lr=1m; μf=500N/rad; μr=450; m=50kg; J=100N·m/rad·s-1; b=50 N·m/rad·s-1;B=40N/m·s-1;v0=10m/s; R=10m.

Here choose α=1 for simulation. β≥max{1,1}, and βis chosen as 20 for simplification.

A platoon with a leading vehicle and 3 following vehicle is simulated in this paper, the leading vehicle travelling along coordinate X at a speed 20m/s. The speed of the vehicles in the platoon is shown in figure 3.

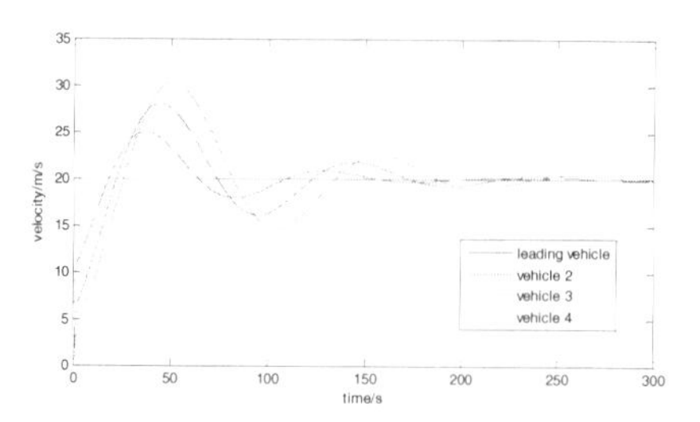

Fig. 3 Platoon velocity

It can be acquired from this figure that after a period of 230 seconds, the vehicles in the platoon become the same speed: 20m/s. The simulated position of the vehicles in the platoon is shown in figure 4.

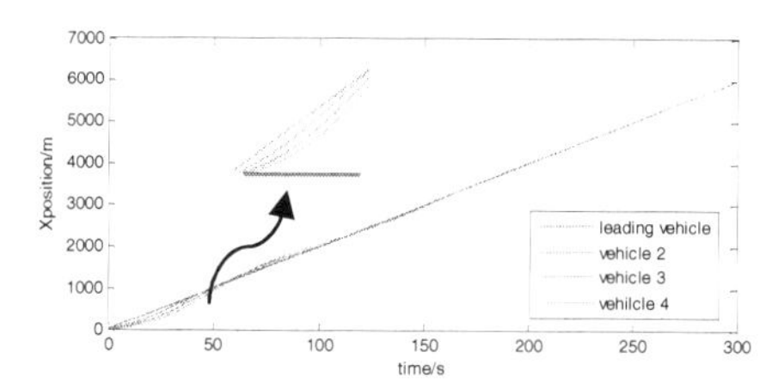

Fig. 4 Platoon position

As can be seen from figure 4, the vehicles in the platoon have different spaces between each other at the very beginning. That is because the leading vehicle has a speed at 20m/s all the time but the following vehicles' start speed is set to zero, so they need a period of time to adjust their speed to 20m/s so as to follow the one in the front.

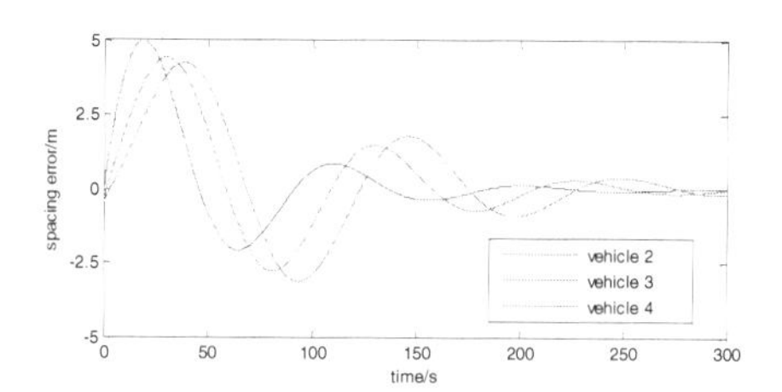

Fig. 5 Spacing error

That is also the reason why it needs a period of time for the spacing error of the 2nd, 3rd and the 4th vehicle to become zero as is shown in figure 5.

5.2 Path Following Task on a Curvilinear Lane

For the second simulation condition, the first vehicle travels along a curvilinear lane at an angle speed about PI/6, two vehicle start to follow the leading vehicle. The XY position of the vehicles in the platoon is shown in figure 6.

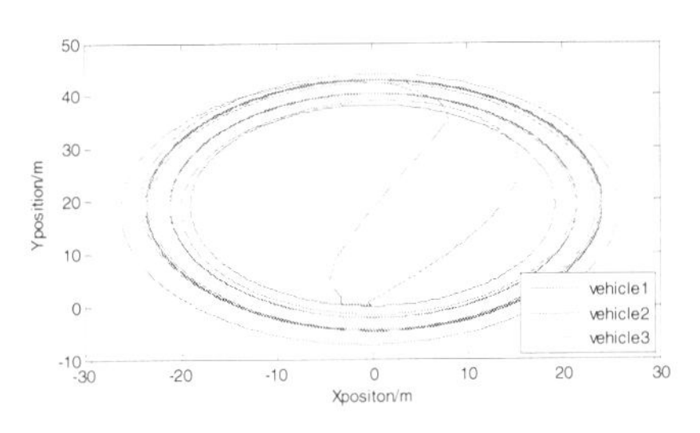

Fig. 6 XY platoon position

As can be seen from figure 6, vehicle 2 and vehicle 3 need some period of time to follow the path which the leading vehicle has passed. That is because at the very beginning the velocity and angle velocity of the following vehicles are equal to zero, they need some time to become stable. The spacing error of vehicle 2 and vehicle 3 is shown in figure7, while the following angle error of them is shown in figure 8.

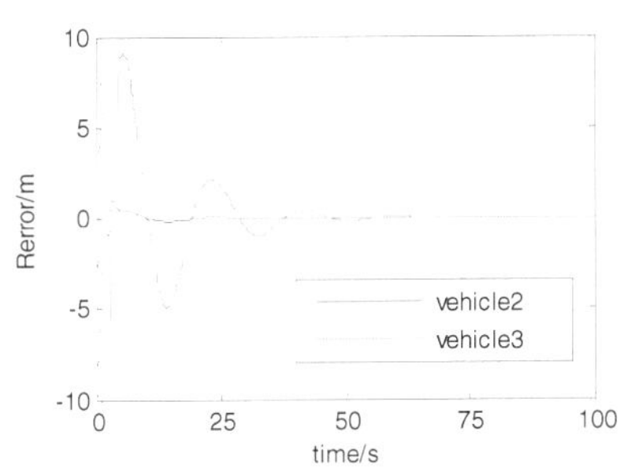

Fig. 7 Spacing error condition 2

It can be seen from figure 7 and figure 8 that after a time of 50 seconds, the speed of the following vehicle becomes stable as well as a time period of 20 seconds by which the following angle become zero. It also can be seen from figure 7 and figure8 and vehicle 3 need more time to become stable than vehicle 2 for the reason that a vehicle can only get information from the one in its front.

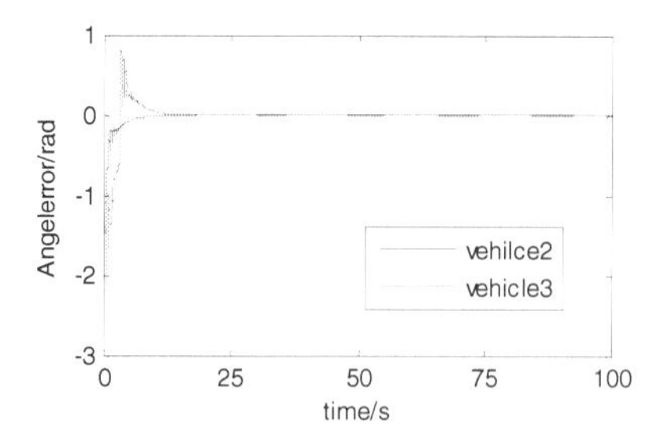

Fig. 8 Following angle error

6 Conclusions

A simplified longitudinal and lateral coupled model of an intelligent vehicle is built in this paper, which can perform rear-wheel-traction and front-wheel-steer maneuver. The intelligent vehicle platoon is composed of a leading vehicle which can travel automatically and some vehicles with CCD camera and laser radar which can get velocity and space information from the one in its front. The controller of the platoon is based on tuned PID algorithm which can be easily be used through hardware such as single chip or DSP. The simulation results show that the controller performs well both when the platoon travelling along a straight lane and a curvilinear one.

References

1. Zhang, Y., Kosmatopoulos, B., Ioannou, P.A., Chien, C.C.: Autonomous intelligent cruise control using front and back information for tight vehicle following maneuvers. IEEE Transactions on Vehicular Technology, 319–328 (1999)
2. Zhang, J., Ioannou, P.: Longitudinal control of heavy trucks in mixed traffic: environmental and fuel economy considerations. IEEE Transactions on Intelligent Transportation Systems, 92–104 (2006)
3. Araeil, J., Heredia, G., Ollero, A.: Global stability analysis of fuzzy Path tracking using frequency Response. Engineering Applications of Artificial Intelligence, 109–119 (2000)
4. Nouvelierea, L., Mammar, S.: Experimental vehicle longitudinal control using a second order sliding mode technique. Control Engineering Practice, 943–954 (2007)
5. Khatir, M.E.: On the Decentralized Control of a Platoon of Autonomous Vehicles, Paper of PhD degree of University of Toronto, pp. 36–39 (2006)
6. Xavier, P.: A Decentralized PID-Based Approach for the Collaborative Driving of Automated Vehicles, Paper of Master's degree of University of Toronto, pp. 44–48 (2009)

An Algorithm of Infrared and Visible Light Images Fusion Based on Infrared Object Extraction

Haichao Zhang, Fangfang Zhang, Shibao Sun, Wen Yang, and Yatao Wang

Abstract. The traditional image fusion methods of Infrared and visible light images neglect differences between the background and the targets, resulting in the poor clarity or week identifiability of the fused image. According to the characters of infrared and visible images, an algorithm based on object extraction is proposed. This algorithm infrared image object is extracted by maximum between-cluster variance, the acquired object information and the background and detail information of visible light image are fusion by NSCT. Thereby improved the specific of visual and effect largely. Finally, the effect of the results is evaluated through objective method. Experimental results indicate that the image fusion obtained by this algorithm possessed the same object as the one in the infrared image, and keeps the same details information as the one in the visible image.

1 Introduction

As a technology of great value in practical uses, image fusion is to integrate multiple image signals into a single image to enhance availability of information and achieve more exact, reliable, and comprehensive description of the images. Fusion of infrared image and visible light image has found wide application for defense purposes. This is attributed to essential differences between infrared image and visible light senses—visible light has ability to represent the background more abundant and legibility than infrared. For example, details of targets that are hard to detect in a visible light image (meaning low visible contract) can sometimes be easily discovered in an infrared image [1].

Haichao Zhang · Fangfang Zhang · Shibao Sun · Wen Yang · Yatao Wang
College of Electronics and Information Engineering,
Henan University of Science and Technology, Luoyang 471003, P.R. China
e-mail: zhcluoyang@tom.com

Z. Qian et al. (Eds.): Recent Advances in CSIE 2011, LNEE 125, pp. 159–165.
springerlink.com

With the development of wavelet theory, wavelet analysis becomes a new research hot spot. Wavelet-based image fusion algorithms [2] reserve and inherit the advantages of the pyramid decomposition algorithms, such as multi-scale, multi-resolution. Because of the non-redundancy, after decomposition based on wavelet theory, the data volume will not expand, but the directional information increases, which is fit to the characteristics of the human vision system. However, the 2-D separable wavelet, generated from the 1-D wavelet, can only capture limited directional information, and thus cannot represent the directions of the edges accurately. In recent years, Do and Vetterli developed a true 2-D image representation method, namely, the contourlet transform(CT) [3], which is achieved by combining the LP[4,5] and the directional filter bank(DFB)[6]. Compared with the traditional DWT, the CT is not only with multi-scale and localization, but also with multi-direction and anisotropy. As a result, the CT can represent edges and other singularities along curves much more efficiently. However, the CT lacks the shift-invariance, exists the frequency aliasing [7], which weakens the effectiveness in image fusion. In 2006, Cunha, et al. proposed an overcomplete transform, namely, the nonsubsampled contourlet transform (NSCT) [8]. The NSCT inherits the perfect properties of the CT, and meanwhile possesses the shift-invariance. Accordingly, a novel image fusion algorithm based on the NSCT is proposed. When the NSCT is introduced to image fusion, more information for fusion can be obtained and the impacts of mis-registration on the fused results can also be reduced effectively. Therefore, the NSCT is more suitable for image fusion.

The character of infrared image and visible light image are not considered in the traditional fused algorithms, resulting in the poor clarity or week identifiability. In this paper, the images are talked individually based on the characters of infrared and visible light images. To intensify clarity and identifiability, the information of visible light image is selected between the simple information of the two images, while the complementary information is fused directly. Fusing the complementary information not only keep the background information, but also improve the clarity. In order to outstanding the image character and improve the clarity, an algorithm based on object extraction is proposed. Several experimental results are employed to demonstrate the great validity and feasibility of the proposed algorithm.

2 Nonsubsampled Contourlet Transform

To get rid of the frequency aliasing of the CT and to enhance the directional selectivity, the NSCT with shift-invariance is proposed in paper [8]. The principle is displayed in Fig.1.

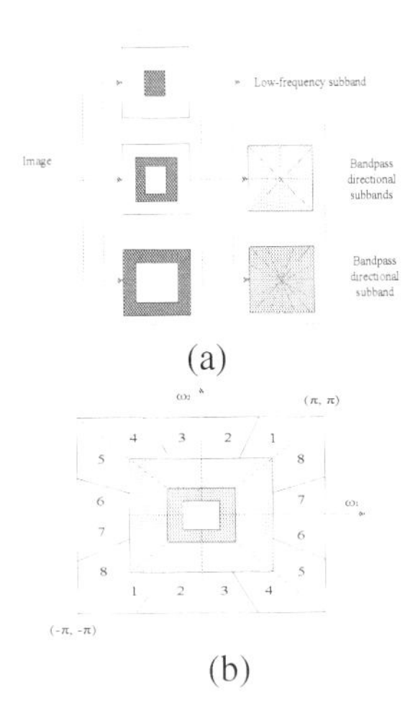

Fig. 1 The nonsubsampled contourlet transform: (a) NSFB structure that implements the NSCT and (b) the idealized frequency partitioning obtained with the proposed structure

The nonsubsampled contourlet transform can thus be divided into two parts that are both shift-invariant: (1) A nonsubsampled pyramid structure that ensures the multi-scale property and (2) A nonsubsampled DFB structure that gives directionality. The process of decomposition is similar to the CT. The NSDFB is obtained by eliminating the downsamplers and upsamplers in the DBF. Since there are no sampling operations in the NSP and the NSDFB, the NSCT has the shift-invariance. Different from the CT, the multi-scale decomposition is achieved not because the Laplacian pyramid transform, but directly by using two-channel non-subsampled 2-D filter banks which meet the precise reconstruction condition. Because the NSP eliminates the sampling operation in the decomposition process, even through the bandwidth of low-pass filter is wider than $\pi/2$, the low-frequency subband cannot cause the frequency aliasing.

The tower filter banks, employed by the NSCT, are two-channel nonsubsampled filter banks. To achieve multi-scale decomposition, conceptually similar to the 1-D nonsubsampled wavelet transform with the a-Trous algorithm, the non-subsampled pyramids (NSP) filters for next level are obtained by upsampler the filters to the previous stage with the sampling matrix $D = 2I = \begin{bmatrix} 2 & 0 \\ 0 & 2 \end{bmatrix}$. The NSPFB is iteratively used. After J-level decomposition, $L+1$ subband images are obtained, all of which has the same size as the input image.

The directional filter banks, employed by NSCT, are two-channel nonsampled fan filter banks. To achieve multi-direction decomposition, the NSDFB is iteratively used. The filter banks used in the level is obtained by sampling matrix D through the filter banks in the previous level. The matrix D is $D = \begin{bmatrix} 1 & -1 \\ 1 & 1 \end{bmatrix}$. If the subband image at one scale is decomposed, 2^l direction subband images are obtained, all of which have the same size as the input image. After j -level NSCT decomposition, one lowpass subband image and $\sum_{j=1}^{J} 2^{l_j}$ (l_j denotes the direction decomposition level at the j -th scale) bandpass directional subband images are obtained.

3 The Proposed Method

The visible light image has an ability to represent the background more abundant and legibility than infrared. The detail of target is hard to detect in the visible light image in the corrosponding region while it can easily be discovered in the infrared image. At present, most image fusion algorithms perform processing on the entire scene; however, infrared images could not obtain detailed background and exact location of target. The main reason for combining visible and infrared sensors is that a fused image construction and unambiguous localization of a target represented in the infrared image with respect to its background represented in the visible light image, so it is reasonable to introduce the technology of infrared object extraction into infrared and visible light image fusion. Specific implement steps are as follows:

3.1 Target Extraction

Generally the target region corresponds to the area with particular property. In order to identify and analyze the target, the corresponding region needs to be extracted from the original image, which is target and background separation. The corresponding target and background regions are obtained through the separation operation on the corresponding infrared and visible light images. The fused effectiveness will be affected by the accurate of image segmentation directly, and then the method of image segmentation is important particularly.

One aim of fusion is to grasp the integral information of the circumstance. Because the infrared images are heat induced, the higher temperature they have, the higher intensity they show in the image. And the infrared images have a character of two-peak histogram, so the image separation is achieved by threshold effectively.

In this paper, the target region in infrared image is obtained by threshold and morphology method at first. Then the visual light "target region" with the infrared target image is obtained by pixels corresponding. The images in Fig.2 show the target extraction of the origin image.

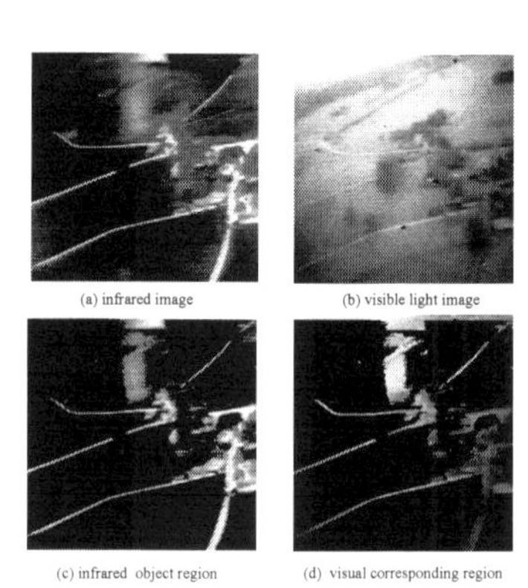

Fig. 2 Target extraction image

3.2 *Target Region Fusion*

The target image is decomposed by NSCT, and then the coefficients can be classified into a low-frequency subband and multiple high-frequency subbands. They are fused separately using different fusion rules. By successively performing inverse nonsubsample contourlet transform to the modified coefficients for all decomposition subbands, the final fused image can be constructed.

3.2.1 Low-Frequent Fusion

As the coefficients in the coarsest scale subband represent the approximation component of the source image, the simplest way is to use the conventional averaging method to produce the composite coefficients. However, it can't obtain fused approximation of high quality. Here the local energy is used to compute the final coefficients. The region energy can be calculated by

$$E_P = \sum_{s=-1}^{1}\sum_{t=-1}^{1}\left(P(i+s, j+t) - P(i,j)\right)^2 \tag{1}$$

The fusion rule is finally defined as

$$\left.\begin{array}{ll} FL(i,j) = AL(i,j), & \left| E_A(i,j) > E_B(i,j) \right| \\ FL(i,j) = BL(i,j), & \left| E_A(i,j) \le E_B(i,j) \right| \end{array}\right\} \tag{2}$$

3.2.2 High-Frequency Fusion

The high-frequency coefficients with large absolute value are considered as sharp brightness changes or salient feature in the corresponding source images, such as the edges, lines, contours and object boundaries. Therefore, the rule of high-frequency selects the max absolute value in the fused image. The fusion rule is defined as

$$\left.\begin{array}{ll} FH(i,j) = AH(i,j), & \left| AH(i,j) > BH(i,j) \right| \\ FH(i,j) = BH(i,j), & \left| AH(i,j) \le BH(i,j) \right| \end{array}\right\} \tag{3}$$

3.3 *Increase Background Information*

Finally we get the fused image by combining the fused target image and the background of the visible light image. Here are the results of two group experiments to certify the proposed method.

4 Experiments Study and Analysis

Originally, the infrared and visible light images have different information. In order to reveal the efforts of different algorithms, three different algorithms are compared. It can be seen that each algorithm has different vision effects. Fig.3(c)-(e) show the fused images using three different type of fused rules. Table I illustrate the performance of the different methods. Different objective standards have different meanings, entropy, average gradient, and the mean cross entropy are selected to evaluate the quality of the different rules.

4.1 The Experiment

In order to prove the feasibility of the proposed fusion rule, the fusion rules of target extraction are applied to NSCT. Fig.3(f) is fused image obtained with NSCT and the proposed method of target extraction, which bears witness to the good performance, feasibility, and effectiveness of the proposed method with the its region segmentation rule. The NSCT is shift invariant, better characteristic in direction and non-subsampled with the sub-bands images of the same size. Composed with Fig.3(c)-(e), Fig.3(f) has better preference and, especially, can effectively describe the target that have plentiful linear or curve contours. Moreover, the proposed image can extract most features of target from original image and preserve most spectrum information. Therefore, Fig.3(f) possesses the best vision effects in comparison with other algorithms.

Table 1 Performance of various Results on processing Fig.3

Method	Entropy	Average Gradient	The mean Cross Entropy
Wavelet fusion	7.4779	3.9879	2.0796
Contourlet fusion	7.4948	4.7514	1.8220
NSCT fusion	7.4953	4.3774	1.8674
The proposed method	7.9386	4.4999	1.4166

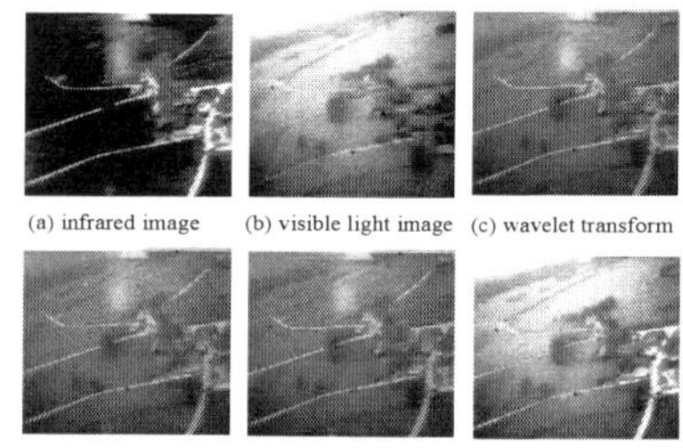

Fig. 3 Experiment

From the run of experiments, the results prove that the proposed method can obtain the best-performance. The results also show that the proposed method can be additionally applied to improve the fusion results. It can be seen that due to the various imaging principle and environment the source images with different modality contain complement information. The proposed method is applied to these images.

5 Conclusions

As the novel tool, NSCT offers better advantages of directionality, localization, anisotropy and multiscale, which cannot be perfectly achieves by traditional multiscale analysis like wavelet. In this paper, a novel method based on the target extraction is proposed. Considered the character of source images and the advantage of NSCT, the background information in the visible light image and the target information in the infrared image are fused by the proposed algorithm.

References

[1] Toat, A., van Ruyven, J.J., Valeton, J.M.: Merging thermal and visible images by a contrast pyramid. Optical Engineering 28(7), 789–792 (1989)
[2] Tao, G.Q., Li, D.P., Lu, G.H.: Study on image fusion based on different rules of wavelet transform. Acta Photonica Sinica 33, 211–224 (2004)
[3] Do, M.N., Vetterli, M.: The Contourlet Transform: An efficient directional multiresolution image representation. IEEE Transactions on Image Processing 14, 2019–2106 (2005)
[4] Burt, P.J., Adelson, E.H.: The laplacian pyramid as a compact image code. IEEE Transactions on Communications 31, 532–540 (1983)
[5] Do, M.N., Vetterli, M.: Framing pyramids. IEEE Transactions on Signal Processing 51, 2329–2342 (2003)
[6] Bamberger, R.H., Smith, M.J.T.: A filter bank for the directional decomposition of image: theory and design. IEEE Transaction on Signal Processing 40, 882–893 (1992)
[7] Liu, Y., Guo, B.L., Wei, N.: Multifocus image fusion algorithm based on contourlet decomposition and region statistics. Image and Graphics, 707–712 (2007)
[8] Cunha, A.L., Zhou, J.P., Do, M.N.: The nonsubsampled contourlet transform: theory, design and applications. IEEE Transactions on Image Processing 15, 3089–3101 (2006)

The Design of Ultra High-Speed Intelligent Data Acquisition System

Xiaolei Cheng, Xiaoping Ouyang, and Fang Liu

Abstract. For the requirement of cosmic rays detecting and high intensity pulsed radiation field diagnosing, an ultra high-speed intelligent data acquisition system (UIDAQ) without CPU and MPU is designed. The indispensable functions could be operated only by one piece of ADC and one piece of FPGA, without the accessing of the "brain" (CPU or MPU) of the whole detecting system or diagnosing system. Then the tests of indispensable functions is introduced, to confirmed that the valid signals could be distinguished and sampled with sampling rate over Gs/s by the UIDAQ system intellectively and intellectively. Including its self-regulation functions, the UIDAQ performs as the "vegetative nervous" of the whole detecting system or diagnosing system.

Index Terms: Ultra high-speed, Intelligent data acquisition, Full digitizing design, Valid signals distinguishing.

1 Introduction

The research of cosmic rays and high intensity pulsed radiation field require to distinguish the valid transient signals lasting only nanoseconds from the invalid signals and record them under hostile measuring environments. The problems mentioned below caused the difficulty in system designing:

1, the signal integrity (SI) would be deteriorated by the conventional complicated intelligent circuits;
2, the necessary Gbytes/s processing rate for intelligent functions will make the charge on the CPU (or MPU) become too heavy to be processed normally and enduringly [1-4].

To solve these problems, an ultra high-speed intelligent system independent of the CPU and the complicated intelligent circuits is required.

Xiaolei Cheng · Xiaoping Ouyang · Fang Liu
School of Nuclear Science and Engineering, North China Electric Power Univ., Beijing, 102206, China

Z. Qian et al. (Eds.): Recent Advances in CSIE 2011, LNEE 125, pp. 167–174.
springerlink.com

2 Principle Analysis

2.1 *Full Digitizing Design*

The design procedures to solve those problems can be described as 'full digitizing design', which means to convert all the external analog signals (including diagnosed signals and controlling signals) to digital signals on board as much as possible. Through the Full Digitizing Design, most of the analog circuits have been converted as digital circuits, which could be integrated in one piece of FPGA. Then the circuit of UIDAQ system could be simplified as shown in Fig.1 [5, 6].

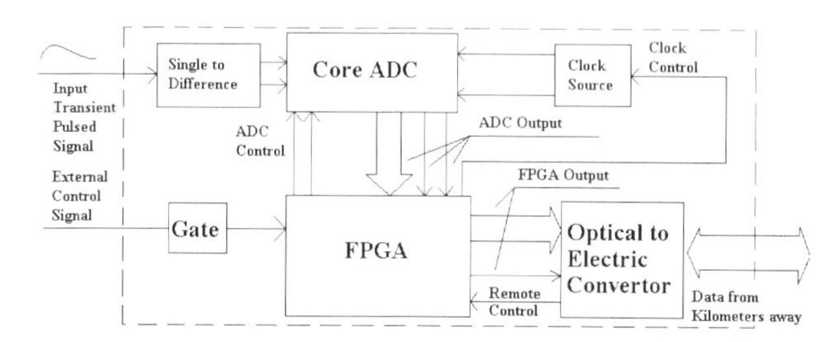

Fig. 1 The simplified circuit of UIDAQ system

2.2 *Intelligent Functions Achieving*

The full digitizing design of main intelligent functions will be introduced orderly:

Automatic Pretrigger - As showed in Fig.2, the transient signal could trigger the sampling process by itself, and its rising edge could be recorded integrally. The operating of data acquisition is showed in Fig.3 [7].

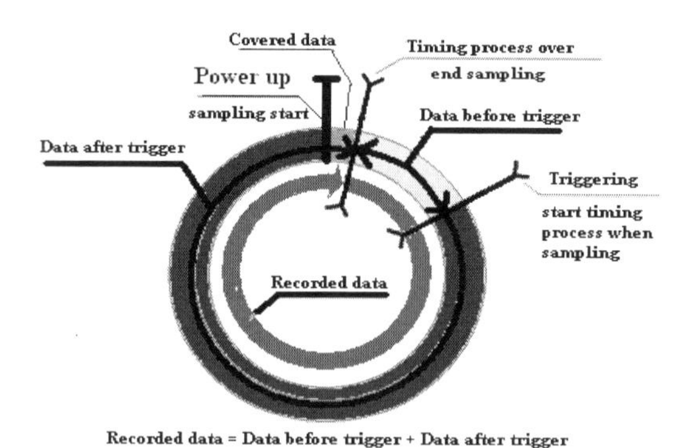

Fig. 2 Digital automatic pre-trigger mode

Delayed External Trigger – Sometimes, the multi UIDAQ systems must be triggered by external clocks to ensure the sampling of each will start in a certain timing sequence. Usually an external clock mastering circuit is needed, which will cost more power and make the whole detecting or diagnosing system more complicated and unstable.

To ensure the stabilization of the system, the multi UIDAQ systems could be started in a certain timing sequence by the only global synchronous clock through the intelligent function called delayed external trigger, which is achieved by the internal logical processing of the FPGA.

If there are n (n=1, 2, 3...) UIDAQ systems and the required trigger time of the sampling are t_1, t_2, t_3...t_n, then this function could delay the time intervals between the arriving of synchronous clock (t_0) and the start of sampling as t_1-t_0, t2-t0,t3-t0...tn-t0, which make the sampling of each UIDAQ systems start in a certain timing sequence (t_1,t_2,t_3...t_n) without any external circuits.

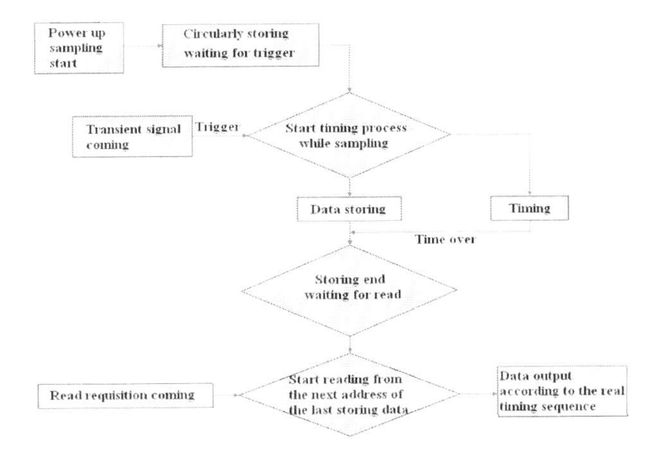

Fig. 3 Flow diagram of data acquisition

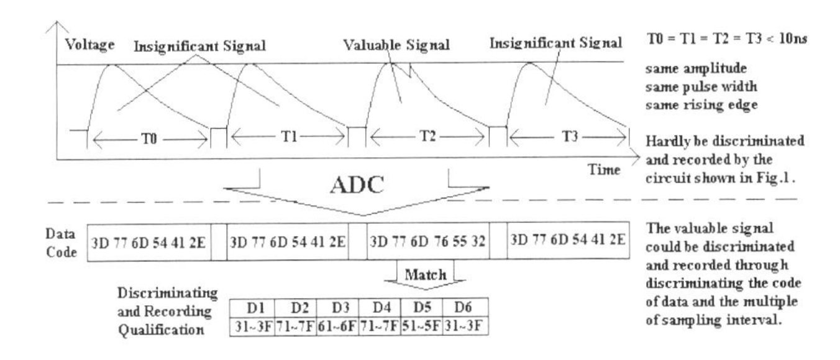

Fig. 4 Certain characteristic discriminating

Valid Signal Distinguishing – The UIDAQ system still would be interfered by the electromagnetic pulse under the normal electromagnetic shielding through the "Back-door coupling", mainly appear as the interfering signals and the baseline drift. As shown in Fig.4, the valid signal could be distinguished from the interfering signals accurately through a concise logical processing in the FPGA of the UIDAQ system.

Auto Baseline Restoring – When the UIDAQ system is operating, the baseline of the input signals are monitoring by the FPGA. If the baseline is abnormal, the internal logical controlling will be triggered and the revisal codes will be sent to the adjusting interface of ADC to adjust back the baseline.

2.3 *Integrated Design of FPGA*

All the intelligent functions must operate parallelly, which usually depends on the processing of CPU or MCU. Through the "Full Digitizing Design" method, that can be achieved by the FPGA as shown in Fig.5 [8].

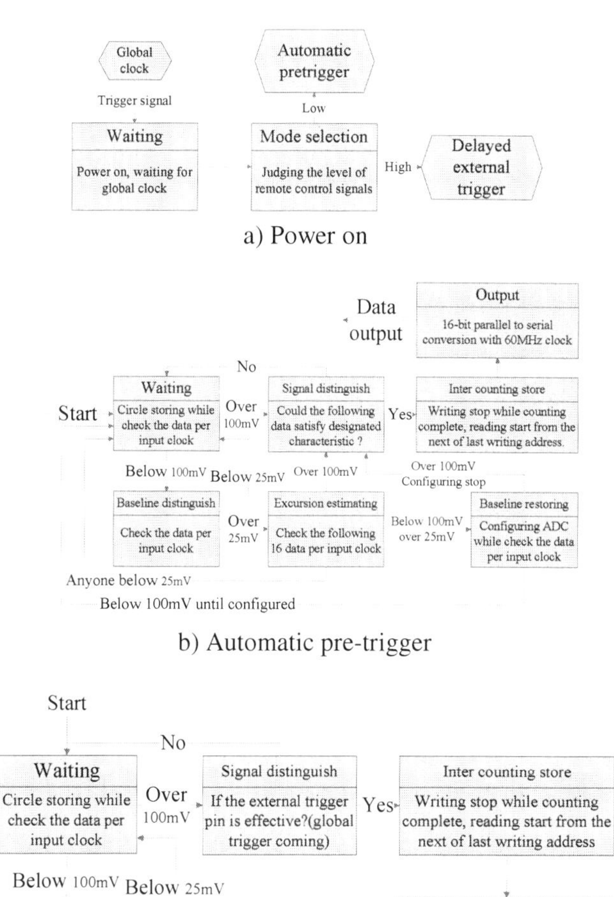

Fig. 5 Integrated design of FPGA

3 Results

3.1 Automatic Pre-trigger

An UIDAQ system has been designed to verify the intelligent functions. It's an 8-bit data acquisition system with 2Gs/s sampling rate[9].As shown in Fig6, the UIDAQ system could be triggered by the valid signal itself without the rising edge missing and the peripheral pre-trigger circuits. That will reduce the instability factor of the whole detecting or diagnosing system.

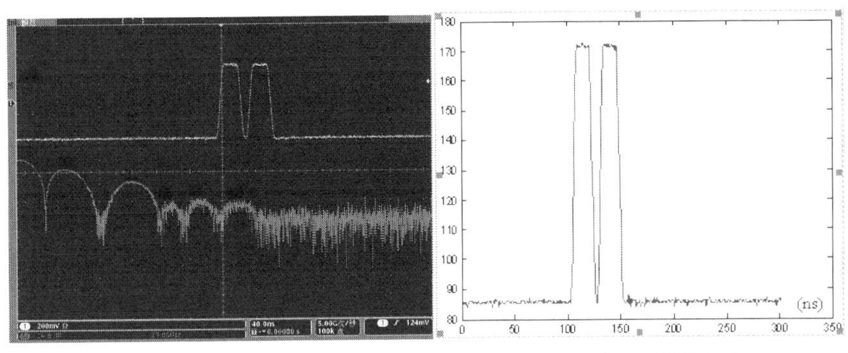

Fig. 6 Test for automatic pre-trigger

3.2 Delayed External Trigger

As shown in Fig.7, the interval between the global clock and the start of sampling could be controlled as it has been required (25ns in this test).

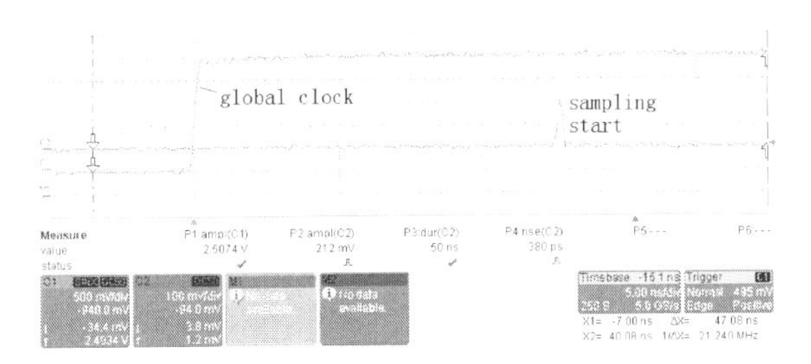

Fig. 7 Delayed external trigger

As shown in Fig.8, a mark will be inserted into the sampling result to mark the temporal relation between the signal and the start of sampling. From that relation, the temporal relation between the signal and the global clock will be figured out.

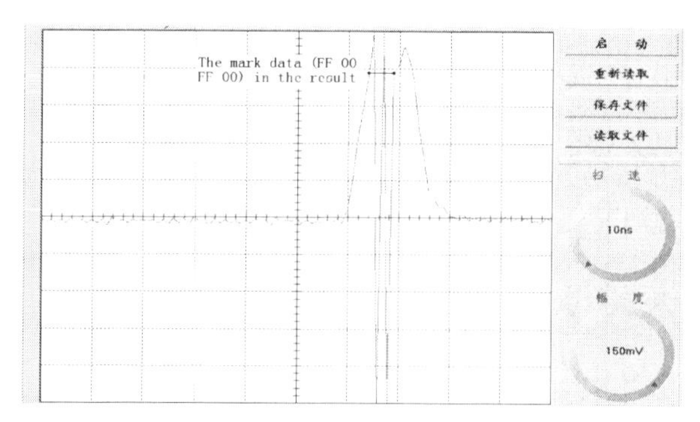

Fig. 8 The mark in sampling result

3.3 Valid Signal Distinguishing

As showed in Fig.9, the sampling of this system will not be triggered by transient interference signals (pulsed width <3ns in this test). It could be triggered only by the valid signals (pulsed width>3ns in this test).

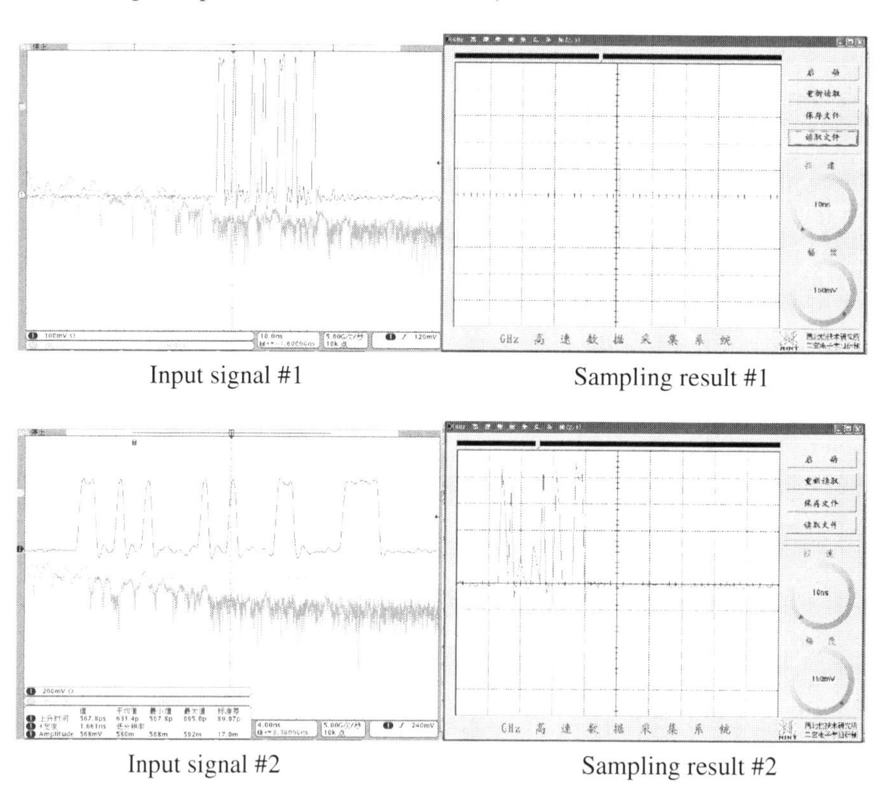

Fig. 9 Test for valid signal distinguishing

3.4 Auto Baseline Restoring

As shown in Fig.10, the baseline of the input signal could be restored by the logical operation in the FPGA of this system.

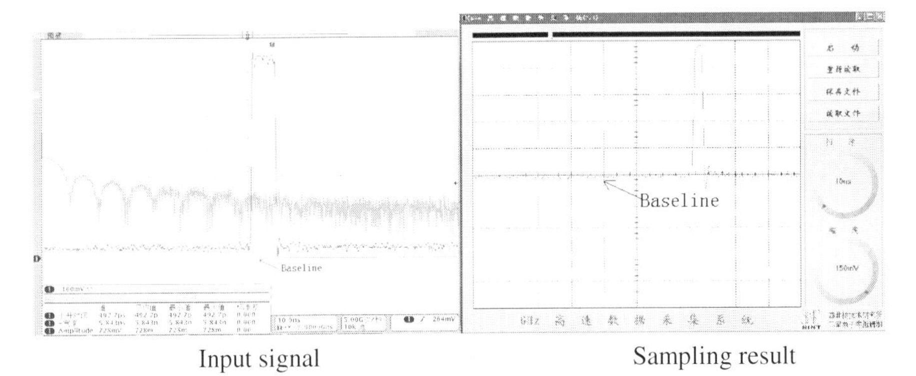

Fig. 10 Test for baseline restoring

4 Conclusions

The UIDAQ system is designed to achieved the requirement of cosmic rays detecting and high intensity pulsed radiation field diagnosing, without accessing the CPU (or MPU) and any other complicated circuits. All the intelligent functions being achieved by only one piece of ADC and one piece of FPGA will reduce the power dissipation and ensure the signal integrity (SI) of the whole system.

Further more, the UIDAQ system operates as "vegetative nervous". That will optimize the system as to free up its "brain" to cut the risk of system halted and extend its useful life.

Acknowledgments. Thanks to the supporting by the National High Technology Research and Development Program of China (2007AA01Z275).

References

[1] Jiao, W.: The Space Exploration, pp. 68–129. Peking University Press, Beijing (2005) (in Chinese)

[2] Ouyang, X.: Detecting Technique of Pulsed Radiation. Chinese Engineering 10(4), 44–55 (2008) (in Chinese)

[3] Ouyang, X., et al.: The System Design and Experiment Method of Detecting System in Pulsed Radiation Field diagnosing. Atomic Energy Press, Beijing (2009) (in Chinese)

[4] Bitossi, M., Cecchi, R., et al.: The Data Acquisition System of the MAGIC-II Telescope. In: IEEE International Workshop on Intelligent Data Acquisition and Advanced Computing Systems: Technology and Applications, Dortmund, Germany, September 6-8 (2007)

[5] Graham, H.J.M.: High-Speed Digital Design (2004)
[6] Brooks, D.: Signal Integrity Issues and Printed Circuit Board Design (2005)
[7] Cheng, X., et al.: The Design of 1Gsps Real-Time Sampling System for Transient Pulsed Signal. In: The 16th IEEE NPSS Real Time Conference, Beijing (2009)
[8] Cheng, X., et al.: An Ultra-high-speed Front-end Intelligent Digital Transmission System for High-intensity Pulsed Radiation Field Diagnosis. In: 2009 IEEE Nuclear Science Symposium and Medical Imaging Conference, Orlando (2009)
[9] Cheng, X.: The Digital Transmission in Pulsed Radial Detecting of Nuclear Explosions, pp. 60–62. The Doctoral Dissertation of Tsinghua University (2010) (in Chinese)

Simulation of Wealth Distribution

Fangfeng Zhang, Tejaswini K. More, and Xuehu Zhang

Abstract. Under what conditions do inequalities within society emerge? To explore this question we modeled the evolution of unequal distributions of wealth and social capital within artificially constructed societies. In order to better understand the causative factors of unequal wealth distribution, we examined and extended the NETLOGO Wealth Distribution Model made by Michael Gizzi, Tom Lairson and Richard Vail. Cursory examination of the data show that increases in all individuals' fields of vision decreases inequality as measured by the Gini coefficient. Analyses of extreme examples of the model are helpful in understanding the most potent factors affecting inequality.

Keywords: inequality, networks, multi-agent, NETLOGO, Gini coefficient, wealth distribution.

1 Introduction

In modern societies equipped with advanced transportation and communication facilities, social interactions amongst people play an important role in determining an individual's social status and potential wealth[1].

In our paper, we are developing several models that are designed to help us understand not just why inequity persists within society, but also how it evolves over time. Such an inquiry is important because it helps explains Pareto's Law[2] in which the rich in society, though few in number, tend to get richer while the poor [many, in comparison] tend to get poorer. Such queries, if fruitful, may prove useful in advising policy decisions, which aim to reduce the negative effects of huge disparities in wealth and power within society[3].

Fangfeng Zhang
School of Information Beijing Wuzi University, Beijing, China

Tejaswini K. More
School of Medicine Yale University, New Haven, CT., U.S.A

Xuehu Zhang
Peking University, Beijing, China
e-mail: zzf_flora@163.com

Z. Qian et al. (Eds.): Recent Advances in CSIE 2011, LNEE 125, pp. 175–179.
springerlink.com

The use of NETLOGO models of social inequity comes out of the recognition of the fact that the notion of "wealth" has several disparate interpretations. In our model, we give consideration to the notion of "wealth" as it refers to physical resources such as food, capital, or money.

In our paper, we try to find the answer of the following wondering: What are the key modulators of wealth inequality within this model? Specifically, we ask:

How does increased population density affect the distribution of wealth?

To what extents do increases in an individual's capacity to see the topology of their environment affect global measurements of inequality?

How do variations in the productivity of the landscape affect the inequality of the system?

2 Methods

The NETLOGO model used in this paper is an adaptation of Epstein & Axtell's "Sugars cape" model. In this model, independent agents navigate a non-homogenous landscape collecting grain ["wealth"] under variable conditions of population growth, field of vision, grain growth capacity of the landscape, distribution of productive landscape, and more. In order to survive, individuals must consume part of their accumulated grain [their "metabolism"]. We modified the rules of the original NETLOGO Wealth Distribution Model. The model we used to simulate wealth distribution was defined by the following set of rules:

The model begins with all individuals having the same amount of wealth.

Each person moves in the direction where the most grain lies.

Each time tick, each person eats a certain amount of grain ["metabolism"].

When their lifespan runs out, or they run out of grain, they die and produce a single offspring.

Wealthy individuals can choose to establish settlements on productive land; the settler gets all of the harvest from the patch.

When settlers die, the patch becomes open for settlement by another individual.

For death by old age capture the old-wealth, transfer it to the new generation of individuals who share 50 / 50 with new siblings.

Initially, all individuals are assigned random values of vision, and metabolism. The maximum values of such variables can be manipulated in the model.

Inequality is measured by calculation of the Gini coefficient.

3 Results

The results of a NETLOGO simulation of an extreme example of a society demonstrating low population [n=2], large fields of vision [max = 15], agents with low resource consumption [metabolism = 1], land with maximum productivity [NUM-GRAIN-GROWN=15, PERCENT BEST-LAND = 25%], equal distribution of life expectancy [all agents live until age 56].

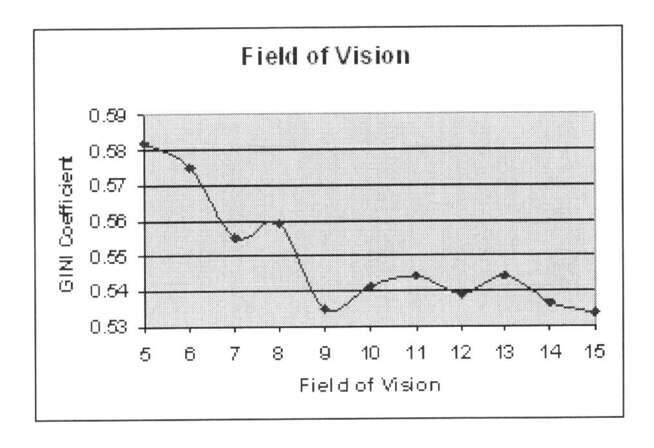

Fig. 1 Relation between GINI coefficient and field of vision

In this experiment, the maximum possible field of vision was sequentially increased. The other conditions [kept constant] were as follows: PERCENT-BEST-LAND =10%; GRAIN-GROWTH-INTERVAL=1; NUM-GRAIN-GROWN=4; NUM-PEOPLE slider = 250; LIFE-EXPECTANCY-MIN=1; LIFE-EXPECTANCY-MAX=83; METABOLISM-MAX = 15.

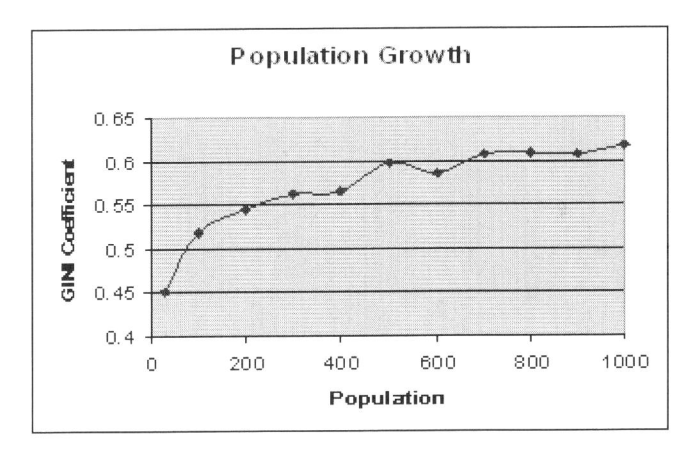

Fig. 2 Relation between GINI cofficient and population

In this experiment, initial size of population was sequentially increased to simulate increasing population density. The other conditions [kept constant] were as follows: PERCENT-BEST-LAND =10%; GRAIN-GROWTH-INTERVAL=1; NUM-GRAIN-GROWN=4; MAX-VISION= 9; LIFE-EXPECTANCY-MIN=1;LIFE-EXPECTANCY-MAX=83; METABOLISM-MAX = 15.

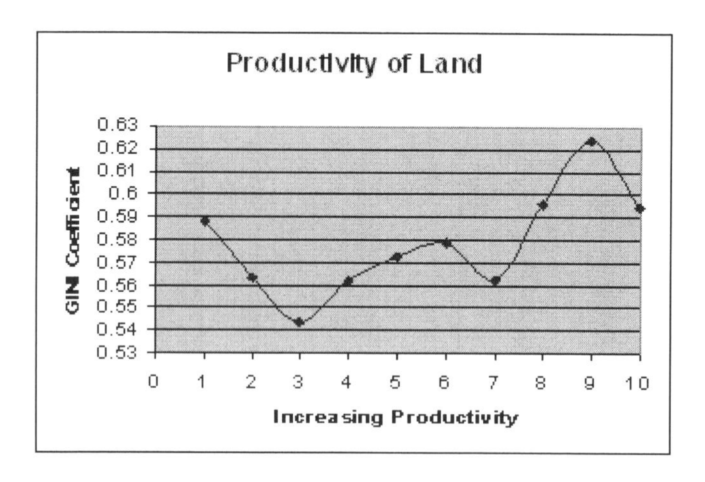

Fig. 3 Relation between GINI coefficient and productivity of land

In this experiment, the NUM-GRAIN-GROWN variable was sequentially increased to simulate increasing productive capacity of the landscape. The other conditions [kept constant] were as follows: PERCENT-BEST-LAND =10%; GRAIN-GROWTH-INTERVAL=1; NUM-PEOPLE= 250; MAX-VISION= 9; LIFE-EXPECTANCY-MIN=1; LIFE-EXPECTANCY-MAX=83; METABOLISM-MAX = 15.

4 Conclusions

Cursory examination of the data show that increases in all individuals' fields of vision decreases inequality as measured by the Gini coefficient.

Increasing population density results in increasing inequality.

The relationship between increasing production of resources and inequality is unclear. It is possible that Simon Kuznets' theory that inequality increases over time, then at a critical point begins to decrease may explain the data obtained.

Analyses of extreme examples of the model are helpful in understanding the most potent factors affecting inequality. Taken together, these experiments suggest that low population, individuals with large fields of vision, agents with low resource consumption, land with maximum productivity, equal distribution of life expectancy, and stable population growth would characterize a society with the lowest measurement of inequality.

The models used in the study are "exploratory models", according to the classification by Prof. Holland. However, we do believe that the simulation results and analysis presented here are highly relevant to the understanding of the urgent social inequality problem that persists over human history. We also acknowledge that our models are really primitive and inadequate. Many important factors affecting social inequality such as geography, heritage, history and money flows are not included.

Acknowledgment. This study was supported by Beijing WuZi University Research Base Program, Grant # WYJD200902 and Funding Project for Academic Human Resources Development in Institutions of Higher Learning under the Jurisdiction of Beijing Municipality (PHR201006129), Funding Project for Excellent Human Resources Training of Beijing (2010D005009000009) and Funding Project for Beijing Wuzi University Young Research (2010XJQN044).

References

1. Axelrod, R., Hamilton, W.D.: The Evolution of Cooperation. Science 211, 1390–1396 (1981)
2. Axtell, R.L., Epstein, J.M., et al.: The Emergence of Classes in a Multi Agent Bargaining Model. In: Durlauf, S., Young, H.P. (eds.) Social Dynamics. MIT Press, Cambridge (2001)
3. Young, H.P.: Individual Strategy and Social Structure: An Evolutionary Theory of Institutions. Princeton University Press, Princeton (1998)

A Study on Human Integration of Audiovisual Spatial Information for Multi-sensor Fusion Technology

Qi Li, Ning Gao, and Qi Wu

Abstract. Much knowledge has been gained on how human brains are activated on audiovisual integration. However, few data were acquired about effects of spatial information in audiovisual integration. The integration of audiovisual spatial information was investigated using behavioral measures in humans in two experiments which were designed to separate out spatial information from audiovisual integration. The results showed the integration of spatial information promoted the behavioral responses, which offered base data for multi-sensor data fusion algorithms and robotics.

1 Introduction

In recent years, multi-sensor data fusion technology have been widely used whether in the military or the civilian areas, however, research in this area is still in its infancy, not yet a unified theory and effective fusion model and algorithm [1, 2]. The human brain is an optimal multi-sensory information processing system which can combine information (scenery, sound, smell and touch, etc.) from different sensory channels (eyes, ears, nose, limbs, etc.) into a unified perception.

Vision and auditory were the two major means of access to knowledge. In daily life, the phenomenon of audiovisual information integration can be often seen. For example, our perceptual system automatically fuses the information deriving from images and sounds when we watch television in order to understand the depicted content. However, how our perceptual system integrates the different sensory information remains unclear. Moreover, we want to investigate the mechanism of humans' audiovisual integration using the behavioral methods and offer some base data for multi-sensor data fusion algorithms and robotics.

Qi Li · Ning Gao · Qi Wu
School of Computer Science and Technology; Brain-computer interface lab, Changchun University of of Science and Technology (CUST), Changchun, China
e-mail: qili.email@gmail.com

Z. Qian et al. (Eds.): Recent Advances in CSIE 2011, LNEE 125, pp. 181–186.
springerlink.com

Audiovisual information contained many elements, such as spatial element, time element, attention element and so on. The audiovisual information integration was a integrated process of various elements of audiovisual information.

Many previous studies have investigated the multisensory audiovisual information integration as a whole, but not focused on a certain element. The behavioral results showed that responses to audiovisual stimuli are more rapid and accurate than the responses to either unimodal (visual or auditory) stimulus [3, 4, 5]. However, it was unclear which element elicited the promoting effect.

Recently, the influence of spatial congruency on audiovisual interaction was investigated in a divided-attention task, in which both auditory and visual information were task-relevant and were requested to attend [5, 6, 7]. The results of these studies indicated that the neural activity in a spatially incongruent condition was inconsistent with that in a spatially congruent condition. Unfortunately, the results did not verify whether the integration of audiovisual spatial information promoted or demoted the behavioral responses.

Two experiments were designed to separate out the spatial information from the audiovisual information and we discussed the effects of spatial information in audiovisual integration in the promoted behavioral responses.

2 Methods

2.1 Subjects

Thirteen healthy volunteers participated in the experiment (males, aged 22-26 years; mean 23.4). All of the subjects have normal or corrected-to-normal vision and normal hearing capabilities. The experiment protocol was approved by the Ethics Committee of Changchun University of Science and Technology (CUST). After a full explanation of purpose and risks of the research, the subjects had gave written informed consent for all of the studies according to a protocol approved by the institutional research review board.

2.2 Stimuli

Subjects were required to conduct two experiments.

(1) Experiment1: Experiment1 contained two stimulus types which are unimodal visual (V) stimuli and bimodal audiovisual (VA) stimuli. Each type had two subtypes of standard stimuli and two subtypes of target stimuli. The V standard stimulus consisted of a checkerboard (5.2 × 5.2 cm, subtending a visual angle of about 5°) for a duration of 40 ms (Figure1 A)). The V standard stimulus was presented on a display at an angle of about 12° from a centrally presented fixation point.

The VA standard stimulus was a V standard stimulus accompanied by a simultaneous auditory stimulus which was presented through two speakers placed on either side of the display. The auditory stimulus consisted a 1600 Hz sinusoidal tone pips, presented at a sound pressure level of 65 dB for a duration of 25ms (with linear rise and fall times of 5ms, Figure2 A)).

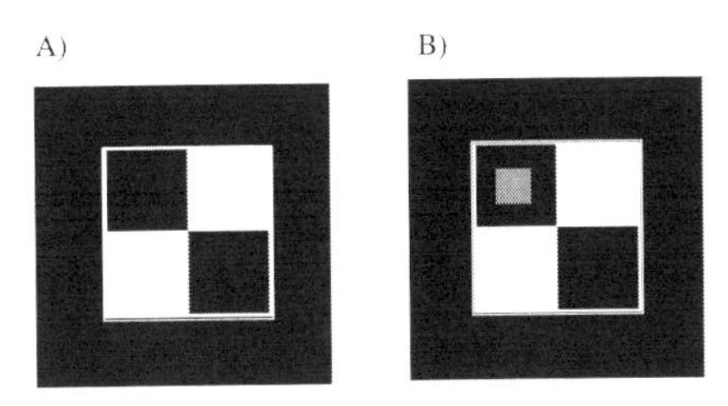

Fig. 1 Examples of visual stimuli. A) Standard visual stimulus; B) Target visual stimulus.

The V target stimulus was very similar to the V standard stimulus but contained one dot in the lower right or upper left (Figure1 B)). The VA target stimulus was a V target stimulus accompanied by a same task-irrelevant simultaneous auditory stimulus. Task difficulty was titrated for each subject based on the V target stimulus so that detection required highly focused attention but could be performed at a correct rate of about 80%, which was accomplished by slightly changing the contrast and / or the size of the dot.

(2) Experiment2: Experiment2 was similar to experiment1. The only difference lied in the VA stimuli in which the auditory stimuli were presented through one ipsilateral speaker with the simultaneous V stimulus as showed in Figure 2 B).

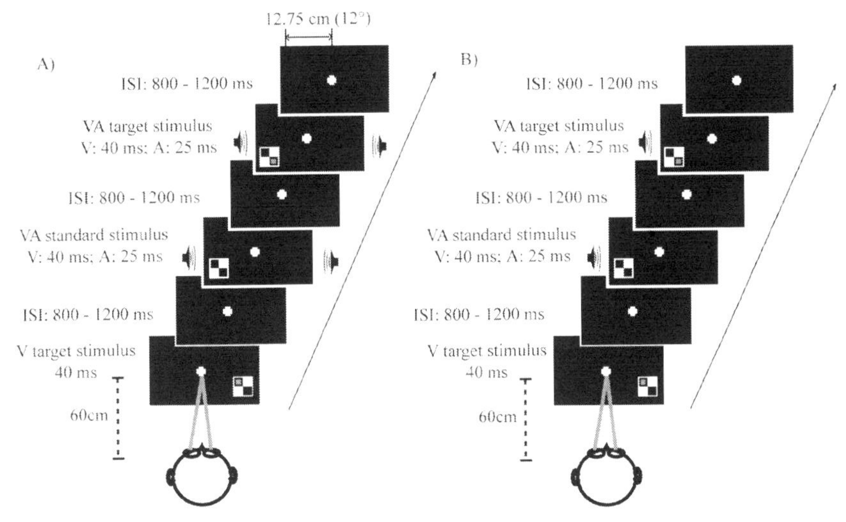

Fig. 2 Experiment arrangement and procedure. A) Arrangement and procedure of experiment1; B) Arrangement and procedure of experiment2.

2.3 *Task and Procedure*

Experiments were conducted in a dimly lit and sound-attenuated room.

In each experiment, there were three sessions. Each session consisted of 80 unimodal V stimuli and 80 multimodal VA stimuli. The frequency of target stimulus was 50%. All the stimuli were randomly presented, and stimuli of each type were presented with equal probability on left or right side. The interstimulus interval (ISI) varied randomly from 800 to 1200 ms (mean ISI 1000 ms). Subjects were required to conduct two experiments. In each experiment, subjects were required to fixate on a centrally presented fixation point to covertly direct attention to visual stimuli while ignoring all auditory stimuli. The subject's task was to detect visual target stimuli on each side as quickly and accurately as possible, responding by pressing the left button of a mouse. At the beginning of each experiment, subjects were given a number of trials for practice in order to ensure that they understood the paradigm and were familiar with the stimuli. Subjects were allowed to take short breaks between sessions.

2.4 Data Analysis

Average Reaction time (RT) and accuracy were computed separately according to stimulus types. The average RT and accuracy of different stimulus types were compared using repeated measure analyses of variance (ANOVA).

3 Results

Fig.3. showed the average RT of V and VA stimuli in experiment1 and experiment2 respectively. The ANOVA showed that there was no difference between the average RT of V target stimuli and that of VA target stimuli in experiment1[F (1, 12) =3.537, P=0.084]. However, in experiment2, the average RT of VA target stimuli was significantly faster than that of V target stimuli [F (1, 12) =15.014, P<.002].

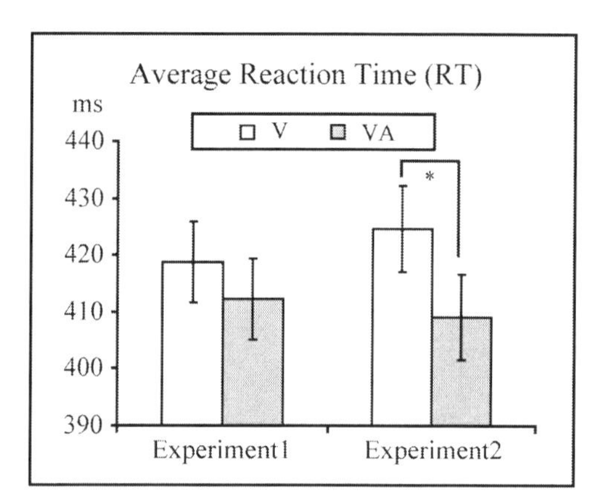

Fig. 3 Average RT of V target stimuli and VA target stimuli in experiment1 and experiment2. Average RT are in millisecond.

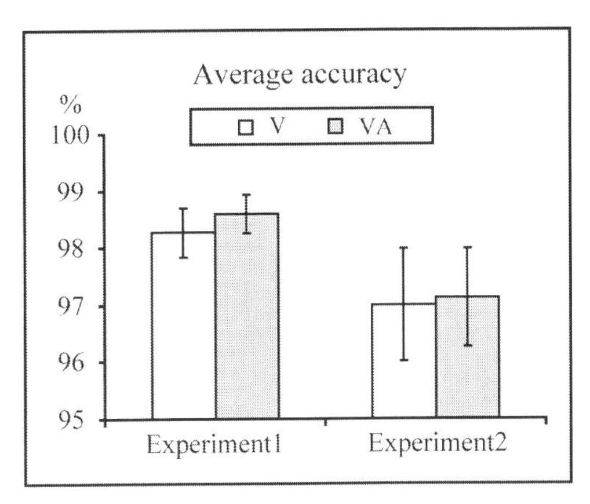

Fig. 4 Average accuracy of V target stimuli and VA target stimuli in experiment1 and experiment2.

Fig. 4 showed the average accuracy of V and VA target stimuli in the experiment1 and experiment2 respectively. The ANOVA revealed that there was no difference between the accuracy of V target stimuli and that of VA target stimuli in experiment1 [$F(1, 12) = 1.000$, $p=0.337$] experiment2 [$F(1, 12) = 0.069$, $p=0.798$].

4 Discussion

Two experiments were designed to separate out spatial information of audiovisual stimuli. In experiment1, the V stimuli were randomly presented on left or right side. Half of these lateral visual stimuli were accompanied by a simultaneous auditory stimulus which was presented through two speakers placed on either side of the display. The location of visual stimulus in audiovisual stimulus was inconsistent with that of auditory stimulus. While subjects detecting VA target stimuli, simultaneously presented auditory stimulus did not offer spatial information. In experiment2, the auditory stimulus of audiovisual stimulus was presented on same side with visual stimulus (left or right). While subjects detecting target stimuli by visual information, simultaneously presented auditory stimulus offered spatial information. So the effects of spatial information in audiovisual integration could be assessed by comparing the results of experiment1 and experiment2.

The results showed that RTs of VA target stimuli were significantly faster than those of V target stimuli. The results were consistent with those of previous audiovisual integration studies and indicated that visual and auditory stimuli were effectively integrated [5-6, 8-10]. However, in experiment1, we did not find the promoting effects. It was indicated in experiment2 that the integration of audiovisual spatial information promoted the behavioral responses.

5 Conclusion

The effects of spatial information in audiovisual integration were investigated in the two experiments which were designed to separate out spatial information from audiovisual integration. Our results showed the integration of audiovisual spatial information promoted the behavioral responses, which offered base data for multi-sensor data fusion algorithms and robotics.

In future studies, we will specify the neuronal mechanism of audiovisual spatial information integration using the event-related potentials.

Acknowledgments. The study is supported by the Jilin Provincial Science and Technology Development Program of China (No.20100171), and the Scientific Research Foundation for the Returned Overseas Chinese Scholars, State Education Ministry.

References

[1] Ernst, M.O., Banks, M.S.: Humans Integrate Visual and Haptic Information in a Statistically Optimal Fashion. Nature 415, 429–433 (2002)
[2] Alais, D., Burr, D.: The Ventriloquist Effect Results from Near-Optimal Bimodal Integration. Current Biology 14(3), 257–262 (2004)
[3] Molholm, S., Ritter, W., Murray, M.M., Javitt, D.C., Schroeder, C.E., Foxe, J.J.: Multisensory auditory-visual interactions during early sensory processing in humans: a high-density electrical mapping study. Brain Res. Cogn. Brain Res. 14, 115–128 (2002)
[4] Teder-Salejarvi, W.A., McDonald, J.J., Di Russo, F., Hillyard, S.A.: An analysis of audio-visual crossmodal integration by means of event-related potential (ERP) recordings. Brain Res. Cogn. Brain Res. 14, 106–114 (2002)
[5] Teder-Salejarvi, W.A., Di Russo, F., McDonald, J.J., Hillyard, S.A.: Effects of spatial congruity on audio-visual multimodal integration. J. Cogn. Neurosci. 17, 1396–1409 (2005)
[6] Gondan, M., Niederhaus, B., Rosler, F., Roder, B.: Multisensory processing in the redundant-target effect: a behavioral and event-related potential study. Percept. Psychophys. 67, 713–726 (2005)
[7] Harrington, L.K., Peck, C.K.: Spatial disparity affects visual-auditory interactions in human sensorimotor processing. Exp. Brain Res. 122, 247–252 (1998)
[8] Giard, M.H., Peronnet, F.: Auditory-visual integration during multimodal object recognition in humans: a behavioral and electrophysiological study. J. Cogn. Neurosci. 11, 473–490 (1999)
[9] Frassinetti, F., Bolognini, N., Ladavas, E.: Enhancement of visual perception by crossmodal visuo-auditory interaction. Exp. Brain Res. 147, 332–343 (2002)
[10] Schroger, E., Widmann, A.: Speeded responses to audiovisual signal changes result from bimodal integration. Psychophysiology 35, 755–759 (1998)

Innovative Web-Based Tool for Safer Transitions of Patients through Healthcare Systems

Ranjit Singh, Raj Sharman, Ashok Singh, Ron Brooks,
Don McLean, and Gurdev Singh

Abstract. Improved quality and safety of care and patient satisfaction can be achieved with s`tructured inter-unit and inter-setting communication with computers. Presented here is a new tool designed to implement a web-based innovative transitions improvement process for creation of situation-aware teams to monitor quality of transitions. The tool is designed to promote the competencies of practice-based learning and improvement, and systems-based practice, as well as the development of a context-sensitive culture of safety. The system is based on visual workflow models of the entities and interactions necessary for reliable and timely transitions from or to any setting. In pilot studies the tool has been well received by staff who found it to be intuitive and user-friendly. We propose that healthcare teams tailor the tool for their unique setting, and continually improve it based on internal evaluation and feedback.

1 Introduction

1.1 Objective

Transitions of patients, especially from one hospital unit to another, are particularly unsatisfactory, undermining patient satisfaction, increasing health care

---

Ranjit Singh · Raj Sharman · Gurdev Singh
State University at Buffalo, NY. USA
e-mail: gsingh4@buffalo.edu

Ashok Singh
Niagara Family Medicine Associates, NY. USA

Ron Brooks
Dendress Corporation, Buffalo NY. USA

Don McLean
Niagara Falls Memorial Medical Center, NY. USA

Z. Qian et al. (Eds.): Recent Advances in CSIE 2011, LNEE 125, pp. 187–194.
springerlink.com

costs and impairing safety-based quality[1-4]. In the USA most of the forty million discharges from the hospital each year occur without timely and reliable communications, presenting a major hazard to patients. The authors' premise is that well structured discharges from hospitals can result only from sound practices of well structured transitions/handoffs within the hospitals. In the US, 19% of Medicare patients are readmitted within 30 days of discharge [5]. World Health Organization's (WHO) expert panel has identified lack of communication and coordination to be the top research priority for developed countries [6,7]. The US Agency for Health Research and Quality, the Joint Commission and the Alliance of Independent Academic Medical Centers, among various others, have therefore made pressing calls for patients to be transitioned with well structured and standardized formats, bearing timely and reliable information [8].

Costs and safety hazards can be reduced with reliable transitions/hand-offs by "breaking down the silos of care" [9]. The director of the US Agency for Health Research and Quality (AHRQ) draws attention to safety issues related to these in her Podcast [10]. This should be seen in light of the fact the USA spends nearly a sixth (16.5%) of its Gross National Product on health care every year. The equivalent average % age expenditure among OECD countries is about 9.

For change improvement to be sustainable it is necessary to empower all the stakeholders at the front-end of the healthcare processes. The essential prerequisite in this study, therefore, was to adopt a bottom-up approach [11-15]. Additionally, designing care around the patients' preferences, values and beliefs can not only improve outcomes but improve patients' satisfaction with the care. This care includes special attention to safe continuity and coordination of care described above. Lack of continuity is transparent to patients leading to dissatisfaction [10].

The general objective, therefore, was to improve quality and safety of inpatient care and patient satisfaction with structured inter-unit and inter-setting communication; bearing timely, reliable and meaningful information for care providers and patients through a bottom-up approach. This work should help us move in the direction of the objectives stated in the recent UK Department of Health's White paper [16] that aims to:

- put patients at the heart of everything the NHS does;
- focus on continuously improving those things that really matter to patients - the outcome of their healthcare; and
- empower and liberate clinicians to innovate, with the freedom to focus on improving healthcare services.

Seeing the huge potential of IT the US Center for Medicaid and Medicare Services (CMS) defines "*meaningful use*" of IT as using an Electronic Health Record (EHR) for the objectives that fall under these general topics:

- Improving quality, safety, efficiency, *care coordination*, population and public health;

- Reducing health disparities;
- Engaging patients and their families; and,
- Ensuring adequate privacy and security protections for personal health information.

1.2 Design

As stated earlier, the Joint Commission has recently called for standardized approaches to transitions communications, including opportunities to ask and respond to questions. Process mapping at the macro- and micro-levels can be a great aid to understanding the risks associated with these transitions. A visual approach [17] is particularly useful for tracking patients' encounters with the healthcare system over time and recording their interactions across the various parts of the system. This approach can help improve continuity of care and reduce the risk of errors that occur during transitions. Therefore, a tool, described below, was designed especially for: (1) creation of *situational awareness* and formation of teams to monitor quality of transitions with the tool, promoting the *competency of practice-based learning and improvement*, (2) creation of visual workflow models of the entities and interactions necessary for reliable and timely transitions for each unit, promoting the *competency of system-based practice*, (3) creation of standardized, SBAR (Status, Background, Assessment and Recommendation) based, prototype formats for transitions for each of the units, (4) evaluation of the format by respective teams and revision of the formats, (5) implementation and measurement of their efficacy followed by continuing format improvement.

This tool allows individual members of a health care team to report cases of unsatisfactory transitions in a way that promotes learning and aids focused development or improvement of transitions formats/tools.

2 Web-Based Transitions Tool

The use of computers allows members of the team to communicate with each other to coordinate efforts in bringing about optimum care for the patient, through a web-based interactive tool. The tool resides on a separate web server of its own with its access point embedded into another master "Home Page". By doing so it rapidly becomes a convenient part of a user's daily workflow. Also by embedding the link into an existing page the tool software already 'knows' who and what site the data is associated with, thus streamlining data entry.

The Status-Background-Assessment-Recommendation System (SBAR) [18] system provides the content and the mode of communication. The tool was developed for transitions not only within the hospitals but also for transitions to-and-from primary care providers. It was designed to encourage and facilitate the cyclic improvement process portrayed in Figure 1.

Figure 2 shows the opening screen of the web-based version with some explanatory notes. This provides general orientation for the participant with the aid of PowerPoint presentation and allows access to instructions for reporting and statement of benefits of so doing. Fig. 3(A) is an example screen for assessing and reporting the quality of transition *to* primary care office *from* another setting such as hospital. Input to questionnaires includes not only SBAR content but also the processes of communication. Space is provided for 'story' telling about the unsatisfactory transition being reported. Fig. 3(B) shows the screen that allows the user to see the compiled results of all the entries by various members of the team over a predetermined period of time. This compiled data, presented to the respondent in a bar chart form (Fig. 3(C)), helps identification of most significant vulnerabilities and aids consensus based prioritization of improvement interventions.

3 Discussion

The tool provides the ability to collect and display trends in errors which occur during a transition. These trends can be collected and displayed both locally and globally via a common SQL based database located at the web server. These facilitate prioritization and appropriate focus based on the frequency and severity of the error/s. In addition, because of the central and global nature of the web, regional, national and worldwide trends can be discovered and addressed not only locally but by a wider group of public health policy makers, provided the latter respect and allow for the unique nature of individual settings. Figure 4 portrays the concept. The acronym PSO refers to Patient Safety Organizations being setup by the US for the purposes of receiving, aggregating and analyzing adverse events reports data. The overall objective of PSOs is to learn from errors and devise interventions to improve safety. They can also help create a National Patient Safety Data (NPSD).

The US Center for Medicaid and Medicare Services (CMS), as stated earlier, is calling for IT use for:

1. Improving quality, safety, efficiency, care coordination, population and public health;
2. Reducing health disparities;
3. Engaging patients and their families; and,
4. Ensuring adequate privacy and security protections for personal health information

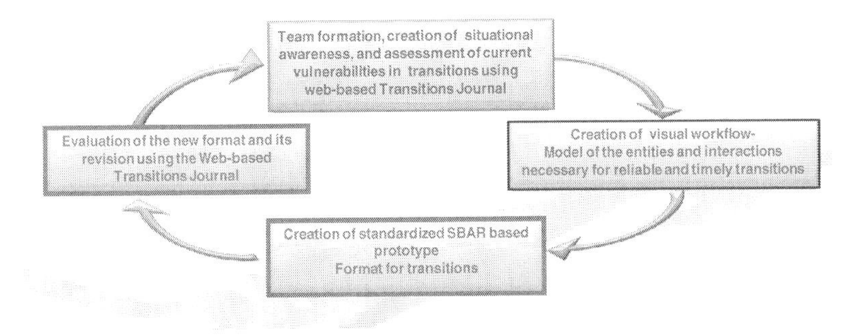

Fig. 1 Cyclic process of monitoring and improving Transitions

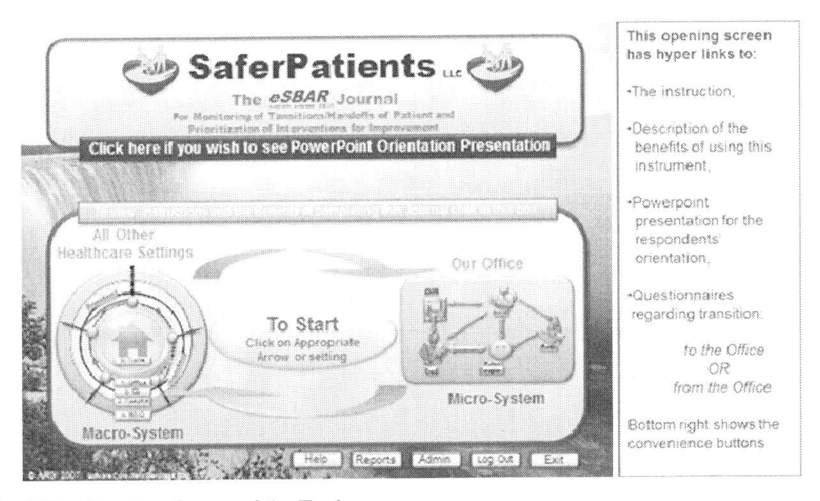

Fig. 2 The Opening Screen of the Tool

Competent transitions make significant contribution to all the above but particularly to the first and the third. Our approach can, therefore, be seen as a contribution to enhancing the "meaningful use of IT".

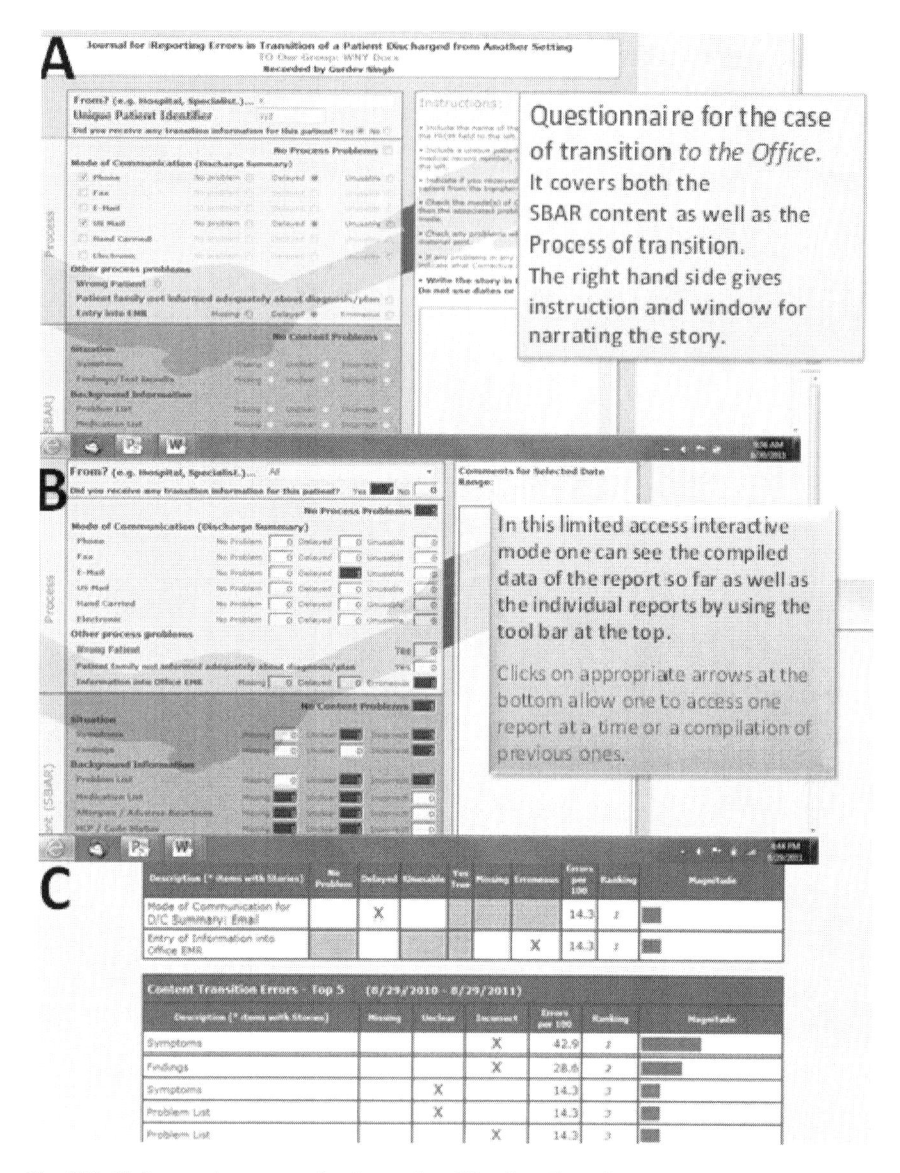

Fig. 3(A-C) Interactive screen for Reporting, Viewing Compiled Data, and Top 10 Errors [19

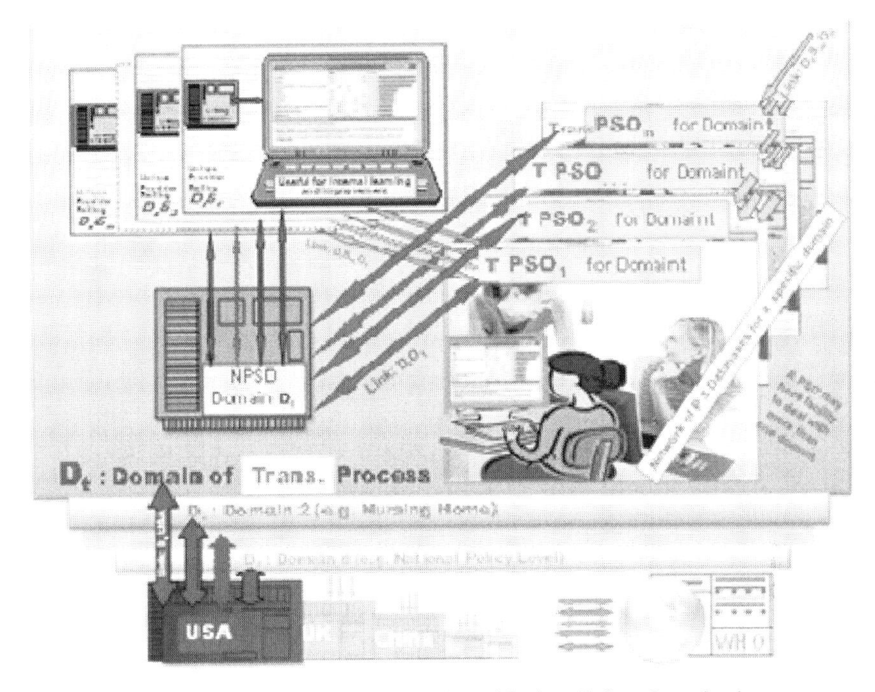

Fig. 4 Concept for TransPSO: a Transitions Oriented Patient Safety Organization

4 Conclusions

The proposed approach using the tool:

1. Provides opportunities for learning and improvement in quality of communication for safe, effective and efficient transitions of care;
2. Instills a systems approach;
3. Promises a very significant potential to help:

   - Improve continuity of care.
   - Improve compliance with important national patient safety goals (e.g. Joint Commission) through improved safety and patient satisfaction.
   - Reduce duplication and unnecessary work, thus reducing costs and increasing efficiency.
   - Encourage team formation and teamwork embedded in context-sensitive culture of safe transitions.
   - Provide mutual emotional and intellectual support for all staff.

The potential of this Transitions Tool as a teaching aid for all healthcare professionals also needs to be explored.

References

[1] British Medical Association. Safe handover: safe patients
[2] Aspden, P., Corrigan, J.M., Wolcott, J., et al. (eds.): Patient safety: Achieving a new standard for care. Institute of Medicine, Committee on Data Stand ards for Patient Safety, Washington, DC (2004)
[3] JCAHO. Patient Safety Goals for 2006
[4] National Quality Forum, Agency for Health research and Quality. Safe Practices for Better Healthcare: A consensus Report (2003)
[5] HMO Workgroup on Care Management. One patient, many places: Managing health care transitions. AAHP_HIAA Foundation, Washington, DC, http://www.ahip.org
[6] WHO. World Alliance for Patient Safety, http://www.who.int/patientsafety/en
[7] Research Priority Setting Working Group of the World Alliance for Patient Safety. Summary of Evidence on Patient Safety: Implications for Research (2008), http://www.who.int/patientsafety/information_center
[8] JCAHO. Sentinel event analysis statistics (1995-2004), http://www.jointcommission.org
[9] Medicare Payment Advisory Commission. Reforming America's Health Care System. Statement before the Senate Finance Committee Roundtable (2009)
[10] Clancy, C.: How to Avoid the Round-Trip Visit to Hospital. Navigating the Health Care System. AHRQ, http://www.ahrq.gov/cosumer/cc/cc060110.html
[11] Singh, R., Singh, A., Taylor, J.S., Rosenthal, T., Singh, G.: Building learning practices with self- empowered teams for improving patient safety. J. Health Manag. 8, 91–118 (2006)
[12] Singh, R., Singh, A., Servoss, T.J., Singh, G.: Prioritizing threats to patient safety in rural primary care. J. Rural Health 23(2), 173–178 (2007)
[13] Singh, R., Naughton, B., Anderson, D.R., Singh, G.: Building Self-Empowered Teams for Improving Safety in Post-operative Pain Management. In: Henriksen, K., Battles, J.B., Keyes
[14] Singh, G., Singh, R., Thomas, E.J., et al.: Measuring Safety Climate in Primary Care Offices. In: Henriksen, K., Battles, J.B., Keyes, M.A., Grady, M.L. (eds.) Advances in Patient Safety: New Directions and Alternative Approaches. Culture and Redesign, vol. 2, pp. 59–72. Agency for Healthcare Research and Quality, Rockville (2008); ; AHRQ Publication No. 08-0034-2
[15] Singh, R., Pace, W., Singh, A., et al.: A Visual Computer Interface Concept for Making Error Reporting Useful at the Point of Care. In: Henriksen, K., Battles, J.B., Keyes, M.A., Grady, M.L. (eds.) Advances in Patient Safety: New Directions and Alternative Approaches, Assessment, vol. 1, pp. 307–320. Agency for Healthcare Research and Quality, Rockville (2008); AHRQ Pub., No. 08-0034-1
[16] House of Commons Committee of Public Accounts. A Safer Place for Patients: learning to improve patient safety. HMO. HC 831, London (2006)
[17] Singh, R., Pace, W., Singh, S., et al.: A concept for a visual computer interface to make error taxonomies useful at the point of primary care. Informat. Prim. Care. 15, 221–229 (2007)
[18] Joint Commission. What does JCAHO expect for handoffs? OR Manager 22(4), 11 (2006)
[19] Singh, R., Roberts, A., Singh, A., et al.: Improving Transitions in In-Patient and Out-Patient Care Using a Paper Or Web-based Journal. Journal of the Royal Society of Medicine (London) Short Report (2011)

On the Design and Experiments of a Fluoro-Robotic Navigation System for Closed Intramedullary Nailing of Femur

Sakol Nakdhamabhorn and Jackrit Suthakorn

Abstract. Closed Intramedullary Nailing of Femur is a frequent orthopedic surgical operation. Traditionally, surgeons are performing the procedure based real-time on fluoroscopic images. Therefore, a great amount of X-Ray is exposed to both patient and surgeon. One of the most difficult tasks is the distal locking process which requires trail-and-error and skill for understanding 2D images generated from 3D objects. This paper describes the overall design and experimental sections of our research series on developing a surgical navigation system using a robot guiding device based on fluoroscopic images. Two approaches with and without employing optical tracking to generate relationship among tools, patient, and imaging system are presented. Moreover, a new technique to interface between surgeon and system is proposed to reduce the problem of hand-eye coordination on most surgical navigation system. This work is built on our previous work on path generation for distal locking process. Our experiment results show a promising solution for guiding the surgeon during the distal locking process in Closed Intramedullary Nailing of Femur.

Keywords: Fluoroscopic Navigation, Surgical Navigation, Robot-Assisted Surgery, Computer-Integrated Surgery, Intramedullary Nailing.

1 Introduction

1.1 Motivation

Closed Intramedullary Nailing or "Closed Nail" is a frequent orthopedic surgical procedure for fixing a long bone, such as, femur or humerus bones. The closed

Sakol Nakdhamabhorn · Jackrit Suthakorn
Department of Biomedical Engineering and Center for Biomedical and Robotics Technology, Faculty of Engineering, Mahidol university,
Sayala, Nakorn Pathom, Thailand
www.BART LAB.org

Z. Qian et al. (Eds.): Recent Advances in CSIE 2011, LNEE 125, pp. 195–203.
springerlink.com

nailing procedure starts with bone fixing, guide-wire inserting, bone canal reaming, nailing inserting and screw locking. Traditionally, surgeon is performing the procedure based on the real-time fluoroscopic images. Therefore, both patient and surgeon are exposed with a large amount of X-Ray exposures [1]. One of the most difficult tasks is the locking process, especially, the distal locking process. This requires a number of trials and errors with high skill to realize the 3D object from the 2D fluoroscopic images. This skill is required for realizing the position and orientation of screw locking path.

This paper describes the overall design and experimental sections of our research series on developing a surgical navigation system using a robotic guiding device based on fluoroscopic images.

1.2 Related Works

Numerous computer-integrated surgical systems [2-3] have been released to assist surgeons in various operations. The goal is to reduce the difficulties in process, to reduce time consuming and to use less of X-ray exposure. Yaniv and Joskowicz [4-5] developed a precise robot to guide in positioning the distal locking intramedullary nail. The system is automatic positioning a mechanical drill guide mounted on a miniature sized robot using a fluoroscopic image. The mean accuracy in vitro experiment is angular error of 1.3 degree and translation error of 3 mm.

HIT-RAOS" [6] is another system developed by Du Zhi-jiang *et al.* The system aims to assist surgeons in several steps on close nailing procedure. Such as, to reposition bone fractures, to guide the surgeon in locking process, to reduce the surgeon's working under C-arm by a tele-operation system. The system consists of patient station, computer control system and surgeon station. At patient site, a repositioning robot was developed. The operation table could adjust its position for locating the patient at different pose. The fluoroscope was modified for tele-operation so the surgeon can control the fluoroscope's position from remote position. In the registration algorithm, a pre-calibrated maker box is attached at the end of guiding robot. The system interacted with user through graphic interface.

An approach for recovery the position and orientation of distal hole's axis was previously proposed by our corresponding author [7-9]. The proposed algorithm required only the area of distal hole's projected image to recover the nail's rotation angle based on hole's area and characteristics of hole's axes. The brief description of the algorithm is explained in section 2.3.

2 Fluoro-Robotic Navigation System

This section covers overall of our navigation system, kinematic analysis and relationships, two navigation approaches with and without optical tracking, a brief description of our path generation algorithm, and guiding system.

2.1 Overall System

Our Fluoro-Robotic Navigation System can be separated into 3 major subsystems. The first subsystem is a distal locking hole recovery subsystem. This subsystem acquires a few X-Ray images from the fluoroscope to recover an axis of distal locking hole's position and orientation which are used to calculate for the path generation. The second subsystem is the mapping 2D X-Ray coordinates into the real 3D world environment coordinates. The third subsystem is the guidance system which is an important system to interact with the surgeon. In this surgical guidance system, we propose a robot-assisting guidance technique to navigate the surgeon to perform drilling the distal locking hole. Figure 1 illustrates the overall navigation system.

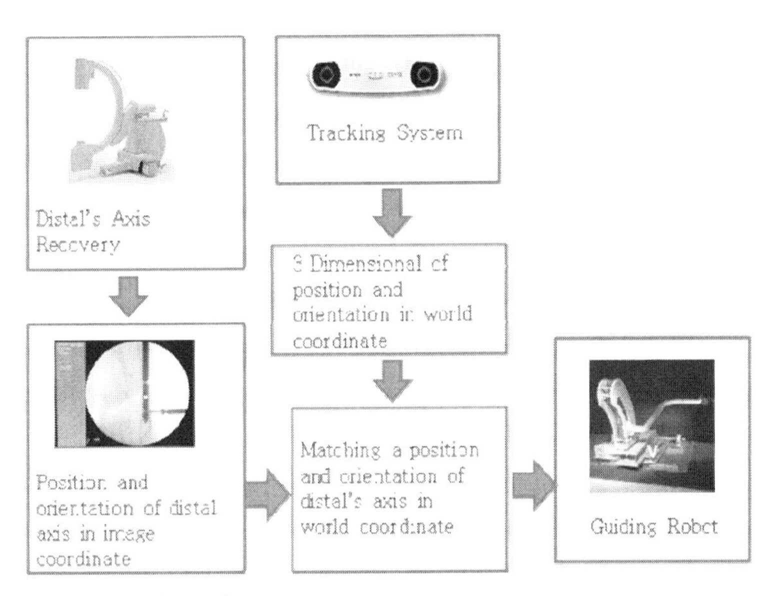

Fig. 1 The diagram of overall system.

2.2 Kinematic Relationships

An important part of the navigation system is the tracking system. This system is to determine the position and orientation of each related object in the procedure. The relationship between each object, such as, drilling tool, patient, imaging system, can be determined with the kinematic relationship. The position and orientation relationships are used in the matching process to register positions of pixels in 2D X-Ray image coordinates into the 3D world coordinates. The two approaches, which are ones with and without optical tracking, are proposed here.

2.2.1 The Approach I: With the Optical Tracking System

An optical tracking system is employed in the first approach to generate a cycle of kinematic relationships between objects. A number of optical markers are attached to; fluoroscopic imaging system, patient bone, and guiding robot. The optical tracking system provides positions and orientations in 3D-space of those three objects related to the optical tracking coordinate.

In order to generate a trajectory path for distal locking process, matching coordinates from different objects are required. The fluoroscopic calibration process is also required to register the position of each pixel in 2D X-Ray images into the position in 3D world coordinates. Therefore, the position and orientation of distal locking hold axis in the real 3D world coordinates can be determined. The homogenous transformations to demonstrate the kinematic relationships among objects are shown in Figure 2.

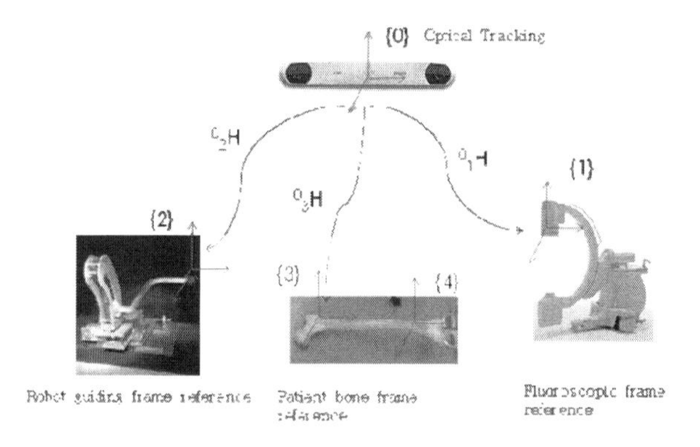

Fig. 2 Transformation diagram of this system

2.2.2 The Approach II: Without the Optical Tracking System

The second approach is employed an X-Ray marker to be attached on robotic guiding device. The shape of an X-Ray marker is designed in a circular shape which is similar to the distal locking hole's shape. The marker is attached to the end-effecter of guiding robot which to be appeared in the view of fluoroscopic image. Therefore, the position and orientation relationships between distal locking hole and the guiding robot can be calculated by comparing the rotation angle of the X-Ray marker with the angle of distal locking hole.

2.3 Recovery of Distal Locking Hole on Intramedullary Nail

The algorithm of the recovery system is based on our previous studies in [7-10]. In this algorithm, only a few images are required to realize the distal locking hole's axis. The algorithm utilizes characteristic information of the intramedullary nail, such as, nail radius, major and minor axes of distal locking hole, and the area of distal locking hole. The algorithm is separated into two iterations. The first iteration is the simulation process. In this process, the nail is rotated at all possible projection.

Then, the entire of simulated nail images are extracted a ratio between major and minor axes of distal locking hole. So, the ratio between major and minor axes in each rotation angle is stored together with an *X-Y-Z* Euler rotational angle in a database. This part has done in pre–operative process. The second iteration is for extracting information from the real-time distal locking hole images. The edge detection and fitting ellipse techniques are applied to find the location and other similar information from distal locking hole. The ratio between major and minor of distal locking hole is then calculated in real-time matching process to recover a distal locking hole's orientation. The cubic curve fitting is applied to find the best fit to the data in database. The curve is called "tool curve" as shown figure 3. The tool curve is used to reversely recover the rotational angle of distal locking hole's axis. This step is operated in intra-operative process. The simple cubic curve fitting equation is shown as equation 1.

$$y = ax^3+bx^2+cx+d \quad (1)$$

Where, y is the result of rotation angle in degree, a, b, c and d is a constant
x is the ratio between major and minor of distal locking hole.

2.4 Robotic Guiding System

The robotic guiding system is a significant proposed part in our surgical navigation system. This system interacts with the surgeon by automatically showing the generated surgical path. Most of navigation systems are usually displayed the guiding information through a monitor by demonstrating animation graphic on the screen. However, in the real situation, the surgeon would prefer to pay more attention on a surgical site more than the screen. Our approach removes the hand-eye coordination problem. The robotic guiding system is a 4 degree-of-freedom robot as shown in figure 4. The robot is placed at the side of the operation table. After the navigation system generates the trajectory path, the robot is then pointing the laser beam to the entry location while path orientation is also given by comparing the drilling tool axis and the laser beam axis.

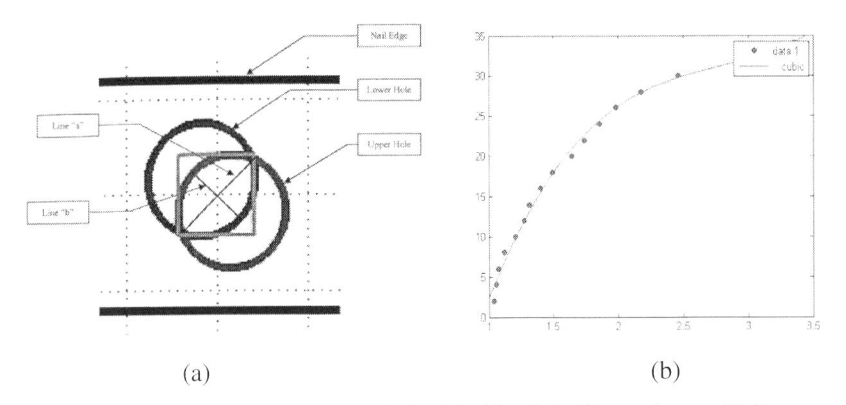

(a) (b)

Fig. 3 (a) A picture show an information of distal locking hole (b) a tool curve [7-9]

In another robot guiding approach, the robot's end-effector is attached with an X-Ray marker. The robot can be designed in a smaller size than the first approach. The robot is imitated the movement of fluoroscopic system as shown in figure 5. The first joint of the robot is use to rotate about Y-axis of nail. The second joint is used to rotate about X-axis along with a nail. The guidance system starts when a fluoroscopic image is captured. The image consists of the X-Ray marker's shape, and the distal locking hole. Then, the distal locking hole axis is recovered at the same time as the recovery of robot's position and orientation. The degree of the rotation between x-ray marker and distal locking hole is determined. So, the robot is then controlled to move to the guiding position and orientation.

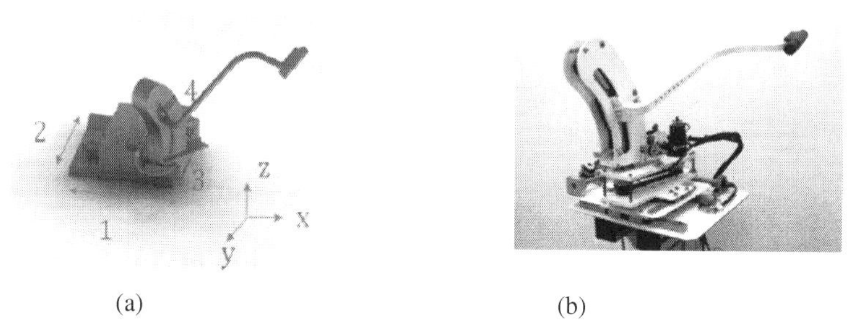

(a) (b)

Fig. 4 (a) A design of guiding robot. (b) a prototype of guiding robot[10]

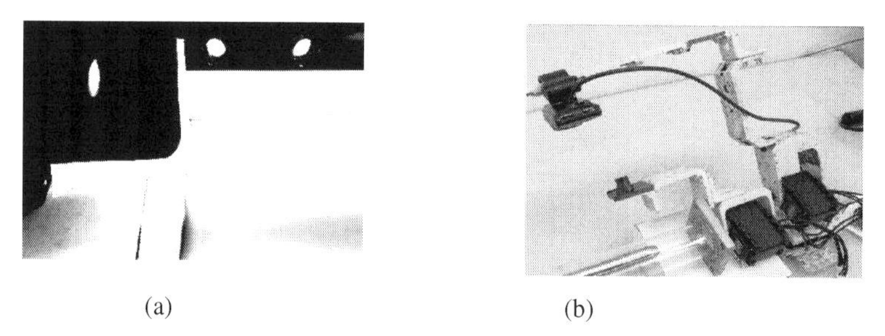

(a) (b)

Fig. 5 (a) A simulate image of distal locking hole and a x-ray marker (b) a model of the second approach.

3 Experiment Setup and Results

3.1 System Simplification and Experimental Set Up

A phantom model and the simulated fluoroscopic imaging system are set up as shown in figure 6. The mimic of fluoroscopic system is created by attaching a small camera on a robotic manipulator driven by servo motor system to imitate the motion of fluoroscopic system. The robot guiding, which is attached with a simulate X-Ray marker, also moves in the similar way as the fluoroscopic system.

The simulated images were captured at 2, 4, 6, 8, 10, 12, 14, 16, 18, 20, 22 and 24 degrees of rotation to make a "Tool curve". In the experiment, simulate fluoroscope is rotated about *X*-axis of distal locking hole, while captures at 2, 4, 6, 8, 10, 12, 14, 16, 18 and 20 degrees of rotation. In addition, the images of a shape of marker are captured at 2, 4, 6, 8, 10 and 12 degree to make another "Tool curve".

In the experiment, the expected result is that the robot manipulator should move to the same axis of distal locking hole. Then, the robot can demonstrate the guiding of the trajectory path to perform the distal locking process.

The graphic user interface program shows the results of the detection and extraction information of distal locking hole as shown in figure 7. The experiment results to recover the distal locking hole's axis is shown in table 1 (a). The average error of the predicted angle is about 0.82 degree. The results of the marker rotation angle are shown in table 1 (b). The average error of the predicted angle is about 0.85 degree.

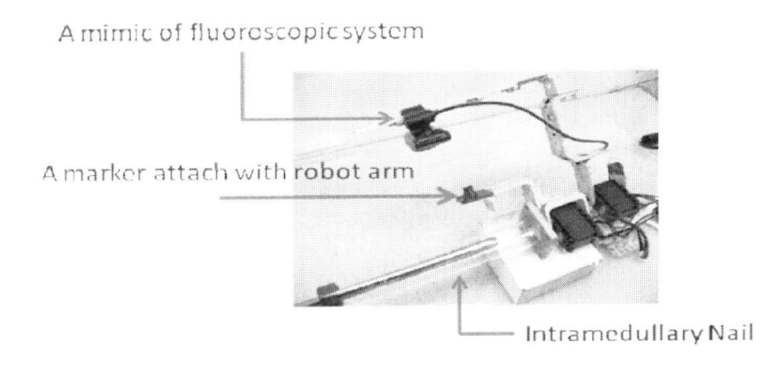

Fig. 6 The component of the experimental set up.

3.2 Experimental Results and Analysis

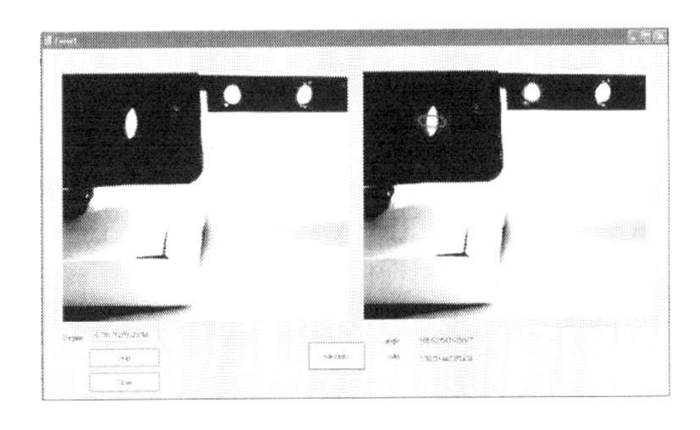

Fig. 7 A program show the recovery of the distal locking hole axis

4 Conclusion and Discussion

This paper has described an overall project on developing an orthopedic surgical navigation system for closed intramedullary nailing of femur. Kinematic based discussion on two proposed approaches with and without optical tracking have been presented briefly. A new concept of robotic guiding device to reduce the hand-eye coordination problem has also proposed. The system has built on our previous work on the algorithm to generate the surgical trajectory based on fluoroscopic images. An experimental set up and results to demonstrate the proposed concept have been demonstrated. The results have shown our method to be a promising solution with a reasonable error less than 0.85 degrees in rotational angle of nail which can be accepted by the orthopedic surgeon who specialize in this case. [This project is supported by the National Metal and Materials Technology Center (MTEC) of Thailand under the Project No. MT-B-50-BMD-14-125-G.]

Table 1 (a) The results of distal locking hole axis' recovery. (b) The results of a prediction angle of the X-Ray marker

(a)

The rotation angle of distal locking hole	The prediction angle
2	3.24
4	3.54
6	4.96
8	7.21
10	9.63
12	11.91
14	14.17
16	16.62
18	19.58
20	22.1

(b)

The rotation angle of a marker	The prediction angle
2	1.50
4	3.48
6	5.47
8	7.43
10	9.18
12	9.82

References

1. Skjeldal, S., Backe, S.: Interlocking medullary nails radiation doses in distal targeting. Arch. Orthopaedic Trauma Surg. 106, 179–181 (1987)
2. Whatling, G.M., Nokes, L.D.M.: Literature review of current techniques for the insertion of distel screws into intramedullary nails. Int. J. Care Injured. 37, 109–119 (2006)

3. Taylor, R.H., Stoianovici, D.: Medical Robotics in Computer-Integrated Surgery. IEEE Trans. Robot. Automat. 19(5), 765–781 (2003)
4. Joskowicz, L., Milgrom, C., Shoham, M., Yaniv, Z., Simkin, A.: Robot-guided long bone intramedullary distal locking concept and preliminary results. In: Proc. Computer Assisted Rodiology and Surgery, pp. 485–491 (2003)
5. Yaniv, Z., Joskowicz, L.: Precise robot-assisted guide positioning for distal locking of intramedullary nails. IEEE Trans. Med. Img. 24(5), 624–635 (2005)
6. Du, Z.-J.,Min-xiu, K., Fu, L.-X., Sun, L.-N.: A Novel Fluoroscopy-Guided Robot-Assisted Orthopaedic Surgery System. In: Int. Conf. on Robotics and Biomimetics, pp. 1622–1627 (2006)
7. Suthakorn, J., U-Thainual, P., Mahaisavariya, B.: An Efficient Algorithm for Recovering Distal Holes' Axes in Intramedullary Nail. In: Proceedings of the 3rd WACBE World Congress on Bioengineering 2007, Bangkok, Thailand, July 9-11 (2007)
8. Suthakorn, J., U- Thainual P.: A New Algorithm for Recovering Distal Hole Pose in Intramedullary Nail. In: Proceedings 5th World Congress on Biomechanics, Munich, Germany, July 29-August 4 (2006)
9. Neatpisarnvanit, C., Suthakorn, J.: Intramedullary Nail Distal Hole Axis Estimation using Blob Analysis and Hough Transform. In: Proceedings 2006 IEEE International Conference on Robotics, Automation and Manufacturing (RAM 2006), Bangkok, Thailand, June 7-9 (2006)
10. Nakdhamabhorn, S., Suthakorn, J.: Design and Development of System Integration for Fluoroscopic Navigation Using Surgical-Guiding Robot. In: World Congress on Computer Science and Information Engineering (CSIE 2011), Changchun, China (2011)

Functional Size Measurement Using Use Case: *From the Viewpoint of Flow of Event*

Xiaomin Zhang and Aihua Ren

Abstract. Due to its abstract principles and being defined not from the viewpoint of development, the COSMIC method is a relative time-consuming job for a developer who is familiar with Use Case to use. To solve this problem, several researchers have mapped the COSMIC method onto Use Case. However, the traditional mapping methods are at the high level of Use Case which is in a large granularity and shortage of details. In this paper we propose the new mapping rules between them. We focus on the viewpoint of Flow of event in Use Case with more details to make the result refined. On the other hand, it can avoid double counting of function point bring by the association between Use Cases. In addition some more detail principles are also provided.

Keywords: funciton point, the COSMIC medhod, Use Case, Flow of event.

1 Introduction

Functional Size Measurement methods aim to provide a technology-independent way of sizing the software. A "Functional Size" is defined in the standard ISO/IEC 14143/1:2007 as "a size of software derived by quantifying the Functional User Requirements" from the view of Actor. The COSMIC method is the most advanced method of measuring a functional size of software, which is developed by the Common Software Measurement International Consortium, the "COSMIC" organization[1].

The common characteristics of the COSMIC method and Use Case[2] are that both of them describe functional requirements from actor's point and both of them are technology-independent. It is feasible to map the COSMIC method into Use Case and it is easier for a developer to count function points from the Use Case.

However, traditional mapping methods are usually at the high level of Use Case which is in a large granularity and shortage of details, and they could not well solve

Xiaomin Zhang
Computer school, BeihangUniversity, Beijing, China
e-mail: prefaye@yahoo.com.cn

Aihua Ren
Computer school, Beihang University, Beijing, China
e-mail: renah@buaa.edu.cn

Z. Qian et al. (Eds.): Recent Advances in CSIE 2011, LNEE 125, pp. 205–210.
springerlink.com

the problem of double counting of function point brought by the dependency between Use Cases. To solve the problems mentioned above, we propose the new mapping rules between them and define the principles in detail. We focus on the viewpoint of Flow of event in Use Case with more details to make the result refined.

2 Releted Works

2.1 Overview of the COSMIC Method

The main principle of the COSMIC method is that the software is decomposed into several layers, and one layer consists of several components, and functional processes are identified from one component and therefore their data movements are defined. There are four types of data movement(entry, exit, write, read), then one CFP (Cosmic Function Point) is defined as the size of one data movement. The primary elements of COSMIC contain functional user, boundary, scope, level of granularity, layer, peer component, functional process, trigger event, data movement, data group, data attribute, etc.

2.2 The Mapping between COSMIC and UML

For a software engineer, it is a little more time-consuming to understand the terms, principles or process of FSM (Function Size Measure) compared to UML modeling. So several papers discuss how to adapt FSM (Function Size Measure) to UML model. For example, Fetcke[3], Bevo[4],and Jenner[5][6], etc. Fetcke[3] kicked off the issue by mapping IFPUG FPA[7] to UML modeling. And Bevo[4], Jenner [5][6], Condori-Fernández[8], Klaas [9], Asma and Hanene[10] adapted COSMIC for the UML diagrams or Use Case in a high level granularity. However, In the following research, most of articles focus on adapting the COSMIC method to other UML diagrams rather than use cases, such as sequence diagram, activity diagram, etc. Since it becomes a tendency to use high-speed, low-cost function point analysis[11], it is time-consuming to identify function point from these diagram, so in our method we identify function point based on Use Case.

3 Mapping Principles

As is shown in table 1, most of the elements in COSMIC can be directly mapped onto Use Case, such as functional user, boundary, scope, granularity, functional process. And several elements in COSMIC could not easy to find similar terms in Use Case such as peer component, data movement, object of interest, data group, data attribute, so we must seek the proper terms in Use Case to map onto the COSMIC method. In addition, there are also several elements which could not map onto Use Case such as layer, etc. In order to explain these mappings in detail, we provide some principles.

Table 1 Mapping between COSMIC and Use Case.

The COSMIC method	Use Case
functional user	actor
boundary	boundary in Use Case
scope	scope in Use Case
level of granularity	Use Case granularity
layer	none
peer component	Use Case package
functional process	flow of event
event	action that trigger the flow of event
data movement	event in Use Case
object of interest	class from Use Case
data group	entity in flow of event
data attribute	attribute of entity

Principle 1: Mapping of functional user
In COSMIC method, functional user sends or receives data from the system which could be an actor, hardware device, or software. In Use Case, actor represents anything that interacts with system[2], and it also could be an actor, hardware device, or software. Thus, it is natural to project functional user to actor.

Principle 2: Mapping of boundary and scope
In COSMIC method, the boundary is defined as a conceptual interface between the software being measured and its functional users. In Use Case, The boundary is defined as a logical boundary between actors and use cases.

In COSMIC method, the scope represents the functional user requirements of the software being measured. While in Use Case, the scope represents the use cases which describe the functional user requirement. Thus, it is easy to map boundary or scope onto Use Case.

Principle 3: Mapping of granularity
In COSMIC method, the same "level of granularity" of the Functional User Requirements means requirements are elaborated in same detail. There is also the concept of "level of granularity" in Use Case, and it is vital to consider granularity when the software is being measured. For example, if you identify function point in the granularity of functional process in COSMIC, then you have to identify function point in the granularity of event flow in Use Case, and this is called "standard measure", or else it is called "early or rapid approximate sizing[12]".

Principle 4: Mapping of layer and peer component
In COSMIC method, A layer is a partition resulting from the functional division of a software system, and there is a "superior/subordinate" hierarchical dependency between the functional services provided by software in any two layers. However, there is not the concept of hierarchical relationship between use cases, and use case is just the description of requirements in the style of text[2].

Use case package[13] that organizes use cases into logical groups can be projected to peer component in COSMIC. One problem is that a data move maybe happened between two peer components which has been ignored when use case.

Principle 5: mapping of functional process
In COSMIC method, a functional process is an elementary component of a set of Functional User Requirements comprising a unique, cohesive and independently executable set of data movements[1]. As mentioned earlier, a flow of event in Use Case describes the sequences of actions. Usually, one functional process can be mapped onto one or more flow of event, in order to map onto functional process, several flow of event have to be amalgamated into one. These flows of event to be amalgamated must be triggered by actor and at least two data movement (one is entry, the other is write or exit) will be identified from them. In addition, the development of every type of flow of event such as basic flow, alternative flow and abnormal flow will increase the size and effort, so each of them could identify function point separately.

Principle 6: mapping of event
In COSMIC method, an event causes a functional user of the piece of software to initiate one or more functional processes. In Use Case, an event causes the use case to be initiated. Thus we take it for granted that the event can be mapped.

Principle 7: identifying function point
In our approach, data movement, object of interest, data group and data attribute can be identified from flow of event. A data group is a distinct, non empty, non ordered and non redundant set of data attributes where each included data attribute describes a complementary aspect of the same object of interest. The subject in the sentence (flow of event) such as actor, software, hardware device often identified as "functional user". The verbs in the sentence often identified as "data movements". The nouns that data movement moved often identified as "data group" and the attribute of the noun often identified as "data attribute".

4 Counting Function Point

4.1 Eliminating Duplicate Counting

The dependency relations between use cases include expansion, inclusion and generalization. We specify that the function point in the independent use cases or dependent use cases should be counted separately because the basic mapping unit is the flow of event not the use case. In the other word, whether it is independent use case or dependent use case, the flows of event it contains at least could form a functional process which can identify function point alone. This approach just concerns on the current use case alone and do not consider the dependency relations between use case, so it avoid the duplicate counting bringing by repeat use of use case.

4.2 Formula of Counting Function Point

$$size(software) = \sum size(useCasePackage_i) \tag{1}$$

$$\begin{aligned} size(useCasePackage) &= \sum size(basicUseCase_i) \\ &+ \sum size(extendUseCase_i) + \sum size(includeUseCase_i) \\ &+ \sum size(generalizeUseCase_i) \end{aligned} \tag{2}$$

$$\begin{aligned} size(useCase) &= \sum size(basicFlow_i) \\ &+ \sum size(altenativeFlow_i) + \sum size(abnormalFlow_i) \\ &+ \sum size(preCondition) + \sum size(postCondition) \end{aligned} \tag{3}$$

$$\begin{aligned} size(eventFlow) &= \sum size(entry_i) + \sum size(exit_i) \\ &+ \sum size(read_i) + \sum size(write_i) \end{aligned} \tag{4}$$

We use formulas above to count the total size of software, $i \in (1, n)$, and n represents the number of the elements which could be use case Package, use case or flow of event, etc.

In our method, the software is first decomposed into use case packages, so we can see from formula 1 that the size of software to be measured equals the total size of use case packages. Because we measure use cases separately, in formula 2 we add all the size of use cases including basic use cases, extensive use cases, use cases of generalization, and inclusive use cases, etc. We identify function point from flow of event, pre-condition and post-condition in a use case, so in formula 3 we add the size of these elements to form the size of use case. The left of formula 4 is actually any item of the set including basic flow, alternative flow, abnormal flow, pre-condition, post-condition, and the right is the sum of data movements including entry, exit, read and write.

5 Conclusion

In this paper we have proposed sizing software with Use Case from the viewpoint of event flow. On the basic of previous works we have defined more refined mapping rules between the COSMIC method and Use Case. And the core and key principle is that we map functional process onto event flow rather than Use Case. On the one hand, it makes the mapping at a low granularity which could produce a more accurate result compared with previous research. In addition, it could eliminate duplicate counting bringing by the associations between Use Cases.

To make this method practical and easy to use, next step will be to develop a sizing tool to support our method. In this way collecting data about functional size and effort then we continue our research on project estimating.

References

1. The COSMIC Measurement Manual (The COSMIC Implementation Guide for ISO/IEC 19761:2003), version 3.0.1 (May 2009)
2. Leffingwell, D., Widrig, D.: Managing Software Requirements A Use Case Approach, 2nd edn. (2004)
3. Fetcke, T., Abran, A., Nguyen, T.: Mapping the OO-Jacobsen Approach to Function Points. In: Proceedings of Tools 23 1997 – Technology of Object Oriented Languages and Systems. IEEE Computer Society Press, Los Calamos (1998)
4. Bevo, V., Levesque, G., Abran, A.: Application of the FFP method to a specification in UML notation: account of the first attempts at application and questions arising (1999); Translated by Jenner, M.: French original 'Application de la méthode FFP à partir d'une spécification selon la notation UML: compte rendu des premiers essais d'application et question, http://www.lrgl.uqam.ca/publications/ (retrieved from May 17, 2004)
5. Jenner, M.: COSMIC FFP 2.0 and UML: Estimation of the size of a system specified in UML – Problems of granularity. In: Proceedings of FESMA-DESMA Conference 2001, Heidelberg, Germany (2001)
6. Jenner, M.: Automation of Counting of Functional Size using COSMIC-FFP in UML. In: Proceedings of 12th International Workshop on Software Measurement, IWSM 2002, Magdeburg, Germany (2002)
7. Function Point Counting Practices Manual, release 4.2, The International Function Point Users Group (2004)
8. Condori-Fernández, N., Abrahão, S., Pastor, O.: Onthe Estimation of Software Functional Size from Requirements Specifications. Journal of Computer Science and Technology 22, 358–370 (2007)
9. van den Berg, K., Dekkers, T., Oudshoorn, R.: Functional size measurement applied to UML-based user requirements. In: Proceeding SMEF Conference (2005)
10. Sellami, A., Ben-Abdallah, H.: Functional size of use case diagrams: a fine-grain measurement. In: 2009 Fourth International Conference on Software Engineering Advances (2009)
11. Jones, C.: A new business model for function point metrics, Version 10.0 (August 2009)
12. Advanced_and_Related_Topics (The COSMIC Implementation Guide for ISO/IEC 19761:2003), version 3.0.1 (May 2009)
13. Armour, F., Miller, G.: Advanced Use Case Modeling Software Systems (2004)

Process-Based Measurement on Airworthy Software

Fen Sun, Yumei Wu, and Deming Zhong

Abstract. According to the characteristics of the airworthy software, this paper provides to measure the software from the development process, which is from the processes of requirement, design, code and test. The software is measured and people respectively get a measure in each stage, then the four measures are integrated together to draw a comprehensive software measure, which completes measuring the software based on the process.

Keywords: ariworthy software, requirement measure, design measure, code measure, test measure.

1 Introduction

Airworthiness refers to the inherent quality of the aircraft that flies safely in the expected environment (including takeoff and landing), and the quality can be durative to be preserved through the proper maintenance. Generally the definition is the state in which the aircraft is fit for the flight. The airworthy software can be understood as the software that meets the airworthy standards or passes the airworthy certification. The airworthy software requires high safety and process dependability, and the characteristic of the airworthy software is achieving the software's safety and ensuring the process dependability through the development processes, thus ensuring the software development processes is essential to the airworthy software, and this paper proposes the measure based on the process, which not only can ensure the software development process, but also gives a comprehensive measure ultimately, which provide objective reference for each stage of the software process and the final quality of the software, and also play a role of supervision and inspection for the software quality.

Fen Sun · Yumei Wu · Deming Zhong
Department of Reliability and System Engineering,Beijing University of Aeronautics and Astronautics, Beijing
e-mail: sunfen1111@163.com, wuyumei@buaa.edu.cn

Z. Qian et al. (Eds.): Recent Advances in CSIE 2011, LNEE 125, pp. 211–216.
springerlink.com

1.1 Requirement Process Measurement

Airworthy software requires the software requirements should be integrity, consistent and correct; system requirements assigned to the software should be traced back to one or more software high requirements; high requirements meet the software requirement standards, and should be verifiable and consistent [1], so accuracy, completeness, consistency, realization, verifiability, traceability of the requirements are chosen as the metrics during the measurement of requirement process. Because the airworthy software requires high safety, the safety must be considered.

In this paper, t_i is denoted as the requirement feature, where i = 1,2, ..., 8; a_i is expressed as the degree of the attention on the requirement feature t_i and can be obtained by averaging and rounding the effective data ,where i = 1,2, ..., 8

Compute the attitudes of assessment staff and round and average them to get the attention of various features. sd_i is the standard evaluation on the requirement feature t_i, where i = 1,2, ..., 8; give the basic requirement of the quality of the requirement feature t_i ,and if find out that the evaluation on the requirement feature t_i has met (not less than)the demand, consider the requirement feature t_i is satisfied.

c_i is the performance evaluation of the requirement feature t_i ,and the positive value indicates the positive evaluation, and negative value indicates the negative evaluation; the greater the absolute value is, the greater the impact is. r_i is a certain requirement.

Start the measure calculations after obtaining the above information. s_i is denoted as the evaluation result of the feature t_i and is the comprehensive evaluate the result various roles (such as customer, project manager, demander, designers, testers, etc.); $s_i = \frac{(c_i - sd_i) \times a_i}{9}$, when $s_i \geq 0$ exists, indicates the positive attitude on the feature t_i , and consider the feature t_i of the requirements is acceptable; $s_i = \frac{(c_i - sd_i) \times a_i}{9}$, when $s_i \leq 0$ exists, indicates the negative attitude on the feature t_i , and consider the feature t_i of the requirements is unacceptable and it needs for further improvement. x_s is denoted as the requirement measure, of which $x_s = \sum_{i=1}^{8} \frac{s_i \times a_i}{\sum_{i=1}^{8} a_i}$, The greater the absolute value of x_s is, the higher the degree of satisfaction or dissatisfaction is.

1.2 Design Process Measurement

The airworthy software requires the low-level requirements and the software architecture should meet the software design standards and should be traceable, verifiable and consistent on the development process [1], so realization, reliability and completeness are chosen as the metrics during the measurement of design process. Because the airworthy software requires high safety, the safety must be considered. The checklist method is used in this paper to complete the measure. Concrete steps are as follows.

Firstly, develop the checklist table. Develop the checklist for the software design according to the concrete software and referring to the software requirement specification, and that is the design measures to be taken as to each requirement.

Secondly, check the software design according to the checklist. Referring to checklist, check the outline design and detailed design and whether all the requirements have been designed. If there are a requirements ignored, mark them out.

Then, determine the metric weights. According to the software design, classify the metrics to determine which design is feasible, which design belongs to reliability design, which design belongs to safety design (design can be operable, and it also can be reliability or safety design), and whether all the software requirements have been covered by all the design. In generally, the weights α_i , of which $\sum_{i=1}^{4} \alpha_i = 1, i = 1,2,3,4$, are done by experts and experts rate each feature based on the measurement results of the software design, and the score is n_i , $n_i \in [0,10]$ $i = 1,2,3,4$.

Finally, according to the formula $s_s = n_1\alpha_1 + n_2\alpha_2 + n_3\alpha_3 + n_4\alpha_4$, get the measurement result of the software design.

1.3 Code Process Measurement

The airworthy software requires the codes realize the low-level requirements and meet the software architecture and the codes are traceable for the design and meet software coding standards on the design process [1], so a measure of coding per metric selected from a consistent coding standards, the main consideration, so portability, reliability, functionality are selected as the metrics of the code process measurement, which are considered mainly the consistency with the coding standards. Because the airworthy software requires high safety, the safety must be considered. Specific measurement steps are as follows.

First, design the defect table, and that is the evaluation principles.

Second, determine the metric weights. Generally the weights are divided into five levels, and 5 is the highest, that is, the highest weight.

Then, check the sample. Sample n copies of the code from all the code of the system to be evaluated, and each code contains m lines of code (if there are extra m lines in the copy, intercept m lines of code from the copy.) Assess the each copy of code which is sampled line by line according to software metric features, identify defects in each copy of code and classify and record them according to metric features. We can get a datum matrix and the number of defects $\omega_i = (\omega_{i1}, \omega_{i2}, \omega_{i3}, \omega_{i4})$,of which, $1 \le k \le 4$ is the final total number of defects of the k-class quality characteristic of the i copy of the code , by assessing the each copy of code which is sampled line by line according to software metric features, identifying defects in each copy of code and classify and recording them according to metric features, and can get the comprehensive defect score of the code is $r_i = \sum_{k=1}^{4} \omega_{ik}\beta_k$, of which, β_k is the metric weight and $\sum_{k=1}^{4} \beta_k = 1$.

Finally, the data which are all passed the assessment to be gotten compose a sample set and the defect value is signed as $b_s = (r_1, r_2, r_3, r_4 ... r_n)$, and by the same we can get the last defect of the software code through calculating the weight authorities, and that is $b_s = \sum_{i=1}^{n} r_i\phi_i$, of which, $\sum_{i=1}^{n} \phi_i = 1$ and ϕ_i is the weight of the i copy of the code. Usually the weights are given by the expert [2]. The average code defect number per 100 lines is $b_{ss} = \frac{100b_s}{mn}$.

1.4 Test Process Measurement

The airworthy software requires the test should be based on software requirements and sufficient [1], so the adequacy of testing is selected as the metric. The ability of exposing the wrong of the test method and the ability of reducing the residual risk of the test are very typical aspects to measure the test method, therefore, these two are also selected as the metrics. Specific measurement method can see Paper 8,9 and 10 in reference [3][4][5].

2 Measurement Based on the Process

In this paper, AHP approach is used to assign the weights φ_i to requirement, design, code, test, and i = 1,2,3,4, and the integrated measure result can be obtained by using the weight algorithm formula $result = x_s\varphi_1 + s_s\varphi_2 + c_s\varphi_3 + b_s\varphi_4$ after the good distribution of the weights.

The final measure of the software can be obtained by using the weight arithmetic after deriving the requirement weight, the design weight, the code weight and the test weight. Consider the overall measure meets the demand, but the measure of some stage may not meet the demand, so determine whether the software ultimately meets the measure demands, not only consider the final measure, but also consider the measures of the stages; only when both meet the demands, it just meets the measure demands. The reference values of the airworthy software measurement results are given, as shown in **Table 1**.

Table 1 The reference values of the airworthy software

assessment	requirement	design	code	test	result
1	$x_s \geq 4$	$s_s \geq 30$	$b_{ss} \leq 1$	$c_s \geq 6$	$\geqslant 35$
2	$x_s \in (0,4)$	$s_s \in (20,30)$	$b_{ss} \in (1,5)$	$c_s \in (4,6)$	(25,35)
3	$x_s = 0$	$s_s = 20$	$b_{ss} = 5$	$c_s \in (3,4)$	(23,25)
4	$x_s \in (-4,0)$	$s_s \in (10,20)$	$b_{ss} \in (5,10)$	$c_s \in (1,2)$	(18,23)
5	$x_s \leq -4$	$s_s \leq 10$	$b_{ss} \geq 10$	$c_s \leq 1$	$\leqslant 18$

1 in the first column in Table 4 shows being very satisfied; 2 shows being satisfied; 3 shows just meeting the requirements; 4 expresses the dissatisfaction; 5 expresses being very dissatisfied.

3 Summarize

According to the characteristics of the airworthy software, start from the software development process to draw a comprehensive measure by selecting the appropriate metrics for each process. This method not only draws the software measure, but also provides an objective reference for the software development process, which can make developers and users understand the situation of the software development process; it also has a good effect on supervising and inspecting the improvement of the software quality, and provides a reference for the assessment of the airworthy software. This method is suitable for the airworthy software or the software with the requirement of high safety, but due to the weight and expert scoring, the quantitative values obtained are with certain subjectivity.

References

[1] RTCA—178B.the software test center of air force eauipment, pp. 26–29
[2] Peng, Y., Wang, S.: Ingrated software quality assessment method based on code review. Software Guide 9(3) (2010)
[3] Xu, Z., Wu, F.: The measure of software test quality. Computer Engineering and Application 21 (2002)
[4] Yu, S., Wu, F.: The method and application of operational profile of interlocking software based on knowledge accumulation. In: Proceeding National Conference on Testing, pp. 291–295 (2000)
[5] Zhao, J., Wang, D., Shi, L.: A measurement and evaluation method on software dependability. Computer Science 35 (2008)
[6] Xie, J., Liu, C.: Fuzzy mathematic method and the application, 3rd edn., pp. 175–191. Indstry Press of Huazhong University of Science, Wuhan

An Improved Algorithm Based on Max-Min for Cloud Task Scheduling

Gao Ming and Hao Li

Abstract. Max-min algorithm is a classic grid computing task scheduling algorithm, but Max-min algorithm for task scheduling under the cloud have many problems. To solve this problems, this paper propose an improved algorithm MMST based on Max-min, which can not only solve the problem short task waiting time is too longer, but also can solve the problem when the number of the task is little, but using too many resources. Finally the algorithm MMST reduces the expenses of cloud services.

Keywords: Max-min, MMST, Improve service quality, Reduce cloud costs.

1 Introduction

Cloud Computing is a new business computing model, which comes from Distributed Computing, Parallel Computing, Grid Computing[1]. The resources of cloud computing are heterogeneous, dynamic and heterogeneous, and the cloud environment is a commercial platform, so how to reasonably use cloud resources and effectively divided cloud computing resources becomes more important.

In the cloud computing, the user is not using resources for free. In the process of the user using cloud computing services, how to minimize the cost and perfectly meet the needs of users, has become the current cloud computing must be addressed. In the traditional grid task scheduling algorithm, Max-min algorithm is a classic batch mode heuristics. Here, we try to introduce Max-min algorithm to the task of scheduling.

Gao Ming · Hao Li
School of Information Science and Engineering
Yunnan University
Kunming, China
e-mail: 29168961@qq.com

Z. Qian et al. (Eds.): Recent Advances in CSIE 2011, LNEE 125, pp. 217–223.
springerlink.com

2 Max-Min Algorithm Application in a Cloud

Max-min algorithm is one of the based grid scheduling algorithm[2,3]. Using the simulation tools Cloudsim, Max-min algorithm was simulated. The simulation parameters are as follows (the number of task represents the length of one task, and the number of computing resources represents the computing power of the resources Mips): a task sequence, t1(200), t2(300), t3(300), t4(500), t5(800), t6(1000);

Case 1: There are only two resources in the cloud r1(300), r2(300);
Case 2: There are enough resources in the cloud, r1(300mips), r2(300), r3(300), r4(300), r5(200), r6 (200), r7 (200), r8 (200);

According to the scheduling algorithm, experimental results are as follow:

Table 1 Max-min algorithm in simulation (case 1)

	Task1	Task2	Task3	Task4	Task5	Task6
Finish time	5.1	5.3	4.3	4.4	2.7	3.3

Table 2 Max-min algorithm in simulation (case 2)

	Task1	Task2	Task3	Task4	Task5	Task6
Finish time	1.0	1.5	1.0	1.7	2.7	3.3

From the simulation results, we can see that when the computing resources is not enough for scheduling, the short task's completion time is always longer, and some times the results cannot be the satisfaction of users. On the other hand, when the computing resources is enough for scheduling, all the task's completion time is always good, but in this case, Max-min algorithm using all six cloud-computing resource from the pool of computing resources that waste a lot of cloud computing resources.

In this case, Max-min algorithm is clearly not well adapted to task scheduling work in cloud computing environment. This paper presents an improved algorithms MMST (Max-min Spare Time) algorithm based on Max-min algorithm.

3 An Improved Algorithm—MMST

In the actual Task scheduling, the user often has the time required to complete the task. In order to complete the task better and make users satisfaction, the time parameters was introduced in the algorithm. At the same time, the face of Max-min algorithm, appear in the actual scheduling of the short task completion time is too long, take up the task when the light goes too many resources and other issues, we propose improvements MMST (Max-min Spare Time) based on Max-min algorithm. Improved algorithm for the MMST is described as follows:

(1) if the task set T is not empty otherwise continue execution (13)
(2) Calculate all the tasks required to the cloud computing resources
(3) $M = \sum_{i=1}^{n} Q_i / T_{DEAD_i}$
(4) Get resources set (R1,R2,...Rn) from resource pool, the resources' computing of set ≥m,and the maximum computing power of resources Ri≥MAX(Ti)
(5) for the task set t of all the task t_i
(6) for the resource set t of all the task r_j
(7) $C_{ij} = tr_{ij} + tw_j$
$$TL_{ij} = T_{DEAD_{ij}} - C_{ij}$$
(8) for the every element of TL_{ij}
(9) Find the task $t_k$which have a shortest waiting time .when the waiting time of tasks was same, priority select the max and fast task
(10) When matching the task t_k with resources, MMST find the resource which the task can be finished in the smallest integer waiting times, and match the resources $r_k$to the task
(11) Take the (t_k, r_k) into the scheduling vector K
(12) Delete the task t_k from T
(13) Update tw, TL_{ij}
(14) scheduling K

The tw_j is an expected time that means resources r_j will complete all tasks in the ready time after tw_j time. tr_{ij} is a expect time that used to represent the finish time resources r_j executive task t_i. T_{DEAD_i} is the maximum waiting time. TL is an $n \times m$ matrix, and it used to record the spare time resources r_j executive task t_i. Vector K is used to store the mapping results, and store each task and resource pair. Each time traversing the TL matrix, Max-min algorithm mapped the task which has the minimum spare time to the resource which has the maximum executive time. Updating the TL matrix and the tw vector, and repeat the mapping process. If there are competing between task $t_i$and t_j, MMST algorithm chose the big task first.

4 Experiment Simulation

4.1 Experiment Environment Cloudsim

University of Melbourne, Australia, the grid project announced the availability of laboratory and Gridbus cloud computing simulation software called CloudSim. It is discrete event simulation package SimJava on the development of libraries, on Windows and Linux cross-platform, CloudSim inherited GridSim programming model to support cloud computing research and development, and provides the following new features: (1) support large-scale cloud computing infrastructure modeling and simulation; (2) a self-sufficient to support the data center, service agents, scheduling and allocation policy platform [4].

4.2 MMST Algorithm Simulation

Because the code is burdensome, this article does not show all code in the algorithm simulation tests. This is only the core code which is the MMST algorithm.

```
  System.out.println("new time");
  for(i=0;i<6;i++)
  {    for(j=0;j<2;j++){
q=cloudlet[i].getCloudletLength()/vms[j].getMips()+c[j];
  double TL=cloudlet[i].getCloudletOutputSize()-q;
  m[i][j]=TL;
System.out.print(TL);System.out.print("   ");}};
  System.out.print("\n");
  int z1=0;int z2=0;double z3=999;
  for(i=0;i<6;i++){
   for(j=0;j<2;j++){
   if(m[i][j]<z3){z3=m[i][j];z1=i;z2=j;}}};
  int max1=0;int max2=0;double max3=0;
  for(j=0;j<2;j++){
   if(m[z1][j]>max3){max3=m[z1][j];
  max1=z1;max2=j;};};
  double th=0;
  th=m[max1][max2];
  c[max2]=c[max2]+th;
  cloudletList.add(cloudlet[max1]);
  broker.submitCloudletList(cloudletList);
  broker.bindCloudletToVm(cloudlet[max1].getCloudletId
(),vms[max2].getId());
  cloudletList.clear();
  cloudlet[max1].setCloudletOutputSize(1000);
```

4.3 Analysis of Experiment Result

1) Lab 1:the resources were enough

t1(200MI,2);t2(300,3);t3(300,3);t4(500,5);t5(800,8);t6(1000,10);

The resources in the resource pool:

r1(300mips);r2(300);r3(300);r4(300);r5(200);r6 (200);r7 (200);r8 (200);

The resource requires through MMST algorithm to predict the resource, $M = \sum_{i=1}^{n} Q_i/T_{DEAD_i}$=100+100+100+100+100+100 = 600 MIPS.

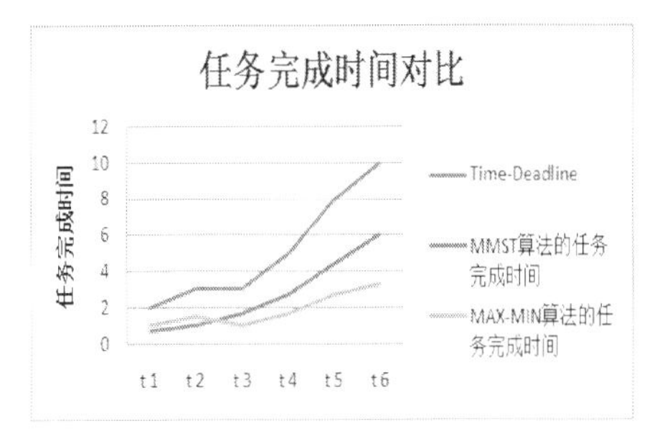

Fig. 1 The compare of task complete time

Figure 1 shows, when cloud computing resources were not enough, it still can be a very good schedule to complete the task before the dead time. When the case of sufficient computing resources, Max-min algorithm can also be effective to ensure that the task be completed before the latest completion time.

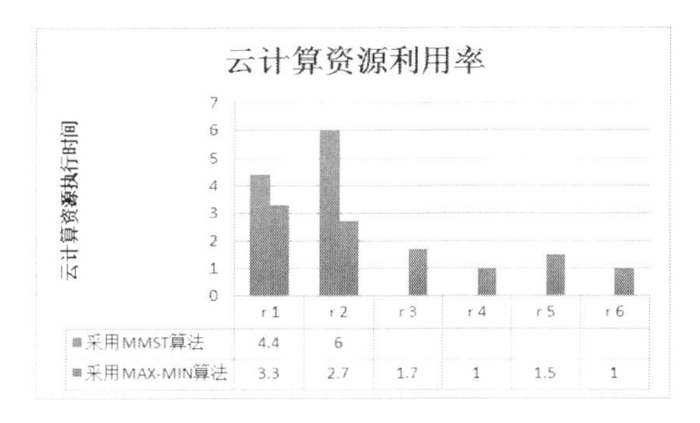

Fig. 2 The compare of resource utilization

Figure 2 shows, when the task queue arrives, due to the different scheduling strategies, MMST algorithm using only two resources R1, R2 and Max-min algorithm called six resources. Assuming the prices of cloud computing resources were 300 $ / s, and the cost of using the MMST algorithm was: $2 \times 6 \times 300 =$ 3600$, the cost of using the Max-min algorithm was: $6 \times 3.3 \times 300 =$ 5940$.we can see that when the resources were enough, the max-min algorithm wasted too many resources that make the resources loading imbalance. But in this case, the MMST algorithm only wasted two resources that make the resources loading balance.

2) Lab 2:the resources were not enough

Changing the last experiment, assuming the resource pool only has r1, r2 two resources and the same task queue.

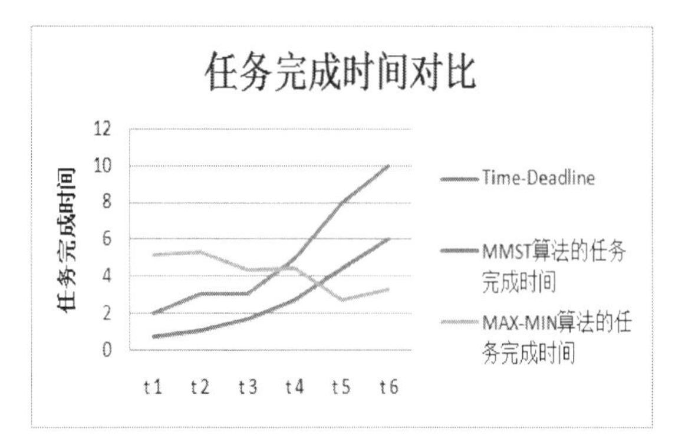

Fig. 3 The compare of task complete time

Figure 3 shows, when the resources were not enough, the max-min algorithm make the short overtime and reduce the quality of service. But in this case, the MMST algorithm can still ensure all the tasks can be done in the dead time.

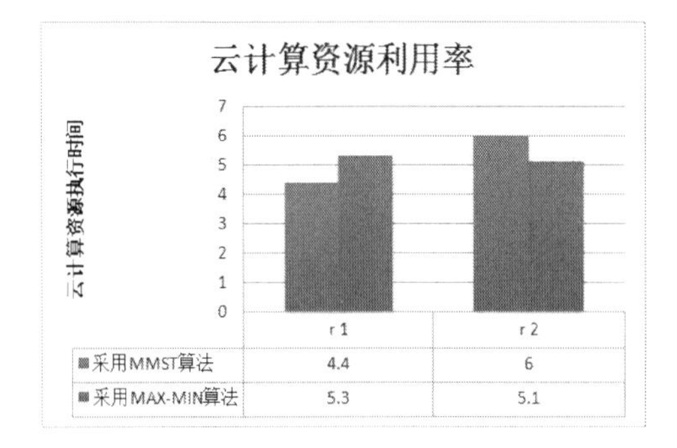

	r1	r2
采用MMST算法	4.4	6
采用MAX-MIN算法	5.3	5.1

Fig. 4 The compare of resource utilization

Figure 4 shows, when the resources were not enough, the max-min algorithm wasted too many resources that make the resources loading imbalance. But in this case, the MMST algorithm only wasted two resources that make the resources loading balance.

5 Summary

Based on the classic grid max-min algorithm, this paper found the max-min algorithm has some problems in a cloud environment. From the experiments on the

Cloudsim platform, this paper found that max-min spare time algorithm is more efficient than Max-min algorithm.In future, we the MMST algorithm can be used in the actual environment. we will continue to improve it in the next step.

Thanks

1. Supported by the National Nature Science Foundation of China 2010 (No.61063044).
2. Supported by Key laboratory in Software Engineering of Yunnan Province under Grant No.210KS05.

References

1. Wang, Q., Wu, Y.: Virtualization and Cloud Computing. IBM (May 2010) (in Chinese)
2. Armstrong, R., Hensgen, D., Kidd, T.: The relative performance of various mapping algorithms is independent of sizable variances in run-time predictions. In: 7th IEEE Heterogeneous Computing Workshop (HCW 1998), pp. 79–87 (March 1998)
3. Freund, R., Gherrity, M., Ambrosius, S., Campbell, M., Halderman, M., Hensgen, D., Keith, E., Kidd, T., Kussow, M., Lima, J., Mirabile, F., Moore, L., Rust, B., Siegel, H.: Scheduling resources in multi-user, heterogeneous, computing environments with SmartNet. In: 7th IEEE Heteroge-neous Computing Workshop (HCW 1998), pp. 184–199 (March 1998)
4. http://baike.baidu.com/view/3652473.html
5. http://baiyunl.javaeye.com/blog/707935

Dynamic-Feedback-Based Connection Pool Framework for Database Cluster

Sui Xin-zheng and Cheng Ren-hong

Abstract. Database Connection Pool has been widely used in database applications, because it not only can efficiently manage the database connections, but also can improve the system performance. When database structure is a cluster, load balancing always should be considered. However, the traditional connection pool statically manages the connections that have already established. It cannot make adjustment according to the changes of environment, which may lead to load imbalance or even overload. To solve this problem, we propose a dynamic-feedback-based connection pool framework which can distribution the workload according to real-time system resource usage of nodes in cluster, and also can achieve the overload prevention. Experiment evaluation has demonstrated the feasibility and effect of this framework.

Keywords: dynamic feedback, connection pool, load balancing, database, cluster.

1 Introduction

Database connections provide a way for client application to access database server. Its establishment requires user authentication, security context configuration and other operations. This will result in a certain communication, memory and other cost. The frequently establishing and releasing database connections will bring great overhead and slow down the whole system. In order to solve these problems, the connection pool technology has been produced. It not only can manage database connections that are expensive and limited, but also can improve the system stability and efficiency.

A cluster comprises multiple interconnected nodes that appear as if they are one server to end users and applications. It can provide high extensibility, availability

---

Sui Xin-zheng
College of Information Technical Science, Nankai University, Tianjin, 300071, China
e-mail: xinzhengs@gmail.com

Cheng Ren-hong
College of Software, Nankai University, Tianjin, 300071, China
e-mail: chengrh@nankai.edu.cn

Z. Qian et al. (Eds.): Recent Advances in CSIE 2011, LNEE 125, pp. 225–231.
springerlink.com

and processing capability at a relative lower price than mainframe solution. Many commercial databases take advantage of this structure. In a cluster system, load balancing should always be considered. However, if the database application uses a connection pool, in which connections are already established and pointing to different cluster nodes. Without special treatment, the distribution of connections is random. It may lead to load imbalanced or even overload.

According to the problem above, this paper presents a dynamic-feedback-based connection pool framework. It can distribute the connections according to the feedback of cluster nodes' system resource usage. More over, if the tasks arrival intensity too high to the node's processing ability, it can also achieve the overload prevention. In the experiment, we use the CPU intensive workload (task) to demonstrate effect of the framework.

2 The Load Balancing Strategies

Due to the task characteristic, node processing ability, data partition and other factors, the cluster nodes are often having uneven load. The nodes with light load can not fully utilize its computing power, while the nodes with heavy load are struggling in resource competition, which will slow down the speed of the entire system. In this case, load balancing must be considered.

Load balancing is an idea of ensuring that each node in the cluster will have approximate same amount of load, to optimize resource utilization, improve system throughput and avoid the overload from happening. Generally, there are two kinds of task distribution strategy to achieve load balancing which are static (like random, round-robin, hash, etc.) and dynamic. The static strategies are relative simple. It cannot response to the changing of load and sometimes has a poor effect. The dynamic strategies distribute tasks based on the current load status of each node. The load can be represented by the task queue length [1], system resource usage (CPU [2], memory [3], disk I/O [4] and network communication [5] [6]).

Usually, the size (running time) of tasks is variable, for example big task may contain many small tasks. Also, the exact resource consumptions are unknown in advance and resources competition may occur between tasks at the running time. In these cases, one plus one does not necessarily equal to two. So, it is impossible to achieve absolute load balancing but an approximate one. Load balancing feature can be implemented in the database layer or the application layer. If a connection pool is used, it can only be implemented in the application layer. In order to achieve load balancing, we adopt the control theory which is mature and widely applied in the computer science [1] [7] [8] [9]. Briefly, the close loop control method we use can be described as that the controller dynamically (periodically) adjusts the related parameters based on the feedback from the system, in order to achieve the desire goal.

3 The Framework of Dynamic-Feedback-Based Connection Pool

The dynamic-feedback-based connection pool framework uses a modular design pattern and resides in the client side. We design it more generic to adapt to more situations. The framework structure is shown in Fig. 1:

1) Task Dispatcher (TD): Based on the DM's decision, the TD dispatches tasks from the TQ to the TEs with the information of which connections to be used.
2) Task Executor (TE): The TE gets the connection from the CP, executes the task and releases the connection back to the CP when the task is finished.
3) Task Queue (TQ): The TQ plays a role in buffering the tasks and is waiting for the TD to process. It runs in first come first serve (FCFS) order.
4) Connection Pool (CP): The CP maintains multiple connections pointed to different nodes in the cluster. It provides two basic functions which are "GetConnection" and "ReleaseConnection".
5) Resource Monitor (RM): The RM gathers system resources usage information from each node in the cluster. Based on different need of DM, different data can be collected.
6) Database Cluster (DC): The DC is composed of multiple nodes. It appears as if one server to the end users and applications. The nodes are interconnected by high speed network and sharing data for each other.
7) Decision Maker (DM): Based on the data collected, the DM determines which node the task should be assigned. The collected data can be resources usage information from RM, task information from the TD or task execution information from the TE.

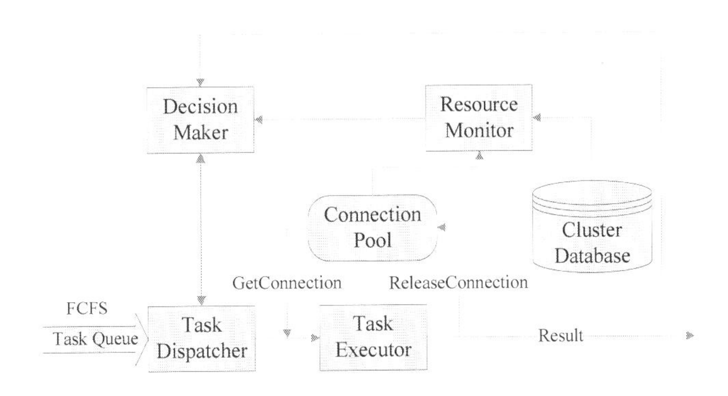

Fig. 1 Dynamic-Feedback-Based connection pool framework

4 Implementation

If the coming frequency of tasks is very high and beyond the nodes' processing ability, the CP maybe dries. There will be a new queue of TEs waiting for connections in front of the CP. The task may wait for a while to get the connection. We should notice that the DM's decision is based on the time when the TD assigns the task, not the time when the TE gets the connection. The period between this two may cause the decision expired. So in this case, tasks should be return back to the TQ when the CP is dry. And also the DM will stop making decisions when the dry of the CP is detected.

Another problem is that when the task dispatching of TD is based on the RM and the sampling period is larger than the frequency of coming tasks. There may be more than one tasks get decision from DM by the same data collected by the RM. This situation will lead to wrong result. So the DM needs to ensure that each task dispatching process is based on the current data. There are two methods to solve this problem. One is to delay the decision making of DM, and hold the task until the next sampling of RM. This will enlarge the response time of the task. Another is to narrow the sampling period of the RM. This may cause more overhead. But the sampling period cannot be too small for nodes to respond to the last arrived task. Generally, the task does not achieve its maximum resource utilization immediately. There is a process from initialization to execution and the resource consumption increasing gradually. So a tradeoff should be made between these cases, S is the sampling period, F is average arrival interval of tasks and R is the system response time:

- If $F >= R$, then $S = R$, tasks should be delayed to the next sampling period.
- If $F < R$, then $F < S < R$, and the DM will ensure that each task assignment will use the current data.

However, the exact resource consumption of tasks is unknown in advance and maybe larger than the rest of system resources. The node maybe overloaded by executing a single new task. It is hard for the DM to make the right decision but the possibility of overload can be lowered. There maybe some tasks have finished and are releasing the resource. So the task distribution strategies should not only consider the current resource consumption, but also need the trend of it. The discussion can be made as follows:

- If the current resource consumption is more than or near the threshold, it should not distribute new tasks but waiting.
- If the current resource consumption is much less than the overload threshold, but the trend is upward, it should not distribute new tasks.
- If the current resource consumption is much less than the overload threshold and the trend is downward, new tasks can be distributed.

5 Experiments and Evaluation

The experiment environment consists of an application client, a database server with two nodes. They are connected by a 100M switch. The specific configuration is shown as follows:

1) Application Client: Windows XP SP3, Lenovo M5100t, AMD PhenomII X4 810 2.6GHz, 3GB RAM, 320GB HDD.
2) Database Server: Solaris 10, Oracle 10g RAC, Lenovo M5100t, AMD PhenomII X4 810 2.6GHz, 4GB RAM, 320GB HDD.
3) Disk Server: Ubuntu 10.04 Server, HP ProLiant ML350 G3, Intel Xeon X4 2.8GHz, 1GB RAM, 72G+36G SCSI.

In order to show the feasibility and effect of the proposed framework, we simply use a CPU intensive task. The task is implemented by a stored procedure that can generate sustained CPU consumption and its sizes are independently random. We assume the arrival process of tasks is a Poisson Process. Let λ as the intensity of the homogeneous Poisson process {N(t), t>= 0}, where λ> 0. The task arrival interval sequence is {Tn, n> 0}. At this point, T1, T2, ... , Tn are independent and obey exponential distribution of parameters λ. Random (0,1) means the uniform random number of interval (0,1). So the task arrival time interval can be expressed as:

$$T_n = -\frac{\ln(Random(0,1))}{\lambda}$$

The CP is initialized by 20 connections for each node and the parameter λ is 0.8. The overload threshold is the CPU utilization rate over 80%. The RM's sampling time is 1 second. We conducted four experiments. The details are described as follow:

- Exp1: The TD randomly distributes connections to the tasks.
- Exp2: The TD distributes connections to the tasks according to the CPU usage of each node.
- Exp3: Same as Exp1 but with extra load (CPU usage is about 50%) on node1.
- Exp4: Same as Exp2 but with extra load (CPU usage is about 50%) on node1.

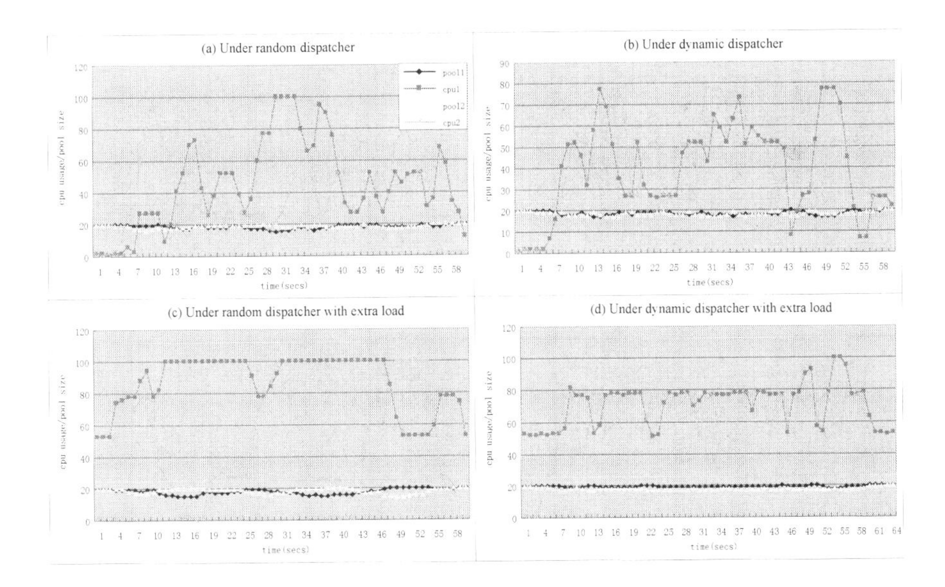

Fig. 2 The resource usage under two different dispatchers.

Fig. 2(a) shows experimental results of Exp1. The cpu1 represents the average CPU utilization rate of the node1. The pool1 means the rest connections of node1 in the CP. It is the same for the cpu2 and pool2. From the result, we can see that the load of the two nodes is not even, and the situation of overload happens more than once. The result of Exp2 is shown in Fig. 2(b). The CPU load of two nodes is more even than Exp1, and no overload happens.

The result of Exp3 is shown in Fig. 2(c). At the very beginning, the CPU utilization rate of node1 is 53% and 2% of node2. With the arrival of tasks, the CPU utilization rate of node1 increased gradually to 100% and maintained for a long time, while the CPU utilization of node2 is not that much. The load on node1 and node2 is not balance at all. The result of Exp4 is shown in Fig. 2(d). The CPU utilizations of the two nodes are mostly less than 80% and the load is more even than Exp3.

6 Conclusion and Future Work

In order to achieve load balancing for database cluster by using a connection pool, we study the workload scheduling strategy, and proposed a dynamic-feedback-based connection pool framework. The experiment results show that the proposed framework can not only balance the load across nodes, but also can prevent the overload happening. However, only one controller (TD) is not very comprehensive for load balancing in cluster database. In the future, we decide to add more controllers to the framework which can adjust more performance parameters based on the feedback of RM. We believe that the adaptation and effect of the framework will be better.

Acknowledgments. This work is supported by the China Academic Degrees & Graduate Education Development Center under contract number CDGDCPGC2009K1.

References

1. Tang, Z., Birdwell, J.D., Chiasson, J., Abdallah, C., Hayat, M.: Closed loop control of a load balancing network with time delays and processor resource constraints. In: Tarbouriech, S., Abdallah, C., Chiasson, J. (eds.) Advances in Communication Control Networks. LNCIS, vol. 308, pp. 245–268. Springer, Berlin (2004)
2. Harchol-Balter, M., Downey, A.: Exploiting process lifetime distributions for load balancing. ACM Trans. Comput. Systems 3(31) (1997)
3. Acharya, A., Setia, S.: Availability and Utility of Idle Memory in Workstation Clusters. In: Proceedings of the ACM SIGMETRICS Conference on Measuring and Modeling of Computer Systems (May 1999)
4. Qin, X., Jiang, H., Zhu, Y., Swanson, D.R.: Dynamic Load Balancing for IO- and Memory-Intensive Workload in Clusters using a Feedback Control Mechanism. In: Kosch, H., Böszörményi, L., Hellwagner, H. (eds.) Euro-Par 2003. LNCS, vol. 2790, pp. 224–229. Springer, Heidelberg (2003)
5. Cruz, J., Park, K.: Towards communication-sensitive load balancing. In: Proceedings of the 21 International Conference on Distributed Computing Systems (April 2001)

6. Bahi, J.M., Contassot-Vivier, S., Couturier, R.: Dynamic load balancing and efficient load estimators for asynchronous iterative algorithms. IEEE Trans. Parallel Distributed Systems 16(4), 289–299 (2005)
7. Yang, G., Zhou, X., Pan, H.: Feedback-based Framework of Adaptive Threads-pool Management. Computer Engineering 21, 65–67 (2006)
8. Shao, G.: Adaptive Scheduling of Master/Worker Applications on Distributed Computational Resources. Ph.D. thesis, Univ. of California, San Diego (May 2001)
9. Lightstone, S., Surendra, M., Diao, Y., Parekh, S.S., Hellerstein, J.L., Rose, K., Storm, A.J., Garcia-Arellano, C.: Control theory: a foundational technique for self managing databases. In: ICDE Workshops, pp. 395–403 (2007)

Loyalty-Based Resource Allocation Mechanism in Cloud Computing

Yanbing Liu, Shasha Yang, Qingguo Lin, and Gyoung-Bae Kim

Abstract. According to the characteristics of the cloud computing environment, A resource scheduling framework which takes into account customer satisfaction and the capacity of cloud platform. The concept of trust is also introduced in the architecture. Simulation shows that the trusted computing based on loyalty can improve the successful transaction rate of the system under the environment of cloud computing, and it can meet the requirements of cloud computing.

1 Introduction

Cloud computing is the hot technology in IT industry nowadays. Because of many unique properties it processes, the resource allocation and scheduling algorithms works in grid computing cannot run effectively in the cloud computing environment[1]. Map-Reduce is a programming model implementation which process and generate large data sets associated[2]. The scheduling model for mapreduce research focuses on applications currently, namely, how to use the model in the existing computing environment [3].

Based on the framework of master-slave and introduction of role-based access control, a scheduling model is proposed in the paper takes full consideration of the trust of node and meets the requirements of different user about service.

Yanbing Liu
Department of Computer Science. Chongqing University of Posts and Telecommunications.
e-mail: Liuyb@cqupt.edu.cn

Shasha Yang
Institute of mobile internet security, Chongqing University of Posts and Telecommunications
e-mail: yangshasha1987@hotmail.com

Qingguo Lin
Institute of mobile internet security, Chongqing University of Posts and Telecommunications.

Gyong-Bae Kin
Department of Computer Education of Seowon University, 361-742, Korea

Z. Qian et al. (Eds.): Recent Advances in CSIE 2011, LNEE 125, pp. 233–238.
springerlink.com

2 Resource Allocation Framework in Cloud Environment

2.1 Resource Allocation Process in Cloud Computing

Map-Reduce is not only a programming model but also a task scheduling model. With the help of MapReduce programming model, services are firstly given to the users whose requirements about resource is urgent and of higher priority, providing more reliability. As the task is completed, the master will evaluate slave according to the performance of the task.

Fig. 1 Resource allocation in cloud computing. the resource scheduling framework in cloud computing provide appropriate resources according to the level of the tasks provided.

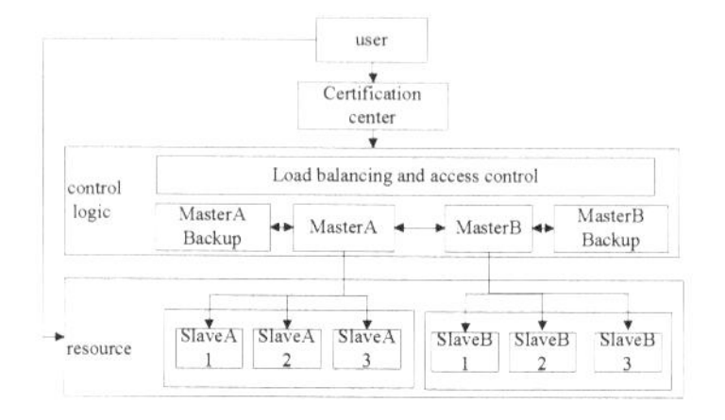

2.2 Resource Scheduling Model in Cloud Systems

First, the system will make classification of the tasks and establish expectations. Then select suitable virtual machine resources based on the feature of tasks and its expectations and have the task work. When the job is finished, the trust evaluation mechanism will calculate the reliability of resources and trust level according to the result of resource allocation, statistics the customer satisfaction and correction model and adjust the level of resources in the resource center.

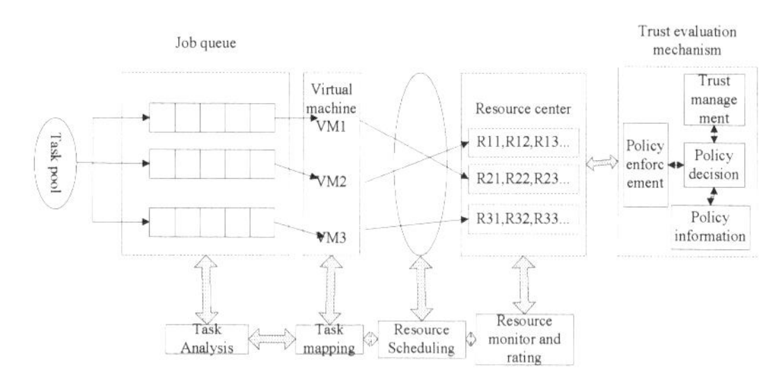

Fig. 2 Resource scheduling in cloud system

3 Turst Calculation Based on Loyalty

3.1 Definition

Definition 1: Trust[4]: when node a assume node b give a strict action as an expected, then a trust b.

Trust has a degree, built on the previous history and can be spread through the recommendation. The article divides trust into direct trust and recommendation trust which is also supposed to be credibility.

Definition 2: Direct Trust: The establishing of trust relationships between is depends on the tractions coming from direct experiences. The trust value can be obtained from the transactions between parties.

Direct trust is direct experience built on the base of interaction of both sides. After each interaction, every node will give a score to its interactive object and submit to the master.

Definition 3: Recommendation trust: Two nodes have no direct transactions, the relationship is established according to the recommendation of other nodes and the valued is based on the assessment of other nodes.

Definition 4: Transaction: Interactions between nodes.

Definition 5: Loyalty: The reliability basis of mutual trust between nodes;

In cloud system, as different users tractions is uncertain, a mechanism providing suitable resource to users according to users' particular expectations is required.

Definition6: The rate of dependency on trust level recommended to master mB by mA is the recommendation of mA, denoted by $Tr_{(mB,mA)}$.

3.2 Resource Allocation Model Based on Trust

In the cloud environment, loyalty based trust accounting framework is shown in figure 3. In the initial operation of the system, all the masters works well and realize the direct trust management to its slaves, that is to say, there is no recommendation trust. As the system is running, masterA happens to be effectiveness, other masters have to take place of masterA and realize the management to salveA. So the first recommendation trust occurs.

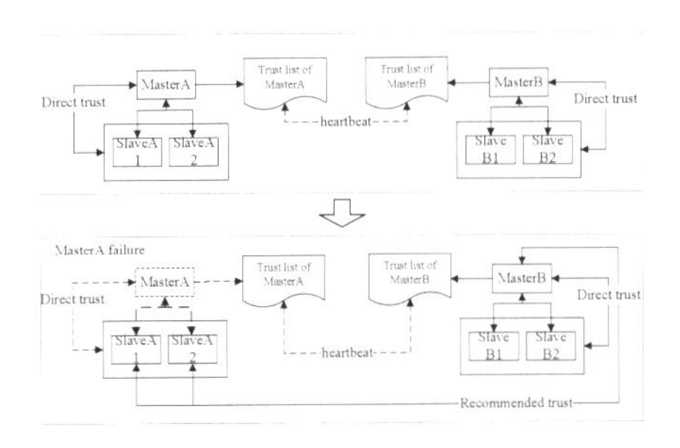

Fig. 3 Turst model in cloud computing

3.3 Loyalty-Based Trust Algorithm

The form of the evaluation value is $r=(m,s,c,i,v)$,where m,s represents the master and slave participating in the ith interaction. v is score m give to s about c (quality of service, honesty). $-1 \le v \le 1$, a negative value indicates dissatisfaction and on the contrary, the positive means satisfied. The evaluation set is $R_D(m,s,c)$.To find out the trust level of the target node; the model needs accumulate related historic evaluations and design an evaluating function according to each evaluation. FIRE model adopt the newly function as the weight of evaluation and give more weight to new assessment.

The framework of loyalty-based trust algorithm is shown as follows:

Algorithm 1 Loyalty-based Trust

Input: Master mA's and mB's historic score about slave

Output: The final evaluation on trust master m gives to slave s.

Initialization, define variables ε_1, ε_2 ; $DTG_{(m,s)}=0$; $Rp_{(m,s)}=0$;

Calculate weight according FIRE model: $W_{Re}(r_k)=e^{-\Delta t(r_k)/\lambda}$ (1)

Node's(s) expectation of the kth interaction: $P_k(m,s)=\dfrac{S_{ms}+1}{S_{ms}+F_{ms}+1}$ (2)

if (recommended trust is existed)

{Calculate the recommended trust $Tt^{k+1}_{(mB,mA)}=(1-\gamma)Tr^{k}_{(mB,mA)}+\gamma\theta$ (3)

if ($I_k(m,s)/I_c(m,s)\ge 1$) $W_{Loy}(r_k)=\dfrac{P_k(mA,s)I_k(mA,s)}{I_c(mB,s)}$ (4)

Else $W_{Loy}(r_k)=1$;

Calculate evaluation of weighting function in the recommended trust:

$W_w(r_k)=Tr(mB,mA)W_{Re}(r_k)W_{Loy}(r_k)$ (5)

Final recommended trust: $Rp_{(mB,s)}=\dfrac{\sum_{r_k\in R_w(mA,s)}W_w(r_k)DTG_{(mA,s)}}{\sum_{r_k\in R_w(mA,s)}W_w(r_k)}$ } (6)

else if(direct trust is existed)

{ if ($I_k(m,s)/I_c(m,s)\ge 1$) $W_{Loy}(r_k)=\dfrac{P_k(mA,s)I_k(mA,s)}{I_c(mB,s)}$;Else $W_{Loy}(r_k)=1$;

Calculate evaluation of weighting function in recommended trust: $W_D(r_k)=W_{Re}(r_k)W_{Loy}(r_k)$ (7)

Final direct trust: $DTG_{(m,s)}=\dfrac{\sum_{r_k\in R_D(m,s)}W_D(r_k)v_k}{\sum_{r_k\in R_D(m,s)}W_D(r_k)}$ } (8)

Based on the direct trust and recommended trust, the ultimate trust of master to salve is determined: $TG_{(m,s)}=\varepsilon_1 DTG_{(m,s)}+\varepsilon_2 Rp_{(m,s)}$ (9)

In the final decision of trust level master to slave, $\varepsilon_1+\varepsilon_2=1, \varepsilon_1>0, \varepsilon_2>0$。

Table 1 Loyalty-based trust algorithm parameters

Parameter	Meaning
ε_1	The proportion of direct trust of overall trust
ε_2	The proportion of recommended trust of overall trust
$DTG_{(m,s)}$	Master m's direct trust to slave s
$Rp_{(m,s)}$	Master m's recommended trust to slave s
$W_{Re}(r_k)$	Evaluation of FIRE model
$\Delta t(r_k)$	the difference between current and evaluation recording time in FIRE model
λ	Parameter about time in FIRE model
$P_k(m,s)$	Node s's expectation of the kth future interaction
S_{ms}	The number of successful times in the previous k-1 transactions
F_{ms}	The number of failure times in the previous k-1 transactions
$I_k(m,s)$	The importance of kth transactions
$I_c(m,s)$	The importance of current transactions
$Tr_{(mB,mA)}$	The recommended trust of master MA
γ	master mB's study factor of recommended trust to master mA,the value depends on the number of interactions between mA and mB.
θ	evaluation of similarity coefficient between master mA and master mB

4 Simulation and Experiment

As a part of cloud environment can be viewed as a special grid, we adopt Gridsim to simulate a cloud computing environment to check the operation of the algorithm. We take the experiment about the malicious behavior-conspiracy. After simulation, the result is shown in figure 4.The method proposed in the paper is provided with the mechanism to discriminate nodes with malicious behavior.

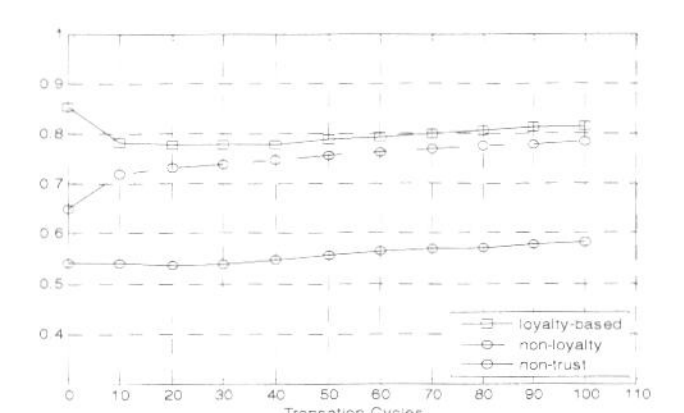

Fig. 4 Owing to the influence of malicious behavior, the successful interaction rate declines in the beginning. While with the increasing of number of interactions, the assessing nodes can play an active inhibition for target nodes on malicious behaviors. So in the latter part of the interaction, a high rate of successful

5 Conclusion

Cloud computing is a distributed computing system built on top of cheap and unreliable hardware. The unreliability of hardware requires that upper software should provide high reliability assurance. By introducing loyalty-based trust mechanism into cloud computing, the paper proposes a new resource scheduling model. The model gives assessment according to the real-time condition of the system and then allocates resources based on the assessment. This kind of dynamic feedback mechanism guarantees the stability of the system and reliability of the services effectively.

Acknowledgments. The work was supported by National Program on Key Basic Research Project (973 Program) of P.R. China (Grant No. 2010CB33470), National Natural Science Foundation of P.R. China (Grant No. 60973160), Project of Chinese Ministry of Education and Chongqing Education Committee(Grant No. 209101 and Kjzh10206), and Natural Science Foundation of CQ CSTC (Grant No. 2009BA2024).

References

1. Sun, R.F., Zhao, Z.W.: Resource scheduling strategy baed on cloud computing. Aeronautical Computing Technique 5(3) (2010) (in Chinese)
2. Dean, J., Ghemawat, S.: MapReduce:simplied data processing on large clusters. In: Proceedings of OSDI 2004: 6th Symposium on Operating System Design and Implemention, San Francisco, CA (December 2004)
3. Xie, J., Yin, S.: Improving MapReduce performance through data placement in heterogeneous hadoop clusters. In: 2010 IEEE International Symposium on Parallel & Distributed Processing, Workshops and Phd Forum, pp. 19–23 (April 2010)
4. Wang, F., Wang, R.C.: Trust evaluation on the base of Gird environment. Computer and Digital Engineering 36(10), 129–134 (2008) (in Chinese)
5. Huynh, T.D., Jennings, N.R.: An integrated trust and reputation model for open multi-agent systems. Journal of AAMAS 13(2), 119–154 (2006)

Power Optimization of Parallel Storage System for Applications with Checkpointing

Yong Dong

Abstract. The power optimization of massive storage system is an important part of the whole high performance computing system power optimization. Checkpoint is an important method keeping system available. In this paper, three different power states are defined for optimizing the power of storage servers. After finishing checkpoint, computing nodes send an instruction to storage server to change its power state. If storage server accommodates several checkpoint image files, it receives several power instructions at the same time which may conflict. State set is used to avoid such conflict and make them work properly. We use power meter to evaluate the effectiveness of power optimization.

1 Introduction

Too high power has been a serious problem of High Performance Computing (HPC) systems. The optimization for storage system power is significant for reducing the power consumption of the whole HPC system.

Checkpoint is an important method to improve system's fault tolerance. It creates separate image file for each process with huge amount of data. These write operations increase the power of storage server greatly. It is necessary and significant to optimize energy consumption of storage server. Checkpoint operations with interval make storage servers for image files not always in working state. When a storage server is idle, we set it to low power state.

The main contributions of this work include:

1. Propose a checkpoint based power state transformation method for optimizing the power consumption of storage servers.
2. Propose a state sets based conflict elimination method to solve the conflict from multiple conflicting power state scaling instructions
3. Verify the correctness and usefulness of our method by experiments.

Yong Dong
Computer School, National University of Defense Science and Technology
e-mail: yongdong@nudt.edu.cn

Z. Qian et al. (Eds.): Recent Advances in CSIE 2011, LNEE 125, pp. 239–244.
springerlink.com

The rest of this paper is organized as follows. Section 2 reviews the related works. Section 3 introduces checkpoint operation based power state transform method and conflict elimination method. Section 4 evaluates our method by experiments followed by conclusion in section 5.

2 Related Works

How to reduce the power of storage system has been a focus. In details, we divide such researches into different kinds to illustrate.1) Modeling the disk power. Effectively modeling the disk power is the basic of optimization, such as research in reference [5,9,10,13]. 2) Using Cache to reduce disk accesses. Qinbo Zhu et al. [14,15] used memory as Cache and proposed two Cache replacement algorithms to reduce power of storage system. MAID (Massive Array of Idle Disks) [2] and PDC [6](Popular Data Concentration) are also avoided the accesses disks through cache.3) Utilizing compiling techniques to obtain more idle time of disks [17,18]. 4) Using DRPM (Dynamic Rotations Per Minute) [7,16] technique to reduce the power of disks. 5) Using disk spin down (shutdown) technique. Disk spin down technique sets those disks which are unused for a long time to spin down state to reduce the energy consumption of disks [3,4,8,11,12,20].

3 Power Optimization of Parallel Storage System

Object storage is popular storage architecture and is widely used in many HPC systems. There are two kinds of storage servers in it: metadata server and object storage server. The metadata of a file is stored in metadata server as well as the content of that file is stored in object storage servers. Utilizing many object storage servers instead of one storage server improves the available storage bandwidth. In this paper, the parallel storage system is divided into two regions in which one of them mainly used for checkpoint data storage. As figure 1 shows, region 1 and region 2 are both composed by storage servers. Region 1 accommodates regular file meanwhile region 2 stores checkpoint images files which we optimize the power consumption of storage servers in it.

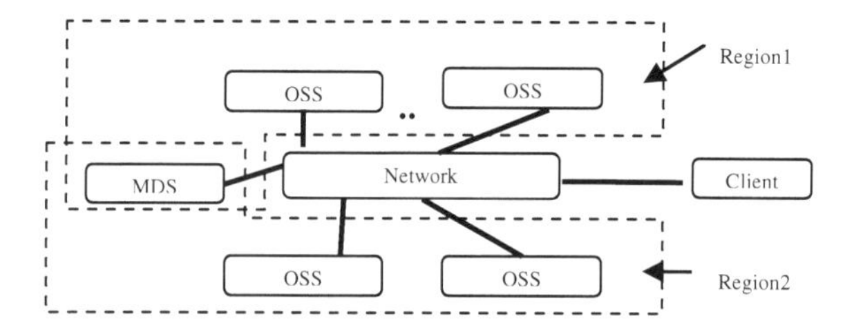

Fig. 1 Architecture of object storage system

The power optimization of storage server can be implemented by scaling processor's frequency and changing power state of storage devices. Modern processor supports different clock frequencies. The higher frequency brings about higher power consumption. Storage device includes a number of disks which are defined with different working states: active, idle and standby. Disk can be spun down for power saving with state transformation from active or idle to standby.

We define three states for storage server: S1, S2 and S3.

S1: It is working state for storage server. Processors in it work with the highest frequency, and storage devices are active for storage requests.

S2: There is no request being served on storage server. Processors and storage devices are all waiting for request.

S3: In this state, processors are scaled down with the lowest frequency and storage devices are spun down.

The values of power with these three states are defined as P1, P2 and P3 which meets P1>P2>P3. Storage server can switch among different states by scaling processor's frequency and storage device's state.

In HPC system, if a job start checkpoint operation, the checkpoint image files are created on storage servers and related data is stored. We modify checkpoint progress for power optimization. When computing node receives checkpoint signal, it create image file firstly. Before related data is written, we get information of the storage server which stores the image file. Next, we send power state transformation instruction to storage server to change its power state to S1. When storage server is changing power state, the computing nodes communicate for reaching consistent global state. After all data is written to image file, another power state transformation instruction is sent to the storage server to change its power state to S3. Storage server will work in low power state for power savings until next checkpoint request arrives.

When the application has a large number of processes or there are multiple checkpoint operations, there are a lot of image files. As the number of storage servers in parallel storage system is constant, one storage server serves several different files which may lead to the conflict of power state instructions, especially for different type of instructions. Assume there are two image files on the same storage server. When an image file is closed, storage server receives a power state scaling instruction to change power state from S1 to S3. But this storage server is storing data to the other file which need set power state to S1. These two power state scaling instructions are conflicting. We propose a state sets based conflict elimination method to solve the conflict from multiple conflicting power state scaling instructions.

Firstly, we construct a set according to power state scaling requests. Whether or not to execute current power state scaling instruction is determined by current set status. When one storage server receives a power state scaling instruction request R, it generates identification I according to the id of the job and the number of that process. For each storage server, all such identifications form a set G. Set G embodies the current working state of this storage server. When storage server receives power state transformation instruction, it make decision whether or not to respond it based on the set G. Figure 2 shows the whole decision process. Here, R_{normal} and R_{down} represent the request setting storage sever to normal power state and the request setting storage server to low-power state, respectively.

```
Storage server receive instruction R
Construct identification I according information from R
if (R is R_nomral)
then
          G=GU {I}
          if (storage server is in power state S3)
          then
                    Scaling power state of storage server to S1
                    Change the power state from S3 to S1
          fi
fi
if (R is R_down)
then
          G=G-{I}
     if (G=Φ)
          then
                    Scaling power state of storage server to S3
                    Change the power state to S3
fi
```

Fig. 2 Flow of conflict elimination method

4 Experiment

The parallel computing system we use in our experiment is composed by 32 computing nodes and 9 storage servers. There are two Intel Xeon 5355 processors in each computing node with 32GB memory. We take one storage server as metadata server, and the other eight ones as data storage servers. In each data storage server, there is one Adaptec 1220 RAID card connected with seven SATA disks. Computing nodes and storage servers are communicated with each other by Infiniband. The parallel file system we use is Lustre. We use YOKOGAWA WT1600 for power measurement which is a high-precision, wide-bandwidth power meter. The benchmark we use is FT and LU in NPB with D class and different number of processes. Table 1 shows the details.

Table 1 Details of benchmarks

Program	Scale	Processes	Size image files
LU	D	8	11.2GB
FT	D	256	112GB

Figure 3 shows power of storage server with program LU. From this Figure, we can see that without the method we proposed, the average power of storage server is about 360W in idle. When computing nodes start writing data, the power of storage server is up to 420W. Figure 3(b) shows that with power optimization method, the average power of storage server reduces to about 310W when there is no data writing, which makes average power saving of 13.8%.

Figure 4 shows the power variety of storage server when there are several image files. It uses FT with data class D. Because there are large amount of data, the time of higher power is much longer. With power optimization method, the average power of storage server is much lower. Similar with figure 3, the average power saving is also almost 14% when storage server is idle.

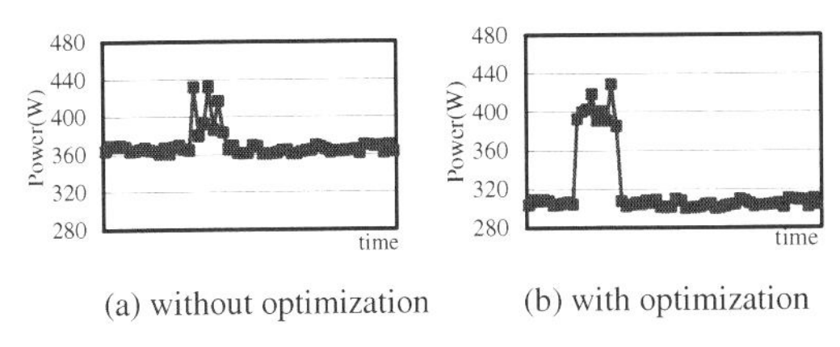

(a) without optimization (b) with optimization

Fig. 3 Power change of storage server with one image file

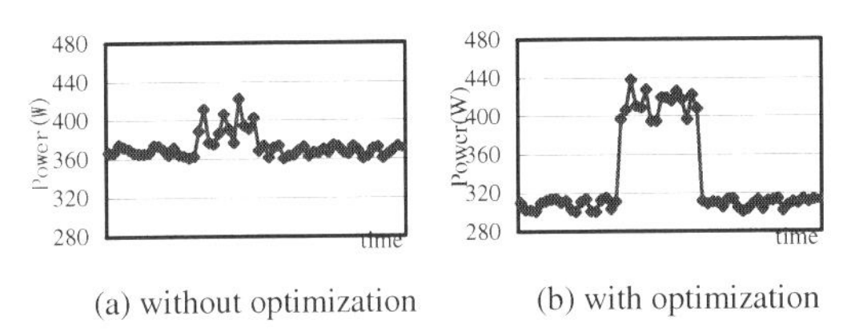

(a) without optimization (b) with optimization

Fig. 4 Power change of storage server with several image files

5 Conclusions

This paper proposes power optimization method for applications with checkpointing It takes advantage of idle time when there is no data writing to change storage server to low power state. In our experiment, we measure power variation of storage server by power meter. Experiment results show that our method is effective which makes power saving of storage server is 13.8% when there is no checkpoint date writing.

Acknowledgments. This work is supported by the National Natural Science Foundation of China under grant No. 60903044.

References

[1] Khargharia, B., Hariri, S., Szidarovszky, F., Houri, M., El-Rewini, H., Khan, S.U., Ahmad, I., Yousif, M.S.: Autonomic Power and Performance Management for Large-Scale Data Centers. In: 21th International Parallel and Distributed Processing Symposium (IPDPS 2007), pp. 1–8. IEEE, Long Beach (2007)

[2] Colarelli, D., Grunwald, D.: Massive Arrays of Idle Disks for storage archives. In: The 2002 ACM/IEEE Conference on Supercomputing (2002)

[3] Li, D., Cai, H., Yao, X., Wang, J.: Exploiting redundancy to construct energy-efficient, high-performance RAIDs. Department of Computer Science and Engineering University of Nebraska-Lincoln (2005)
[4] Li, D., Wang, J., Varman, P.: Conserving Energy in Conventional Disk based RAID Systems. In: The 3rd International Workshop on Storage Network Architecture and Parallel I/Os 2005, Saint Louis, Missouri (2005)
[5] Helmbold, D.P., Long, D.D.E., Sconyers, T.L., Sherrod, B.: Adaptive disk spin-down for mobile computers. Mobile Networks and Applications 5(4), 285–297 (2000)
[6] Pinheiro, E., Bianchini, R.: Energy Conservation Techniques for Disk Array-Based Servers. In: The 17th Interantional Conference on Supercomputing, pp. 66–78 (2004)
[7] Carrera, E.V., Pinheiro, E., Bianchini, R.: Conserving Disk Energy in Network Servers. In: The 17th International Conference on Supercomputing, pp. 86–97 (2003)
[8] Douglis, F., Krishnan, P.: Adaptive Disk Spin-Down Policies for Mobile Computers. Computing Systems 8(4), 381–483 (1995)
[9] Ganger, G., Worthington, B., Patt, Y.: The DiskSim Simulation Environment Version 2.0 Reference Manual, `http://www.ece.cmu.edu/ganger/disksim/`
[10] Zedlewski, J., Sobti, S., Garg, N., Krishnamurthy, A., Wang, R.: Modeling hard-disk power consumption. In: The 2nd USENIX Conference on File and Storage Technologies (2002)
[11] Li, K., Kumpf, R., Horton, P., Anderson, T.: Quantitative Analysis of Disk Drive Power Management in Portable Computers. In: The USENIX Winter Conference, pp. 279–291 (1994)
[12] Ca, L., Lu, Y.H.: Joint Power Management of Memory and Disk. In: Proceedings of the Conference on Design, Automation and Test in Europe, vol. 1, pp. 86–91. IEEE Computer Society (2005)
[13] Greenawalt, P.: Modeling power management for hard disks. In: The Conference on Modeling, Analysis, and Simulation of Computer and Telecommunication Systems (1994)
[14] Zhu, Q., Shankar, A., Zhou, Y.: PB-LRU: A Self-Tuning Power Aware Storage Cache Replacement Algorithm for Conserving Disk Energy. In: The 18th Annual International Conference on Supercomputing, ICS 2004. ACM, Saint Malo (2004)
[15] Zhu, Q., David, F.M., Devaraj, C.F., Li, Z., Zhou, Y., Cao, P.: Reducing energy consumption of disk storage using power-aware cache management. In: 10th International Symposium on High Performance Computer Architecture (2004)
[16] Gurumurthi, S., Sivasubramaniam, A., Kandemir, M., Franke, H.: DRPM: Dynamic Speed Control for Power Management in Server Class Disks. In: 30th Annual International Symposium on Computer Architecture (ISCA 2003), San Diego, California, pp. 169–180 (2003)
[17] Son, S.W., Chen, G., Kandemir, M.: Disk Layout Optimization for Reducing Energy Consumption. In: ICS 2005. ACM, Boston (2005)
[18] Son, S.W., Kandemir, M., Choudhary, A.: Software-Directed Disk Power Management for Scientific Applications. In: The 19th International Parallel and Distributed Processing Symposium (2005)
[19] Top 500 list, `http://www.top500.org`
[20] Lu, Y.H., Chung, E.Y., Simunic, T., Benini, L., Micheli, G.: Quantitative Comparison of Power Management Algorithms. In: The Design Automation and Test in Europe, DATE (2000)
[21] Yokogawa Electric Corporation. WT1600 Digital Power Meter, `http://tmi.yokogawa.com/products/digital-power-analyzers`

Research on Type-Safety Parallel Update

Zhang Shi and Jiang Jian-Min

Abstract. In distributed system, interaction makes program harder to do update dynamically. In this paper, a parallel update calculus with multi-version classes is established with the goal of understanding the underlying foundations of update parallel program, for the purpose of understanding how to best build reliable parallel updatable programs. The calculus formulates as an extension of calculus FJ (Featherweight Java) with multi-versions class, and extension of supporting communication between processes. This calculus can be used as the base of parallel updatable program.

1 Introduction

Dynamic software updating (DSU) enables running programs to be updated with new codes and data without interrupting their executions. But it is hard to realize in parallel system. In parallel system, interaction introduces many uncertainties while updating. What happens when updating, and how to guarantee type safety during updating? In this paper, a formal approach should be developed to set a foundation for implementers of parallel DSU. We introduce a multi-versions classes update calculus for Java parallel software. In this calculus, classes with different version can coexist, which permit program interacting with others regardless of updating simultaneously. The calculus is formulated as an extension of FJ (Featherweight Java)[1] with multi-versions classes.

2 A Calculus of DSU

In this section, a formal calculus for parallel update with multi-versions classes is defined. Syntax, semantic and type system of the calculus should be described. We also focus on some properties of the calculus at the end of the section.

Zhang Shi · Jiang Jian-Min
Dept. of Computer Science and Engineering, Fujian Normal University, China
e-mail: {shi,jjm}@fjnu.edu.cn

Z. Qian et al. (Eds.): Recent Advances in CSIE 2011, LNEE 125, pp. 245–251.
springerlink.com

2.1 Syntax

We formalize our results on the base of FJ, through extending with an update operator, multi-versions classes and two special methods for communication in parallel system. Syntax is defined in the following.

$v ::= \text{new } C(\overline{v}) \mid \text{new } C(\,)$ $\qquad$ $L ::= \text{class } C \text{ extends } D\ \{\overline{C}\ \overline{f};\ K\ \overline{M}\}$

$K ::= C(\overline{C}\ \overline{f})\{\text{super}(\overline{f});\ \text{this}.\overline{f} = \overline{f};\}$ $\qquad$ $M ::= C\ m(\overline{C}\ \overline{x})\ \{\text{return } e;\}$

$e ::= x \mid e.f \mid e.m(\overline{e}) \mid \text{new } C(\overline{e}) \mid (C)\ e \mid e.\text{send}(e,e) \mid e.\text{accept}(e,x) \mid e.\text{update}(C,L)$

$CT = CT \cup \{(C^n, \text{class } C \text{ extends } D^m\{\ldots\})\}$

$P ::= (CT,e)$ $\qquad$ $PS ::= P \mid PS|P$

The meta-variables A,B,C,D and E range over class names, v over values, f and g over field names, m over method names, x over variables, d and e over expressions, L over declarations, K over constructor declarations which can have different arguments, M over method declarations. We write $\overline{f}$ as shorthand for possibly empty sequence $f_1,\ldots,f_n$ (and similar to $\overline{M}$), $\overline{C}$ as shorthand for possibly empty sequence $C_1^{n1},\ldots,C_n^{nk}$. C^m means class C with version number m. If we write C without version number in typing rules, it represents class C with any version.

Send and Accept are for communicating. Send and Accept require two parameters, and the first one for channel. update(C, L) represents dynamic update class C with new declaration L in running program. New declaration of class C will be added to CT through adding version number, and it needn't waste time on updating old object to conforming new declaration.

Class table CT is a mapping from C^n, a class name with version number, to class declarations L'. L' represents class declaration in running program. A program is a pair (CT,e) of class table and expression. PS is a set of programs running in parallel system. Expression e can be object creation, method invocation, field access, casting, variable, send method, accept method and class update.

For typing and reduction rules, some auxiliary definitions are needed, which are show in the following.

Class declaration lookup: $\qquad CT(C^n) = \bullet \quad \text{if } \forall n (C^n, L') \notin CT$

$CT(C^n) = L' \quad \text{if } (C^n, L') \in CT \qquad CT(C) = CT(C^m) \quad m = \text{maxversion}(CT, C)$

Method MaxVersion: $\text{maxversion}(CT,C) = \max\{n \mid C^n \in CT\}$

Field lookup: (with $CT(C^n) = L'$)

$$\textit{field}(\text{Object}) = \bullet \qquad \frac{L' = \text{class } C \text{ extends } D^m\ \{\overline{C}\ \overline{f};\ \overline{K}\ \overline{M}\} \quad \textit{fields}(D^m) = \overline{D}\ \overline{g}}{\textit{fields}(C^n) = \overline{D}\ \overline{g},\ \overline{C}\ \overline{f}}$$

Method lookup: (with $CT(C^n) = L'$)

$$\textit{mset}(\text{object}) = \{\text{update}\} \qquad \frac{L' = \text{class } C \text{ extends } D^m\ \{\overline{C}\ \overline{f};\ \overline{K}\ \overline{M}\}}{\textit{mset}(C^n) = \overline{M} \cup \textit{mset}(D^m)}$$

Method type lookup:

$$\frac{L=\text{class C extends D } \{\bar{C}\ \bar{f};\ \bar{K}\ \bar{M}\}\quad B\ m(\bar{B}\ \bar{x})\{...\}\in \bar{M}}{\mathit{mtype}(m,C)=\bar{B}\rightarrow B}$$

$$\frac{L=\text{class C extends D } \{\bar{C}\ \bar{f};\ \bar{K}\ \bar{M}\}\quad m\notin \bar{M}}{\mathit{mtype}(m,C)=\mathit{mtype}(m,D)}$$

Method body lookup: (with $CT(C^n)=L'$)

$$\frac{L'=\text{class C extends } D^m\ \{\bar{C}\ \bar{f};\ \bar{K}\ \bar{M}\}\quad B\ m(\bar{B}\ \bar{x})\{...\}\in \bar{M}}{\mathit{mbody}(m,C^n)=\bar{x}.e}$$

$$\frac{L'=\text{class C extends } D^m\ \{\bar{C}\ \bar{f};\ \bar{K}\ \bar{M}\}\quad m\notin \bar{M}}{\mathit{mbody}(m,C^n)=\mathit{mbody}(m,D^m)}$$

Class declaration lookup: If class C with version number n is defined in CT, then $CT(C^n)$ returns declaration of C^n, otherwise it return null. If class doesn't specify a version number, then CT(C) returns last declaration of C.

As to other definitions about lookup, it first looks up definition in class C. If no exist, it looks up in its parent class.

2.2 Computational Semantics

The reduction rules include congruence and computations. They are shown as following. We briefly comment on some computations in the definition.

Congruence:

If P and Q are variants of alpha-conversion then $P\equiv Q$.

RP-COMMUTATIVITY $P|Q\equiv Q|P$ RP-ASSOCIATIVITY $(P|Q)|R\equiv P(Q|R)$

RP-PARALLEL $\dfrac{P\rightarrow P'}{P|Q\rightarrow P'|Q}$ RC-FIELD $\dfrac{e_0\rightarrow e_0'}{e_0.f\rightarrow e_0'.f}$

RE-INVK-ARG $\dfrac{e_i\rightarrow e_i'}{\text{new } C_0(\bar{v}).m(v_1,...,v_{i-1},e_i,...)\rightarrow \text{new } C_0(\bar{v}).m(v_1,...,v_{i-1},e_i',...)}$

RC-NEW-ARG $\dfrac{e_i\rightarrow e_i'}{\text{new } C(v_1,...,v_{i-1},e_i,...)\rightarrow \text{new } C(v_1,...,v_{i-1},e_i',...)}$

RC-CAST $\dfrac{e_0\rightarrow e_0'}{(C)e_0\rightarrow (C)e_0'}$ RC-EXPRESSION $\dfrac{e_0\rightarrow e_0'}{(CT,e_0)\rightarrow (CT,e_0')}$

RC-UPDATE $\dfrac{e_0\rightarrow e_0'}{e_0.UPDATE(C,L)\rightarrow e_0'.UPDATE(C,L)}$

Reduction relation is of the form P→P', read "program P reduce to program P' in one step." We write →* for the reflexive and transitive closure of →. In most of the situations, CT doesn't change during reduction. So we abbreviate P→P' to e→e' excepting RC-EXPRESSION and R-UPDATE. Through combining with RC-EXPRESSION with e→e', P→P' can be achieved. Send and accept have the same reducing rules with other methods.

RP-COMMUTATIVITY and RP-ASSOCIATIVITY indicates parallel programs agree with commutative and associative.

RP-PARALLEL: Program P can reduce to P' without affecting other programs.

Computation:

$$\text{R-FIELD} \quad \frac{fields(C) = \bar{C}\ \bar{f}}{(\text{new } C(\bar{v})).f_i \rightarrow e_i} \qquad \text{R-CAST} \quad \frac{C <: D}{(D)\ (\text{new } C(\bar{v})) \rightarrow \text{new } C(\bar{v})}$$

$$\text{R-INVK} \quad \frac{mbody(m,C) = \bar{x}.e_0}{(\text{new } C(\bar{v})).m(\bar{v}) \rightarrow [\bar{v}/\bar{x}, \text{new } C(\bar{v})/\text{this}]e_0}$$

$$\text{R-UPDATE} \quad \frac{\begin{array}{c}L\text{=class } C \text{ extends } D\{...\}\\ L \text{ UPDATABLE TO } CT(C) \quad n>maxversion(CT,C)\end{array}}{\begin{array}{c}(CT, \text{new } C_0(\bar{v}).update(C,L)) \rightarrow (CT \cup (C^n, L'), \text{new } C_0(\bar{v}))\\ L'\text{=class } C \text{ extends } D^m\{...\} \text{ with } m=maxversion(CT,D)\end{array}}$$

$$\text{R-COM} \quad \frac{\begin{array}{c}P \xrightarrow{send(new\ C(\bar{v}_1),new\ D(\bar{v}_2))} P' \quad Q \xrightarrow{accept(new\ C'(\bar{v}_3),x)} Q'\\ (CPORT)\ \text{new } C(\bar{v}_1)\text{=}(CPORT)\ \text{new } C'(\bar{v}_3) \quad P|Q \text{ COM OK}\end{array}}{P|Q \rightarrow P'|[\text{new } D(\bar{v}_2)/x]Q'}$$

R-FIELD. When class C has field f, the field access operator reduces fields to its expression,. It also require "new C($\bar{e}$)" be reduced to "new C($\bar{v}$)" first.

R-CAST. Casts an object to a special class, classes must conform to C<:D.

R-INVK. This reduction rule reduces method invocation to its definition expression, and it replaces parameter variables with parameter values and an object.

R-UPDATE. During updating class, it adds L', new declaration of C, into CT, and assign version number m to class C with m>maxversion(CT,C). L' treats the newest declaration of D as its superclass. Value "new C($\bar{v}$)" will not make any change in the form. As to L, it must be UPDATABLE TO last declaration of C.

R-COM. Communicating reduction requires that program P can be reduced to P' through send operator and programs Q can be reduced to Q' using accept operator. Send and accept must have the same channel, that is $(CPORT)\ \text{new } C(\bar{v}_1)\text{=}(CPORT)\ \text{new } C'(\bar{v}_3)$, and also requires P|Q COM OK which defined in typing judgment. After communication, variable x in Q will be replace by value "new C($\bar{v}_2$)".

2.3 Type System

The Typing rules for expression, method declarations, class declarations and updating are in the following.

Subtyping: $C <: C \quad \frac{C <: D \quad D <: E}{C <: E} \quad \frac{\text{class } C \text{ extends } D\{...\}}{C <: D}$

Typing: Here Γ range over partial functions from identifier x to type T, and CT range over partial functions from class names C to class declarations. CPort is a special type used to represent channel.

$$\text{T-VAR}\quad \Gamma \vdash (CT,x):\Gamma(x) \qquad \text{T-FIELD}\ \frac{\Gamma \vdash (CT,e_0):C_0^{\ n} \quad fields(C_0^{\ n}) = \bar{C}\ \bar{f}}{\Gamma \vdash (CT,e_0.f_i):C_i^{\ m}}$$

$$\text{T-INVK}\ \frac{\Gamma \vdash (CT,e_0):C_0^{\ n} \quad mtype(m,C_0^{\ n}) = \bar{D} \rightarrow C^m \quad \Gamma \vdash (CT,\bar{e}):\bar{C} \quad \bar{C}<:\bar{D}}{\Gamma \vdash (CT,e_0.m(\bar{e})):C^m}$$

$$\text{T-NEW}\ \frac{fields(C^n) = \bar{D}\ \bar{f} \quad \Gamma \vdash (CT,\bar{e}):\bar{C} \quad \bar{C}<:\bar{D} \quad n=maxversion(CT,C)}{\Gamma \vdash (CT,new\ C(\bar{e})):C^n}$$

$$\text{T-UCAST}\quad \frac{\Gamma \vdash (CT,e_0):D^n \quad D<:C}{\Gamma \vdash (CT,(C)e_0):C^m \quad m=maxversion(CT,C)}$$

$$\text{T-DCAST}\quad \frac{\Gamma \vdash (CT,e_0):D^n \quad C<:D \quad C \neq D}{\Gamma \vdash (CT,(C)e_0):C^m \quad m=maxversion(CT,C)}$$

$$\text{T-UPDATE}\quad \frac{\Gamma \vdash (CT,new\ C_0(\bar{v}).e_0):D^m \quad L\ UPDATABLE\ TO\ CT(C)}{\Gamma \vdash (CT,new\ C_0(\bar{v}).update(C,L).e_0):D^n \quad n=maxversion(CT,D)}$$

$$\text{T-SEND}\quad \frac{\begin{array}{c}\Gamma \vdash e_0:C_0^{\ m} \quad mtype(send,C_0) = (C_1,C_2) \rightarrow C^n \\ \Gamma \vdash e_1:C_1^{\ n1} \quad \Gamma \vdash e_2:C_2^{\ n2} \quad C_1^{\ n1}<:CPORT\end{array}}{\Gamma \vdash e_0.send(e_1,e_2):C^n}$$

$$\text{T-ACCEPT}\quad \frac{\begin{array}{c}\Gamma \vdash e_0:C_0^{\ m} \quad mtype(accept,C_0) = (C_1,C_2) \rightarrow C^n \\ \Gamma \vdash e_1:C_1^{\ n1} \quad \Gamma \vdash x:C_2^{\ n2} \quad C_1^{\ n1}<:CPORT\end{array}}{\Gamma \vdash e_0.send(e_1,x):C^n}$$

An environment Γ is a finite mapping from variables to types, written “x:C“. The typing judgment for expression has the form “ $\Gamma \vdash (CT,e):C$ “. The typing rules are syntax directed, with one rule for each form of expression, except T-CAST. In the following, we explain some typing rules.

T-UPDATE. This typing rule requests class declaration L UPDATABLE TO CT(C). if C=D, Then the program’s type will be D^n with n=maxversion(CT, D). If it’s UPDATABLE and $C \neq D$, then the program’s type will not change.

T-SEND. In order to send object, it require send method has type $(C_1,C_2)\rightarrow C^n$, e_0 has type $C_0^{\ m}$, two parameters have type C_1 and C_2. It also requires $C_1^{\ n1}$ is the subclass of CPORT. The result is of type C^n. Classes with different version have the same inherit relation. So for T-SEND, method’s arguments can have different versions.

T-ACCEPT. During accept object, it require accept method has type $(C_1,C_2)\rightarrow C^n$, e_0 has type $C_0^{\ m}$, two parameters have type C_1 and C_2. It also requires $C_1^{\ n1}$ is the subclass of CPORT. The result is of type C^n. Classes with different version have the same inherit relation. So for T-ACCEPT, method’s arguments can have different versions.

Typing judgments, which include Method typing, Class Typing, Updatable typing and communication typing, are shown as follows

$$\text{T-METHOD}\quad \frac{\begin{array}{c}\bar{x}:\bar{C},\ this:C^n \vdash e_0:E_0 \quad E_0<:C_0 \quad CT(C^n)=class\ C\ extends\ D^m\{...\} \\ \text{if } mtype(m,D^m)=\bar{D} \rightarrow D_0, \text{then } \bar{C}=\bar{D} \text{ and } C_0=D_0\end{array}}{C_0\ m(\bar{C}\ \bar{x})\{return\ e_0;\}\ OK\ IN\ C^n}$$

$$\text{T-CLASS}\quad \frac{\begin{array}{c}CT(C^m)=\text{class } C \text{ extends } D^n\{\overline{C}\ \overline{f};\ K\ \overline{M}\} \quad \mathit{fields}(D^n)=\overline{D}\ \overline{g} \\ K=C^m(\overline{D}\ \overline{g},\ \overline{C}\ \overline{f})\{\text{super}(\overline{g});\ \text{this}.\overline{f}=\overline{f};\} \quad \overline{M}\ \text{OK IN}\ C^m\end{array}}{\text{class } C \text{ extends } D^n\{\overline{C}\ \overline{f};\ K\ \overline{M}\}\ \text{OK}}$$

T-UPDATABLE

$$\frac{\begin{array}{c}L=\text{class } C \text{ extends } D\{\overline{C}_1\ \overline{f}_1, K\ \overline{M}_1\} \quad CT(C)=\text{class } C \text{ extends } D^m\{\overline{C}_2\ \overline{f}_2, K\ \overline{M}_2\} \\ \text{for all } f_i \in \overline{f}_2,\ \text{there exist } f_j \in \overline{f}_1 \ \text{ with } f_{2i}=f_{1j} \text{ and } C_{1i}<:C_{1j} \\ \text{for all } M_{2i} \in \overline{M}_2,\ \text{there exist } M_{1i} \in \overline{M}_1 \ \text{with same type and name}\end{array}}{L\ \text{UPDATABLE TO}\ CT(C)}$$

T-COM

$$\frac{\begin{array}{c}P=(CT,e_0.\text{send}(e_1,e_2)) \quad Q=(CT',e_3.\text{accept}(e_4,x)) \quad \Gamma_P \vdash (CT,e_1):C^{n1} \quad \Gamma_Q \vdash (CT,e_4):C'^{n2} \\ \Gamma_P \vdash (CT,e_2):D^m \quad C=C' \quad n1=n2 \quad D^m<:\Gamma_Q(x) \text{ or } (\Gamma_Q(x)=D^n \text{ and } n \geq m\ )\end{array}}{P|Q\ \text{COM OK}}$$

T-MATHOD. The typing judgment has the form M OK IN C^n. It uses the expression typing judgment on the body of the method, where free variables are the parameters of the method with their declared types. In case of overriding, if a method with the same name is declared in the superclass, then it must have the same type between them.

T-CLASS. The typing judgment has the form L' OK. It checks whether the constructor applies super to the fields of the superclass and initializes the fields declared in this class or not, and whether each method declaration in the class is ok or not.

T-UPDATABLE. This typing judgment requires that all of CT(C)'s fields $\overline{f}$ still exist in the fields of new declaration $\overline{g}$, and the corresponding type is subtype relation, that is $\overline{f} \in \overline{g}$ and $C_i<:D_i$. The judgment of updatable also requires that all methods exist in CT(C) must be exist in new declaration with the same return type and parameter type. The implementation of the method can be changed. Of course, the new class can add some new methods and fields into the L with name different with those existed.

T-COM. This judgments for communicate has the form "P|Q COM OK". It requires that they communicate with the same channel type, and variable x in accept function must be superclass or older version than D^m. When D^m isn't newer than $\Gamma_Q(x)$ or is the superclass of $\Gamma_Q(x)$, then they may cause error when x invoke methods or access fields that are not exist in D^m. Through this judgment, we can make sure that communication can be made in parallel system, and it will not cause errors in program Q.

2.4 Properties

In this section, type soundness is shown. We prove such a result: if the program is well typed and it reduces to a normal form, and the updating conforms typing judgment defined above, then program result is either a value of subtype of the original expression's type, or an expression that get stuck at a downcast, or a value of class C with new declaration. Proves are not presented in this paper.

Theorem 1 (Subject Reduction). If $\Gamma \vdash (CT,e):C^m$ and (CT, e)→(CT', e') then $\Gamma \vdash (CT',e'):D^n$ for some D<:C or CT(C^n) UPDATABLE TO CT(C^m) with C=D and n>m.

Theorem 2 (Type Soundness). If $\varnothing \vdash (CT,e):C^m$ and $(CT,e) \vdash (CT',e')$ with e' a normal form, then e' is a value v with $\varnothing \vdash (CT,v):D$ and $D<:C$, or a value v with $\varnothing \vdash (CT,v):C^n$ and CT(C^n) UPDATABLE TO CT(C^m) and n>m, or an expression containing $(D)\ new\ C(\overline{e})$ where $C<:D$.

3 Related Works

There are some DSU researches related to parallel system. A formal framework for studying on-line software version change was described in [2]. It gives a general definition of validity of on-line sufficient conditions for ensuring validity for a procedural language. [4] represents an upgrade system that permit updating at class. It permits object of different versions coexist through inheriting from abstract interfaces that cannot change across versions. This constraint necessarily limits program evolution. [3] provides a simple formalizing mode for reasoning about dynamically updatable program.

4 Conclusions

In this paper we present a parallel update calculus with multi-versions classes for updating class dynamically. The calculus express communication between updatable programs without interrupting program's running. Multi-versions classes can coexist in the running programs.

References

1. Igarashi, A., Pierce, B., Wadler, P.: Featherweight Java: A Minimal Core Calculus for Java and GJ. In: Meissner, L. (ed.) OOPSLA 1999, pp. 132–146. Denver, CO (1999)
2. Gupta, D., Jalote, P., Barua, G.: A Formal Framework for On-line Software Version Change. IEEE Transactions on Software Engineering 22(2), 120–131 (1996)
3. Bierman, G., Hicks, M., Sewell, P., Stoyle, G.: Formalizing Dynamic Software Updating. In: Proceedings of the Second International Workshop on Unanticipated Software Evolution (USE) (April 2003)
4. Gilsi, H., Robert, G.: Dynamic C++ classes- A lightweight mechanism to update code in a running program. In: USENIX Annual Technical Conf. (1998)

The Research on Cloud Resource Pricing Strategies Based on Cournot Equilibrium

Bo Wang, Hao Li, Yuanyuan Miu, and Bing Kong

Abstract. The ultimate aim of cloud computing is to sharing resources and services, therefore, how to management and assign resources effectively become the most important part of cloud computing. The introduction of economics can be reflecting resource supply and demand dynamic change primly. This paper put forward on the basis of improving economic cloud calculation model based on the economic grid models, and develops the pricing strategy of bottom resource based on the Cournot equilibrium theory, so as to realize maximization of the resource participants' benefit.

Keywords: Cloud Bank, initial price, Cournot equilibrium.

1 Introduction

Cloud Computing (Cloud Computing) is a distributed processing (Distributed Computing), Parallel Processing (Parallel Computing) and Grid Computing (Grid Computing) development, or that these commercial implementations of the concept of computer science [1]. The core idea of cloud computing is virtualization, one or more virtual data center resources, the service provided to users in the form of rent. In the cloud, virtual physical resources, virtual architecture, and virtual middleware platform and business applications are based on service provision and use of the form [2]. Cloud computing resources together to form a unified data center to provide services outside, due to cloud resources, cost-effective, high reliability,

---

Bo Wang · Hao Li · Yuanyuan Miu · Bing Kong
School of Information Science and Engineering
Yunnan University
Kunming, China
e-mail: bcai.wang@foxmail.com

Z. Qian et al. (Eds.): Recent Advances in CSIE 2011, LNEE 125, pp. 253–259.
springerlink.com

application distribution, scalability, flexibility and other characteristics, the user's use of resources and resource availability are constantly changing. Therefore, the introduction of economic models to guide resource management and allocation of resources by the resource prices to reflect supply and demand changes.

Cloud computing business model common are IaaS, PaaS and SaaS, the contents of this paper is mainly achieved in IaaS mode. The current academic has made a variety of economic cloud computing model (Cloud value chain reference model[3], a decision model[4]), but the two models are for service level, not specifically related to the underlying resource pricing strategy. In this paper, the existing cloud computing resource management model is proposed based on a market-based mechanism for resource management approach to Cournot equilibrium as the theoretical basis to achieve a cloud computing model to quantify the underlying resources and pricing.

2 Cloud-Computing Model Based on the Economics of Research and Investigation

Cloud computing and distributed computing, grid computing has some similarities, are on the network through the integration of distributed resources to provide powerful computing ability. Cloud computing business model more suitable for industrial development, and therefore received broad industry support, can develop rapidly. Economic cloud model is currently involved are:

(1). Cloud value chain reference model [3]: the United States and Germany and some other scientists in the extensive analysis of existing business model cloud, cloud services, stakeholder relations, market allocation and the value of the structure of the basis, then presents the cloud of the value chain reference model. The proposed model can adapt to the cloud computing market a wide range of business services under the program.

(2). "Great cloud" BC1.0 [9]: developed by China Mobile and China's second session of the Conference on cloud computing released. The release of the "big cloud" BC1.0 has a distributed file system, distributed mass data storage, distributed computing framework, cluster management, cloud storage systems, flexible computing systems, parallel data mining tools and other key functions. And now China Mobile has built 1,000 servers, 5,000 CPU Core, 3000TB storage size of "big cloud" laboratory.

The cloud computing model to introduce economic concepts from different angles, and thus solve the problem of resource allocation and resource control. As cloud computing environments have a certain complexity of computing resources offered by a corresponding feature, so this bank in the original mesh [5] model, based on its improved and the introduction of the Cournot equilibrium theory, and further study modeling to quantify the economic cloud computing resources and pricing.

3 Based on the Underlying Economics of the Cloud Bank Pricing of Resources

In the cloud computing environment, resource providers are defined as producers. In order to attract consumers and resources to maximize the use of idle resources, they will provide a competitive service strategy; application requests the user is defined as the consumers, in their needs within the time limit, hoping to use the least resources to meet the cost to service needs. So supply and demand balance is the most basic problem to be solved, the price response to resource information to a large extent, the strategy is based on the reasonable price of supply and demand.

Resource providers such as CPU, memory, storage, and other physical resources, to complete a specified service (Figure 1). Resources to provide the difference between the resources and therefore to develop a specific evaluation strategy, also need to consider the interaction between resource providers, and ultimately give a preliminary pricing of the various resources and information together with the resource description record in the CIS. Through virtualization of physical resources to form different types of resources registered to the UDDI.

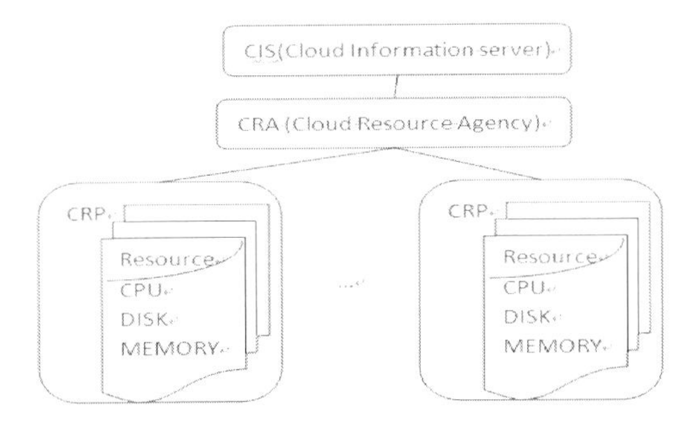

Fig. 1 Cloud Resource provider (CRP) logical structure

Resources to provide specific resources need to be an abstract description of the first, so the price is measured using specific resources. For each category of resources calculated according to the pricing strategy of the basic price of each resource. Because there is competition between resource providers the relationship, in order to maximize the interests of all resource providers to ensure that the premise of price equilibrium, the Cournot model is introduced to solve this problem.

3.1 Cournot Equilibrium Theory

Different economic model determines a different way of price determination of resources and services. In this model, based on the cloud bank, recorded in the CIS in the initial price of the underlying resources, resources provided by the interaction between the decision and has nothing to do with consumers. Monopoly characteristics of the model is only taking into account the resources of each party, the price of resources by the resource providers the interaction between the decision strategy between resource providers as to achieve price equilibrium. Allocation of resources under the equilibrium price is usually the most reasonable and effective.

Definition 1: Cournot equilibrium (Courtnot equilibrium) in a market economy conditions, the participants are known to each other's production, will bring them to determine their most profitable output, then output will not change any of the participants [8].

Cournot equilibrium solution is a Nash equilibrium of production. Nash equilibrium is considered to be in the interests of all participants, while maximizing self-interest is maximized. Therefore, according to Cournot equilibrium theory, in cloud computing environments to maximize each participant's own interests.

Definition 2: Cournot equilibrium in cloud computing cloud computing environment, resource providers are known to all participants in the amount of resources per unit time in the case, choose to make available resources in order to maximize their own interests.

Resources in order to provide resources to the greatest degree of use of resources between different interests are interrelated, the relationship between the game, the game's final result is that the participants achieve their own interests in various resource maximize at the same time, the optimal use of resources.

3.2 Pricing Resource Pools

Taking into account work already made, that is a measurement of each of the resources provided by a single resource prices. Based on the price, the introduction of the concept of equilibrium, and finally reach a preliminary resource prices. Traditional Cournot equilibrium only take into account the two players, so the cloud computing environments need to be extended in order to achieve maximum benefits to meet the purpose of multiple participants. This article first gives the following assumptions:

(1) cloud computing environment with multiple resource providers, each provider has multiple resources available for, and the new provider at any time and resources to join and exit.

(2) resource providers for market share among competing with each other, assuming that their mention in the known providers of other resources provided under the premise of the availability of resources to determine their own, so to maximize their own interests.

(3) resource providers is no difference between the products, but also that different consumers choose different resources get the same effect.

That the i-th number of resource providers of resources, the i-th individual resource providers the price of resources, total resources provided by the cost price of C, there are multiple participants in the case of multiple resource prices set is $C = \{C_1, C_{2,} \dots C_n\}$, the i-th resource cost price; the

$$C_i = A_i * cost\_price_i \quad (1)$$

In the cloud, the interaction between resource providers, so the basic price is based on the need to find resources for the initial price of P, P is a function of total resources, where A is total resources, $A = \sum_1^n A_i$,

Each resource provider i think they provide resources to the total impact of prices, the impact is:

$$\frac{\partial P}{\partial A_i} = \frac{dP}{dA} \cdot \frac{\partial A}{\partial A_i} = \frac{dP}{dA} \quad (2)$$

Resources are i number of resources, single resource price P (A), so the cost for the revenue function as:

$$\pi_i = P(A) * A_i - C_i \quad (3)$$

To make the resource provider i to maximize the benefits of (5) of the derivative to get

$$\frac{d\pi_i}{dA_i} = A_i * \frac{dP}{dA} + P(A) - \frac{dC_i}{dA_i} = 0 \quad (4)$$

That is simplified to,

$$A_i = \frac{\acute{C}_1 - P}{\acute{P}} \quad (5)$$

At this point (5) on the left of the resource providers that i have decided to provide the resources, then their interests are maximized. Affected by supply and demand as prices change, so i with each resource provider to provide the resources for. In (5), if known P (A) the specific function of the type, you can find the profit-maximizing resource provider i the amount of resources to be provided.

Taking into account the existence of competition between the various participants in the interaction between the first specific consideration of resource providers the relationship between i and j. In (5) where on the derivative, i and j drawn between the slope of reaction function, there are n resource providers, including $j \neq i$ and $j \in \{1,2, \dots, n\}$..After simplification was:

$$\frac{dA_i}{dA_j} = \frac{(P - \acute{C}_i)\ddot{P} - \acute{P}^2}{2\acute{P}^2 - \ddot{C}_i\dot{P} - (P - \acute{C}_i)\ddot{P}} \quad (8)$$

Result is a linear function on, so there, but the price function P (A) can not determine its form, the game process is the balance of M () or non-existent equilibrium. If P (A) is a linear function on, you can make $P(A) = \sum_1^n \alpha_i * A_i + K$, which is the balance parameter, the parameter K that price and the relationship between the amount of resources. There are $\ddot{P} = 0$, then (9) can be simplified into:

$$\frac{dA_i}{dA_j} = -\frac{\acute{P}^2}{2\acute{P}^2} = -\frac{1}{2} \quad (9)$$

At this point both the response function shown in Figure 2.

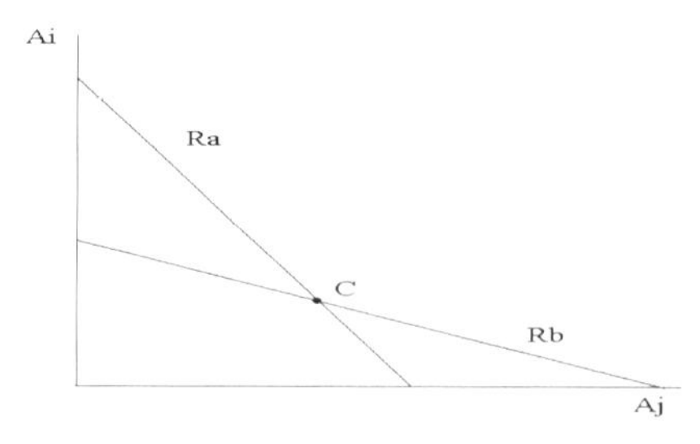

Fig. 2 Cournot equilibrium response curve

The first step is to solve the equilibrium assumption is symmetric equilibrium, so there are (), where each participant faced the same conditions.

At this point there is: $A_i = \frac{\acute{c}-P}{\acute{P}} = \frac{cost\_price_i - \sum_1^n \alpha_j * A_j + K}{\sum_1^n \alpha_j}$,

So this time, $A_i = \frac{cost\_price_i + K}{2\sum_1^n \alpha_j}$ into solving P (A) was $P(A) = \sum_1^n (\alpha_i * \frac{cost\_price_i + K}{2\sum_1^n \alpha_j}) + K$,.

4 Summary and Further Work

In this paper, the original mesh model based on the bank itself, based on characteristics of cloud computing cloud computing proposed to improve resource management and scheduling model. First, for each resource provider, considering the resources provided by the factors (CPU, disk, etc.), gives a basic price of resources solving methods; as cloud computing environment with multiple resource providers, between different resource providers there is competition, the goal is to maximize their own interests, so the final record in the CIS in resource prices also need to consider the environmental factors, which introduces a Cournot equilibrium, find the function of each participant's interests, to consider to the case of multiple participants, given the interests of the participants to maximize the resources should be provided when the amount of formula, and ultimately arrive at an initial price of resources.

But the cloud computing resources with the task matching the price to pay many of the issues or strategic issues such as open, we will model the existing cloud bank to further discussion.

Thanks

1. Supported by the National Nature Science Foundation of China 2010 (No.61063044).
2. Supported by Key laboratory in Software Engineering of Yunnan Province under Grant No.210KS05.

References

1. Zdnet, cloud computing and grid computing (October 14, 2010) (in Chinese), http://cloud.it168.co/a2010/0410/871/000000871762.shtml
2. Lenk, A., Klems, M., Nimis, J., et al.: What's inside the cloud? An architectural map of the cloud landscape. In: Proceedings of the 2009 ICSE Workshop on Software Engineering Challenges of Cloud Computing (CLOUD 2009), Vancouver, Canada, May 23 (2009)
3. Mohammed, A.B., Hwang, J.: Cloud Computing Value Chains: Understanding Businesses and Value Creation in the Cloud. In: Economic Models and Algorithms for Distributed Systems, pp. 187–208. Birkhaser Verlag, Basel (2009)
4. Lee, I.: A Model for Determining the Optimal Capacity Investment for Utility Computing. In: Economic Models and Algorithms for Distributed Systems, pp. 187–208. Birkhaser Verlag, Basel (2009)
5. Li, H., Tang, G., Lu, J., Yao, S.: The QoS Resource Quantification Based on the Grid Banking Model. In: Proceedings of the First International Conference on Networked Digital Technologies (NDT 2009) Ostrava, The Czech Republic, Ostrava, July 29-31, pp. 548–553 (2009) ISBN:978-1-4244-4615-5; (published by IEEE)
6. Buyya, R.: Economic-based distributed resource management and scheduling for Grid computing, in Thesis, Monash University (2002)
7. Wolski, R., Plank, J.S., Bryan, T., Brevik, J.: G-commerce: Market Formulations Controlling Resource Allocation on the Computational Grid. In: Prceedings of the 15th International Parallel & Distributed Processing Symposium, IEEE Computer Society, Washington, DC (2001)
8. Rasmusen, E.: Games and information. Blackwell Publishers (1997)
9. eNet, china mobile put fword BC1.0, http://www.enet.com.cn/article/2010/0521/A20100521657286.shtml (October 10, 2010)

Embedded Software Test Model Based on Hierarchical State Machine

Shunkun Yang, Guangwei Zhang, Qingpei Hu, and Hong Kong

Abstract. A test model based on hierarchical state machine is proposed for embedded software testing in this paper. This state machine model shares the same characteristics of object-oriented methodology, such as encapsulation, delegation, and polymorphism. With consideration of model-based embedded software testing, the states and events of the model are analyzed. Also, a test state machine is designed and implemented, illustrated with a practical application example.

Keywords: Embedded software, Software testing, Hierarchical state machine.

1 Introduction

Embedded software has been widely adopted in industrial applications, such as electronics, aviation, and spaceflight. For these critical applications, reliability is a basic requirement. If the software fails, terrible consequences will be caused, which would lead to huge loss of economic profit of human lives. To ensure the reliability of software, testing is always taken as the main practical approach. Also, testing for embedded software has drawn more and more attention for the critical roles they play in industrial systems.

Shunkun Yang
School of Reliability and System Engineering, Beihang University, Beijing, China
e-mail: ysk@buaa.edu.cn

Guangwei Zhang
Department of Global Business Services, IBM, Beijing, China

Qingpei Hu
Academy of Mathematics and Systems Science, Chinese Academy of Science, Beijing
e-mail: qingpeihu@amss.ac.cn

Hong Kong
Department of Quality, Safety, and Environment, Vestas Wind Systems, Beijing, China

Z. Qian et al. (Eds.): Recent Advances in CSIE 2011, LNEE 125, pp. 261–267.
springerlink.com

As a test strategy and method, state machine is widely used for test case generation and selection of test inputs. Finite state machines(FSMs) and extended finite state machines(FSMs) are commonly used in test sequence derivation [1~2]. The FSM model has been widely studied and many methods are available for the purpose if test data generation[3~4]. Generally, most embedded software could be tested in a way based on state machine, as they totally or partially present some behavior based on states.

Especially, model-driven embedded software test method replaces physical devices with models to input test data and to receive feedback data from the target system [5].

The motivation for our work is the observation that there are many state transitions and event handlings in these models. In order to manage these states and the whole testing process, hierarchical state machines (HSM) is proposed to construct the structure frame of test model. With HSM based testing model, test state could be configured and controlled according to test requirement in a convenient way, and also the test that based on state could also be carried out easily. In the following sections the state machine and state graph is firstly introduced. Then the HSM based software testing models is proposed, with comprehensive analysis of the model under testing environment. Also, a HSM is designed and implemented based on a test state machine for embedded software with incorporation of the theory of model driven software testing and state machine.

2 Test Model

Modeling is an effective approach for embedded software testing. Models pass test data to Software-Under-Test (SUT) and receive feedbacks through interfaces and related services to enable closed-loop test. The most important features of test model are interchangeability, real-time performance, and concurrency. Test model is the key point in many aspects, such as improving the construction efficiency of test environment and expanding the test ability. However, SUT and its external environment are so complex that it brings need for a more general structure, which could easily construct the test model and support code generation. Considering this, we introduce state machine theory, and combine it with test model frame to build a test state machine model for embedded software to satisfy these requirements.

In order to build our test state machine model, we firstly refer to Moore state machine: define all the states of test models and treat related actions as part of the state definition. This could make the structure of test state machine clearer, and also make it easier to test SUT through controlling the model state. Secondly, hierarchical state machine is used to abstract common behavior of some states to build the behavior of public states. This is intended to reduce the explosive growth in the number of states and transitions to describe more complex system through reuse.

2.1 Model States and Events Analysis

Many embedded systems totally or partially present behavior based on states. These behaviors have direct influence on state and performance of embedded software. Understanding the states of test model is the basis for automatic software system test. Table 1 list several states of test models that are common in a test process. Periodic state, event state, conditional state, and failure state are nested in model running state. All of them constitute a hierarchical nesting state in which the nested child states (periodic state, .etc.) reuse the behaviors of its father state. Once the differences from father state are defined, all the public events could be handled by father state. Then tasks could be added to child state, which enable the test model to implement all kinds of behaviors.

Table 1 State of test model

Model state		Detail description
Initial state		Model initialize variables
Waiting state		Model is waiting to receive event
Running state	Periodic state	Model is running periodic task
	Event state	Model is running event task
	Conditional state	Model is running conditional task
	Failure state	Model is running error task
Locking state		Malfunction happens to model , without any action
Terminal state		Model is ending, quit test process

During the test process of embedded software, things that could influence the change of test model state could be divided into two groups: interior event and exterior event. Most interior events are related to tasks, and they could transmit model to target state to run specific tasks. Interior events include model task, clock, and semaphore. Exterior events mainly mean the test instruction that is sent to test models. They control the tasks and states of models in the form of function-instructions. Exterior event is the main means for testers to control test model states through test state machine during test process. State transition of test model could be seen in Fig 1.

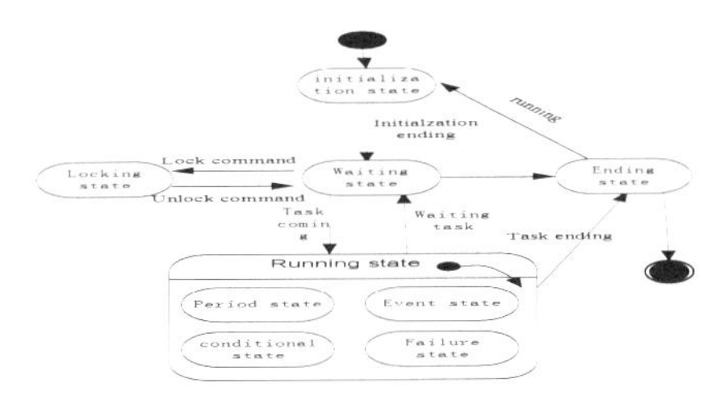

Fig. 1 State transitions of test model

2.2 Model Architecture

The test state machine model should not only implement model state scheduling, but also should satisfy the requirement of real-time performance, expansibility and robustness. When building test state machine model, the most important factor is real-time performance, because if the test input/output cannot meet the real-time requirement, the test results are meaningless. Expansibility could enable test state machine model to add states and events more easily and this is also the guarantee of the improvement of test efficiency and reusability of test model; Robustness should also be satisfied that it could response to events and finish state transitions timely when the explosion of model states happened and evade the dead situation of test state machine at some abnormal condition.

In order to make the structure of test state machine model clear and could function independent, Modular and hierarchical design method is used to design the structure of the test model which can also achieve good expansibility. Test state machine model is divided into several sub-modules. When some new states, tasks and events are needed to be added, the only thing needs to do is to add new definition into related sub-module; The scheduling module is in charge of the whole running process of test state machine model to ensure its real-time performance during the test process. On the other hand, in order to reduce the number of modules linked with state machine and exterior event, some specific interfaces are used to define the states and behaviors of models to enable testers to define and control the state and task of the test model.

The detail structure of the test state machine can be seen in Fig 2, including six modules: initial module, schedule module, register module, state module, event module, and task module. When a test process starts, initial module initializes the model variable and related resource to set the model at waiting state. When an event arrives, the event will be parsed by event module, and then the schedule module passes the event information to register module which send the target state of the event back to schedule module, schedule module distributes the target state to state module which is in charge of state transitions. Then, schedule module let task module to run related tasks.

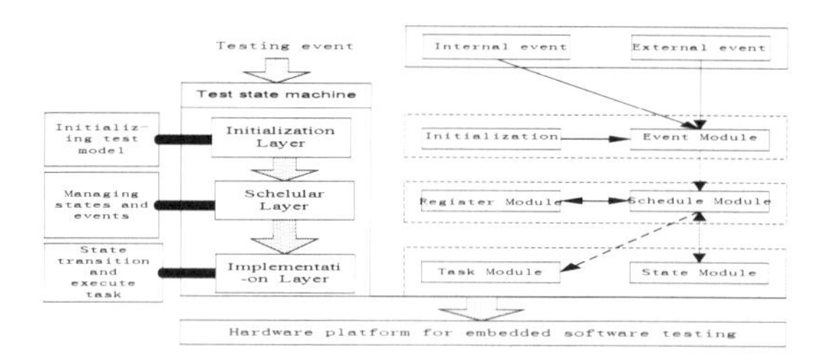

Fig. 2 Architecture of HSM based test model

2.3 Model Implementation

The core of the test state machine model is the schedule module (see Fig 3). The interfaces between schedule module and other modules are very simple, which could help to reduce coupling. The implementation of test state machine model is focusing on three aspects: 1) supporting behavior inherited in hierarchical state; 2) using state entry and exit action to realize guaranteed initialization and cleanup; 3) supporting specified state model through class inheritance.

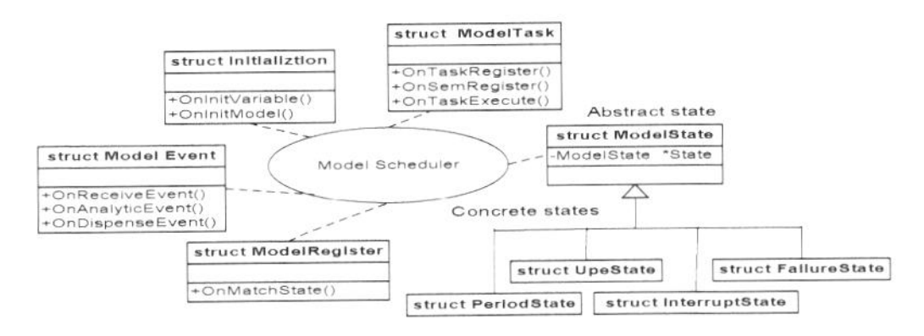

Fig. 3 Implement of test state machine

Test model is implemented by C language to guarantee its real time property. Just as the hierarchical state described, each state has its own behaviors or behaviors inherited from its father states. So how to use C to realize encapsulation, inheritance and polymorphism is the key problem. In fact every test model is a class with attributes and functions. Attributes could be defined by structure, while the functions could be called by the function pointer variable. Each function is a parameter that is represented by a structure pointer. There are many inherited relationships between child state and father state in hierarchical states, in which the purpose is to embed the attribute of father state into child state to enable the instantiation of child state could call the methods of father state.

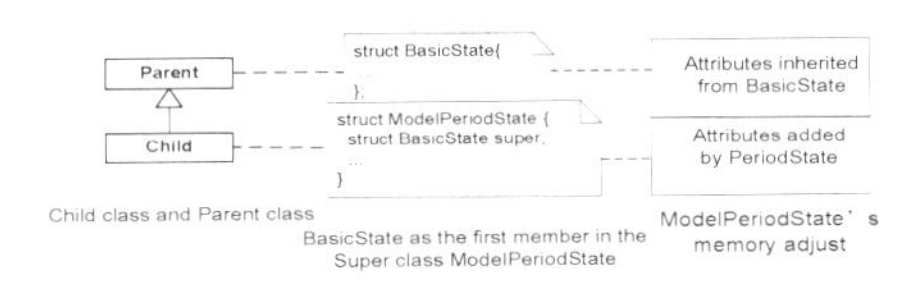

Fig. 4 Inheritance of test state machine

The method described in Fig 4 can only be applied in case of single inheritance, because the class which has more than one father could not use all the attributes of their father class. Inheritance can add more responsibilities to the function of constructor and de-constructor, because each instance of child state has an embedded instance of father state. The constructor of child state has to call the constructor of father state explicitly to initialize the controlling parts of father state. In order to avoid potential relativity, the constructor of father state should be called before all the attributes have been initialized.

On the base of inheritance implementation, polymorphism could also be realized. When child state inherited from father state, functions in different child states can be called by the pointer of father state, so according to the difference of public events, father state will handle them correspondingly. In the test state machine model, when model enters into the running state, the running state could not only handle public events, but also call the functions of child state through the information of schedule model to execute different actions of models to realize the polymorphism.

3 Application Examples

In order to validate the feasibility of the test state machine model, we applied it in a practical airplane flight control software testing project, and construct a test model to simulate the navigation system which interacting with the flight control system under the practical environments. The navigation system model has all the states we discussed before. After initializing, it stays at the waiting state waiting for the coming events; while in the running state, it keeps sending the periodic navigation data to flight control system according to the rest requirement. Fig 5 represents the state graph of navigation system under the control of test state machine.

Now, we show how the test state machine model will react when the navigation system fails during the software testing of flight control system, as shown in Fig 6. After the initializing process, the navigation system enters into waiting state. When testing instruction arrives, it will be parsed by the event module and sent to schedule module which will distribute the event to register module. After acquiring the target state of navigation system and finishing state transitions with the help of state module, the schedule module will activate related tasks of the current state by the task module. We send failure instruction of navigation system model to test state machine which will switch the periodic state of the navigation system model to failure state, and let it stop sending any data to flight control system. At this point, the reaction of the software of the flight control system could be observed.

According to this example, it could be seen that test model based on state machine is able to switch states easily and could enable testers to carry out more complex test.

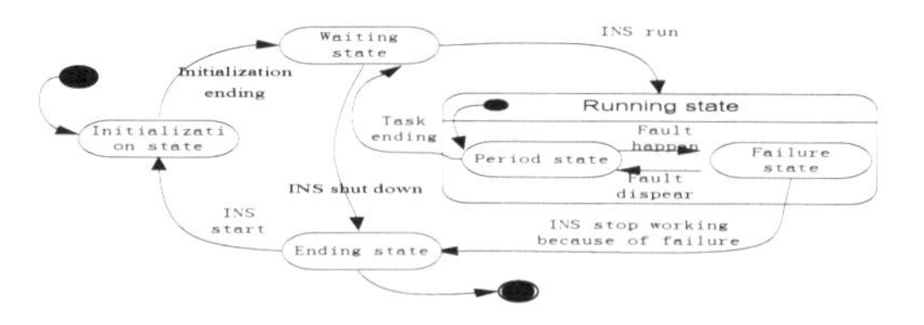

Fig. 5 State graph of navigation system model based on HSM

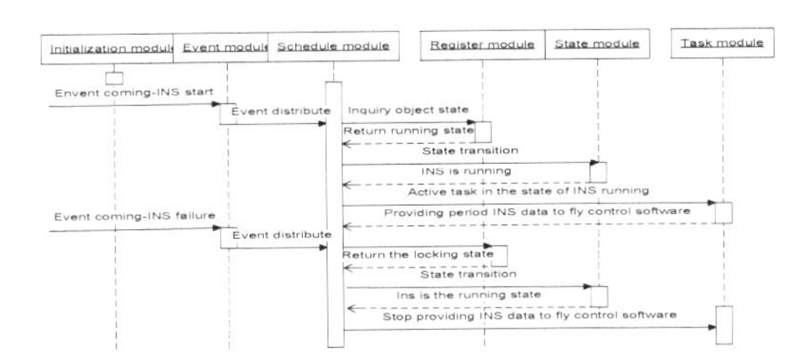

Fig. 6 Sequential graph of flight control test model

4 Conclusions

The test model based on hierarchical state machine is proposed in this paper. Through this mechanism not only real-time performance, expansibility and robustness could be enhanced, but also make it easy to implement and greatly improve the advantage of test model and test efficiency. This is an interesting topic and preliminary study is conducted in this paper. Further comprehensive studies can be explored in the future. More hierarchical states could be supported and applied in different test environment to fulfill the test requirement of complex embedded software testing. Also, the performance comparison with other model-based testing approach could be studied through the practical testing case.

References

1. Subramaniam, M., Guo, B., Pap, Z.: Using Change Impact Analysis to Select Tests for Extended Finite State Machines. In: 7th IEEE International Conference on Software Engineering and Formal Methods, pp. 93–102. IEEE Press, Hanoi (2009)
2. Samek, M.: Practical Statecharts in C/C++ Quantum Programming for Embedded Systems. CMP Books, Elsevier (2002)
3. Mathaikutty, D.A., Ahuja, S., Dingankar, A., Shukla, S.: Model-driven Test Generation for System Level Validation. In: Proceedings of the IEEE High Level Design Validation and Test Workshop, pp. 83–90. IEEE Press, Irvine (2007)
4. Weißleder, S.: Influencing Factors in Model-Based Testing with UML State Machines: Report on an Industrial Cooperation. In: Schürr, A., Selic, B. (eds.) MODELS 2009. LNCS, vol. 5795, pp. 211–225. Springer, Heidelberg (2009)
5. Gargantini, A., Riccobene, E., Scandurra, P.: A Model-driven Validation & Verification Environment for Embedded Systems. In: Proceeding of International Symposium on Industrial Embedded Systems, pp. 241–244 (2008)

RTEMS OS Porting on Embedded at91r40008 Platform

Wu Kai and Li Fang

Abstract. RTEMS embedded operating system is a very excellent real-time operating system, it's easy to be cut and transplanted. For having a good real-time character and stability, RTEMS can be widely applied in various fields. Take ev40 development board for example, the paper analyzes the main startup process in RTEMS, detail steps of RTEMS embedded operating system is a very excellent real-time operating system, it's easy to be cut and transplanted. For having a good real-time character and stability, RTEMS can be widely applied in various fields. Take ev40 development board for example, the paper analyzes the main startup process in RTEMS, detail steps of the RTEMS porting to at91r0008,it has a certain significance for application and development of the RTEMS operating system.

Keywords: **R**TEMS, embedded, at91r400081.

1 RTEMS Operating System

RTEMS is the abbreviation of Real Time Executive for Multiprocessor Systems, it is an RTOS open source of real-time embedded operating system. RTEMS system has excellent real-time, high stability, and fast speed. The performance of RTEMS equally matches VxWorks. RTEMS kernel is based on preemptive micro-structure, multi-processor, and the core code is written by C / C++ languages, for this, its portability is well. It was first used in the U.S. defense system. RTEMS has a very wide range of applications in aerospace, military, and civilian areas.RTEMS provides a variety of API interfaces. Besides RTEMS API itself, it also provides API interfaces of POSIX. The API enables software designers to get rid of the underlying details of the multi-tasking and multi-processor control. RTEMS also offers memory management, message management, mutex

Wu Kai · Li Fang
Key Laboratory of Power Electronics &Electric drives Institute of Electric Drives Institue of Elctrial Engineering, Chinese Academy of Science
Beijing, China

Z. Qian et al. (Eds.): Recent Advances in CSIE 2011, LNEE 125, pp. 269–274.
springerlink.com

management, GDB debugger and a series of API. The API allows users to concentrate on developing applications, which improve efficiency greatly.

2 RTEMS Startup Process Analysis

Understanding the RTEMS operating system initialization process during startup is necessary for transplanting OS. According to the need of our board, we look for start codes which the hardware need, modify the codes according to the hardware.

RTEMS initialization process is shown in Figure 1, including the steps of defining the interrupt level, masking all interrupts, initializing the BSP (Board Support Package), and initializing the data structure and other major processes.

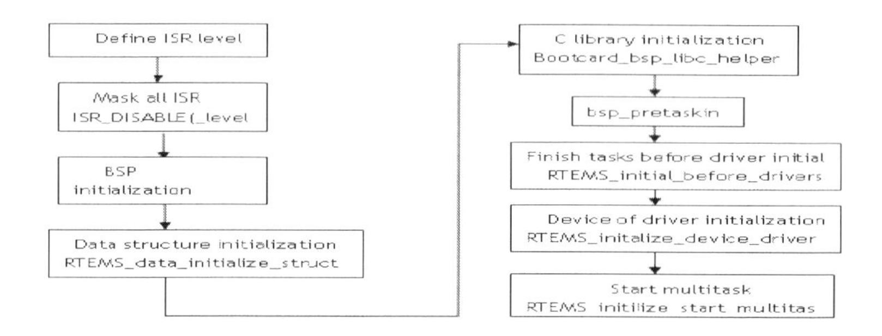

Fig. 1 RTEMS initialization

Besides BSP initialization part, there are four functions need to be focused on:

RTEMS_initialize_data_structures
RTEMS_initialize_before_drivers
RTEMS_initialize_device_drivers
RTEMS_initialize_start_multitasking

1)RTEMS_initialize_data_structures(*configure_table) initializes the data structure ;ensures consistency between the various data structures; initializes the user initialization task list, which contains all types of required information of RTEMS system (single-processor / multi-processor). The data structure contains the most important parts of the RTEMS such as operating space required, the size, starting address and number in each tick of the subtle, the number of ticks corresponding to each time slice and so on.

2)RTEMS_initialize_before_drivers (void) completes tasks before the stage driver initialization, which mainly concentrate on loading API_extension_list's all nodes,. If it is configured as multi-processor systems, it will also create the creation job of the MPCI communication server

3) RTEMS_initialize_device_drivers (void) loads all the device driver. If it is configured as multi-processor systems, it will also initialize the MPCI layer. Its internal execution is divided into four functions.

4)RTEMS_initialize_start_multitasking(void) initializes the system's multi-tasking environment, selects a thread inheriting the priority of this thread itself to start multi-tasking.

3 Steps of Porting RTEMS to at91r4000

When the RTEMS being transplanted to a new CPU, we must configure the development environment and provide for board support package (BSP) of appropriate version. BSP of the system is directly related with the hardware independent layer, which is the basis for embedded real-time system transplantation. As the key of the porting process, BSP's main function is to shield the hardware, provide the operating system driver and hard driver.

The main functions BSP achieved include:

Target board hardware initialization, special in the initialization CPU for the entire system, providing the underlying hardware support;

Providing for the operating system device drivers and system interrupt service routine;Defining operating system functions, and providing a real-time multitasking operating environment for the software system;

Initialization of the operating system, getting ready for the OS's normal running;

This paper's development environment is based on EV40 target board. Development environment is consisted of host-based virtual machine under Ubuntu Linux operating system, cross-GDB debugger which connects to GDBsever under Windows environment through network. The target board connects to the GDBsever with J-link. And at the same time, the Linux system is required to install RTEMS cross compiler and debugging environment. The tool-chains are provided on the official website.

Detailed steps are as follows:

1) Select the nearest BSP template for reference. Modify the source code, if the template is close to the target board, it can save a lot of time.
2) Create new folders in the directory, referring to the chosen template. Create a new profile BSP.cfg for BSP configuration.
3) Modify the makefile file and the link files
4) Modify basic informations such as the startup code, interrupt vectors and stack to be established. Modify information relevant to target board and target CPU information.
5) Write the related device drivers required;
6) Compile the BSP, RTEMS kernel libraries;
7) Tailor the RTEMS according to different applications, debug and build the operating system.

4 More Steps in Transplantation RTEMS BSP

1)Modify Makefile.am and configure.ac:
The two files are the dependent files needed by autotools.We can modify these two files according to the exist subdirectory.

```
  Modifying  Makefile.am:
  include_HEADERS = include / bsp.h
  include_HEADERS = include/tm27.h
  nodist_include_HEADERS = include / bspopts.h
  console_SOURCES = console / uart.c .. / .. / shared /
console.c
  abort_SOURCES = .. / shared / abort / abort.c,
      Modify configure.ac:
  AC_INIT            ([rtems-c-src-lib-libbsp-arm-ev40],
[_RTEMS_VERSION], [rtems-bugs@OARcorp.com])
```

2) Modify make/custom/ev40.cfg directory
According to the added folders ,copy any cfg files, modify RTEMS_CPU, RTEMS_CPU_MODEL, RTEMS_BSP_FAMILY and other information, documentation, generation of the path, the compiler options:

```
 CPU_FLAGS =-mcpu = arm7tdmi
 FLAGS_OPTIMIZE_V =-o2-g
    ... ...
```

3) Modify the linker script
Linker script provides the size of the stack, the environment variable size of the kernel. Modify the linker script as follows:

```
  In the file c/lib/libbsp/ev40/startup/linkcmd
 MEMORY {
sdram: ORIGIN = 0x30000000, LENGTH = 64M
        }
 Link address in the memory and stack space of the kernel:
 _sdram_base  =  DEFINED  (_sdram_base)?  _sdram_base:
0x30000000;
 _sdram_size = DEFINED (_sdram_size)? _sdram_size: 64M;
 _irq_stack_size    =    DEFINED    (_irq_stack_size)?
_irq_stack_size: 0x40000;
... ...
```

4) Modify the startup code
If you load the kernel using the uboot , some major work such as start the chip can be given to uboot. Main function of RTEMS kernel' startup code is used to initialize the stack and bss segment, and clear the mmu cache and so on.

Set the stack for each mode are as follows

```
mov r0, # (PSR_MODE_SVC | PSR_I | PSR_F)
msr cpsr, r0
mov r0, # (PSR_MODE_IRQ | PSR_I | PSR_F)
msr cpsr, r0
ldr r1, = _irq_stack_size
ldr sp, = _irq_stack
add sp, sp, r1
... ....
```

5) Configure and compile RTEMS library
After the completion of the above steps, we can start the RTEMS configuration. The same with other types of operating systems, the operating system configuration is completed through setting a series of macro definitions .These macro RTEMS are related with global variables and constants, such as RTEMS configuration table, CPU dependency information table, the system initialization task table, and the user initialization task lists. Condefs.h includes the operating system configuration templates, which can be pre-compiled. For example, RTEMS_configuration_table is a table needed for the kernel, we can configure it to embrace the following aspects:

```
CONFIGURE_EXECUTIVE_RAM_WORK_AREA,
CONFIGURE_EXECUTIVE_RAM_SIZE,
CONFIGURE_MAXIMUM_USER_EXTENSIONS
CONFIGURE_MICROSECONDS_PER_TICK,
CONFIGURE_TICKS_PER_TIMESLICE,
CONFIGURE_IDLE_TASK_BODY
```

6) Compile the RTEMS BSP
Compile the operating system BSP commands following these steps:

```
~$:/ Configure-target = arm-RTEMS, prefix = / opt /
RTEMS / RTEMS-disable-posix disable-networking-enable
RTEMSbsp = "ev40"
@ ~ $: Make
@ ~ $: Make all
```

7) Compile and debug the example routine and display.The example routine is as follows:

```
RTEMS_task init_task (
RTEMS_task_argument ignored
                    )
{
printk ("main: monitor init \ n");
RTEMS_monitor_init (0);
RTEMS_capture_cli_init (0);
RTEMS_task_delete (RTEMS_SELF);
}
# Define CONFIGURE_APPLICATION_NEEDS_CONSOLE_DRIVER
```

```
# Define CONFIGURE_APPLICATION_NEEDS_CLOCK_DRIVER
# Define CONFIGURE_APPLICATION_NEEDS_TIMER_DRIVER
# Define CONFIGURE_MAXIMUM_TASKS 30
# Define CONFIGURE_MAXIMUM_DRIVERS 15
......
```

Compile the routine using the makefile rules provided by RTEMS itself. We should modify the makefile according to the size of the file under the include path, file name and header files.

After packaged by uboot, download the file directly to the flash to run. Run command Loady 0xfe080000 and bootm in uboot,. The results are shown in Figure 2

```
   Data Size:      127072 Bytes = 124.1 kB
   Load Address: 00010000
   Entry Point:  00010000
   Verifying Checksum ... OK
OK
## Transferring control to RTEMS (at address 00010000) ...

*** HELLO WORLD TEST ***
Hello World
*** END OF HELLO WORLD TEST ***
```

Fig. 2 Hello World routine Results

5 Summaries

This paper describes detailed steps of how RTEMS operating system is ported to ev40, and gives the example routine of transplantation. RTEMS is good for its high real-time, high reliability and good portability. It is being paid more and more attention nowadays, RTEMS operating systems support almost all the processor structures widely.The maturity and development of the porting technology in the ARM would promote the development of RTEMS, break the monopoly of other commercial real-time operating systems, and greatly facilitate the embedded system development.

References

1. Yun, Y., Li, Y.-J.: Ported to RTEMS BSP SAILING S698 processor development. Computer Engineering and Applications 45(26), 60–62 (2009)
2. Xu, D., Dong, W., Ning, D., Yongbing, S., Gangming, J.: GUI of RTEMS oriented system design level. Microelectronics and Computer 24(6), 42–45 (2007)
3. Liu, L.: RTEMS CPU Utilization Research and Implementation. Popular Science 108(8), 41–455 (2008)

Embedded Mobile Internet Supervision System for Food Traceability Supplied to Hong Kong

Xianyu Bao and Shaojing Wu

Abstract. An embedded mobile Internet supervision system was developed for the whole process of food traceability, in order to ensure food safety supplied to Hong Kong. The modulated design of each functional unit and Intel X86 architecture gives a compact 205mm×115mm×22m unit with data process and control by both Atom Z530 and SCH US15W, memory, power management, LVDS input/output, RFID data collection, and mobile Internet. Therefore, the system could work independently as a hand-held supervision device or directly tied to the cross-border vehicles, suitable for the border inspection, record access and browsing, and peccancy vehicles positioning and tracking.

1 Introduction

The construction of public information platform of Shenzhen-Hong Kong food supply chain marks the significant improvement of the information publicity and supervision of exported food safety. The platform also makes a higher requirement of controls over source and supply chain of export food, including plant and animal breeding, production, processing, transportation, clearance, package and distribution. Thus, it is urgent to develop an intelligent mobile supervision system which can locate food, trace the origin anytime and anywhere, to replace the traditional manual way. Since 2009, Tsinghua University and Shenzhen Academy of Inspection and Quarantine have been collaboratively developing an embedded mobile Internet supervision system for food traceability exported to Hong Kong, which has achieved some initial results in mobile Internet devices , the platform of fruit and vegetable exported to Hong Kong [2], and virtual gates of ports [3].

In this paper, the RFID and mobile Internet technologies are adopted to implement the whole process traceability and supervision of various nodes in supply chain of exported food. Besides, the continuous maturity of Intel X86 architecture and the rapid development of Atom processing technology provide an effective way to implement embedded mobile supervision system.

Xianyu Bao · Shaojing Wu
Shenzhen Academy of Inspection and Quarantine, Guangdong, China

Z. Qian et al. (Eds.): Recent Advances in CSIE 2011, LNEE 125, pp. 275–280.
springerlink.com

2 Working Principle

As Fig. 1 illustrates, the system is composed of two core Intel chips, which construct a CPU + Bridge X86 architecture including data processing based on Atom Z530[4], peripheral interface control based on SCH US15W[5], memory, power management, LVDS, RFID data collection, and mobile Internet.

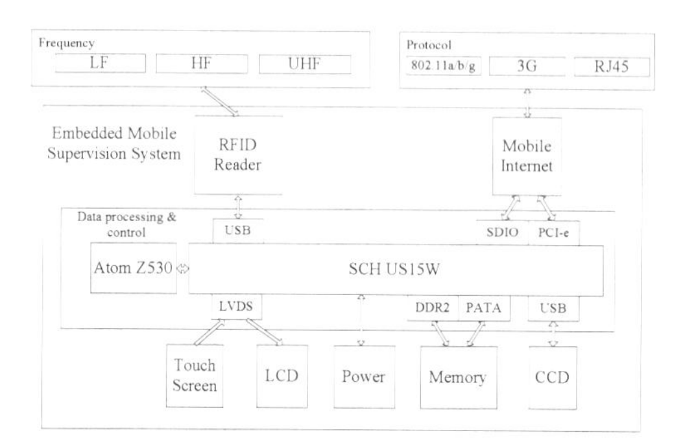

Fig. 1 The embedded mobile supervision system

Its specific working principle is as follows. Data processing and control module handles realtime data computation and the control of peripheral interface circuit. Touch screen input module and RFID data collection module can be used to write food related information to RFID tags attached to food's package in realtime. Meanwhile, the information is, respectively, stored into local and remote supervision databases via mobile Internet module (including 802.11 b/g/n、3G or RJ45). When inspectors supervise on site, they can fill the supervisal records online by the system. When irregularities are discovered, they can take photos on site via CCD camera as evidence. When clearance, border inspectors use the system to retrieve RFID tag information to compare inspected information with the one in databases, which could achieve fast clearance.

The system adopts tablet PC-wise design method. It also runs Windows XP, which is compatible with most application software. Another advantage is that RFID data collection module with LF, HF, UHF frequencies is changeable.

3 Division of Functions and Hardware Design

3.1 Data Processing and Control Module

As a core module of the embedded mobile supervision system, take the low power consumption, computation capability, peripheral logic, protection, storage, mobile Internet communication into consideration, Atom Z530 (441 balls, 13×14mm) and SCH US15W (1249 balls, 22×22mm) from Intel are chosen. They could achieve data transmission through high speed 400/533MT/s FSB.

Atom Z530 focuses on real time data processing, which employs the 45nm high-K metal gate transistor manufacturing process and advanced consumption management technology. Its frequency is up to 1600MHz, Thermal Design Power (TDP) as low as 2W (The TDP of mainstream portable PCs is 35W) [6]. SCH US15W chipset equipped with GMA 500 graphics card possesses 3D image processing power, 720p, 1080i HD video hardware decoding functions. What's more, it integrates with most I/O interfaces of PCs, such as PCI-e、 USB、 SDIO、 LVDS、 PATA, etc.

3.2 Memory Module

The voltage of the kernel of memory chip is 1.8V and it can extend up to 2GB DDR2 533MT/s. When motherboard is on power, it can load data from user specified hardware, CD-ROM or other external storage. Meanwhile, rapid data swamp can be implemented between memory and bridge chip by the 64 SM_DQ data signal pins and 8 SM_DQS strobe signal pins. Memory module can choose DDR2 DIMM slot standard suitable, market well-known brand memory.

Hard disk makes use of PATA interface standard, with data transmission speed up tot 133MB/s. In our system, we choose Silicon's SST85LD1008M SSD with 8G capability and can be extended to 64GB. Its package format is LBGA.

3.3 LVDS Module

LVDS module consists of touch input circuit, video control circuit and LCD display circuit. Touch input circuit transmit its sensed position information (X_RIGHT, X_LEFT, Y_UP, Y_DOWN) to the bridge chip after the processing by the conversion chip TSC2007IPW. The TSC2007IPW implements the location of touch position information and instruction execution. Video control circuit consists of bridge chips and four pairs of interconnected data signal differential lines in LVDS LA_D[0:3], among them, one pair of differential clock signal line LA_CLK completes synchronous transmission of video data together. The output of LCD display of video signal is achieved by using LVDS cable to connect LVDS interface and LCD.

3.4 RFID Data Collection Module

There are two parts of jobs that RFID data collection module performs. On one hand, it outputs information of RFID tag to LCD display after its processing by data processing and control module. On the other hand, it writes the touch inputted data into RFID tag after its processing by data processing and control module. The module utilizes two half-duplex RF communication chips ADF7020[7] to achieve above functions. It also uses miniUSB interface to connect to SCH US15W chip directly. It configures its receiving and sending modes and working frequencies by the peripheral control circuit. In receiving mode, it proceeds to baseband processing after the signal selection, amplification, demodulation and other processing. In sending mode, it modulates and amplifies baseband signal, and then proceeds to send the signal via generated carrier.

3.5 Mobile Internet Module

One feature of the embedded mobile supervision system is that it can connect to the Internet anytime anywhere. The system provides both wireless and wired connection ways to achieve that, including 802.11 b/g/n, 3G, and RJ45 Ethernet interface. All data collected by RFID data collection module is sent to supervision databases via this mobile Internet module, in order to record data and retrieve data for comparison when necessary.

3.6 Power Management Module

The system compasses processor, bridge chips, memory, hardware, touch screen, LCD, PCI-e, SDIO, USB and other kinds of digital and analog circuits, which have different requirement of power. In order to make all these modules perform well but also minimize power consumption to a longer battery usage, we design 20 groups of power between0.9~5V, with considerations of different modules' refined requirement and power conversion efficiency.

4 Design of Touch Screen Input Software

The developed system adopts full-touch input to replace traditional keyboard and mouse input. To handle the touch location, early method is adopting simple registration of touch area or increment of the icon size to achieve a better effect of touch. Recent years, the mainstream method is to display touch area on predefined regulated location of touch screen. In this way, users can just place the stylus onto the area they want to touch in their each operation and then click the area [8]. But for devices running Windows XP, using touch screen to browse websites, drag and drop windows, click icons which are close to the edge and other fine human-computer interactions, higher requirements are demanded on the press information of touch screen.

The system uses seven inches four-wire resistive touch screen. First, define the central position coordinates (x_0, y_0), then divide this touch screen into A, B, C, D four equal size area. Its characteristic is partitioning big screen into small screen, and then adjusting position delicately and independently to every small screen. Suppose touch screen controller supports N bits A/D conversion (8 bits and 12 bits for TSC2007IPW chip), then the central position coordinates (x_0, y_0) is equivalent to $(2^{N-1}-1, 2^{N-1}-1)$.

Due to the nonlinear characteristics of the touch screen, there are different adjusting ways to edge and non-edge areas, elaborating as follows:

(1) Edge adjustment

A area: if $x<10$, then $x=10$; if $y>2^N-7$, then $y=2^N-7$.

B area: if $x>2^N-7$, then $x=2^N-7$; if $y>2^N-7$, then $y=2^N-7$.

C area: if $x>2^N-7$, then $x=2^N-7$; if $y<7$, then $y=7$.

D area: if $x<10$, then $x=10$; if $y<7$, then $y=7$.

(2) Non-edge adjustment

A area: if $x<2^{N-1}-1$, then $x-=9\times(2^{N-1}-1-x)/(2^{N-1}-1)$;
if $y>2^{N-1}-1$, then $y+=17\times(y-2^{N-1}+1)/(2^{N-1}-1)-\tau$.

B area: if $x>2^{N-1}-1$, then $x+=7\times(x-2^{N-1}+1)/(2^{N-1}-1)$;
if $y>2^{N-1}-1$, then $y+=17\times(y-2^{N-1}+1)/(2^{N-1}-1)-\tau$.

C area: if $x>2^{N-1}-1$, then $x+=7\times(x-2^{N-1}+1)/(2^{N-1}-1)$;
if $y<2^{N-1}-1$, then $y-=25\times(2^{N-1}-1-y)/(2^{N-1}-1)+\tau$.

D area: if $x<2^{N-1}-1$, then $x-=9\times(2^{N-1}-1-x)/(2^{N-1}-1)$;
if $y<2^{N-1}-1$, then $y-=25\times(2^{N-1}-1-y)/(2^{N-1}-1)+\tau$.

τ is correction value, $\tau\in[0,9]$. In non-edge adjustment, there is no correction value for X-axis, because in fact its size is slightly bigger than that of Y-axis. If the system supports gravity sensing, the adjustment is according to the rotation direction of gravity sensing and updates area in real time, while the adjustment of X-axis and Y-axis and correction value remain the same.

5 Performance

The system is completed with adoption of modularized function unit and stack assembly architecture, and integration all previous modules into a space of length 205mm ×width 115mm × height 22mm.

The performance of the system are as follows.

Screen: 7 inches LCD screen with full touch input capability; network: supporting wireless protocol 802.11a/b/g, 3G and wired network RJ45; operating system: Windows XP; weight: 980g (including battery); volume: length 205mm × width 115mm × height 22mm; standby time: 150 minutes with standard 1850mA battery; stability: 3*24 hours o, 3DMark05 stable with continuous normal tests.

6 Conclusion

According to the requirements of the whole process supervision of all node of supply chains in exported food, this paper developed an embedded mobile supervision system based on Intel Centrino Atom processor technology. Through adoption of X86 architecture and modularized function units design conceptions, the system integrates data processing and control, RFID data collection, mobile Internet and some other modules into a space of length 205mm × width 115mm × height 22mm, and it also realizes a highly efficient software design way of touch screen input based on Windows XP operating system. The system is already applied to various nodes in supply chains of exported food in Shenzhen.

References

[1] Bao, X.Y., Lu, Q., Zheng, W.M.: Design of embedded mobile Internet device based on Atom. In: CACIS, pp. 91–95. USTC Press (2010)

[2] Lu, Q., Wang, X., Liu, S.Y.: Research on RFID supervision technology for vegetables supplied to Hong Kong. Journal of Plant Qurantine, 32–34 (2009)

[3] Lu, Q., Ding, X.Y., Chen, X.: Design and realization of virtual gate system based on RF technology. Journal of Modern Electronics Technique, 139–144 (2010)

[4] Intel corporation: Intel ATOM processor Z5×× series (Rev. 005) (2008), http://download.intel.com/design/processor/specupdt319536.pdf

[5] Intel corporation: Intel system controller hub (Rev. 2.0) (2008), http://download.intel.com/design/processor/specupdt/364236.pdf

[6] Gerosa, G., Curtis, S., Addeo, M.: A sub-2 low power IA processor for mobile internet devices in 45nm high-k metal gate CMOS. IEEE Journal of Solid-State Circuits, 73–82 (2009)

[7] Analog Devices, Inc.: ADF7020-high performace ISM band FSK/ASK transceiver IC (2001), http://cn.alldatasheet.com/datasheet-pdf/pdf/83646/AD/ADF7020.html

[8] Sharp corporation: Input device and a touched region registration method. China Patent no. 03136826 (2003)

Abstract Mechanisms of GEF and Techniques for GEF Based Graphical Editors

Deren Yang, Min Zhou, Fen Chen, and Jianping Ma

Abstract. The Graphical Editor Framework (GEF) is a popular framework built to develop rich graphical editor for models. But being lack of detailed documentation, it is difficult to use GEF in practice. In this paper, the principle, composition and main classes (interfaces) of GEF were analyzed. The three levels of abstract mechanisms of GEF were studied. The implementing techniques for GEF based graphical editors were analyzed. And the implementation of a graphical editor instance was described. The study is helpful to implement GEF-based graphical editor.

Keywords: raphical Editor Framework, Principle, Model, Abstract Mechanism, Implementing Techniques.

1 Introduction

The Graphical Editor Framework (GEF for short) can be used to develop graphical editor for models, and put model to graphical environment for editing [1]. GEF is completely application neutral and can be used to build but not limited to activity diagram, GUI builder, class diagram editor, state machine, workflow editor, and even WYSIWYG text editor, etc. The study of abstraction mechanisms of GEF and its implementing techniques are useful to guide related application software development.

The main contributions of this paper are as follows. First, GEF's three levels of abstraction, namely abstraction mechanism based on MVC architecture, and design-oriented and implementation-oriented were analyzed. Secondly, GEF implementing technology, such as key techniques, methods and related operations were studied. Thirdly, a graphical editor instance was described.

The organizational structures of this paper are as follows. The related research background is briefly described in part 2; and GEF principle, composition

Deren Yang · Min Zhou · Fen Chen · Jianping Ma
School of Science, Ningxia Medical University, Yinchuan China
e-mail: ydr@tom.com

Z. Qian et al. (Eds.): Recent Advances in CSIE 2011, LNEE 125, pp. 281–287.
springerlink.com

and functions are analyzed in Part 3; and abstraction mechanisms of GEF are studied from three levels in Part 3 also; and implementing technique of GEF is analyzed in Part 4; and finally conclusions and future research directions are given in Part 5.

2 A Brief Overview of Related Researches

GEF is widely used in developing graphical tools, There are a lot of related researches, such as used for Multi-Agent system development [2] [3]. However, most of these researches only focused on how to use GEF in implementation. In fact, as an infrastructure, GEF has relatively complex abstraction mechanisms, and is lack of related research, such as in literal [4] only implementation level of abstraction was discussed. Therefore, in order to better realize its value, such as for software process modeling, it is necessary to study GEF abstraction mechanism and to explore its implementing process step by step.

3 GEF's Abstraction Mechanisms

In GEF, Model and View can communication only through Controller, with which couple problem is solved, and by request, edit strategy and command mode burden of Controller is reduced, and so interdependence of components is minimized.

GEF's principle is as follows:

(1) Editor accepts operation, and converts to request which will be submitted to Controller;
(2) Controller accepts request which will be forwarded to Edit Policy Manager;
(3) Edit Policy Manager establishes command for request;
(4) Editor executes Command and modifies Model;
(5) Controller monitors Model's change and notifies View to update;
(6) View refreshes display.

GEF framework supports interaction caused from mouse and keyboard, and displays model graphically. It offers two plug-ins: Draw2d plug-in is a lightweight graphical component system, and provides layout and display graphics rendering tools, and Gef plug-in adds editing function to Draw2d. The plug-ins are run-time library [1].

3.1 MVC-Based Abstraction

GEF uses MVC architecture (seeing Figure 1), and with it some important graphical editing functions, such as Drag, undo/redo, move, delete, change size of graph, can be achieved.

The necessary conditions for Model are as follows. First, Model must contain all data user wants to modify. Secondly, Model can not hold a reference to its View, and it can not understand the rest of editor. Thirdly, Model must notify its changes to Controller.

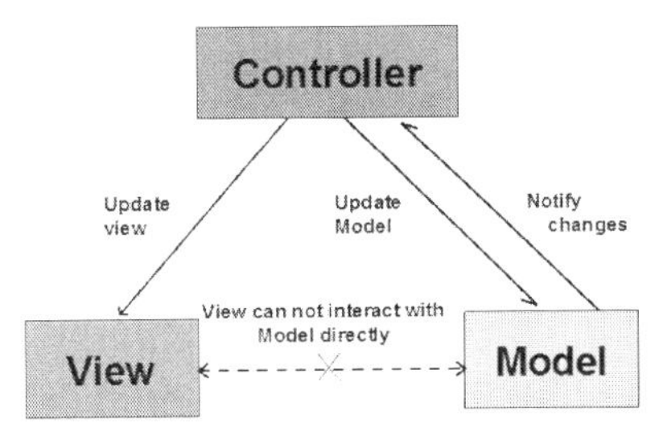

Fig. 1 MVC-based abstraction

The necessary conditions for View are as follows. First, View can not hold data which is not defined in Model. Secondly, View can not hold any reference to Model. Thirdly, View do not understand Controller, and not participate controller logic in any way.

Model and View can not refer each other, and their contact only depends on Controller. Controller is a bridge between them, and it is the core component of GEF.

The advantages of this structure are as follows. Model is used to store data, and View is used to display data, and Controller is used to bind Model and view logic. Therefore, Model is concise and well-defined, and roles of the structure are clearly separated, and it facilitates design and implementation.

3.2 Design-Oriented Abstraction

In design level, in order to refine above MVC abstraction appropriately, user and editor should be taken into account, which is conducive to system analysis and design, and its logic is shown in Figure 2.

Editor is a place for Controller to show View. Its role is presenting model graphically, and it also provides edit, echo, tool tips and other functions.

By selecting and clicking editor's operations, Users can produce events.

Controller (EditParts) is responsible for commenting event, and converts it to model's operation, and is in charge of maintaining View. Each model object and its view have a dedicated controller, and the controller is responsible for collecting model information, setting its graph properties. Controller involve in editing model process, namely using edit policy (EditPolicies) to handle editing tasks.

In order to let Controller know model's changes, Model must implement a notification mechanism. Controller is registered as event listener for corresponding model object to listen for change notification, and to notify View to update in time according to model's new state.

View is any object visible to user, which is built with Draw2d. GEF provides support to Draw2d graphics (Figures) and tree (Treeitems). The former is used to display editing area, and the latter is used to show structure according to outline.

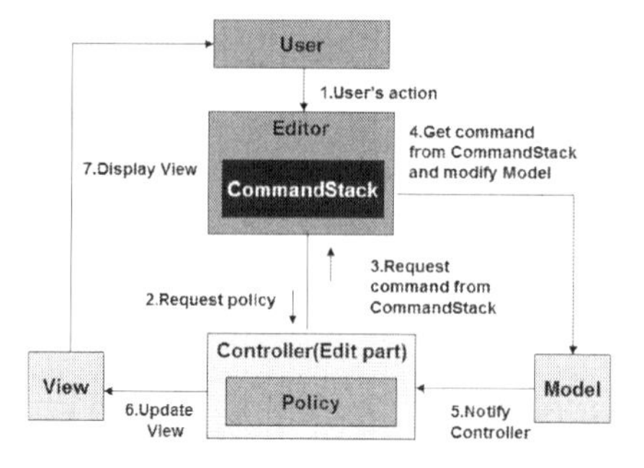

Fig. 2 Design-oriented abstraction

3.3 *Implementation-Oriented Abstraction*

From developers' perspective, it is necessary to further refine Editor and Controller in Figure 2, such as to separate command stack and command from Editor, to separate edit policy from Controller, which is conducive to programming.

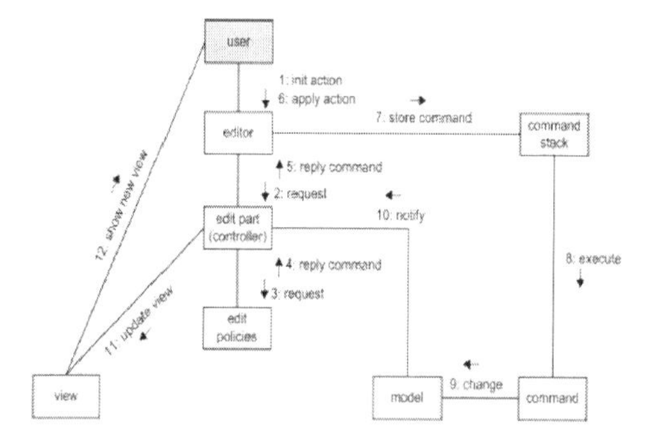

Fig. 3 Implementation-oriented abstraction

User submits request to editor, editor manages command stack by Editing domain, and every command of the stack be executed by execute() to modify model, then the request generated is passed to Controller of the model. A series of edit policies are installed in Controller, and each policy has its corresponding commands, and a strategy exists and can handle the request, then related change is notified to editor, and view refreshes accordingly, and finally a renewed view is displayed to user. Figure 3 illustrates this abstract Mechanism [4].

4 Techniques for GEF Based Graphical Editors

To implement a GEF-based editor, the following three questions must be addressed. First, some mechanisms must be implemented to automatically build and display view defined for model when editor being opened. Secondly, some mechanisms must be implemented to update view after model is changed immediately. Thirdly, some mechanisms must be implemented to capture operation caused by graphical user interface, and to convert it into model's change.

4.1 Key Techniques

User's editing operation is converted to Request which is called Role in GEF. Role is realized by edit policy, and main function of edit policy is to create Command according to Request, and Command will directly operate model. Some policies are configured for each Controller, and user's operate to Controller will be transferred to corresponding edit policy. Its advantage is that controllers can share a number of repetitive operations.

GEF uses EditPartViewer class to implement editor. Model is created by Plain Old Java Objects (POJOs). Controller is created by EditPartFactory. View is created in editor, and is deployed through configureGraphicalViewer(), and so model and controller is connected. Model is shown by calling method initializeGraphicalViewer().

4.2 Implementing Method

To implement menu in user interface, it need call two classes and its methods in Actions package.

First, run() of DiagramAction class is called. After opening dialog box in Diagram menu, buildActions() of DiagramActionBarContributor class calls addRetargetAction(), and RetargetAction button icon is displayed in main menu or toolbar.

Secondly, declareGlobalActionKeys() calls getAction() of toolBarManager.Add, and editor will map Action to corresponding RetargetAction, and enabling RetargetAction of menu and toolbar to have Action's function.

In Commands package, add/remove connection command, and add/delete/move node command, and direct edit commands, and arrow connection command, delete command and create command exist. In every Command, execute, undo/redo and other methods exists used to execute commands.

In EditPartWithListener class, EditPart needs to be registered in model in order to monitor model, and registering and removing listener by calling activate() and deactivate().

PartFactory class is used to connect Model and Controller. Among them, editPartForElement() is used to capture Controller of Model, and createEditPart() is used to connect model and controller.

4.3 Operational Steps and Methods

Operational steps and main methods are as follows:

(1) User operates graphical user interface;
(2) Operation creates an event, and the event is captured by editor with firePropertyChange(), and editor passes the event by calling buildActions() to current tool (active tool);
(3) Current tool interprets event queue as Request, and submits Request to Controller by calling performRequest(), and Controller passes Request to edit policy by calling createEditPolicies(), and edit strategy corresponds request by calling getXXXCommmand(), i.e. calling command from CommandStack. Every edit strategy has its corresponding commands, and the commands used to implement request and to modify the specified model;
(4)Controller listeners model's changes by calling firePropertyChange(), and using propertyChange() to call refreshVisuals() to refresh View.

4.4 An Implementation Instance of Graphics Editor

GEF-based software can be developed as Eclipse plug-in or independent software. If it is developed as standalone software, it can be implemented as a Rich Client Platform (RCP).

An instance of a graphical editor is shown in Figure 4.

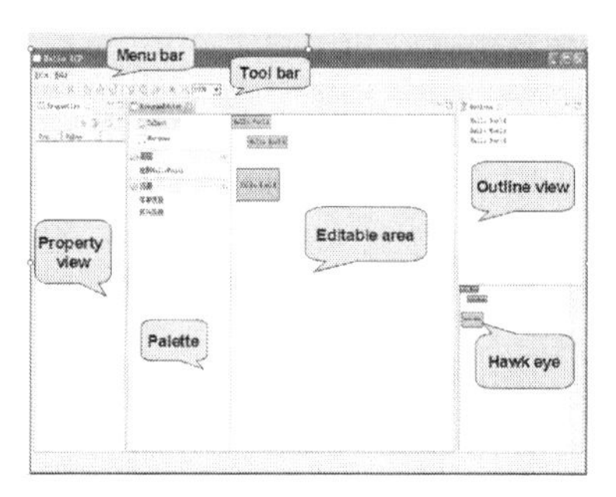

Fig. 4 An interface of editor instance

The implementation steps are as follows:

(1) Create a model as XXXModel;
(2) Create a controller XXXEditPart corresponding to the model;
(3) Create editPartFactory, in which model and controller are connected;
(4) Create GraphicalViewer in editor, and display graphic.

GEF-based editor is created in Rich Client Applications (RCP). gef.tutorial.step source package contains following six built-in classes: Application class, Perspective, ApplicationActionBarAdvisor, ApplicalicationWorkbenchAdvisor, ApplicationWorkbenchWindowAdvisor and Acrivator.

Following packages are added in code package: view-related class gef.tutorial.step.ui; model-related class gef.tutorial.step.model; controller-related class gef.tutorial.step.editparts; command-related class gef.tutorial.step.commands; policy-related class gef.tutorial.step.policies; editor related class gef.tutorial.step.actions; static variables related to class gef.tutorial.step.helper.

5 Conclusions and Future Research

GEF provides a visual editor framework for model. Generally, Model is based on POJOs and it is difficult to manage, the alternative is Eclipse Modeling Framework (EMF), which provides a flexible mechanism for managing model. Currently, Eclipse community studies Graphical Modeling Framework (GMF), combining GEF and EMF, and making GEF editor to use EMF-based model.

Based on GMF framework, author's future research is agent-oriented software process methodology and software paradigm mechanisms and related applications.

Acknowledgements. The research is supported by project of National Nature Science Foundation of China (81160183), Natural Science Foundation of Ningxia (NZ10103, NZ11105) and special talents project of Ningxia Medical University (XT200913).

References

1. Eclipse Project group (2010).GEF Project, http://www.eclipse.org/gef/ (accessed September 10, 2011)
2. Hubner, J.F., Boissier, O., Kitio, R., et al.: Instrumenting multiagent organisations with organisational artifacts and agents: giving the organisational power back to the agents. Journal of Autonomous Agents and Multi Agent Systems 20, 369–400 (2009), doi:10.1007/s10458-009-9084-y
3. Fortino, G., Garro, A., Mascillaro, S., et al.: ELDATool: A Statecharts-based Tool for Prototyping Multi-Agent Systems. In: Proceedings of Workshop on Objects and Agents, pp. 14–19 (2007)
4. Buchmann, T., Dotor, A., Westfechtel, B.: Model driven development of graphical tools: Fujaba meetsgmf. In: Proceedings of the 2nd International Conference on Software and Data Technologies (ICSOFT 2007), pp. 425–430 (2007)

An Implementation on Instruction Design Ontology

Dongqing Xiao, Muyun Yang, Sheng Li, and Tiejun Zhao

Abstract. Current AID (automated instruction design) platforms are criticized for incapable of transfer domain knowledge to the user intelligently owing to the lack of detailed and machine readable description of instruction design. We investigate how to apply the ontology to solve the problem. Based on former works including IMS Learning design and EML, we employ protégé to build instruction design ontology in OWL-DL (ontology web language–descriptive language). We improve the implement of activities to support ordinal relationship reasoning. We also discuss the clarity, coherence and expandability of the implementation for building one sub-class PBL (Problem-based learning). The experiments confirm our method as a feasible solution for AID semantic description.

Keywords: AID, Semantic Model, EML, OWL.

1 Introduction

Growing interest in AID (automated instruction design) motivates the maturity of related theories and the development of some application platforms. However, the majority of platforms stay in the form of template writing or instruction design knowledge management database, failing to transfer knowledge intelligently to ordinary teachers who lack professional instruction design knowledge. The bottleneck is how to build a detailed, standardized and machine readable description of instruction design, i.e. the instruction design ontology. The chief challenge in this issue is how to extract a consonant complete set of conceptions from the various instruction modes and describe relations among conceptions in a specific machine language.

To the best of our knowledge, related works to our paper include the semantic model of instruction design [5] and learning design proposed by IMS (instructional management system) in UML (uniform model language). However,

Dongqing Xiao · Muyun Yang · Sheng Li · Tiejun Zhao
School of Computer Science and Technology,
Harbin Institute of Technology, Harbin, China
e-mail: {dqxiao,ymy,tjzhao}@mtlab.hit.edu, lisheng@hit.edu.cn

Z. Qian et al. (Eds.): Recent Advances in CSIE 2011, LNEE 125, pp. 289–295.
springerlink.com

these models are not perfect. The semantic model of instruction design just suits well-structured learning process for cognitive objectives and meanwhile IMS learning design suffers semantic loss for adopting UML.

Based on previous research, we modify some concepts for whole harmonization, and present techniques to interpret concepts and relationships among core concepts.

2 Ontology

Ontology Theory. Ontology is usually referred as a formal representation of knowledge by a set of concepts within a domain and the relationships between those concepts [2]. In other words, ontology applies to knowledge representation and information modeling. As for the instruction design ontology, it attempts to reach a consensus of instruction design knowledge and translate knowledge into a machine readable way for further application, e.g. teaching diagnosis and teaching strategy decision etc.

Description Language. Here, we choose sub-language OWL-DL (ontology web language description language) to describe the whole ontology and SWRL (semantic web rule language) to support the reasoning process for its *well-defined syntax, well-defined semantic, good support for reasoning, abundant descriptions, the convenience for expression* .

Ontology Editor. We use protégé1 (a free, open source ontology editor and knowledge-based framework) for ontology coding, and its visualization and navigation plug-ins will facilitate coding and the deducted result presentation. In addition, it provides good connector for external semantic reasoner: Facet2.

3 Instruction Design Construction

This paper adopts Mike Ushold & Michael Gruninger's ontology construction methodology—skeletal methodology.

Demand Analysis. The user consists of the ordinary teacher, the instruction designer. Through the introduction of instruction design ontology, it is expected that computer can offer some advices (such as the recommendatory sequence of activities or strategy) based on the existing knowledge description when the user needs. To achieve this goal, instruction design should include conceptual descriptions of instruction practice and instruction strategy.

Ontology Capture. First, instruction process itself can be viewed as a system. The system goal is to bring out several learning objectives. The components of this system include the role, the learning context and the instructional material. These components interact in order to achieve the goal. Thus far, we exact some important concepts in instruction design. By reference to the descriptive

---

[1] Available at `http://protege.stanford.edu/`

[2] Available at `http://www.cs.man.ac.uk/~horrocks/FaCT/`

framework in EML, we amended the deployment of concepts and added some concepts (subject for resource management; activities for instructional process description, etc).

3.1 Core Concepts Capture and Coding

Hierarchy of core concepts. We begin core concepts capture and coding with the hierarchy of core concepts. Instruction design: {learning objectives, components, learning context, subject} Component: { role, activities}.In the light of B·S·BLOOM [7], learning objectives in cognitive domain include six levels: knowledge, comprehension, application, analysis, synthesis, evaluation. IMS assumes that learning objectives are parallel. However, this assumption is contrary to the general view and proved to be incorrect in reasoning process. Inspired by the idea that learning is advancing process and so are learning objectives, we organize them as shown in Fig.1.

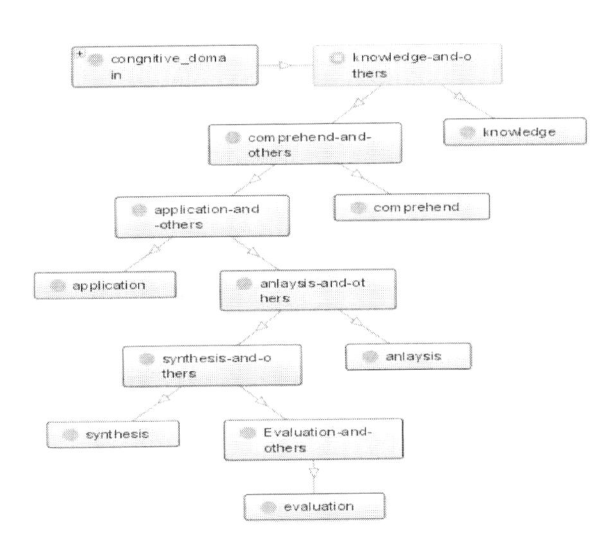

Fig. 1 Organization of learning objectives

IMS suggests that activities consist of three components, learning activity, supporting activity and activity structure. Instead of treating activity structure as subclass of activity, we take activity structure as a concept linked to specific practice. To represent sequence and selection of activities in activity structure, we adopt recursive definition.

E.g.1: *Minimal activity structure* AS_n

$$(A_{n-1}(A_n))$$

E.g.2: *Bigger activity structure* AS_i

$$(A_i,(AS_{k+1}))$$

This method of definition reserves semantic meaning about activity sequence as rich as possible. Through machine learning we can get the dependency of activity pairs which serve to reference of potential activities and ordinal relationships when ordinary teachers conduct learning design.

To realize the above-mentioned definition, we introduce two relationships (*has-direct-content, has-minstructure*).

For example, after translating into OWL, the definition of minimal structure equals "Has-direct-content only An-1 and *has-minstructure* (*has-direct -content only* A_n)".

The above-mentioned relationships can retain information as much as possible when it is used to collect certain instruction design cases with great value. However, it is impractical to build so detailed relationships for numerous activities. Yet, the user would like to get more recommendatory activities. This situation gives rise to the introduction of a looser ordinal relationship. HereSo we build some rules for looser ordinal relationships deduction.

The protégé produces the rule set in the following form utilizing SWRL.

$$entity(?x) \wedge ((\neg / op(?y,?x)) \wedge / \vee (\neg / op(?z,?x))) \overset{*}{\rightarrow} (?z,?x) \text{ (OP: object property)}$$

Let AS be the set of activity structure, A be the set of activities, and R be the set of relations between AS and A. Among the following relationships, f1 and g1 are stricter while others are looser.

f1: has-direct-content *f2: has-content*
g1: has-minstructure *g2: has-structure*
P: prior to *N: next to.*
A fraction of rule:

$$f1(AS, A) \Rightarrow f2(AS, A) \quad (1)$$

$$g1(AS_k, AS_l) \Rightarrow g2(AS_k, AS_l) \quad (2)$$

$$\exists AS_l \in \{AS_l \mid f2(AS_l, A_j) \wedge g2(AS_k, AS_l)\} \Rightarrow f2(AS_k, A_j) \quad (3)$$

$$\exists AS_l \in \{AS \mid g2(AS_l, AS_m) \wedge g2(AS_k, AS_l)\} \Rightarrow g2(AS_k, AS_m) \quad (4)$$

$$\exists AS_l \in \{AS \mid \neg g2(AS_i, A_i) \wedge f2(AS_i, A_j)\} \Rightarrow P(A_i, A_i) \quad (5)$$

$$P(A_i, A_i) \Rightarrow N(A_i, A_i) \quad (6)$$

According to this rule set, the inference engine can deduct some relationships among activities from learning design case collection.

Supporting activity. A support activity can optionally be connected to a role. It means that the activity is repeated for every member in the supported role (learner/staff). Since the well-known learning activities, Support activities are typically performed by staff members (e.g. tutors) to support learners. In some pedagogical models however, learners can support learners (peer to peer teaching). It is also possible that staff members support staff members. When the optional role-ref element is set, it is expected that the support activity will act for every single user in the specified role(s). That is: the same support activity is repeated for every user in the role(s). When the role-ref is not available, the support activity is a single activity (like the learning-activity). We realize some instances such as monitoring, assessing the performance, group organization, problem-evoking, and test.

Integrating the Existing Ontologies. There are numerous complicated multi-relationships among existing ontologies. Because of adoption of *UML (one kind of second-order predicate logic descriptive language)*, IMS learning design has no other choice to transform every *pluralistic relationship* into many binary relationships. It causes loss of semantic meanings and expressing brevity while it goes against the user's convenience. Overcoming the inconvenience of UML, OWL translates them into properties. In ontology editor, the N-gram building mode is just for muti-relationship.

Then the following two figures show the pluralistic relationships for domain concepts.

activities		
use the tools	Class	tools
use-the resource	Class	learning-resource
involved the roles	Class	component
has-objectives	Class*	objectives
name	String	
...		

Fig. 2 Attributes of activities

roles		
has number	Integer	
use the tools	Class	tools
in the contest	Instance	context
use-the resource	Class	learning-resource
act	Class	component
...		

Fig. 3 Attributes of roles

4 Instruction Design Ontology Validation

We conduct the experiment in two phases. In the first phase, we discuss the clarity, coherence and extensibility of the whole instruction design ontology by building the classical instructional mode PBL (problem-based learning). Then we try to deduct the sequence of activities involved in PBL and analyze the difference between deducted result and the real one.

PBL is a student-centered instructional strategy, in which students solve problems collaboratively and reflect on their experience. According to barrows' model, PBL contains several activities: 1. group organization, 2. new problem proposing, 3. Getting consensus, 4. Learning activities, 5. Present solutions, 6. Reflection.

We choose some concepts' capture and coding to show how to build PBL ontology.

Role: students and teachers or experts. **Learning objectives:** Precisely to solve the concrete problem, meaning to get basic knowledge. In other words, PBL seeks knowledge beyond getting basic knowledge in B·S·BLOOM's systemic classification. Its expression: not (has-objectives only Knowledge) has-objectives min 2 cognitive-domains.

The coding process demonstrates instructional design ontology's clarity (without misunderstanding or ambiguous concept) and expandability (we build PBL ontology in the framework of learning design ontology without any change).

With the help of visualization plug-in and external semantic reasoner, i.e. Facet 3.1, protégé can get activity sequences as shown in Figure.4 for given selection of activities without sequence. The brown lines stand for "has prior to "while the green ones stand for "has next to".

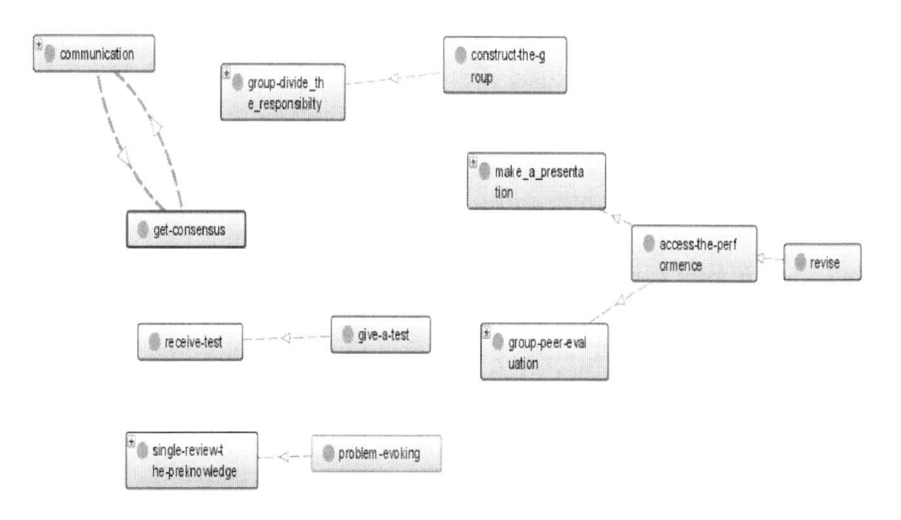

Fig. 4 The result of ordinal relationships reasoning

As shown in Figure.4, the deduction sequence is in accordance with the real one, however not a complete one and fails to link up all activity. Though such, it can provide quite a few information for activity arrangement. We think the cause of sequence discontinuity is the existence of concept incoherence or the scarcity of ordinal relationship.

5 Conclusion and Future Work

We explore how to apply ontology theory to knowledge representation and information modeling in learning design. First, we exact core concepts and organize them into harmonious framework by reference to EML and IMS learning design with nuanced modifications. And, we introduce two types of relationship to represent activity sequence which make machine deduction available. Then, we develop the whole learning design ontology adopting OWL.

We test the clarity, coherence, and extensibility of the whole instruction design ontology by building one classical instruction mode PBL. The result demonstrates the ontology is robust for detailed instruction mode. And we observe that the deducted result about activity sequence is in accordance with real ones though there are several interruptions in sequence.

In future work, we plan to extend our learning design ontology to incorporate more learning design rule sets and cases, which provide the widen and deepen domain knowledge.

Acknowledgments. This work is supported by Supported by the National Natural Science Foundation of China under Grant No. 60736014 and the National High Technology Development 863 Program of China under Grant No. 2006AA010108.

References

1. Perez, A.G., Benjamins, V.R.: Overview of Knowledge Sharing and Reuse Component: Ontologies and Problem Solving Methods. In: Stockholm, V.R., Benjamins, B. (eds.) Proceedings of the IJCAI99 workshop on Ontologies and Problem Solving Methods, KRR 1999, pp. 1–15 (1999)
2. Guarino, N.: Formal Ontology in Information Systems. In: Proceedings of FOIS 1998, Trento, Italy, June 6-8, pp. 3–15. IOS Press, Amsterdam (1998)
3. Wei, S., Lu, Q., He, K., Li, Y.: The Semantic Model of AID And Its Realizing Methods. Open Instruction Research 15(6) (2009)
4. EML (instruction modeling languag), `http://celstec.org/content/learning-networks-professionals`
5. Cui, G., Ren, X., Zhang, H.: SMID: A Semantic Model of Instructional. In: 2009 First International Workshop on Ed Global Learning Consortium, IMS Learning Design Information (2009)
6. Global Learning Consortium, Inc. IMS Learning Design Information Model (January 20, 2003), `http://www.imsglobal.org/learningdesign/ldv1p0/ldv1p0speclicense.html`
7. Paul, R.: Critical thinking: What every person needs to survive in a rapidly changing world, 3rd edn. Sonoma State University Press, Rohnert (1993)

Application of DXF in Developing Vectorgraph Edition System for Laser Carve

Yan Zhao, Hongyi Gu, and Ying Che

Abstract. AutoCAD is best tool to edit vectorgraph, it can save vector graphs in file with extension of Drawing Interchange Format （DXF） . This paper provide a solution that generate data for laser carve by analyzing DXF file, which get help from AutoCAD's powerful vector editing function and provide a new way to develop similar software. In this article, not only the format of DXF file and the illustration of algorithm about reading vector data from DXF file are introduced , but also the suggestion about sampling points from vector graphs by the given precision . At last, the developing trend of this technique and constructive methods to develop the similar software are proposed.

Keywords: DXF, Vectorgraph, Laser Carve, AutoCAD.

1 Summarize of DXF

Computer has been widely applied in graphics. There are two kinds of graphics, bitmap and vectorgraph. The vectorgraph has many merits, for example, occupies little space, isn't distorted after being enlarged and so on, that makes people favorite it very much. AutoCAD of AutoDesk company is most famous, it is very remarkable and the format of it's DXF file has been industry standard, many graphic edition system can load and edit DXF file.

2 Structure of DXF File

There are three important concepts of DXF file structure, section, group code and group code value.

（1） Section: DXF range features of graphics by section, one section store one type of features. There are five sections in each DXF file.

Yan Zhao · Hongyi Gu · Ying Che
Changchun University of Science and Technology, China

Z. Qian et al. (Eds.): Recent Advances in CSIE 2011, LNEE 125, pp. 297–301.
springerlink.com

(2) Group Code: DXF uses group code to describe properties of graphics, each property of a graphic is corresponding to one group code.

(3) Group Code Value： Group code shows the type of property and group code shows the value of property.

You can get help from home page of AutoDesk for more information for free.

3 Solutions of Developing Vectorgraph Edition System for Laser Carve

3.1 Common Solution

Now, almost all companies who product laser carve machine develop vectorgraph edition software by themselves, which cost much source of human beings and money. This solution has some problems, such as, long period of development and frequently updates.

3.2 New Solution

A vectorgraph edition system must have these major functions. First, providing a platform to create and edit vectorgraph. Second, offering an application program to sample points from vectorgraph according to proper precision. Third, developing a driver program to control laser carve machine. People who have rich experience in developing software must agree that the first part cost much work. If we can take advantage of AutoCAD's strong points on editing vectorgraph, we just devote to develop the other two parts.

AutoCAD provide several formats of file to save vectorgraph including DXF, it's an open format. DXF file is composed of ASCII, you can use notepad of Windows Operating System to open and read it. We can read all information of vectorgraph from DXF file easily as we do AutoCAD ourselves.

For example, we can draw a line in AutoCAD, save it in file with ".dxf" extension. Next we need to make a computer program, load this file, read the line's start point and end point, sample all points for laser carve according to proper precision, and then call driver program of laser carve machine to carve this line.

This solution has some advantages as follow:

(1) It makes use of AutoCAD's powerful functions on editing vector graphic, and need not develop similar software ourselves, that will shorten the development period and reduce the cost.

(2) The algorithm is steady, that will reduce the cost to maintain the software.

(3) Easy to widespread. AutoCAD is one of some software which should be mastered in many colleges and universities as a curriculum. Compared with the platform developed by company producing laser carve machine, the worker came from colleges and universities prefer to AutoCAD. This solution brings you more opportunities to catch user.

4 An Example

This section will show an example to describe the whole process of developing a vectorgraph editing software by using the DXF file.

4.1 Edition and Saving of Vectorgraph in AutoCAD's Environment

Open AutoCAD's editing environment, draw a line and a circle , save work in file with name of "Test.dxf" .Please pay attention to the extension of file

4.2 Get Information of Vectorgraph from DXF File

1. Enumerate graphics from DXF file

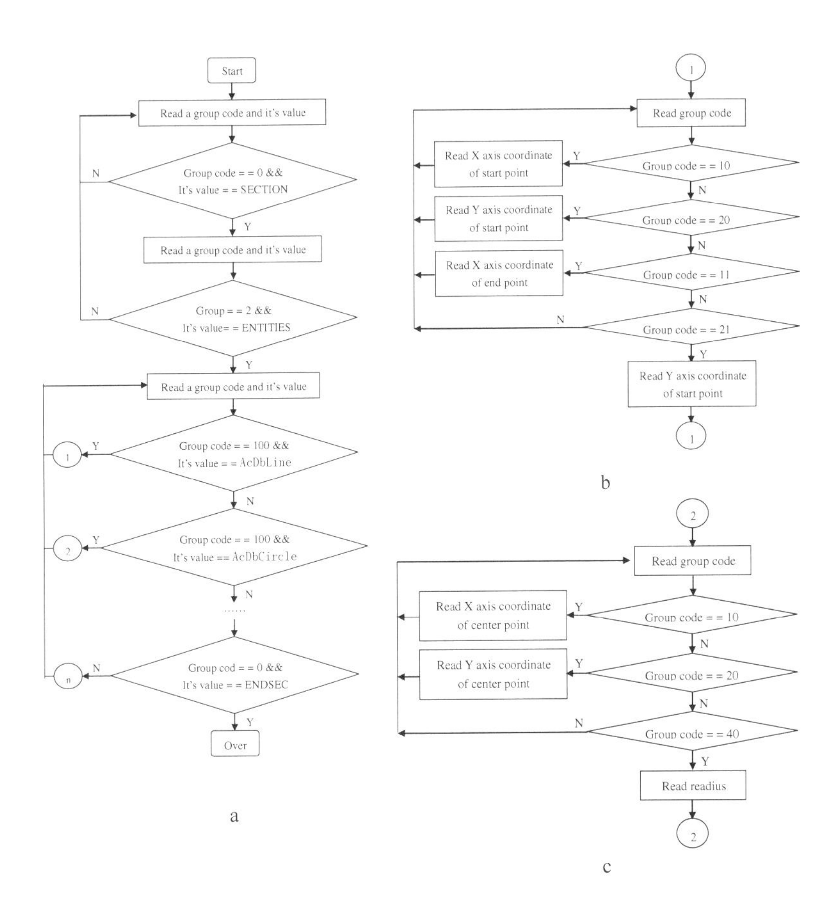

Fig. 1 Flow chart of enumerate graphics in DXF file

4.3 *Sampling Points for Laser Carve*

Before sampling points from the vectorgraph, we must determine the sampling precision. The precision is determined by size of facular, the bigger of the facular size, the lower of the precision.

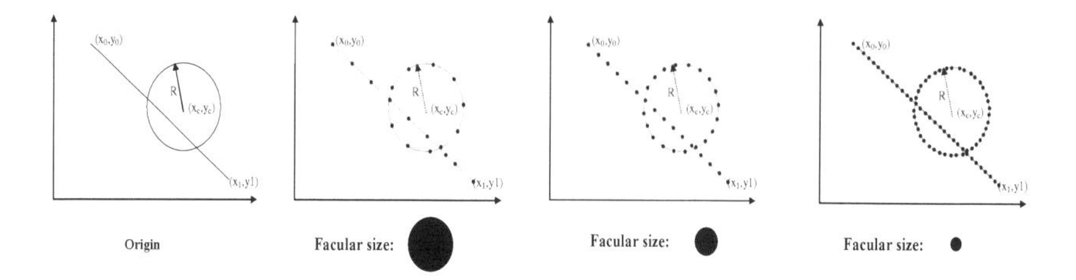

Fig. 2 Sampling precision map facular size

Treat all kinds of graphic as Bezier, that make us use one arithmetic to sample points from all kinds of graphic.

Example:

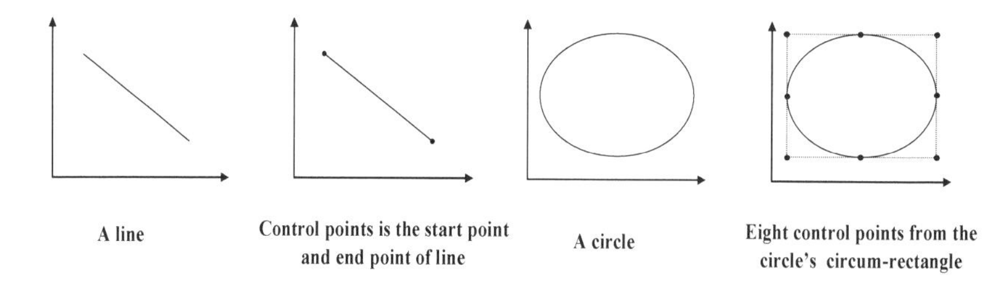

Fig. 3 Choose control points of graphics

From this example, please take care of that it is important to select control points according to every kind of graphic, and it's a difficulty of designing software.

5 Use VBA of AutoCAD

AutoCAD integrates VBA development environment and provides abundant built-in objects for user to create、 edit graphics and get any information of them, and we can use it to promote our strategy as follow.

(1) Develop a COM component named LaserCarve, expose interface named ILaserCarve, create function named Carve with three parameters, graphic type, amount of control points and an array contain control points. The function Carve

should sample points from vectorgraphics for laser carve according to proper precision, and call driver of laser carve machine, transfer an array contain laser carve points to it, control the machine to work.

(2) Open VBA development environment in AutoCAD platform, create a procedure named EnumEntity.

Call built-in objects' method to enumerate graphics and get their property. Save our work in a file with name and path of "e:\ EnumEntity.dvb".

(3) Create a toolbar on AutoCAD platform named LaserCarve.

Add a button on the tool named ExcuteCarve, and set it's macro to call the program created in step two.After these work, we can create and edit graphic on AutoCAD platform and click the button to complete laser carve.

This strategy has some advantages:

(1) Integrate laser carve function into AutoCAD platform and easy to use.

(2) COM is language irrespective program development techniques and easy to update.

(3) Its easier to get graphics' information by built-in objects then any other method.

References

1. Zhang, F., Zheng, X., Lu, L.: AutoCAD VBA second development, pp. 176–193. Qinghua University Press, Beijing (2006)
2. Cottingham, M.: Mastering AutoCAD VBA, pp. 35–102. Electronics Industry Press (2001)
3. Sun, L.: Computer Graphics, pp. 119–202. Harbin Industrial Technology University Press (2006)
4. Zhang, G.: AutoCAD2000 VBA Development Technology, pp. 236–257. Qinghua University Press (2000)
5. MacGeorge, G.: Start automating AutoCAD today with basic VBA. Inside AutoCAD (2006)
6. Shirley, P., Peters, A.K.: Fundamentals of Computer Graphics (2002)
7. Research and Development of Engineering Graphics Intelligent Examination System. In: Proceedings of the 8th China-Japan Joint Conference on Graphics Education (2007)

Gantt Chart Generation Technology Based on Web Applying in Manufacturing Execution System

Yadong Fang and Laihong Du

Abstract. In order to resolve the problem of graphic display for production dispatching result in the process of manufacturing execution system application, three methods of gantt chart generation based on web are disposed. Firstly, the core classes of JFreeChart component is introduced, and ChartDataSet class is constructed to realize accession of gantt chart data. Further more; the generation instance of gantt chart by JFreeChart is introduced. Secondly, the output steps of gantt chart for plant job dispatch are discussed, and effect figure by ChartDirector is illustrated. Lastly, the method of gantt chart for plant job dispatch by java servlet technology is introduced, and the generation steps and instance of gantt chart are discussed.

1 Introduction

Gantt chart is largely used to express the arrangement of production plan in production scheduling. In gantt chart, bar charts indicate the work and time, which shows the task situation clearly. Gantt chart is a effective tool to solve the workshop production scheduling problem, it shows the work has been done and the work to be done on the time horizontal axis, the people and equipments undertake the work are shown on the vertical axis, so that the complex data of production scheduling algorithm are displayed with intuitive graphical visualization and quantification, and the workshop plan arrangement situation is clear at a glance, therefore gantt chart is an effective tool for workshop managers to understand the overall situation and arrange the schedule. With the deepening of MES applications, the needs of remote production scheduling in discrete enterprise under

Yadong Fang
Electro-mechanical Engineering College of Xi'an Technological University,
Xi'an, Shaanxi, 710032, China
e-mail: fangyadong@xatu.edu.cn

Laihong Du
School of Administration Xi'an University of Finance and Economics

Z. Qian et al. (Eds.): Recent Advances in CSIE 2011, LNEE 125, pp. 303–308.
springerlink.com

Intranet and Internet environment are more and more, which urgently required solving the problem of graphic display and interaction based on web in remote production scheduling.

ILOG Corporation which is merged by IBM designed very advanced software engine including c++ and Java software components for business rules, optimization and visualization. ILOG gantt software components is a gavtt components based on MVC , it can build rich Gantt charts in Microsoft Visual Studio .NET and Borland C# Builder , achieve interactive adjustment of job-shop scheduling based on Web, but it can not edit production plan by moving the gantt charts with mouse, Java chart engine JFreeChart is open source project of open source code site SourceForge.net, it can be deployed on the J2EE application server by JFreeChart.jar and Jcommon.jar , gantt chart based on Web can be generated with its API, it can produce hot tip, but does not support interactive operations. Chart Director has rich graphical component libraries, development platform supports various Web application, allows users to control and customize chart details, generated image map support tool tips and mouse interaction with the graphical control, however, user need to buy the appropriate module, and the technical details are fully packaged, no interactive Gantt applications are reported. In this paper, to meet the demand of production scheduling graphic display in networked manufacturing process, three methods of gantt chart generation are introduced emphatically.

2 The Generation of Gantt Chart Based on JFREECHART

JFreeChart is a Java project of an open source site –SourceForge.net. It is mainly used to develop a variety of charts, such as pie charts, bar, line, area, distribution, mixing diagrams, Gantt charts etc. It can deploy a J2EE application server by packaging component application Jar files, and through its application programming interface we can finish the generation of Gantt chart based on Web.JFreeChart is mainly consisted of org.jfree.chart and org.jfree.data.The former is mainly concerned about the graphic itself and the later concerned about data of graphical display. Org.jfree.chart package mainly includes class: ChartFactory, ChartUtilities, JFreeChart, renderer, plot, axis, etc. Org.jfree.data package mainly includes class: gantt, category, etc. JFreeChart engine itself provides a factory class used to create different types of chart objects. ChartUtilities class can output completed Chart object to the PrintWriter object.

This paper builds Java classes- ChartDataSet to achieve Gantt data acquisition, which class diagram is shown in Figure 1, Gantt chart of key equipment occupancy shown in Figure 2. AccessDatabase class is used to operate MES database to obtain manufacturing tasks and equipment, team and other dynamic information. Key Equipment sort is used to determine key equipment of the process of shop scheduling dispatching. The following is the detailed descriptions about properties and methods of ChartDataSet class.

1. Construction method set ChartDataSet, creating a sery of task object -s1 through class Task Series, and key equipment class key Equipment Sorting will create a instantiated object, by which equipment returns a variable number of

key equipment, and then construct a new array taskInfoSpt [][] combined with database query method of database operations encapsulated class Access Database. The array is constructed by task number, end time and start time. It can be checked out according to two requirements: critical equipment number and the scheduled task status.
2. After the time array was built, we use the method of separation of string class StringTokenizer and the method of string processing such as substring, lastIndexOf and indexOf to deal with the string of the start time and end time. Then, the year, month, day and hours are separated from the time format and saved to the corresponding time array according to the requirements of processing date.
3. We create data which is needed in gantt chart with the help of the Task, TaskSeries and askSeriesCollection classes. First, we create the task object which contains the data in the array of beginTime[][] and endTime[][] with the Task classes. Then, we add object t2, t3, t4 and t5 created by Task classes with the help of the object s1 created by TaskSeries.
4. Create a collection object of task series set using the TaskSeriesCollectionof class which can realize gantt classified data set interface. At this time, the object collection will add tasks series object s1.Then, the collection object is retuned by the method setChartDataSet. In this way, we can get the data needs to be displayed by Gantt chart.

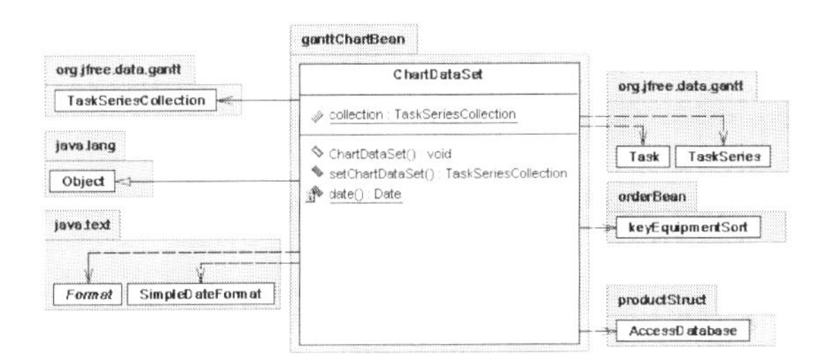

Fig. 1 Get the class diagram of the gantt chart data

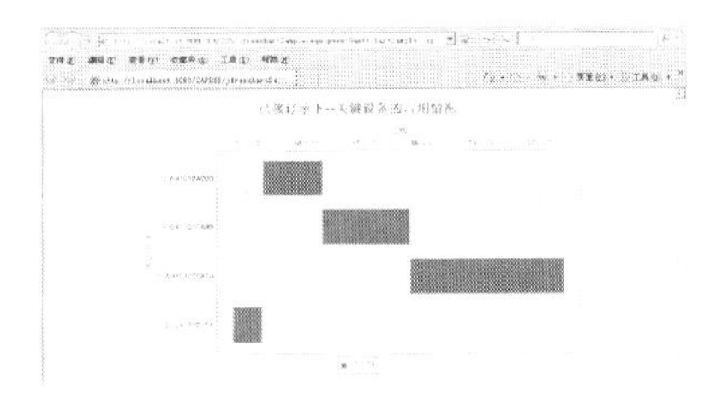

Fig. 2 JFreeChart gantt chart of the key equipment take up situation

3 Method of Gantt Chart Generation Based on ChartDirector

ChartDirector control use convenient, quick, flexible, have powerful function, and strong interactivity. In web server and embedded application development, as a ideal tool, it have some characteristic, such as rich component libraries of chart graphics, beautiful figures, widely development platform support, support many web application, have the object which based on an API(application programming interface) and allows users to control and custom chart details, all above made it can design satisfaction chart for users. ChartDirector have many classes, such as GetSessionImage.class, XYChart.class, Chart.class, DataSet.class, Axis.class, DrawArea.class.Etc.Fig.3 is a gantt chart example for job shop scheduling results, the main steps to create gantt charts output about workshop scheduling problem by ChartDirector are as follow:

1. Put ChartDirector classes pack file ChartDirector. jar to the directory that the Web application class library Tomcat4.1 in the path Tomcat 4.1 \ webapps \ ROOT \ Web-INF\ the lib, and set corresponding web program configuration file of ChartDirector.
2. Call this control class and through the JavaBean technology called ant colony algorithm to solve the single objective workshop scheduling problem class ACOforJSP and processing equipmentforJSSP class. The two classes are translated from Java source code themselves.
3. Call processing equipment class EquipmentforJSSP method to create a device array, and store the data of the array in a label array named labels, it signify the task gantt charts y-coordinate represents.
4. Setting the task index number, start date, end date and color that each progress bar corresponds to in the Gantt chart. Call the ACOforJSSP method to create a new array sammple[][].In the task index number stored in the array taskNo []array sammple[][] of fifth columns by 1 numerical value. So it can correspond to the label array device tag number. And then a new array sammple[][] in the third and fourth columns respectively stored to the start date and end date of the array startDate[]array endDate[], then the new array sammple[][]in the sixth columns representing each of the progress of the sixteen hex color value stored in an array of colors colors[].
5. Create a Gantt chart object chart, the size of 700 x 400 pixels, the background color for the 0xffcccc, the border color for 0x000000, 3D border width is 1 pixels. Such as XYChart C = new XYChart (700, 400, 0xffcccc, 0x000000, 1).
6. Arrange a Gantt chart overall title, fonts, font and background color.
7. Setting related property of the rectangular drawing area in the position (140, 55), and setting the date of horizontal scale and display format, setting horizontal scale value.
8. Add a variety of color layer to the progress bar of the Gantt chart and the width of the progress bar degree. Setting related property of the illustrations (Legend) in the position (140,320).
9. Generating pictures. Firstly, generate figures through the method of makeSession (request, "chart1") and output to the Session variable chart1, and then generate hot tip for the pictures.

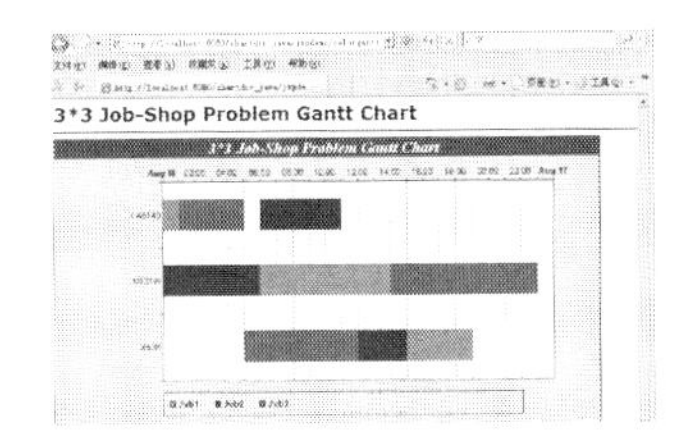

Fig. 3 Job shop scheduling ChartDirector gantt chart

4 Generation Method of Gantt Chart Based on Java Servlet

The third way is write Java classes directly. It inherits the HttpServlet classes in the bag servlet, and the gantt chart will be chalked out in the page of your browser by adding code processing HTTP request in the method of doGet and doPost. The job shop scheduling gantt chart effect is shown in Fgure 4, and which is generated by the servlet classes. The specific procedure is as follows:

1. Set the output document MIME type for "image/PNG", and then, create gantt chart canvas size by instantiation BufferedImage class, finally,get the graphical environment objects.
2. Set the titles of gantt charts, draw background and write the name of coordinate transformation using the setColor, fillRect, drawRect and drawString method in the Graphics classes.
3. Draw gantt chart grid lines using setColor, drawString and drawLine method in the Graphics classes.
4. Draw specific gantt charts and coordinate transformation coordinates scale using the same method.
5. Instantiation ServletOutputStream classes, and coded image using the write method in ImageIO classes.
6. The Java classes are compiled, and deployed to Web server. At the same time, add the following code to the deployment of file web.xml.

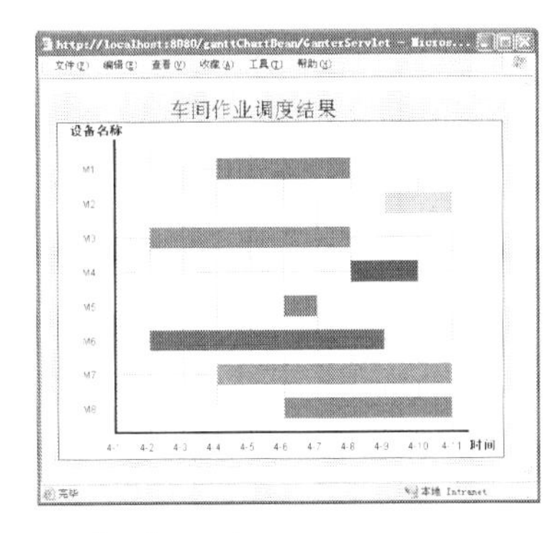

Fig. 4 Job shop scheduling servlet Gantt chart

5 Conclusion

Three display methods of production scheduling Gantt chart that supporting the Web application are proposed during the needs of manufacturing execution system scheduling are analyzed in this paper. In the application process of discrete manufacturing execution system, the above method can solve the basic production scheduling graphics display problem. But because all of the three methods export gantt chart to the page in the form of pictures, so they only support simple image hot operation, and the mouse action response mechanism is weak. Rely on the mouse drag and click operation to realize the adjustment of the task using JavaScript language response page elements action will became one of the hot research topic in the MES.

Acknowledgments. The research work is supported by Shanxi Provincial Programs for Technology Research and Development under Grant No.2010K01-076, namely "The Research and Application of Digital Manufacturing Execution System".

References

1. IBM ILOG Production, http://www.ilog.com/ (cited February 15, 2009)
2. JFreeChart, http://www.jfree.org/index.html (cited February 17, 2009)
3. ChartDirector, http://www.advsofteng.com/product.html (cited February 18, 2009)

Legacy System User Interface Reengineering Based on the Agile Model Driven Approach

Yen-Chieh Huang and Chih-Ping Chu

Abstract. As technologies are continuously evolving, it has become almost mandatory to update and redesign systems to be in flux. Organizations can obtain the most cost-effective and valuable solutions by transforming their legacy assets. Reengineering the Legacy System generally aims to lessen the cost of application maintenance and make the systems more up-to-date and efficient. Thus, we provide an agile reengineering methodology for a Legacy System to smoothly evolve to a new system, and simultaneously, the Model-Driven Approach and Service Oriented Architecture (SOA) are introduced into the reengineering process. The aim of this methodology is to use static and dynamic analysis to model the Legacy System interface. The final output model can assist the model transformation tool to rapidly generate a specific language interface and data service, and then join the UI and data service under SOA. This agile reengineering methodology is a new development process in system reengineering.

1 Introduction

The evolution of information technology is the latest challenge facing today's enterprise, and due to the rapidly changing world, the Legacy System needs to evolve. Most small and medium enterprises are using Legacy Systems rather than the latest technology, such as a high level user interface and a more effective database. Beside, most legacy systems lack support documents or even the original system source code [1]. Agile processes are characterized by considerably less emphasis on analysis and design than almost all other modern life-cycle models.

Yen-Chieh Huang
Lecturer, MeiHo University
23, Ping Kuang Rd., Nei Pu Hsiang, Ping Tung 912, Taiwan
e-mail: p7894121@mail.ncku.edu.tw

Chih-Ping Chu
Professor, National Cheng-Kung University
No.1, University Road, Tainan 701, Taiwan
e-mail: chucp@csie.ncku.edu.tw

Z. Qian et al. (Eds.): Recent Advances in CSIE 2011, LNEE 125, pp. 309–314.
springerlink.com

Implementation begins much earlier in the life-cycle rather than using detailed documentation [2]. MDA can be divided into four phases: Computation Independent Model (CIM), Platform Independent Model (PIM), Platform Specific Model (PSM), and Code [3, 4].

In this research, we build an agile model-driven reengineering methodology to reengineer a legacy system. Besides, we chose Flex as the specific platform. In the process of agile reengineering, we focus on building a standard process to model the legacy system interface and to produce the system model by UML. First, we construct a CIM. Due to the specific platform we chose, we don't emphasize PIM and PSM mapping. We refine CIM for the PSM. These models will be part of the new system documentation. Next, we develop a model transformation tool. This tool parses the PSM model and proceeds with the model transformation operation to immediately generate the Flex interface and PHP data service. Then, we use Flash 4 Builder IDE connecting the UI and PHP data service and other web services under SOA. Finally, we generate a new system mock-up interface with basic functions for demonstrating to clients. The clients give feedback and add into the next iteration cycle. This is a new development process in system reengineering.

2 Related Works

Reengineering means to reorganize and modify existing systems to enhance them or to make them more maintainable [5]. The reengineering method includes the user interface compatibility, database compatibility, transition support, system interface compatibility, training and so on [6]. Bisbal et al. further classified the system evolution as including wrapping, migration, and redevelopment [7]. In this research, we apply legacy system migration to a new platform or environment, and the system functionality as well as the data will remain unchanged and be termed migration [8].

As the name implies, AMDD is the agile version of Model Driven Development (MDD). MDD is an approach to software development where extensive models are created before the source code is written. A primary example of MDD is the OMG's MDA standard. With MDD, a serial approach to development is often taken, which is quite popular with traditionalists. AMDD creates agile models which are barely adequate to drive your overall development efforts.

A service-oriented architecture is essentially a collection of services. These services communicate with one another. The communication can involve either simple data passing or two or more services coordinating some activity. Some means of connecting services to one another is required. The SOA means a service consumer sending a service request message to a service provider, and then the service provider returns a response message to the service consumer. The request and subsequent response connections are defined in some way understandable to both the service consumer and service provider. A service provider can also be a service consumer [1].

3 Research Model

We propose an agile and iterative reengineering methodology based on MDA and SOA. Procedure for each cycle:

Step1: Legacy system interface modeling to CIM (analysis requirement).
Step2: Constructing PIM according to CIM.
Step3: Constructing PSM according to PIM.
Step4: Code generation

The modeling architecture is illustrated in Fig.1. In the legacy system we only have limited information to reengineer. We focus on the legacy system that only exists as an executable system without any source code or documentation. In this case, we can only obtain information from the user interface. First, we derive CIM from UI, refine it to PIM (optional), and finally refine it to PSM. In PSM, we need to derive the Robustness class diagram and sequence diagram describing the full system model. Further, we need to model the user interface component and layout style so we will be able to shorten the adaptation period of the user after converting it to a new system. Note, due to the specific platform Flex, we determined it unnecessary to construct a PIM model.

The Robustness class diagram includes the boundary, control, and entity classes. The system is comprised of a User Interface, Control Logic and Entity Datafile.

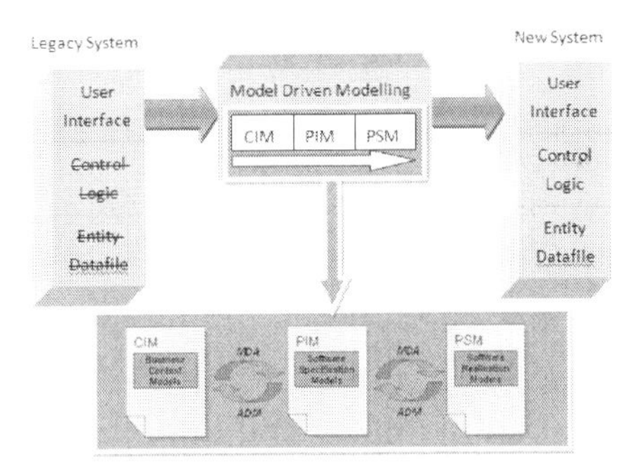

Fig. 1 Model Transformation Architecture

4 CIM Dynamic Analysis and Static Analysis

The CIM describes the business context and business requirements. In addition, we integrate and modernize existing legacy systems according to new business needs in this phase. In the CIM modeling phase, we process dynamic and static analysis. The final outcome of this phase is the use case, activity diagram and Interface Glossary Table. In the following explanation we use a legacy system from National Cheng Kung University Chinese Language Center Management, as shown in Fig.2.

This system is a data centric management system. This is a system that lost maintainability due to the lack of source code, documentation and author. The business logic of the language center has changed so this system needs to evolve. We will discuss a systematic approach (dynamic analysis and static analysis) for improving the legacy interface and follow-up modeling.

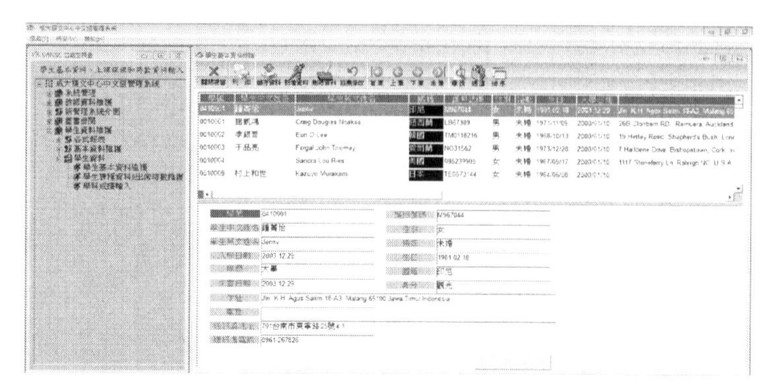

Fig. 2 Legacy system interface example case-NCKU Chinese Language Center Management System

The purpose of dynamic analysis is to analyze the state transition of the user interface and the use case of the system. The state transition model can be completely presented using an activity diagram, which is one kind of navigation model. We can identify states between activities in the activity diagram. Each state is a group of frames and each frame is a group of UI components. The purpose of static analysis is to model the interface static structure. In the static modelling phase, we identify the state of the user interface. Next, we need to identify what content will be placed into the state. We cut all user interfaces' states into fragments. The fragment rules include tow methods, one is frame by frame, and the other is function by function.

5 PSM Modelling

We chose the Flex Platform in this research, so we will not discuss PIM in this paper. We use the outcomes of the CIM phase. The outcomes are the use case, activity diagram and Interface Glossary Table to further refine the PSM. PSM describes the system model with respect to the chosen software technology platforms Flex. The main outcome of this stage is the Flex Robustness class diagram and sequence diagram. The top–down approach is used in this stage. We construct the Robustness class diagram following boundary, control, and entity classes. The stereotype uses << >> to mark the elements. The sequence diagram interprets the State transition sequence (review the activity diagram), messages exchange and function called. The sequence diagram is useful during the implementation phase, and we can review the sequence diagram to perform UI Data binding. The classes of boundary, control, entity, use case, and sequence diagram are three interlocking aspects that all influence one another. In this stage we complete the Robustness class diagram and use case, so we can continue to derive the sequence diagram. Three of these diagrams must keep refining and adjusting to maintain consistency.

6 Model Transformation Tool

Most MDA-compliant CASE tools can only generate partial code and method declaration. Therefore, they lack practicality. We narrow the scope and optimization of the applications for the data centric system. In most data centric cases, there are many operations controlled by the database, for example *Login, Insert, Update, Delete*, and *Search*. The tool we developed utilizes these operations. The tool work uses Flash 4 Builder IDE. The IDE can easily process UI data service binding. In the transformation process, the tool requests some information. There are SQL statements, inputs, outputs type and information for connecting to the Mysql database. The tool will encapsulate these operations in PHP data service format, which can be called to obtain a real return. It greatly improves the source code coverage. In addition, the tool will generate a Flex MXML interface file and embed the ActionScript code to control the state transition. Eventually, we will able to use the Flash 4 Builder to proceed with UI data service binding. Flash 4 Builder IDE is also able to import and bind the WSDL and Http web services to form different combinations of services under SOA.

Finally, after completing data binding, we use Flash 4 Builder to compile the code to a single executable SWF flash file. This file is embedded in a normal html page and executed by Flash player, which is cross platform illustrated in Fig.3.

Fig. 3 New system in Flash

Here, we complete the first iteration cycle. At this stage, the result can be shown to the client for feedback. The feedback will be added to the model to generate another result for the client in the second iteration cycle.

7 Conclusions

This research outlines the steps and procedures to implement a Model-Driven approach to modeling a legacy system. Dynamic analysis and static analysis are used to analyze the Legacy System interface and model for PSM. Eventually, we use a developed assistant tool to automatically generate the Flex interface and data service in PHP. Finally, two results are connected in a Flash 4 IDE environment. This research has thus described a complete agile and iterative reengineering process.

The main contributions of this study are as follows:

- Legacy system agile reengineering methodology uses MDA.
- Agile iteration and increment method use in the reengineering process.
- SOA model can rapidly provide composite services.
- Construction of the modeling procedure to systematically analyze the legacy system interface to derive from the system model.
- Creating Flex PSM stereotype profiles.
- Development of the reengineering process from the legacy system to Web 2.0 RIA in Flash.

References

1. Almonaies, A.A., Cordy, J.R., Dean, T.R.: Legacy System Evolution towards Service-Oriented Architecture. In: Proceedings of the 10th European Conference on Software Maintenance and Reengineering (2006)
2. Schach, S.R.: Object-Oriented Classical Software Engineering, 7th edn. McGraw-Hill (2007)
3. Kleppe, A., Warmer, J., Bast, W.: MDA Explained: The Model Driven Architecture^{TM}: Practice and Promise. Addison Wesley (April 2003)
4. Koch, N.: Classification of model transformation techniques used in UML-based Web engineering. IET Software 1(3), 98–111 (2007)
5. Sommerville, I.: Software Engineering, 8th edn. Addison Wesley (2006)
6. Gowthaman, K., Mustafa, K., Khan, R.A.: Reengineering Legacy Source Code to Model Driven Architecture. In: Proceedings of the Fourth Annual ACIS International Conference on Computer and Information Science (ICIS 2005) 2005E, pp. 1–6 (2005)
7. Bisbal, J., Lawless, D., Wu, B., Grimson, J.: Legacy Information Systems: Issues and Directions. IEEE Software 16(5), 103–111 (1999)
8. Albanese, C., Bodhuin, T., Guardabascio, E., Tortorella, M.: A toolkit for applying a migration strategy: a case study. In: Proceedings of the Sixth European Conference on Software Maintenance and Reengineering 2002, March 11-13, pp. 154–163 (2002)

Research and Implementation on the AJAX Tag Framework Based on J2EE*

Yang Xiao-jie, Weng Wen-yong, Su Jian, and Lu Dongxin**

Abstract. Ajax, because of its asynchrony and new experience to the Web users, is becoming more and more popular in the information system. This paper not only provides the Ajax Tag Framework X-DWR TDF(X-DWR Tag Developed Framework) which is based on J2EE and expanded the DWR, but also gives detailed description of its implementation and frame structure. In a word, it provides a simple and efficient way for people to develop the Ajax's applications based on J2EE

Keywords: J2EE, Ajax, Tag, Framework.

1 Introduction

Recently, Ajax became more and more popular in the developers. It can be applied in the UI development of the application systems based on WEB, and give users a new operation experience. Ajax is the fusion of many technologies and has an incomparable advantage in the user experience, response ability, network independence, easy deployment, easy system upgrade, rational utilization of local resources and so on. At present, many Ajax frameworks have been used in various applications. However, due to the lack of uniform standard, the applicable scope and development mode of these frameworks is different. Generally speaking, there are two widespread frameworks. One is based on the pure JavaScript and the other

Yang Xiao-jie · Weng Wen-yong · Su Jian
College of Computer and Computing Science Zhejiang University City College Hangzhou, China
e-mail: {YangXiaojie,WengWenyoug,SuJian}@zucc.edu.cn

Lu Dongxin
College of Computer Science and Technology Zhejiang University Hangzhou, China
e-mail: dongxin6.lu@gmail.com

* This work is supported by Zhejiang Provincial Education Department Research Project (No.Y20083064) and Science Innovation Project of HangZhou University Key Lab (No.20080433T01).

** Corresponding author.

Z. Qian et al. (Eds.): Recent Advances in CSIE 2011, LNEE 125, pp. 315–321.
springerlink.com

one is based on the server. These two kinds of frameworks have a big difference, and they both have own advantages and disadvantages. However, they have in common is that the high requirements for developers. This limits the scope of the application development of Ajax. In order to expand the scope of applications, reduce the development requirements for Ajax application and make full use of the advantages of these two frameworks, we should integrate them. This paper integrates these two frameworks and provides a new application development framework(X-DWR Tag Developed Framework or X-DWR TDF). First, this paper will introduce the mainstream technologies in current. Then, the structure and realization of X-DWR TDF will are introduced in detail.

2 J2EE Framework and AJAX Technology

J2EE defines the standard for developing multi enterprise applications. The J2EE platform simplifies enterprise applications by basing them on standardized, modular components, by providing a complete set of services to those components, and by handling many details of application behavior automatically, without complex programming. The two most important problems, bottom components and presentation layer, in the current development of enterprise applications solved through this platform. The application development based on J2EE uses the standard system structure, component and the loose design. So, as in [1], it can conceal many details and easy to be used and upgraded. Taglib (tag library), which is using the J2EE framework and based on Java Bean, is a kind of extension of the JSP Tag. Generally speaking, the custom tag is mainly used for hiding objects, managing submission of HTML form and accessing database or other enterprise service, such as E-mail and directory operations, etc, as in [2]. Generally, the developers of custom tag are the programmers which are very familiar with Java programming, data accessing and enterprise service accessing. So, the custom tag let HTML designers don't have to focus on the complex business logic, and could only focus on the web page design. What's more, it not only rational divides the developers and users of tag library, but also packages repeated work and greatly improves the productivity and makes the tag library can be used in different projects. Therefore, it reflects the idea of software reuse very well. Now many mainstream J2EE frameworks provide their own tag library, such as Struts.

Ajax, which full name is "Asynchronous JavaScript and XML", is a technology which is used to develop interactive web applications, as in [3]. It uses the object of XMLHttpRequest to do asynchronous data query and retrieval, and uses XML as data transmission protocols. Then, combine DOM (Document Object Model) and JavaScript to dynamically display and interact. In this way, we could just update the datum which users are interested in. This not only saves the limited bandwidth resources and improves the response speed and user experience, but also reduce the server load pressure. For a typical WEB request, the traditional WEB application allows users to fill in a form. When the form was submitted, application will send a request to the WEB server. Server receives and processes the form, and then returns a new page. In this way, users must wait until the new page back. Because the most HTML code of these two pages is often the same, this method is waste in bandwidth. In contrast, Ajax application can only send or access the necessary data to the server. It uses SOAP or other web service protocols based on XML. Because the client uses JavaScript and asynchronous mode to treat the response from the server, so the users don't have to wait the response from the server and can simultaneously do other things, after sending the request. When the server finished this treatment, the client automatically updates this datum.

Along with the Ajax gradually mature, some simplified Ajax application development frameworks also began to appear. Currently, the popular Ajax frameworks are as follows:

DWR: DWR is an open source code and uses Apache licensed protocols in its solutions. It contains a Java library in the server, a DWR servlet and JavaScript library.

GWT: Web Toolkit (GWT) is a framework. It supports to use java to develop and debug Ajax applications.

JSON-RPC-Java: JSON-RPC-Java is a WEB application Middleware. It is based on JSON (JavaScript Object Notation) and allows DHTML to directly access remote Java applications.

Struts AjaxTags: Struts AjaxTags is an application framework. It added the Ajax (Asynchronous Javascript+XML) technique support in the existing Struts HTML tag library.

AjaxAnywhere: AjaxAnywhere converts the existing JSP components and simply divides the Web page into several areas, in order to realize the Ajax application framework.

These Ajax frameworks, according to their functions and realization methods, can be divided into pure JavaScript library and server Ajax library. They both have their own unique advantages and application scopes:

Pure JavaScript library: It provides the core, remote script.
Advantages:
1. Can handle a variety of server-side languages;
2. Can better separate client-side code and server-side code;

Defects:
1. Need developers to serialize the objects returned from the server;

Server Ajax library: It generates the Ajax's codes in server-side.
Advantages:
1. Can minimize the handled JavaScript code (various types of library is slightly different);
2. Can map the local server objects, as such as database records, to the JavaScript equivalents;

Defects:
1. Bind the JavaScript code to server-side languages too close;

3 X-DWR TDF Framework Introduction

X-DWR TDF framework is based on J2EE and integrates two mainstream Ajax frameworks. Moreover, it is appropriate expanded DWR by defining a group of tags, to simplify the Ajax's development. This framework abstracts and classifies the mainstream Ajax's application problems, and packages the controlling events, form events and callback responses in the form of Tag. In this way, it solves some cross-browser issues. If application developers use these special tags, parameters and events will be dynamically bound, in the page initialization. When a page event is triggered, DWR engine will communicate with the server, and return the processing results from the server and transform them into the objects which can be recognized by JavaScript. Then, the callback function, which is designated by the event, will treat this response. This method, which tag and specific page events were binding only in the runtime, not only effectively improves code's modification, simplicity and readability, but also significantly reduces the complexity in development and is more convenient to extend Ajax applications in original systems. X-DWR TDF framework also provides some flexibility. According to the actual needs, the developers could program JavaScript to process the special requests and responses. In addition, combining with the JS kit, we can provide abundant page expression and new operation experience for users.

4 X-DWR TDF Structure and Realization

Framework structure through abstracting, according to the rules of coupling and cohesion, separates the different function operations and maintains the semantic unification. Then, it uses one mechanism to manage these operations or operation sets. Its purpose is to create some reusable components, to reduce the complexity of the application system development and effectively improve software system's quality in expansibility, maintainability, testability and stability. X-DWR TDF structure is shown in figure 1.

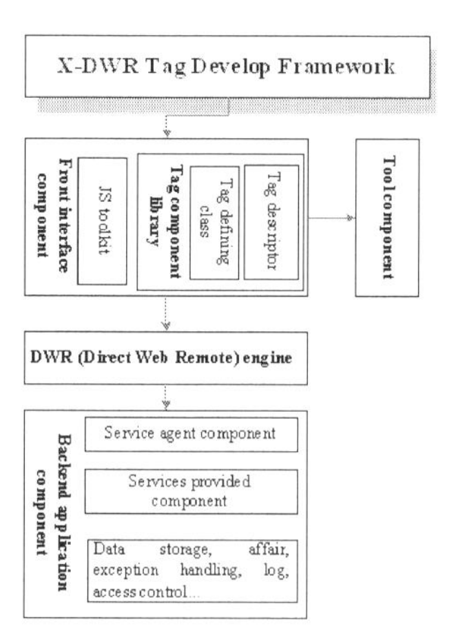

Fig. 1 X-DWR TDF structure

The X-DWR TDF framework was designed strictly in accordance with the MVC's three-layer structure, and was built in according to the standard requirements of J2EE. It mainly contains four parts: front interface component, tool component, DWR engine, and backend application component. The front interface component and tool component belongs to the display layer, DWR engine belongs to the control layer, and backend application component belongs to the business logic layer. X-DWR TDF framework and J2EE framework achieves a seamless integration. They use common durable layer to finish data access.

A.Front interface component:

This component contains JS toolkit and tag component library. The JS toolkit provides a set of useful JavaScript functions. They can be called by the front interface component, at run time.

This set of functions not only deal with the simple WEB client business, such as form validation, dynamic control of the DOM, the data formatted, accessing local resources, etc., but also achieve a certain degree of cross-browser support and reduce the complex programming for WEB developers. The tag component library provides various tags to develop Ajax applications in WEB client. According to abstracting, modeling and defining the common logics, this paper described these tags. Then, these tags were classified by their respective functions and construct into different Ajax application tags. According to the principle of information hiding, these tags also provide standard interfaces for users. When the page initializes, according to these user's configuration parameter, the tags will be bound with the special page components or form events, and then finish the special Ajax request. In this way, the original application is very easy to add new Ajax needs, and the Ajax business design and interface design can be separated. The method, which is only binding in the initialization, provides support for the normative of code and the expandability of applications. All of these are not to increase much redundancy for the system, and will not affect the complexity of the system's release and deployment.

In the process of realizing the Ajax application, using the front-end interface component can greatly reduce the difficulty of programming. Based on using these tags, developers don't need to understand the Ajax principle and carry out complex JavaScript programming to complete the corresponding functional requirements. It is helpful to improve the usability of Ajax framework.

B.Tool component:

This program is used to define tag, and provides some common data structure and processing mode. Its purpose is to reduce the coupling between the tags, simplify the extending of tag function and the adding new tags, and improve the flexibility of framework design.

C.DWR engine:

DWR can serialize object and transform the JAVA object into the JavaScript object. This make WEB developer handle objects become more convenient. DWR has a Servlet, which maps to a specific file (such as: DWR.xml). According the configure information in this specific file, this Servlet initialize the server's methods, objects and their attributes which is need public to the Web customers, and create relevant JavaScript agent classes. Using DWR as the realization framework of Ajax can avoid the injurer caused by the dependent injecting of codes to original framework. Because, when began to build Ajax demands, it only need to add the DWR libraries into the original framework.

DWR belongs to the server-side Ajax framework, and can map the local server objects, such as record set of database, into the JavaScript equivalents. Through using the parameter of a particular method to specify a function, then callback the function when the data returns, it solved the problem in combining the asynchronous characters of Ajax and the synchronous characters of calling normal Java methods. Before the data return, client still can continue doing relevant logic. This can guarantee the continuity and comfort in the user's operations, as in [4].

D.Backend application component:

Backend application component is a server component, and it is also the core of this Ajax application framework. It is responsible for handling business logics and application services. The service agent component announced its public business methods. On the one hand, this way clears the contents of the service. On the other hand, this way protects the ability of the security control of the system access and refuses unauthorized services. The service provided component is focused on business processing and returns results to the caller. This component integrates with the J2EE business layer component and uses the data model and data access objects (DAO), which is the same with persistence layer, to complete the necessary data access logic. It also completed the specific operations, such as services, logs, exception handling, and access control and so on, by dint of the original mechanism of the J2EE. However, all of these operations could do without prior consideration in the J2EE framework; the backend application component provides the expansibility of them.

5 Application

X-DWR TDF is a development framework of WEB layer. It not only can greatly improve the efficiency of development, but also can enhance the system's ability in readability, modifiability and expansibility. This paper used the example, which is similar with Google Suggest in their function, to appear the effect of using X-DWR TDF. The result is shown in table 1:

Table 1 The compare in using X-DWR TDF or not

Content		pure JavaScript Ajax	server Ajax	Using X-DWR TDF
demands for programmer	principle of Ajax	familiarity	understand	understand
	Coding ability with JavaScript	skilled, familiarity	General	Not required
amount of code	presentation layer	20-30 lines	10-20 lines	1 line
	logic layer	30-40 lines	10-20 lines	10-20 lines
serialize object		need	not need	not need

6 Conclusion

J2EE framework is one of the mainstream enterprise frameworks, at present. Considering its high development ability, good flexibility and good platform, it has been adopted by most of the enterprises. Along with the diversification of user needs, Ajax applications are increasingly widespread. This paper referenced two types of current mainstream Ajax framework, and provided a new Ajax framework based on J2EE, that is X-DWR TDF(X-DWR Tag Develop Framework). Its purpose is to adapt to more Ajax's demands in the development process of J2EE project, and to simplify the Ajax's realization of these demands and provide more efficient development mode for programmers. This framework also exist some shortage and need to continue to improve. First, because needs to loading tags in each initialization, so if large amounts of JavaScript were encapsulated in tags, this will reduce the execution speed of page. Second, tags can be restricted by parameters, so event called will also been restricted.

References

1. Li, H.-Z., Liu, X.-L., Guo, D.-Q., et al.: The discuss of J2EE architecture. Journal of the Hebei Academy of Sciences 20(3), 152–156 (2003)
2. Zhang, Y., Ma, Y.-X.: Introduce of the custom tag libraries in JSP. Modern Electronic Technology 27(24), 100–101 (2007)
3. Tan, R.-S.: Ajax technology and its development prospect. Heilongjiang Science and Technology Information 11S, 87 (2007)
4. Wang, N.-H., Jin, X.: Ajax technology and its realization of DWR framework. Automation Technology and Application 26(12), 92–94 (2007)

The Design and Realization of Core Asset Library on the Data Processing Domain

Xinyu Zhang and Li Zheng

Abstract. Data Processing Domain (DPD) is becoming increasingly important in the software industry, for which Software production line provides an important support. Core asset library construction is the major task on software product lines. This paper describes the design and implementation of the Core asset Library, researches the key technology and provides the information for software companies.

Keywords: Core asset library, Software reuse, Data Processing.

1 Introduction

With the development of science and technology, each industry produces more and more data. Data Processing Domain (DPD) is becoming increasingly important in the software industry. Moreover, with the Complexity of Data Sources and the variety of users' requirements, data processing presents challenges for realization of software product lines. As noted, a software product line is a set of software-intensive systems sharing a common, managed set of features that satisfy the specific needs of a particular market segment or mission and that are developed from a common set of core assets in a prescribed way in place [1].

This paper investigates the product line on the data processing domain and introduces the design and realization of core asset library in details. Meanwhile it explores key technical skills related to core asset library. This paper is organized as follows. Section 2 presents the background of this work. Section 3 is the core of the

Xinyu Zhang
Computer and Information Management Center, Tsinghua University, Beijing100084, China
e-mail: zxy.tsinghua@gmail.com

Li Zheng
Department of Computer Science and Technology, Tsinghua University, Beijing100084, China
e-mail: zhengli.ts@gmail.com

Z. Qian et al. (Eds.): Recent Advances in CSIE 2011, LNEE 125, pp. 323–327.
springerlink.com

paper and presents the system architecture and the design of core assets library as well as the design and implementation of core-asset-component. Section 4 concludes this paper and presents our future work directions.

2 Background

Core asset library construction is the major task on software product lines domain engineering. Software assets consist of components, tools and related documentation. All assets are managed by core asset library. Tools relating to assets library include development tools for component integration, core assets library management tools and component test tools. First of all, the development staff develops a new component on the basis of requirements through development tools for component integration. Second, the new component will be tested by component test systems and evaluated according to trustworthiness proves [2]. In addition, it will be distributed to the core asset library by means of core asset library management tools. The development staff will further be responsible for the search, download, and maintenance of documentation and version. Integrated Development Platform is able to access and use the new component through the general component interface.

3 Design

3.1 System Structure Design

Figure 1 displays the architecture of core asset library. The core asset library system comprises core asset, core asset management tool, code component development tool and code component test tool [3,4].

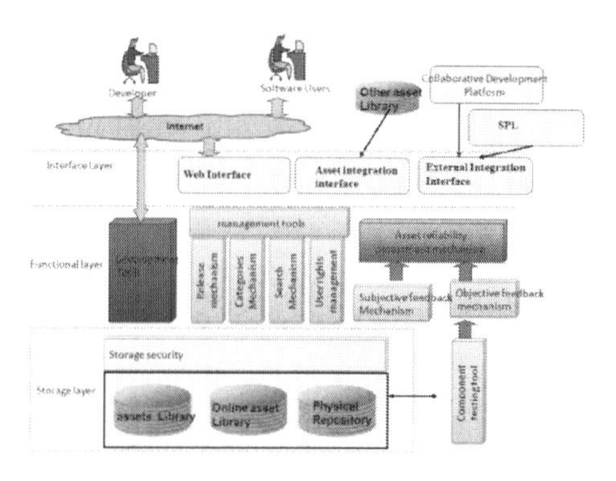

Fig. 1 System structure of the Core asset library

3.3 Code Component Development Tools

The code component development tools mainly provide newly-built components and functions of editing, compiling and unit test. On the software product line platform for data processing, data interaction between components is through XML mainly. Therefore the input and output parameters of methods that the component provides should be in XML. For general use of components, the input and output parameters are all defined by XML string data. As for components' each method, it normally contains 3 tasks: to parse XML input parameters, to execute the business logic and to encapsulate XML results. After new compilation, test for simple units should be performed. Components should possess various properties that can be configured to satisfy all kinds of requirements in product line assembly.

3.4 Code Component Test Tools

The code component test tools are the test tools for product lines. They provide objective evaluation of assets' trustworthiness adopting the B/S structure. They also combine with core asset library management systems and can supply black box test function for all code components inside the system. The test tools can graphically edit test data and process xml test data.

The main functions of code component test tools are as follows:

1) To display component list for choice.
2) To display the function list of appointed components;
3) To display notation of the selected function or description of input & output;
4) Users can enter XML data as function input;
5) To call selected function based on user input used as entry parameters;
6) After that, present output of the function;
7) Provide description for safety test, performance test and reliability test.

3.5 The Trustworthy Assurance Mechanism

The trustworthy software possesses trustworthy properties with the domain requirements. Taking the Software Trustworthiness Classification Specification proposed by the first seminar of the Trustworthy Software Tools and Integration Environment program supported by the National High Technology Research and Development Program of China (863 Program) into consideration in the data processing domain oriented software product line, we hold that data processing domain component should bear the following dependable features[5,6,7,8]:

- Correctness: It is the probability of realization of user expected functions. This is the lowest and basic requirement for software systems.

- Reliability: It is the probability of failure-free software operation during a specified period of time in a specified environment.
- Security: It is the probability of software to provide and safeguard privacy, integrity, availability and validity.
- Real time: It is the probability of software to response and output within a prescribed time.
- Maintainability: It is the probability that the software product can be modified. The modification may mean to correct or improve the software so that the software can adapt to the changes in the environment, requirements and functions in the specification.
- Survivability: It is the probability of software to continue to function during a natural or man-made disturbance or during the occurrence of failures and to restore full service within the prescribed time.

In order to realize the measurement of trustworthy properties, we suggest the following trustworthy assurance framework demonstrated as the trustworthiness tree model:

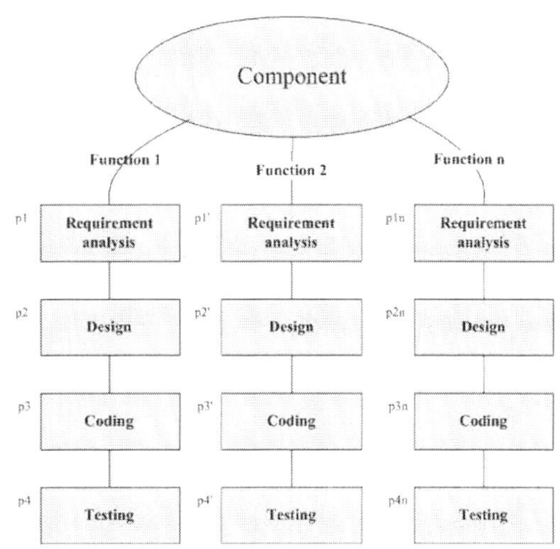

Fig. 2 Trustworthiness tree model

During the software development, there are four major links calling for trustworthy proofs. They are the requirement phase, the design phase, the encoding phase and the testing phase. The final trustworthiness value is calculated by evaluating the trustworthiness value at every stage.

4 Conclusion

In the field of core asset library research, attempts have been made on its stored content, management and measurement, and few have touched on the issue of its realization process. This paper made a detailed introduction to the ingredients and key technical skills of core asset library based on data processing domain, and provided a significant guidance for software product line development.

The future task is to explore the Method of the Measurement for Component Trustworthiness and to make Core asset library provide trusted component.

The national science and technology support plan 2008BAH29B01.

References

1. Frangois, C., Roger, C.: A Product Line engineering practices model. Science of Computer Prograrnrning 57, 73 (2005)
2. Ding, J., Liu, Y.: The Evolution of the Core Asset Library in the Software Product Line, vol. (1), pp. 81–83. Shanxi Institute of Education
3. Medvidovic, N., Dashofy, E.M., Taylor, R.N.: Moving architectural description from under the technology lamppost. Information and Software Technology 49(1), 12–31 (2007)
4. Dashofy, E.M.: Supporting Stakeholder-driven, Multiview Software Architecture Modeling. Ph.D. thesis, University of California, Irvine (2007)
5. TRUSTIE Group (TRUSTIE-STC V 2.0) V2.0 (June 2009)
6. TRUSTIE Group (TRUSTIE-STE V 2.0) V2.0 (June 2009)
7. Liu, X., Sun, H., Bing, Liu, C., Yang, Y., Wang, H., Li, X.: Software Trustworthiness Evidence Framework Specification(TRUSTIE-STEV2.0) (2009)
8. NASA. Software Safety NASA Technical Standard. NASA-STD-8719. 13A (September 15, 1997)

The Development and Realization of 3D Simulation Software System in Mobile Crane

Xin Wang, Youguo Liang, Rumin Teng, and Shunde Gao

Abstract. Considering mobile cranes, a general solution of the 3D simulation software system is presented, and the kinematics math model of mobile cranes is built. Then the simulation system including crane oriented modeling, crane action simulating and crane collision is developed based on Ogre and Bullet. Finally, the usability of our simulation system is proved with an example of mobile crane.

Keywords: mobile crane, crane configuration, 3D simulation.

1 Introduction

With computer graphics and virtual reality technology to further development, 3D simulation technology has been widely used in construction machinery field, not only in the design process, but also in the process of using the crane sale[1]. For familiar product features, performance and usage, foreign product manufacturers were developed crane attendant software for crane configuration and 3D simulation, such as the German company Liebherr, Demag and Manitowoc Company with the launch of cart software. These features of the software is based on two-dimensional wireframe image technology, a combination of arm selection and simulation operation simple process, while providing the necessary information of the crane shown in Figure 1 Lieberr's vehicle screenshots. But up to now has not developed such a vehicle software, and with the increase of domestic products, to help customers quickly

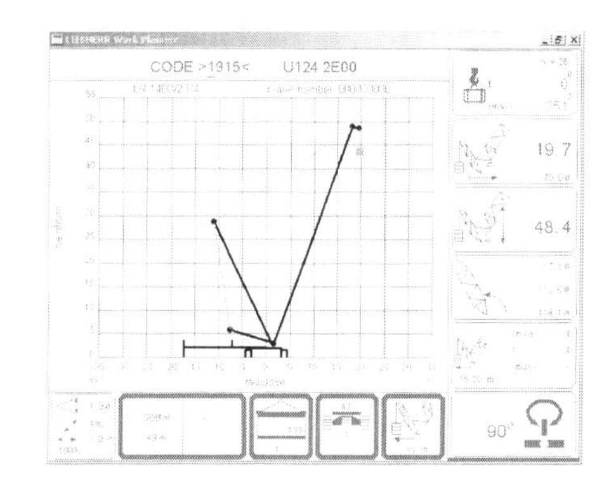

Fig. 1 Liebherr's attendant software system

Xin Wang · Youguo Liang · Rumin Teng · Shunde Gao
School of Mechanical & Engineering Dalian University of Technology, Dalian, Liaoning, China

Z. Qian et al. (Eds.): Recent Advances in CSIE 2011, LNEE 125, pp. 329–335.
springerlink.com

accomplish more tasks in this vehicle software development is imperative. With computer graphics and virtual reality technology rapid development more realistic visual effects, more comprehensive data information software system is possible.

This article will focus construction crane in the research and analysis on the basis of product characteristics, to establish the system framework and the crane.kinematics model by opening the engine to achieve a variety of essential functions, and ultimately developed a sophisticated and practical vehicle software.

2 Research and Design the System Framework

System's goal is to achieve operating cranes interactive simulation software system. Crane configuration subsystem, Extended engine subsystem, 3D modeling subsystem, Human-computer interaction subsystem,Simulation subsystem, Simulation output subsystem components, the structure of the framework shown in Figure 2.

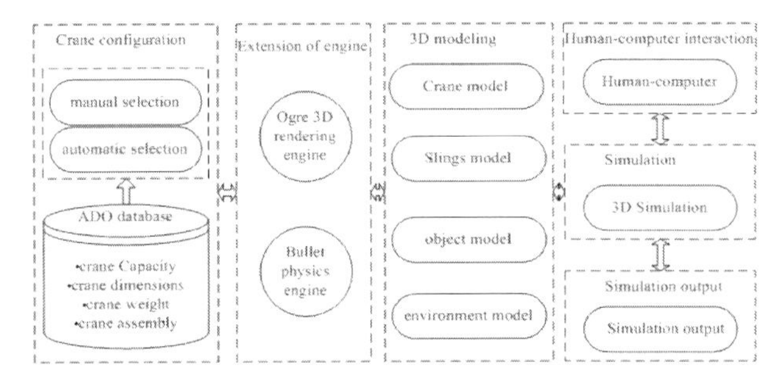

Fig. 2 Construction cranes operating simulation software system framework

Subsystems and the main functions are as follows:

1. Crane configuration subsystem. Crane configuration including manual selection and automatic selection of cranes working condition. Selection process through the back-end database crane dimensions, weight, center of gravity data to calculate the corresponding mathematical model, using ADO database to find lifting performance data, filter out to meet the operational requirements of the operating conditions.

2. Extension of engine subsystem. Including extension of Ogre 3D rendering engine and Bullet physics engine. Ogre provides a convenient quick method of modeling, rational and efficient organization and management of the scene, combined with crane features and operating requirements, so that the system needs to adapt. Bullet physics engine based on the use of the intermediate plug OgreBullet scene model to achieve a precise collision detection and minimum distance calculation.

3. 3D modeling subsystem. Including model building cranes, slings modeling, object modeling operations and operating environment modeling. while building a model crane collision , motion control systems and control system status information.

4. Human-computer interaction subsystems. Keyboard and mouse control provides two ways to control the crane of the action. Can modify Crane parameters, operating parameters of the object, the operating environment parameters and the parameters of sling.

5. Simulation subsystem. Man-machine interface for receiving input information and generate a variety of command and control of each module is operating normally coordinate the data exchange module.

6. Simulation output subsystem. Including the crane status data output, data output ground pressure, collision detection output, the output of the minimum distance calculation and alarm output.3 Key Technologies

3 Key Technology

3.1 System Mathematical Model

Crane configuration and 3D simulation subsystem have to build mathematical models of cranes, the main solution to the boom or jib angle and the lifting height is calculated for different operating conditions are different methods of calculation[2].

1. Mode of operation for the main boom, need to calculate main boom elevation and lifting height.

$$A = \arctan(\frac{d}{c}) + \arccos(\frac{b-a}{\sqrt{c^2+d^2}}) \quad (1)$$

$$H = \sqrt{(c^2+d^2)-(b-a)^2} \quad (2)$$

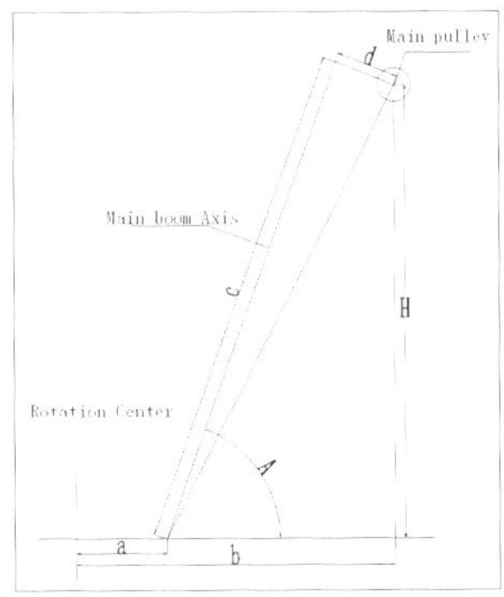

Fig. 3 Main boom structure

2. mode of operation for the fixed jib, need to calculate main boom elevation fixed jib elevation and lifting height.

$$A = A_1 + A_2 + A_3 \quad (3)$$

$$C = A - B \quad (4)$$

$$H = \sqrt{h^2-(b-a)^2} \quad (5)$$

Formula:

$A_1 = \arctan[d/(c+e)]$;

$A_2 = \arccos\frac{(c+e)^2+d^2+h^2-f^2-g^2}{0.5\sqrt{(c+e)^2+d^2}h}$;

$A_3 = \arccos\frac{b-a}{h}$;

$h = [(c+e)^2+d^2+f^2+g^2 - 2\sqrt{(c+e)^2+d^2}\sqrt{f^2+g^2}\cos D]^{1/2}$;

$D = 180 - B - \arctan(g/f) + A_1$.

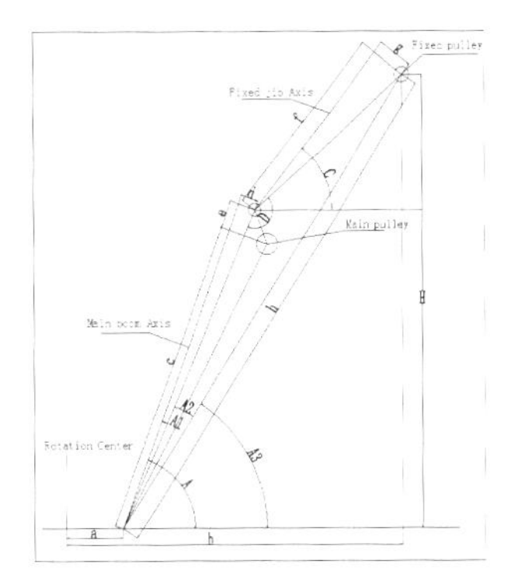

Fig. 4 Fixed jib structure

3. mode of operation for the luffing jib, need to calculate luffing jib elevation and lifting height.

$$C = \frac{\arctan(g/f) + \arccos M}{\sqrt{g^2 + f^2}} \tag{6}$$

$$H = \sqrt{(c+e)^2 + d^2}\,\sin(A - A_1) + \sqrt{g^2 + f^2 - M^2} \tag{7}$$

Formula:

$$M = b - a - \sqrt{(c+e)^2 + d^2}\,\cos(A - A_1);$$

$$A_1 = \arctan\frac{d}{c+e}.$$

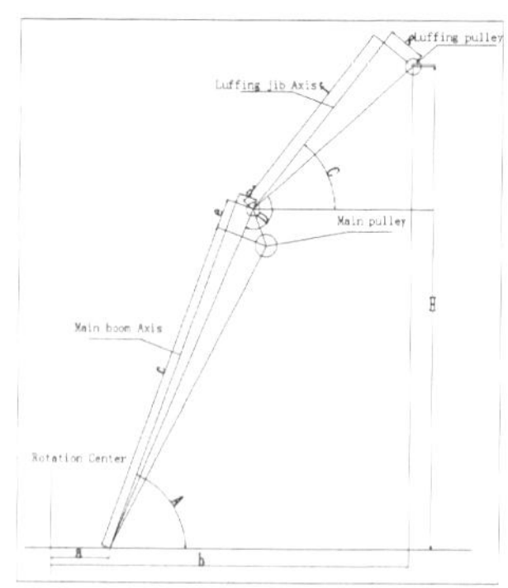

Fig. 5 Luffing jib structure

3.2 3D Modeling

3D modeling is an important part, to complete the mechanical components of cranes, slings, job objects, and operating environment. 3D modeling based on Ogre, hierarchical scene tree structure, interactive modeling and geometric modeling 3DSMAX combined with Bullet physics engine, an effective and realistic 3D model[1].

For example, the standard main boom boom system modeling, use of ADO database technology to read the standard back-end database storing data dimensions boom boom assembly and standard relational data, as shown in Table 1, and corresponding data processing, such as the 12HB treated as data 0m and model information HB.mesh, followed by processing into the system to identify the final processing of information, as shown in Table 2.

Table 1 Standard main boom assembly table

Conditions	Boom length	Boom assembly
HB	48	12HB+6H+6H+12PH+12HT
…	…	…

Table 2 Mesh file assembly table

Conditions	boom length	Boom assembly
HB	48	HB. mesh+H. mesh+H. mesh+P H. mesh+HT. mesh Distance relative to base section：0m, 12m, 18m, 24m, 36m
…	…	…

The standard main boom assembly shown in Figure 6, 48m main boom model shown in Figure 7.

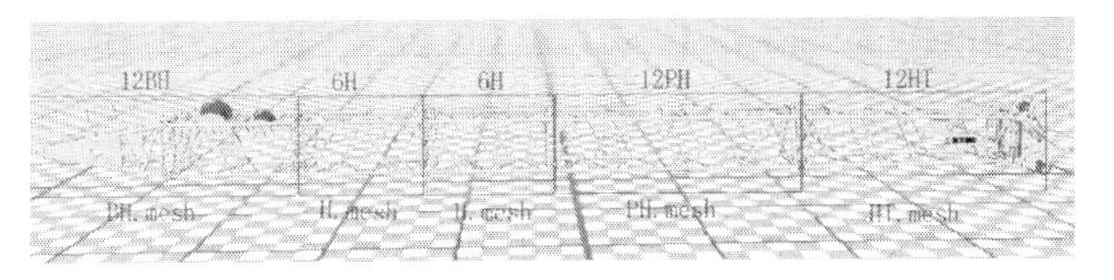

Fig. 6 Standard main boom assembly

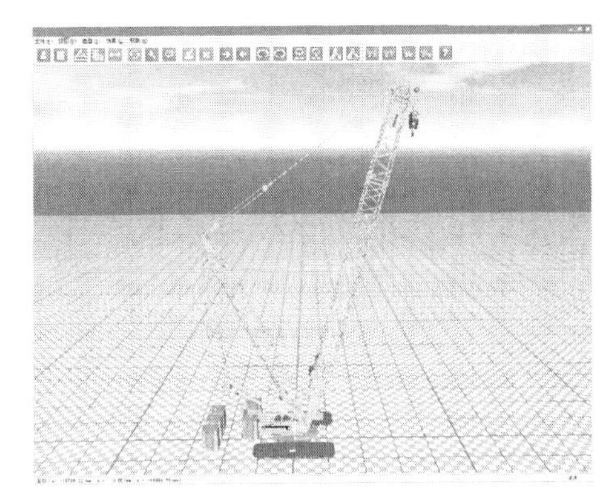

Fig. 7 48m main boom crane model

3.3 3D Simulation

1. Crane motion simulation process. Crane movements are walking machine, turntable rotation, boom luffing, jib luffing and upgrade from the basic movement and sport by their composite. Because the system uses to organize the tree-level model of the scene[4], so by modifying the parent node of each node in the coordinate system of the position vector and orientation matrix can be achieved, such as walking cranes, slewing, luffing, upgrade from a variety of crane bodies action, thus achieving the object translation work, lifting, luffing and other simulation exercise.

In the implementation process, the use of hierarchical tree modeling scene crane parts for model management. The use of top-down tree structure to represent the various components of cranes, which can effectively describe the various components of the crane 3D spatial relationships, but also very natural movement associated with the various components to achieve[5].

2. Simulation of collision detection. In this paper, AABB bounding volume hierarchy methods and the Bullet physics engine, collision detection to provide the interface combine to form a broad / narrow two-stage collision detection method.

AABB bounding volume hierarchy method, the use of simple shapes bounding box will be wrapped up complex geometric objects, collision detection during the first intersection between the bounding box test; if the bounding box intersection, then the geometric precision of the collision between the object detection. Obviously, AABB bounding volume hierarchy tree method for judging the intersection of two geometric objects is not very effective. System scene collision detection between objects in the calculation, including the bounding volume and bounding volume intersection test between the geometry and the geometry and the intersection between the test, which detects the intersection between the bounding volume also contains sub-scene intersection test between the sub-scene and the intersection between geometry and the geometry test the intersection between the detection. Narrow phase collision detection is also known as precise collision detection, bounding volume intersection test is unable to determine whether two objects intersect when the precise and complex intersection test[6]. Bullet physics engine to use the interface mesh body collision detection and accurate collision detection geometry, simple to implement operations, overhead, and improved collision detection speed.

3.4 Simulation Example

To a company 450 ton mobile crane as an example the use of the software system. Operating conditions in the operating range of 20m, the demands for improvement in re-180t of the cylindrical object, lifting height of 25m. Selection of parameters of Figure 8 corresponds to lifting performance enumerable table, Figure 8 crane operating conditions for the selection interface.

According to operational requirements, select the following conditions: conditions SHB, boom length 48m, radius 24m, rated load 198t, Superlift counterweight 250t, Superlift radius 12m, which created 3D crane model shown in Figure 7.

Rigging according to actual situation, job objects, operating environment parameters, and generate three-dimensional image, and then operating on the simulation, the customer moves through the keyboard to achieve a variety of crane control, shown in Figure 9.

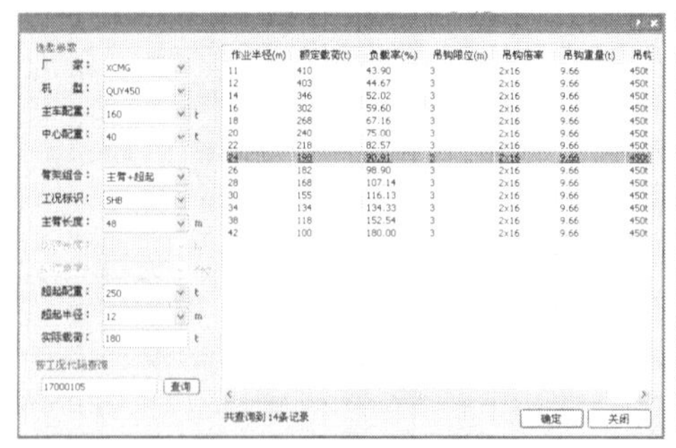

Fig. 8 Crane configuration

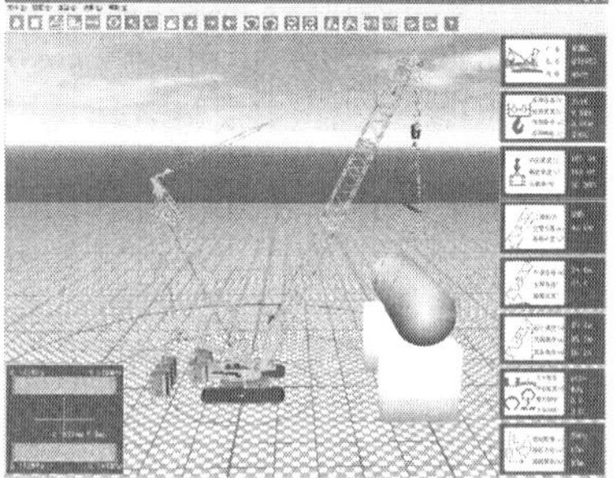

Fig. 9 3D Simulation

4 Conclusion

Considering mobile cranes, Ogre 3D rendering engine and Bullet physics engine is built based on a 3D simulation system operating framework to achieve, crane configuration, 3D modeling, 3D simulation, collision detection and status information display function, through mobile cranes instance, that the job simulation results realistic intuitive, clear and friendly interface, full-featured.

References

[1] Zhang, J., Gang, Z., Rui, Z., et al.: Based on virtual reality simulation system for shipborne special crane. Computer Engineering and Applications 16, 174–178 (2002)
[2] Wu, D., Lin, Y., Wang, X., Wang, X.: For program demonstration of the mobile crane hoisting system design simulation. Journal of System Simulation 20(5), 1187–1191 (2008)
[3] Bo, Y.: OGRE graphics rendering engine based on the visual simulation technology research and implementation. University of Electronic Science and Technology, Chengdu (2006)
[4] Lin, Y.: Based on three-dimensional engine simulation system of lifting (Master thesis). Dalian University of Technology, Dalian (2008)
[5] Maria, A.: Introduction to Modeling and Simulation. In: Andradottir, S., Healy, K.J., Withers, D.H., Nelson, B.L. (eds.) Proceedings of the 1997 Winter Simulation Conference (1997)
[6] Madera, F.A., Day, A.M., Laycock, S.D.: A Hybrid Bounding Volume Algorithm to Detect Collisions between Deformable Objects. In: Second International Conferences on Advances in computer-Human Interactions, ACHI 2009. Inst. of Eles. and Eles. Eng. Computer Society, Cancun (2009)

Modeling Web Applications for Software Test*

Min Cao and Haiqiang Li

Abstract. The chief task of software test based on model, which is an efficient way for testing the Web application, is building an abstract model of the Web application waiting for test. One problem with current Web application test technologies is the lack of tools for modeling the whole Web software. Specifically, there is the lack of support for describing Web application from the view of action and function. This paper is concerned with providing the support for development and test of Web application. The presented novel model, named CBTOWADM, abstracts the Web application as a tree based on its system function and business process. CBTOWADM not only simplifies the design and development of the Web application, but also acts as the model middleware for software test. We describe the basic model definition and the application in software test of CBTOWADM.

1 Introduction

To ease the difficulty of web application testing, a plethora of automated tools and testing frameworks are now available for different aspects of testing. However, most of the existed test tools, including checking tool of link, validation of HTML, catching and roll-back tools, security testing tools, work load testing tools, and so on, are static certification and measuring tools and do not support the function testing of the web application [1]. Model-based testing method of web application provides the support of automatic test case generation for testing function of the

---

Min Cao
School of Computer Engineering & Science, Shanghai University
email: mcao@shu.edu.cn

Haiqiang Li
School of Computer Engineering & Science, Shanghai University
email: lhqmaillove@163.com

* This research is partially supported by Shanghai Leading Academic Discipline Project (Project Number: J50103).

Z. Qian et al. (Eds.): Recent Advances in CSIE 2011, LNEE 125, pp. 337–342.
springerlink.com

web application. The chief task of the software test based on model is building the abstract model of the web application.

Currently, the approaches of building the model of web applications in common use are generally describing the web pages and the relations between different web pages through UML or entity-relation graph, while the dynamic behavior and actions of the web application are modeled by using decision-making or state chart. These approaches seldom model web applications from the view of their behavior, action and function, let alone present functional testing scheme [2]. A. Andrews and J. Offutt address the problem of automatically testing web applications at the system level. They use a hierarchical approach to model potentially large web applications. The approach builds hierarchies of Finite State Machines (FSMs) that model subsystems of the web applications, and then generates test requirements as subsequences of states in the FSMs. These subsequences are then combined and refined to form complete executable tests. The constraints are used to select a reduced set of inputs with the goal of reducing the state space explosion otherwise inherent in using FSMs [3]. But the test of the interconnection or interaction between components and component composition is ill-considered in this approach.

This paper focuses on modeling the web application and the approach of software test based on model. The research concerns the decomposition of the web application. After that, the components constitute the web application and the relations between these components are analyzed. Thus, it is possible to model the web application from the view of system function and component interconnection.

2 Research on Modeling Web Applications

Generally speaking, according to the business process and system function, a web application can be divided into different parts called *module*. Each module can be divided into different smaller module, which is called sub module. Each module or sub module calls *component* or components to realize its function. The presented CBTOWADM(Component-Based & Tree-oriented Web Application Development Model) combines this kind of thought to development method based on component. It uses a logic tree, whose root represents the whole web system, to model a web application. One non-terminal node can either be a module, a sub module or be a called component. The leaf nodes of the tree generally bind components.

One component can be called by one or more module / sub module, thus the architecture of a web application may exchanged from a logic tree to a logic graph. In order to maintain the characteristics of the logic tree, CBTOWADM allows more than one nodes to bind to a same component.

According to above thought, CBTOWADM is defined as a 5-tuple (T, f_m, f_c, f_n, I), where T is a logic tree, I is a finite interface-level set of operations predefined on the tree T, and f_m, f_c, f_n are all mappings. The concrete definitions are as follows.

Definition 1: *T* is a logic tree represents a web application. Its extension includes 1) A finite set of R is introduced to represent the relations between a tree node and its parent. Each $rel \in R$ is method call, inherited relation, and so on, and each *rel* can own a value. 2) Each node of a tree can bind to a component. 3) All the nodes of a tree can be different types, such as a module node means a business process, and a component node binds to a concrete component, etc.

Definition 2: f_m: $M \rightarrow N$, is a mapping from nodes of a tree to function modules of a web application. N is a finite set of nodes of a logic tree except the root. M is a finite set of function modules or business processes which compose a web application. f_m allows the user to map a node of a tree to a module of a web application.

Definition 3: f_c: $C \rightarrow N$, is a mapping from nodes of a tree to components. N is a finite set of nodes of a logic tree except the root. *C* is a finite set of component entities, each of which contains description, implementation, and such useful information of the component. f_c allows the user to map a node of a tree to a component.

Definition 4: f_n: $Na \rightarrow N$, is a mapping from nodes of a tree to navigation tree or sub navigation tree of a web application. A web application can behave itself as a navigation tree, and a logic tree or sub tree can be mapped to this kind of navigation tree. *Na* is a finite set of navigation trees that compose a web application. N is a finite set of nodes of a logic tree except all the leaf. f_n allows the user to map a non-terminal node of a tree to a navigation tree of a web application.

Definition 5: *I* is a finite interface-level set of operations predefined on a logic tree. It provides a complete set of API extended from the basic operations of a logic tree.

A component of CBTOWADM can be a dynamic link library, a web page, the logic function of a web page, a class, a web service, a database or a table of a database, an image file, etc. The dependencies of components can be static or dynamic. In order to describe the software architecture of a web application, and develop the web application by using the described software architecture, CBTOWADM provides the support of component description and component dependency description [4].

3 Application of CBTOWADM in Software Test

CBTOWADM can be used either when developing a web application by assembling the existed components or when testing a web application through the model-based testing approaches and tools.

3.1 Model Abstract of Web Application

In the procedure of software test of the web application, some of the web applications are constructed according to certain model, such as UML or FSM, and abide the rules of the model strictly. However, there are web applications which are developed without any model. Therefore, it is difficult to test these web applications constructed ad arbitrium. CBTOWADM settles this problem well. It abstracts the model of the web application, as shown in Figure 1.

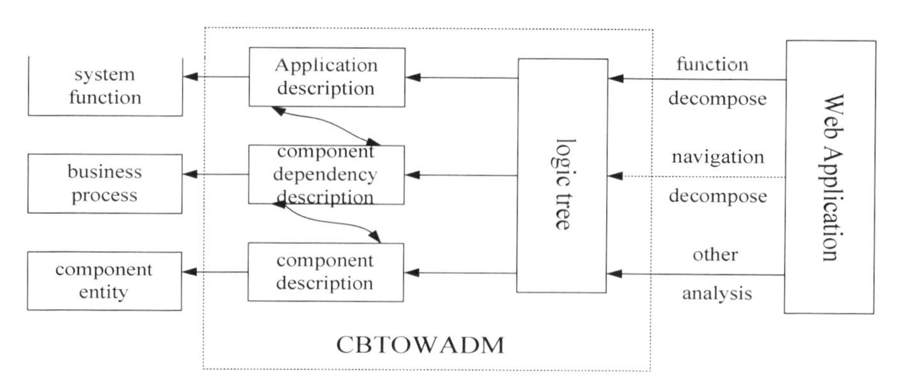

Fig. 1 CBTOWADM model of the web application

First, the function of the web application which is waiting for test is analyzed and abstracted. The web application is divided into different modules from the view of the software architecture, the relations and interconnections between different modules are analyzed. Then, according to the approach provided by CBTOWADM, the decomposed modules of the web application are abstracted as components. The relations and interconnections between modules are abstracted to component dependencies. The component and component dependencies are described according to the rules of CBTOWADM. Based on these component descriptions and component dependency descriptions, the web application is abstracted and described as a logic tree. After that, the decomposed components, business process, system function, and so on, are matched to component description, component dependency description, and logic tree, respectively. Last, the components are analyzed and decomposed concretely when necessary, and the static or dynamic component dependencies are found.

The above procedure decomposed a web application step by step from the view of system function. Both the components and component dependencies are described, as well as the whole web application is abstracted. Thus, not only is the skilled programmer or software designer able to describe the software architecture of the web application, the decomposed components and the component dependencies, but also the ecumenical business operator who is familiar with system function and business process instead of software development technique is able to describe the web application and component dependencies easily. Of course, the detailed analysis and model construction of components is usually delivered to skilled software developer.

3.2 Apply CBTOWADM When Testing the Web Application

Figure 1 illustrates the procedure of modeling an existed web application. That is, the CBTOWADM model of an existed web application is abstracted according to above steps. The next step for testing the web application is transforming the CBTOWADM model to an appropriate model, such as UML or FSM. Then the test cases are produced automatically. Figure 2 hints the application of CBTOWADM in software test approach and test tool.

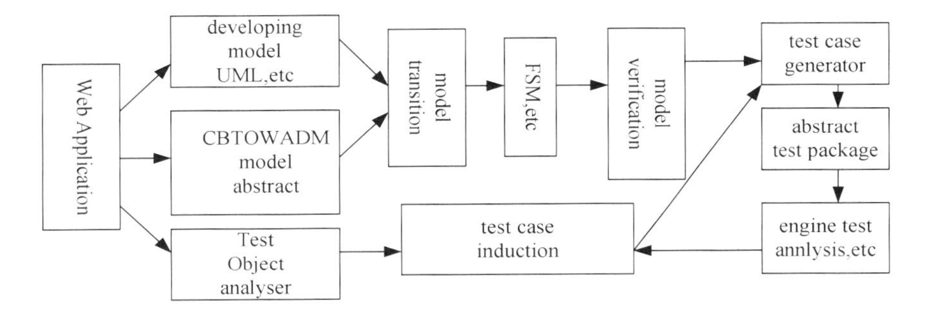

Fig. 2 Web application test system based on model

The web applications, which are developed according to certain model, such as UML, can be described by using its model directly. The other web applications are abstracted and described by using CBTOWADM. Then, the CBTOWADM or UML description of the web application is abstracted as an appropriate model, which is suitable for software test, through model transition. The abstracted model is described by formal or half-formal approach, such as FSM. After that, the transformational model is tested and verified. Combining the verified model with the induction of test case engendered by test analyser, the test case generator produces abstract test package. The abstract test package is tested by test runtime, which is constituted of test engine, analysis engine, and so on. The generated test run path can be used to optimize and extend the test case induction. The above procedure is executed repeatedly, and at last, the test result of the web application is verified.

4 Conclusions

CBTOWADM describes the software architecture of the web application by using the logic tree structure. It not only simplifies the development and assembly of the web application, but also realizes the non-gap connection between requirement, design, development, and assembly of a web application. CBTOWADM can also abstract the function model of existed web application for software test. The future work includes developing tools of model verification and automatic model transformation.

References

1. Hower, R.: Web Site Test Tools and Site Management Tools. Software QA and Testing Resource Center (2002), `http://www.softwareqatest.com/qatweb1.html`
2. Li, L., Miao, H., Chen, S.: Test Generation for Web Applications Using Model-Checking. In: SNPD 2010, pp. 237–242 (2010)
3. Andrews, A., Offutt, J., Alexander, R.: Testing Web Applications by Modeling with FSMs. Software Systems and Modeling 4(3), 326–345 (2005)

4. Ye, F., Zhao, W., Peng, X.: A Component-Assembling Based Approach to Web Application Development. Journal of Nanjing University (Natural Science) 41(Z1), 405–410 (2005) (in Chinese)
5. Li, Q., Cai, H.: Web Application Based on Plug-ins Structure. Computer Science 36(4A), 331–333 (2009) (in Chinese)
6. Valderas, P., Pelechano, V.: Introducing requirements traceability support in model-driven development of web applications. Information and Software Technology 51(4), 749–768 (2009)

A Fully Concurrent Garbage Collector

Delvin Defoe, Morgan Deters, and Ron K. Cytron

Abstract. Over the past two decades, the number of memory-managed programming languages has proliferated. Such languages use uniprocessor memory management techniques to allocate and reclaim storage on behalf of programs. Multicores and other forms of multiprocessor systems have evolved to become common place. However, deployed memory management techniques largely have not taken advantage of the benefits of multiprocessor systems. We present a fully concurrent garbage collector that offers negligible synchronization cost and leverages the power, speed, and potential of multicores and other multiprocessor systems. This collector makes its best effort to avoid suspending application threads as many concurrent (and other *on-the-fly*) collectors do. Instead, it retrieves requisite data opportunistically from each thread to compute its view of the heap and perform garbage collection work. Our collector is an efficient, high performance garbage collector that yields features that are desirable for multiprocessor real-time systems.

1 Introduction

Garbage collected languages like Java, Matlab, Python, PHP, and C# are continuing to be developed and deployed. Multicores and other forms of multiprocessor

Delvin Defoe
Dept. of Computer Science and Software Engineering, Rose-Hulman Institute of Technology, Terre Haute, IN 47803 USA
e-mail: defoe@rose-hulman.edu

Morgan Deters
Computer Science Department, New York University, New York, NY 10012 USA
e-mail: mdeters@cs.nyu.edu

Ron K.Cytron
Dept. of Computer Science and Engineering, Washington University in St. Louis, St. Louis, MO 63130 USA
e-mail: cytron@cse.wustl.edu

Z. Qian et al. (Eds.): Recent Advances in CSIE 2011, LNEE 125, pp. 343–363.
springerlink.com

systems are also evolving at an alarming rate and are becoming common place. However, garbage collection techniques have not kept up to truly allow these languages to leverage the processing power of these super fast parallel processing systems.

Traditional garbage collection techniques were limited in many respects, assuming a list structure for data, or a single-threaded uniprocessor architecture. Uniprocessor systems are no longer the *state-of-the-art*, and the most efficient garbage collector is no longer a *stop-the-world* collector that executes in a single thread when all mutators[1] are suspended.

The field of garbage collection has grown tremendously in an effort to keep up with increasingly complex computing technology. Incremental collectors that interleave their execution with mutator code in a more fine-grained manner have evolved. However, they are constrained in the architecture they target, namely uniprocessor systems. Although this interleaving allows both the collector and mutators to make incremental progress during the intervals when each executes, on a uniprocessor architecture only one thread, the collector or a mutator, can execute at a time. There is no support for parallel execution of multiple mutators or concurrent execution of collector and mutators.

To address these limitations, concurrent collectors and parallel collectors emerged. Concurrent collectors perform garbage collection work, in a single thread, in parallel with mutator execution. These collectors allow multiple mutators to run at the same time, leveraging the speed and processing power of multiple processors. Parallel collectors use several threads to do their work, thereby allowing both the collector and mutators to leverage multiple processors.

Many papers written on concurrent collection and many implementations of concurrent collectors present this collection technique as one that does not scale well with the number of mutators. They employ a snapshot approach that suspends all mutators at the same time, typically at the start or end of a garbage collection (GC), to compute a unified view of the heap. Since mutators can only be suspended at *safe points* [4], the duration of a suspension grows linearly with the number of mutators.

In order to truly take advantage of the capabilities of multicores and other multiprocessor systems, *on-the-fly* techniques [15, 4] that do not suspend all mutators at the same time during a collection are needed. Most *on-the-fly* collectors suspend mutators one at a time, as needed, to compute their view of the heap. Bacon *et al.* [4], and Levanoni and Petrank [15] have each presented an *on-the-fly* collector; however, there are limitations with their approaches. The collector described in this paper improves upon these *on-the-fly* collectors by opportunistically avoiding mutator pauses and reducing the number of mutator synchronization points per collection.

[1] A mutator is an application thread. It is typically called a mutator because the collector perceives it as a nuisance that simply mutates the heap.

1.1 Contributions

In this paper we present a fully concurrent garbage collector that makes its best effort to never pause, but run in parallel with mutators at all times. The four main features of our collector can be summarized as follows.

- Reference counting garbage collection is the approach employed in this collector. To reduce reference counting overhead, our collector lazy updates the reference count of an object by accurately accounting only for pointers in the heap. Section 3 describes how reference count updates are computed in a lazy fashion.
- This is an *on-the-fly* collector that transacts with each mutator on an individual basis, as needed, to compute its view of the heap. Mutators are not suspended during a transaction, as is the case for other *on-the-fly* collectors. Our collector uses the computed view to update the reference counts of heap objects, discover dead objects and reclaim the storage that they occupy.
- Our collector minimizes synchronization overhead between the collector and mutators. Mutators use a short, non-blocking, atomic write barrier to make changes to the heap. The write barrier requires no synchronization between mutators. Moreover, the collector does not transact with a mutator when it is modifying the heap.
- Garbage collection work is performed in cycles and our collector uses a minimal number of per-mutator transactions per collection to compute its view of the heap. These transactions are also short and non-intrusive. We describe this process further in Section 2.2.

In sections 5 and 6 we compare our collector to another *on-the-fly* collector, namely Levanoni and Petrank's collector [15]. It is that collector that inspired our work. We also suggest a few implementation approaches and discuss associated issues in Section 4. These are the main contributions of this paper.

2 Collector Overview

The essence of our collector is that it is an *on-the-fly* reference counting garbage collector that computes a sliding view of the heap in cycles, while it provides reduced synchronization with mutators and performs reference counting in a lazy fashion. In this section we describe some of these features in greater detail. For convenience we refer to this collector as a fully concurrent garbage collector (FCGC).

2.1 Computation of Sliding View

Unlike collectors that require a consistent view of the heap to perform GC accurately, FCGC requires only a "fuzzy" picture of the heap. A consistent view is derived typically from a real or virtual snapshot of the heap taken at the beginning, middle, or end of a collection. Given such a view of the heap, most concurrent collectors are able to determine which heap objects are *live* (or reachable) and which heap objects are *dead* (or unreachable). They use such liveness information to reclaim the storage of dead objects. While a snapshot facilitates collection, computing a snapshot is costly. The collector usually suspends all mutators at the same time to compute the snapshot. During the suspension, all processors except the one on which the collector is executing are idle. To overcome this wasting of processor cycles, FCGC uses the notion of a sliding view [15]. A sliding view is an inexact view of the heap that is computed during a collection when the collector transacts with all mutators one at a time, as needed, during a time interval, say $[t_1, t_2]$. During this interval, each mutator contributes to the computation of the view by exchanging state information with the collector. Such information is collected in each mutator's local buffer when it modifies references to heap objects (from other heap objects). The collector subsequently uses the accumulated information to update the heap reference count of each affected heap object and to reclaim the storage of objects with zero heap reference counts that are not pointed to from the root set.

2.2 Cyclic Execution

FCGC executes garbage collection in cycles. During a collection, it engages each mutator in transactions called *soft handshakes* or simply handshakes [15]. A handshake is a synchronization point between the collector and the mutators in which the collector interacts with each mutator on an individual basis, as needed. Typically, in *on-the-fly* collectors that execute garbage collection in cycles, a per-mutator interaction entails suspending the mutator, retrieving its buffer, and subsequently resuming it. Each mutator receives an empty buffer before it is resumed. A subtle yet significant difference between the way our collector transacts with mutators is worth noting. FCGC makes its best effort to avoid suspending mutators. Instead, is uses a primitive *try-retry* mechanism (see Section 4.4) to effectively retrieve the requisite buffered information from each mutator. It retrieves such information only when the mutator is not using its buffer. FCGC uses only two such transactions while other *on-the-fly* collectors use at least four handshakes [15].

2.3 Reduction of Synchronization

FCGC is a non-intrusive collector. It uses a non-blocking write barrier that requires no synchronization between mutators. Mutators may execute concurrently in the write barrier without interfering with each other. The write barrier is short, performs minimal locking, and executes quickly. Interactions between the collector and mutators are also short and on a per-mutator basis. Each mutator is engaged twice per collection in a non-blocking handshake as described in Section 2.2. Thus, the synchronization between the collector and mutators is low.

2.4 Lazy Reference-Counting

An added benefit of FCGC is that it updates the reference count of each heap object lazily—it does not keep account of changes to pointers in the stack or registers [8]. The classical referencing-counting collector [6] and many of its derivatives adjust object reference counts for every pointer assignment. FCGC only adjusts reference counts of heap objects affected by assignments in successive collections. We illustrate this deferred reference counting [8] approach with an example. Let ptr be a pointer field in an object, say obj. Assume that during the previous collection $obj.ptr = O_1$. Suppose during the current collection the following assignments are made in this order:

$$obj.ptr = O_2,\ obj.ptr = O_3,\ \ldots,\ obj.ptr = O_n.$$

At the end of the current collection we only decrement the reference count of object O_1 and increment the reference count of O_n. Not only is this approach lazy, it also saves significant overhead. The classical collector would adjust reference counts as follows:

$$O_1.rc--,\ O_2.rc++,\ O_2.rc--,\ \ldots,\ O_n.rc++,$$

a total of $2(n-1)$ adjustments where $n > 1$. $O_n.rc$ means rc is the reference count of object O_n. These $2(n-1)$ adjustments are summarized in only two adjustments in FCGC.

3 Collector Details

FCGC is an *on-the-fly* reference-counting garbage collector whose ideal target is a multi-threaded or multiprocessor system. The collector may execute in its own dedicated hardware thread or on its own dedicated processor. Alternatively, it can share processing elements with mutators. In order for the FCGC algorithm to work correctly, instructions in the write barrier and code that instantiates new objects must not be reordered by the compiler. These procedures must also be executed atomically. This sequential memory consistency is not uncommon since most platforms offer a consistent view of memory for reads and writes made to the same address. For systems that do not guarantee sequential consistency, a few simple modifications can be made to the algorithm [15] to get it to work on those systems. For the remainder of our discussion, we assume that the underlying hardware architecture guarantees sequential memory consistency and that the write barrier and the new object allocation code can execute atomically.

3.1 Mutator Role in Garbage Collection

Mutators affect GC by instantiating new objects in the heap and storing references to heap objects in pointer fields. When a mutator instantiates a new object obj, the heap reference count of obj is initialized to zero because the collector has not yet determined which pointer references obj and for how long. If at the end of the current collection there are pointers that reference obj, its reference count is adjusted. Objects with a reference count of zero are candidates for GC at the end of a collection. As such, to prevent newly instantiated objects from leaking memory, the addresses of these new objects are placed in a table that the collector later examines for collation. Such a table is called a zero-count-table (ZCT) and each mutator has one. Figure 1 details the instantiation routine and one of the ways in which ZCT's are used.

Procedure **Object**: **Instantiate**(*size*: **Integer**)
    // new object obtained from allocator
1. O = allocate(*size*)
2. $ZCT_i = ZCT_i \cup \{O\}$
3. return O

Fig. 1 Instantiates new object.

Procedure **Update**(*s:* **Pointer**, *new:* **Object**)
1. Object *old* := **read**(*s*)
2. **if** $\neg Dirty_0(s) \wedge \neg Dirty_1(s)$ **then**
3. $Buffer_i := Buffer_i \cup \langle s, old \rangle$
5. *// dual dirty flags*
    $Dirty_{d_i}(s) :=$ **true**
6. **write**(*s*, *new*)
7. **if** $Snoop_i$ then
8. $Roots_i := Roots_i \cup \{new\}$

Fig. 2 Write barrier uses dual dirty flags. i is the mutator index and $d_i \in \{0,1\}$ indexes the dirty flag in use for the current collection.

Each mutator T_i is also given a local buffer $Buffer_i$ to store $\langle s, old \rangle$ tuples that the collector uses to compute its sliding view of the heap and to update the heap reference counts of affected objects. In a $\langle s, old \rangle$ tuple, s denotes a pointer that receives an assignment during the current collection and *old* represents the address of the last object that s referenced during the previous collection. Mutator T_i buffers $\langle s, old \rangle$ only if it is the first mutator to assign to s during the current collection. Each pointer s is associated with a pair of dirty flags, one of which is raised when s is first updated during the current collection. Updates are done with the write barrier depicted in Figure 2. One dirty flag is used for even-numbered collections and the other is used for odd-numbered collections. The existence of two dirty flags per pointer avoids dirty flag collisions, which can be confusing to the collector. While other approaches use additional handshakes to address these collisions [15], FCGC avoids them by using two dirty flags. This approach works regardless of the number of mutators and the number of collections.

Each mutator T_i has a field $d_i \in \{0,1\}$ that indexes the dirty flag it is allowed to manipulate based on the collection in which it is executing (even or odd). For example, when T_i is executing in an even-numbered collection it uses the write barrier to modify dirty flag $Dirty_0(s)$. However, when it is executing in an odd-numbered collection it uses the write barrier to modify dirty flag $Dirty_1(s)$. Initially, for all mutators T_i, d_i takes on the value $k_0(\text{mod } 2)$, where k_0 is the number of the first collection. d_i is subsequently updated as part of the first handshake of each collection—see Figure 3. Although lines 9 and 10 in Figure 3 present the possibility of an infinite loop, we use thresholding to break out of the loop. For each mutator T_i, if the collector fails to swap its buffer after a certain number of attempts, the collector will pause the mutator to retrieve its buffer. Buffer swapping is almost a constant time operation (just a few pointer manipulations), so a mutator does not have to be suspended for a long time. After a successful swap, the mutator is resumed.

Procedure **InitiateCollection**
1. **for each** thread T_i **do**
2. $Snoop_i :=$ **true**
3. $buffer\_swapped_i :=$ **false**
4. // HS_1 begins
   for each thread T_i **do**
5. **while** $buffer\_swapped_i ==$ **false**
6. // retrieve T_i's local buffer ignoring
   // duplicate pointer information
   $Hist_k := Hist_k \cup Buffer_i$
7. // give T_i an empty local buffer
   $Buffer_i := \emptyset$
8. // set the dual dirty flag index
   $d_i := (k+1) \textbf{mod } 2$
9. **if** buffer swap successful **then**
10. $buffer\_swapped_i :=$ **true**

Fig. 3 InitiateCollection uses dual dirty flags per pointer. i is the mutator index and k is the collection number.

Note, the instructions in the procedure in Figure 1 or the write barrier in Figure 2 must not be reordered by the compiler and both these routines must be executed atomically. Failure to enforce this restriction can result in the incorrect behavior of the collector. The write barrier utilizes a *snooping* mechanism to determine whether T_i is involved in the computation of a *sliding view* [15]. This feature can be explained as follows. A sliding view of the heap is computed during a collection over the interval $[t_1, t_2]$. This can be perceived as a view of the heap that slides in time. While that view is being computed, pointers are updated by receiving references to different objects. The snooping mechanism is used to ensure that objects that receive new references during the current sliding view computation are not reclaimed during the current collection. Such objects are considered *roots* for the current collection and are logged in per-mutator local root buffers, $Roots_i$. These local root buffers are consolidated during the next handshake. Section 3.2.3 describes the consolidation step more fully.

3.2 *The Collector*

We now describe the main functions of the FCGC collector. Pseudocode for each key function is provided.

3.2.1 Initiating a Collection

The collector begins a collection by enabling the snooping mechanism described in Section 3.1. The collector then executes in the first handshake, HS_1—see Figure 3. During HS_1, the collector interacts with each mutator on an individual basis. The details of this interaction is described in Section 2.2 and Section 3.1. After retrieving local buffer $Buffer_i$ from each mutator T_i, FCGC consolidates $Buffer_i$ into a global per-collection history buffer list, $Hist_k$, for the current collection. $Hist_k$ actually contains a list of buffer addresses where each address points to a local buffer $Buffer_i$ that was just retried from mutator T_i. For convenience, we refer to $Hist_k$ as a global list of all the tuples in the retrieved buffers $Buffer_i$. The tuples in $Hist_k$ are subsequently used to update the reference counts of heap objects and to reclaim objects whose heap reference counts fall to zero.

3.2.2 Resetting Dirty Flags

The tuples in $Hist_k$ were actually buffered during the previous collection, *i.e.*, prior to handshake HS_1 of the current collection. The pointers listed in $Hist_k$ are exactly the pointers that were updated at least once during the previous collection. These pointers need to have their dirty flags reset so the collector can determine whether they are modified in successive collections. This action is necessary to summarize reference-count updates—see Section 2.4. Only the objects that are affected by pointer updates in successive collections should have their reference counts adjusted (incremented or decremented). Moreover, in the write barrier, a mutator can buffer a pointer only if its dirty flag ($Dirty_0(s)$ or $Dirty_1(s)$) is not dirty. The routine to reset dirty flags is detailed in Figure 4.

Note that resetting of dirty flags is very specific. Only one set of dirty flags gets reset, *i.e.*, the flags raised before HS_1 of the current collection. The set of flags modified since HS_1 are not inadvertently reset, so they do not need to be restored. The collector should have exclusive write access to $Dirty_{k\,(\mathrm{mod}\,2)}(s)$ when it resets

```
Procedure ResetDirtyFlags
1.    for each ⟨s, old⟩ ∈ Hist_k do
2.        Dirty_{k (mod 2)}(s) := false
```

Fig. 4 Procedure **ResetDirtyFlags** uses dual dirty flags. k is the collection number.

it. This should not be a problem since mutators that concurrently attempt to mark s as dirty will use the other dirty flag to do so. When a mutator *checks* the dirty status of a pointer, as is done in the write barrier in Figure 2, both dirty flags are checked. Checking both flags is necessary to determine whether the pointer is dirty from the current collection or from the previous collection. Again, this should not pose a problem because when a mutator checks the dirty status of a pointer, it is only doing a read on the dirty flag that the collector resets. It does not write that flag.

Prior to using the procedure in Figure 4 to reset dirty flags of pointers in collection k, for all s in the system, if

$$Dirty_{k\,(\mathrm{mod}\,2)}(s) = true$$

then $s \in Hist_k$. After resetting dirty flags of pointers in collection k, for all $s \in Hist_k$,

$$Dirty_{k\,(\mathrm{mod}\,2)}(s) = false.$$

3.2.3 Consolidating Local Buffers

The collector engages mutators in a second handshake, HS_2. During HS_2, the snooping mechanism is disabled; mutator local root buffers are consolidated into a global per-collection root buffer; new objects are consolidated into a global per-collection ZCT, and mutator local buffers are consolidated in a global per-collection history buffer for the next collection. In the same vein that $Hist_k$ was described in Section 3.2.1, these global per-collection buffers are actually lists of addresses, where each address refers to a local buffer that was just retried from mutator T_i. This global history buffer consists of pointers updated since HS_1.

Consolidation simply amounts to retrieving the aforementioned per-mutator buffers, storing their addresses in global per-collection buffers, and returning to each mutator empty per-mutator buffers to log new information until the next retrieval. Consolidation is performed by procedure **Consolidate** in Figure 5 and serves as HS_2.

The local state of a mutator in this context, denoted by $State_i$, refers to all pointers (to heap objects) that are immediately available to the mutator. This includes pointers from the stack, registers, and global variables. Only the mutator and the collector have access to $State_i$. The same is true for $Roots_i$, defined in Section 3.1.

3.2.4 Adjusting Heap Reference Counts

FCGC now has enough information to adjust the heap reference count (rc) of each heap object pointed to by the pointers in $Hist_k$. Objects logged as the 'old values'

```
Procedure Consolidate
1.  local Temp := ∅
2.  Roots_k := ∅
3.  for each thread T_i do
4.     while buffer_swapped_i == false
5.        Snoop_i := false
6.        // retrieve snooped objects
          Roots_k := Roots_k ∪ Roots_i
7.        // give T_i an empty Roots_i buffer
          Roots_i := ∅
8.        // copy thread local state and ZCT
          Roots_k := Roots_k ∪ State_i
9.        ZCT_k := ZCT_k ∪ ZCT_i
10.       ZCT_i := ∅
11.       Temp := Temp ∪ Buffer_i
12.       if buffer swap successful then
13.          buffer_swapped_i := true
14. // consolidate Temp into Hist_{k+1}
    for each ⟨s, old⟩ ∈ Temp do
15.    if s ∉ Handled then
16.       Handled := Handled ∪ {s}
17.       Hist_{k+1} := Hist_{k+1} ∪ {⟨s, old⟩}
```

Fig. 5 Procedure consolidates all the per-mutator local information into per-collection buffers. Like HS_1, this procedure makes its best effort to not suspend mutators. A mutator will only be suspended if the collector exceeds a threshold number of attempts to retrieve the mutator's local buffer.

of pointers in $Hist_k$ have their *rc* fields decremented. Objects referenced by pointers in $Hist_k$ have their *rc* fields incremented. Figure 6 details how these reference count updates are done. By simply reading the contents of $Hist_k$, it is not always feasible to identify objects whose reference counts need to be incremented. Consider, for example, a pointer $p \in Hist_k$ with a raised dirty flag. The address of object obj, which is the 'old value' of p, is not in $Hist_k$ because p was logged by a mutator after it completed HS_1. The address of obj is thus logged in either a mutator local buffer or $Hist_{k+1}$. In order to identify objects such as obj and correctly adjust their reference counts, additional processing of pointers such as p is required.

Pointers such as p are referred to as *unresolved pointers* because the collector has not yet resolved the objects to which they point. Unresolved pointers are logged in global per-collection buffer $Unresolved_k$ so they can be processed further when the mutator local buffers and $Hist_{k+1}$ are examined. Mutator buffers are read asynchronously by procedure **ReadBuffers** in Figure 7. FCGC does not clear these buffers, instead it collates their content in another per-collection buffer called $Peek_k$. $Hist_{k+1}$ is then read by procedure **ReadHistory**, in Figure 8, and added to $Peek_k$. $Peek_k$ is subsequently used to determine which objects need to have their *rc* fields incremented. Procedure **IncRC** in Figure 9 increments the *rc* fields of such objects.

```
Procedure AdjustRC
1.   Unresolved_k := ∅
2.   for each ⟨s, old⟩ ∈ Hist_k do
3.     curr := read(s)
4.     if ¬Dirty_0(s) ∧ ¬Dirty_1(s) then
5.       curr.rc := curr.rc + 1
6.     else
7.       Unresolved_k :=
           Unresolved_k ∪ {s}
8.     old.rc := old.rc − 1
9.     if old.rc = 0 ∧ old ∉ Roots_k then
10.      ZCT_k := ZCT_k ∪ {old}
```

Fig. 6 Procedure adjusts rc fields of heap objects identified by pointers in $Hist_k$. It uses dual dirty flags in checking the dirty status of each s. Care must be taken to ensure that if s appears in $Hist_k$ multiple times, it is only processed once.

```
Procedure ReadBuffers
1.   Peek_k := ∅
2.   for each T_i do
3.   // copy buffers without duplicates
       Peek_k := Peek_k ∪ Buffer_i
```

Fig. 7 Procedure reads mutator local buffers without clearing them.

```
Procedure ReadHistory
1.   // copy Hist_{k+1} without duplicates
       Peek_k := Peek_k ∪ Hist_{k+1}
```

Fig. 8 Procedure reads history buffer of next collection and adds it to $Peek_k$.

```
Procedure IncRC
1.   for each ⟨s, old⟩ ∈ Peek_k do
2.     if s ∈ Unresolved_k then
3.       old.rc := old.rc + 1
```

Fig. 9 Procedure increments rc field of objects identified by unresolved pointers.

```
Procedure ReclaimGarbage
1.  ZCT_{k+1} := ∅
2.  for each object obj ∈ ZCT_k do
3.    if obj.rc > 0 then
4.      ZCT_k := ZCT_k − {obj}
5.    else if obj.rc = 0 ∧ obj ∈ Roots_k then
6.      ZCT_k := ZCT_k − {obj}
7.      ZCT_{k+1} := ZCT_{k+1} ∪ {obj}
8.  for each object obj ∈ ZCT_k do
9.    Collect(obj)
```

Fig. 10 Procedure determines which objects are garbage collectible and collects them with procedure **Collect**.

3.2.5 Reclaiming Storage of Garbage Objects

After the reference counts of heap objects are adjusted, FCGC ventures into reclaiming the storage occupied by garbage objects. Garbage objects are heap objects with heap reference counts of zero that are not marked as roots. An object is marked as a root object if it is in $Roots_k$. Objects with pointers in $Hist_{k+1}$ that become garbage are not collected at this time. Their collection is deferred to the next collection since their pointers were last updated during the current collection. Furthermore, they became garbage after the current collection started. Thus, there is potential for a small amount of floating garbage. However, this floating garbage does not linger since it is collected quickly—in the next collection. The procedures responsible for reclaiming garbage objects are **ReclaimGarbage** in Figure 10 and **Collect** in Figure 11.

```
Procedure Collect(obj: Object)
1.   local DeferCollection := false
2.   for each pointer s ∈ obj do
3.     if ¬Dirty_0(s) ∧ ¬Dirty_1(s) then
4.       DeferCollection := true
5.     else
6.       val := read(s)
7.       val.rc := val.rc − 1
8.       write(s, null)
9.       if val.rc = 0 then
10.        if val ∉ Roots_k then
11.          Collect(val)
12.        else
13.          ZCT_{k+1} := ZCT_{k+1} ∪ {val}
14.  if ¬DeferCollection then
15.    return obj to general purpose allocator
16.  else
17.    ZCT_{k+1} := ZCT_{k+1} ∪ {obj}
```

Fig. 11 Procedure that collects garbage objects. It uses dual dirty flags in checking dirty status of each s.

4 Implementation issues

A number of implementation design decisions need to be made in implementing a collector like FCGC. Here, we explore the space of possible high-performance implementations and discuss our proof-of-concept implementation in the GNU Compiler for Java (GCJ) version 4.1.0, which comes bundled with GCC [11].

4.1 Buffer Representation

There are many possibilities for buffer representation in implementing an *on-the-fly* reference counting garbage collector. The performance of the write barrier is affected by the selection of a buffer representation, as is the method with which the collector swaps buffers with mutators.

We choose to represent buffers as per-mutator arrays of log entries. We maintain two such arrays for each mutator, used alternately in collections: while the mutator logs entries in one array, the collector operates on the other. During the first handshake of a collection, the mutator is instructed to switch to the other array and the collector is guaranteed exclusive access to the previous. In this way, buffer-swapping is almost a constant-time operation, just a few pointer manipulations. The avoidance of mutator blocking is an essential advantage of our collector.

4.2 Maintaining Dirty Flags

We consider two possible ways to store dirty flags for pointers in objects. As discussed in earlier sections, space must be reserved to store two dirty flags for each pointer in an object. In our implementations of FCGC, we use a bit for each dirty flag.

4.2.1 Store Dirty Bits in Pointers

Pointers to Java objects are typically word-aligned, so on 32-bit systems the lower two bits of a pointer are effectively unused (they are always zero). They can thus be used to store the two dirty flags.

This approach has several advantages. There is no object size overhead for dirty flags, and the dirty flags for a given pointer are easy to find. However, where hardware support does not exist, the compiler must arrange to mask out these two bits on every pointer access. This is a constant-time operation, but it increases code size and degrades performance due to the large number of such operations. In systems with specialized memory addressing hardware that automatically masks out some parts of dereferenced pointers, this is an ideal approach to storing dirty flags.

4.2.2 Store Dirty Words in Objects

If dirty flags are not stored directly in pointers, they must be stored elsewhere. They could be placed in an associated data structure and objects could point to them. This correspondence would necessarily be one-to-one, and would require the overhead of an additional pointer per object. We find this unsatisfactory, though we admit that such an implementation might be simpler than our approach. We do not consider that approach further in this article.

The approach we used in our implementation stores dirty flags in objects. We pack 32 bits of dirty flags into a *dirty word* and sprinkle such dirty words throughout each object, as needed. Each dirty word stores dirty flags for 16 pointers.

There are severe constraints on the placement of dirty words. Type substitution in object-oriented languages (using a pointer-to-`B` as if it were a pointer-to-`A`, where `B` is a subclass of `A`) requires that the runtime memory layout of a subclass contains, as a prefix, a valid object of superclass type. Strictly speaking, this object "prefixing" is not necessary for type substitution; indeed, in the presence of multiple inheritance (in languages that support it), another approach is necessary, involving adjustment of the `this` pointer by a statically-known offset. However, in Java implementations, and for the first inherited type in languages supporting multiple-inheritance, the method of object prefixing described is generally used. Therefore, once a dirty word is allocated for a class, that word must exist at the same offset in all subclasses.

If we limit ourselves to solutions with as few extraneous, unused bits of dirty-field information as possible ("leftovers" from the final dirty word), this can be achieved in one of two reasonable ways. First, we could place dirty words above the object pointer as depicted in Figure 12(a): object fields are laid out as usual, with superclass fields appearing first; dirty words are laid out at negative offsets from the object base pointer, with superclass dirty words appearing closer to the base and subclass fields appearing at lower addresses. This is an attractive solution. It is simple, and maintains compatibility with traditional object layout schemes.

The approach has two drawbacks. First, a pointer is separated by some distance from its associated dirty word. They almost certainly sit on a different cache line, and may even reside on different pages (if the allocator carelessly allocates such objects). Further, such a layout complicates implementation of some object-oriented languages (especially implementations of multiple-inheritance languages like C++) in which information of varying length is already laid out above the object pointer, depending on the type of the object or the kind of type inheritance employed.

Another approach is to sprinkle dirty words among the fields of object as depicted in Figure 12(b). In this approach, dirty words are placed for every 16 fields of pointer type in the object, with a dirty word appearing before all fields it represents. Unused portions of dirty words in superclass layouts are used for subclass pointer fields before allocation of new dirty words. The first dirty word of each object is actually only partially used for dirty-field information so that a 4 bit heap reference count can be accommodated. We chose to use four bits for the reference count (although all 32 bits could be used) because most objects are typically referenced less than 32 times

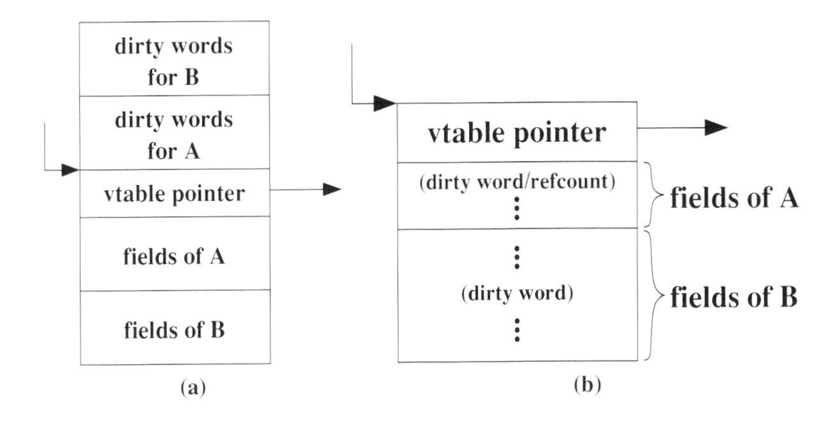

Fig. 12 Two possible object layouts. **(a)** Placement of dirty words *above* object pointer. **(b)** Placement of dirty words among the fields of an object.

during their lifetime [9]. Thus the first (partial) dirty word holds dirty information for 14 fields of pointer type.

We remark that since we use only four bits to hold the reference counting field of an object, stuck reference counts will occur. In such cases, we use a novel *on-the-fly* mark-sweep collector [3] occasionally to either reclaim such objects or to restore their reference counts. We will also use the *on-the-fly* mark-sweep collector to reclaim cyclic data structures, since reference counting collectors cannot reclaim such objects.

4.3 Pointer Modifications in Log Entries

Buffer entries in our FCGC are represented by the 4-tuple:

$$\langle ptr, dirty\_word, dirty\_bit, oldvalue\rangle$$

The dirty word associated with a pointer is given. The dirty bits associated with the pointer are also given.

This approach has the effect of using four words for the size of each buffer entry. Although this overhead seems a bit much for a pointer update, it is not unreasonable as buffer space is recycled between collections. One way to keep buffer overhead low is to use the pair $\langle ptr, oldvalue\rangle$ as described in Section 3. But doing so requires storing dirty flags in the unused bits in the pointers themselves—an approach that is potentially more expensive for architectures without specialized hardware support for masking out dirty bits, as noted in Section 4.2.

Another point that is worth noting is the possibility for buffer overflow. Each buffer is a fixed size array. Should that size be exceeded for a mutator, the last entry in the buffer should be a pointer to a linked list that acts as a cache for additional buffer space. Of course, the size of the buffers is a parameter that can be used to configure the collector.

4.4 A Non-blocking Write Barrier

Care must be taken in data structure selection and development of the write barrier. The write barrier must be safe, yet non-blocking.

In our implementation, the write barrier does not perform any locking, and the collector may perform operations in parallel. Some of these operations, during handshakes, may be unsafe and may lead to the collector having an incorrect view of the mutator's state. This potential *unsafety*, however, is rare, and more pertinently, it is detectable in the collector. When detected, the collector is able to undo and repeat the handshake until a safe handshake is performed or until a threshold number of failed attempts to perform a handshake with a mutator has been exceeded. *A mutator scarcely loops in the write barrier*. One is paused only for a brief moment if the collector fails a threshold number of attempts to perform a handshake with it. Otherwise, any collector performance degradation will generally be invisible to the mutators. They may in certain circumstances observe a longer collection phase and thus, a longer time before they are provided a clean buffer.

This design eliminates (for the most part) mutator blocking, but effectively causes the collector to block until a safe handshake can be performed. We have removed the overhead of scheduling decisions, and even that of unnecessary system calls. What we provide is a primitive *try-retry* mechanism that effectively takes the place of a heavier lock.

5 Preliminary Results

We report measurements we collected from running an instrumented version of the collector to generate statistics for measurable features of interest. This version of the collector is not accompanied with the *on-the-fly* mark-sweep collector we referenced in Section 4.2. Our primary goal here is to study the behavior of the write barrier and the handshaking mechanism. We used two simple Java programs to explore these routines.

	With no collector	With the FCGC collector	overhead	% overhead
Time (s)	15.91	16.91	1.0	6.29 %

	clock tick count	Time (s)	% overhead
Handshake Time	205332	8.581×10^{-5}	0.009 %
Write Barrier Time (logging)	1082804796	0.453	45.253 %
Write Barrier Time (no logging)	403531284	0.169	16.865 %

Fig. 13 (Top) Overhead for executing SortNumbers 100 times with no collector and with our collector. (Bottom) Overhead distribution for FCGC when executing SortNumbers 100 times with the collector enabled. The total time for each operation for the 100 runs of the application is recorded.

- **SortNumbers:** This single threaded application sorts numbers using selection sort.
- **SimpleThreads:** This application illustrates simple interactions between two concurrent threads.

The experiments were performed on a Dell Precision 530 workstation with two physical Xeon 2.4 GHz CPUs with Intel's hyper-threading that supports two threads per CPU; 512MB physical memory and 2GB of swap space; the Debian distribution of LINUX with kernel 2.6; a heap size of 32MB, and clock tick of 2392.791 MHz (as reported in `/proc/cpuinfo`). This is equivalent to 0.418 nanoseconds per tick.

5.1 Collector Overhead

To estimate the collector overhead, we ran SortNumbers 100 times with the collector and 100 times with no collector. For each run, 1000 randomly generated doubles were sorted and the results were printed to the console. The results from this experiment are given in Figure 13 (top).

The overall overhead associated with the collector seems small–6.29% from this experiment. This low overhead may be attributed to the fact that the system's configuration is ideal for the number of threads in the application and the collector combined—the Dell system has two physical processors while the application and the collector together have a total of two threads.

To gain insight into the overhead distribution in FCGC, we used the x86 assembly language instruction `rdtsc` to count the number of clock ticks for each collector operation we wanted to measure, namely the write barrier and the handshaking mechanism. Figure 13 (bottom) records the measurement for FCGC's overhead distribution.

There is a significant difference between the handshake overhead and the write barrier overhead. This is expected, since the collector runs far fewer collections than mutators update heap pointers. The overhead also differs between the two paths through the write barrier—in one path the mutator logs buffer entries and in the other path it does not. To better understand these differences, we performed further experiments.

5.2 Investigation of Overhead

We ran SortNumbers 100,000 times and for each run, we timed and counted the number of handshakes and the number of write barrier executions for each path through the write barrier (logging entries and not logging entries). We also ran the same experiments with SimpleThreads to help us determine whether the overhead distribution differs when multiple threads are involved. We compare the mean time per handshake with the mean time per write barrier execution for each path through the write barrier to determine whether such comparison accounts for the differences in overhead distribution. Figure 14 summarizes our results.

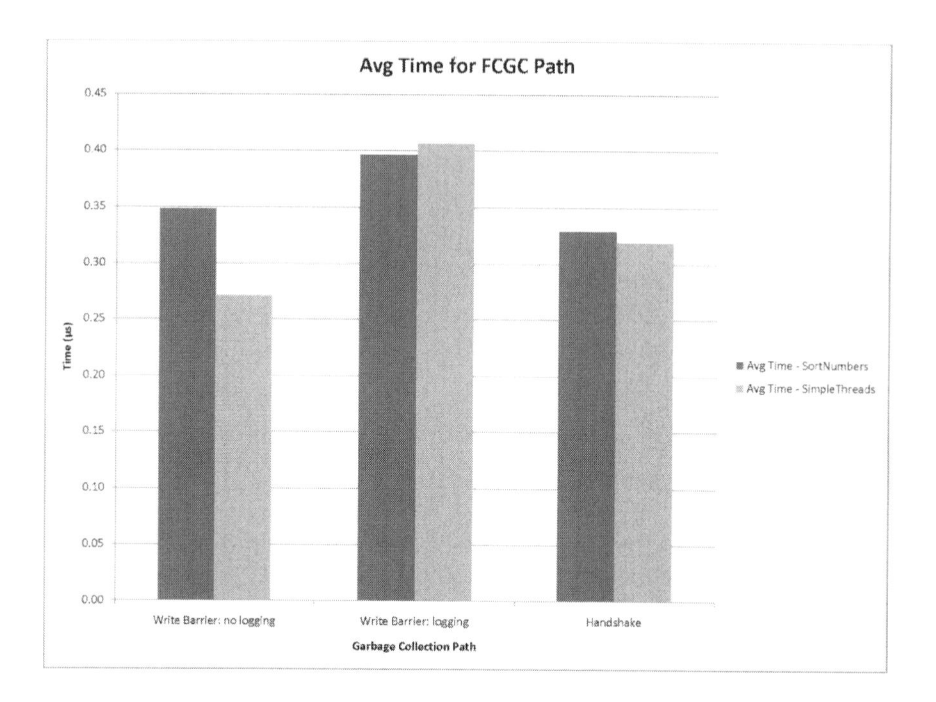

Fig. 14 Average time cost for an operation, observed for the 100,000 runs of SortNumbers and SimpleThreads when the collector is enabled and the operation is executed.

Our preliminary results suggests that the non-blocking write barrier (see Sections 4.4) and the handshaking mechanism described in Section 2.2 pay dividends. We notice that the handshake mechanism runs fast. There are two possible reasons why it runs so fast. First, mutators never blocked. Suspending a mutator can be expensive especially if system calls are made to pause and resume it. Moreover, blocked mutators make no progress while suspended and throughput suffers. The other reason is that the collector uses no heavy locks or other heavy synchronization mechanism to perform a handshake. Instead, it uses the *try-retry* mechanism described in Section 4.4. Although the collector may momentarily block on a handshake, we did not observe such blocking in our experiments. We used a counter to keep track of the number of times the collector failed to perform a handshake with a mutator. Hence, no threshold was exceeded and no mutator needed to be paused. Levanoni and Petrank [15] designed, implemented, and tested a similar *on-the-fly* collector and concluded that the write barrier is fast. Figure 14 suggests that our handshaking mechanism can be just as fast and takes virtually constant time. This observation coincides with our handshake mechanism—a few pointers are modified to swap buffers between the collector and mutators instead of copying individual entries from mutator buffers to global per-collection buffers.

6 Related Work

6.1 A Comparison of FCGC and LPC

Levanoni and Petrank designed and presented an *on-the-fly* reference counting garbage collector for Java[15], which we denote LPC for short. We noted in Section 1.1 that LPC inspired our work. Our collectors are similar in several respects and different in other respects. We compare the write barrier and handshaking mechanism of both collectors.

The write barriers in FCGC and LPC are very similar. Their only difference lies in how they update dirty flags. In FCGC, one of two dirty flags is updated at a time, whereas in LPC, only one dirty flag is used. Consequently, the overhead associated with the write barrier execution in each of these collectors is comparable.

We noticed from our preliminary results that the handshaking mechanism in FCGC executes in virtually constant time; however, this is not the case in LPC. In an LPC handshake, the collector suspends each mutator in turn, copies its buffered entries to a global consolidated history buffer, then resumes the mutator. Scalability is a concern since mutators can only be suspended at safe-points. FCGC handshakes do not suffer from this concern. Moreover, FCGC uses two handshakes per collection while LPC uses four. The fact that FCGC's handshakes are few, short, and non-blocking, coupled with the non-mutator-blocking write barrier, these features are insightful in thinking about how to design multiprocessor real-time garbage collectors.

6.2 Other Related Work

Reference-counting garbage collection dates back to the early days of *list processing* [6] and it has been adopted subsequently by many systems [12, 7, 1, 18].

Deferred reference counting (DRC) [8] overcomes some of the overhead of reference counting [14, 13] by accurately accounting only for pointers in the heap. Objects cannot be collected in DRC when their heap reference counts drops to zero, because stack cells, registers, and global variables must be checked for references. Our collector employs DRC by adjusting reference counts only after the second handshake.

Limited-field reference counting restricts the size of a reference count to a few bits, which usually suffices for most objects. If a limited-field reference count overflows, a tracing collector (e.g., an *on-the-fly* mark-sweep collector [3]) or some other algorithm is used to restore it. FCGC is such a collector, using four bits to hold each object's reference count in the object's header.

The list of *concurrent* and *multithreaded* collectors is extensive [17, 2, 7, 10, 16]. One shortcoming of many of these collectors, as presented in the literature, is their need to suspend all mutators to compute a snapshot of the heap. *On-the-fly* algorithms [4, 5, 15] avoid suspending multiple mutators at the same time. FCGC is based on LPC [15], which uses the notion of a sliding view instead of a snapshot.

7 Conclusions and Future Work

We presented an *on-the-fly* reference-counting garbage collector, FCGC, that targets a multi-threaded, multiprocessor environment. FCGC is based on a similar collector from Levanoni and Petrank [15]. FCGC is a fully concurrent garbage collector that makes its best effort to never pause, but run in parallel with mutators at all times. It uses lazy updates to adjust the reference counts of heap objects; its write barrier is non blocking and requires no synchronization between mutators. FCGC performs collection work in cycles and uses as few as two per-mutator transactions (handshakes) per collection to compute its view of the heap. These features make FCGC a non-intrusive collector.

In addition to the small number of handshakes that FCGC offers (two versus four in LPC), it makes its best effort to never pause mutators, unlike other collectors (e.g., LPC) that suspend each mutator in turn to perform a handshake. Instead, it uses a *try-retry* mechanism is which the collector loops until it succeeds in performing a handshake or some pre-defined threshold of failed attempts has been reached. The result is a reduction in synchronization cost and temporal overhead per collection.

We have implemented and tested a proof-of-concept version of the collector. We presented our observation in Section 5 and noted that both the write barrier and handshake are fast—they execute in almost constant time. We plan to implement FCGC in a commercially available Java Virtual Machine so we can perform a more thorough analysis of its behavior. We also plan to investigate the possibility of reducing the synchronization between the collector and mutators further to exploit other possible benefits.

References

1. Aho, A.V., Kernighan, B.W., Weinberger, P.J.: The AWK Programming Language. Addison-Wesley (1988)
2. Appel, A.W., Ellis, J.R., Li, K.: Real-time concurrent collection on stock multiprocessors. SIGPLAN Notices 23(7), 11–20 (1988)
3. Azatchi, H., Levanoni, Y., Paz, H., Petrank, E.: An on-the-fly mark and sweep garbage collector based on sliding views. SIGPLAN Notices 38, 269–281 (2003)
4. Bacon, D.F., Richard Attanasio, C., Lee, H., Rajan, V.T., Smith, S.: Java without the coffee breaks: A nonintrusive multiprocessor garbage collector. In: SIGPLAN Conference on Programming Language Design and Implementation, pp. 92–103 (2001)
5. Bacon, D.F., Rajan, V.T.: Concurrent Cycle Collection in Reference Counted Systems. In: Lee, S.H. (ed.) ECOOP 2001. LNCS, vol. 2072, pp. 207–235. Springer, Heidelberg (2001)
6. Collins, G.E.: A method for overlapping and erasure of lists. Communications of the ACM 3(12), 655–657 (1960)
7. DeTreville, J.: Experience with concurrent garbage collectors for Modula-2+. Technical Report 64, Digital Equipment Corporation Systems Research Center (August 1990)
8. Peter Deutsch, L., Bobrow, D.G.: An efficient incremental automatic garbage collector. Communications of the ACM 19(9), 522–526 (1976)
9. Dieckmann, S., Hölzle, U.: A study of the Allocation Behavior of the SPECjvm98 Java Benchmarks. The MIT Press (2001)

10. Doligez, D., Leroy, X.: A concurrent generational garbage collector for a multi-threaded implementation of ML. In: Conference Record of the Twentieth Annual ACM Symposium on Principles of Programming Languages, SIGPLAN Notices, pp. 113–123. Association for Computing Machinery (1993)
11. Free Software Foundation. GCC, the GNU Compiler Collection - GNU Project - Free Software Foundation, FSF (2007), `http://gcc.gnu.org/`
12. Goldberg, A., Robson, D.: Smalltalk-80: the language and its implementation. Addison-Wesley Longman Publishing Co., Inc., Boston (1983)
13. Hartel, P.H.: Performance Analysis of Storage Management in Combinator Graph Reduction. PhD thesis, Department of Computer Systems, University of Amsterdam, Amsterdam (1988)
14. Jones, R., Lins, R.: Garbage collection: algorithms for automatic dynamic memory management. John Wiley & Sons, Ltd. (1996)
15. Levanoni, Y., Petrank, E.: An on-the-fly reference counting garbage collector for Java. In: Proceedings of the 16th ACM SIGPLAN Conference on Object Oriented Programming, Systems, Languages, and Applications, pp. 367–380. ACM Press (2001)
16. O'Toole, J.W., Nettles, S.M.: Concurrent replicating garbage collection. Technical Report MIT–LCS–TR–570 and CMU–CS–93–138, MIT and CMU (1993); Also LFP94 and OOPSLA93 Workshop on Memory Management and Garbage Collection
17. Steele, G.L.: Multiprocessing compactifying garbage collection. Communications of the ACM 18(9), 495–508 (1975)
18. Wall, L., Schwartz, R.L.: Programming Perl. O'Reilly and Associates, Inc. (1991)

A Metamodel for Internetware Applications

Zhiyi Ma and Hongjie Chen

Abstract. In the Internet environment, software has the characteristics such as open, dynamic evolution, and hard-controlled. Aiming at the characteristics, this paper presents a metamodel for modeling Internet software. The metamodel includes a set of modeling concepts based on CBDI Meta Model for SOA Version 2.0. The metamodel is used to model Internet software in the phases of analysis, design, and deployment, and to model the evolution of Internet software, including runtime self-adjustment.

Keywords: Metamodel, Internet Software, Service Innovation.

1 Introduction

Internet has already developed into the significant infrastructure in the modern information society. How to realize the share and integration of a variety of resources in the open, dynamical and hard-controlled Internet environment has become an important challenge to the software technology [1].

The software entities in the resources mentioned above are often developed and managed independently, and collaborate each other in different network sites. The behaviors of these entities need to be regulated automatically with the change of the running environment and application requirements, and the applications consisting of these entities also need to be statically adjusted and dynamically evolved. Thus, the requirements for the software entities are that they should be autonomous, cooperative, evolutionary and context-aware. Due to the fact that the software entities in one network site can provide diverse services, such entities should be multi-objectives. Moreover, the software entities should be reliable due to their complexity. Naturally, the applications consisting of the software entities also need to have the above features.

Zhiyi Ma · Hongjie Chen
Software Institute, School of Electronics Engineering and Computer Science, Peking University
Key laboratory of High Confidence Software Technologies (Peking University), Ministry of Education, Beijing, China
e-mail: {mzy,chenhj}@sei.pku.edu.cn

Z. Qian et al. (Eds.): Recent Advances in CSIE 2011, LNEE 125, pp. 365–373.
springerlink.com

The paper defines a software entity that has such features as an Internetware. An Internetware is of cooperative and self-adaptable in the Internet environment. An Internetware application consists of a set of relative Internetwares. The self-adaptability of an Internetware application embodies in its run phase. First the Internetware application needs to meet the user-predetermined objectives, and then the relative Internetwares exert their self-adaptability to meet the original or new objectives with the changes of the running environment and application requirements. The latter may lead to change the architecture of the application

Due to the features of Internetware applications, the traditional software modeling languages are not suited to model Internetware applications in the complex Internet environment. Therefore, the paper presents a metamodel for modeling Internetware applications.

2 Overview of the Internetware Application

Internetware is a core concept in this paper for sharing and integrating a variety of software in the open, dynamic evolution and hard-controlled Internet environment. Figure 1 illustrates how to develop and maintain Internetware applications.

An Internetware application includes an Internetware application model and a component model for implementing Internetwares. A set of Internetwares forms an Internetware application, which runs in the Internetware running platform in the Internet environment. Internetwares are implemented by the components, which runs in the component running platform in the Internet environment. Both platforms have the ability of real-timely showing their running states and behaviors, and also have the ability of helping the software running on them real-timely to show their internal states and behaviors. On both platforms, an Internetware application can exert its self-adaptability to adapt to the change of the running environment and application requirements. In Figure 1, an evolving Internetware application is represented using the Internetwares circled in a cloud-shaped graph and the relative components.

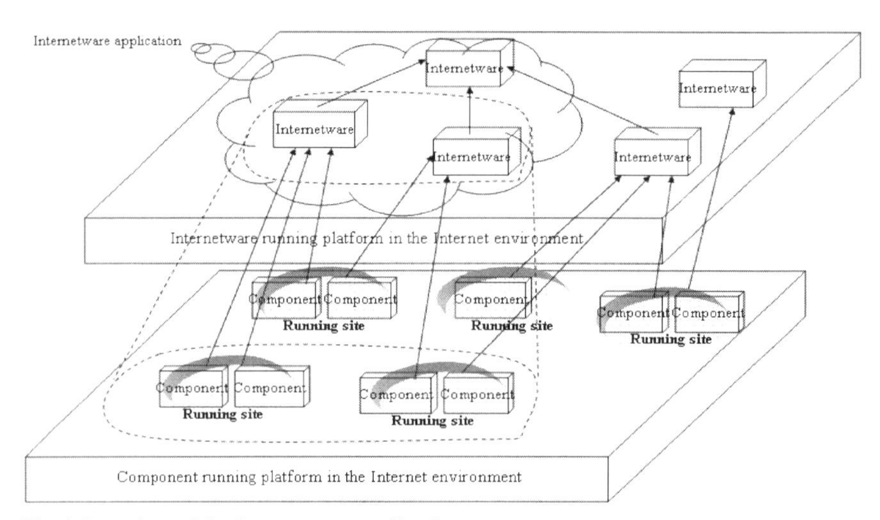

Fig. 1 Overview of the Internetware applications

3 Internetware Metamodel

An Internetware is a software entity that has cooperativity and adaptability in the Internet environment. It is equipped with the features of autonomy, cooperativity, context-aware, multi-objective, and reliability. The concept service is completely defined in the SOA metamodel (CBDI-SAE) [2] issued by CBDI organization. Service has the features of cooperativity and multi-objective. The paper extends its definition, i.e., adding autonomy, context-aware, and reliability to it. Figure 2 shows the package structure diagram of the Internetware metamodel based on CBDI-SAE.

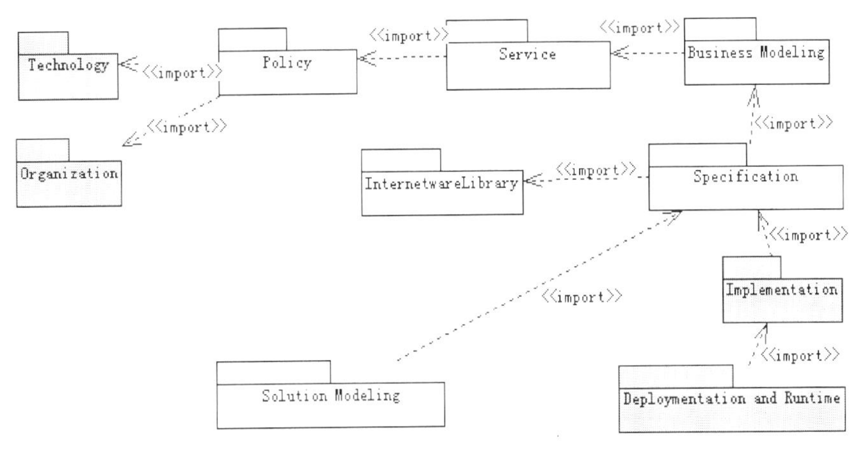

Fig. 2 Package structure diagram of the Internetware metamodel based on CBDI-SAE

In Figure 2, except *InternetwareLibrary*, the others are the original structure of CBDI-SAE, and the packages without filled color are the extended packages and added packages1.

3.1 Extension of the Business Modeling Package and Solution Modeling Package

Both packages are used to build business models. An important concept in the Business Modeling package is *business service*, which is provided by activities in business processes. A business service realizes definite business objectives, namely meeting definite function requirements, quality requirements and constraints. The paper extends concept *BusinessObjective*, See Figure 3.

BusinessObjective is divided into *ServiceObjecitive*, *FunctionObjecitive*, and *BehaviorObjecitive*. *FunctionObjecitive* is the relatively complete requirement that one or several business services meet, and also can embody the internal and external

[1] In the paper, the elements without filled color are the extended and added ones.

interaction of the applications. *ServiceObjecitive* is used to organize *FunctionObjecitives* and show the high-level ability of business services. Each *FunctionObjecitive* is embodied by detailed behaviors. Such behavior is defined as *BehaviorObjecitive*. That is, a set of relative behavior objectives focus on what they provide to meet the interaction performed by one *FunctionObjecitive*. The above high-level objectives are the abstract of the low-level objectives and the low-level objectives are the realization of the high-level objectives.

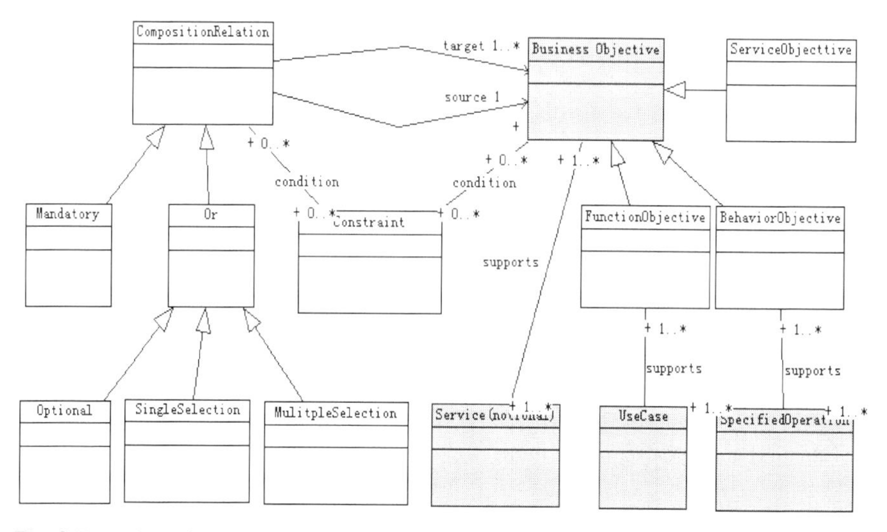

Fig. 3 Extension of Solution Modeling package and Business Modeling package

CompositionRelation is divided into *Mandatory* and *Or*. *Or* includes *Optional*, *SingleSelection*, and *MulitpleSelection*. *SingleSelection* means to choose one from some options. *MulitpleSelection* means to choose several from some options.

CompositionRelation can be applied between *ServiceObjecitives* (that is, a *ServiceObjecitive* can be divided into *sub-ServiceObjecitives*), between one *ServiceObjecitive* and one or several *FunctionObjecitives*, between *FunctionObjecitives*, and between one *FunctionObjecitive* and one or several *BehaviorObjecitives*. However, it can not be applied between *BehaviorObjecitives* (that is, a *BehaviorObjecitive* can not be divided). Constraints can apply to the above objectives and relations between the objectives in aspects of functions and non-functions.

Solution Modeling package is used to build requirement models. *UseCase* and *SpecifiedOperation* in the package support *FunctionObjecitive* and *BehaviorObjecitive*, respectively.

The extended *Business Modeling* package increases its ability of modeling business objectives in different abstract levels. The extended *Solution Modeling* package can be used to build the relations between *ServiceObjecitives* and obtained requirements. Both extended packages are the foundations for building the other models of Internetware applications.

3.2 Extension of Service Package and Specification Package

The *Service* package is be used to describes the basic information of services, i.e., the information for originally recognized services and relations between the services. We add a Meta-attribute has *Autonomy* into the meta-class *Service* to indicate whether a service is autonomy (without showing the metamodel here).

The *Service* package is be used to describes the basic information of services, i.e., the information for originally recognized services and relations between the services. We add a Meta-attribute hasAutonomy into the meta-class Service to indicate whether a service is autonomy (without showing the metamodel here).

Specification package is be used to specify services. Figure 4 shows the extension of the package.

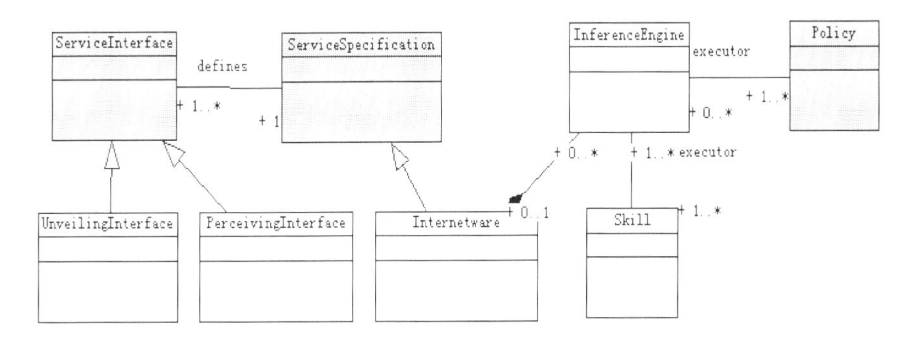

Fig. 4 Extension of Specification package

The added UnveilingInterface and *PerceivingInterface* are sub-meta-classes of the interface *ServiceInterface*. *UnveilingInterface* is used to outward show the necessary internal information of an Internetware and *PerceivingInterface* to perceive running environment. Added *Internetware* is a sub-meta-class of *ServiceSpicification* and has an added meta-class *InferenceEngine* as its part. *InferenceEngine* can reason according to the defined policies. Added *Skill* is used to adjust the Internetware own behavior according to the reasoning result. Internetsoftware defined like this has reactivity and autonomy.

3.3 Added InternetwareLibrary Package

Because Internetwares are diversiform and the providers of Internetwares often locate in the different geographical positions, it is necessary to model the management (such as classification and release) of Internetware contracts. A metamodel is presented for specifying the structure of service contract library systems and their interconnection in [3]. The metamodel is a UML 2.0 profile and can be used to model the inner structure of Internetware contract library systems and the topology structure between Internetware contract library systems.

3.4 Extension of the Policy Package

The UML Profile QoS&MF [4] issued by OMG gives the metamodel of the framework describing the requirements and properties of quality of services, and the metamodel describing the ability of fault tolerance. QoS&MF may combine with *Policy* package of CBDI-SAE to specify the policies of Internetwares.

(1) Quality of service

Policy package of CBDI-SAE is used to define the business rules and technology rules (namely policies) in service-oriented computing. Each policy consists of options, and each option consists of assertions. This package is also used to specify the effect of the policies, relations between policies, and the organizations and roles affected by policies. For modeling Internetware applications, *Policy* package should include a quality metamodel. Figure 5 shows the extension of *Policy* package in quality.

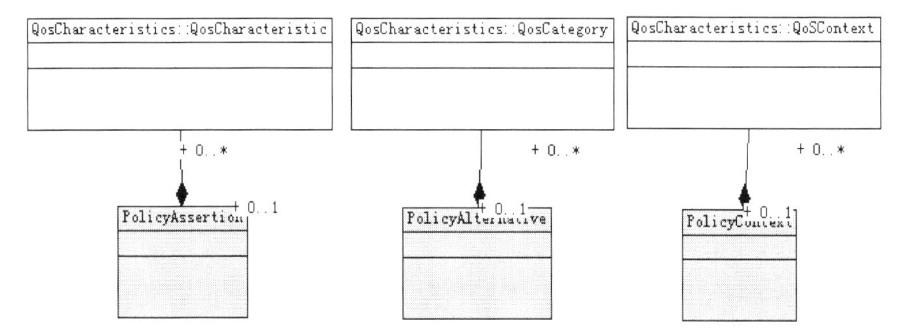

Fig. 5 Extension of Policy package in quality

QoSCharacteristic in the *QoSCharacteristic* package of QoS&MF is used to describe quality of service, and *PolicyAssersion* in CBDI-SAE is used to describe policy assertions. Hence, *QoSCharacteristic* is defined as a part of *PolicyAssersion*. *QoSCategory* in *QoSCharacteristic* of QoS&MF is used to organize *QoSCharacteristic*, and *PolicyAlternative* in *QoS&MF* is used to describe alternative policy options and organize *PolicyAssersions*, thus *QoSCategory* is defined as a part of *PolicyAlternative*. Since *PolicyContext* in CBDI-SAE is used to define the policy context and includes quality of service, it includes *QosCharacteristics::QoSContext* in QoS&MF.

(2)Fault tolerance mechanisms

Policy package of CBDI-SAE should have the ability of building fault tolerance models. The fault tolerance metamodel in QoS&MF is used to extend *Policy* package in the paper. Figure 6 shows the extension result.

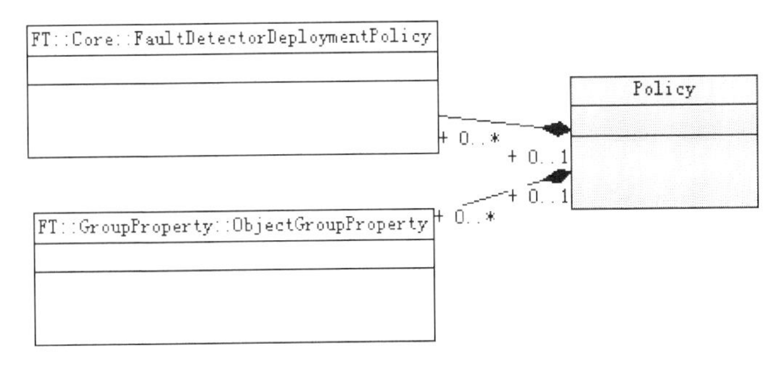

Fig. 6 Extension of Policy package in fault tolerance

FaultDetectorDeploymentPolicy in the *FT::Core package* of QoS&MF is used to describe how to monitor software defects. *ObjectGroupProperty* in the *FT::GroupProperty* package of QoS&MF is used to model fault detection, consistency check, and defect treatment. Therefore, *Policy* of *Policy* package in CBDI-SAE includes them.

Because ServerObjectGroup in FT::Core of QoS&MF may be an element of any kind needing fault tolerance treatment, Service, ServiceSpecification, AutomationUnit, BusinessSereviceSubject, and Node in CBDI-SAE are all its sub-meta-class (without showing relations between them here).

4 Related Work

Soo Ho Chang points out that it is necessary to model the variability of services and interface mismatch handling, and gives a suggestion for dynamic match [10], but does not give a concrete modeling language. The K-Component framework emphasizes the separation evolution logic and business logic of architecture [8]. The self-evolution of the framework mainly embodies the replacement of components that only have local views, and thus this confines the ability of self-evolution. The Arch-Ware [7] extends π calculation for modeling architecture and then supports the development of evolutional software. The degree of formalization of Arch-Ware is higher, but it lacks the support of the self-evolution environment.

The intelligence Agent technology has been used to research on the development of the adaptable software [5, 6]. The technology is effective for developing the software applications that have high intelligence behavior. However, there are still some problems about the integration of it and the current software development methods. For example, Maze [11] is the influential Agent-oriented analysis and design methods, and uses the parent-child relation between objectives to identify Agents; in fact, there are other relations between objectives. It is difficult to build a software method compatible with the current mainstream software technologies based on the Agent technology [12].

Lu Jian et al. prsent the preliminary conceptual frame and basic features of Internetwares, and commentate related technologies. Further they present an environment-driven model for Internetwares based on Agent technology and ontology technology, and builds some prototype systems [1]. Mei Hong et al. present a software architecture centric development method that emphasizes software components as basic entities and reflective middleware as the running platform [9]. The paper presents the detailed metamodel for Internetware applications based on services besides components.

5 Conclusion

The paper extends CBDI-SAE and combines it with QoS&FT and the service contract library metamodel presented by us. They, which are extended in UML 2.0, and the extension in this paper form a modeling conceptual system as the metamodel for Internetware applications. The metamodel provides the systematic support for analyzing, designing, implementing, deploying, and evolving (including self-adjusting runtime) Internetware applications.

There are some open issues to be addressed in the metamodel for developing Internetware applications. Especially, the intelligence degree of the self-adaptability of the Internetware applications needs to be improved, together with other reasoning mechanisms. Furthermore, we will pay further attention to research on the reliability of Internetwares.

Acknowledgments. The work supported by the National Basic Research Program of China under Grant No.2011CB302604, and the Scientific Research Foundation for the Returned Overseas Chinese Scholars, State Education Ministry.

References

1. Lu, J., Ma, X., et al.: Research on the Internetware Oriented Environment Driven Model and the supporting technology. Science in China Series E 38 (2008)
2. CBDI. Meta Model for SOA Version 2.0. Everware-CBDI Inc. (2007)
3. Ma, Z., Zhou, Y.: Modeling Distributed Service Contract Library Systems Based on UML. Acta Electronica Sinica 35(8) (2006)
4. OMG. UML Profile for Modeling Quality of Service and Fault Tolerance Characteristics and Mechanisms. OMG ptc/04-09-01 (2004)
5. Chang, Z., et al.: Component Model and Its Implementation of Internetware Based on Agent. Journal of Software 15(5), 1113–1124 (2008)
6. Lu, J., et al.: Research on the Internetware Model Based on Agent. Science in China Series E: Information Sciences 35(12), 1233–1253 (2005)
7. Oquendo, F., Warboys, B., Morrison, R., Dindeleux, R., Gallo, F., Garavel, H., Occhipinti, C.: ARCHWARE: Architecting Evolvable Software. In: Oquendo, F., Warboys, B.C., Morrison, R. (eds.) EWSA 2004. LNCS, vol. 3047, pp. 257–271. Springer, Heidelberg (2004)
8. Dowling, J., Cahill, V.: Self-Managed decentralized systems using k-components and collaborative reinforcement learning. In: Proceedings of the 1st ACM SIGSOFT Workshop on Self-managed Systems, pp. 39–43. ACM Press, New York (2004)

9. Mei, H., et al.: A software architecture centric engineering approach for Internetware. Science in China Series F: Information Sciences 49(6), 702–730 (2006)
10. Chang, S.H.: A Systematic Analysis and Design Approach to Develop Adaptable Services in Service Oriented Computing. In: Proceedings of IEEE SCW, pp. 375–378 (2007)
11. DeLoach, S.A.: The MaSE Methodology, In Methodologies and Engineering for Agent System. Kluwer Academic Publishing, New York (2003)
12. Lu, J., Ma, X., et al.: Rresearch and Development of Internetwares. Science in China Series E: Information Sciences 36(10), 1037–1080 (2006)

Design and Implementation of Intermediate Representation and Framework for Web Applications

Tomokazu Hayakawa, Shinya Hasegawa, and Teruo Hikita

Abstract. By the spread of Rich Internet Applications (RIAs), Web applications are becoming more convenient to use. Although there are many RIA technologies such as Ajax, JavaFX, Flex and Silverlight, by far there seems no common technique for specifying RIAs. The absence of the technique is considered inconvenient, especially when using RIAs in business. Hence, current Web applications heavily depend on the technology they use. Therefore, when the technology becomes obsolete, developers have to redevelop their applications by using other RIA technologies. Transforming an existing RIA to another through an intermediate representation is one of the solutions for it, but few attempts seem to have been made by this method. In this paper, we realize this, especially focusing on UI, by using an intermediate representation and a framework. The intermediate representation we propose here is in an XML format and provides an easy way of specifying a RIA for developers. The UI information is categorized into three parts: an widget part, a style part and a behavior part, which are held in a DOM tree, a CSSOM-like tree and an abstract syntax tree for ECMAScript, respectively. Moreover, the framework provides a default implementation in Java so that developers can easily extend it, if necessary. We show that the proposed method can solve the problem through an example of transforming a Web application from DHTML to JavaFX.

1 Introduction

By the spread of the Internet, Web applications are widely used nowadays. However, they were sometimes inconvenient because they only provided limited user interfaces, so that Rich Internet Application (RIA) technologies are proposed to enhance them. Although RIAs consist of many technologies such as Ajax, JavaFX[9], Flex[1] and Silverlight[7], there is no common technique for specifying a RIA.

Tomokazu Hayakawa · Shinya Hasegawa · Teruo Hikita
Meiji University, 214-8571, Japan
e-mail: {t_haya,s-hase,hikita}@cs.meiji.ac.jp

Z. Qian et al. (Eds.): Recent Advances in CSIE 2011, LNEE 125, pp. 375–384.
springerlink.com

Consequently, Web application developers have to choose one of these technologies, and they need to redevelop them when the chosen RIA technology becomes obsolete. Thus, there is a strong need to solve the problem of the dependency among RIA technologies, especially when using them in business.

This paper is intended to solve the problem, especially focusing on UI, by using an intermediate representation and a framework that aim at transforming an existing RIA to another RIA format through a vendor-neutral format. We have called the system "*Web-IR*". Although several studies have been made on transforming legacy Web applications, there seem to have been few studies on transforming a RIA to another.

The key ideas of our present work are (1) the widget and the style information of a Web application are held in a DOM tree and a CSSOM-like tree, and its behavior information is held as an AST (Abstract Syntax Tree), and (2) the framework provides a default implementation in Java and developers can easily change the behaviors of the transformation by extending corresponding classes, if necessary[11, 12].

The rest of this paper is organized as follows. Section 2 describes related studies and technologies on transforming Web applications. In section 3 we introduce an example of the transformation from DHTML to JavaFX by our proposed method. Section 4 describes the design of the intermediate representation. Section 5 presents the design and the implementation of the framework. Section 6 describes conclusion and further tasks.

2 Related Studies and Technologies

2.1 Related Studies

Yu Ping, Kostas Kontogiannis and Terence C. Lau[10] proposed a framework which transforms existing Web applications to equivalent ones based on MVC model. Alessandro Marchetto and Filippo Ricca[6] proposed a methodology which converts existing Java applications to Web services. Piero Fraternali, Sara Comai and Alessandro Bozzon[4] proposed a model-driven approach and realized an automatic code generation of RIAs. However, they did not aim at transformation among RIAs.

2.2 Related Technologies

XUL (XML User Interface Language)[8] and XForms[15] are XML-based languages that aim at specifying UI information of a Web application. They are similar to our intermediate representation in that they consist of core technologies of Web applications such as HTML, CSS and JavaScript, but are different from ours in that they are not designed to specify Web application behaviors.

Velocity[2] and XSLT[16] are kinds of template engines. They are similar to our framework in that they replace symbols (strings) or tags (elements) in a user-defined template and construct an output, but are different from our framework in that they cannot easily handle script languages such as JavaScript.

3 Example of Transformation between Web Applications

Figures 1-4 show the results of the transformation performed by the framework of ours, where a page in HTML is transformed into JavaFX. As figures 1-2 indicate, there are three widgets on the page, a button, a text and an image, which are commonly used in modern Web applications. We can see that the appearances of the UI, such as spacing and the notify message, are slightly changed before and after the transformation. This is because each RIA technology has a different policy of layout for its UI. In each of these, a click event is dispatched to the button, and the result of a clicking is shown in figures 3 and 4.

Figures 5-7 show the source codes of the sample input HTML application, the transformed intermediate representation and the transformed output JavaFX application, respectively. The latter two are obtained by our system. The description language of the input and the intermediate representation is JavaScript, and the one of the output is JavaFX Script.

The process of the transformation is in two steps: (1) from an input to the intermediate representation and (2) from the intermediate representation to an output.

In figure 5, the identifier *b1* is used to reference the element of the button. In figure 6, the identifier *b1* in the intermediate representation is used to identify the alert button. Finally, in figure 7, the identifier *b1* is used to declare a variable. In all the three cases, the identifier *b1* is used to reference the button in each script.

Fig. 1 Sample (HTML) Application

Fig. 2 Transformed (JavaFX) Application

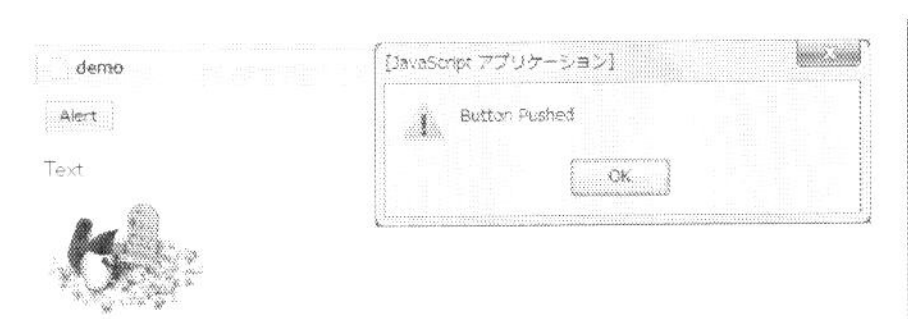

Fig. 3 Sample (HTML) Application after Clicking

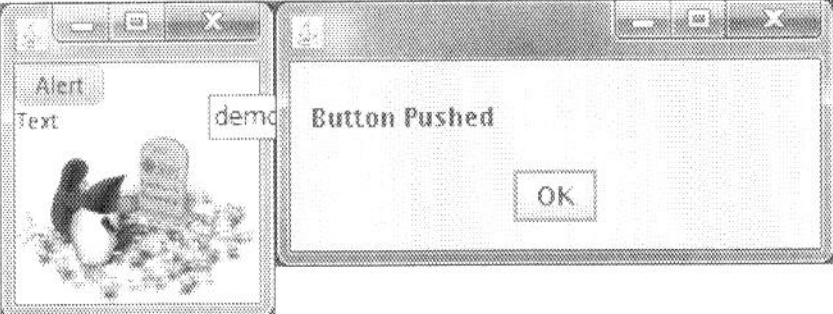

Fig. 4 Transformed (JavaFX) Application after Clicking

4 Intermediate Representation

When designing the intermediate representation, we have chosen XML[13] and basic Web technologies such as HTML, CSS and JavaScript as description languages in order to maximize its versatility. Our intermediate representation consists of three parts: an widget part, a style part and a behavior part (Figure 8). They correspond to HTML, CSS and JavaScript of a Web application, respectively. Although it currently supports a small subset of elements that many RIA technologies commonly support, developers can use them easily without detailed knowledge.

```
<html>
    <head>
        <script type="text/javascript">
            function init(){
                var b1 = document.getElementById("b1");
                b1.addEventListener(' click' ,func1,false);
            }
            function func1(){
                alert("Button Pressed");
            }
        </script>
    </head>
    <body onload="init();">
        <div>
            <input id="b1" type="button" value="Alert"/>
            <p>Text</p>
            <img src="SunRIPsmall.png" width="120" height="80"/>
        </div>
    </body>
</html>
```

Fig. 5 Source of Sample (HTML) Application

```
<application>
    <widget>
        <button id="b1" value="Alert"/>
        <text>Text</text>
        <image src="SunRIPsmall.png" width="120" height="80"/>
    </widget>
    <style></style>
    <behavior>
        function init(){
            b1.addEventListener('click',func1,false);
        }
        function func1(){
            alert("Button Pressed");
        }
    </behavior>
</application>
```

Fig. 6 Source of Intermediate Representation

```
var b1: Button;
var stage: Stage = Stage{
    scene: Scene{
        content: VBox{
            content: [
                b1 = Button{
                    text: "Alert";
                    onMouseClicked: function(e:MouseEvent): Void{
                        Alert.inform("Button Pressed");
                    }
                },
                Text{content: "Text";},
                ImageView{
                    image: Image{
                        url: "{__DIR__}SunRIPsmall.png";
                        width: 120;
                        height: 80;
                    }
                }
            ]
        }
    }
}
```

Fig. 7 Source of Transformed (JavaFX) Application

```
<?xml version="1.0" encoding="UTF-8"?>
<application>
    <widget>
        <!-- UI Information -->
    </widget>
    <style>
        <!-- Style Information -->
    </style>
    <behavior>
        <!-- Behavior Information -->
    </behavior>
</application>
```

Fig. 8 Skeleton of Intermediate Representation

4.1 Widget Part

The widget part has <*widget*> as its root element, and has other elements as its children to hold the information of UI widgets, such as a textbox and a password box. Table 1 shows the elements which can be used currently in our system.

4.2 Style Part

The style part has <*style*> as its root element, and has other elements as its children to hold the information of styles such as font sizes and font colors. In our system, the input style information (e.g., CSS) is transformed into an equivalent XML, so that no information will be lost in the transformation. However, the reproducibility of the style information depends on how it is supported in the RIA technology of the output.

4.3 Behavior Part

The behavior part has <*behavior*> as its root element, and has other elements as its children to hold the information of the behaviors of an application, such as a key pressed event and a mouse pressed event. Table 1 shows the supported mouse and keyboard events. An important point here is that both the widget and the style parts use XML elements, but the behavior part uses the ECMAScript[3] in order to reduce the size and improve readability. The another reason that we use ECMAScript is that it is commonly used by Web developers, and it has simpler grammars in compared with the other programming languages. However, our system currently provides a limited support for it, so that some high-level features such as *prototype* are not yet supported.

4.4 Extensibility

If developers want to add new UI elements or style elements to the intermediate representation, they can simply add the elements into it with a few tasks described in section 5.5.

5 Framework

5.1 Overview

Our framework uses Java as its development language, and it supports (1) transforming a Web application into the intermediate representation and (2) transforming the intermediate representation into another RIA (Figure 9). Also, it uses the design patterns[5], such as Abstract Factory and Visitor, to improve reusability and extensibility. In addition, we have chosen basic Web technologies such as XHTML, CSS and JavaScript, as description languages to improve usability. A key point of our system is that the information among the transformation is held and passed by the DOM trees (Figure 10), so that the average Web developers can understand and use the system easily.

5.2 *Classes and Interfaces*

Figure 11 shows the class diagram of the framework, where concrete classes are omitted. The roles of the each interface are described in the following subsections.

5.2.1 Application Interface

The interface named *Application* represents a Web application. In our current system, there are three subinterfaces named *AjaxApplication*, *IRApplicaiton* and *JavaFXApplication*. Each of them represents an Ajax application, the intermediate representation and a JavaFX application, respectively.

Table 1 Widgets and Events in Intermediate Representation

Element Name	Event Name
button	onclick
radiobutton	onmousedown
image	onmouseup
label	onkeypress
checkbox	onkeydown
textbox	onkeyup
passwordbox	
combobox	
listbox	

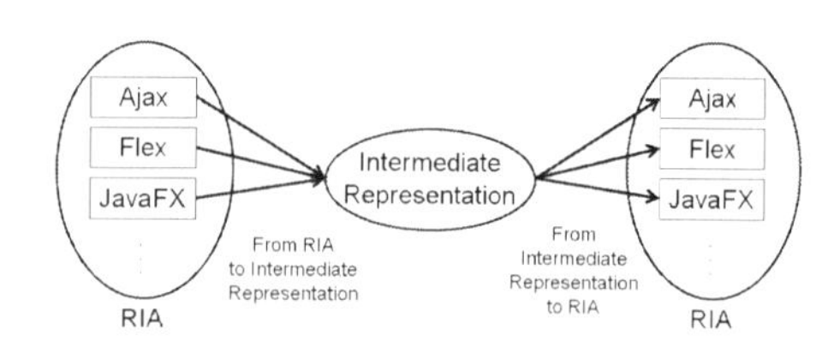

Fig. 9 Overview of Framework

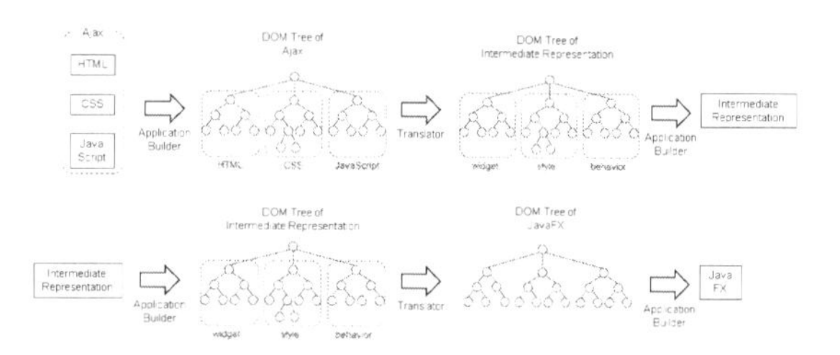

Fig. 10 Process of Transformation

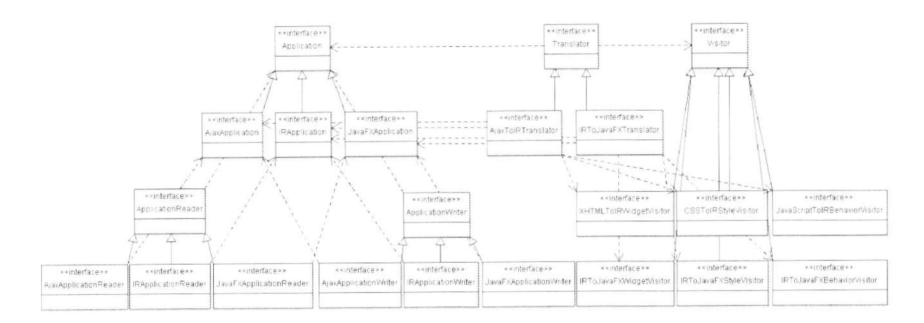

Fig. 11 Class Diagram of Framework

5.2.2 ApplicationReader Interface

The interface named *ApplicationReader* represents a reader which reads an application from a source file and returns an instance of the interface *Application*. Currently, our system has three subinterfaces named *AjaxApplicationReader*, *IRApplicationReader* and *JavaFXApplicationReader*. Each of them reads an Ajax application, the intermediate representation and a JavaFX application, respectively.

5.2.3 ApplicationWriter Interface

The interface named *ApplicationWriter* represents a writer which writes an application to a destination file. Currently, three subinterfaces named *AjaxApplicationWriter*, *IRApplicationWriter* and *JavaFXApplicationWriter* are provided.

5.2.4 Translator Interface

The interface named *Translator* represents a translator which translates Web applications, and the translator consists of three visitors which have roles to transform widgets, styles and behaviors, respectively. Currently, there are two translators named *AjaxToIRTranslator* and *IRToJavaFXTranslator*. Each of them translates an Ajax application to the intermediate representation, and the intermediate representation to a JavaFX application, respectively.

5.2.5 Visitor Interface

The interface named *Visitor* represents a visitor which walks around DOM trees and processes each element of the trees. Each time an element is found in the tree, the visitor is called back and this process is repeated recursively. Finally, the output is produced. The current system provides six default visitors named *XHTMLToIRWidgetVisitor*, *CSSToIRStyleVisitor*, *JavaScriptToIRBehaviorVisitor*, *IRToJavaFXWidgetVisitor*, *IRToJavaFXStyleVisitor* and *IRToJavaFXBehaviorVisitor*.

5.3 Method of Transformation

The process of the transformation is as follows.

5.3.1 Construct DOM Tree from Input

An input file is parsed and transformed into a DOM tree. As stated above, the intermediate representation consists of three parts, so does the DOM tree. For example, if an input is HTML application, the tree parts correspond to XHTML, CSS and JavaScript, respectively.

5.3.2 Replace Elements

The constructed DOM tree is transformed according to the rules initially specified in the framework. As a DOM tree consists of three parts, three visitors are used in the transformation. Since our framework is designed to use the visitor pattern, so that developers can easily change the transforming behaviors.

5.3.3 Output DOM Tree

The transformed DOM tree is written to a file in an application specific format.

5.4 Implementation

5.4.1 Input

Our system can accept a Web application as an input which uses the technologies shown in table 2. Also, table 4 shows the modules of XHTML[14] supported by the framework.

5.4.2 Output

Our system produces a JavaFX application as an output, and table 3 shows the UI controls of JavaFX which can be used in the output.

5.5 Extensibility

If developers want to change the behaviors of the transformation or add new elements into the intermediate representation, they are achieved by simply extending the corresponding classes and overriding the corresponding methods since our framework is well-designed by employing the design patterns.

Table 2 Used Technologies

Technology	Version
XHTML	1.1
CSS	2.1
JavaScript	1.5
XML	1.1
Java	1.6
DOM	Level 2

Table 3 Supported JavaFX Controls and Corresponding Elements

JavaFX Control	IR Widget
Button	button
CheckBox	checkbox
ChoiceBox	combobox
Label	label
ListView	listbox
PasswordBox	passwordbox
RadioButton	radiobutton
TextBox	textbox

Table 4 Supported XHTML Modules

Module Name	Support Status
Structure Module	partial
Text Module	yes
Hypertext Module	partial
List Module	partial
Object Module	no
Presentation Module	yes
Edit Module	partial
Bidirectional Text Module	no
Forms Module	partial
Table Module	partial
Image Module	yes
Client-side Image Map Module	no
Server-side Image Map Module	no
Intrinsic Events Module	partial
Meta Information Module	no
Scripting Module	partial
Stylesheet Module	yes
Link Module	yes
Base Module	no

6 Conclusion

By using the intermediate representation and the framework, developers can generalize their RIAs and transform them into another RIAs easily. Hence, our goal, to reduce the dependency between Web applications and underlying RIA technologies, is achieved to some extent. On the other hand, it still remains work to enhance the capability of our framework to transform more complicated behaviors of Web applications. The improvements of the capability and supports for other RIA technologies are planned to be implemented in the near future.

References

1. Adobe: Flex, `http://www.adobe.com/products/flex/`
2. Apache Software Foundation: Velocity, `http://velocity.apache.org/`
3. Ecma International: ECMAScript Language Specification 3rd edn. (1999), `http://www.ecma-international.org/publications/standards/ecma-262.htm`
4. Fraternali, P., Comai, S., Bozzon, A.: Engineering Rich Internet Applications with a Model-Driven Approach. ACM Transactions on the Web 4(2) (2010)
5. Gamma, E., Helm, R., Johnson, R., Vlissides, J.M.: Design Patterns: Elements of Reusable Object-Oriented Software. Addison-Wesley Professional (1994)

6. Marchetto, A., Ricca, F.: Transforming a Java Application in an Equivalent Web-Services Based Application: Toward a Tool Supported Stepwise Approach. Web Site Evolution (2008)
7. Microsoft: Silverlight, `http://www.microsoft.com/silverlight/`
8. Mozilla Foundation: XUL, `https://developer.mozilla.org/En/XUL`
9. Oracle: JavaFX, `http://javafx.com/`
10. Ping, Y., Kontogiannis, K., Lau, T.C.: Transforming Legacy Web Applications to the MVC Architecture. In: Eleventh Annual International Workshop on Software Technology and Engineering Practice (2004)
11. W3C: CSSOM, `http://dev.w3.org/csswg/cssom/`
12. W3C: Document Object Model (DOM) Level 2 Core Specification, `http://www.w3.org/TR/DOM-Level-2-Core/`
13. W3C: Extensible Markup Language (XML) 1.1, 2nd edn., `http://www.w3.org/TR/xml11/`
14. W3C: Modularization of XHTML, `http://www.w3.org/TR/xhtml-modularization/`
15. W3C: XForms, `http://www.w3.org/TR/xforms/`
16. W3C: XSLT, `http://www.w3.org/TR/xslt`

Design and Implementation of Market Management System Model Based on the Branch and Bound Algorithm and Internet of Things

Yang Liu and Wenxing Bao

Abstract. To solve the problem of the market information on halal, the collection mode of the market information, the trace of the distribution of the goods, we propose a model , based on the Branch and Bound Algorithm and Internet of Things. According to the situation of Ningxia halal market , we design and implement this model. This system fulfills the function of gathering the market price, warehouse management, optimization of distribution and the trace of goods distribution. The system meets the requirements of halal markets and improves the efficiency of storage and transportation and the halal market competition.

Keywords: halal, Livestock product, market management, Branch and Bound Algorithm.

1 Introduction

With the rapid development of informatization, the information revolution bring about great changes to our country. Most of the developed countries pay more attention to the technology of using information to improve the efficiency of the market management. This has become a major trend of the market-managing development in the world. In our country, the reform and development of the market management has achieved outstanding results after reform and opening up. Therefore, local government at several levels think highly of the construction of market management system.[1] However, due to the weakness of the current market supervision and exchange of information management, the market is still disorder, especially the informatization of the halal market management . In addition, with China's entry into WTO, international markets bring about a powerful influence and pressure to the domestic markets and related industries. Therefore,

Yang Liu · Wenxing Bao
School of Computer Science and Engineering North University for Ethnics
Yinchuan Ningxia, P.R. China
e-mail: {nlgliuyang,bwx71}@163.com

Z. Qian et al. (Eds.): Recent Advances in CSIE 2011, LNEE 125, pp. 385–391.
springerlink.com

in order to consolidate the status of the market management, implement market supervision and cxchangc of information and enhance competitive forces of markets and related industries of our country, we need get down to the informatization of halal market instantly.

Ningxia is the only hui nationality autonomous region in China. Halal is not only the preponderant farm products of Ningxia, but also one of the eleven preponderant farm produces in China. So the Halal of NingXia has certain advantages in culture and resource . [2]

Although Halal of NingXia has some characteristics, its scale is small. The relative lack of market information on Halal markets restrains the further development of halal.To solve these problems and satisfy the requirements of market information on Halal markets, this paper develops a halal market-managing system; based on the Branch and Bound Algorithm and Internet of Things. The deatails asr as follows: the system realizes the the collection of the storage, the transportation and the sale of halal, builds halal market management database, and apply EAN/UCC-128 and GPRS, making use of the Branch and Bound Algorithm to optimize the storage and the distribution. This supports the informatization management of halal market.

2 Overall Structure and Key Technology of the system

Based on the analysis of the requirement on information on Halal markets, we propose a solution about halal market management using the Branch and Bound Algorithm and Internet of Things. For the warehouse of the Halal markets, we establish records about the storage of halal products through pasting up barcode on the box, and record the information about these products when they are sent in,transferrd, or send out of the warehouse. When the products are sent out of the warehouse, we print EAN/UCC-128(a trace code) on the packing box and put the data into the central server. Then the consumer can trace the information of the products by message or network. At the same time, researchers can gather the information of the selling price from various sales markets by PDA, then send the information to the central server with GPRS, and establish the price information file of the halal products. The overall structure of the system is given in Fig. 1.

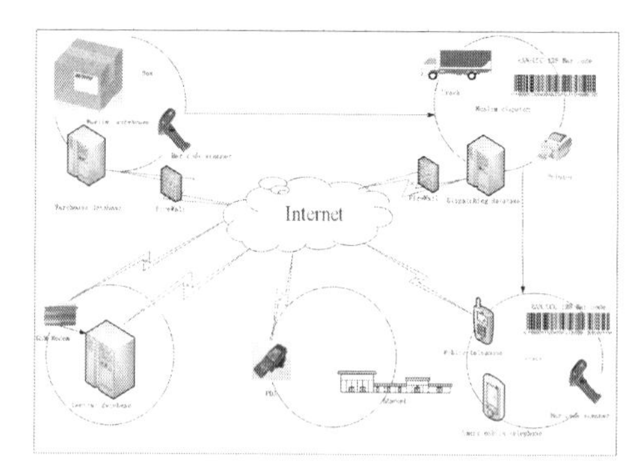

Fig. 1 Overall Structure of System

2.1 The Branch and Bound Algorithm

The basic idea of the Branch and Bound Algorithm is to search all the possible solution space from the optimization of the problem which is constrained by some conditions. During the implementation, we put all the possible solution space into a less subset , calculate the upper limit or lower limit of each subset. When the value of those subsets surplus the possible solutions, we set no further branch. Thus, many subsets will not be considered, and the search range will be manified. When all the possible solutions are in the bounds of the subsets, we can find the feasible solution. Therefore, in general, the algorithm can ensure to find the optimal solution.[3]

In view of the optimal path in the system is traveling salesman problem(TSP), we describe the problem: finding Hamilton which has the least weight in a complete graph. Assuming G=(V,E), V={1,2,...,n} is a vertex set, E is a edge set, and each edge e=(i, j) ∈E has a non-negative weight w(e), what we need is to find a Hamilton C whose total weight $w(c) = \sum_{e \in E(c)} w(e)$ is the least.

Setting d_{ij} as the distance between city i and city j, $d_{ij}>0$, $d_{ii}=0$, $i,j \in V$, and

$$x_{ij} = \begin{cases} 1, (i, j) \text{ is in the optimal path} \\ 0, \text{other} \end{cases} \tag{1}$$

Then the mathematical model of TSP can be also described as following linear programming form:

$$MinZ = \sum \sum d_{ij} x_{ij}$$
$$st \begin{cases} \sum_{j \neq i} x_{ij} = 1, i \in V \\ \sum_{i \neq j} x_{ij} = 1, j \in V \\ \sum_{i,j \in S} x_{ij} = |S| - 1, S \subseteq V \\ x_{ij} \in \{0,1\} i, j \in V \end{cases} \tag{2}$$

Here, |S| denotes the number of vertex in set S; the first two constraint means each vertex with exactly one in-edge and one out-edge; and the third constraint is to guarantee that there is no sub-loop.

The application of the Branch and Bound Algorithm can be describe as following: [4].

- Setting up an adjacency matrix which indicates a weighted path diagram, as Fig.2, the weight record the distance between two vertexes;

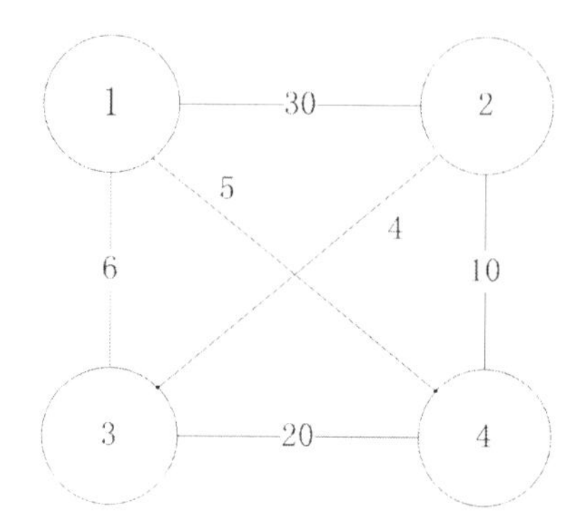

Fig. 2 Vertex Weighted Graph

- Establishing a min-heap to indicate the priority queue of the living node;
- Calculating the least spending of initial edge of each vertex;
- If the weighted graph has no initial edge ,then this algorithm is over; else ,initializing the algorithm with the least calculated spending of initial edge;
- Using "while" to accomplish the extension of inner node of the arrangement tree;
- When a leaf node of the arrange tree become current extension node, the algorithm is over.

2.2 Design of the Trade Code Called EAN/UCC-128

EAN-128 transfers the information into barcode, based on the defining standard of code EAN/UCC-128. It adopts the logic of 128 codes and has features of integrity, compactibility, connectivity and high reliability. According to the information of logistics demand and the existing experience, we choose EAN UCC-128 as the trace code of the logistics and adopt the corresponding application identifier code which is regulated in GB / T16986-2003 《EAN·U CC system application identifier》 [5],as Fig. 3.

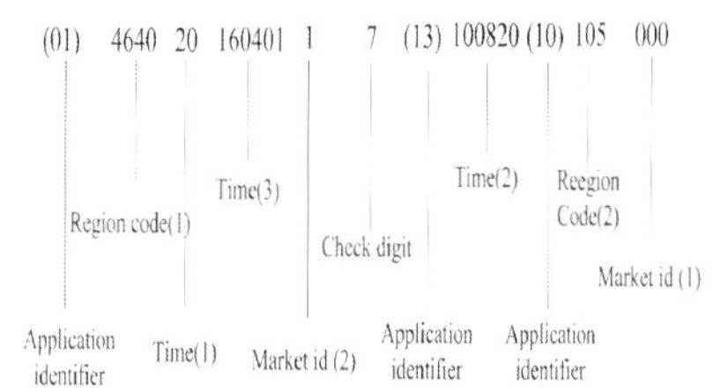

Fig. 3. Encoding Structure of EAN/UCC-128

According to the characterizes of different codes,we use the identifiers of trade account AI (01) to identify logistics identified code, the identifiers of date of the products AI (13) to indentify the date of the products ,and the identifiers of the source entity reference code AI (01) to identify the serial number of the market which the logistics starts. Check digit is check bit, Region code(1) and Region code(2) shows the area code of the market, and they are conformed by GB/T2260-

1999,for example 640105 indicates xixia district of NingXia YinChuan . Market id(1)and Market id(2) indicate the code of the market in fig 3.the market code is "0001", Time(1)、Time(2) and Time(3) indicate the exact time of the distribution ,in fig 3 which is 16:04:01 2010/8/20 .

2.3 GPRS

The GPRS is abbreviation of general packet radio service, based on the network of GSM (Global System for Mobile Communication. It is called the 2.5 generation of the mobile communications technology, and combines wireless communications with Internet. The remote data acquisition system based on GPRS is connected through wireless IP interface mode provided by GPRS wireless communications network system of the China Mobile, which supervise and manage the real time data information of industry and civilian to implement the gathering, the delivery and the unified management and control of the remote data. GPRS has the characteristics of high speed and low cost. Its transmission speed can be achieved 171.2kb/s. Compared with wired communication, GPRS has more merits such as being flexible in organizing a network, easy extension, low pay, easy to maintain, and high cost performance.[6]

3 System Function Model and Implementation

Through investigation and analysis, the management system of halal market is divided into 4 sub-systems in this study: warehouse management system of halal, market distribution system of halal, traceability system of distribution information of halal and market information collection system of halal , as Fig.4.

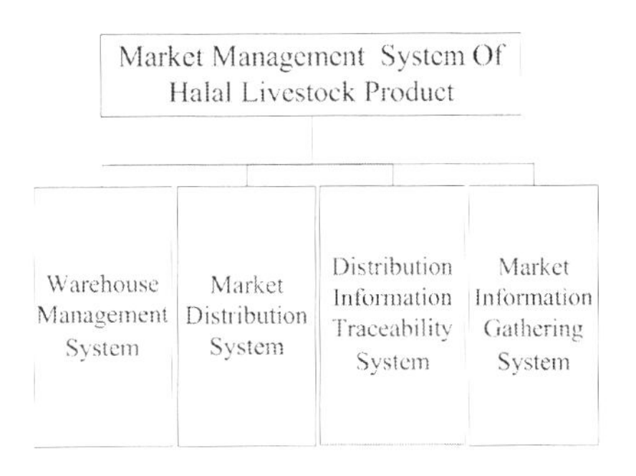

Fig. 4 Architecture of System Function Module

Warehouse Management System of Halal: Using EAN/UCC-128 bar code technology, database technology and computer network technology, the system carries out the electronic management of stocking, transferring, delivery and storage optimization based on halal storage information, so as to monitor all aspects of warehousing and make it scientific and reasonable.

Market Distribution System of Halal Using EAN/UCC-128 bar code technology, database technology and computer network technology, the system carries out the electronic management of distribution of halal based on delivery warehouse information of the halal. By analyzing the distance and other factors, the system proposed a optimal path for the traveling salesman problem.

The text description can be expressed as follows: Given a set of N cities and the direct distance between each two ,we find a closed loop, making each city exactly once and only once, and the total travel distance the shortest. Using the language of mathematics, finding a loop$T=(t_1,t_2,\cdots,t_n)$, make the following objective function the smallest:

$$f(T)=\sum_{i=1}^{n-1} d(t_i,t_{i+1})+d(t_n,t_1) \tag{3}$$

In the above formula (2), t_i means denotes city code, valued from 1 to n ,and $(t_1,t_2,\cdots,t_n)$can be seen as a permutation on n; $d(t_i,t_j)$ denotes the distance between t_i and t_j .For the symmetric TSP, it has $d(t_i,t_j) = d(t_j,t_i)$. At the same time, the system monitors distribution environment and officers in the process of halal distribution , to ensure that products meet the requirements of halal food, and also lay the foundation for the traceability of distribution information.

Distribution Information Tracking System Of Halal Using SMS Cat and Mobile Communication Technology ,the system has carried out that when receiving the query message ,it automatically search the database according to logistics distribution signal , and send the distribution information to the query phone .enquiry officers can also visit the website for information inquirements. As Fig.5.

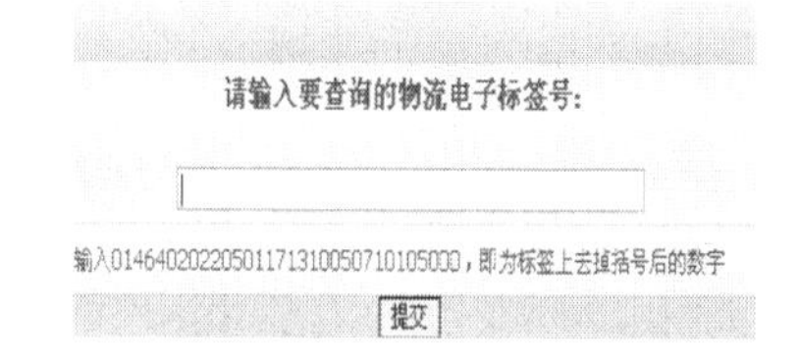

Fig. 5 Distribution Information Tracing System

Market Information Gathering System Of Halal

In support of GPRS technology, through Tcp/Ip protocol and asynchronous transfer mode, the system has achieved data transmission between servers and handheld devices in any direction using PDA. The market information gathering staff can send the collection information price to server at any time. as Fig.6.

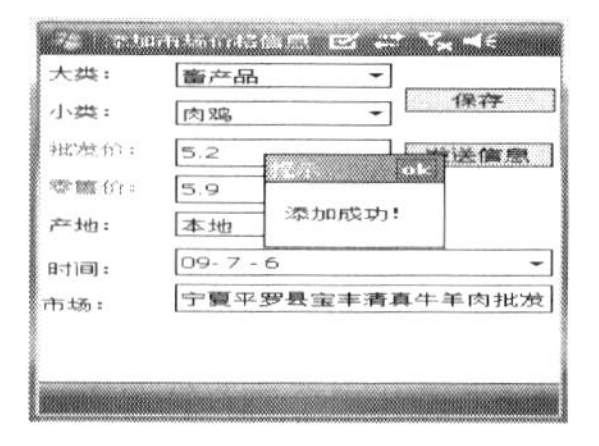

Fig. 6 Price Information Collection System

Software Development Environment
On the basis of the warehousing , distribution, information traceability and collection of halal , The system chooses cross-platform Java language to develop warehouse management system , market distribution system and distribution information tracking system ,selects c# language to develop market information gathering system, and use SQL Server 2005 CE for embedded database. It selects Windows XP as the computer operation system , and all use SQL Server 2005 for database.

4 Results and Analysis

Using halal market as study object, the research sets up key information database of halal market management ,based on the branch bound algorithm and the Internet of Things technology. Furthermore, this paper realizes a system of halal warehouse, distribution, logistics information traceability, market information gathering and many other functions.

Halal industry is the advantageous and characteristical industry of Ningxia national economy. It is a common consensus of government and enterprise to develop market management of halal vigorously, and form a range of market management system of the halal imperatively. Methods, products and application demonstrations of results adopted in this study are beneficial to market management of halal. It will improve the competitiveness of halal in halal market and high consumption area on home and overseas markets.

Acknowledgments This work is supported by the National Key Technology R&D Program of China under Grant NO. 2007BAD33B03.

References

1. Yan, H.: Development of Information System of Construction Market Sypervision and Business Management Based on Web. In: Proceedings of 2005 International Conefrence on Construction & Real Estate Management, pp. 566–569 (January 2005)
2. Ge, Z., Tian, L.: Thoughts on the supply chain management on Islamic beef and mutton industries in Ningxia. Journal of Agricultural Sciences 29(4) (2008)
3. Baiduencyclopedia (2010), `http://baike.baidu.com/view/1304851.htm?fr=ala0_1`
4. Wang, X.D.: Design and Analysis of Computer Algorithm, 3rd edn., pp. 225–229. Tsinghua University Press (2006)
5. Chunling, Y., Bing, Z., Songhao, G., Rui, W., Zhaoyu, H.: Design and Implement of Traceable Label of Vegetable Produce Applied in Vegetable Quality and Safety Traceability System. Food Science 28(07), 572–574 (2007)
6. Liu, J.: Remote Data Acquisition System Base on GPRS. Electronic Engineering & Product World. Design Field 3, 99–101 (2008)

An Efficiency Optimization Strategy for Huge-Scale Data Handling

Congqi Xia and Yonghua Zhu

Abstract. This paper discussed the efficiency optimization of handling huge scale data and analyzed the relationship of the factors and the efficiency. The cost of the whole access procedure has been analyzed as well. It proposed overhead shifting strategy to optimize the performance of handling these problems. The main idea of this overhead shifting method is to reduce the critical factor—communication overhead. It shifts the logic process code to the client. This strategy deals with the dirty data, which is the result of itself, are by implement a state notification protocol. The protocol notifies the client whether its data becomes dirty so that the client can easily decide how to behave. An experiment was done to prove the superiority of the new strategy. The results showed that this overhead shifting strategy improves most accesses and operations and makes the user experience being satisfied.

1 Introduction

Twenty years have passed since the ERP System was first introduced. Enterprise Application Suite is a new name for formerly developed ERP systems which includes almost all segments of the business by using ordinary Internet browsers as thin clients. Though traditionally ERP packages have been on-premise installations, ERP systems are now also available as Software as a Service.1

Yet the efficiency and the user experience of the EAS become the choke point when dealing huge-scale data. In this paper, the main purpose is to analyze the overhead of the whole process when calling a service and figure out an approach to optimize it.

Congqi Xia
School of Computer Engineering and Science Shanghai University,
No.149 Yan Chang Road Shanghai, China
e-mail: silverbelial@gmail.com

Yonghua Zhu
Computing Center of Shanghai University Shanghai University, Bldg D,
Shang Da Road Shanghai, China
e-mail: zyh@shu.edu.cns

Z. Qian et al. (Eds.): Recent Advances in CSIE 2011, LNEE 125, pp. 393–398.
springerlink.com

2 Related Work

The common way to optimize the performance are to 1) optimize the algorithm, 2) adds triggers and stored procedures and 3) use connection pool. These methods had little effect on the huge-scale data.

With the advent of high speed networks, the future communication environment is expected to comprise a variety of networks with widely varying characteristics. The next generation multimedia applications require a transfer of a wide variety of data such as voice, video, graphics, and text which have been widely varying access patterns such as interactive, bulk transfer, and real-time guarantees. Traditional protocol architectures have difficulty in supporting multimedia applications and high-speed networks because they are neither designed nor implemented for such a diverse communication environment.2

Despite the advantages of the network protocol described in 2, it cannot be applied to every business software since it is too expensive as a hardware supplement is needed.

3 Analysis of the Cost

Considering the time cost of a single access, it's easy to separate the whole procedure into four parts:

1. Rendering Cost— This kind of time cost can be considered as a static factor. Even though it ascents when the amount of the data enlarges, it depends on only the language you use to program the page, which means there are few ways to reduce this type of overhead.
2. Communication Overhead— Communication overhead is a variable factor and changes uncontrollable.
3. Logic Process Cost— Logic process overhead is often affected by the overhead of the server. It varies from time to time because the amount of calling of the services changes all the time.
4. Database Operation Cost— Database overhead can also be considered as a static factor. This kind of overhead can be reduced by programming techniques such as triggers and store procedures.

A set of experiments has been done to figure out the relationship between those factors and costs.

The first experiment is focused on the relation between data amount and overload. The numerical result is shown in Table. 1. To analyze the characteristics, the numerical result is turned to the line chart as Fig. 1. The overload of the server can affect the performance a lot. Another experiment was done to reveal the relation between time cost and server overload. The numerical result is shown in Table.2 and the line chart is shown in Fig. 1.

Table 1 The relation between Data and Overloads

Data Amount (KB)	DB Overhead (ms)	Logic Overhead (ms)	Communication Overhead (ms)	Data Amount (KB)	DB Overhead (ms)	Logic Overhead (ms)	Communication Overhead (ms)
100	15	47	219	600	31	282	12484
200	47	78	735	700	172	2062	20578
300	62	78	1813	800	63	1921	25765
400	62	78	2969	900	63	2062	34328
500	47	234	7359	1000	78	3594	45313

Table 2 The relation between Time Cost and Server Overhead

Threads	The Latest Response (ms)	Threads	The Latest Response (ms)	Threads	The Latest Response (ms)
100	5218	400	21028	700	36921
200	9723	500	26723	800	41892
300	16021	600	32432	900	47234

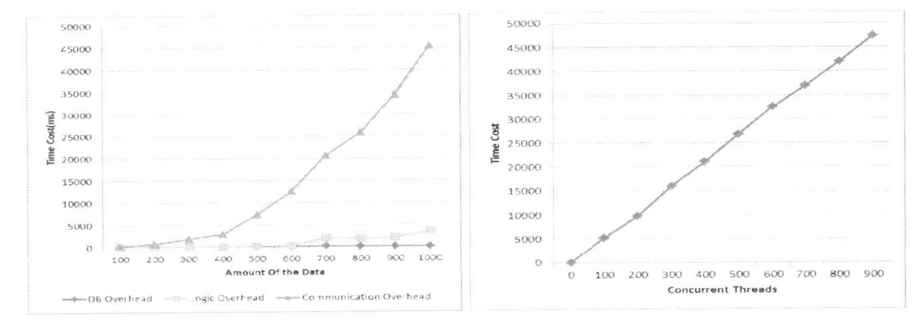

Fig. 1 The relation between Data/Overheads and relation between Time-Cost/Server-Overhead

According to the experiment results stated in the previous paragraph, following conclusions can be summarized: 1) page rendering and database operation overhead can be treated as static factors. 2) communication overhead has an exponential relationship with the amount of data. 3) the logic process overhead has a logarithmic relationship with the amount of data. 4) the logic process overhead has a linear relationship with the server overhead.

4 Strategy of Performance Improving

To improve the performance of the system, some factors must be reduced or some relations shall be changed.

4.1 Primary Solution

The requirement of accessing the server should be satisfied regardless the amount of the overhead it leads to. The communication overhead becomes the critical cost in the whole accessing procedure, since it increases exponentially with the increment of the data amount.

The problem is inevitable. However, if saw another way, the best solution exists where the biggest flaw stands. Fig. 2 describes the normal software structure involving the Internet. By researching the traditional structure, ways to solute this problem can be found from these aspects: 1) reduce the communication overhead which is a time killer when the problem scales grew huge; 2) reduce server overhead which makes the logic process unbearable slow. It's obviously noticeably, the Internet part makes the communication part inevitable and the logic procedure is all accomplished on the server.

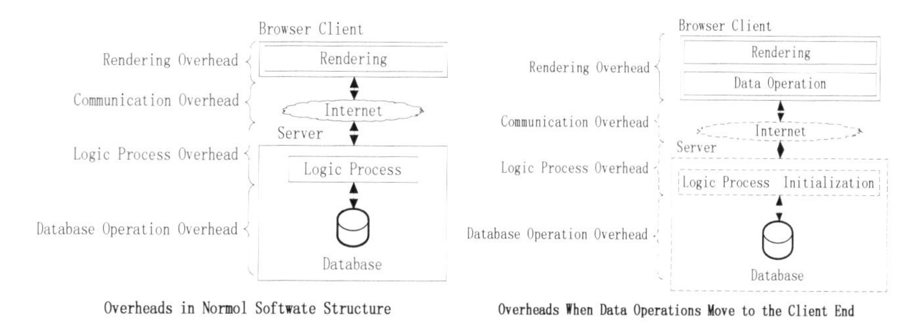

Fig. 2 Overheads in Normal Structure and when Data Operations Move to the Client End

An overhead shifting strategy is proposed in this paper. As the new structure demonstrated in Fig. 2, the logic process is partly moved to the client, rapidly reduce the times of the communication and relieve the server from heave overhead.

In this circumstance, the large-scale overheads occur only when the clients get the data for the first time and synchronize the data. Variable d refers to the data amount. Variable c refers to the time cost in each period. Variable f refers to the coefficient factors in each relationship. And variable s refers to the static factors. The following equations exist in traditional structure:

$$c_{render} = s_{render}, c_{database} = s_{database}, c_{communication} = f_{communication} \bullet d^{n}, c_{Logic} = f_{Logic} \bullet \log_{n'} d$$

$$\begin{aligned} c_{sum} &= c_{render} + c_{communication} + c_{Logic} + c_{database} \\ &= s_{render} + s_{database} + f_{communication} \bullet d^{n} + f_{Logic} \bullet \log_{n'} d \end{aligned} \quad (1)$$

And the statements shown in (2) exits in the new structure.

It's obvious that the new structure has much better time-cost. However it has some side effects on the system, like dirty data.

$$
\begin{aligned}
s_{render} &= \begin{cases} s_{render} & \text{,when initializing} \\ s_{render} + f_{render} \bullet d & \text{,when operating} \end{cases} \\
c_{communication} &= \begin{cases} f_{communication} \bullet d^{n} & \text{, when initializing} \\ 0 & \text{, when operating} \end{cases} \\
c_{Logic} &= \begin{cases} f_{Logic} \bullet \log_{n'} d & \text{, when initializing} \\ 0 & \text{, when operating} \end{cases} \\
c_{database} &= \begin{cases} s_{database} & \text{, when initializing} \\ 0 & \text{, when operating} \end{cases}
\end{aligned}
\tag{2}
$$

$$
c_{sum} = \begin{cases} s_{render} + s_{database} + f_{communication} \bullet d^{n} + f_{Logic} \bullet \log_{n'} d, & \text{when initializing} \\ s_{render} + f_{render} \bullet d & \text{, when operating} \end{cases}
$$

4.2 Advanced Solution Avoiding Side Effects

To fix the new-raised problem, a protocol has been established to describe the latest changes of the database. The main idea to solve this problem is to monitor the client whether there are changes in the database related to the page and what have been changed. This idea can be presented in the following steps: 1) initialize the data at the server; 2) get the data through the normal routine; 3) store the data in a particular data structure; 4) operate the data in the client; 5) repeat step 4 until the whole work is finished; 6) submit the data; 7) handle dirty data.

Yet there are different kinds of demand on this kind of problem. In some projects, user operation can ignore the others and just get his/her data into the database, which means the latest submitter will get his/her data into the database. Otherwise, user will get failure notification when he is not the first submitter to submit changes. All the operations will be abandoned to assure that there are no dirty data in the database.

A new protocol with more humanity was introduced to avoiding extreme practices, which makes the user to decide how the system will behave. They can choose to ignore the others, or to abort operations have just been done. And the administrator can configure this as well.

Considering the dirty data handling, the formulation of the Time Cost shall be changed as well. Let $l_{protocol}$ refers to the length of the protocol message. Let c'_{sum} refers to the cost when taking the overwrite tactic (ignore the others' change) and let c''_{sum} refers to the cost when talking the dirty data checking tactic (abandon the formal operation), following equations will be founded:

$$c'_{sum} = c_{sum} \tag{3}$$

$$
c''_{sum} = \begin{cases} s_{render} + s_{database} + f_{communication} \bullet d^{n} + f_{Logic} \bullet \log_{n'} d, & \text{when initializing} \\ s_{render} + f_{render} \bullet (d + l_{protocol}) + f_{communication} \bullet l_{protocol} + f_{Logic} \bullet l_{protocol}, & \text{when operating} \end{cases}
\tag{4}
$$

It's easy to prove that (5) is always positive, which means the overhead shifting strategy is less sensitive about the increment of amount of data.

$$\frac{d(c_{old} - c_{new-worst})}{dA} = nf_{communication} d^{n-1} + \frac{1}{d \ln n} - f_{render} \tag{5}$$

To demonstrate how good this overhead shifting strategy works, another experiment was done. The average costs are listed with the caparison of the normal value in Table.3. It's crystal that the overhead shifting strategy works perfectly.

Table 3 The efficiency of overhead shifting stratedy

	Normal	Initialization	Operation
Time cost (ms)	5218	5121	172

5 Summary

As describe previously, the strategy proposed in this paper can easily improve the performance of the EAS without any changes about the main logic procedure.

By shifting the logic processing part to the client, it can make full use of the browser client and make sure that the users have the best experience.

Used in the project, this page-logic-shifted scheme has been proved to improve the efficiency of web projects.

References

1. Wikipedia, the free encyclopedia, Enterprise resource planning (2010), http://en.wikipedia.org/wiki/Enterprise_resource_planning
2. Jain, P.K., Hutchinson, N.C., Chanson, S.T.: Protocol architectures for delivering application specific quality of service, pp. 313–320. IEEE (1995)
3. Cormen, T.H., Leiserson, C.E., Rivest, R.L., Stein, C.: Introduction to algorithms, 2nd edn. The Massachusetts Institute of Technology Press (2002)

Impact on Chunk Size on Deduplication and Disk Prefetch*

Kuniyasu Suzaki, Toshiki Yagi, Kengo Iijima, Cyrille Artho, and Yoshihito Watanabe

Abstract. CAS (Content Addressable Storage) systems reduce total volume of virtual disk with deduplication technique. The effects of deduplication has been evaluated and confirmed in some papers. Most evaluations, however, were achieved by small chunk size (4KB-8KB) and did not care about I/O optimization (disk prefetch) on a real usage. Effective disk prefetch is larger than the chunk size and causes many CAS operations. Furthermore, previous evaluations did not care about ratio of effective data in a chunk. The ratio is improved by block reallocation of file system, which considers access profile. Chunk size should be decided by considering these effects on a real usage. This paper evaluates effectiveness of deduplication on a large chunk of CAS system which considers the optimization for disk prefetch and effective data in a chunk. The optimization was achieved for boot procedure, because it was a mandatory operation on any operating systems. The results showed large chunk (256KB) was effective on booting Linux and could maintain the effect of deduplication.

1 Introduction

Content Addressable Storage (CAS) becomes popular method to manage disk image for many instances on virtual machine [1,2,3]. In CAS systems, data is managed by certain size of chunk, and it is addressed not by its physical location but by a name derived from the content of that data (secure hash is used as a unique name usually). CAS system can reduce its total volume by data deduplication which shares same content chunks with a unique name.

Most CAS systems use 4 or 8 KB chunk size. The size fits to traditional block size of file system. It facilitates the abstraction of block device. However, it was too small from the views of effectiveness of disk access. If 4KB chunk is managed by a

Kuniyasu Suzaki · Toshiki Yagi · Kengo Iijima · Cyrille Artho
National Institute of Advanced Industrial Science and Technology, 1-1-1 Umezono, Tsukuba, Ibaraki, Japan 305-8568

Yoshihito Watanabe
Alpha Systems Inc.., 6-6-1 Kamidanaka, Nakahara-ku, Kawasaki, Kanagawa, Japan 211-0053

Z. Qian et al. (Eds.): Recent Advances in CSIE 2011, LNEE 125, pp. 399–413.
springerlink.com

SHA-1 digest (20 Byte), 1TB virtual disk is expressed by 5GB (5GB/4KB x 20 byte SHA-1 digests) SHA-1 digest. The volume increase is 0.49% but management overhead is not trivial. Larger chunk is desired from the view of management.

On current operating systems, high I/O throughput is obtained by block access optimization, especially disk prefetch. Disk prefetch is a widely used technique to reduce the number of I/O by reading extra blocks and keeping them in a memory as cache. Many small CAS chunks per a disk prefetch are against the optimization, because it causes many searches of chunk in CAS system.

Locality of reference (access locality) is important on CAS system, because scattered (i.e., fragmented) data decreases effective data in a chunk and increases the number of chunks for read requests. Locality of reference is increased by block reallocation of file system, which considers access profile. High locality of reference expands window size of disk prefetch, because it increases cache hit ratio and reduce number of access.

However, large chunk reduces effect of deduplication. Furthermore, larger chunk reduces ratio of effective data in a chunk, because locality of access is limited in general case. We have to adjust chunk size to balance effects of I/O optimization and deduplication on a real usage.

In this paper, the size effect on deduplication and disk prefetch are evaluated on LBCAS (Loopback Content Addressable Storage) [4] which manages each chunk with a file and reconstructs loopback block device. The optimization for access profile is applied for boot procedure, because it is mandatory operation on any operating system. We estimate chunk size impact on deduplication and disk prefetch.

This paper is organized as follows. Section 2 reviews related works and makes clear the features of CAS systems. Section 3 introduces issues on CAS systems: retrieve overhead and alignment. Section 4 describes the implementation detail of LBCAS. Section 5 analyses the effect of disk prefetch on LBCAS. Section 6 evaluates the effect of deduplication and the affinity of disk prefetch. Section 7 discusses directions for future work, and section 8 summarizes our conclusions.

2 Related Work

This section describes features of CAS systems. The features are compared with LBCAS which is used in this paper.

Venti is a pioneer of CAS system, developed for Plan9 data archive[5,6]. Venti is a block storage system in which chunks are identified by a collision-resistant cryptographic hash. A chunk of CAS is created when it is a first appear data in the CAS system. The chunks are stored in a data-base of CAS system, which retrieves data by the hash. The chunks are not removed and read many times. The feature is called WORM (Write Once Read Many). A chunk is self-verifying with its hash digest and can keep data integrity in network. The LBCAS has same features but each chunk is saved in a file with SHA-1 file name because file abstraction is easy to handle. It does not require special data-base.

Current CAS system is divided into two categories with fix or variable length chunk. Foundation[8] use fix length and DeepStore[8] and NEC-Hydra[9] use variable length. Variable length chunk can find effective deduplication, but it causes internal fragmentation on the data-base. From the view of management overhead, fixed alignment is easy to handle. LBCAS uses fix length for maintenancebility.

The effects of deduplication on several OS images were evaluated in [2, 3]. They showed features of deduplication from many aspects: on a single OS image, among related OS images, and between updated OS images. The results showed advantage of deduplication, but they were evaluated on small chunk size (from 512B to 8KB). This paper evaluates effect of larger chunk size considering access optimization on a real usage.

For high throughput, chunk aggregation was introduced for distributed CAS systems [10]. In the paper[10], 8KB chunks were aggregated into larger container object (4MB) and showed efficiency on the network. Paper [11] used same technique and showed 100MB/sec throughput. However, they used special methods for packing and did not care the access pattern on client OS. Our proposal uses general block reallocation technique to aggregate data, which considers access profile.

3 Issues on CAS Systems

CAS system is a virtual block device, which has its own problems. The hash collision is the most popular problem, but there are some other problems related to chunk size. In this section we summarize these problems.

3.1 Issues on CAS Systems

Effects of deduplication appear in many aspects. The effects are categorized in 3 types; intra, inter, and update [3]. Intra is deduplicated chunk which appears in a disk image. Inter appears among disk images. When different OS images have same chunks, they are treated as Inter. Update appears between updated disk images. Update compares updated disk image with previous one. It causes high deduplication, because most parts of disk are not changed in an update.

Intra deduplication makes a profit for each VM instance, because a VM can reduce the total access to a disk image. The most intra deduplications are occupied by zero-filled chunks in general. Unfortunately, they are not effective for a VM instance, because they are not read from an OS. Other intra deduplications reduce the number of access and contribute the performance on a real situation.

Inter and update deduplications make a profit for servers, because consumption of physical storage can be reduced. Fortunately, normal update changes a part of disk image and the remainder are treated as deduplication.

3.2 Retrieve Overhead

Recent disk prefetch mechanism makes larger access than 4KB in order to hide disk access latency. The extra data is saved to a memory as cache. CAS system may cause multiple retrieving of chunk per a disk prefetch. It is against the optimization of disk prefetch. In order to hide the multiple retrieving, CAS system should use suitable chunk size.

However, disk access size varies with hit ratio of cache. Constant access size is suitable for CAS systems. We should re-consider the disk prefetch mechanism.

3.3 *Access Alignment*

CAS systems with fixed chunk size have access alignment. If an access crosses over the alignment, the access has to take some chunks. If the chunk size is small, there are a lot of crossovers. It increases the overhead of retrieving on a CAS system. The behavior differs from physical block device. We should decide a chunk size under the consideration of alignment and access size.

4 LBCAS

LBCAS is a kind of CAS system which was called HTTP-FUSE CLOOP in [7]. It was developed for Internet block device but it is used as a local CAS system. The data chunks are saved to files with their SHA-1 file names and works as a part of loopback block device. The abstraction level is high and the overhead is not trivial. However, file abstraction is easy to handle, because we can utilize many technology of contents deliver (e.g., file sharing, proxy, etc).

LBCAS is made from an existing block device. The block device is divided by a fixed chunk size and each data saved to a small block file. Saved data are also compressed. The compression reduces traffic but it causes internal fragmentation in a file system, because normal file system manages a file with 4KB data block. The same problem causes on a data-base of CAS system which uses fixed size of record and compression. The loss is described in later section.

Each block file has a name of SHA-1 of its contents. The address of block files is managed by mapping table. The mapping table has the relational information of physical address and SHA-1 file name. Figure 1 shows the creation of block files and mapping table file "map01.idx". Block files are treated as network transparent between client and server. A virtual block device (loopback file) is reconstructed with the mapping table file on a client. Client's local storage acts as a cache. The files are measured with SHA-1 hash values when they are mapped to virtual disk. It keeps the integrity of block device.

When an access is issued to a virtual disk on a client, the relevant block file is downloaded and the data is mapped to the virtual disk. The block files are used when it is required, and the file can be removed on a client if needed. Namely, the only necessary blocks files are used on demand.

LBCAS has two level of cache to prevent redundant download and decompression, which is called "storage cache" and "memory cache". Storage cache saves the downloaded block files at a local storage. It eliminates the download of same block file. If the necessary block files are saved at storage cache, the LBCAS works without network connection. The volume of storage cache is managed by water mark algorithm of LIFO in current implementation. The latest downloaded block files are removed when the volume is over the water mark, because aged block files might be used for boot time. Memory cache saves the uncompressed block file at the memory of LBCAS driver. It eliminates decompression when same block file is accessed in succession. Memory cache saves 1 block file and the coverage is the size of block file. It should be coordinated with the disk prefetch of existing OS.

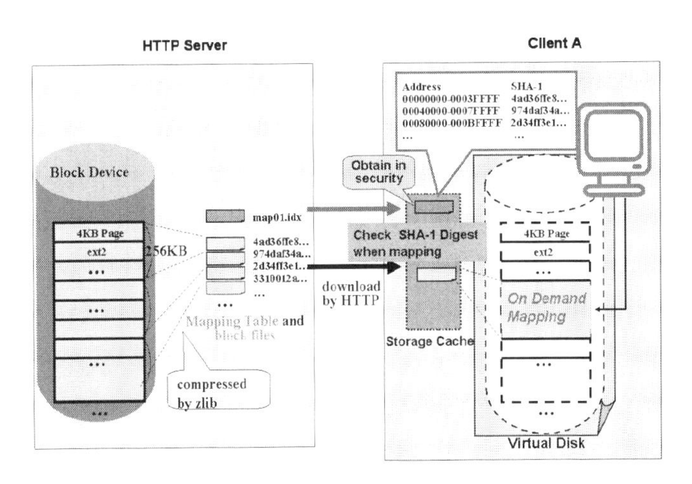

Fig. 1 Server and client for LBCAS.

5 Disk Prefeatch

Most operating systems have the function of disk prefetch to reduce I/O operation. Disk prefetch reads extra data from a block device and saves them to memory as cache. They hide latency of I/O when next read requests hit the cache. When page cache does not hit, the next readahead shrink the window size. The suitable window size achieves efficient usage of cache memory and I/O request, which depends on the locality of reference.

Some techniques of disk prefetch are developed on kernel level and user application level. Block reallocation of file system is considered a part of disk prefetch technique, because it increases locality of reference and improves hit ratio of cache. These techniques are closely related to chunk size problem on CAS.

5.1 Kernel Level

When a read request is issued, kernel reads extra data and saves them to main memory as cache. It reduces the number of I/O operation and hides the I/O delay. The function in Linux kernel is called "readahead" [12,13]. The window of readahead is extended or shrank by the profile of cache hit and miss-hit.

Figure 2 shows the action of readahead on LBCAS. When a readahead operation is issued, some block files are downloaded and mapped to the virtual block device. When chunk size is small, many small chunks per a disk prefetch are used and reduce the effect of readahead.

This figure also shows the effect of storage cache of LBCAS. When a same block file is required sequentially, the block file is stored on the memory cache of LBCAS and the decompression is eliminated.

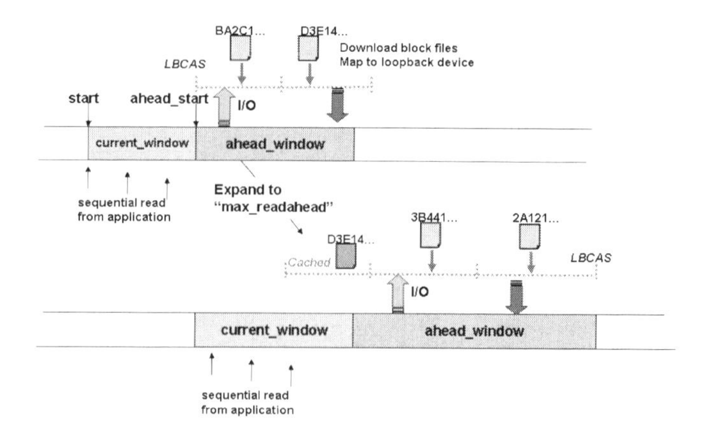

Fig. 2 Relation between LBCAS and readahead.

5.2 User Level

Disk prefetch technique is also developed on user level. For example, libprefetch[14] is a prefetch library to reduce the seek overhead. Especially, Linux kernel has the system call "readahead" from 2.4.13, which populates page cache with data of a file. The name is same to the kernel level "readahead" but the action is different. Whole data of a file is populated on page cache before the file is read.

Some Linux distributions have a tool to utilize the readahead system call to speed up the boot process. The files requested at boot time are listed at"/etc/readahead/boot", and the data of the files are populated on the page cache in advance at boot time. Unfortunately, user level readahead requires a lot of memory because it saves whole data of files. The total memory size has to be balanced. Ubuntu has a tool called "preload" to take a log of files requested at boot time and arranges the file list for readahead.

5.3 Block Reallocation

Most file systems have defragmentation tools to reallocate blocks of file system. It is said to increase the locality of reference and make quick access. The tools, however, reallocate blocks from the view of continuation of file or expansion of spare space. Quick access is a side effect of continuation of file. In order to solve the problem, "ext2/3optimizer" was developed [15]. Ext2/3optimizer takes the profile of accessed blocks of ext2/3 and reallocates the blocks in line.

Figure 3 shows the block image on ext2/3 file system. Left figure indicates before ext2optimizer and right figure indicates after Ext2/3optimizer. Ext2/3optimizer changes pointers of data blocks of inode only. It aggregates the data blocks at the head of device to increase locality of reference. The other structure of ext2/3, namely meta-data of ext2/3, is reserved. The reallocation is based on real access profile and the cache hit ratio keeps high. As a result, the window size of readahead can keep large. On the LBCAS, effective data in a chunk is increased and the number of necessary block files is reduced.

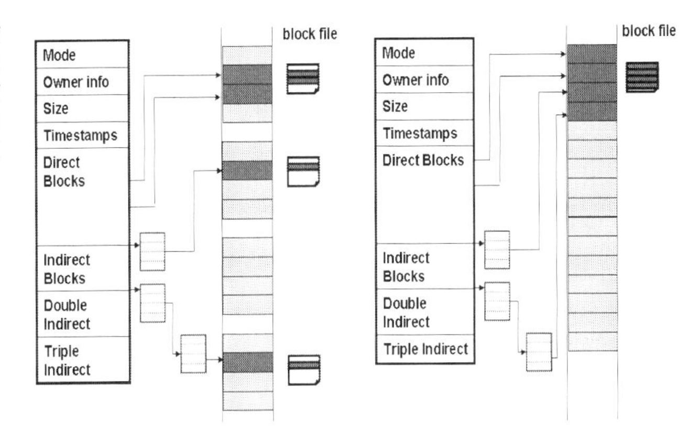

Fig. 3 Reallocation of ext2/3optimizer and effect on block file of LBCAS. Red blocks are accessed on a real usage.

6 Evaluations

This section describes the results of deduplication on some chunk sizes and performance on disk prefetch.

6.1 Deduplication on Larger Chunk

The effects of deduplication on LBCAS are evaluated on Debian GNU/Linux and Ubuntu. Debian (Lenny) and Ubuntu (9.04) were installed on 8GB virtual disk of LBCAS. The chunk size is changed from 4KB to 512KB. We also evaluated Windows XP with some chunk sizes. The results are eliminated by limitation of space but we got same features on them.

1) Intra Deduplication

The left of Table 1 shows the effect of intra deduplication. The deduplicated chunks are categorized by "non-zero data chunk" (NonZero) and "zero filled chunk" (Zero). The Zero was expressed by 1 block file and covered more than 50% in any cases. However, the intra deduplication for NonZero was little (less than 5 %). The effect of NonZero was reduced as long as chunk becomes large, but the reduction of Zero was not affected severely. The intra deduplication was dominated by Zero.

The right half (3 columns) of Table 1 shows the volume usage. Amount of native compressed chunk indicates the volume which does not consider internal fragmentation. Used volume in file system indicates the volume which includes internal fragmentation on ext3. Disk waste rate shows the ratio of data in a 4KB ext3 block, which is caused by internal fragmentation. The internal fragmentation is caused by compression on LBCAS. 4KB chunk caused huge internal fragmentation, because the chunk size is the same as the block size of ext3. All block files have internal fragmentation with compressed chunk data. The disk waste rate indicates true ratio of compression. Other chunk sizes also affected by the internal fragmentation by compression, but disk waste rate is improved because most compressed chunk are exceeded 4KB. The same problem is caused on a data-base of CAS system with irregularly compressed chunk. The larger chunk mitigated this fragmentation problem, especially more than 64KB chunk showed less than 10% loss on 4KB block of ext3.

Table 1 Intra Deduplication on 4KB-512KB chunk and Disk Usage. Upper is Debian and Lower is Ubuntu.

Chunk size Debian	Duplicated Chunk (shared) (%)		Non-duplicated (unique) Chunk (%)	Amount of native compressed chunk (MB) **A**	Used volume in File System (MB) **B**	Disk Waste Rate (fragmentation) **A/B**
	NonZero	Zero				
4KB	4.26	68.2	27.5	929.5	2568.4	0.36
8KB	2.68	67.9	29.4	932.5	1606.7	0.56
16KB	1.68	67.5	30.9	940.2	1296.4	0.72
32KB	0.82	66.7	32.5	945.6	1104.9	0.83
64KB	0.21	65.6	34.2	945.2	1039.2	0.9
128KB	0.02	64	36.0	941.5	991.1	0.95
256KB	0	61.8	38.2	937.9	964.1	0.97
512KB	0	59.1	40.9	936.0	949.8	0.98

Chunk size Ubuntu	Duplicated Chunk (shared) (%)		Non-duplicated (unique) Chunk (%)	Amount of native compressed chunk (MB) **A**	Used volume in File System (MB) **B**	Disk Waste Rate (fragmentation) **A/B**
	NonZero	Zero				
4KB	2.63	69.0	28.4	940.1	2578.7	0.36
8KB	1.08	68.5	30.4	914.9	1639.9	0.58
16KB	0.53	68.0	31.5	892.4	1247.9	0.73
32KB	0.16	67.2	32.6	874.4	1055.7	0.86
64KB	0.01	66.1	33.9	860.4	954.7	0.91
128KB	0	64.5	35.5	852.1	901.3	0.95
256KB	0	62.2	37.8	848.0	874.2	0.97
512KB	0	58.9	41.1	846.3	860.1	0.99

2) Inter Deduplication

Table 2 shows the effect of inter-deduplication between Debian and Ubuntu. The effect was 10% at 4KB chunk but it also reduced as chunk become larger. Unfortunately 4KB was too small considering internal fragmentation problem. Storage server does not need to care about inter deduplication except zero filled block, because the coverage of zero filled block was more than 50 % (Table 1).

Table 2 Inter deduplication between Debian and Ubuntu

LBCAS size	Debian block file	Ubuntu block file	Same block file	Sharing Debian (%)	Sharing Ubuntu (%)
4KB	603,341	612,967	65,457	10.85	10.68
8KB	316,848	322,442	14,042	4.43	4.35
16KB	165,159	165,764	2,884	1.75	1.74
32KB	86,186	85,670	474	0.55	0.55
64KB	44,947	44,427	85	0.19	0.19
128KB	23,599	23,263	24	0.10	0.10
256KB	12,527	12,388	11	0.09	0.09
512KB	6,705	6,742	4	0.06	0.06

3) Update Deduplication

The effect of update-deduplication was also evaluated on Debian and Ubuntu respectively. They were updated every week for security.

Table 3 shows the ratio of reused chunks which indicates deduplication between updates. The ratio was high in any chunk size, because most parts of disk images were not changed. It means the deduplication is effective on updates at any chunk size.

Table 3 Reuse (%) of deduplication Debian and Ubuntu.

Reuse (%) on Debian	4KB	8KB	16KB	32KB	64KB	128KB	256KB	512KB
29/05/2009-05/06/2009 Package 22.94MB	94.1	91.3	88.7	87.0	84.8	81.5	76.7	69.1
05/06/2009-12/06/2009 Package 4.32 MB	98.0	97.3	96.9	96.5	96.0	95.3	94.4	92.6

6.2 *Effect of Disk Prefetch*

We compared the effect of ext2/3optimizer and user level readahead (system call readahead) on LBCAS. Ext2/3optimizer and user level readahead were applied on guest OS (Ubuntu 9.04, Linux kernel 2.6.28, ext3 file system) on KVM virtual machine (version 60). The Ubuntu used 1.98GB volume in 8GB LBCAS.

1) Access Pattern of Disk Prefetch

The access pattern of boot procedure was investigated. We confirmed the characteristics of access pattern and locality of reference and applied ext2/3optimizer on it. From after, we refer user-level readahead as "u-readahead" in order to distinguish it from kernel level readahead.

a) Block Reallocation: ext2/3optimizer

Figure 4 shows the data allocation on ext3, which is visualized by DAVL (Disk Allocation Viewer for Linux) [16]. The left figure shows the original data allocation, and right figure shows the data allocation optimized by ext2/3optimiser.

In the figure, green plots indicate the allocation of meta-data of ext3 which was aligned at the right edge. We confirmed that ext2/3optimizer keeps the structure of ext3. The blue plots indicate the contiguous allocation of data block of file and the yellow plots indicate the non-contiguous allocation. We confirmed that ext2/3optimizer reallocates non-contiguous data at the head of virtual disk. It was the result that ext2/3optimizer exploited the profiled data blocks and aggregated them to the head of the disk. As a result, ext2/3optimizer increased fragmentation from the view of file. DVAL showed that normal ext3 had 0.21% fragmentations but the ext3 optimized by ext2/3optimizer had 1.11%. The relocation, however, was good for page cache. The coverage of readahead was expected to keep large and occupancy of block file of LBCAS would be high.

Figure 5 shows the access trace of the boot procedure. The x axis indicates the physical address and y axis indicates the elapsed time. The red "+" plots indicate the access on the normal ext3 and the blue "X" plots indicate the access on the ext3 optimized by ext2/3optimizer. The figure showed that the accesses to the normal were scattered. The locality of reference was not good and the effect of page cache and the occupancy of block file of LBCAS would be low. On the other hand, the access to the ext2/3optimizer increased the locality of reference, because the most accesses were the head of disk. The rest spread accesses were the meta-data and the volume was little.

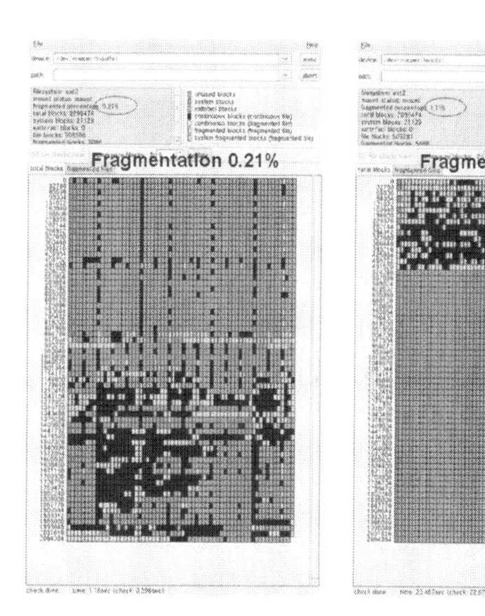

Fig. 4. Visualization of data-allocation on ext3 (left is normal and right is ext2/3optimizer) by DAVL.

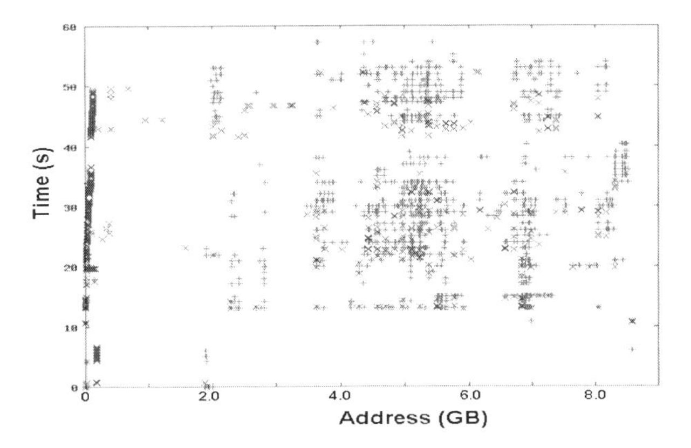

Fig. 5 Access trace of boot procedure (RED "+" indicates normal and BLUE "X" indicates ext2/3 optimizer.)

b) User Level readahead (system call readahead)

Ubuntu has the mechanism to populate the page cache with files required at boot time. The files are described at "/etc/readahead/boot" and "/etc/readahead/desktop". The former file listed 937 files and the total volume was 54.1MB. The latter file listed 281 files and the total volume was 25.0MB. The listed files were not all files required boot time. Ubuntu 9.04 required 2,250 files (203MB) and the half of them were populated on the page cache by u-readahead before they were truly required.

c) Effect of Readahead

Figure 6 shows the frequency for each readahead coverage size on normal, u-readahead, and ext2/3optimizer. The figure shows that ext2/3optimizer reduced the small I/O requests. As a result, the frequency of I/O request was reduced to 2,129 from 6,379 and the coverage of readahead was changed to 67KB from 33KB. The total I/O was 140MB and 208MB on ext2/3optimizer and normal respectively. The I/O request was 2 times wider and the frequency of I/O request was 1/3. The effect of frequency was not the inverse of magnification of I/O. The results indicated that the locality of reference was much improved.

On the other hand, u-readahead showed same tendency with normal. The small requests were reduced and the big requests were increased a little bit. The total I/O of u-readahead was increased to 231MB from 208MB of the normal. The coverage of readahead was expanded to 41KB but it was small than ext2/3optimizer. The result came from that the u-readahead could not decrease the small I/O, which was caused by the locality of reference. The frequency of I/O was 5,827, which was less than normal 6,379, although the total I/O was increased. The results indicated that ext2/3optimizer was much effective than u-readahead from the view of disk prefetch readahead.

Table 4 Volume transitions at each processing level. The upper table indicates the volume transition on guest OS.The bottom table indicates the volume transition on LBCAS.

	Normal	u-readahead	ext2/3opt
Volume of files (number, average)	203MB (2248 Av:92KB)		
Volume of requested chunks	127MB		
Volume of required access which includes the coverage of Readahead (average number of access and size of readahead)	208MB **(6,379 Av:33KB)**	231MB **(5,827 Av:41KB)**	140MB **(2,129 Av:67KB)**

Chunk size	Downloaded size MB (Uncompressed size MB), Occupancy %		
64K	86.1(247), 51.5%	93.4(272), 46.9%	55.3(144), 88.7%
128K	96.8(290), 43.9%	104(315), 40.3%	55.3(149), 85.3%
256K	114(358), 35.5%	123(386), 35.0%	55.6(159), 80.0%
512K	144(474), 26.9%	153(508), 25.1%	55.6(176), 71.8%

2) Effect of Chunk Size on Real Access Pattern

Table 4 shows the volume transitions at each processing level. The upper table shows the total volume requested from transferred OS on three levels: (1) the volume of files which opened by the boot procedure, (2) the chunk volume which is purely required by the boot procedure, and (3) the volume accessed to LBCAS (it includes redundant data covered by readahead). The bottom table shows the status of LBCAS for each chunk size: the volume of downloaded block files, the volume of uncompressed block files, and the occupancy of effective data in the LBCAS.

From the result, we know that the purely used chunk was 63% (127MB/203MB) of volume of opened files at boot time. It meant that 37% was not used and it caused inefficient access request of readahead. The readahead for normal ext2 required 208MB access to the LBCAS. The result shows the 81MB (208MB - 127MB) was redundant access. The u-readahead made much worse and 104MB was redundant access. The problem was solved by ext2/3optimizer significantly. The readahead for ext2/3omitimizer required 140MB. The ext2/3optimizer made 67% better than the normal.

The bottom table shows the status of LBCAS. We confirmed that downloaded files were less than 56MB at any LBCAS size on ext2/3optimizer. However, the normal of 512KB chunk requires 144MB, which is 1.67 much larger than 64KB chunk (86.1MB). It was caused by bad locality of reference. On ext2/3opitimizer, the occupancy was almost same on any LBCAS size but it was decreased from 51.5% at 64KB to 26.9% at 512KB on normal. The result indicated that block reallocation was necessary for LBCAS.

Table 5 shows the frequency of each function of LBCAS for normal, u-readahead, and ext2/3optimizer. I/O requests were issued by guest OS and the frequency was independent of LBCAS. The rest columns indicated the function of LBCAS. The number of uncompress is summation of the number of download and storage cache. The summation of uncompress and memory cache is the total used files on the LBCAS.

Table 5 Frequency of functions of LBCAS. Upper, middle and lower tables show the normal, u-readahead and ext2/3optimizer case, respectively. "Requests" indicates the number of I/O issued by guest OS. The rest columns show the frequency of each function of LBCAS. "Files per request" indicates the frequency of downloads for files per a request.

Chunk size Normal	Requests from guestOS (R) (Av:33KB)	Download (D)	Storage Cache of LBCAS (S)	Uncompress (U) =(D)+(M)	Memory Cache of LBCAS (M)	Files per Request (F) (R)= (1)+(2)+(3) (U)+(M)= (1)+(2)*2+(3)*3
64K	6,338	3,958	1,663	5,621	3,647	(1) 4,148 (2) 1,450 (3) 740
128K	6,381	2,321	1,729	4,050	3,793	(1) 4,919 (2) 1,462
256K	6,379	1,435	1,748	3,183	3,908	(1) 5,667 (2) 712
512K	6,395	948	1,769	2,717	4,019	(1) 6,054 (2) 341
u-read-ahead	(Av:41KB)					
64K	5,825	4,344	1,172	5,516	3,626	(1) 3,537 (2) 1,259 (3) 1,029
128K	5,834	2,526	1,200	3,726	3,761	(1) 4,181 (2) 1,653
256K	5,827	1,544	1,179	2,723	3,908	(1) 5,032 (2) 804
512K	5,822	1,015	1,172	2,187	4,023	(1) 5,434 (2) 388
ext2/3opt	(Av:67KB)					
64K	2,165	2,296	626	2,922	1,311	(1) 941 (2) 380 (3) 844
128K	2,148	1,189	593	1,882	1,398	(1) 1,116 (2) 1,032
256K	2,129	634	576	1,210	1,409	(1) 1,639 (2) 490
512K	2,132	353	517	870	1,520	(1) 1,874 (2) 258

The results showed storage cache and memory cache worked well. Especially the two caches were effective on large chunk size. The frequency of storage cache and memory cache were more than the frequency of download and decompression.

7 Discussions

Security of CAS system is important, especially when it is used in the Internet. The integrity of contents is ensured in many CAS systems, because the secure hashed names are used for verification. However, confidentiality is not ensured. First solution is to use secure file system on CAS. In this style, CAS system has to consider affiliation of access pattern and compression algorithm because it does not know the detail of secure file system. Convert encryption is used in [17] and secret sharing is used in [18] for CAS systems. These security mechanisms are included in CAS systems and access pattern is not changed. The type of implementation can reuse the optimization on file system.

The data blocks are reallocated in order to be in line according to the access profile. It results in keeping large coverage of readahead at the boot procedure. It reduces the boot time but the data blocks are fragmented from the view of file. Unfortunately, the optimization is too tight and it would not fit to another access pattern. If the reallocated data blocks are used in another application, the access pattern cannot get large coverage of readahead. However, boot procedure is special and several files are used at boot procedure only. We have to estimate the special files and its ratio, which are not used for other applications.

8 Conclusions

This paper showed chunk size impact on deduplication and disk prefetch for CAS system. Most CAS systems assumed small chunk size (4KB-8KB) to get the effect of deduplication. However, the size of disk prefetch was larger, when it was optimized by block reallocation based on access profile. The effective chunk size has to be decided to balance the effect of deduplication and I/O optimization.

Experimental measurement was achieved on LBCAS which saves chunk data in a file with SHA-1 file name. The LBCAS image was optimized for reallocation for boot procedure. The results showed the small chunk size was inefficient in many case even if it could get the merit of deduplication. Larger chunk could mitigate the performance degradation with the help of reallocation of block on a file system and constant large disk prefetch. It reduced the boot time of Linux on virtual machine KVM with small traffic.

This paper introduced some aspects of optimization on CAS system. Chunk size was an additional factor and the optimization had to consider it for each real situation.

References

1. Tolia, N., Kozuch, M., Satyanarayanan, M., Karp, B., Bressoud, T., Perrig, A.: Opportunistic use of content addressable storage for distributed file systems. In: Proceedings on USENIX Annual Technical Conference (2003)
2. Liguori, A., Hensbergen, E.C.: Experiences with Content Addressable Storage and Virtual Disks. In: Workshop on I/O Virtualization, WIOV 2008 (2008)

3. Jin, K., Miler, E.L.: The Effectiveness of Deduplication on Virtual Machine Disk Images. In: The Israeli Experimental Systems Conference, SYSTOR 2009 (2009)
4. Suzaki, K., Yagi, T., Iijima, K., Quynh, N.A.: OS Circular: Internet Client for Reference. In: Large Installation System Administration Conference, LISA (2007)
5. Quinlan, S., Dorward, S.: Venti: a new approach to archival storage. In: Proceedings on The Conference on File and Storage Technologies, FAST (2002)
6. Lukkein, M.: Venti analysis and memventi implementation, Master's thesis of University of Twente (2008)
7. Rhea, S., Cox, R., Pesterev, A.: Fast, inexpensive content-addressed storage in Foundation. In: USENIX Annual Technical Conference (2008)
8. You, L.L., Pollack, K.T., Long, D.D.E.: Deepstore: An archival storage system architecture. In: Proceedings 21st International Conference on Data Engineering, ICDE (2005)
9. Dubnicki, C., Gryz, L., Heldt, L., Kaczmarczyk, M., Kilian, W., Strzelczak, P., Szczepkowski, J., Ungureanu, C., Welnicki, M.: HYDRAstor: A Scalable Secondary Storage. In: USENIX Conference on File and Storage Technologies, FAST (2009)
10. Eaton, P., Weatherspoon, H., Kubiatowicz, J.: Efficiently binding data to owners in distributed content-addressable storage systems. In: Proceedings on Security in Storage Workshop, SISW 2005 (2005)
11. Zhu, B., Li, K., Patterson, H.: Avoiding the Disk Bottleneck in the Data Domain Deduplication File System. In: Proceedings of USENIX File and Storage Technologies, FAST (2008)
12. Wu, F., Xi, H., Xu, C.: On the design of a new Linux readahead framework. ACM SIGOPS Operating Systems Review 42(5) (July 2008)
13. Wu, F., Xi, H., Li, J., Zou, N.: Linux readahead: less tricks for more. In: Proceedings of the Linux Symposium (2007)
14. VanDeBogart, S., Frost, C., Kohler, E.: Reducing Seek Overhead with Application-Directed Prefetching. In: USENIX Annual Technical Conference (2009)
15. Kitagawa, K., Tan, H., Abe, D., Chiba, D., Suzaki, K., Iijima, K., Yagi, T.: File System (Ext2) Optimization for Compressed Loopback Device. In: 13th International Linux System Technology Conference (2006)
16. DAVL (Disk Allocation Viewer for Linux), `http://sourceforge.net/projects/davl/`
17. Storer, M.W., Greenan, K., Long, D.D.E., Miller, E.L.: Secure Data Deduplication. In: Proceedings of the 4th ACM International Workshop on Storage Security and Survivability, StorageSS (2008)
18. Douceur, J.R., Adya, A., Bolosky, W.J., Simon, D., Theimer, M.: Reclaiming space from duplicate files in a serverless distributed file system. In: Proceedings of International Conference on Distributed Computing Systems, ICDCS (2002)

Management Information Ontology Middleware and Its Needs Guidance Technology

Yugang Zhu, Kaijun Chen, Xingming Guo, and Yong He

Abstract. We must solve some questions such as low-cost, high efficiency, flexibility, robustness of life and etc. about development and application on management information system in current time. The management information ontology triple, a 3-elements (business-table-predicate) set, can effectively describe the separation and filtration between business and functions, realize truly the management functions from abstract to subtle, while ensuring that complex business can be defined and guided-implemented on-line. Therefore, middleware technology and its needs guidance technology based on management information ontology can effectively reduce the time in the system requirements and analysis and design, shorten the phase in the database design and simplify the document written request, then improve the current software engineering methods.

Keywords: management information system(MIS), ontology middleware, needs guidance, software engineering.

1 Introduction

The object-oriented technology describes the problem what will be solved as an object class, and the associated properties, events and methods, who can be mapped relevantly in real management activities, characterization, driven mechanism and its process. Whether this technology is applied to management information systems (MIS) successfully or not depends on the particle size

---

Yugang Zhu · Kaijun Chen
Zhejiang Technical Institute of Economics, Hangzhou,China

Xingming Guo
Zhejiang Technical Institute of Economics, Zhejiang University, Hangzhou, China(310018)
e-mail: gu0xm30@sohu.com

Yong He
School of Biosystems Engineering and Food Science, Zhejiang University,
Hangzhou, China

Z. Qian et al. (Eds.): Recent Advances in CSIE 2011, LNEE 125, pp. 415–421.
springerlink.com

required to solve the problem[1] because too small size is not conducive to software reuse, and too large size is difficult to encapsulation or inheritance.

It is a good idea that the common characters about the management information processing is analyzed and then the application system development methodology is built by reusing the component library and using the data dictionary[2]. Therefore the essential management features must be excavated sufficiently to the general elements of management, business, data flow, batch and real-time process in order to breakthrough in manage information systems software engineering.

It is a breakthrough in recent years that the ontology theory and middleware technology are used to solve the software engineering problems in development of MIS[3]. According the ontology theory, the requirement of management business can be summarized as a series of concepts and theirs inferences, and then middleware technology can solve the pragmatic realization and inference calculus[4], so that all the problems will be solved such as the information implementation of management business needs, the universal and general requirements in description and processing of management information.

2 Management Information Ontology

2.1 *Management Information System Triple S*

The MIS in any fields can be described an ontology triple: S={U， T， P}

Among above, U is a description set of knowledge and needs in the field, or named as business, T is a set of the two-dimensional relational tables, referred to as the table, P is a set of predicate calculus logic for T not for U, or known as function.

The significance of the triple model of MIS model is to achieve the separation between a MIS business and functional. In the triple S, the two-dimensional relational table T is a carrier of management information, which constitutes an infinite loop of the range set, U is a direct feature reflection in which the MIS can be used in specific business, which constitutes an infinite no-loop of the range set, according the two-dimensional relations definition, P is a finite set based on two-dimensional relational operations, which can achieve various business management information needs through the static description for table T (definition) and dynamic processing (algorithms).

2.2 *Table T*

An ontology concept which can fit all management information processing should be:

- be able to unambiguously describe any specific management information object;
- be able to map the two-dimensional relational database;
- to facilitate needs communication between users and developers;
- can be called, guided to execute online.

According above, the two-dimensional relational tables set T is defined as a core concept of management information ontology(MIO), in order to avoid ambiguity and facilitate communication, the table T is defined as follows:

1)Irregular form (three-dimensional form): A table in MIO can be described as three-dimensional form, known as irregular form, pages, rows, columns are denoted as !,?,: respectively. General cross statistical two-dimensional table with horizontal line and vertical column can be called the regular table, their rows, columns are denoted as ?,: respectively. Regular table is a simplified special case of irregular table, it omits the page dimension! .

2)Current table: the table for the is known as the current table, denoted {!,?,:}, other table is named for table non-ongoing process currently, denoted ABCDXXXXXX [!,?,:], where ABCD is the table code, XXXXXX refers to the table period.

The table period can be expressed with relative representation, for example, 01,02 are expressed last month, previous month, or with absolute representation, and OFOGAA is expressed daily for 7/11/2006.

So, Set T can be expressed as T={t|{!,?,:}, ABCDXXXXXX [!,?,:],{?,:}, ABCDXXXXXX [?,:],!,?,: are All natural numbers, ABCD is arbitrary table code, XXXXXX is arbitrary table period}. Above definition can simplify predicate calculus operations greatly, for example, "line 8 is assigned to 0" can be expressed as:"{8:}=0" simply instead of:

Update [tablename] Set [fieldname1]=0,[fieldname2]=0,...Where [line number] =8

2.3 Predicate Calculus P

The predicate calculus P is the inference regulation in the ontology triple S, and its inference object is table known by definition, moreover, the inference regulation is finite. The ontology triple S has common predicate calculi:

- Create and maintain (eg. table description)|M
- table write |W
- assignment |F
- obtain data correspondingly |B
- sort, classification statistics |P
- delete line(s) |Z
- translate code |D
- table input |N
- table send and receive |R
- sum, count |S
- table union |U
- insert line |I
- obtain code |C

The corresponding component library can be developed for the above predicate calculi and called online or guided to execute by ontology middleware according with requirements.

2.4 Field Description Concept Set U

The management business needs have unlimited growth, so even if the same industry and business MIS will be faced with changes that occur over time or in demand, and these changes will be often shown first (or finally) in the language description way. It can be inferred that the field description set U is an infinite no-loop set.

However, with the definition of triple and its features of the separation between business and functions, unlimited changes and uncertain business in the needs can be solved by limited table forms and their predicate calculus sequence.

For next requirement example:

According with business need, May 28 business meeting decided to change the deficit invoice A reflecting the return of goods into the dedicated refund bill B, invoice shall not be used in deficit. While the follow-up daily C, monthly D, yearly E statistics were added "return" column (column 9) to accurately reflect the statistical results. This needs to change must be completed in the MIS before May 31 deadline to be reflected in the monthly statistics.

Ontology reasoning can:

[Create]table B, [Condition]only>=0, [Maintain]table A, [Condition]only>=0, [Maintain]Table C, D, E, [Add]column 9, [Maintain]predicate calculus in table C, [New] obtain data from table A by the step | U, [Maintain]predicate calculus in table D, [New]obtain data from table A by the step | U, [Maintain]predicate calculus in table E, [New]obtain data from table A by the step | U.

3 Guided By Needs Online

3.1 Executing Priciple of Middleware Guided by Needs Online

A MIO middleware should be guided by the management business needs online, when the business changes, it can translate the customer's natural language into the ontological structural description language directly (requires mapping feature without other tools turning into a structural language). With the quick definition on the middleware platform, the platform invokes dynamically, interprets and executes the definition, eventually shows the desired results. If users would get, the changing and maintenance of requirements would have been completed. The schematic principle is shown in Figure 1.

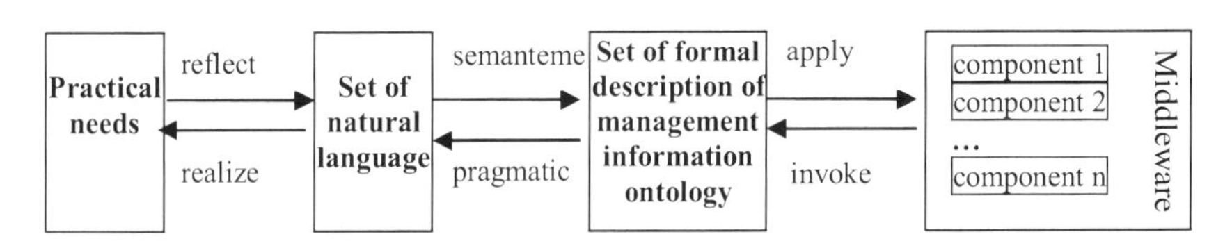

Fig. 1 The executing principle diagram of middleware based on MIO and requirement function components

3.2 Function Architecture of Middleware

Middleware needs to build a flexible functional modules architecture which must be divided into at least three layers, the first layer is the application layer, the second layer is the meta-data dictionary layer, and the third layer is the data warehouse layer as it is guided by the ontology requirements of management information. Applications programs must build a component library corresponding to predicate calculus for dynamic call, interpretation and execution. Figure 2 shows a prototype middleware platform based on the ontology triple of the MIS.

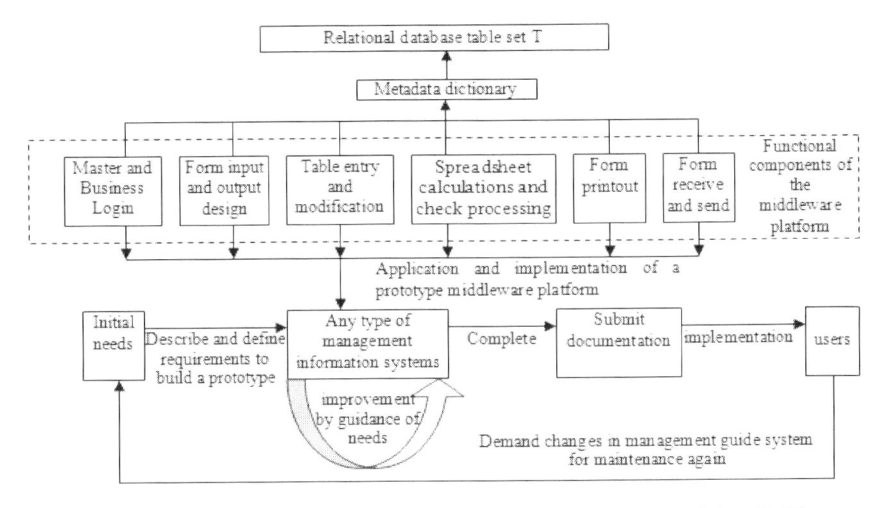

Fig. 2 A prototype middleware platform module based on the ontology triple of MIS

3.3 Development Flow of Middleware Guided by Needs

User's management needs have nothing to do with whether middleware platform can achieve or not, so the first thing is to judge the user's demand for implementation of the ontology middleware under the guidance of needs, to decide whether to call middleware components to achieve a demand directly, or to improve the middleware before it can be called. When the demand can not be implemented by the existing component library, we must separate the functions from the business, strip the relationships between them, add new ontology terms or semantic predicates, form a new semantic description system and its corresponding functional components, and eventually describe and implement the new requirements. Please pay attention to that the newly formed semantic description system and its components must be compatible with the original ontology system, coordinate to a comprehensive component library, and digest to the table set which has nothing to do with specific areas of business. See Figure 3.Middleware should upgrade in evolution type, rather than abandonment type.

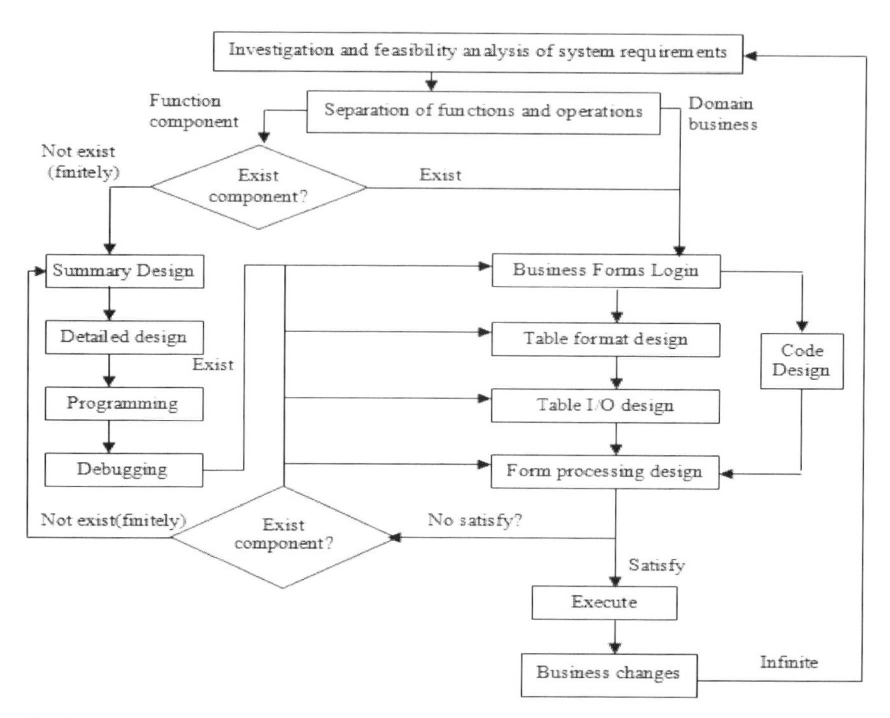

Fig. 3 A development flow based on middleware of MIO guided by needs

4 Conclusion

Because of the finite feature of the predicate calculus and infinite loop of the table set, the ontology middleware function and its development and maintenance must tend to a stable saturation value, while the infinite no-loop diversity of needs in description of domain business can be achieved with the help of the table set and predicate calculus. The practice shows that the process of operation and maintenance is very simple and effective to define needs by guidance, it can effectively reduce the time in the requirement investigation and system analysis and design, shorten the phase of the database design, simplify writing documents requirements, and even authorize users to complete their own requirement. Therefore, the MIO middleware based on the technology has unlimited potential applications.

Thanks

This article is supported by the Eleventh Five-Year National Science and Technology Support Project (2006BAD10A09), Information technology research in agricultural enterprises as the main body in rural areas.

References

1. Gong, H., Qiu, X., Xue, Y., Qian, L.: Layered Software Construction. Computer Engineering and Applications 5, 135–138, 232 (2003) (in Chinese)
2. Wang, X.: An Application System of New Exploitation Means Based on Requirement Commonness,Data Dictionary and Reshipment Component. Application Research of Computers 75, 99–101 (2003) (in Chinese)
3. Guo, X., Guo, T., Zhang, S.: Model of middleware framework in demand for management information ontology. Journal of Zhejiang University (Engineering Science) 2008 42(8), 1286–1293 (2008) (in Chinese)
4. Guo, X., Guo, T., Liu, G., Zhang, S.: Construction of predicate calculus finite set based on the requirement of management information ontology. Journal of Zhejiang University (Science Edition) 36(4), 401–407 (2009) (in Chinese)

Software Development of EVMIS Color LED Display

Zeming Jiang, Chen Cheng, and Xin Chen

Abstract. Industrial color LED display EVMIS is the Industrial Control Equipment integrated GPS, GPRS and SMS. It is composed of two circuit boards by the upper and lower. The upper board, which is developed by the tool QT is mainly for display and data reception use. The software is mainly for showing the functions of the color LED display. That is, it will display data on the color LED display via hardware port response or by reading serial ports, GPS, GPRS and CAN. The lower board programs, which are developed by the tool KEIL are the control by relay and automatic sleep, it will regularly send GPRS data to the server.

Keywords: Industrial color LED display, EVMIS, Linux, QT, AT command.

1 Introduction

Embedded Control System has already been penetrated widely into various aspects of our daily life, industrial control and business application. In the fields of industrial automatisation, various apparatuses and meters have been used in industrial automation, aviation and aeronautics, communication and transportation. The use of Embedded Control System has been increasing.

With the development of Embedded Control System and increasing demands from a wide spectrum of various industries, there is an increasing need for an approach that provides more straight-forward display of current data in those human-computer Interaction systems, for example, the motor vehicle navigation,

Zeming Jiang
Department of Electrical Engineering, Beijing Jiaotong University, Beijing, China

Chen Cheng
Department of Electrical Engineering, China University of Mining and Technology (Beijing), Beijing, China

Xin Chen
Department of Embedded Control, Beijing CSR Times IT Co., Ltd, Beijing, China

Z. Qian et al. (Eds.): Recent Advances in CSIE 2011, LNEE 125, pp. 423–428.
springerlink.com

MRI equipment, patient monitoring system, vehicle entertainment platform, health care system and wireless sensor system. Industrial color LED display undoubtedly offers the ideal solution to straight-forward display of data related to industrial control and operation. In addition, there is an aspiration from a number of domestic engineering vehicle and electric-vehicle producers for the advent of domestically made Industrial color LED display with user- friendly interface and reliable functionality.

The author started software development for wireless color LED monitoring devices since March 2009. The development is conducted on upper board consisting of 7-inch display with already transplanted LINUX and relevant software package and the lower board consisting of LPC2368.

The upper board, which is developed by the tool QT, is mainly for display and data reception use. The software is mainly for showing the functions of the color LED display. That is, it will display data on the color LED display via hardware port response or by reading serial ports, GPS, GPRS and CAN.

The lower board programs, which are developed by the tool KEIL are controlled by relay and automatic sleep, it will regularly send GPRS data to the server.

2 Summary of the Structure

The color LED display consists of two layers (as illustrated in Figure 1).

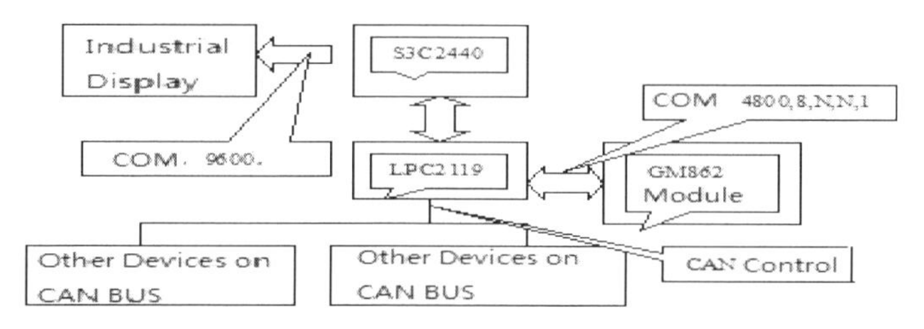

Fig. 1 The structure diagram

2.1 Instruction of the Lower Layer

The main modules of the lower board are LPC2368 and GM862. NXP LPC2368 is a high-performance 32-bit RSIC microprocessor based on ARM7TDMI-S. It is equipped with Thumb command expansion, on chip integration: 512KB Flash supports ISP and IAP, 58K SRAM. The basic frequency is as high as 72MHz, on chip integration: crystal oscillator, 4MHz RC crystal oscillator, PLL advanced Vectored Interrupt Controller, 10/100M Ethernet with DMA, a Full-Speed USB2.0 Device interface, two CAN 2.0B interfaces, a general DMA controller, four UART interfaces, a full-function Modem interface, three I2C serial interfaces, three SPI/SSP serial interfaces, an I2S interface, a SD/MMC memory card interface, six 10-bit ADCs, a 10-bit DACs, four 32-bit capture/comparison clock, watchdog timer, PWM module supporting three-phase motor control, real-time clock with alternative back-up battery, the General Purpose IO and etc.

2.2 *Introduction of the Upper Board*

The upper board is S3C2440 controller chip, mainly responsible for a set of functions including image display and operating system functioning. The operating system is a simplified version of Linux and the graphics program is composed by Qt Creator.

3 Structure of the Bottom Board Program

Figure 2 is the frame chart to illustrate the function of the bottom board.

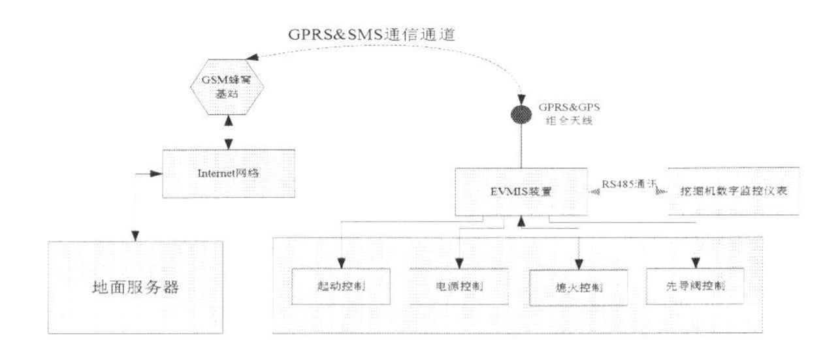

Fig. 2 Frame Chart of the Function

As one of the core components of the whole EVMIS system, the main functions of the bottom board include: GPS data collection; vehicle data collection; GPRS data communication; vehicle locking control etc.

3.1 *Main Circuit on Bottom Board*

Main circuit on bottom board is illustrated in figure 3. In the overall design plan, the guiding principle of the design is based on design reliability in engineering machinery application, power efficiency and cost control, on the condition that the intended functional requirements can be satisfied.

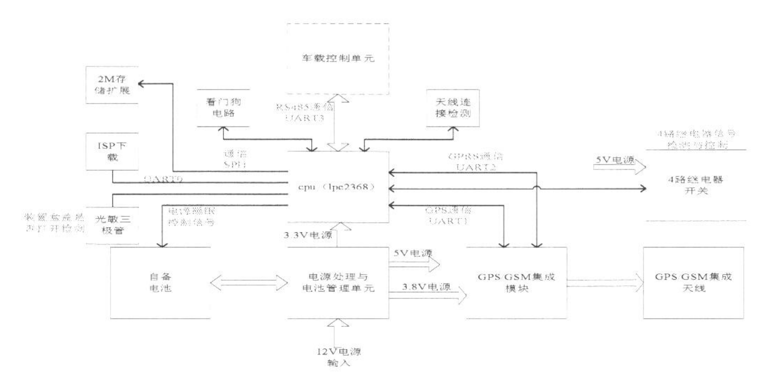

Fig. 3 Schematic diagram of the bottom board

3.2 Communication Linkage

The AT command, ruled by ETSI, is used for the serial port control among communication modules. It is in accordance with the related regulation of GSM. It is AT command that achieves the linkage between LPC2368 and GM862. Below is the AT command being used.

Under the circumstance of good communication conditions of the terminal and the ground receiving station, GPRS is used. Following is the process for the main chip LPC2368 to startup and link GM862 module by using GPS through the AT commands. Like: startup : AT<cr>;Set the baud rate of the module: AT+IPR=115200<cr> and etc.Log off GPRS online mode by +++.

3.3 Reception of Commands

Regularly, for a complete monitoring device, its function is supposedly not limited to monitoring equipment information on general line, sending data to host-computer or guest-computer on the monitoring system. It should be able to receive commands sent from host-computer or guest-computer on the monitoring system that are converted and forwarded in the form of CAN or serial information through terminal, which perform the act of locking engineering vehicles or electronic vehicles. However, as GPRS is a form of unsecured communication method, parity, header and trailer need to be incorporated into the information sent and information is required to be returned to the guest-computer after receiving usable information, otherwise, the guest-computer would keep sending.

4 The Program Structure of Upper Board

4.1 GPS Data Manager

Function. It is mainly to receive the data sent by the serial ports and then display. On the left side, it is a terminal of the serial ports to display the data received from the lower board, including the debugging information. On the right side, it is to display the decryption of the data which is encrypted before. It includes the year, the month, the date, the hour, the minute, the longitude and the dimensionality.

Under the system of Linux, its control on the hardware is actually the read-write of the document of the hardware and the device. It is relatively easier. However, the difficulty bumps up if it is needed to make a graphical interface and able to control the hardware by using QT.Therefore, the programming methods of multithreading and clock are introduced to the writing of programs.

4.2 Multithreading

QT provides support to the thread. Its basic forms are: a thread independent from the platform, the event transfer of the thread safety mode, and a overall QT storage mutex which allows you to invoke QT methods from different thread.

The most important one is QThread. In other words, to start up a new tread is to execute a new QThread::run(). There would be cases that two threads would like to access the same data. Under this circumstance, it is extremely necessary to protect the data. With this purpose, this program also includes a QMutex. A thread can lock on mutex and once it is locked, other threads can not lock it up any more. All such attempt will be blocked until the mutex is released.

4.3 Serial Port Control

The way to control serial ports is basically the same in QT and LINUX, except that when using the multithread program, it needs to quote the header file so as to make QT go deep into Linux system and operate the serial ports.

```
#include   <stdio.h>      /*standard input or output*/
#include   <stdlib.h>     /*standard function library*/
#include   <unistd.h>     /*Unix standard function*/
#include   <sys/types.h>
#include   <sys/stat.h>
#include   <fcntl.h>      /*document control*/
#include   <termios.h>    /*PPSIX terminal control*/
#include   <errno.h>      /*error number*/
```

Having set up like this, serial port data can be captured through reading document. In QT, it needs the command of close(); to close down the serial ports document.

4.4 Button Control

The way to operate button is also to read and write the absolute address. It is much easier comparing to the serial ports. The current value of the button can be obtained by reading 0xE3000002. It is quite similar to operate the rotary encoders. Two values are obtained from reading 0xE3000000, then subtracting one from the other and going through the filter to get the change of the rotary encoders. Illustrated as Figure 4.

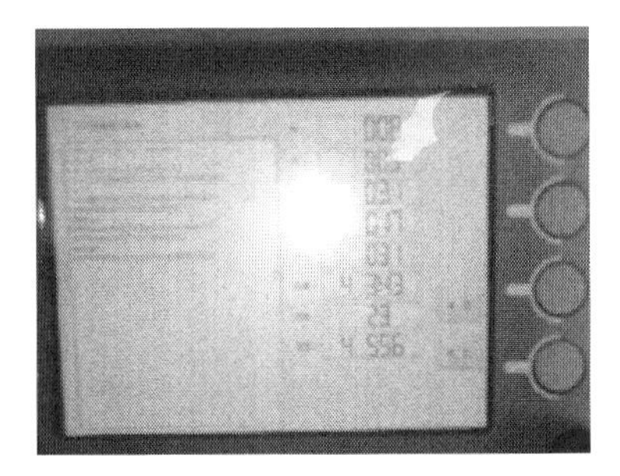

Fig. 4 Push Button Panel

4.5 Dial Display

The serial port data received is displayed vividly in the form of a dial, a numeration table and a progress bar, as shown in Figure 5.

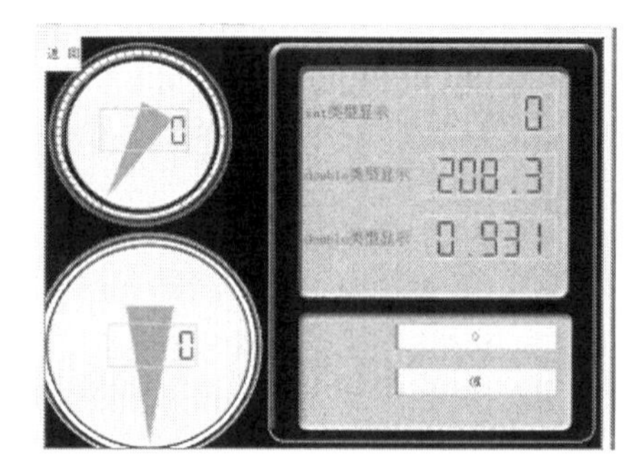

Fig. 5 Dial Display

References

1. Ni, J.: Qt and Linux Operating System Window Design, 1st edn. Publishing House of Electronics Industry, Beijing (2006)
2. Sharma, K., Kabir, M.J.: Red Hat Enterprise Linux 3, pp. 181–183. Tsinghua University Press, Beijing (2006)
3. Gopalan, K.: Real-time Support in General Purpose Operating System, ECSL Technical Report TR92, Experimental Computer System Labs, Computer Science Dept., Stony Brook University, Stony Brook (2001)
4. Mao, D.-C., Hu, X.-M.: Linux Kernel Source Code Analyzing. Chekiang University Press, Hangzhou (2001)
5. Nakamura, S.: Recent Status of Nitride Based LED and Full Color Displays. In: The 3rd Int. Display Workshops, pp. 37–40 (1996)
6. Zheng, W., Buhlmann, P., Jacobs, H.O.: Sequential shape-and-solder-directed self-assembly of functional microsystems. Proc. Natl. Acad. Sci. U.S.A 101, 12814–12817 (2004)

The Definition and Implementation of a Framework for Web-Based Configurable Software

Junjie Wu, Yonghua Zhu, and Huaiyang Zhu

Abstract. As existing web-based application has shortcomings of long development life cycle, low reuse rate and poor expansibility causes a great number of problems for the developers and users. Based on these problems, a framework for web-based configurable software is presented in this paper. This framework is composed of three layers and seven main modules. Under the guidance of this framework, a web-based configurable software was implemented. Finally, the efficiency and the scalability of this framework are proved by a practical application.

1 Introduction

As Internet is growing tremendously, web-based applications are becoming popular. But most of current Web-based systems have a long development life cycle, low reuse rate of the function and poor expansibility. Because modifying and producing a large number of web pages costs a long time, web-based project development life cycle lasts for a long time [1]. Furthermore, most of the web pages are static and there are so many complicated functions, as a result, they are hard to be applied to

Junjie Wu
School of Computer Engineering and Science Shanghai University,
No.149 Yan Chang Road Shanghai, China
e-mail: wjj7221834@gmail.com

Yonghua Zhu
Computing Center of Shanghai University Shanghai University, Bldg D,
Shang Da Road Shanghai, China
e-mail: zyh@shu.edu.cns

Huaiyang Zhu
Shanghai Rongheng Information Technology Ltd,
Floor 13 Bldg 3 No.28 Lane 91 Eshan road Shanghai, China
e-mail: zhu.william@163.com

Z. Qian et al. (Eds.): Recent Advances in CSIE 2011, LNEE 125, pp. 429–434.
springerlink.com

different environments and user requirements. So, to make the architectural evolutionary and extensible, versioning capabilities [6] should be developed.

Configuration [2] is the process of completing a specific task in a work with the tools and methods provided by the application software. Configurable software takes advantage of the feature of configuration, treating object as operating entity to encapsulate and package. Configurable software can save development cost, especially for reconstruction and updating of the application. Configurable software, flexible and scalable[7], has strong adaptability to change with demands, providing good human-computer interaction and high reliability, decreasing repeated work, improving productivity, and shortening the development life cycle of software.

This paper combines the concept of configuration with Web-based software applications and presents the definition and implementation of framework for web-based configurable software, which is simple and convenient for user and logic designers. According to the framework, a hierarchic concept of Functional-Levels (FL) is proposed, which divides the web-based configurable software into three layers and seven modules. Then, implementation of the web-based configurable software is presented as an example to illustrate its procedure. The framework for web-based configurable software in this paper aims to facilitate the implementation of web application. Also, it will give users some hints when attachment or design errors take place.

2 Related Work

2.1 Application Model of the Configurable Software

Configuration is the combination of module or several objects, while it also can be viewed as the evolution of one object. Version [4] is a state of a generic object at a particular time or an evolution of a generic object after a set of changes, but the versions of the generic object share something in common, such as an interface or substitutability in use. When a time stamp in the Configuration model changes, it need two time stamps to form an effectively period to determine an effectively time, as an effectively version. The effectively version can be compared with the prior versions, looked for the different points from the others and to make some planning and adjustments for the change which may happen in the future. In the model of Configuration, component can be viewed as a generic object. Considering the different property of a generic object, versioning can be divided into two concepts: variant and revision [3][4], which has parallel and consecutive versions of the generic object, respectively. The concept of variant and revision form a set of versions, as illustrated in Fig. 1.

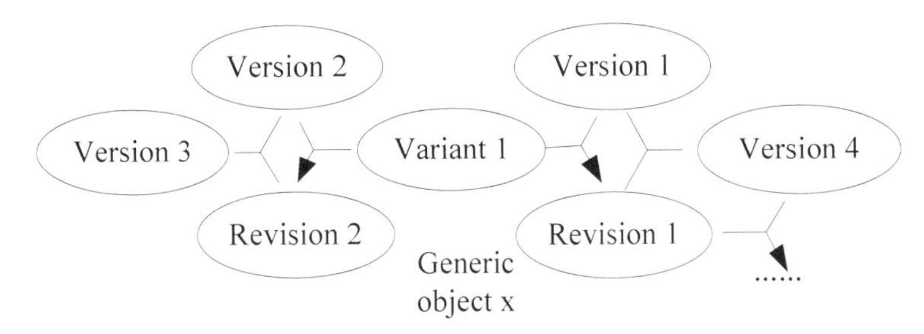

Fig. 1 Generic object with versions, variants and revisions

2.2 Existing Products

Existing web development tool platforms, like ASP.NET, are difficult for users to use because of its complex architecture, while existing configuration web software is not all-purpose, such as DWINS[6].

3 Framework for the Web-Based Configurable Software

Configurable software provides flexible configuration and rapid development tools of building application for users. The framework for the web-based configurable software can be viewed as an object:

$$S = \{C, F, D, L, U, I, A\} \quad (1)$$

S refers to the web-based configurable software; C refers to Component Lib; F refers to Component Factory; D refers to User Interface Designer; L refers to Interface Definition Language; U refers to Control Unit; C refers to User Interface; A refers to Data; Fig.2 shows this framework.

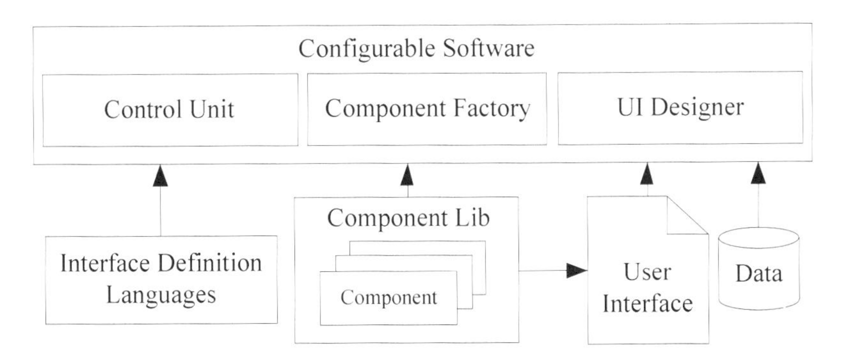

Fig. 2 Framework for the web-based configurable software

3.1 Component Lib and Component Factory

Component lib is a container for components, which can be viewed as a collection of components. Components [2] can be divided into three types: the interface component, the implementation component and the sub-configuration component. These three kinds of components can be linked and combined. Throughout the life cycle, there will be many changes. But the configurable software does not need to change a lot, especially when the implementation component changes without modifying the internal function interface. In practical applications, the modification of the implementation components happens more frequently than user interface. Component factory is charged for governing and classifying the components in the system. It provides a universal interface to all the components.

3.2 User Interface Designer(UI Designer)

UI designer is a graphical configuration tool, whose functions can be listed as below:

- Designing and modifying the user interface by using components which already exist in the component factory.
- Changing the location, size and other properties of the interface component;
- Binding data to the implementation component;
- Establishing the logical relationship with other components;
- Converting user interface into Interface Definition Languages.

3.3 Interface Definition Language (IDL) and Control Unit(CU)

Interface Definition Language (IDL) [5] is used to describe the context of an application. While in the web-based configurable software, IDL is used to accurately describe the context of user interface, the type and properties of each component and the logical relationships between the components. In this framework, the interface of the interface component is specified in an IDL.

Control Unit (CU) is used to parse the IDL, access to the component information, establish relationships between the components, load the entity of component and form the final user interface. When the component is implemented, CU can access the data needed by the component and update persistent data and intermediate results. Furthermore, it can also transfer data between the components.

3.4 User Interface(UI) and Data

User interface is the container of components. Unlike the ordinary static web pages, UI in this framework can be described by the Interface Definition Languages and generated by the CU.

Data will be displayed by the components when the application is running. It also can be the persistent information or intermediate results between the components and server.

4 Implementation of the Web-Based Configurable Software

Based on this framework, web-based configurable software was implemented according to the procedure.

- Design the UI: According to the user demands, components can be picked up from the component factory, created by users, or assembled by the existing components in the component factory. In the UI Designer, components should be selected from the component factory, whose properties can be modified to meet different user demands. Fig.3 shows how the UI designer seems like. After the UI designed, IDL, accurately describing the entire UI can be generated.
- Generate the UI: According to the IDL generated by the UI designer, CU can establish the relationships between the components. Then it will load the entity of components to form the final UI according to the information from IDL. The Fig.3 shows the final UI generated. Tree-based data structure can be built to describe the relationship of components, which makes unified management and loading of components to be possible.

Fig. 3 UI designer and User interface (UI)

- Run-time: When the UI is loaded, the CU will request for data needed from the server. As the server gets the request, it will pull data from database and maintain the persistent information. Then the server sends response to the CU. The CU reads data from the response and binds it with the relative components. When user interacts with the UI, the state of UI will change, that is, the component data or the persistent data will change. As the components aware of the changes, they will send requests to CU to change the intermediate data or the behavior of UI.

5 Conclusions and Future Work

The web-based implementation of configurable software has been proved to improve the efficiency of the web project development. It can greatly improve the scalability and largely shorten the development life cycle of web project.

This framework has been applied in a project of Shanghai Rongheng Information Technology Ltd. However, our framework is still open to improvement and we will also attempt to improve the presentation of existing framework. Interface Definition Language(IDL), platform-independent, should provide accurate description of the page. That is, different page in different languages and platform can be generated by the same IDL, such as VB, ASP, etc. We need to provide more complete IDL in the direction of expansibility to improve the existing framework.

References

1. Hagen, C.J., Brouwers, G.: Reducing software life-cycle costs by developing configurable software. In: Aerospace and Electronics Conference, pp. 1182–1187 (1994)
2. Wheater, S.M., Little, M.C.: The Design and Implementation of a Framework for Configurable Software. Configurable Distributed Systems, 136 – 143 (1996)
3. van der Hoek, A.: Configurable Software Architecture in Support of Configura-tion Management and Software Deployment. Software Engineering, 732–733 (1999)
4. Modelling Configurable Products and Software Product Families. In: Software Configuration Workshop (2001)
5. OMG, Common Object Request Broker Architecture and Specification, OMG Document Number 9 1.1 2.1
6. Balasubramaniam, D., Morrison, R., Greenwood, R.M., Warboys, B.: Flexible Software Development: From Software Architecture to Process. Software Architecture, 14 (2007)
7. Development of Scalable Service-Oriented Grid Computing Architecture, Wireless Communications. Networking and Mobile Computing, 6006–6009 (2007)

Dynamics Simulation of Engine Crankshaft Based on Adams/Engine

Min Cheng, Jiping Bao, and Qiankun Zhou

Abstract. Using professional module Adams/engine build the model for 491 engine and running dynamic simulation, and feting the graphical analysis of crankshaft torsion vibration,besides,finding the Torsion vibration characteristics of the crank train.The research provides the basic theoretical for the optimization of the crankshaft,and has a certain significant for the improvement of the whole engine.

1 Introduction

In our daily life, engine is one of the most common power machinery, and the further develop and enhance of its manufacturing technology has the great significance to social development. The engine crankshaft is the main part of the engine, it works under the impact, torsion, bending stress and torsion stress[1]. Traditional crankshaft analysis[2] is based on the analysis of each component, calculate their inertia force generated by rotating and reciprocating inertia force and pressure of synthesis gas after the outbreak of the solution, the process is very tedious. With the develop of computer and information technology, virtual prototyping technology is widely used. Using simulation software ADAMS (Automatic Dynamic Analysis of Mechanical System) analyze connecting- rod, it can simplify the forces as cylinder gas pressure, reciprocating and rotary inertial force, which will create pressure on crankshaft connecting-rod [3].

Min Cheng
Beijing Forestry University, Jiping Bao
e-mail: chengm001@163.com

Jiping Bao
Beijing Forestry University, Qiankun Zhou
e-mail: baojiping@bjfu.edu.cn

Qiankun Zhou
Beijing Forestry University
e-mail: 113233245@163.com

Z. Qian et al. (Eds.): Recent Advances in CSIE 2011, LNEE 125, pp. 435–439.
springerlink.com

491 engines are widely used in our various models; the engine speed and torque are low. If the performance is enhanced to 6000rpm,the increasing of the speed will bring the original model a range of issues, the most important problem is whether the vibration characteristics match with the original model. In this paper, the use of specialized software ADAMS/engine for the 491 engine crankshaft for torsion vibration analysis, getting the torsion vibration characteristics,providing a theoretical basis for the improving of engine, and it also has a certain significance for the improve of the 491 engine.

2 Basic Theory of Crankshaft Torsional Vibration

In practice, it was discovered that when the engine reaches a certain speed ,operation becomes uneven and with a long time operation, the crankshaft will have the risk of fracture, when the speed increase or decrease, it will improve the situation, or even disappear.,Therefore, this is not due to engine imbalance, A lot of theoretical and experimental studies have shown[4] that it is caused by the occurrence of a significant crankshaft torsional vibration ,the more the number of cylinders, crank the longer, the more serious phenomenon, which is crankshaft torsional vibration.

3 Crank Modeling

The multi-body dynamics model of Crankshaft torsional vibration mainly includes a flexible body model, piston, connecting-rod, flywheel and damper, besides, it also contains the connection for connecting the various parts and the outside force. This paper using ADAMS/engine working on 491 engine model, define global variables, including culinder,piston,crankshaft and connecting-rod.the model define automatically crank kinematics and mutual movement relationships.

3.1 491 Engine Parameters

First, using by 491 engine manual inspection and calculation of incentive-related parameters, as specified in table 1:

Table 1 491Engine Parameters

Type	4-stroke, water-cooled, inline
diameter* Stroke (mm)	91*86
Compression ratio	2.237
Rated power/speed	76/4300
Max torque/speed	193/2000`2600
Idle speed (r/min)	750`850
Crank main journal (mm)	58
Neck axis length (mm)	33.8
Piston diameter (mm)	23
Piston Length (mm)	71.4

3.2 The Import of the Crank and Modeling

Start with Adams/engine run Standard Interface->file->open->assembly, open the model cranktrain_i4_ass.cranktrain_i4,modify the parameters and get the 491 model as picture1.

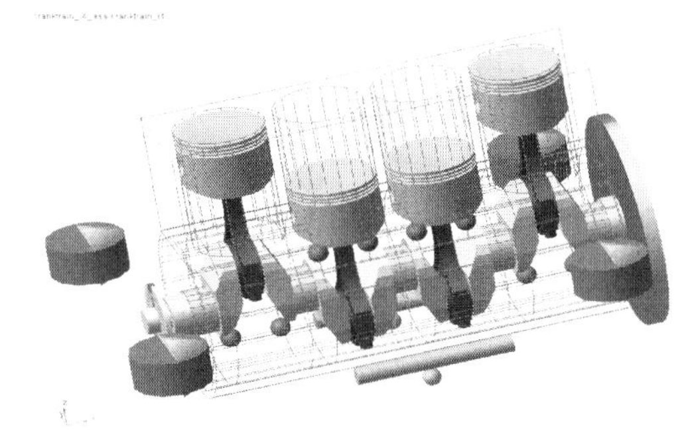

Fig. 1 491engine model

4 491 Engine Crankshaft Torsion Vibration Simulation and Analysis

The 491 engine crank model import Adams / engine interface, its torsional vibration analysis (CRANK_CONCEPT-analysis), getting the engine torsional vibration characteristics whose speed is between 1000rpm-6000rpm. (Higher than the rated speed of 4300rpm)

4.1 491 Engine Crankshaft Torsional Vibration Analysis Simulation

After importing 491 engine, proceed as follows:

1. Modify the model of the crankshaft, right click mouse click and select editable, Paying attention to modify the crank to Torsional.
2. Applied gas pressure, choose Adjust-> Gas Force-> ues_gas_force_1 documents, which is shown in Figure 2.

fev_diesel_001 - 记事本

文件(F) 编辑(E) 格式(O) 查看(V) 帮助(H)

(XY_DATA)
{crank_angle <angle> pressure <bar>}

曲轴转角 (°)	压强 (帕)							
0	0.56759	0.58025	0.56882	0.55956	0.55742	0.55651	0.54859	0.55049
0.5	0.66808	0.68069	0.66831	0.65852	0.65529	0.65606	0.64645	0.6504
1	0.76318	0.77495	0.76215	0.75258	0.74829	0.75082	0.74007	0.74608
1.5	0.8499	0.86002	0.84739	0.83883	0.83361	0.83778	0.8266	0.83447
2	0.92571	0.93347	0.92165	0.91481	0.90889	0.91442	0.90359	0.91291
2.5	0.98874	0.99356	0.98318	0.97869	0.97235	0.97876	0.96915	0.9793
3	1.03785	1.03932	1.03099	1.0293	1.02285	1.02956	1.02201	1.03222
3.5	1.07267	1.07058	1.06487	1.06624	1.05998	1.0663	1.06161	1.07101
4	1.09365	1.08798	1.08538	1.08983	1.08401	1.08923	1.08809	1.09577
4.5	1.1019	1.09284	1.09376	1.10108	1.09586	1.09932	1.10227	1.10737
5	1.09917	1.08709	1.09183	1.10161	1.09701	1.09816	1.10553	1.10731
5.5	1.08766	1.07307	1.08183	1.09347	1.08937	1.08786	1.09976	1.09767
6	1.06987	1.05337	1.06624	1.07901	1.07516	1.07085	1.08714	1.08086
6.5	1.04839	1.03063	1.04763	1.06073	1.05669	1.04973	1.07005	1.05953
7	1.02575	1.00738	1.02843	1.04106	1.03627	1.02711	1.05083	1.03634
7.5	1.00421	0.98585	1.0108	1.02222	1.01605	1.00537	1.0317	1.01376
8	0.98567	0.96785	0.99651	1.00609	0.99785	0.98661	1.01457	0.99395
8.5	0.97155	0.95468	0.98682	0.99412	0.98311	0.97244	1.00093	0.97863
9	0.9627	0.94706	0.98245	0.98721	0.97282	0.96394	0.99181	0.96895
9.5	0.95942	0.94517	0.98355	0.98575	0.96747	0.96165	0.98773	0.96548
10	0.96147	0.94866	0.9898	0.98962	0.96711	0.96551	0.98868	0.96823
10.5	0.96817	0.95675	1.00041	0.99822	0.97135	0.97497	0.99421	0.97663

Ln 37, Col 11

Fig. 2 Gas Pressure

3. Setting shock absorber damping vibrations properties, right-click with the mouse, select modify-> editable, and click "For CRANK CONCEPT only", then confirm.
4. Connecting rod and piston with default properties, then select Simulate-> Crank Train Analysis -> CRANK_CONCEPT_Analysis, the speed set to 1000 ~6000rmp, interval speed is set to 50rpm, click ok,then analysis.

4.2 The Result of 491 Engine Crankshaft Torsional Vibration

Through simulation analysis, the work can get the crankshaft torsional vibration characteristics, as shown in Figure 3 - Figure 6: Figure 3 Deflection angle for the crankshaft speed between 1000rpm - 6000rpm; Figure 4 the maximum shear stress for crankshaft speed between 1000rpm - 6000rpm, Figure 5 the maximum torque for the crankshaft speed between 1000rpm - 6000rpm, Figure 6 safety factor for the crankshaft speed between 1000rpm - 6000rpm.

According to experience, maximum torsion angle for low-power engine is generally less than 3 °. It is known from Figure 3 that Deflection angle is getting higher while the crankshaft speed increases, when the crankshaft speed works below 4450rpm the maximum Deflection angle is 0.3 ° which is within reasonable limits; Otherwise, the crankshaft speed gets to 4900rpm, the Deflection angle gets to 0.39 °, even when speed is 6000rpm, the Deflection angle is 0.43 °, which are all higher than 0.3 °, That means the angle is over 491 engine crankshaft's maximum torsion angle.

It can be drawn from Figure 4, when the crankshaft speed is higher than 4500rpm, torque values significantly increased, it will speed up the crankshaft wear, increasing the rotation no uniformity, the risk of breaking crankshaft gets higher.

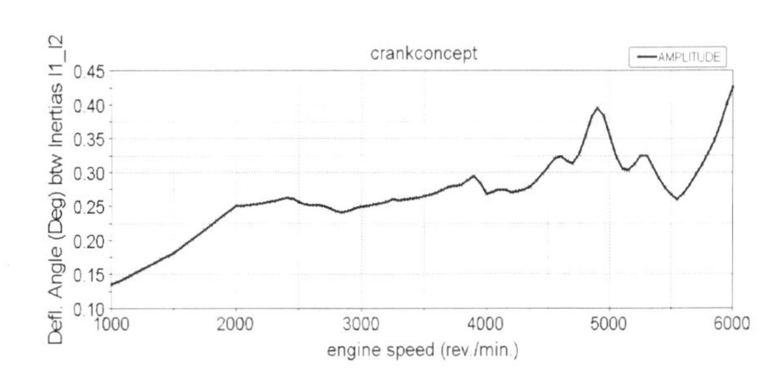

Fig. 3 Deflection Angle

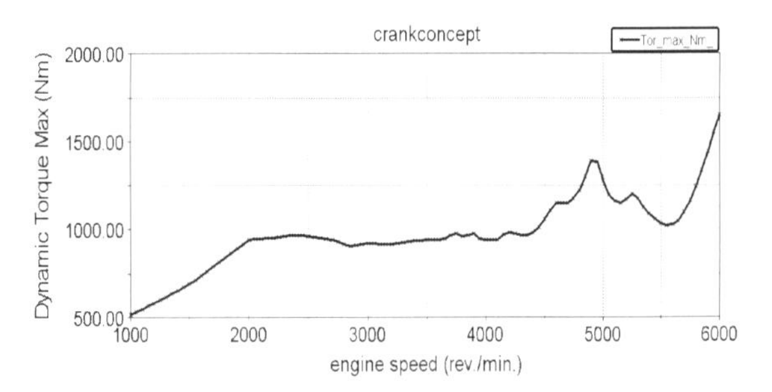

Fig. 4 Maximum power crankshaft torques

Figure 5 shows the safety factor of crank pin when the crankshaft works. Safety factor is defined as:

$$SF = \sigma_D / \sigma_V \tag{1}$$

σ_D means the equivalent alternating stress σ_V means the fatigue limit stress

Figure 5 includes the safety factor for all the crankshaft chamfer the between 1000rpm - 6000rpm.It can be seen from the figure, the value is higher than the limit value of 1.25 in their work which is within the scope of security.

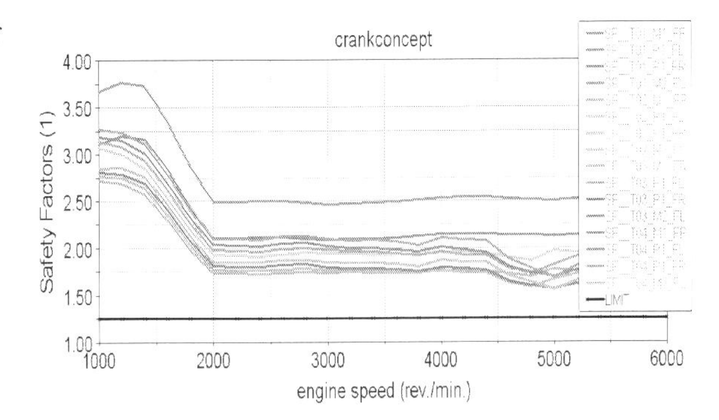

Fig. 5 safety factor of crankshaft chamfer

5 Conclusions

By modifying the parameters of the Adams / engine template model cranktrain_i4_ass to get mechanical analysis of crankshaft torsional vibration, to obtain crankshaft torsional vibration characteristics which works between 1000 - 6000rpm. It can be seen from the figure, when the crankshaft speed above 4200 ~ 4500rpm, the pressure is significantly more than the original strength of the crankshaft .in order to improving the 491 engine performance, the material improving should be considered or improved by strengthen the crankshaft structure. And then enhance the strength of the crankshaft, which can keep its normal work even when its speed promote to 6000rpm.

References

[1] Chen, J.-W., Zhang, Z.-W., Liu, R.-C., Xue, L.-Q., Sun, H.: Dynamic Analysis of Crank Linkage Mechanism Based on ADAMS/View. Tractor&Farm Transporter 36(5) (October 2009) (in Chinese)

[2] Chen, L., Du, Y., Zhang, Y., Teng, F.: Modeling and Simulation of Crankshaft-link-piston based on Adams. Mechanical Transmission, 60 (2010) (in Chinese)

[3] Chen, L., Zhang, Y., Ren, W.: Dynamic Analysis of Mechanical Systems and Application Tutorial for Adams. Tsinghua University Press, Beijing (2005) (in Chinese)

[4] Yang, L.: Internal Combustion Engine Design. Journal of Agricultural Machinery (1981) (in Chinese)

Finding Abnormal Behaviors of Object-Oriented Programs Based on Cumulative Analysis

Xuemei Liu and Shuangmei Liu

Abstract. In this paper, several cumulative analysis methods that are used in fault localization based on program behaviors were analyzed and compared. Because these methods did not take into account the features of object-oriented programs, and thus cannot be properly used in fault localization in object-oriented program. This paper proposed that object behaviors could be described by using the object life behavior model; then object-oriented program behavior model was constructed on the basis of life behavior model; the similarity of program behaviors was indicated by using sequence pattern. Based on the cumulative analysis of program behaviors, the method of finding abnormal behaviors of object-oriented programs was proposed. The model of finding abnormal behaviors was also constructed. The method of finding abnormal behaviors based on cumulative analysis and its implementation procedure were depicted in detail. Afterward, this method has been used in the research of fault localization and has been very effective. This work is supported by the National Natural Science Foundation of China under grant No. 60603039.

Keywords: Object life behavior model, Fault localization, Cumulative analysis; Abnormal behavior.

1 Introduction

Fault localization is an important step in software debugging. An efficient method of fault localization could accelerate the process of debugging and improve the quality of software[1].

Many fault localization methods are based on program analysis techniques. In static analysis, there are many program analysis techniques, such as program flow

---

Xuemei Liu
Department of Software and Information Management, Beijing City University
Beijing, China

Shuangmei Liu
College of Science and Information, Qingdao Agriculture University
Qingdao, Shandong, China

Z. Qian et al. (Eds.): Recent Advances in CSIE 2011, LNEE 125, pp. 441–452.
springerlink.com

diagram, dependency relationship[2], program slice[3], program belief[4], etc. In dynamic analysis thcrc arc also many tcchniqucs, such as log analysis[5], debugger, dynamic slice[6], etc. Neither static nor dynamic analysis takes into account the complete program behaviors. However, the variance analysis based on program behaviors could find suspicious patterns (or abnormal behaviors) and locate faults by comparing the difference among program behaviors.

The variance analysis based on program behaviors could extract program behaviors from many executing processes to find the differences between them by comparing success behaviors and failure behaviors of a program. These differences between program behaviors were abnormal behaviors. Then the modules that might contain faults could be inferred. Thus, finding and analyzing abnormal behaviors is an effective strategy for fault localization.

2 Related Works

The cumulative analysis for fault localization based on program behaviors has been researched extensively, and had a good effect in fault localization.

Jones[7,8]proposed that the faults could be located by using code coverage and testing information visualization, and then developed a support tool, that is, Tarantula. Its central idea was that the coverage state of every statement could be recorded during executing test cases, and inferred no matter the current test case became invalid or not. Giuseppe[9] gathered the executing track of a program that executed successfully and ineffectively, and the results could be shown by using method invocation trees. Analyzing the method invocation trees by using frequent pattern mining algorithms aimed at finding the differences between method invocation trees that executed successfully and ineffectively, which was the method invocation subtree, also named discriminative pattern. The discriminative pattern described the possible faults by comparing and analyzing the context of the execution.

Dallmeier[10] tracked and recorded the trace of method invocation of every object, which was then analyzed to obtain sequence sets by using the sliding window algorithm. By comparing the sequence sets corresponding to every class in the process of success execution and failure execution, the arrangement of classes which were likely to contain faults could be obtained, and then it could help users to detect classes in terms of the priority, the real faults could be found.

However, none of the above-mentioned methods takes into account the features of object-oriented programs, nor do they describe the object behaviors, thus they could not be properly used in fault localization for object-oriented programs. In order to solve the fault localization of object-oriented programs, object behaviors should be described by using the object life behavior model. Based on the model, program behaviors could be described by using sequence patterns. The patterns were analyzed by using variance analysis. The model and algorithm of finding suspicious patterns of object-oriented programs were designed. The research of fault localization could go on based on it.

3 Construct the Behavior Model of Object-Oriented Programs

This paper proposed that a class should be the elementary unit of analyzing program behaviors because the class abstractly described a set of objects that had common features. When executing a program, every class could instantiate many objects. The section of the class behaviors consisted of the behaviors set of the instantiated objects of a class. When locating faults, the section of class behaviors could be obtained after analyzing every object behaviors and gathering them based on Life Behavior Model. Analyzing the section of all classes behaviors could benefit for fault localization.

Definition 1: Life Behavior Model

$LBM(o) := <e1,e2,..ei,\ldots,en>, 1 \leq i \leq n$; $ei: = [signature, t], ei \in EI(o) \cup EC(o)$;

$signature := invocation@site\$thread$; $EI(o) := EIE(o) \cup EIR(o)$;

$EC(o) := \{e|e=[invocation@site\$thread, t]\}$; $EIE(o) := \{e|e=[invocation>@site\$thread, t]\}$

$EIR(o) := \{e|e=[invocation<@site\$thread, t]\}$

n was the number of events happening in the life cycle of the object o. EI(o) showed the called events of the object o. EC(o) showed the calling events of the object o. The event e was described as [signature, t]; where 'signature' described the concrete meaning event e; 't' described what time e happened.

'signature' was formally defined as 'invocation@site$thread', while invocation and site were the methods of the object o; '@' showed that site called invocation; '$' showed the condition of method invocation was thread. EIE(o) was the event set of Invocation Entry. EIR(o) was the event set of Invocation Return. '>' and '<' were use to distinguish the different events of EIE(o) and EIR(o).

LBM roundly recorded the object behaviors in the life cycle which was based on the method invocation between objects, as well as the context and time message created by the method invocation from the standpoint of objects. Describing program behaviors based on LBM could properly characterize the program behaviors message, and then could be the basis of subsequent variance analysis.

Definition 2: Class Behavior Model

CBM(c):={LBM(o)|o is instantiated from c}. When executing a program, the class behavior was a set of all objects behaviors of the same class. That is, CBM was a set of LBM of all instantiated objects.

Definition 3: Program Behavior Model.

PBM(p):={CBM(c)| c is an executed class}. The executed class was the class that had at least one instance object and the LBM of the object was larger than o when executing a program. But PBM(p) was incomplete. It only described the behaviors from dynamic execution and was a section of the program behaviors.

4 Model for Finding Abnormal Behaviors Based on Patterns

In terms of Definition 3, the PBM based on LBM contained abundant program behaviors message at runtime. If variance analysis and comparison were made directly by using of them, a large amount of analyzed date could be brought, which did not benefit for acquiring targeted variance behaviors.

Thus, analyzing program behaviors was based on patterns that were fragments of program behaviors emerging at certain frequencies, then the common features could be extracted. The program execution behaviors were firstly analyzed by using pattern mining algorithm. The message was then abstracted as patterns. By analyzing patterns, the abnormal behaviors of the program could be obtained, and the faults of the program could be located in terms of complete behavior message[11].

Based on the program behavior model[12], both the success patterns and the failure patterns of program behaviors were extracted and the abnormal behaviors could be found by using analysis of variance. These abnormal behaviors of the program were suspicious patterns. In this paper, the object of analyzing patterns in fault localization was a class. The class behavior was defined by instantiated objects behaviors in running program. The suspicious pattern could found by analyzing the class behaviors. The program behaviors related to the suspicious pattern could also be analyzed. This would help locate the faults and calculate the possibility of faults.

The framework of finding abnormal behaviors based on patterns was shown as Figure 1.

4.1 Preparing Data

The program being analyzed could run successfully or fail. The data of executing track for the successful run and failed run could be gathered separately. The program behaviors model would be constructed. To acquire a behavior model of a class, the Lifetime Behavior Model (LBM)[13] of every object could be traversed to acquire sequence composed of called methods of objects by aggregation. The sequence set of called methods of a class could be acquired.

4.2 Extracting Pattern

Based on the rules of extracting pattern, the pattern could be extracted by analyzing the sequence set of called methods of a class. Because recursion and loop existed in the codes, many partially analogical substrings might emerge in the called methods. The sequence set should be preprocessed to reduce the complexity of the set and remove iterative elements. The patterns acquired at last could be more effective. At the same time, sequence normalization could make it easier to process programs. Based on the sequence set that had been normalized and preprocessed, the pattern that was satisfying the rules would be acquired by the algorithm of extracting pattern.

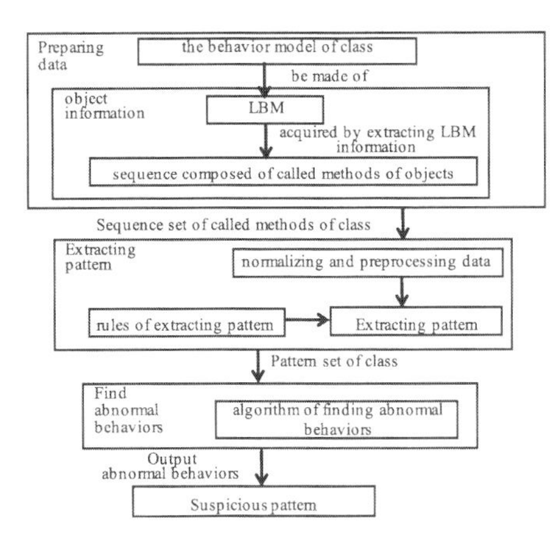

Fig. 1 The framework of finding abnormal behaviors based on patterns

4.3 Finding the Abnormal Behavior

The abnormal behaviors could be acquired by analysis of variance which was done in the set of success patterns and failure patterns, which were suspicious patterns. The abnormal behaviors were caused by faults and emerged mainly in failure patterns and rarely in success patterns. Based on the abnormal behaviors, the reasons of causing faults could be found easily. Thus, in this paper the algorithm of finding abnormal behaviors was proposed, which was the basis of static reasoning.

5 Algorithms of Extracting Patterns

5.1 Sequence Normalization and Pre-procession

Because recursion and loop existed in the codes, many iterative substrings which had same semantic information would emerge in the called sequence, which could increase the length of the sequence being analyzed and the workload of subsequent analysis and computation. Thus, before pattern extraction, preprocessing would be executed to remove iterative substrings. The algorithm was shown as Algorithm 1.

Algorithm 1 Algorithm of preprocessing sequence

```
  GeneralSequencePreProcess(sequence)
  Input:sequence;//the sequence that will be analyzed
  {
  check=checkExist(sequence,clcutSequence);
  If(check=TRUE)//If the sequence has been processed,
           // the result will be acquired directly
{cutSequence=findSequence(clcutSequence,sequence;)
else//To process sequence by calling reduceSequence
{cutSequence=reduceSequence(sequence);
putSequence(sequence,cutSequence,clcutSequence)
}}}
```

The input was the sequence of the called methods to be analyzed. If the sequence had been preprocessed, the result cutSequence would be acquired by retrieving Hash clcutSequence, using the sequence as a keyword. If the sequence had not been processed, preprocessing would be executed by calling reduceSequence, which would search the largest iterative substring of the sequence, record its location, and remove it from the sequence. The result would be stored into clcutSequence.

5.2 Definition of the Pattern Standard

The pattern could be obtained on the basis of the pattern standard. In general, the standard was an n-tuple whose information was different according to different problems. The information comprised by the standard of extracting patterns was as follows:

1) *The minimal pattern length. The length of a pattern ptn was len (ptn).*
2) *The maximal pattern length.*
3) *The minimal support. The support was the total number of times a pattern emerged in a sequence set. The minimal support was the minimal number of permission. This was necessary because it could make the pattern more significant. The support of a pattern ptn was supp(ptn).*
4) *The maximal interval. It supported to discretely analyze the pattern.*
5) *The minimal score of a pattern. The score of a pattern could be computed. If the score was lower than the minimum, the pattern would be eliminated.*

According to the feature of the data to be analyzed, the minimal pattern length is 2 and the maximal pattern length was uncertain. In practice, longer pattern would be computed until no pattern could be found to satisfy the standard. The standard of extracting patterns defined in this paper was c={minLen, maxInterval, minSupp}, in which, minLen=2, manInterval=1, minSupp=3. The standard that every ptn of the pattern set must be satisfied with is Len(ptn)≥minLen. The ptn became an element's substring of SC (the set of called methods) after no more than maxInterval operations were inserted into the ptn, and supp(ptn)≥minSupp.

5.3 Algorithm of Finding Patterns

The algorithm was made of two parts, the sub-algorithm of finding the initial pattern whose length was 2, and the sub-algorithm of finding the pattern whose length was greater than 2.

6) *Finding initial patterns. The initial pattern was the pattern whose length was 2. According to the standard of patterns, the initial pattern could be obtained by traversing the set of called methods. The algorithm was shown as Algorithm 2.*

Algorithm 2 Compute the initial pattern (part)

```
InitPtnMining(currClass,c,currSeq)
Input:currClass;//the class that will be analyzed
Input:c;//the standard of extracting patterns
Input:currSeq;//the sequence that will be analyzed
Output:initPtn;//the set of initial patterns
{for(i=1;i<currSeq.size()-c.maxInerval-1;i++)
 {for(j=i+1;j<i+c.maxInterval+1;j++)
  {esKey=new ESKEY(currSeq,i,j);
  esValue=esInit.get(esKey);
  check=checkExist(esValue);
  if(check=TRUE)
   {tempptn=new cPtn(esKey,currLen);
addSupp(tempptn,1);
utPattern(tempptn,initPtn);
}
  else addSupport(initPtn,esValue)
  }}
 Return initPtn;
 }
```

When the initial pattern of length 2 was extracted, each sequence of the set of called methods needed to be analyzed orderly. In Algorithm 2, every sequence currSeq could be analyzed, but it was only a part of the algorithm to compute the initial pattern. The elements prior to currSeq.size()-c.maxInterval-1 could be analyzed. The method of processing the rest of elements was the same with above-mentioned algorithm.

In Algorithm 2, after analyzing the sequence currSeq, the candidate patterns that were satisfied with maxInterval and minLen could be extracted; their support would be computed and stored. In the whole algorithm, every sequence would be analyzed orderly, the support of existing patterns would be increased, and then the new pattern would be stored. After analyzing every sequence, the initial pattern whose length was 2 will be obtained, which satisfied the standard of extracting patterns.

Defining the patternVec aimed at contributing to finding the pattern whose length was greater than 2. patternVec[i] was composed of patterns whose first element was i. For example, the pattern {2,3} was stored in patternVec[2]. Patternvec was built during computing the initial pattern. The longer pattern would be computed by patternVec based on the initial pattern.

7) *Finding the pattern whose length was greater than 2. The pattern whose length was 3 could be computed on the basis of the pattern whose length was 2. By analogy, the pattern whose length was n+1 could be computed on the basis of the pattern whose length was n. When the pattern being analyzed was ptn_1, the new pattern ptn_3 would be computed to every element ptn_2 of patterVec[ptn1[len(ptn_1)]-1], ptn_3=ptn_1+ptn_2[len[ptn_2-1]]. If supp(ptn_3)> minsupp, ptn_3 was a pattern, by analogy, the longer pattern could be computed. The algorithm was shown as Algorithm 3.*

Algorithm 3 Finding the pattern whose length was greater than 2

```
continuedPtnMining(cPtn)
Input;cPtn//the set of patterns whose length are n
{currPtn=getPtn(nPtn);
curtail=getTail(currPtn);
tailVec=getItem(ptnTail,headTail);
check=checkExist(tailVec);
if(check=TRUE)
  {for each item in tailVec
   {cPtn=getPtn(tailVec);
   TailStr=getTail(cPtn);
   newStr=head+tailStr+";";
   check=checkExist(newStr);
   if(check=FALSE)
   {putPtn(newStr,++cndNum);
   putPtn(cndNum,new cPtn(newStr,currLen);
   }
else continue;
}}}
}
```

In Algorithm 3, the pattern whose length was n+1 was possible to be found by traversing the set of patterns whose length was n, which would be stored into pattenrVec. Based on the standard of patterns, the pattern set of the class could be computed by analyzing the sequence set of called methods, and the abnormal behaviors would be found.

6 Finding Suspicious Patterns Based on Variance Analysis

Definition 4: Alphabet. It was the set of all methods of classed in the program that would be analyzed, marked as M.

Definition 5: SO (the Sequence of Called Methods of Object o). SO was the sequence composed of events of o-method which existed in EI(o). SO(o)=<o-m1,…,o-mi,…,o-mk>, $\forall o\text{-}m_i, \forall o\text{-}m_{i+1}$, $\exists T_i < T_{i+1}$

Definition 6: Pattern. Pattern was defined as $pattern = \{m_i m_j ... m_n \mid m_i, m_j ... m_n \in M\}$. Pattern was a segment of sequence that was composed of elements of M satisfied with the standard of patterns. A set of patterns was PTN, whose element was ptn. FPTN was the set of failure patterns and SPTN was the set of success patterns.

Definition 7: Abnormal Behavior. Abnormal behavior was defined as the difference between successful running and failure running of a program. Although abnormal behavior did not always cause failure, it could help to locate the faults. Abnormal behavior was indicated by the suspicious pattern of a class.

Definition 8: Suspicious Pattern DPTN. When a ptn satisfied the following standards, the ptn could be defined as a suspicious pattern, 1) the ptn only emerged in the set of failure patterns; 2) the support of ptn in the set of failure patterns was no less than λ times of the support in the set of success patterns.

S(ptn, F)=Support[ptn in FPTN] showed the support of ptn in the set failure patterns. S(ptn, S)=Support(ptn in SPTN) showed the support of ptn in the set of success patterns. If S(ptn,S)=0 and S(ptn,F)>0, ptn was a suspicious pattern. Otherwise, set dm(ptn)= S(ptn,F)/ S(ptn,S) as the suspicious degree of ptn(dm(ptn)>0). If dm(ptn)≥λ, ptn was a suspicious pattern. If a pattern simultaneously emerged in the set of failure patterns and the set of success patterns, the analysis method of the pattern whose length was 2 was different from that of the pattern whose length was greater than 2.

8) *The pattern whose length was 2. When the length of ptnk was 2, if dm(ptnk)≥λ, ptnk was a suspicious pattern. λ was the threshold of the suspicious degree which could be computed by experiments.*
9) *The pattern whose length was greater than 2.*

Algorithm 4 Compute the pattern whose length was greater than 2

```
calDptn(ptnm)
Input: ptnm,//the pattern whose length is greater than 2
{PTNK=GetSubPtn(ptnm,n-1) //to obtain the //sub-pattern of ptnm whose
length is n-1
For each Ptnk in PTNK
{check=checkIsDPTN(Ptnk)
If(check=TURE){
              setFlg(true);
exit;
       }}
if(!getFlg){
      if (dm (ptnm) ≥λ)){
            dptnm=new DPTN(ptnm)
     }}}
```

In Algorithm 4, the sub-pattern whose length was n-1 could be computed firstly by analyzing the pattern whose length was greater than 2. If a sub-pattern had been found as a suspicious pattern, it would be inserted into the set of suspicious patterns and the analysis was finished. Otherwise, the suspicious degree of ptnm was analyzed and it would infer whether it was a suspicious pattern. According to the length, the pattern simultaneously emerged in FPTN and SPTN could be analyzed by Algorithm 4.

This paper was not the first to propose to use analysis of variance in fault localization. In many existing research, analysis of variance had been used in the model based on the nearest neighbor method and fault localization based on invariance[14]. What made it different in this paper was that analysis of variance could be made continuously to find the suspicious pattern. Because the abnormal behaviors were often caused by faults, the common feature of program behaviors were indicated by patterns, the abnormal behaviors could be found by computing patterns. By analysis of variance the suspicious pattern would be acquired, which was the beginning of fault localization.

The data flow diagram of sub-module to find suspicious patterns was shown as Figure 2.

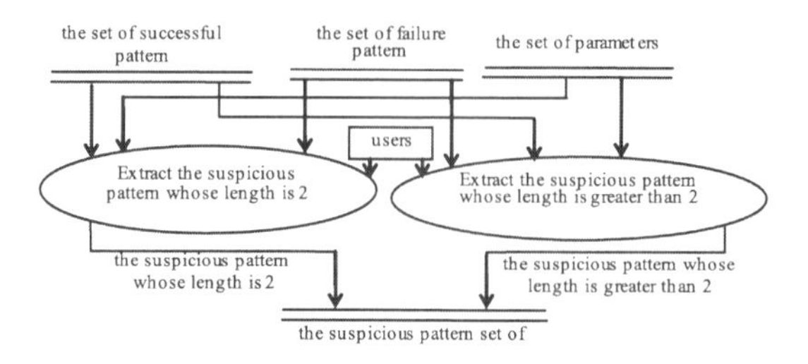

Fig. 2 The diagram of sub-module to find suspicious pattern

For each class to be analyzed, its suspicious pattern could be found by analysis of variance based on the set of success patterns and failure patterns. The suspicious pattern set of a class was composed of the suspicious pattern whose length was 2 or greater than 2, which was the input of fault reasoning.

7 Case Study

Based on the algorithm of finding suspicious patterns, the BBN for fault reasoning could be constructed, whose nodes consisted of the suspicious patterns and the callers of the methods that constituted the suspicious patterns, through which the faulty modules could be found and the probability for each module containing the faults could calculated.Two sets of experiments (There were 11 experiments) were done by selecting seven pieces of real software, which aimed at verifying the effectiveness of this method. The first set of software consisted of DataTest, ThreadTest, FileChooserDemo, Maze, DecimalToBinary, and AutoMachine. The second set was WSDL.

The controlled experiment was done in the first set, and the non-controlled experiment was done in the second set. For the controlled experiment, the faults were embedded in the program. The data of failure execution could be obtained by testing the software with embedded faults. For the non-controlled experiment, the runtime data of the developing software that contained faults was gathered as failure data.

Some index data of eleven experiments about fault localization was shown in Table I. The data demonstrated that this technique could achieve an average accuracy of 0.761 and an average recall of 0.737. This fault localization technique was very effective and had high practical value.

Table 1 Relative Index of Program Performance

Software	Precision	Recalll
DataTest	0.83	1
ThreadTest	0.83	1
DecimalToBinary2	0.78	0.88
DecimalToBinary1	0.71	0.71
Maze	0.83	0.71
FileChooserDemo	0.75	0.67
AutoMachine1	0.75	0.64
AutoMachine2	0.72	0.62
AutoMachine3	0.72	0.58
AutoMachine4	0.7	0.55
WSDL	0.75	0.75
average	0.761	0.737

In this paper, an algorithm of finding suspicious patterns was proposed. By variance analysis and backward reasoning of the program behavior relative to failed and successive execution, the modules that possibly included faults were acquired to help programmers reduce the program elements that they need to consider in debugging.

However, for the running programs, the method invocation sequence was only one aspect of the object behaviors. The program behaviors shown by it only covered part of the program behaviors. Some other elements of program behaviors could not be depicted by the model defined in this paper, such as data flow. Thus, the other faults such as computing expressions and error processing in a method could not be located by the method mentioned in this paper. In subsequent research we will take into account other methods to solve the problem.

References

1. ISO, IEEE Standard Glossary of Software Engineering Terminology,ANSI/IEEE Std 729. IEEE, New York (1983)
2. CODESURFER[EB/OL]. US GrammaTech, Inc. 2007, `http://www.codesurfer.com/` (February 6, 2009)
3. Weiser, M.: Program slicing: Formal, psychological and pracical investigation of an automatic program abatraction method. PhD dissertion. University of Michigan, Ann Arbor, Michigan (1979)
4. FINDBUGS[EB/OL], Maryland: Lesser GNU Public License (2006), `http://findbugs.sf.net/` (December 4, 2008)
5. CHAINSAW[EB/OL] Oliver Burn (2002), `http://logui.sourceforge.net/` (December 10, 2008)
6. Margaret Ann Francel. Fault localization through execution traces. PhD dissertion. Georgia Institute of Technology, USA (2002)
7. Jones, J.A., Harrold, M.J., Stasko, J.: Visualization of Test Information to Assist Fault Localization. In: 24th International Conference on Software Engineering (ICSE 2002), pp. 467–477. ACM, Orlando (2002)
8. Jones, J.A.: Fault Localization Using Visualization of Test Information. In: The 26th International Conference on Software Engineering (ICSE 2004), pp. 54–56. ACM, EICC (2004)

9. Di Fatta, G., Leue, S., Stegantova, E.: Discriminative Pattern Mining in Software Fault Detection. In: The 3rd International Workshop on Software Quality Assurance (SOQUA 2006), 14th ACM Symposium on Foundations of Software Engineering (ACM SIGSOFT 2006), pp. 62–69. ACM, Portland (2006)
10. Dallmeier, V., Lindig, C., Zeller, A.: Lightweight Defect Localization for Java. In: Gao, X.-X. (ed.) ECOOP 2005. LNCS, vol. 3586, pp. 528–550. Springer, Heidelberg (2005)
11. Abdelwahab, H., Timothy, C.L.: An Efficient Algorithm for Detecting Patterns in Traces of Procedure Calls. In: IEEE First International Workshop on Dynamic Analysis (WODA 2003), Portland, Oregon, pp. 33–36 (2003)
12. Agrawal, R., Srikant, R.: Mining sequential patterns. In: The 11th International Conference on Data Engineering, pp. 3–14. IEEE Press, Taipei (1995)
13. Wu, J., Jia, X., Liu, Y., Li, G.: Java Object Behavior Modeling and Visualization. In: International Conference on Software Engineering Advances (ICSEA 2006), Washington, DC, USA, pp. 60–65 (2006)
14. Mohammad, E., Eleni, S.: Analysis of Web-usage Behavior for Focused Web Sites: A Case Study. Journal of Software Maintenance, 129–150 (2004)

Proof Automation of Program Termination

Stefan Andrei, Kathlyn Doss, and S. Kami Makki

Abstract. Total program correctness includes partial program correctness and program termination. These two subproblems are undecidable for general programs. As the predominant part of total correctness, the partial program correctness was studied intensively by the research community. Nevertheless, proving termination for an arbitrary program is challenging. This paper describes an automated method that tests for termination of modulo-case functions by conducting a mathematical induction proof. The system takes a modulo-case function and builds an execution trace tree from its inverse. Then, a linear polynomial is formed to capture the level of the executed trace tree for which the termination problem holds. Using the cases of the original function, an inductive proof is automatically generated. If a function successfully terminates, its runtime is computed based on the polynomial.

Keywords: termination analysis, induction proof, modulo-case function.

1 Introduction

Termination represents one of the key topics in computer science. When a program has finished computation of a given input it has terminated. Termination for any arbitrary program can be difficult to determine. In fact, in 1936, Turing showed that the termination problem, also known as the *halting* problem, was undecidable [7]. The truth of any nontrivial statements about arbitrary functions is undecidable, as stated in Rice's theorem [5]. Although there does not exist a general solution applicable for all programs, termination can be sometimes solved on an individual basis [8].

Total correctness includes partial correctness and termination analysis. According to [6], termination analysis is a challenging research topic in both theory (mathematical logic, proof theory) and practice (software development, formal methods). Most of the program termination tools cannot automatically use

---

Stefan Andrei · Kathlyn Doss · S. Kami Makki
Lamar University, Beaumont, Texas, 77710, U.S.A.
e-mail: `{Stefan.Andrei,Kathlyn.Doss,Kami.Makki}@lamar.edu`

Z. Qian et al. (Eds.): Recent Advances in CSIE 2011, LNEE 125, pp. 453–459.
springerlink.com

the size-change termination technique of [4] because the recursive calls may not retain the same parameters throughout. Meanwhile, the affine-based SCT of Anderson and Khoo, [1], does not directly show that this function terminates [2]. Also, systems like the Omega Calculator [3] were not able to assess the termination of all of the example functions that this system was able to confirm. This paper describes an automated technique for obtaining the complete proof of functional program termination without human intervention. The language describing a program is irrelevant as it is easy to translate a program in a different language. For example, Microsoft Research developed a static analysis tool, Terminator [10], to prove termination of Windows device drivers written in C. The technique presented in [2] is systematic, in that it shows termination of modulo-case functions with the designer's assistance. By transforming the method from [2] into an entirely automated algorithm, this current work is able to generate a complete proof using mathematical induction for the modulo-case functions.

Section 2 provides preliminaries for the termination algorithms. Section 3 describes the *hybrid exponential normal form*, a combination of an exponential and a first-degree polynomial. Section 4 shows our experimental results of our implementation tool. Section 5 is the conclusion and future work.

2 Preliminaries

We consider the nontrivial class of modulo functions that describes the recursive calls as a sequence of function compositions of $f()$ [2]. The numerical normal form for functions having one parameter is F: N → N, given by:

where $n'_0, \ldots, n'_{m-1}$ are constants, the guards $x = n_0, \ldots, x = n_{m-1}$ may be also called terminating conditions. This paper considers the termination problem for programs like this:

```
while (x ≠ n0 and ... and x ≠ nm-1) x = F(x);
```

Note that $\beta_1(x), \ldots, \beta_{p-1}(x)$ are boolean expressions that may contain binary operators including *modulo* and are called non-terminating conditions. Functions $f_0(x), \ldots, f_{p-1}(x)$ are invertible functions. The function f: $A \to B$ is an invertible function if there exists g: $B \to A$ such that $f \circ g = 1_B$ and $g \circ f = 1_A$, where 1_A and 1_B represent the identity functions over A and B, and $\circ$ represents the function's composition. Function $f_1()$ below is in numerical normal form:

$$f_1(n) = \begin{cases} 0 & n=0 \text{ or } n=1 \text{ or } n=2 \\ n/3 & n=0 \pmod 3 \text{ and } n>2 \\ 2n-2 & n=1 \pmod 3 \text{ and } n>2 \\ n+1 & n=2 \pmod 3 \text{ and } n>2 \end{cases}$$

In $f_1()$ above, $n = 0$, $n = 1$, and $n = 2$ are all terminating cases. Non-terminating cases contain *modulo* operators: $n \equiv 0 \pmod 3$ and $n > 2$, $n \equiv 1 \pmod 3$ and $n > 2$, and $n \equiv 2 \pmod 3$ and $n > 2$. The expressions on variable n are: $n / 3$, $2n - 2$, and $n + 1$. Note that n and m are integers, and $n \equiv m \pmod p$ is shorthand for there exists an integer r such that $n - m = p * r$.

The execution trace tree of function $F()$ is denoted $ETT(F)$. Such a tree is a pair (V, E), where V is the set of nodes or vertices, and E is the set of arcs known as edges. Note that the root remains unlabeled. Direct descendents of the root are terminating cases of the function, and all other nodes will be labeled with naturals. The arcs represent a solution from $F(x) = y$ as going from node y to node x.

We associate to any y and k, which are positive integers, a finite p-ary tree with root y, having k levels, denoted $ETT_k(y) = (V_k, E_k)$. Each node v belonging to V_k, except the root, is labeled with a natural number, denoted by $label(v)$. The node v may have 1, 2, ..., p descendants depending on its label x. That is for all i belonging to $\{0, \ldots, p - 1\}$ whenever $\beta_i(f_i^{-1}(x))$ holds, then v has the descendant labeled by $f_i^{-1}(x)$. If all the paths from the root are finite, then the termination problem can be easily solved. Namely, for the inputs that are labels of $ETT(F)$, the termination problem holds, whereas for the rest of them, the program does not terminate [2].

The modulo-case functions considered in this paper have the general pattern:

$$F(x) = \begin{cases} 0 & x = 0 \text{ or } x = 1, \\ a_0 \cdot x + b_0 & x \equiv 0 \pmod{p}, \\ \cdots & , \\ a_{p-1} \cdot x + b_{p-1} & x \equiv p-1 \pmod{p}. \end{cases}$$

The $ETT(F)$ is built using the inverse of the affine functions from the $F()$ function. Based on the modulo-case function, any arbitrary node v may have up to p descendants each. "That is, for all i belonging to $\{0, \ldots, p - 1\}$ whenever $(x - b_i) / a_i \equiv i \pmod{p}$, then v has a descendant labeled by $(x - b_1)/a_1$" [2]. For example $f_1^{-1}() = \{(m, 3m) \mid m \geq 1\} \cup \{(m, (m + 2) / 2) \mid m \geq 3 \text{ and } m \equiv 0 \pmod{6}\} \cup \{(m, m - 1) \mid m \geq 4 \text{ and } m \equiv 0 \pmod{3}\}$. The $ETT(f_1)$ is first constructed by labeling the root's immediate descendants with the termination cases 0, 1, and 2. The following levels' nodes are created based on which conditions their immediate ancestors met. Figure 1 shows the first five levels of $ETT(f_1)$.

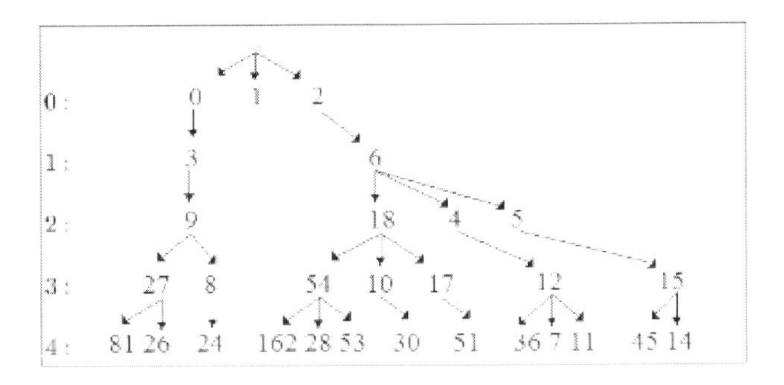

Fig. 1 The first five levels of $ETT(f_1)$.

We refine the termination method of [2]. The first step is to determine $s = \max_{i = \{0, \ldots, p-1\}} \{[1/a_i], 1\}$ and $\varphi(k) = s^k$ if $s > 1$ or k if $s = 1$, where $[x]$ represents the integer ceiling of x.

The number s is used to define $\varphi(k)$, the finite integer set, $\{0, 1, \ldots, \varphi(k)\}$, that appears in the proof. The second step is to inverse the function $f()$, and build the *ETT(f)*. The *ETT()* will be generated up to the smallest level where every label from the first set $\{0, 1, \ldots, \varphi(k)\}$ and then for the following set $\{0, 1, \ldots, \varphi(k+1)\}$. A polynomial, P, is then formed such that $P(k)$ is the smallest level of *ETT(f)* needed for a given set $\{0, 1, \ldots, \varphi(k)\}$ are labels of *ETT(f)*. Algorithm **A** shows steps of the termination method:

```
1.  Find the inverse function f^-1(), ETT(f), s, φ()
2.  d = 1; polynomialFound = false;
3.  while (true) {
4.    while (polynomialFound == false) {
5.     Find the polynomial P of degree d such that P(k) = sl_k, for all k belonging
       to {1, ..., d + 1}, where sl_k represents the smallest level such that {0, ..., φ(k)}
       labels ETT(f);
6.      if (sl_{d+2} ≤ P(d + 2)) polynomialFound = true;
7.      else d++;
8.      if (d > d_max) return 'Polynomial up to degree d_max was not found' ;
      }
9.    result = CheckInductionStep();
10.   if (result == true) {
11.    Let S be the set of naturals for which the induction step holds;
12.    if (S == N) return 'Yes' and runtime = φ^-1(P(k));
13.    else return 'Yes' for n belonging to S and runtime = φ^-1(P(k)); and
                   'No' for n belonging to S and runtime = infinity;
14.     polynomialFound = false; d++;
      }
}
```

Theorem 1. [2] Let $f()$ be a modulo-case function, and d_{max} a positive integer as the maximum degree for the polynomial. Algorithm A will provide:

- If $f()$ is terminating, then return '*yes*' and an estimation of running time;
- If $f()$ is not terminating, then return '*no*' and a domain constraint;
- Otherwise, return 'Polynomial up to degree d_{max} was not found' ■

The `CheckInductionStep()` method is the most challenging step from Algorithm A, as there is no algorithm for general termination problem [5]. Dealing with symbolic proofs represents an important area in the symbolic computation community. To the best of our knowledge, there is no fully automated algorithm able to generate proofs by mathematical induction for general programs. Next section presents a canonical form of terms useful to automatically generate the termination proof.

3 The Hybrid Exponential Normal Form

The key issue in automatically generate a proof based on mathematical induction lies in proving the induction step without an "explosion" of terms and cases. We successfully identified a large class of terms that include exponential terms and linear degree polynomials useful for proving the induction step for nontrivial modulo-case function termination. The below definition introduces the canonical form of this new class of terms, called the *hybrid exponential normal form*.

Definition 1. We say that a *hybrid exponential*, *HE*: $\mathbf{N} \times \mathbf{R} \rightarrow \mathbf{N}$, is in *normal form* if there exist a, b, and c real constants, n and k positive integers such that $HE(d, r) = a * n^{k+d} + b * r + c$. ■

This hybrid exponential normal form is not just important being an addition between an exponential and a linear degree polynomial, but it is closed under term substitution (shown in Definition 2), multiplication by a scalar, and addition by a constant.

Definition 2. We say that a hybrid exponential, *HE*: $\mathbf{N} \times \mathbf{R} \rightarrow \mathbf{N}$, is obtained by *term substitution* from the hybrid exponential, *HE'*: $\mathbf{N} \times \mathbf{R} \rightarrow \mathbf{N}$, given by $HE'(d, r) = a * n^{k+d} + b * r + c$, if there exists a hybrid exponential, *HE''*: $\mathbf{N} \times \mathbf{R} \rightarrow \mathbf{N}$, such that $HE(d, r) = a * n^{k+d} + b * HE''(d, r) + c$. ■

The other two operations over hybrid exponentials are much simpler to be defined than the term substitution. Let us consider a hybrid exponential, *HE*: $\mathbf{N} \times \mathbf{R} \rightarrow \mathbf{N}$, where $HE(d, r) = a * n^{k+d} + b * r + c$, and a scalar s, a real number. Then the multiplication by s leads to the hybrid exponential *HE'*: $\mathbf{N} \times \mathbf{R} \rightarrow \mathbf{N}$, given by $HE'(d, r) = a * s * n^{k+d} + b * s * r + c * s$. Similarly, the addition by a constant ct, a real number, leads to the hybrid exponential *HE''*: $\mathbf{N} \times \mathbf{R} \rightarrow \mathbf{N}$, given by $HE''(d, r) = a * n^{k+d} + b * r + c + ct$. Even if the above forms clearly indicate that a hybrid exponential is closed under multiplication by a scalar and addition by a constant, the closure under term substitution needs more elaborated calculations (Lemma 1).

Lemma 1. The hybrid exponentials are closed under term substitution. ■

Next section describes some of the most important implementation details, as well as the experimental results for some nontrivial modulo-case functions.

4 Experimental Results

Our implementation deals with the termination proof, by considering linear polynomials. The `CheckInduction Step()` method is implemented in Java programming language. The data structures used to store the *ETT* include two array lists, one for the numbers generated and one for the corresponding level of each generated number. More details about our tool can be found in our unpublished technical report [9].

Our Java programming language implementation makes use of the hybrid exponential normal form as a canonical way to represent the expressions for the termination proof, that is, $a * n^{k+d} + b * r + c$. According to Lemma 1, any hybrid exponential is closed under term substitution. Obviously, any hybrid exponential is also closed under addition of a constant, multiplication of a scalar. This means that by performing any of the mentioned operations to a general term, we obtain a new term of the same form. This is important because when the proof is being performed the expression will keep the same form no matter what degree the expression grows to be.

The tree $ETT(f_1)$ has been built and its first five levels were illustrated in Figure 1. The number s was determined to be 3, and $\varphi(k) = s^k$, since $s > 1$. When k is 1, the set to be found is {0, …, 3}, and when k is 2, the set is {0, …, 9}. These sets are generated by $ETT(f_1)$ levels 1 and 4, respectively. From this information we can find a linear polynomial that satisfies $P(1) = 1$ and $P(2) = 4$, that is, $P(k) = 3 * k - 2$. We initially test the induction step for the set {0, …, 27}, and check that by level 7 (that is given by $P(3) = 7$), all of the $ETT(f_1)$ labels from the new set have been generated. The coefficient of our polynomial is used to determine if termination can be proved using induction. The number of steps of the induction proof should not exceed the coefficient in the number of times that the function must be reapplied to any case of the function. Using our implementation, the termination problem holds.

5 Conclusion and Future Work

Our method generates the termination proof of a nontrivial class of programs, the modulo-case functions. The algorithm used to perform the induction proof uses symbolic expressions to evaluate the function. Our implemented tool can provide a written proof indicating termination of the function, or a partial proof stating the polynomial needs to be considered at a higher degree.

We plan as future work to extend the class of functions for which this automated method is able to provide a proof for termination.

Acknowledgments. This work was partially supported by National Science Foundation, REU site, grant #0851912.

References

1. Anderson, H., Khoo, S.C.: Affine-Based Size-Change Termination. In: Ohori, A. (ed.) APLAS 2003. LNCS, vol. 2895, pp. 122–140. Springer, Heidelberg (2003)
2. Andrei, S.: Termination Analysis by Program Inversion. In: Proceeding or the Ninth 2008 Symposium on Numerical Applications of Symbolic Computation (SYNASC 2008), pp. 26–29. IEEE Computer Society, Timisoara (2008)
3. Kelly, W., Maslov, V., Pugh, W., Rosser, E., Shpeisman, T., Wonnacott, D.: The Omega Library interface guide. Technical Report CS-TR-3445, CS Dept., University of Maryland, College Park (March 1995)

4. Lee, C.S., Jones, N.D., Ben-Amram, A.M.: The Size-Change Principle for Program Termination. In: Conference Record of the 28th Annual ACM SIGPLAN-SIGACT Symposium on Principles of Programming Languages, pp. 81–92. ACM Press (2001)
5. Rice, H.: Classes of Recursively Enumerable Sets and their Decision Problems. Transactions of the American Mathematical Society 89, 25–29 (1953)
6. Termination Portal, Termination Portal (December 2009), `http://termination-portal.org/wiki/Termination_Portal`
7. Turing, A.: On Computable Numbers, with an Application to the Entscheidungs Problem. Proceedings of the London Mathematical Society 42, 230–265 (1936)
8. Weimer, W., Forrest, S., Le Goues, C., Nguyen, T.V.: Automatic Program Repair with Evolutionary Computation. Communications of the ACM 53, 109–116 (2010)
9. Doss, K., Andrei, S., Makki, K.: An Automatic Induction Proof for Program Termination Analysis, Technical Report, No. 2, Lamar University, pp. 1–11 (August 2010)
10. Cook, B., Podelski, A., Rybalchenko, A.: TERMINATOR: Beyond Safety. In: Ball, T., Jones, R.B. (eds.) CAV 2006. LNCS, vol. 4144, pp. 415–418. Springer, Heidelberg (2006)

Testing of Loop Join Points Using AspectJ

Mutum Zico Meetei, Anita Goel, and Siri Krishan Wasan

Abstract. Testing is an important phase during the development of software. Testing is performed at different levels of abstraction – unit, integration and system levels. In this paper, we use Aspect-Oriented Programming (AOP) using AspectJ, for automated testing of Java program at unit level. This paper focuses on testing the execution details of loop join point. We use variables of type static for the testing of loop join point. We observed that using an aspect (a modularizing unit) improves the overall understanding of loop join points, makes the crosscutting concern explicit and reduces code tangling.

Keywords: Unit testing, Loop Join Point, AOP, AspectJ.

1 Introduction

Unit testing is a process of testing the individual subprograms, subroutines, or procedures in a program [4]. A class is the unit of testing in object-oriented software. Testing of a class involves testing of the methods defined in the class, and their impact on the state of the object. The state of the object needs to be observed both before and after the execution of the method. The encapsulation feature of object-oriented software hides the data structure, posing difficulty in accessing the state of the object, which is essential for verifying the correctness of processing and for error diagnosis.

In this paper, we use AspectJ [6, 1], an implementation of Aspect-Oriented Programming (AOP) [7] for Java [11] programming language to perform testing at unit level. AOP encapsulates the crosscutting concerns in an *aspect* - a modular unit of crosscutting implementation. A "join point" defines an event in a system where a crosscutting concern can be joined with the core concern at compile time or run time. Many points in the execution flow of the program can act as join

Mutum Zico Meetei · Siri Krishan Wasan
Department of Mathematics, Jamia Millia Islamia, New Delhi, India

Anita Goel
Department of Computer Science, Delhi University, New Delhi, India

Z. Qian et al. (Eds.): Recent Advances in CSIE 2011, LNEE 125, pp. 461–467.
springerlink.com

points. AspectJ deliberately exposes only a subset of all possible join points like constructor/method call, constructor/method execution, field set/get points, object/ class initialization and exception handling points. AspectJ doesn't expose *for* loops or *if* statement [9]. This paper focuses on testing the loop join points.

Here, we propose a technique for unit testing of *for* loops and *if* statement in Java program. We compare the execution details when the variables are defined as static type and non-static type. We are able to trace the execution details of the loop join points when the variable is of static type. Defining an a*spect* has helped us, as it weaves with the Java code, leading to clean and non-tangling code. We use eclipse [1] framework, for supporting AspectJ code.

In this paper, Section 2 presents a brief definition of AOP. Section 3 discusses unit testing. Section 4 discusses the loop join points. Section 5 describes the aspects for loop join points. Sections 6 discuss the related work. Section 7 states the conclusion.

2 Aspect-Oriented Programming

AOP is a new programming paradigm, which is recently popular for the use of modularizing concerns that cross-cut the basic functionality of a program [8]. Typical crosscutting concerns include logging, testing, performance optimization and design pattern. AOP provides a language mechanism that explicitly captures crosscutting concerns, which makes it possible to crosscut other concerns in a modular way. The most popular AOP is AspectJ - an aspect oriented extension to Java. It introduces some new constructs such as *introduction, join point, pointcut, advice* and *aspect,* for better handling of crosscutting concerns.

AspectJ supports join points for method execution, method call and field access, but, *for* loops and *if* statement are also join points. AspectJ cannot intercept the iterations of *for* loop. However, our technique is able to trace the internal execution details of the loop join points, such as *for* loops and *if* statement.

3 Unit Testing

Unit testing requires testing of each unit (module) of a program in order to verify that the detailed design for the unit has been correctly implemented. The intention for unit testing is to manage the combined elements of testing, since attention is focused initially on small units of the program. There are two types of unit testing – specification-based testing (black-box testing) and program-based testing (white-box testing). Our technique focuses on conditional coverage - testing loop statement and conditional statement. Testing *for* loops and *if* statement are important during unit testing, as they make path testing difficult due to the significantly increased number of possible paths.

3.1 Unit Testing Using Aspects

Over the years, there exists a number of testing tools for Java programs. However AOP provides a new challenge and approach to the existing tools such as Junit and Jtest. The advantage of using AspectJ is that, it addresses the problem of separation of concerns in programs, which is well suited to the unit testing. Our testing approach uses AspectJ for observing the internal execution details during testing of java program. The execution consists of the following details-

- Class being executed
- Method call or execution of the class
- Line number of the codes
- Control flow of the loops and *if else* statements

The idea behind our testing technique is simple; we use an *aspect* to observe the internal details of execution during testing. Aspect is defined without affecting the source code during software development and is woven with the source code during compilation.

4 Loop Join Points

A join point is an identifiable execution point in a system. Join points in AspectJ include method call, constructor call, method call execution, constructor call execution, field get, field set, exception handler execution, class initialization, and object initialization.

Initialization of class with static variables is used to perform class-level crosscutting. Class initialization join points represent the loading of a class including the initialization of the static parts. Object initialization join pint is defined when a dynamic initializer is executed for a class. This join point occurs only in the first called constructor for each type in the hierarchy. Unlike class initialization that occurs when a class loader loads a class, object initialization occurs when an object is created. Constructor call execution join points are similar as the method call join points except that it deals with constructors. This join point is triggered before the constructor starts execute. It is defined when the constructor is called on the object and occurs after constructor call join point. Field set join points are defined when attributes associated with objects are written. However initialization of constant fields is not join points. The field's write access join point encompasses the constructor call execution; field's read access encompasses reading the field as part of creating an object.

A loop join point includes *for* loop and *if* statement. Loops do not have arguments as that of other join points such as method calls and executions; they often depend on contextual information to which a programmer may want to access. Every loop statement has a unique entry point and exit point; *before* advice can be woven with the entry and the *after* advice with the exit point. However the *around* advice may not apply to loop join point.

5 Aspect for Loop Join Points

In this section, we present our aspect code for checking loop join points. Aspects are composed of point cuts, advice and intertype declarations. Aspects on high level can be dividing into two types - development aspect and production aspect. The development aspect is used during initial development to support testing, debugging and performance tuning.

Our technique uses aspect for testing loop join points. For this, we define the variables to be type *static*. We use *privileged* aspect for accessing the class private/ protected attributes and methods. This kind of access is required for the testing at unit level; **privileged aspect** autotrace{ Advice code }. We illustrate our testing technique with an example of a simple Java program showing loop join points-*for* and *if* statements, as shown in Fig. 1.

```
 4 public class firsttest {   public static int  s, i ;
 6 public  static void main(String arg[]){. . . . . . .
10        int n = in.nextInt (); int   k ;
12        for (i= 0; i<n; i++){  s=+i ; //Line number 13
14        System.out.println( "the  sum is "+s); }
16        if (i  >5){
17        k=+i;
18        i++;  System.out.println ( "sum!   = "+k); }
```

Fig. 1 Example of Java program using *for* and *if* statements

For checking preconditions we use *before* advice, that would execute before the specified pointcut executes. The *before* keyword exposes variable names (types) that can be used in the advice code. We define level number L3, within the advice body so as to identify the execution detail at unit level. The *after* keyword is used for the postcondtions. The advice code template for checking preconditions and postconditions for loop join points is shown in Fig. 2.

```
Pointcut PL2():!within ( autotrace)&&!execution(* *.*(..))
&& !execution(* .new (..)); . . . . . . . . . . . . .
before():PL2(){printUnitLevel(thisJoinPoint);}. . . .
after ():PL2 ()  {    printUnitLevel (thisJoinPoint);  }
```

Fig. 2 AspectJ codes for Precondition and Postcondition

The pointcut *PL2 ( )* specifies, using a method signature, that calls of any method on any instance of class is intercepted. The *after* advice is used to intercept a method after the execution or calls. In the advice code; *!within(autotrace)* specifies that the version of method calls is not within the aspect *autotrace*.

5.1 Tracing Loop Join points

Our technique uses tracing aspect for observing the internal execution details during testing of loop join points. We use a set of interfaces - *thisJoinPoint, thisJoinPointStaticPart, thisEnclosingPointStaticPart* to display the hierarchy of execution of class, parameter values of methods entry/exit, and dynamic and static information of join points. The code template for the interface used for unit testing are shown in Fig. 3.

```
private void printUnitLevel (JoinPoint jp ){ println ( "L3/: " +jp);
StringBuffer argst =new StringBuffer (" L3/ARGS:\t");
Object []args =jp.getArgs();
for (int length =args.length , i = 0 ;i <length ;  ++i){
argst.append (" [" + i +  "] ="  + args[i]); }println (argst);
SourceLocation SL =jp.getSourceLocation();
println  ("  L3/Source  :    "  +SL.getFileName  ()  +  ";Line  no:"
+SL.getLine())
```

Fig. 3 Code template for interface

Our technique uses a simple aspect for testing the internal execution details at unit level. During unit testing, the output and input of the methods and the impact of the method execution on the state of the object is observed. This includes observing the loop join points. We are able to trace execution details of loop join points when the variable is defined as static type. We compare the internal execution details of the loop join points when the variables are defined in static and non static type. The level number *L3* indicates the testing at the unit level. Fig. 4 shows the output of *for* loop and *if* statement when the variable is defined as static type. Fig. 5 shows the execution details of the Java program defined in Fig. 1.

```
L3/Source :  firsttest.java;Line no:12
L3/call(StringBuilder  java.lang.StringBuilder.append(int))
L3/Source :  firsttest.java;Line no:14
L3/get(int firsttest.i)     L3/ARGS:
L3/Source : firsttest.java;Line no:16
L3/get(int firsttest.i)      L3/ARGS:
L3/Source :  firsttest.java;Line no:17
L3/set(int firsttest.i)    L3/ARGS: [0] =17
L3/Source :  firsttest.java;Line no:18
L3/call(java.lang.StringBuilder(String))
```

Fig. 4 Internal execution details of loop join point in static type

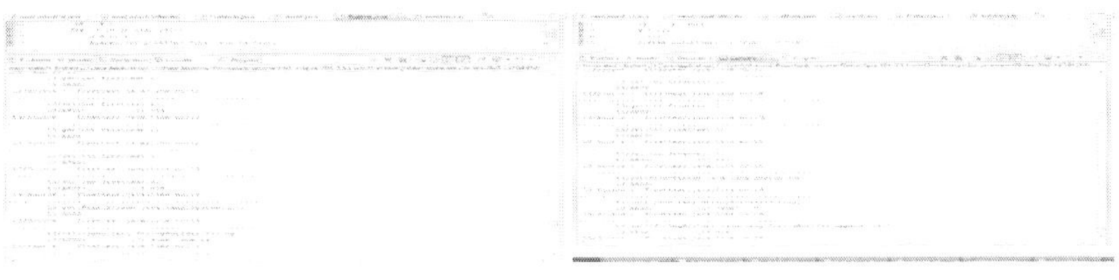

Fig. 5 Output of execution details of loop join points using AspectJ

5.2 Code Coverage

All the methods in Java program defined in Fig. 1 are traced, which help us to generate coverage of method, class, loop join points and inheritance. The coverage report can be viewed as in percentage coverage of the line numbers and loop join points traced by the aspect. The report helps us to gather details of the executions efficiently when the variable is defined in static type. Table 1 shows our test coverage report.

Table 1 Coverage at Method, Class and Loop Join Points.

Static Type		*Non-Static Type*	
Uncovered Line no:	*Uncovered Loop Join Points*	*Uncovered Line no:*	*Uncovered Loop Join Points*
Line no:5 , 11	*None*	*Line no:7,11,12,13,16, 17,18*	*for (i=0; i<n; i++), if (i>5)*
coverage: 85.71%	*%coverage: 100%*	*%cover-age: 50%*	*%coverage : 0%*

6 Related Works

Harbulot and Gurd [5] present a loop join point model which demonstrates the need for a more complex join point in AspectJ. However this research lacks analysis of infinite loops that contain Boolean conditions. They present LoopAJ that provides a point cut for selecting loop join points, and allows parallelizing the execution of the specified loop body. Xie et al. [12] proposed a synchronized block join points. The contribution of their work is to enable selecting synchronized statements as join points.

Delamare et al. [2] present Vidock-a tool for impact analysis of aspect weaving on test cases. They perform a static analysis of Java program that identifies that subset of test cases impacted by the aspect weaving. Seyster et al. [10] present a framework for GCC – a widely used compiler infrastructure to invoke each join point. It helps to customize the inserted instrumentation and to leverage static analysis. They show its applications over heap visualization, integer range analysis and code coverage. Farhat et al. [3] presents two categorization of non functional requirement and show how each type requires a different testing strategy using aspects. They extrapolate techniques of testing functional requirements and suggest complementary methods to test NFRs of object-oriented system using aspects.

7 Conclusion

In this paper, we present a testing technique for testing the loop join points in Java program using AspectJ. We describe tracing the loop join points using AspectJ.

This is possible when the variable is defined of static type. It facilitates to observe the internal execution details of the loop join point, such as *for* loop and *if* statement. We also compare testing coverage when the variable is defined both as static as well as non static type.

References

[1] AspectJ (2010), http://www.eclipse.org/aspectj/
[2] Delamare, R., Munoz, F., Baudry, B., Traon, Y.L.: Vidock: a Tool for Impact Analysis of Aspect Weaving on Test Cases. In: International Conference on Testing Software and Systems. IFIP, Brazil (2010)
[3] Farhat, S., Simco, G., Mitropoulos, F.J.: Using aspects for testing nonfunctional requirements in object-oriented systems. In: Proceedings of the IEEE Southeast Con., Concord, NC, USA, March 18-21, pp. 356–359 (2010); ISBN: 978-1-4244-5854-7
[4] Myers, G.J.: Art of Software Testing, 2nd edn. John Wiley & Sons,Inc., New Jersey (2004)
[5] Harbulot, B., Gurd, J.R.: A join point for loops in aspect. In: AOSD: Proceedings of the 5th International Conference on Aspect Oriented Software Development, pp. 63–74. ACM, New York (2006); doi:10.1145/1119655.1119666
[6] Kiczales, G., Hilsdale, E., Hugunin, J., Kersten, M., Palm, J., Griswold, W.G.: An Overview of AspectJ. In: Lee, S.H. (ed.) ECOOP 2001. LNCS, vol. 2072, pp. 327–353. Springer, Heidelberg (2001)
[7] Kiczales, G., Lamping, J., et al.: Aspect-Oriented Programming. In: Aksit, M., Auletta, V. (eds.) ECOOP 1997. LNCS, vol. 1241, pp. 220–242. Springer, Heidelberg (1997)
[8] Meetei, M.Z., Goel, A., Wasan, S.K.: An overview of Logging with Aspect Oriented Programming. Accepted for publication in WCSE, Xiamen, China (May 19-21, 2009)
[9] Laddad, R.: AspectJ in Action: Practical Aspect Oriented Programming, 2nd edn. Manning Publication, USA (2009)
[10] Seyster, J., Dixit, K., Huang, X., Grosu, R., Havelund, K., Smolka, S.A., Stoller, S.D., Zadok, E.: Aspect-Oriented Instrumentation with GCC. In: Barringer, H., Falcone, Y., Finkbeiner, B., Havelund, K., Lee, I., Pace, G., Roşu, G., Sokolsky, O., Tillmann, N. (eds.) RV 2010. LNCS, vol. 6418, pp. 405–420. Springer, Heidelberg (2010)
[11] Sun Microsystems, Java 2 plateform standard edition v 1.4.2 API specification (2010), http://java.sun.com/j2se/1.4.2/docs/api/
[12] Xi, C., Harbulot, B., Gurd, J.R.: Aspect-oriented support for synchronization in parallel computing. In: PLATE 2009: Proceedings of the 1st Workshop on Linking Aspect Technology and Evolutin, pp. 1–5. ACM, New York (2009); doi:10.1145/1509847.1509848

The Research of Determining Class Testing Sequence

Dan Liu, Junhui Zhang, Hongyu Zhai, Li Li, Liu Liu, and Xiaohui Yang

Abstract. Determining class testing sequence is an important part of Object-orient-ted integration testing. It influences the object class bunch of testing very much.T-his article mainly narrates a method of determining class testing sequence, using t-he object chart of class to determine class's in and out degree and considers the cl-ass's relation of Dynamic dependence and indirect dependence . By comparing the in and out degree of class, the sequence is decided.

1 Introduction

In the object-oriented programming, the interdependence of is extremely close, is unable to carry on the class testing radically in the translation incomplete program. Therefore Object-oriented integration testing usually be carried on after the translation completed. And The object-oriented program has the dynamic characteristic while the control of program so that only to do integration testing based on black box's on the entire translation program. But the traditional integration testing is through various of integrated strategy to complete various functional module, generally may happens after the subprogram translation completes in the situation. Therefore the traditional integration testing was not already suitable for the object-oriented integration testing.

In the object-oriented integration testing ,class testing sequence is a very important aspect. Until now there have many researches of class of testing sequence's determination. In literature [1-2] proposed one method of the use object relational graph determination class testing sequence. But the method has some shortcutting. And in literature [3] changes the shortcoming of literature [1-2] by using the method of union the design pattern and Object relational graph. But object-oriented pattern relational graph testing sequence production method existence certain insufficiency which can't vivid performance of the class's dependence. Only analysis the class's static relation but lack the dynamic class dependence

Dan Liu · Junhui Zhang
School of Computer Science and Technology, Changchun University of Science and Technology, No. 7089WEIXING Road, Changchu, Jilin, 130022, P.R. China
e-mail: ld_1983@hotmail.com, zhang_junhui@126.com

Z. Qian et al. (Eds.): Recent Advances in CSIE 2011, LNEE 125, pp. 469–474.
springerlink.com

analysis. Literature [4] regards as the object chart an oriented graph, using its out and in-degree carries on the analysis, but there are also some shortcomings. This article proposes one method to determine class testing sequence through unifies the above two methods.

2 Basic Definition

The object relational graph is an oriented graph which takes class as a node, the dependence of classes as side. There are three basic relations in Object-oriented such as Inherits, association, aggregation. Above all is direct dependence. But possibly there has the indirect dependence and the dynamic dependence in the object-oriented model. Literature[5] had pointed out explicitly the dynamic dependence's relations so-called dynamic dependence may infer from the static relations. If class Y is the service class of class X .Therefore in the program is executing, class X and all direct either the indirect subclass dynamic relies on Y and all direct or the indirect subclass.

Definition 1. Supposes G =<V,E> is an oriented graph corresponds to an object relational graph, V stands for the apex, and E stands for dependence relations between classes in oriented graph. Classes Set is $T=\{T_1, T_2, \ldots, T_n\}$.

Definition 2. The dependence of Integration testing class is in oriented graph direct to all class of dependence set: $C_I = \{ C_I\ 11 ,\ldots C_I\ 1n ,\ldots C_I\ 21 ,\ldots C_I\ 2n ,\ldots C_I\ ij ,\ldots C_I\ nn \}$, $C_I\ ij= 1$ stands for class T_i depends on class T_j; and $C_I\ ij= 0$ stands for class T_i and class T_j have no relations.

Definition 3. The being depended relations of Integration testing class is be directed to all class of dependence set: Co = { Co 11 ,...Co 1n ,...Co 21 ,...Co 2n ,...Co ij ,...Co nn } , $C_I\ ij= 1$ stands for class T_i depends on class T_j; and $C_I\ ij= 0$ stands for class T_i is not depended by class T_j .

Definition 4. The set which depends class Ti is: $CI(Ti) = \{ CI\ ij \mid CI\ ij= 1, Ti \in T, CI\ ij \in CI \}$, The class set which class Ti depends is : $Co\ (Ti) = \{ Co\ ij \mid Co\ ij= 1, Ti \in T, Co\ ij \in Co \}$.| CI(Ti) | which is in-degree named as divI(Ti) stands for the count of the class which is depended by class Ti(Ti ∈T). | CI(Ti) | which is out-degree named as divO(Ti) stands for the count of the class which is depend class Ti(Ti ∈T).

Definition 5. Some class T_i in oriented graph if $divI(T_i) = 0$. And this class is named independent class. This class has higher priority than other classes.

Definition 6. M is the set of class testing sequence. $M = \{M_1,M_2,\ldots,M_j\ldots,M_n\}$. If $i < j$ class M_i has higher priority than class M_j. $1 < i < j < n$.

3 The Method of Producing Class Testing Sequence

3.1 Main Step

Step1: The class testing sequence's point is to eradicate ring circuit in classes. If the sum of class A's out-degree and in-degree is max to the sum of class B's out-degree and in-degree, class A has higher priority than class B. Therefore we can delete the relation of class A to class B while using Stub(B)instead of class B.

Step2: Extracting the out-degree of all classes. If some class's out-degree is 0,the class is main class which has higher priority. This kind of class judging it's in-degree, if the class's in-degree is higher, it has higher priority and put it into set M. If its in-degree is the same, we sort them by according to its indirect dependence relation and the dynamic dependence relation.

Step3: Deleting the class in set M. If other classes depend above class, setting this class's out-degree from 1 to 0. Repeat step2 until set M is null.

3.2 The Example of Producing Class Testing Sequence

Class set T ={A, B,C,D,E,F,G,H,K,L}
(1) Fig. 1 is an example of object relational graph. There are ring circuit in oriented graph. Firstly we must reduce the ring circuit. In Fig. 1: There is ring circuit between class F and class H . The sum of class F is :

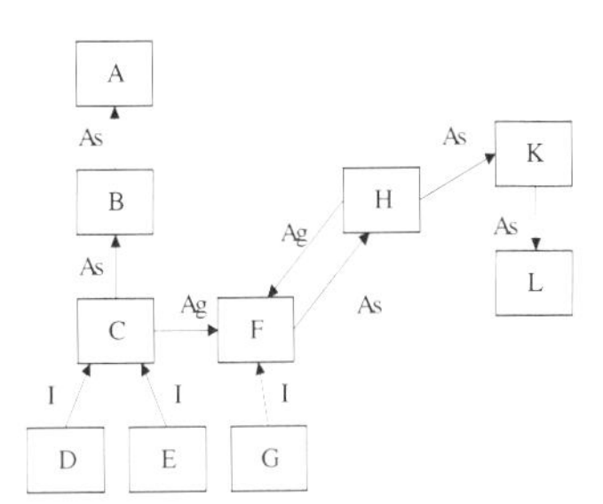

Fig. 1 Object relational graph example

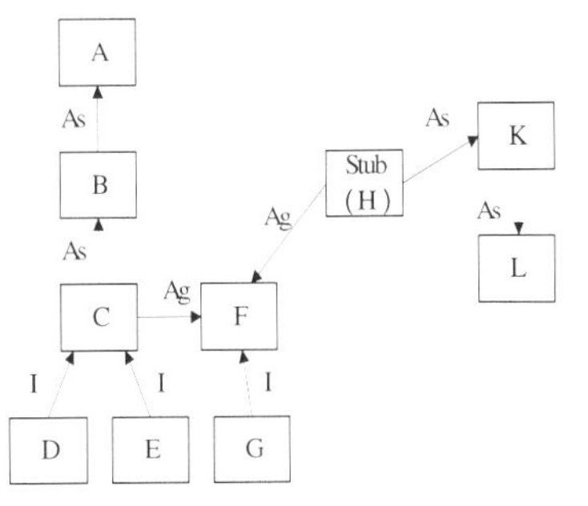

Fig. 2 Eradicates ring circuit's object relational graph

$$C_I(F)\ |+|\ Co\ (F)\ |=3+1=4, \quad (1)$$

The sum of class H is:

$$|CI(H)\ |+|\ Co\ (H)\ |=1+2=3, \quad (2)$$

because 4>3s,class F has higher priority than class H, deleting the depended relation between class F and class H, using Stub(H)instead of class H so that the relational graph is Fig. 2. From Fig. 2 we can know the non-direct dependence and dynamic dependence relation using the following Table 1 and Table 2 stands for the dependence of classes.

Table 1 Non-direct dependent and dynamic dependence relations of classes

Class name	Non-direct dependent relations	Dynamic dependence
C	A	G
D	B	G,F
E	B	G,F
Stub(H)	L	

(2)From Table 1: Co(A)= Co(L)= Co(F)= 0 , and $C_I(A) = C_I(L)=1$, $C_I(F)= 3$. Class F's priority is higher. The non-direct dependent relations is the same which depended class A and class L. So that we can put the two classes into the set M , M = M = {F, A, L}. And deleting class F,A, L form relational table while we get Table 3.

Table 2 Direct dependent relations of classes

	A	B	C	D	E	F	G	Stub(H)	K	L	Co
A	0	0	0	0	0	0	0	0	0	0	0
B	1	0	0	0	0	0	0	0	0	0	1
C	0	0	0	0	0	1	0	0	0	0	2
D	0	0	1	0	0	0	0	0	0	0	1
E	0	0	1	0	0	0	0	0	0	0	1
F	0	0	0	0	0	0	0	0	0	0	0
G	0	0	0	0	0	1	0	0	0	0	1
Stub(H)	0	0	0	0	0	1	0	0	1	0	2
K	0	0	0	0	0	0	0	0	0	1	1
L	0	0	0	0	0	0	0	0	0	0	0
C_I	1	1	2	0	0	3	0	0	1	1	

Table 3 Relational table of deleting class A,L

	B	C	D	E	G	Stub(H)	K	Co
B	0	0	0	0	0	0	0	0
C	1	0	0	0	0	0	0	1
D	0	1	0	0	0	0	0	1
E	0	1	0	0	0	0	0	1
G	0	0	0	0	0	0	0	0
Stub(H)	0	0	0	0	0	0	1	1
K	0	0	0	0	0	0	0	0
C_I	1	2	0	0	0	0	1	

(3)From Table 3:Co(B) = Co(K)= Co(G) = 0 ,C_I(B) = C_I(K) = 1. Class B is indirect depended by class D, E so that class B's priority is higher than class K.C_I(G)= 0 .Class G has the lowest priority. M={ F,A, L,B,K, G}. Deleting B,K,G from Table 3 while we has Table 4.

Table 4 Relational table of deleting class B,K,G

	C	D	E	Stub(H)	Co
C	0	0	0	0	0
D	1	0	0	0	1
E	1	0	0	0	1
Stub(H)	0	0	0	0	0
C_I	2	0	0	0	

(4)From Table 4:Co(C) = Co(Stub(H))= 0 ,C_I(C) =2, C_I(Stub(H)) =0 .Class C's priority is higher than class H so that M={ F, A, L,B,K,G,C,Stub(H)}.We get Table 5.

(5)From Table 5,Co(D) = Co(E) = 0,C_I(D) = C_I(E) = 0 and through Table 1. The result is M={ F,A,L ,B,K,G,C,Stub(H),D,E}.

(6) According to Table 1 we can get the set that M={F,A,L,B,K,G,C, Stub(H),D,E}which equals M={ F,A,L,B,K,G,C,H,D,E}.

Table 5 Relational table of deleting class C, Stub（H）

	D	E	Co
D	0	0	0
E	0	0	0
C_I	0	0	

4 Concluding Remark

Class testing sequence's determination is an important question in an object-oriented integration testing .This article is proposed in the object-oriented pattern relational graph testing sequence production method and based on the complex network object-oriented integration testing method, considering the non-direct dependence and the dynamic dependence relations in classes, but possibly has certain insufficiency in the class testing sequence's determination's efficiency, next step will carry on the correction.

References

[1] Kung, D.C., Gao, J., Hsia, P., et al.: Class firewall, test order, and regression testing of object- oriented programs. Journal of Object-Oriented Programming 8(2) (1995)

[2] Tai, K.C., Daniels, F.J.: Test order for inter- class integration testing of object-oriented software. In: Proceedings of the CompsAC 1997-21st International Computer Software and Applications Conference, pp. 602–607 (1997)

[3] Gao, J., Qi, L.: The method based Object-oriented pattern relational graph of testing sequence production. Shanghai Normal University (Natural Sciences) 28(3) (1999)

[4] Li, L.: Object-oriented integration testing based on the complex network. Shanghai Second Polytechnic University(20090708)

[5] Li, C.: Object-oriented integration testing based on expansion object relational graph. Anhui Broadcast Film and Television Professional Technology College (2008); 1009-3044(2008)14- 20938- 02

Twice Rewritings to Reduce Test Case Generation with Model Checker*

Ye Tian, Ying Chen, and Hongwei Zeng

Abstract. Generating test cases with a model checker provides an effective means to perform test automation. However, there are usually much redundant calls to the model checker such that degrade the performance of test case generation, and also the generated test suit is often much redundant. In this paper, we propose a reduction approach to test case generation by using property rewriting technique. The approach includes twice rewritings: one is for eliminating redundant test goals represented in LTL properties, and the other for eliminating redundant test cases. A simple example, withdrawing money from ATM, is employed to illustrate our approach.

1 Introduction

Model checking has gained a great popularity recently in the field of software testing for reason of generating test cases automatically. However, the executions of test sets take up CPU time and memory [1]. Besides, once a new test started, the additional cost in resetting the initial state increased for several activities were invoked in this course [1, 2]. Therefore we are confronted with the problems of massive test cases and high cost in model checking executions. Emphasis is placed on the reduction of test case generation by applying the method rewriting to property sets which we will investigate in our paper.

Ye Tian · Ying Chen · Hongwei Zeng
School of Computer Engineering and Science
Shanghai University
Shanghai 200072, China

* This work is supported by National Natural Science Foundation of China under grant No. 61073050, the Natural Science Foundation of Shanghai Municipality of China under Grant No.09ZR1412100, and Shanghai Leading Academic Discipline Project, Project Number: J50103.

Z. Qian et al. (Eds.): Recent Advances in CSIE 2011, LNEE 125, pp. 475–481.
springerlink.com

The process of using model checker to test the software can be described as follows: The model constructed from specifications can probably be an FSM/EFSM or DFD which illustrates the relations between states and guard conditions of transitions. Properties are usually characterized by LTL/CTL syntax to adapt model checker [3]. The model checker checks whether the model satisfies the properties or not. A counterexample shows a single available computation path along the model of the system which reflects the relations between each state [8] such that a test case can be constructed. To trap the model checker to generate a counterexample, we adopt the trap property [7] which is the negation of a property describing a test requirement. The main theme of this paper is to reduce the properties and the test cases. According to test cases generated previously we can optimize the test sets with certain methods such as using test-tree [4]. However, our main strategy of optimization is that omit the properties or test cases which are covered by the new generated test sequences. Whereas there is little evidence and comparison to say achieving the best test suit for there is no standard to weigh the level "best" with different techniques in complex circumstances [9].

Next section, the concept of rewriting [6] involving possibility knowledge is investigated. In section 3, we demonstrate a simple example with EFSM to yield test cases with the model checker SMV [8], we propose an approach to the reduction of the test cases and the properties. Finally, we draw the conclusions and put forward our future work.

2 Preliminaries

The model checker SMV provides a language for describing the models and can check the validity of LTL/CTL formulas on those models. In general, test cases generated automatically by model checker can be a mass. We considered two aspects: 1) reduce dynamically property sets to cut down the cost and time in calling model checker. 2) Reduce the number of test cases in the test suit [5].

The model checker is called to generate a test case for a property, then the generated test case is utilized to rewrite other properties , if the rewriting result for some property ϕ is false, we conclude that ϕ is covered by the test case and it can be deleted from the property sets.The definition of the state rewriting is introduced below, where ϕ is the objective property intended to be rewritten. s indicates the current state of the test case to evaluate the property.

$$(G\phi)\{s\}=\phi\{s\}\wedge G\phi \tag{1}$$

$$(\phi_1\wedge\phi_2)\{s\}=\phi_1\{s\}\wedge\phi_2\{s\} \tag{2}$$

$$(\phi_1\vee\phi_2)\{s\}=\phi_1\{s\}\vee\phi_2\{s\} \tag{3}$$

3 Test Case Generation

As an example, withdrawing money from ATM specified in EFSM is used to demonstrate the feasibility of applying rewriting method to test generation. There are four states in this simple ATM model which can be seen in figure 1.

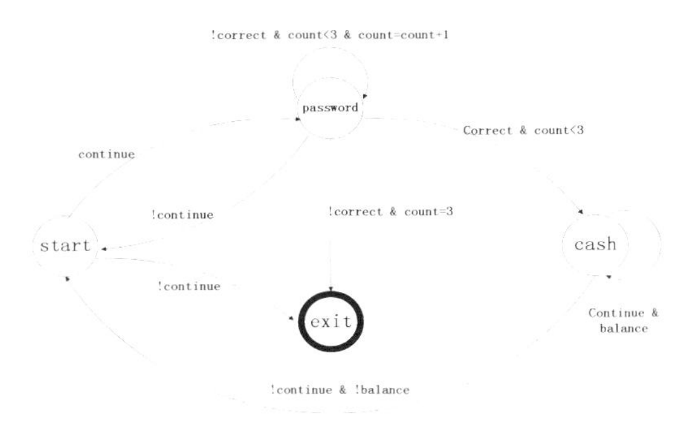

Fig. 1 When entering the wrong password exceed 3 times, it has to fail to the *exit* state. We use guard condition *count* to record the times of entering password and *balance* to determine whether the credit card is overdraw. *Continue* is a Boolean value that represents whether the client wants to continue the operations in ATM. *Correct* indicates the right password.

To generate test cases in accordance to a given test object, we first create suitable trap properties. For a transition $s_1 \xrightarrow{g} s_2$,its corresponding trap property can be represented in the formula $G(state = s_1 \,\&\, g \rightarrow X(state != s_2)$, saying that there is not any transition from s1 to s2 in case of the guard condition g holding.

On the principle of trap property definition above, we formulate the trap property based on LTL as G(state = start & continue ->X (state! = password)).Trap properties depicted for the whole model are listed in table 1.

Table 1 The trap properties given by the model

	Trap property
P1.	G (state = password & !continue ->X (state !=start))
P2.	G (state = start &!continue ->X (state !=exit))
P3.	G (state = start & continue ->X (state != password))
P4.	G (state = password & !correct & count <3 ->X(state !=password))
P5.	G (state = password & !correct & count =3 ->X (state !=exit))
P6.	G (state = password & correct & count <=3 ->X (state !=cash))
P7.	G (state = cash & continue & balance ->X(state !=cash))
P8.	G (state = cash & (!continue | !balance) ->X (state !=start))

Table 2 shows the details of the test sequences with evaluated values and states generated from the 3 trap properties.

Table 2 Parts of the test cases generated from p1 to p4

property	continue	correct	count	balance	state
p1	1	0	0	0	start
	0	0	1	0	password
	0		1		start
p2	0	1	0	0	start
	0		0		exit
p3	1	0	0	0	start
	0		1		password

By using the model checker, 8 test cases are created as counterexamples to show violations. Before the utilization of the rewriting, the result shows above indicates inexplicitly that there may exist some redundant or similar test cases and the rewriting should be introduced to advance test suit.

4 Twice Rewritings

In this section, an algorithm for test reduction using twice rewritings is developed. The algorithm of the twice rewritings is shown in figure 2. The first time rewriting starts from property 1 to the last, when the rewriting result comes to false and the corresponding property is deleted, the time and the cost of calling SMV will be cut down. Then the second time rewriting is called to minimize the test suit on the presupposition of changeless coverage simultaneously.

```
  //The arithmetic of the first time rewriting:
  firstrewriting( p[],c[])
{
       i=0;
  while ( p [i] )
      {
   call SMV to generate test case c[i];
        for ( i+1 ; i<=n ;i ++)
          {
             rewriting ( p[i+1] );
             if (rewriting ( p[i+1] )=false)
             delete p[i+1];
          }
                i++;
      }
  }
//The arithmetic of the second time rewriting:
   secondRewriting( p[],c[])
{
       j=n;
  while( c[j] )
     {
        for ( j ; j>0 ; j-- )
          {
            rewriting ( p[j-1] );
         if (rewriting ( p[j-1] )=false  & p[j-1])
           delete c[j-1];
          }
                j--;
     }
}
//show the result:
   While (c[j])
      {
        Show c[j];
        j++;
       }
```

Fig. 2 Shows the algorithm of twice rewritings. Two arrays p[i], c[j] are constructed to characterize the sequences of the properties and test cases. *i* and *j* indicate the index of the property and a test case respectively. Finally the algorithm output the reduced test Suit.

4.1 First Time Rewriting

In the first time, we choose one of the properties as a testing property to first generate a rewriting test case to rewrite all the other properties which we called objective properties. There is no need to monitoring all the properties to generate test cases at the first time, for the method rewriting will help us reduce some duplicated or redundant objective properties in first turn and the next testing property may probably have been already deleted in previous process.

In this way, we are able to save the time and cost in executing process. Model checker generate the test case by monitoring testing property p1 and we record that test case c1 for rewriting. The rewriting result shows that there is no relation between property 1 and 2 otherwise we can get the false result. Then we choose the next objective property p3 (G (state = start $\wedge$ continue ->X (state $\neq$ password))) and use the same method to rewrite it with the test case c1 generated by testing property p1. The process can be easy to understand.

G (state = start $\wedge$ continue ->X (state $\neq$ password)) $\{s_0\}$
= ((state=start $\wedge$ continue ->X(state $\neq$password))$\{s_0\}$ $\wedge$ ϕ)
=(((state=start$\wedge$continue)$\{s_0\}$->X(state$\neq$password))$\{s_0\}$$\wedge$$\phi$)
= (true -> (state $\neq$password) $\wedge\phi$)=(state $\neq$password) $\wedge\phi$)

The second state s_1 is continue = 0, correct = 0, count = 1, balance = 0, state = password. Then keep on rewriting:

$\phi_1$$\{s_0\}$ = ((state $\neq$password)$\wedge\phi$) $\{s_1\}$
= ((state $\neq$password) $\{s_1\}$ $\wedge\phi\{s_1\}$)
=false $\wedge\phi\{s_1\}$
=false

In s_1 state, state $\neq$password is false, the whole rewriting is false. It is proved that objective property ϕ is covered by the test sequence. We can draw the conclusion that the testing property p1 covers property p3 so that it can be eliminated from the property sets. Then we keep on rewriting p4, p5, p6, p7, p8 with the testing property p1 and none of these properties is reduced in this rewriting turn.

4.2 Second Time Rewriting

Our main goal is to reduce the test cases, the reduction of properties in first time rewriting is obviously not enough to meet our demand of optimization, so that we will do a second time rewriting to further reduce the test suit. For fear of the influence of the first time rewriting, we rewrite these test cases reversely from the last test case generated by p8. All the properties and the test cases are showed in table 2. Our strategy is that first we use c8 to rewrite other properties from p7 to p1.

During the process of rewriting, it can be found that rewriting p6 with c8 can obtain the result false, which means test case generated from p8 can fully cover the objective property p6, so does p4 with c5.

Table 3 Twice rewritings results compared with the original result

property	Trap property	TC	sequence
p1	G (state = start & continue ->X (state != password))	c1	<start,password,start>
p2	G (state = start &!continue ->X (state !=exit))	c2	<start,exit>
p3	~~G (state = password & !continue ->X (state !=start))~~	c3	
p4	G(state=password & !correct & count <3- >X(state !=password))	c4	~~<start,password,password>~~
p5	G (state = password & !correct & count =3 ->X (state !=exit))	c5	<start,password,password,password,password ,exit>
p6	G (state = password & correct & count <3 ->X (state !=cash))	c6	~~<start,password,cash>~~
p7	G (state = cash & continue & balance ->X(state !=cash))	c7	<start,password,cash,cash>
p8	G (state = cash & (!continue | !balance) ->X (state !=start))	c8	<start,password,cash,start>

Ultimately come to the conclusions in table 3: c8 covers { p8, p6}, c7 covers { p7, p6 }, c6 covers { p6}, c5 covers { p5, p4} , c4 covers {p4}, c2 covers{ p2}, c1 covers {p1 }. Indeed only with c8, c7, c5, c2, c1 can already satisfy the need of full coverage.

5 Conclusions

This paper defines a notion of rewriting to fulfill our requirements in optimizing test case generation. First time we rewrite the properties to reduce some redundant properties to cut down the cost of calling model checker SMV, from this aspect, we shorten the time and the consumption of the SMV implementation. Second time, we rewrite the properties reversely to reduce some duplicated or redundant test cases to avoid the test case explosion. We still confront with the problem of how to efficiently deal with the extension in real models with our rewriting method. That is also an optimization to these test cases which we will discuss in the future.

References

[1] Hamon, G., Moura, L., Rushby, J.: Generating efficient test sets with a model checker. In: 2nd IEEE International Conference on Software Engineering and Formal Methods (SEFM), Beijing, China (September 2004)
[2] Hierons, R.M.: Using a minimal number of resets when testing from a finite state machine. Information Processing Letters 90, 287–292 (2004)
[3] Huth, M., Ryan, M.: Logic in Computer Science: Modelling and Reasoning about Systems. Cambridge University Press (2000)

[4] Zeng, H.W., Miao, H.K., Liu, J.: Specification-based Test Generation and Optimization Using Model Checking. In: Proceedings of First Joint IEEE/IFIP Symposium on Theoretical Aspects of Software Engineering (TASE 2007), pp. 349–355. IEEE Computer Society Press, Shanghai (2007)
[5] Hong, H.S., Cha, S.D., Lee, I., Sokolsky, O., Ural, H.: Data Flow Testing as Model Data Flow Testing as Model Data Flow Testing as Model Checking. In: Proceedings of the 25th International Conference on Software Engineering (ICSE 2003), Portland, Oregon, USA, (May 3-10, 2003)
[6] Fraser, G.: Issues in using model checkers for test case generation. Journal of Systems and Software 82(9), 1403 (2009)
[7] Kandl, S., Kirner, R., Puschner, P.: Automated Formal Verification and Testing of C Programs for Embedded Systems., Proceedings of the 10th IEEE International Symposium on Object and Component-Oriented Real-Time Distributed Computing (ISORC 2007) (2007); 0-7695-2765-5/07
[8] Meolic, R., Fantechi, A., Gnesi, S.: Witness and Counterexample Automata for ACTL. In: de Frutos-Escrig, D., Núñez, M. (eds.) FORTE 2004. LNCS, vol. 3235, pp. 259–275. Springer, Heidelberg (2004)
[9] Fraser, G., Gargantini, A.: An evaluation of specication based test generation techniques using model checkers. In: Proc. of Testing: Academic & Industrial Conference Practice and research techniques (TAICPART 2009), pp. 72–81. IEEE Computer Society (2009)
[10] Legeard, B., Peureux, F., Utting, M.: Controlling test case explosion in test generation from B formal models. The Journal of Software Testing, Verification and Reliability 14(2), 81–103 (2004)

Unit Testing Memory Management with Microsoft Pex ⋆

Xiaoyu Liu, Liang Zhou, Xiangpeng Zhao, and Hongli Yang

Abstract. Memory management is a function of operating system. Unit testing is used to validate the correctness of programs. *Pex* is a Parameterized unit test tool, which can create a small test suite with high code coverage. The paper describes how to use *Pex* to test a simulation program of memory management.

1 Introduction

With the rapid development of software industry, software development grows more and more complexity. In order to improve the efficiency of software development and software reliability, software testing has been paid attention by industry and academy, increasingly. Currently, software testing is the only effective way that uses to verify whether the software has arrived what they expected. Unit testing is the first step in testing procedures, the importance of it is self-evident.

Unit testing tests the single part of a program. Usually, a unit is a method or a function in a program. Unit testing could detect many problems which other test methods can not detect. Unit testing also uses to verify whether the program is agreed with the specification. *Pex* [1, 2] is a new tool that helps in understanding the behavior of. Net code, debugging issues, and in creating a test suite that covers all corner cases – fully automatically. It is developed by Microsoft.With automatic

Xiaoyu Liu · Liang Zhou · Hongli Yang
College of Comp. Sci., Beijing Univ. of Tech.
e-mail: t_t@emails.bjut.edu.cn, liangzhou@email.bjut.edu.cn, yhl@bjut.edu.cn

Xiangpeng Zhao
Microsoft ATC Beijing, China, e-mail: xiazhao@microsoft.com

⋆ Supported by open foundation of State Key Laboratory of Computer Science, Institute of Software, Chinese Academy of Sciences (No. SYSKF1008), Subject and Postgraduate Education Construction Project of Beijing Municipal Commission of Education.

Z. Qian et al. (Eds.): Recent Advances in CSIE 2011, LNEE 125, pp. 483–488.
springerlink.com

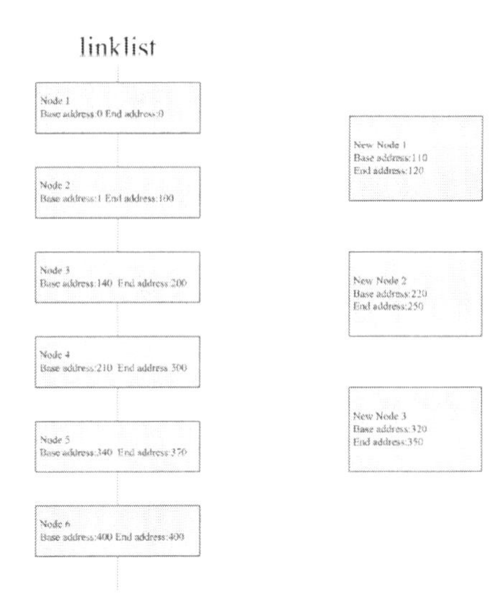

Fig. 1 Status of new pages

detecting, *Pex* can analyze a program and generate the necessary test cases. Using *Pex* may help developers to alleviate the workload of the program development.

Developers often seek to test individual components in isolation, to make testing more robust and scalable. A common approach is to use dummy implementations. Moles is a tool to generate dummy implementations. In this paper we use an example of a memory management program to introduce the benefits of unit testing with *Pex*.

2 The Memory Management Algorithm in C#

Memory management is a very important function of the operating system. To study memory management algorithms, we have implemented a small simulation program using C#, which only includes very a few core features such as adding and querying a page. Note that our main purpose is to show how we can use *Pex* to help testing complicated algorithms, and the algorithms *per se* is not important.

2.1 The Algorithm of Simulation Memory Management

The program uses a doubly-linked list to simulate a main memory. Nodes of the list express existing pages of main memory. Each node of the list have a base address and a end address. So, every node have a range of address and free-frames represented by the range of address between two page nodes. When a node is needed to be added in the list, the program will determine whether the new node is included in existing nodes. If the now node is isolated, the program will add the now node

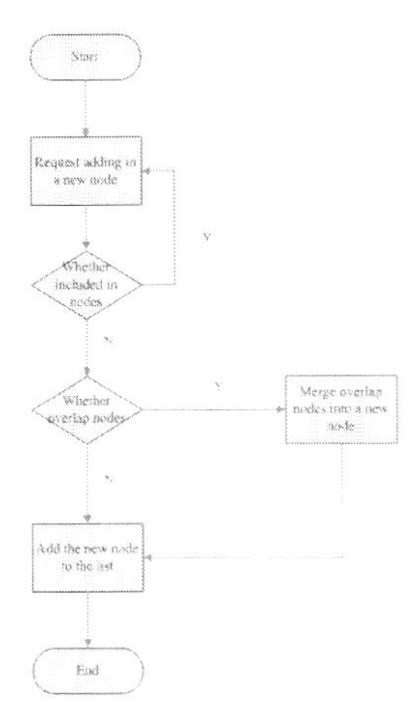

Fig. 2 Flow-diagram

into the list. If the new node included in any existing nodes, the program do nothing. If the new node overlaps any existing nodes, the program merge overlapped nodes into a new node. Figure 1 shows these status of new nodes. The algorithm of the program is showed by Figure 2.

2.2 *Sample Code*

Because of space limitations, we just put two methods of program as example to explain the process of the C# code. The first method is *Add*. The second one is the method *SearchVA*.

2.2.1 Add Method

Add is a public method of the *VARangeListSimulator* class. It has an object of the *SimpleVarange* class as a parameter. Its function is adding in a node to the list. The location of the node relies on addresses of it. Adding in different nodes to the list may lead the program to execute different branch of paths, because of the following code in the program:

```
ulong newRangeBA = rBA.InRange ?rBA.Prev.Va-
-lue.BaseAddr : r.BaseAddr;
ulong newRangeEA =rEA.InRange ? rEA.Prev.Va-
-lue.EndAddr : r.EndAddr;
```

The following code is the core of the program code, which is the *Add* method.

```
public void Add(SimpleVARange r)
{
   var rBA = SearchVA(r.BaseAddr);
   var rEA = SearchVA(r.EndAddr);

   if (rBA.Prev == rBA.Next &&rBA.Next== rEA.Prev &&
                              rEA.Prev == rEA.Next)
   {
       return;
   }
   else
   {
        ulong newRangeBA =rBA.InRange?
        rBA.Prev.Value.BaseAddr : r.BaseAddr;

        ulong newRangeEA = rEA.InRange?
        rEA.Prev.Value.EndAddr : r.EndAddr;

        var newSimpleRange = new LinkedListNode<Simple-
        -VARange>(SimpleVARange.CreateSimple-
        -VARangeByAddr(newRangeBA, newRangeEA));

        SimpleVARanges.AddBefore(rBA.Next,
        newSimpleRange);

        DeleteRanges(o=>o !=newSimpleRange.Value&&Simple-
        -VARange.Overlap(newSimpleRange.Value, o));

        MergeRangesIfConsecutive(newSimpleRange.Previous,
        newSimpleRange);
        MergeRangesIfConsecutive(newSimpleRange,
        newSimpleRange.Next);
   }

}
```

3 The Unit Testing with Pex

Microsoft *Pex* is an automated test input generator that uses dynamic symbolic execution to check if the software under test agrees with its specification. This automation makes software development more productive and software quality increases. *Pex* produces a small test suite with high code coverage from parameterized unit tests. We test the simulation program with *Pex*. Because *Pex* can automatically analysis the tested code, we just write the testing methods of the testing program.

3.1 Unit Testing of Add

The testing program generates a simulation object of the *SimpleVARange* class called *r*. The simulation object is generated by the moles. It makes the tested code independent. The testing program also generates an object of the *VARangeListSimulator* class that called target. *Pex* selects different return values of range for the mock object *r* as parameters. *r* will added to the list by the *Add*. And the test program verifies the *Add*.

We also add two nodes into the object target that in order to test different scenarios.

When the new node and existing nodes overlap each other, the program will generate a new node, and delete old nodes from the list. The parameter of this code is an object of other class. Therefore, the tested code is not completely independent.

There are three different status of parameters that makes program produce different branch. *Pex* selects the new input by the status of path coverage of *Add*. It will be reflected in test results. In addition, some boundary value may lead the program to throw Exception. As this is a simulation program of the Memory management, we set these Exception throw by the tested method are not considered as wrong.

Test Code of method add:

```
public void Addtest(ulong x,ulong y) {
    PexAssume.IsTrue(x<=y);
    var target = new VARangeListSimulator();
    //generate an object of VARangeListSimulator

    var r1 = new LinkedListNode<SimpleVARange>(
    SimpleVARange.CreateSimpleVARangeByAddr
    (8 * 1024, 16 * 1024));

    var r2 = new LinkedListNode<SimpleVARange>
    (SimpleVARange.CreateSimpleVARangeByAddr
    (24 * 1024, 64 * 1024));

    target.SimpleVARanges.AddAfter
    (target.SimpleVARanges.First,r1);

    target.SimpleVARanges.AddAfter(r1,r2);

    //add two memory block nodes to target

    var r = new MSimpleVARange();
    r.BaseAddrGet = delegate()
    {
        return x;
    };
    r.EndAddrGet = delegate()
    {
        return y;
    };
    //r is a simulator object  of  SimpleVARange  class
    target.Add(r);
}
```

Pex automatically executes the program. It chooses the value of the test suit according to nodes that already existed in the list. *Pex* selects some boundary value of error-prone, and chooses the input to make the program executes different paths. The Figure 3 is the result of the program.

The results shows: First, *Pex* randomly selects a set of value (16384,155649) for testing. It means that the range of the adding node is (16384,172033). Then *Pex* record the result of the path branches, so that remaining pathes of the branches can be tested. In the Figure 3, *Pex*also selects other three values. In addition, *Pex*use the value group (0,0) to test add a new node with the zero size. The simulation program does not deal with it.

Finally, *Pex* also tests the Exception that causes by the boundary value of *Int*64. This Exception is easily overlooked by traditional unit test.

Pex Exploration Results - stopped - 6 tests, 41 runs- 208/214 dynamic coverage

ulatorTest.Addtest(UInt64 x, UInt64 y) Views Ask A Question

Review bold issues: Global Events | All Tests | All Events | 2 Static Field Stores

	x	y	Summary/Exception	Error Message
1	16384	155649		
2	8193	40960		
3	8193	12288		
4	0	0		
5	24574	ulong.Max...	InvalidOperationException	Sequence contains no matching element
6	8190	ulong.Max...	InvalidOperationException	Sequence contains no matching element

Fig. 3 Testing result of Add

	Not Covered ...	Not Cove...	Covered (Blocks)	Covered (...
VARangeListSi...	45	23.20%	149	76.80%
.ctor()	0	0.00%	10	100.00%
<DeleteRan...	0	0.00%	3	100.00%
Add(class ...	0	0.00%	46	100.00%
DeleteRang...	0	0.00%	15	100.00%
MergeRange...	0	0.00%	28	100.00%
SearchVA(u...	2	9.09%	20	90.91%
IsValid(cl...	4	12.90%	27	87.10%
<DeleteRan...	3	100.00%	0	0.00%
DeleteRang...	17	100.00%	0	0.00%
SearchRang...	19	100.00%	0	0.00%

Fig. 4 Code coverage of Add

The results from the path coverage can be seen at the Figure 4 *Add* is 100% covered. Notes that the dynamic coverage is not 100%, because *Pex* just fully cover the *Add*.This shows the path of the tested code branches has been fully tested.

4 Conclusion

Using *Pex* and Moles, developers can design parametric unit testing code to test the correctness of programs.In this experiment, *Pex* can automatically analyze the code of these methods, and generate a group of efficient test suites for testing. It shows that the use of *Pex* can reduce some work of writing unit test code. *Pex* also finds a number of exceptions which can not be easily found with traditional unit test. By using of *Pex* , developers can focus more into the implementation of function, rather than the preparation of unit test code. With the development of software, *Pex* will be more important.

Acknowledgement. Thanks Kang He, Chen Deng and Fan Yu for their helpful comments.

References

1. Pex, `http://research.microsoft.com/en-us/projects/Pex/`
2. Tillmann, N., de Halleux, J.: Pex - White Box Test Generation for NET. In: Beckert, B., Hähnle, R. (eds.) TAP 2008. LNCS, vol. 4966, pp. 134–153. Springer, Heidelberg (2008)

White-Box Test Case Generation Based on Improved Genetic Algorithm

Peng Wang, Xiao-juan Hu, Ning-jia Qiu, and Hua-min Yang

Abstract. Some intermittent or transient failures are particularly difficult to diagnose in highly complex and interconnected systems. This paper focuses on the use of genetic algorithms for automatically generating software test cases. In particular, this research extends a newly improved genetic algorithm, which adopts back propagation algorithm for local fine-tuning in the final link, and speeds up access to the best population. The various approaches offer opportunities for performance improvements that make these techniques more scalable for realistic applications.

1 Introduction

Software test, the important means for software quality assurance, intends to reveal the potential problems fast and early in software applications, encourages developers to solve these problems as quickly as possible and provide customers with high-quality software products. At present, the software test accounts for 50% of the total cost [1-3]. Test cases are not only important tool for achieving software testing effectively but also the key to ensure the effectiveness of software testing. Hand-written testing is heavy workload, long testing cycle and large testing loopholes. It makes automatic generation of testing cases which could become the major research issue. Because not it only reduces the total cost of software development, but also significantly it improves software reliability and shortens the development cycle of software application products.

Efficient test case generation method can generate high-quality and low-data test cases, reducing software total development costs. Therefore, an effective test case generation research method is practical significance. Automatic generation of test cases is a process that looked for a group of testing data to meet the given criteria in

Peng Wang · Xiao-juan · Hu · Ning-jia Qiu · Hua-min Yang
Department of Computer Science
Changchun University of Science and Technology
Changchun, Jili Province, China
e-mail: wangpeng@cust.edu.cn

Z. Qian et al. (Eds.): Recent Advances in CSIE 2011, LNEE 125, pp. 489–495.
springerlink.com

a data field. In recent years, the problem of which is generating the test case has transformed into the path search problems [4-5]. As the test case generation is an undecidable problem, accounting for the measured size and complexity of procedures, the general search algorithm has been extremely limited. General algorithm has a high degree of robustness and good global search capability as well as a very clear advantage in dealing with uncertainty search problems, so it has been used to solve these problems. However, in fact, it cannot be guaranteed convergence to global optimal solution, and not be suitable for local search; it is only close to the global optimal solution. So that the convergence rate is significantly decreasing, diversity is also to fast reduce [6]. In order to overcome these deficiencies, we combined the genetic algorithm and back propagation algorithm (called "BP algorithm" for short). In the initial stage, we introduced BP algorithm to optimize the newly species that based on combination of variation operators in genetic algorithm, reducing the number of iterations and speeding up accessing to the best path [7]. Currently, the test case generation can approximately be divided into two categories, white box test case generation method and black box test case generation method. This paper only discusses white box test case generation method.

2 Test Case Generation Based on GA and BP Algorithm

2.1 Basic Principles

The genetic algorithm is developed based on the survival of the fittest principle. It is a group optimization search algorithm which combines the theory of natural selection and population chromosomes randomly exchange mechanism. The chromosome of population is targeted for the major object and encoded, and selection, chiasma and variation operations on the encoded search space. Selection, chiasma and variation are known as the genetic manipulation among these operations. Parameter setting, population initializing, parameter encoding, fitness function and genetic operation setting are the five elements of genetic algorithm [8-10].

Genetic algorithm simulates the process of biological evolution and gene manipulation in computer, and not demands specific information of the object. The fitness function is almost limitless. Neither it is requires continuous function, nor it requires differentiable function. The new population contains many individuals with high fitness function value, which are screened out by the set fitness function and a series of genetic operations. It includes large amounts of information of former generation, and introduces new individuals which are better than the previous generation individuals. The average value of group fitness is rising up to meet the certain optimized criteria by this cycle.

The automatic test case generation method, through the optimized search algorithm, intends to search out the best cases that meet the requirements of software testing in the large test case space. Because of the complexity of the problem, there are still many problems to be solved.

2.2 Basic Ideas of Genetic-BP Algorithm

1. The population takes the real number coding scheme. Each connection weight is denoted by a real number, so the distribution of the connection weights can be used a set of real numbers to express. The new initial population Q whose individual number is N, does not exist the same individual.

2. The setting of fitness function and error function.

3. Determine whether the entire target path is covered by current population, or whether the error number is less than the given value, or whether the number of test time is larger than the given number. If you meet one of above conditions, it will put out the optimized group, and end the calculation, otherwise, it will determine the sub-target coverage path, and go to the next step.

4. Evaluating various weighs. The choice of individuals to inherit is based on the fitness function value. The higher fitness function value, the higher probability is copied to the next generation.

5. According to the individual fitness function values in the current population, we adjust the chiasma, variation probability and finalize.

6. Set the initial position, chromosome number of chiasma and variation operation, and then Execute the genetic evolutionary operation, at last, the new population Q1 is generated.

7. Recording the maximum, minimum and average value of the progeny population to fit function. In order to maintain species diversity, the maximum deviation record is remained between the offspring data and the parent data.

8. Setting the current offspring generation as the parent population of the next generation. The test Number increased by 1 and return to the third step to judge.

2.3 The Construction of the Fitness Function

The fitness function is selected based on the effectiveness of the guidance for the direction of optimization parameters in the search space, so that the genetic algorithm can gradually approach the optimal parameters, to instead of misconvergence and local optimum. The fitness function of this algorithm can be used the following equation (1):

$$f(x) = \begin{cases} C^{*}-g(x) & g(x)<C \\ 0 & \text{Other cases} \end{cases}$$

$$\begin{cases} C^{*} = \| g_{m}(x) - E[g(x)] \|_{2} + E[g(x)] \\ g_{m}(x) = \max\{ g(x) \} \end{cases} \tag{1}$$

The parameter gm is the maximum in current input space. The value of E[g(x)] is the average value of N objective function. The parameter C* is the summation of E[g(x)] and the European norm of the difference between gm and E[g(x)]. It not only can be to guarantee the non-negative fitness function value, but also to make the value varying with individual in the input space. The genetic manipulation is more reasonable than before with the restraint of the fitness function, because the results are not consistent by different technicians to set up for the same objective in genetic algorithms.

2.4 Improve and Optimize the Operators

The traditional chiasma and variation operator are fixed. They damaged the genes with higher fitness function values, and slowed the search speed. The new operators modify these shortcomings, as follows:

$$P_c = \begin{cases} k_1(f_{max} - f'_c) / (f_{max} - \bar{f}), & f'_c \geq \bar{f} \\ k_3, & f'_c < \bar{f} \end{cases}$$

$$P_m = \begin{cases} k_2(f_{max} - f) / (f_{max} - \bar{f}), & f \geq \bar{f} \\ k_4, & f < \bar{f} \end{cases} \quad (2)$$

In the equation (2), $f'c$ is the larger fitness function value in the parent group before chiasma; f is the fitness function value of individual; and $k_1=k_3=1$, $k_2 = k_4= 0.5$.

The individual is called high-quality individual when its fitness function value is higher than the average fitness function value; if its probability of Pc and Pm take a smaller number, the convergence of genetic algorithm will be strengthened. If the Pc and Pm probability of individuals is larger, whose value of fitness function is lower than the average value of fitness function, the genetic algorithm will not fall into local solutions.

2.5 Factors of Affecting the Test Case Generation

The test case generation algorithm has made a lot of improvement and optimization; however, there are still some factors that can not be accurately estimated and computed. Summarized as follows:

1. Test Environment Configuration.
Different hardware and software configuration have great impact on the enforcement of the test cases in different test environment, such as software configuration, operating system and database configuration, etc.

2. The Extent of Understanding of Requirements
From the bug source analysis report, most bugs are due to the inadequate or inappropriate understanding of the requirements. Therefore, to enhance the understanding of requirements analysis is necessary.

3. Staffing
The result is different for different personnel to complete for the same project. In particular, the experience and quality of software developers and testers are the key factors.

4. Bug Number
The number of bug which obtain the algorithm operation has a big gap compared the number of bug found later. There are many unexpected issues during the software implementation.

5. User Satisfaction
When the user acceptance testing finished, there will be some user feedback views or bugs. Those issues are the omission of the existing test cases. According to the feedback, we should modify and re-perfect the testing case.

3 Experiments

The problem of differentiating the shape of triangle contains a clear and complex logic, so it is widely used in the software test case generation as an example. The genetic algorithm has an advantage in search capabilities and time efficiency compared with the adaptive genetic algorithm (called "AGA" for short) through the analysis of the iterations and time of right triangle test case generation. Set the population size from 100 to 300, and run ten times for each species. We record the number of iterations and running time to find the optimal solution to analyze. The population size represents the total number of test sub-path in Table 1. The number of iterations and the consuming time of Genetic -BP algorithm are lower than those of the AGA algorithm from the data of Table 1, so the Genetic-BP algorithm can significantly improve the efficiency of generating test cases.

Table 1 The iterations and time of right triangle test case generation

Population size	The Best iteration of GABP	The Worst iteration of GABP	The Best iteration of AGA	The Worst iteration of AGA	Time of GABP	Time of AGA
300	2	9	12	255	14.8	420.7
280	2	8	10	153	14.2	395.8
260	2	8	4	168	13.3	370.5
240	1	8	15	170	12.9	328.6
220	2	9	3	125	12.1	305.8
200	2	12	3	160	12.5	240.6
180	3	10	4	154	11.9	290.1
160	6	11	12	350	11.7	267.8
140	6	13	9	500	12.1	375.2
120	5	16	9	460	10.9	227.1

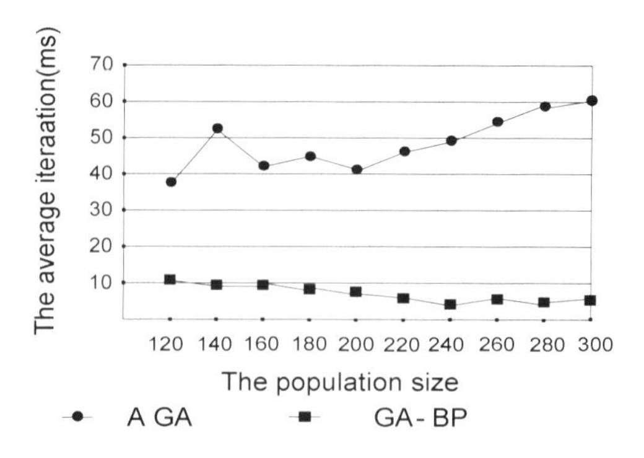

Fig. 1 The average value of iteration for right triangle test case generation.

In Fig.1, the horizontal axis is the population size, and the vertical axis is the time of generating the optimal solution which is the average running time of ten times.

The experimental data show that the test case generation based on the Genetic-BP algorithm has obvious advantages with the same parameters.

4 Summarizes

This paper presents a new test case generation method which combine genetic algorithm with BP algorithm. We use BP algorithm to enhance the search capabilities in areas of test cases after the genetic algorithm generate test cases. So it can overcome falling into local optimum easily and improve efficiency.

References

[1] Tsoulos, I.G., Lagaris, I.E., Gen, M.: An Enhanced Genetic Algorithm for Global Optimization. Computer Physics Communications 61(19), 2925–2936 (2008)

[2] Michael, C.C., McGraw, G.E., Schatz, M.A.: Generating Software Test Data by Evolution. IEEE Trans. on Software Engineering 27(12), 1085–1110 (2001)

[3] Berndt, D.J., Watkins, A.: Investigating the performance of genetic algorithm-based software test case generation. High Assurance Systems Engineering 4(13), 261–262 (2004)

[4] Berndt, D., Fisher, J., Johnson, L.: Breeding software test cases with genetic algorithm. System Sciences 02(01), 6–9 (2003)

[5] Yao, Y.: New test case generation method based on genetic algorithm. Computer & Digital Engineering 231(1), 18–21 (2009)

[6] Xia, Y., Liu, F.: Automated software test data generation based on immune genetic algorithm. Computer Applications 28(3), 723–725 (2008)

[7] Li, K.-S., Dai, Z.-H.: An improved BP algorithm based on evolutionary algorithm. Microcomputer Information, 23–25 (2010)

[8] Wang, L., Yang, X.-H.: Application of back propagation artificial neural networks in bug analysis. Journal of Computer Applications 30(1), 153–155 (2010)
[9] Du, S.-Q.: A novel algorithm of optimizing neural network weights and its application. Journal of Northwest University for Nationalities 30(76), 27–31 (2009)
[10] Erdogmus, D., Principe, J.C.: Generalized information potential criterion for adaptive system training. IEEE Trans. on Neural Networks 13(5), 1035–1044 (2002)

A Formal Method for Analyzing Trust Chain of Trusted Computing Platform

Yasha Chen and Yu Sun

Abstract. A formal method is used to analyze the integrity of trust chain in trusted computing platform. Based on Biba's security model, components of the platform are tagged with different security level when system boots. With the satisfaction of "read-up, write-down", information flow analysis mechanism is used to describe the security threats between different components. An evaluation model is built and realized.

1 Introduction

With the information security practice, people gradually realize that most of the security risks are from the computer, so it is necessary to ensure information security of computer itself. This must be achieved by taking measures from chips, hardware architecture and operating systems etc, which result the basic idea of trusted computing.

Trust equals to reliability plus security approximately[1]. Trusted computing system is able to provide system reliability, availability, information security and behavior security. The idea of trusted computing derives from community. The basic idea of it is to establish a trust root in the computer system, and then to create a chain of trust. The lower layer measures the higher layer and then trusts it.

Above all, the root of trust extends the trust to entire computer system as to ensure the trust of the computer system.

In 1999, IBM, HP, Intel, Microsoft and other well-known IT companies had launched the TCPA (Trust Computing Platform Alliance). In 2003, TCPA was

Yasha Chen
Department of Electrical and Information Engineering, Naval University of Engineering, Wuhan, China
e-mail: cys925@yahoo.com.cn

Yu Sun
School of Computer, Beijing University of Technology, Beijing, China
e-mail: syking@163.com

Z. Qian et al. (Eds.): Recent Advances in CSIE 2011, LNEE 125, pp. 497–502.
springerlink.com

reorganized as the TCG (Trusted Computing Group). Meanwhile, all kinds of applications supporting trusted computing are widely used, such as Microsoft's VISTA Bit-Locker technology. At home and abroad, the research on chain of trust is also in full growing vigorously, such as the OpenTC, NGSCB, TXT technology, etc[1].

At the same time, the profit-driven manufacturers who are considered to promote the development of trusted computing are not really good for the development of trusted computing. Currently, trusted computing products in large numbers, but, there is often a lot of controversy to the real key technologies, such as the trust chain. Therefore, it is necessary to analyze the chain of trust technical, and to identify defects of specification.

2 Biba Model

The TCG specifications, such as chain of trust, are closely following the security models and requirements. The core idea of chain of trust is integrity measurement, which aims to ensure the integrity of different components while system starts up. Biba [2-5] model is a powerful tool to describe the integrity property.

2.1 Definitions of Biba Model

Biba model defines additional markers (integrity label, il) property for the integrity of the subject and the object, which are prerequisites property for access control. Biba model uses the following terms:

- S: the set of subjects, its members can be expressed as s, the subject generally refers to the process;
- U: the set of legal users of the system, the user who launches the process is the owner of the process;
- O: the set of objects, its members can be expressed as o, most of the objects have capability of storing data, the subject can convert to object when the former is called by other subjects;
- I: the set of integrity tag;
- il: $S \bigcup O \Rightarrow I$, is the integrity property of subject or object, the integrity property of the subject s and object o are expressed as $il(s)$ and $il(o)$ respectly;
- $\leq$ (or leq): the relationship between two integrity tags, the relationship on the set of integrity tag constitutes a lattice, if x and y satisfy the relationship x $\leq$ y, then it is can say that x is not bigger than y;
- o: the access when subject reads from an object, which can be expressed as "soo";
- m: the access when subject writes data to an object, which can be expressed as "smo";
- i: the access when subject invoke an object, which can be expressed as "s1is2"

2.2 *Introduction of Biba Model*

The BLP model only addressed the issue of confidentiality of information without involving integrity issues; its presence in the integrity of the definition has some defects. BLP model does not take effective measures to restrict unauthorized modification of information, which could make illegal, unauthorized tamper. Considering these factors, Biba defines information integrity level as confidentiality level defined by BLP as to not allow the access from low-level process to high-level process, which means that any subject only could write information down to object whose securiy level is lower than the former.

The main two properties of Biba model are depicted as figure 1.

- a. no "write-up", which makes the object with high-level integrity is made by subject with high-level integrity, which ensures that the object with high-level integrity can not be modified by process with low-level integrity.
- b. no "read-down", which makes any subject can not read (or execute) the data with low-level integrity.

Biba model can be expressed with a partial order.

- ru, the read operation, if and only if SC(s) ≤SC(o).
- wd, the write operation, if and only if SC ≥SC (o).

BLP model and Biba model complement each other. Biba model focuses on information integrity other than confidentiality. The chain of trust of TCG uses integrity measurement technology to ensure secure boot. Therefore, Biba model is suite for studying the transferring of chain of trust.

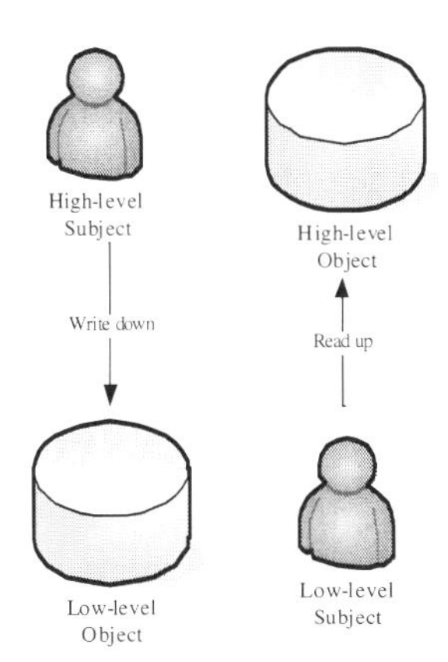

Fig. 1 Biba Model

3 Analysize of Chain of Trust Based on Biba Model

3.1 Security Level of Chain of Trust

Firstly, this paper defines the security level of chain of trust as followings:

- The highest security level: TBB, including CRTM and TPM.
- The higher security level: POST Code, Option ROMs.
- The high security level: IPL.
- The low security levlel: OS-Loader.
- The lower security level: ACPI table.

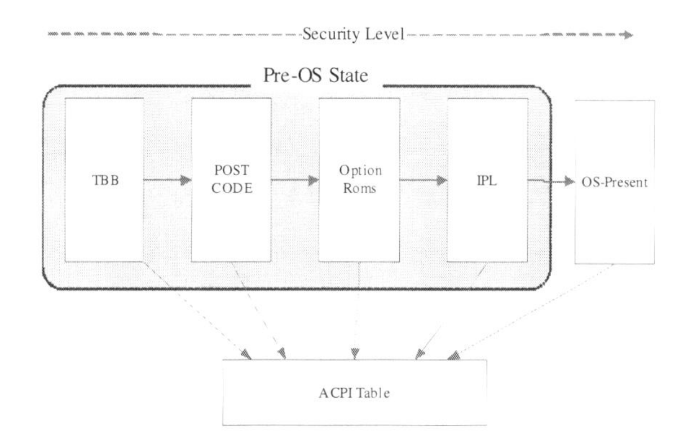

Fig. 2 The security level of chain of trust

3.2 Information Flow Analyze for Chain of Trust

The information flow includes four kinds of operations during transferring of chain of trust, which is depicted as figure 3.

- High-level subject measures integrity of low-level object.
- High-level subject writes data down to low-level object, which means that event log is created.
- Low-level process accesses high-level process, such as IPL accesses the interface, INT 1A, of CRTM.
- Low-level process access high-level process, such as IPL updates PCR of TPM by calling interrupt interface.

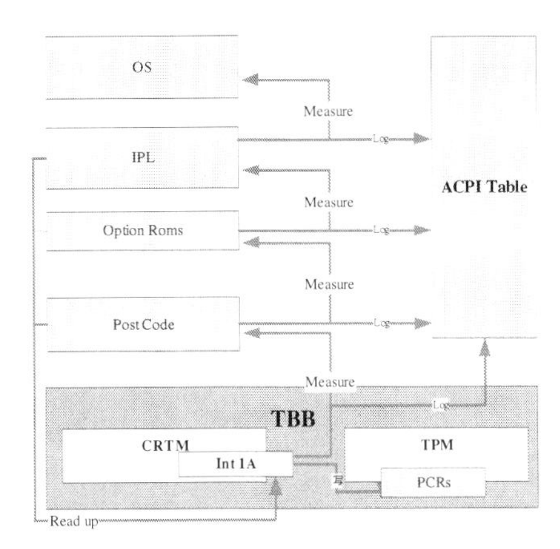

Fig. 3 Data flow analyze for chain of trust based on Biba model

This paper lets the second operation and the fourth operation as "write" operation, and lets the first operation and the third operation as "read" operation. The bootstrap flow of chain of trust is as follows.

- a. The platform starts up, and TBB has the highest level integrity. TBB begins to establish chain of trust by calling INT 1A interrupt which has low-level integrity.
- b. Another highest level subject, CRTM, measures (read down) from the code and data of POST, Option ROMs which are low-level objects. The low level process accesses (read up) the interrupt interface of INT 1A of CRTM and updates PCRs of TPM, and updates the event logs of ACPI table, and then transfers to next component.
- c. IPL continues to measure the low level objects, and accesses (read up) the INT 1A.

As to transferring of chain of trust, it is necessary that high-level process measures low-level process other than in turn. When low-level process wants to access high-level process, it only has read-up permission, and could not have any write operation to high-level process. The Biba model could restrict chain of trust into "read-up" and "write-down" access control policy, but in reality, the chain of trust exists operation of "write-up" such as low-level process accesses the interface of CRTM for updating certain PCRs of TPM, which is allowed in the specification of chain of trust and TPM. It is obvious that these operations violate the rules of Biba model, which is depicted as figure 4.

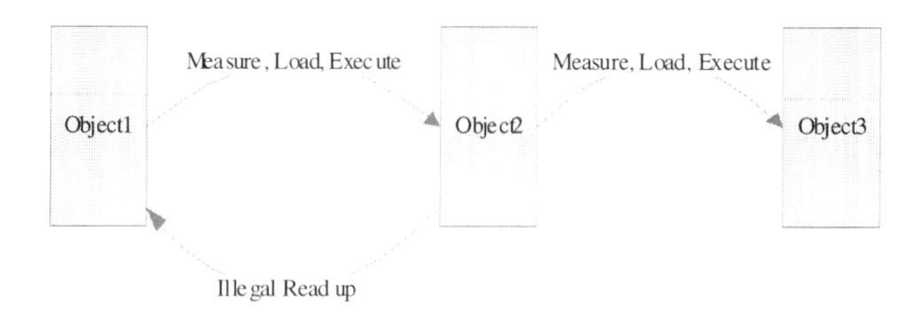

Fig. 4 The illegal read-up of chain of trust

Therefore, there is still a gap between TCG specifications of chain of trust and Biba model. When the low-level process had been measured by the high-level process, the former could access (write-up) the latter without any authorization, which damages the integrity of high-level and could let attacker or malicious code utilize it. Therefore, TCG specification although defines the transferring rules of chain of trust, but it do not ensure the integrity of high-level component.

4 Conclusion

This paper firstly describes the TCG specification, and analyses the integrity and security of chain of trust based on Biba model. The implement of chain of trust do not conform to Biba model and there is integrity risk because of no write-up policy, which needs to protect high-level measured subjects. This is the study content of next step.

References

1. Zhang, H., Luo, J., et al.: Development of Trusted Computing Research. Journal of Wuhan University (Natural Science Edition) 18(5), 1–3 (2006) (in Chinese)
2. Biba, K.J.: Integrity considerations for secure computer systems. USAF Electronic System Division, Hanscom Air Force Base, Tech. Rep.: ESD-TR-76-372 (1977)
3. Mayfield, T.: Integrity in automated information system. National Computer Security Center, Technical Report, 79–91 (1991)
4. Joint Technical Committee 1. ISO/IEC 15408-2 (1999)
5. Loscocco, P.A., Smalley, S.D., Muckelbauer, P.A., et al.: The inevitability of failure: The flawed assumption of security in modern computing environments. In: Proc. The 21st National Information Systems Security Conf., pp. 303–314. National Computer Security Center, Crytal (1998)

A Novel Method for Searching Services of Cloud Computing

Tianji Wu, Kui Xiao, and Kui Yang

Abstract. While the number of software services grows rapidly, service retrieval are severely hampered by the growth. To overcome this limitation, several methods have been proposed to improve the service retrieval mechanisms. In this paper, we propose a novel method for searching services of cloud computing by exploiting the semantics and ontology technologies. Key tags are extracted from software documents, and are used as metadata of the service resources. Ontologies of the resources are constructed with the tags to facilitate the searching process.

Keywords: service retrieval, tag, ontology, cloud computing.

1 Introduction

Today, cloud computing as a new business model grows rapidly all over the world. Saas (Software-as-a-Service) is the first type of cloud computing, and these services are attracting more and more attention worldwide. Salesforce.com is a leader of the Saas market. It created a concept, Saas CRM, and has made a success in providing online CRM services. Iaas (Infrastructure-as-a-Service) is another type of cloud computing. Amazon Web Services provides virtual servers with unique IP address and blocks of storage on demand. Paas (Platform-as-a-Service) is the third type of cloud computing. Google Apps is one of the most famous Paas providers. Developers can create applications with the provider's APIs[1].

Tianji Wu
Digital Department of library, Huazhong Agricutural University, Wuhan, China

Kui Xiao
School of Computer and Software, Wuhan Vocational College of Software and Engineering, Wuhan, China

Kui Yang
Modern Educational Technology Center , Huazhong Agricutural University
e-mail: yangk@mail.hzau.edu.cn

Z. Qian et al. (Eds.): Recent Advances in CSIE 2011, LNEE 125, pp. 503–508.
springerlink.com

People may want to know how to find out the exact services that they need. They could make use of search engines, such as Google and Baidu. But it is not an efficient way for the services retrieval. According to the tags provided by the developers, users could capture the common information of software services that they need. Generally speaking, people do not know the exact names of software services they want. Therefore, the traditional way of service retrieval do not work well for users. Understanding the meaning of users' requirements is very important for the retrieval. Researchers want to address the problem with semantics and ontology technologies.

According to the most cited definition of the Semantic Web literature, an ontology is an explicit specification of the conceptualization of a domain [2]. Guarino clarifies Gruber's definition by adding that the AI usage of the term refers to an engineering artifact, constituted by a specific vocabulary used to describe a certain reality, plus a set of explicit assumptions regarding the intended meaning of the vocabulary words" [3]. An ontology is thus engineered by - but often for members of a domain by explicating a reality as a set of agreed upon terms and logically-founded constraints on their use[4].

In this paper, a novel method is proposed for searching services of cloud computing by exploiting the semantics and ontology technologies. The details of the method are presented in Section 2, then in Section 3 we show the results of the experiment. Some related work is discussed in Section 4. Finally, we conclude with a summary and a discussion of future work.

2 Approach

In this section, we give an overview of how to find out the exact software services that users want. The service retrieval includes two steps: (1)mining tags from software documents(subsection 2.1); (2) constructing ontologies with the tags(subsection 2.2).

2.1 Mining Tags from Software Documents

The web pages could be found out by making use of its metadata, such as keywords. The metadata is the "primary key" of web pages. However, software services on the internet do not own these "primary key" to facilitate the service retrieval, and they are published with programs and documents. Software documents describe the software architecture, functions, and operation details. We mine key tags from the software documents, then construct ontologies on top of the key tags, that are the "primary key" of software resources on the internet.

[5] introduced a novel method for mining tags from software documents. The approach can be described in the following six steps:

1. Term extraction: firstly, lexical analysis is performed to parse the software documents, after which terms (noun) are extracted from the documents.
2. Term similarity computation: the similarity between terms is calculated.

3. Document similarity computation: similarity of documents is calculated on top of the numbers of the similar terms shared in the documents.
4. Software similarity network construction: a software similarity network is constructed, where the nodes represent software services, edges denote similarity relationship between software services, and the weights of the edges are the similarity degrees between software services.
5. Community Detection: a hierarchical clustering algorithm is used to cluster software in the constructed software similarity network.
6. Tag mining: the characteristic terms in each of software communities are mined, and the result is stored in the form of ontology. The process for mining tags is shown in Fig 1.

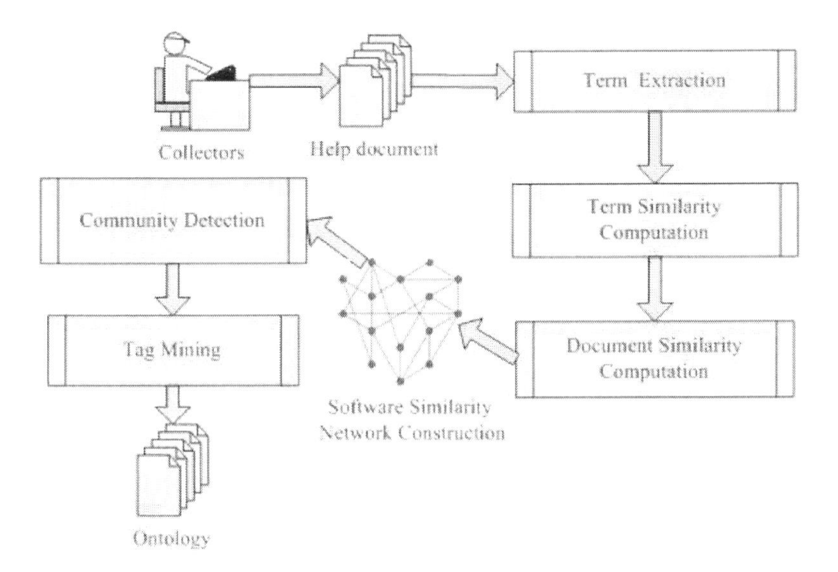

Fig. 1 The process for mining tag knowledge

2.2 *Constructing Ontologies with Tags*

Ontologies can be used as the "primary key" of service resources. Based on the technologies of ontology matching and semantic search, people could find out the exact services that they need.

[6] proposed a new method for constructing ontologies with tags. First, on top of online vocabulary resources, such as Wikipedia and WordNet, we transfer the tags into concepts. Second, the concepts can be mapped with the elements of ontologies which have been published on the internet. Then there may be some relationships between these concepts, because the elements are probably connected with each other. Third, the developer communities could validate these relationships, and add new relationships on demand. Using this semi-automated way, we turn the tags into ontologies.

3 Experiment

In the following, we demonstrate the experiment process and analysis the experimental data from Alisoft.com. Our first step was Mining tags from software documents. Second, tags were used to construct ontologies to facilitate the searching of the software services from Alisoft.com.

In our experiment, more than 100 software documents were collected beforehand. Then we extracted terms from documents, and transferred term into regular noun phrase (NP). Next, we computed the similarity between NPs. On top of the similarity of NPs, we calculated the similarity between the software documents. The document similarity is the software similarity. Then we construct software similarity network (S2N), Fig 2 shows the S2N of the experiment. In the network, node represents software, and edge represents similarity between softwares. In addition, each edge has a weight. After that, the softwares were divided into several categories with clustering algorithm. Category was called software community. By a threshold, the tags of each community were selected from NPs.

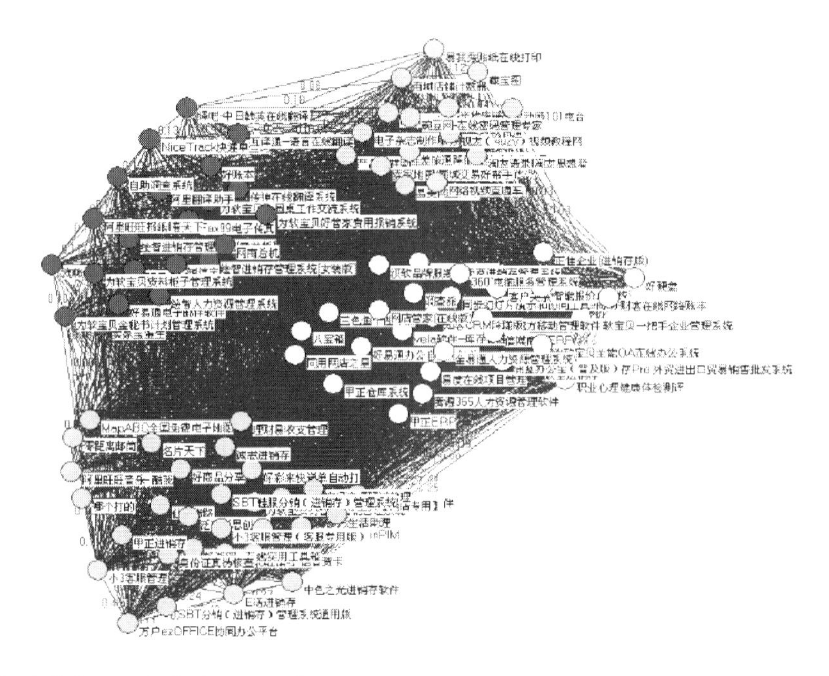

Fig. 2 Software similarity network

The following keywords are tags extracted from a software community:

{统计 流量统计 统计软件 批量修改 批量处理 批量处理软件 批量上传 店铺 淘宝店铺 自动上架 均匀分布 橱窗推荐 评价}

First of all, these tags could be viewed as concepts. Using the online vocabulary resources, we defined the hierarchical relationships between concepts. Secondly, the concepts were mapped with elements of ontologies on the internet. Relationships between the elements were transferred into the connections between concepts. Finally, the model were validated by users, and some additional but necessary relationships were added to the model too. Fig 3 shows the ontology model of the tags.

4 Related Work

A folksonomy is a system of classification derived from the practice and method of collaboratively creating and managing tags to annotate and categorize content; this practice is also known as collaborative tagging, social classification, social indexing, and social tagging [7]. Some people focus on constructing ontology on top of tags of folksonomy. [8] investigates a new approach to integrate an ontology on top of a folksonomy. They describe an application that filters del.icio.us keywords through the WordNet hierarchy of concepts, to enrich the possibilities of navigation. However, there is a drawback in the approach. WordNet concepts include only formal English words, therefore they could not describe all the tags that provided by common users. Some tags are new words created by common users, some tags are compositions of old words and so on. These tags could not be found in the WordNet or dictionaries.

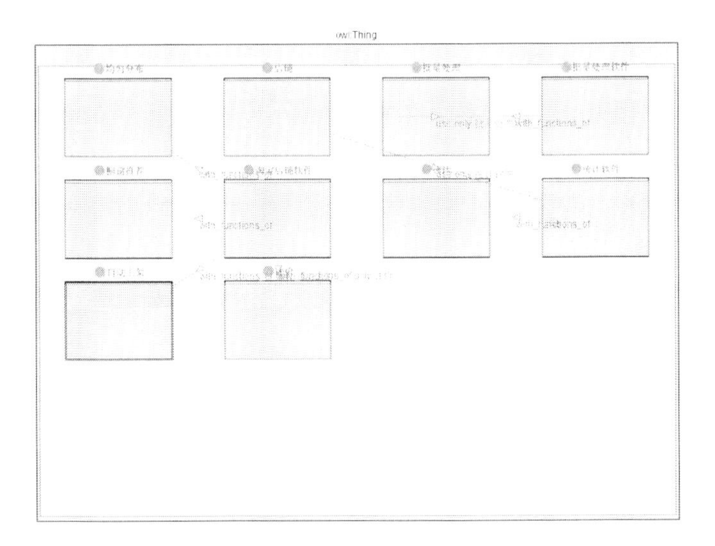

Fig. 3 Jambalaya of ontology

5 Conclusion and Future Work

Generally speaking, traditional retrieval way could not help people find out software services effectively, because search keywords do not describe the essence of a software service. In this paper, tags were extracted from software documents firstly. Then ontology was constructed to depict software services to facilitate the service retrieval. Semantic information added to software services made the searching process more intelligent.

As future work, it would be also interesting to improve the ontology integration, and to make the process full-automated. Furthermore, we hope semantics search could act better in the service retrieval.

Acknowledgments. This work is partially supported by Educational Reform and Realization Plan Of Huazhong Agricutural University (2007C44). We would like to thank Li Qin at State Key Laboratory of Software Engineering, Wuhan University for useful conversations.

References

1. Cloud Computing From Wikipedia, http://en.wikipedia.org/wiki/Cloud_computing
2. Gruber, T.R.: Towards Principles for the Design of Ontologies Used for Knowledge Sharing. In: Guarino, N., Poli, R. (eds.) Formal Ontology in Conceptual Analysis and Knowledge Representation. Kluwer Academic Publishers, Deventer (1993)
3. Aberer, K., Cudré-Mauroux, P., Catarci, A.M.O(e.) T., Hacid, M.-S., Illarramendi, A., Kashyap, V., Mecella, M., Mena, E., Neuhold, E.J., De Troyer, O., Risse, T., Scannapieco, M., Saltor, F., Santis, L.d., Spaccapietra, S., Staab, S., Studer, R.: Emergent Semantics Principles and Issues. In: Lee, Y., Li, J., Whang, K.-Y., Lee, D. (eds.) DASFAA 2004. LNCS, vol. 2973, pp. 25–38. Springer, Heidelberg (2004)
4. Mika, P.: Ontologies Are Us: A Unified Model of Social Networks and Semantics. In: Gil, Y., Motta, E., Benjamins, V.R., Musen, M.A. (eds.) ISWC 2005. LNCS, vol. 3729, pp. 522–536. Springer, Heidelberg (2005)
5. Qin, L., Li, B., Pan, W.-F., Peng, T.: A Novel Method for Mining SaaS Software Tag via Community Detection in Software Services Network. In: Jaatun, M.G., Zhao, G., Rong, C. (eds.) Cloud Computing. LNCS, vol. 5931, pp. 312–321. Springer, Heidelberg (2009)
6. Van Damme, C., Hepp, M., Siorpaes, K.: Folksontology: An Integrated Approach for Turning Folksonomies into ontologies. In: Proceedings of the ESWC Workshop Bridging the Gap between Semantic Web and Web 2.0
7. Folksonomy From Wikipedia, http://en.wikipedia.org/wiki/Folksonomy
8. Laniado, D., Eynard, D., Colombetti, M.: Using WordNet to turn a folksonomy into a hierarchy of concepts. In: 4th Italian Semantic Web Workshop Semantic Web Applications And Perspectives, SWAP 2007 (2007)

A Quantitative Management Method of Software Development and Integration Projects

JiangHong Shu,Yong Duan, and Fang Wang

Abstract. The evaluation of software is always the software corporation's concern. The paper focuses on software projects and makes the classification for projects. For different software types, we set up the different project stages in which we propose and define the evaluation projects and combine "Red, Yellow, Green" light system with the score scheme instead of only using "Red, Yellow, Green" light system. The paper proposes and establishes the visualized and effective quantitative evaluation method and achieves the effective management and evaluation for software projects.

Keywords: software projects, evaluate, quantitative management.

1 Forward

As the development of software industry, how to effectively evaluate the software projects has attracted more and more attention of all software companies gradually. The object of software projects management is the software engineering project, the involved range of which covers the whole software engineering process. In order to make sure that the software projects will be successful, the key point is to make proper arrangement and execution on the working rang of the software, possible risks and required resources and so on.

Based on conclusion of success and failure in past tasks, the writer makes quantitative sorting of effective evaluation in the software project management, develops a set of effective quantitative evaluation standards for the software projects, through a visualized marking method of "Red light, Stop; Green, Go; Yellow, Warn", marks the stage of the project and the current stage status to obtain visualized warning result. On this basis, the writer combines "Red, Yellow, Green" light system with the score scheme to finally obtain the quantitative evaluation result through calculation.

JiangHong Shu · Yong Duan · Fang Wang
Shandong Hi-Speed Information Engineering Co., Ltd

Z. Qian et al. (Eds.): Recent Advances in CSIE 2011, LNEE 125, pp. 509–514.
springerlink.com

2 Preparation of Project Evaluation

During the actual software project management process, as it involves many human factors, all kinds of risks of the projects are caused, as a result, it is difficult to follow uniform standards regularly during the project execution process. On one hand, the fact is that it is difficult to follow even there are uniform standards and requirements; on the other hand, it has relatively high requirements on management personnel if management is executed according to experience. In order to solve the contradiction of above two facts, it is necessary to sum up experience to get common experience and standards on the basis of experience, thus to achieve the purpose of uniformly following.

2.1 Organizational Mode and Team Building

In order to smoothly develop the management for the software projects and achieve effective results, it is required that the whole company should have coordinated action.

The management and evaluation for the software projects are closely related to the organizational mode and team building for the projects. The traditionally universal organizational mode is "project manager" management mode , the classically organizational structure is shown as the following figure:

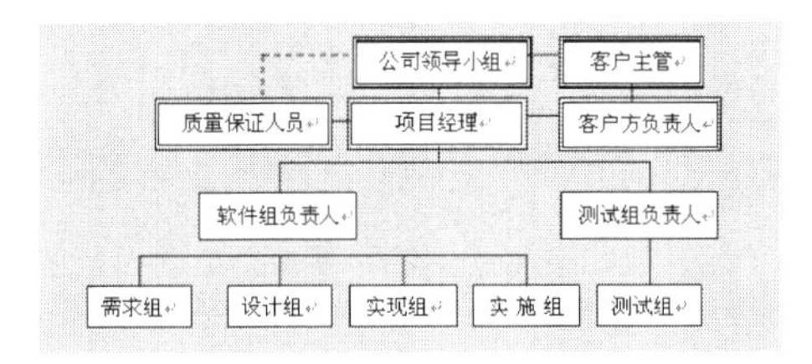

Fig. 1 Traditionally organizational structure for projects

In order to perform effective quantitative evaluation for the whole project, based on the traditionally organizational structure, a special "project evaluation team" is increased, which performs quantitative evaluation for the relative software projects in the company, informs the corresponding company (department) leaders, project management departments and project teams with the final results of evaluation, and select relative results of evaluation to issue within the company. The improved organizational structure is shown as the following figure:

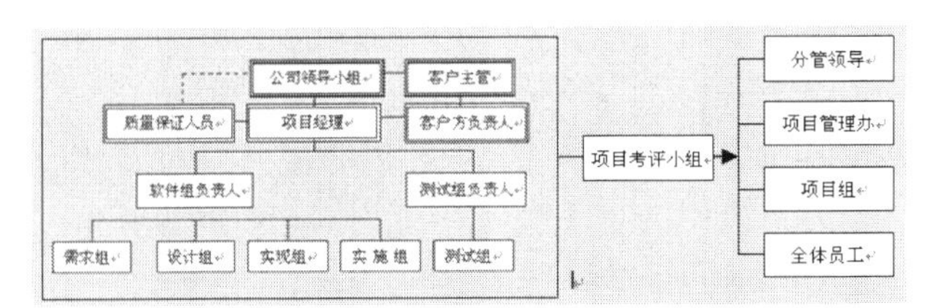

Fig. 2 Improved organizational structure for projects

2.2 Definition of Project Types

During the management for the software projects, besides standard software projects there are also non-standard software projects, such as projects with less or no code programming, that is to say, software projects taking implementation as main subject; meanwhile, there are also independent software maintenance projects.

First, we will define the types of projects. We can classify the software projects into the following types according to common types of software projects:

a) Software development projects :Software development projects are software projects including the whole process of demand analysis, design, coding, testing and the implementation, the main working contents is software development.
b) System integration projects :System integration projects are software projects taking the application implementation of software as the main working content. A small quantity of code development may be included, but not the main working part.
c) Support service projects :Maintenance service projects are software projects taking application maintenance of software as the main working content. Fundamentally, the development of software codes is not included.

As the particularity of support service projects, this text mainly describes the quantitative evaluation method for the software development projects and system integration projects.

2.3 Stage Division of Projects

a) Relative stage division of the software development projects

合同文本	●
项目计划	
需求分析说明书	●
评审报告	●
概要设计说明书	●
数据库设计说明书	●
详细设计说明书	●
评审报告	●
更改评审表	●
源代码	●
测试计划	●
测试用例及大纲	●
测试缺陷	●
测试报告	●
用户手册	●
培训计划	●
安装光盘	●
维护记录	●
客户确认报告	
客户验收报告	●
项目总结报告	●
	94

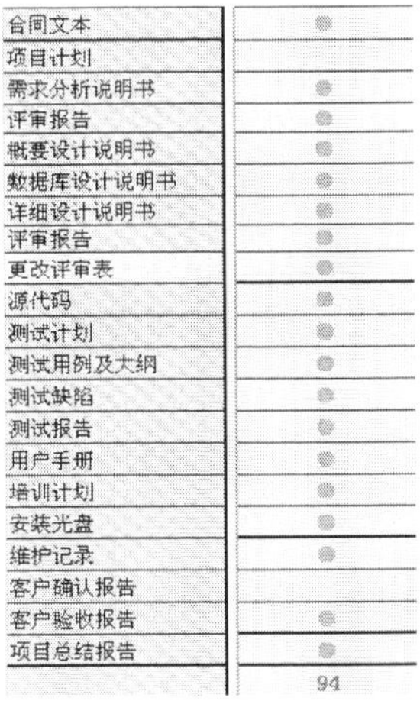

Fig. 3 Relative stage division of the software development projects

b) Relative stage division of the system integration projects

合同签订	合同文本	●
准备阶段	项目计划	●
	施工进度计划	●
	开工申请	●
设备到货	设备不开箱检验记录单	
	设备开箱检验记录单	
系统联调及测试	系统联调计划	
	系统联调测试报告	
	机电完工安装调试检验报告	
	设备安装后测试记录表	●
用户培训	用户培训计划	
	培训教材	
	培训记录单	
试运行	系统试运行报告	●
交工验收	交工验收报告	
	项目总结报告	

Fig. 4 Relative stage division of the system integration projects

3 Evaluation Standards and Evaluation Process for Projects

Chairman Mao said "everything will be successful if you prepare it in advance, otherwise it will be failed" in *On Protracted War*, that is, you cannot get victory in the war without planning and preparation in advance. This well explains the importance of planning. The same principle applies to here, after the software projects are classified and the project process is divided into stages, if there is no corresponding project planning, the evaluation of the whole software project cannot be executed in reference, just like a piece of worthless paper.

3.1 Formulation of the Project Evaluation Standards

During formulation of the project evaluation standards,for each assessment item, different prompt statuses should be defined, respectively:

●	Yellow: Prompt
●	Red: Warning
●	Green: Normal
●	Gray: Uninvolved

Based on setting red, yellow and green lights for prompt and warning, this text combines Red, Yellow and Green colors with the corresponding scales and obtains the result of quantization through calculation.

- The corresponding score of the item with green light is x, that is, the full score of each evaluation item is x;
- The corresponding score of the item with yellow light is y, y < x;
- The corresponding score of the item with red light is z, z < y;

So, the quantitative evaluation result = $$\frac{\sum_{i=1}^{a} ax_i + \sum_{j=1}^{b} by_j + \sum_{k=1}^{c} cz_k}{nx_i} * 100$$

Wherein, n=a+b+c z<y<x

For example, it is supposed that x=5, y=2, z=0, that is to say, the corresponding scores of the assessment items are green light 5, yellow light 2 and red light 0.

During assessment, it is necessary to calculate the total assessment items according to the stage of the project, evaluate each assessment item, mark red, yellow and green lights according to the results of assessment, calculate the total assessment scores according to red, yellow and green lights, and compare the assessment scores with the total scores of the total assessment items to obtain the comprehensive score finally, that is, the final result of quantization.

Above contents will be described by an example below:

Such as Fig. (4) Relative stage division of the system integration projects,during the whole evaluation, n=5, a=3, b=2, c=0

According to the calculation formula for quantitative evaluation:

$$\frac{\sum_{i=1}^{a} ax_i + \sum_{j=1}^{b} by_j + \sum_{k=1}^{c} cz_k}{nx_i} * 100$$

The comprehensive score (round) is:

Score for evaluation=(5*3+2*2)/25*100=76

That is, the comprehensive score of the project in this month is 76.

Through this evaluation method, for software development projects and system integration projects, by red, yellow and green light system, the presence or absence and quality of working products of each project in each stage can be intuitively and comprehensively understood; meanwhile, the comprehensive score is obtained through the quantitative calculation combined with the score scheme, further the quantitative evaluation result is obtained and issued within the company through relative modes. This contributes to compare the ranking situation of the project in the whole department and in the whole company, and finally promotes the good progress of the project.

3.2 *Project Evaluation Process*

After the types of projects are determined, the project plans are made and the project evaluation standards are formulated, we can assess and evaluate one project. The evaluation process for projects will be described below.

a) Sign the project contract
b) Determine the type of the project
c) Make a project plan according to the project contract.
d) mark the uninvolved assessment item with "gray light".
e) Assess the project at the end of each month according to the project evaluation standards, in conjunction with the current progress situation of the project, and obtain the comprehensive score.

4 Conclusion

During the management of projects, as there are human factors, it is determined that there are many possibilities and difficulties due to these possibilities during the execution of the projects. Managing and evaluating projects by a "Red, Yellow and Green lights" system are attempts of a new management and evaluation mode. A visualized and effective management and evaluation method for the software projects is provided and applied in practice, and has got some achievements.

References

[1] Shaw, D.G.: Team Measurement and Rewards: How Some Companies Are Getting It Right. HRP 18(3) (1995)
[2] Chen, J., Xu, X.: The Research of R&D Professional's Career Development Path and Title Evaluation. Science Research Management 20(3), 34–39 (1999) (in Chinese)
[3] Computer Software Engineering Specification Compilation of National Standards. Standards Press of China, Beijing (1998) (in Chinese)
[4] McGarry, J.: PSM. China Machine Press, Beijing (2003) (in Chinese)
[5] Zheng, R.: Software Engineering (Senior). Tsinghua University Press, Beijing (1999) (in Chinese)
[6] Humphrey, W.S.: Software Process Management. Tsinghua University Press, Beijing (2003) (in Chinese)

Large Vocabulary Continuous Speech Recognition of Uyghur: Basic Research of Acoustic Model

Muhetaer Shadike, Li Xiao, and Buheliqiguli Wasili

Abstract. In this paper, we analysis the characteristics of Uyghur languge, introduce components of speech recognition system, design a program to calculate trigram, bigram and unigram language models based on the corpus of the entire text of the novel Qum basqan sheher, then estimating acoustic models uses unigram and bigram language models calculated above, particularly with some probability theory for acoustic modeling. And discuss some potential problems of the probabilistic speech recognition approaches.

Keywords: Uyghur, n-gram language model, acoustic models, HMM, speech recognition, corpus.

1 Introduction

The most widely used elementary acoustic units in Large Vocabulary Continuous Speech Recognition (LVCSR) systems are phone based, where each phone is represented by a Markov chain with a small number of states.

Phone-based models offer the advantage that recognition lexicons can be described using the elementary units of the given language, and thus benefit from many linguistic studies. It is of course possible to perform speech recognition without using a phonemic lexicon. Compared with larger units, small subword units reduce the number of parameters, enable cross-word modeling, facilitate porting to

Muhetaer Shadike
The Xinjiang Technical Inistitute of Physics & Chemistry, Urumchi, China

Li Xiao
Graduate School, Chinese Academy of Sciences, Beijing, China

Buheliqiguli Wasili
Xinjiang Education Institute, Urumchi, China
e-mail: muhtar_xjzb@163.com

Z. Qian et al. (Eds.): Recent Advances in CSIE 2011, LNEE 125, pp. 515–521.
springerlink.com

new vocabularies and, most importantly, can be associated with back-off mechanisms to model rare contexts.

A given HMM can represent a phone without consideration of its neighbors or a phone in a particular context. The context may or may not include the position of the phone within the word, and word-internal and cross-word contexts may be merged or considered as separate models. Different approaches are used to select the contextual units based on frequency or using clustering techniques or decision trees; and different context types have been investigated: single-phone contexts, triphones, generalized triphones, quadphones, and quinphones, with and without position dependency.

Acoustic model training consists of estimating the parameters of each HMM. The most popular approaches make use of the maximum likelihood criterion, ensuring the best match between the model and the training data. Since the goal of training is to find the best model to account for the observed data, the performance of the recognizer is critically dependent upon the representatives of the training data. Speaker independence is obtained by estimating the parameters of the acoustic models on large corpora containing data from a large speaker population. There are substantial difference in speech from male and female talkers arising from anatomical difference and social ones. It is thus common practice to use separate models for male and female speech in order to improve recognition performance, which in turn requires automatic identification of the gender.

Uyghur language is a Turkic language spoken primarily in the Xinjiang Uyghur Autonomous Region, China, mainly by the Uyghur people. Like many other Turkic languages, Uyghur language displays vowel harmony and agglutination, lacks noun classes or grammatical gender, and is a left-branching language with Subject Object Verb word order. It is widely accepted that Uyghur language has three main dialects, the Central dialects, the Southern dialects and the Eastern dialects, all based on their geographical distribution. Each of these main dialects has a number of sub-dialects which all are mutually intelligible to some extent.

The Central dialects are spoken by 90% of the Uyghur-speaking population, while the two other branches of dialects only are spoken by a relatively small minority. Vowel reduction is common in the northern parts of where Uyghur is spoken, but not in the south.

In Uyghur and other Turkic languages, the basic syllabic structure is centered on a vowel (V), which may be preceded by, or followed by a consonant (C). The following combinations illustrate typical Turkic syllabic structures: V, VC, CV, and CVC. Each syllable must contain a vowel. Note that, due to the great number of loan words, the following syllabic structures with consonant clusters have been adopted into Uyghur: VCC, CVCC, CCV, and CCVC. With the exception of a few loan words, word stress in Uyghur falls onto the last syllable [7].

In this paper, we shall not look at full-blown speech recognition systems. We want to concentrate more on the theories for acoustic modeling than on particular applications. In sections 2, brief describes the components of speech recognition system include concept of front-end processing, decoder, acoustic model and language model. Section 3 outlines of probability theory for acoustic modeling. Sections 4 estimate the acoustic model of Uyghur based on *Qum Basqab Sheher* corpus. Section 5 present some potential problems of the acoustic modeling approach.

2 Components of Speech Recognition System

Today's most speech recognizers are based on statistical techniques, with Hidden Markov Models being the dominant approach. The main components of a speech-recognition system are the front-end processor, the decoder, the acoustic models, and the language models [4]. The speech-recognition problem can be formulated as the search for the best hypothesized word sequence given an input feature sequence.

A sequence of acoustic feature vectors is extracted from a spoken utterance by a front-end signal processor. We denote the sequence of acoustic feature vectors by $x=(x_1+x_2,...,x_T)$, where $x_t \in X$ and $X \subset R^d$ is the domain of the acoustic vectors. Each vector is a compact representation of the short-time spectrum. Typically, each vector covers a period of $10ms$ and there are approximately T=250 acoustic vectors in a 10 word utterance. The spoken utterance consists of a sequence of words $v=(v_1,v_2,...,v_N)$. Each of the words belongs to a fixed and known vocabulary V, that is, $v_i \in V$. The task of the speech recognizer is to predict the most probable word sequence v' given the acoustic signal x. Speech recognition is formulated as a Maximum a Posteriori (MAP) decoding problem as follows:

$$v' = \arg\max_v P(v|x) = \arg\max_v \frac{p(x|v)P(v)}{p(x)} \tag{2.1}$$

where we used Bayes' rule to decompose the posterior probability in (2.1). The term $p(x/v)$ is the probability of observing the acoustic vector sequence x given a specified word sequence v and it is known as the acoustic model. The term $P(v)$ is the probability of observing a word sequence v and it is known as the language model. The term $p(x)$ can be disregarded, since it is constant under the max operation.

a. Front-end processor

The front-end processor compresses the speech signal into a sequence of parameters that will be used to recognize spoken words. The front-end typically operates on short frames (windows) of the speech signal, performing frequency analysis and extracting a sequence of observation vectors of frequency parameters (or acoustic vectors) $x=(x_1+x_2,...,x_T)$, one for each frame. Many choices exist for the acoustic vectors, but Mel-frequency-warped cepstral coefficients (MFCC) have exhibited the best performance to date.

b. Decoder

The central component of speech recognition system is the decoder, which is based on a communication theory view of the recognition problem; it tries to extract the most likely sequence of words $v=(v_1,v_2,...,v_N)$ given the set of acoustic vectors x.

c. Acoustic Models

The acoustic model is usually estimated by a Hidden Markov Model (HMM), a kind of graphical model that represents the joint probability of an observed variable and hidden (or latent) variable [1]. In order to understand the acoustic model, we now describe the basic HMM decoding process. By decoding we mean the calculation of the *arg* max_v in (2.1). The process starts with an assumed word sequence v. Each word in this sequence is converted into a sequence of basic spoken units called phones using a pronunciation dictionary. Each phone is represented by a single HMM, where the HMM is probabilistic state machine typically composed of three state (which are hidden or latent variable) in a left-to-right topology. Assume that Q is the set of all states, and let q be a sequence of states, that is $q=(q_1,q_2,...,q_T)$, where it is assumed there exists some latent random variable $q_t \in Q$ for each frame x_t of x. Wrapping up, the sequence of words v is converted into a sequence of phones p using a pronunciation dictionary, and the sequence of phones is converted to a sequence of states, which in general at least three states per phone. The goal now is to find the most probable sequence of states.

d. Language Models

Language models are used to estimate the probability of a given sequence of words, $P(v)$. The language model is often estimated by n-grams [6], where the probability of a sequence of N words $(v_1, v_2, ..., v_N)$ is estimated follows:

$$p(v) \approx \prod_t p(v_t | v_{t-1}, v_{t-2}, ..., v_{t-N}) \tag{2.2}$$

where each term can be estimated on large corpus of written documents by simply counting the occurrences of each n-gram. Various smoothing and back-off strategies have been developed in the case of large n where most n-grams would be poorly estimated even using very large text corpora.

3 Probability Theory for Acoustic Modeling

Formally, the HMM is defined as pair of random processes q and x, where the following first order Markov assumptions are made:

1. $P(q_t/q_1,q_2,...,q_{t-1})=P(q_t/q_{t-1})$;
2. $p(x_t/x_1,...,x_{t-1},x_{t+1},...,x_T,q_1,...,q_T)=p(x_t/q_t)$.

The HMM is a generative model and can be thought of as a generator of acoustic vector sequences. During each time unit (frame), the model can change a state with probability $P(q_t/q_{t-1})$, also known as the transition probability. Then, at every time step, an acoustic vector is estimated with probability $p(x_t/q_t)$, sometimes referred to as the emission probability. In practice the sequence of states is not observable; hence the model is called hidden. The probability of the state sequence q given the observation sequence x can be found using Baye's rule as follows:

$$p(v) \approx \prod_t p(v_t | v_{t-1}, v_{t-2}, ..., v_{t-N}) \tag{3.1}$$

where the joint probability of a vector sequence x and a state sequence q is calculated simply as a product of the transition probabilities and the output probabilities:

$$p(x,q) = P(q_0)\prod_{t=1}^{T} P(q_t|q_{t-1})p(x_t|q_t) \tag{3.2}$$

where we assumed that q_0 is constrained to be a non-emitting initial state. The emission density distributions $p(x_t/q_t)$ are often estimated using diagonal covariance Gaussian Mixture Models (GMM), also called Continuous Density HMM (CDHMM), for each state q_t, which model the density of a d-dimensional vector x as follows:

$$p(x) = \sum_i \omega_i N(x; \mu_i, \sigma_i) \tag{3.3}$$

where $\omega_i \in R$ is positive with $\sum_i \omega_i = 1$, and $N(x; \mu, \sigma)$ is a Gaussian with $\mu_i \in R^d$ and standard deviation $\sigma_i \in R^d$. Given HMM parameters in the form of transition probability and emission probability, the problem of finding the most probable state sequence is solved by maximizing $p(x,q)$ over all possible state sequences.

4 Estimating Acoustic Models

In order to estimate the acoustic model of particular word sequence, we must count the number of occurrences of particular kinds of word sequences. Let's take an example. Suppose our corpus is the entire text of the novel *Qum Basqan Sheher (The City Flooded with Desrt, Memtimin Hoshur, 1996)*, there are 58284 word tokens. In the *Qum Basqan Sheher* corpus, *'jahan'* occurs 63 times, *'kezdi'* occurs 43 times and *'jahan kezdi'* occurs 40 times. Now we interested in knowing the probability that *'jahan'* come after *'kezdi'*. This would be the acoustic model that *'kezdi jahan'*. Let j denote *'jahan'*, and k represent *'kezdi'*. It is known that

$P(j) = 63/58284 = 0.0010809$,
$P(k) = 43/58284 = 0.0007378$,
$P(k/j) = 40/58284 = 0.0006863$.

These are our language models [2] that based on *Qum Basqab Sheher* corpus. The additional information we have about *'kezdi jahan'* is that *'kezdi'* already recognized, and we use this to estimate the acoustic model of *'jahan'* that it comes after *'kezdi'*, $P(j/k)$.

Baye's rule [5] can be derived from the definition of conditional probability (3.1). Writing this in terms of our events, we are interested in the following probability:

$$P(j|k) = \frac{P(k,j)}{P(k)} \tag{4.1}$$

where $P(j/k)$ represents the acoustic model that *'jahan'* comes after *'kezdi'*, and k is the event that *'kezdi'* already recognized. Using the Multiplication Rule (3.2), we can write the numerator of (4.1) in terms of event k and our language model that the *'jahan'* comes after *'kezdi'*, as follows

$$P(j|k) = \frac{P(k,j)}{P(k)} = \frac{P(j)P(k|j)}{P(k)} \tag{4.2}$$

Now we use (4.2) able to calculate the $P(j/k)$ from the given information:

$$P(j|k) = \frac{P(j)P(k|j)}{P(k)} = \frac{0.0010809 \times 0.0006863}{0.0007378} = 0.0010055$$

5 Potential Problems of the Probabilistic Approach

Although most state-of-the-art approaches to speech recognition are based on the use of HMM, they have several drawbacks [1], some of which we discuss hereafter.

- Consider the logarithmic form of (3.2),

$$\log p(x,q) = \log P(q_0) + \sum_{t=1}^{T} \log P(q_t|q_{t-1}) + \sum_{t=1}^{T} \log p(x_t|q_t) \tag{5.1}$$

there is a known structural problem when mixing densities $p(x_t/q_t)$ and probabilities $P(q_t/q_{t-1})$: the global likelihood is mostly influenced by the emission distributions and almost at all by transition probabilities, hence temporal aspects are poorly taken into account. This happens mainly because the variance of densities of the emission distribution depends on d, the actual dimension of the acoustic features: the higher d, the higher the expected variance of $p(x/q)$, while the variance of the transition distributions mainly depend on the number of states of the HMM. In practice, one can observe a ratio of about 100 between these variances; hence when selecting the best sequence of words for a given acoustic sequence, only the emission distributions are taken into account. Although the latter may well be very well estimated using GMM, they do not take into account most temporal dependencies between them.

- While the EM algorithm is very well known and efficiently implemented for HMM, it can only converge to local optima, and hence optimization may greatly vary according to initial parameter settings. For CDHMM the Gaussion means and variances are often initialized using K-Means, which is itself also known to be very sensitive to initialization.

- Not only is EM known to be prone to local optimal, it is basically used to maximize the likelihood of the observed acoustic sequence, in the context of the expected sequence of words. Note however that the performance of most speech recognizers is estimated using other measures than the likelihood. In general, one is interested in minimizing the number of errors in the generated word sequence. This is often done by computing the Levenshtein distance between the expected and the obtained word sequences, and is often known as the word error rate. There might be a significant difference between the best HMM models according to the maximum likelihood criterion and the word error rate criterion.

Hence, throughout the years, various alternatives have been proposed. One line of research has been centered around proposing more discriminative training algorithms for HMM. That includes Maximum Mutual Information Estimation (MMIE), Minimum Classification Error (MCE), Minimum Phone Error (MPE) and Minimum Word Error (MWE). All these approaches, although proposing better training criteria, still suffer from most of the drawbacks described earlier.

References

[1] Keshet, J., Bengio, S.: Automatic Speech and Speaker Recognition: Large Margin and Kernek Methods. A John Wiley and Sons, Ltd., Publication (2009)
[2] Martinez, W.L., Martinez, A.R.: Computational Statistics Handbook with MATLAB. Charman & Hall/CRC (2002)
[3] Coleman, J.: Introducing Speech and Language Processing. Peking University Press (2010)
[4] Rayner, M., Carter, D., Bouillon, P., Digalakis, V., Wirén, M.: The Spoken Language Translator. Peking University Press (2010)
[5] Smith, K.J.: Finite Mathematics. Brooks/Cole Publishing Company, Monterey (1985)
[6] Shadike, M., Xiao, L., Wasili, B.: Large vocabulary continuous speech recognition: basic research of trigram language model. In: ICMCE (2010)
[7] Zakir, H.A.: Introduction to Modern Uighur. Xinjiang University Press (2007)

A Workflow Engine to Achieve a Better Management of Enterprise Workflow

Wei Li and Lihong Chen

Abstract. Due to traditional workflow engines focus on different aspects, there are two aspects of deficiency. On the one hand, lacking of a unified modeling standards make it inconvenient to build up and analyze enterprise workflows; On the other hand, lacking of rationality analyzing capacities and decision supporting capacities make it could not support closed-loop management of enterprise workflow very well. Against these deficiencies, a new solution is put forward in this paper. In this solution, workflow modeling references to ARIS (Architecture of Integrated Information System) specifications. Rationality analyzing mechanism and runtime exception warning mechanism are provided.

Keywords: Workflow Engine, Closed-loop Management, Workflow Modeling.

1 Introduction

So far, different workflow products possess their suite of models, workflow definition languages, and APIs[1]. The most widely used workflow management platforms (e.g. JBPM 3.0, FlowMark), didn't have enterprise workflows managed in integrated information architecture, which leads to poor portability. Capacities of reasonable constraints detecting were not involved either, managers have to search in the large log file to find problems exist in workflow models after workflow disruption or infinite loop processing had occurred, which should be time-consuming and inefficient. Address these issues, why don't we build up effective mechanisms in workflow engines to archive the goal of efficient modeling and stable operating of enterprise workflows?

2 Engine Architecture

Workflow engine introduced in this paper is composed of workflow modeling and analyzing module, workflow controlling module and decision supporting module.

Wei Li · Lihong Chen
North China Electric Power University
e-mail: liwei@ncepu.edu.cn, lihongchensrc@yahoo.cn

Z. Qian et al. (Eds.): Recent Advances in CSIE 2011, LNEE 125, pp. 523–528.
springerlink.com

Workflow modeling and analyzing module provides the GUI for business workflow designers to build up enterprise workflow models. And then analyze these models and corresponding information configured on them. Finally, generates knowledge and rules for enterprise management and store them in the relationship database in the form of standardized data objects.

Workflow controlling module establishes engine logic according to the knowledge and rules described in the database, generates proper algorithm to perform task assignment and resource allocation.

Decision supporting module is responsible for giving recommendations for optimization of irrational workflows. On the one hand, it analyzes the rationality of workflow model while modeling, in order to ensure the workflow model and engine logic saved in relationship database are correct. On the other hand, it monitors operational status of each workflow instance. Operational status that is quite inconsistent as it is described in the knowledge base will be recorded in the database as runtime exceptions. Administrator can look over these exceptions at any time and make corresponding decisions. These strategic decisions will be saved in the database as new knowledge.

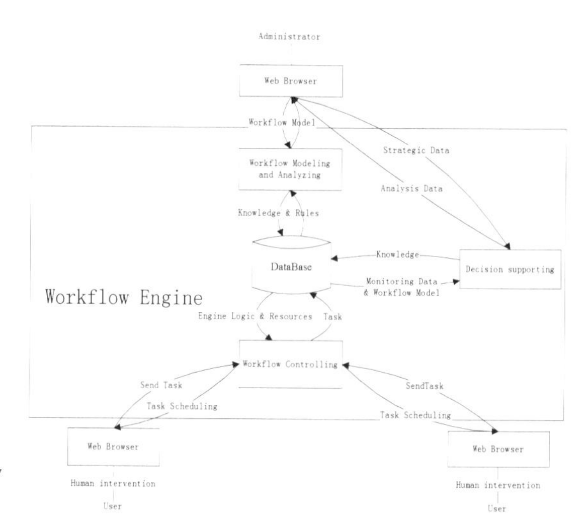

Fig. 1 Architecture of workflow engine

3 Workflow Modeling

{N,T,R,V} make up an integral workflow model[1]. N and T build up the skeleton of enterprise workflow, R and V is the information configured on the skeleton.

3.1 Nodes-N

Nodes include {Start Node, End Node, Task Node, Node, Fork Node, Join Node, Workflow Interface}.Taking its specificity into account, we emphasize on workflow interfaces here.

Workflow Interfaces are the references of workflow models. As it is shown in figure 2, the workflow interface that references to workflow model B is called the inflow, outflow of workflow model A in the case of C, D respectively. In special, workflow A is the sub-workflow of B in the case of E. In converse, workflow B is the sub-workflow of A in the case of F.

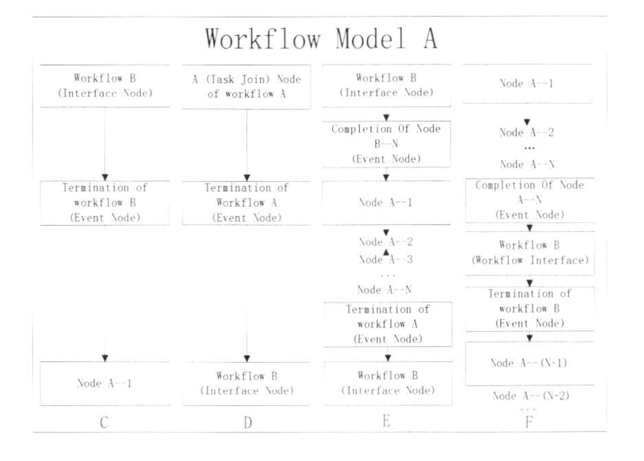

Fig. 2. Different kinds of workflow interfaces

3.2 *Transitions-T*

Transitions include {Conditional Transitions, Simple Transitions}.

Conditional transitions stand for the workflow engine will trigger the successor only if the workflow context or the specific data in database fulfill the conditions configured on conditional transitions.

Simple Transitions stand for the engine will trigger the successor unconditionally with the completion of predecessor.

3.3 *Rules-R*

Rules are available to generate engine logic. Abide by these rules; engine logic decides when to trigger the correct node by judging workflow context during the life cycle of workflow instance.

3.4 *Variables-V*

Variables are arguments that available for match rules configured on engine logic, it can be both workflow context-related data and specific data stored in the database.

4 Workflow Controlling

Workflow controlling performs task scheduling and resource allocating according to the engine logic. In order to have a better control of workflow, the concept of token should be introduced here. Token indicates the current running node of a workflow instance, only the node holds the token has the right of execution. Simultaneously, token maintains context sensitive data of each workflow instance which makes it more convenient to make transition decisions. In addition, taking the cost of passing tokens into account, token of sub-workflow instance inherits from parent workflow instance.

5 Decision Supporting

Capacities of rationality checking, timeout detection and infinite loop detection are included in decision supporting module, which makes this engine more efficient and robust than others in general.

5.1 *Rationality Checking*

Consistency validation of workflow interface is an indispensible means to avoid workflow disruption. Marking the workflow interface references to workflow model 'A' as 'AI', and the workflow interface reference to workflow model 'B' as 'BI'.

1. If 'BI' occurs in 'A' as an inflow interface, 'AI' should be occurs in 'B' as an outflow interface.
2. In special, if 'B' is a sub-workflow of 'A', 'AI' should be occurs in 'B' both as inflow interface and outflow interface.

According to interface validating rules, Figure 3 illustrates workflow model 'B' that matches the workflow model 'A' shown in figure 2. Case 'G', 'H', 'I', 'J' matches case 'C', 'D', 'E', 'F' respectively.

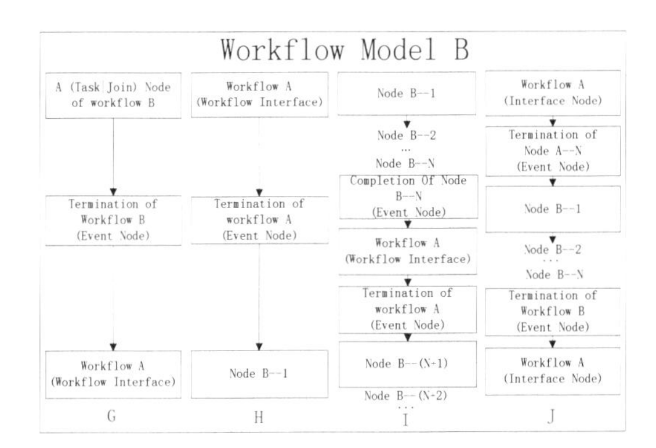

Fig. 3 Interface consistency validating rules

5.2 *Timeout Detection*

Timeout means a business operation costs much more time than it usually needs. Timer starts to work once a task instance is constructed. Marking the usually need time of business processing as 'UT' , time had gone by as 'T', total time ever spent on this task as 'S' , total number of this task ever been instantiated as 'N', longest time it cost that we can tolerate as 'M' . Following case may occur with time going:

1. Task be finished when T≤UT. Then UT = (T+S)/N.
2. Task be finished when UT<T≤M. Then UT = (T+S)/N. Besides, the corresponding timeout action configured on this task will be executed, e.g. the corresponding document will be sent to the direct leader of the original actor if it is an audit-related business processing.
3. Task hadn't been finished until T>M. Information about this task instance will be logged in the database as timeout exception. Administrator can make strategic decisions by checking these exceptions.

5.3 Infinite Loop Detection

If incorrect workflow models were established or inappropriate engine logic were configured on these workflow models, a phenomenon that workflow is running in a sub-process recursively and cannot be driven to any other nodes through self-control may occur. This phenomenon is called infinite loop in this paper.

Both infinite loops caused by simple transitions in a single workflow model and among several workflow models that makes up an integral workflow can be detected while modeling. The detection process can be divided into two steps: First, generate the corresponding controlling graph for each integral workflow. Second, output all the cycles constituted by simple transitions according to the controlling graph. Algorithm to generate controlling graph is described below:

```
Algorithm GWG.
Input<&G,models>//G:empty graph, workflow models
Output<&G>        //controlling graph.
Begin
If(exist{ model1,model2}&& model1 is sub-workflow
    model2&&model1 is (inflow/outflow) of model2).
  Model1 and model2 make up an infinite loop.
  Exit.
End If.
Loop   i←0 to models.size()-1.
    If(models.get(i) has a start node)
        Call GCG. arguments: <&G,m>.
    End If.
End Loop.
End.
```

```
Algorithm GCG .
Input: <&G, w> // G :controlling graph
//w: a workflow model which has a start node
Output: <G>    // G: controlling graph
Begin
List list = list of outflow interfaces of w.
Loop   k←0 to list.size()-1.
If (!(list.get(k) references to parent workflow
    of w) && !(Edge(w,list.get(k)∈G.edges))
      Add Edge(w,list.get(k)) to G.edges.
      Call GCG. arguments:<&G,list.get(k) >.
End If.
End Loop.
End.
```

Detection of infinite loops caused by conditional transitions will be delayed to runtime. If a sequence of nodes be instantiated repeatedly more than 'N' times in the same workflow instance, then the workflow instance will be suspended, corresponding information will be logged in the database as runtime exceptions.

6 Conclusions

In this paper, a new solution is represented to solve two major problems exist in traditional workflow engines. Workflow modeling specifications makes it easy to build up standardized enterprise workflows; Decision supporting capacity make it supports close-loop management of enterprise workflow better.

References

[1] Li, J.: Research and development of web services-based workflow platform. Jiangsu University (2008)
[2] Feng, S.: Workflow platform based on EJB for power plant. North China Electric Power University (2010)
[3] Huan, M.: Research of web services-based lightweight workflow engine. Harbin Engineering University (2007)
[4] Li, H.: EJB 3.0 classic getting start. Tsinghua University Press (2008)
[5] Tretola, R., Barber, S., Erickson, R.: Advancedflex programming. Posts&Telecom Press (2008)

Software Project Process Models: From Generic to Specific

Hao Wang, Xuke Du, and Hefei Zhang

Abstract. Organization Standard Software Process models are based on the assumption of similarity between projects - either cross or within a given business section. The research describes a validated reference metamodel, based on an empirical study on the commonality of different projects from various business areas. Based on the metamodel, we suggest a tailoring approach and tools for design and generation of the specific project process model.

1 Introduction

In an organization, multiple standard reference process models may be needed to address the needs of different application domains, lifecycle models, methodologies, and tools[2]. The organization's set of standard processes contains process elements (e.g., a work product size-estimating element) that may be interconnected according to one or more reference models. Since one model may not be appropriate for all situations, the organization may adopt different reference models.

Process tailoring is making, altering, or adapting a process description for a particular end[2]. For example, a project tailors its defined process from the organization's set of standard process models to meet the objectives, constraints, and environment of the project. That is, tailoring is the act of adapting a standard software process to meet the needs of a specific project[5].

Nowadays, process tailoring is regarded as a mandatory task implemented in the process plan phase, especially when the organization wants to comply with international standards such as ISO or CMMI. However this important activity is usually

Hao Wang · Xuke Du
Beijing University of Posts and Telecommunications, Beijing, China
e-mail: wanghao@bupt.edu.cn, duxuke@yahoo.cn

Hefei Zhang
BCI Technology Inc., Beijing, China
e-mail: service@bcitech.com.cn

Z. Qian et al. (Eds.): Recent Advances in CSIE 2011, LNEE 125, pp. 529–537.
springerlink.com

carried out following ad-hoc approach with neither guideline nor systematic method. Furthermore, to date very few research has been done in this area[10, 12, 13, 3, 11].

In this paper we present a new approach by defining organizational reference metamodel and based on the generated organizational reference process model which can be used as the process tailoring basis for specific projects. Based on our research, a tool called BeyondPLM is implemented to demonstrate the approach.

This paper is organized as follows: Section 2 present a validated process metamodel. In Section 3 we introduce the method of design and generation of organization reference process models. Specific project process model generation approach is introduced in detail in Section 4. We conclude with an outlook on next steps in Section 5.

2 The Reference Metamodel

The proposed metamodel is a well-organized model that combining the process elements in software processes based on the software development lifecycle, including roles, activities, artifacts, standards, guidelines, document templates, checklists, tools, instantiations and historical cases etc. It is also used as domain model to construct the process management tool - BeyondPLM. The proposed metamodel is represented as follows in Figure 1. Before we can render how this metamodel is used to generate the organization reference process model and serve process tailoring, we need to dive into the metamodel first, to explore what exactly constitutes the process model and how they make it.

In an organization, software process assets typically may include workflows, templates, standards, guidelines, checklists, tools, instantiations and historical cases, etc. In our approach, the process assets are boiled down into following process elements, they are "LifeCycleModel", "Process", "Role", "Activity", "Artifact", "Tool", "Metric"and so on, which are organized around "LifeCycleModel" element. The elements, except role, activity, artifact and process, can be regarded as "Support"element.

When we use this metamodel as domain model to build the process managment system - BeyondPLM, this model can be used both in the organization reference process model definition part and specific project process tailoring part. The reason that why are these two parts rather than some other parts, is for such an idea: the organization reference process models offers a series of predefined standard processes needed for software process according to different lifecycle models, and the project process tailoring part supports process tailoring for a specific project corresponding to a suitable reference process model selected from the reference process models; on the other hand, the two parts may be carried out by different departments or roles in practice, for example, one is PMO (Project Management Office), while the other is PM (Project Manager)[8]. Obviously, the metamodel is the basis for the two parts. A further discussion of the relationship between the two parts will be carried on later while going through Section 3 and Section 4.

3 Design and Generation of Organization Reference Process Models

Consider object diagrams in the lifecycle part. Organized around "Lifecycle", four key domain objects build an organic process system through all assets found in software processes being closely integrated and workflows being established orderly. A simple mapping between typical organizational software process assets and key domain objects is shown in Figure 2.

Standards, templates, checklists, instantiations and historical cases of organizational software process assets are categorized into "Support" as "Support" elements for "Role", "Activity" and "Artifact"; workflows of the software processes are mainly implemented by the organization of "Activity" elements, together with "Role" elements who are responsible for the "Activity" and "Artifact" elements which are produced due to the implementation of "Activity"; guidelines concerning process tailoring are partially laid on the elements definition of "Role", "Activity" and "Artifact", while partially laid on "Support" acting as a support.

In such process system, "Lifecycle" serves as a unique identifier to a defined process. One "Lifecycle" is always binding with a "Role" object, an "Activity" object, an "Artifact" object and a "Support" object, and the "Lifecycle" has a one-to-many relationship with them, that is, one "Lifecycle" element has multiple "Role" elements, multiple "Activity" elements, multiple "Artifact" elements and multiple "Support" elements. When creating a lifecycle, guidelines on how to distinguish it

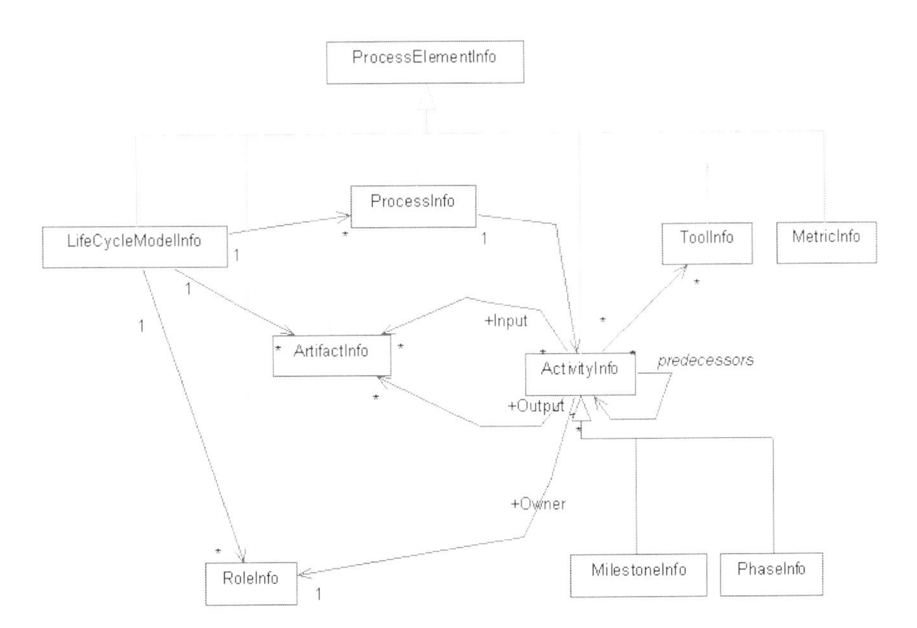

Fig. 1 The Metamodel for Software Process Management

as the proper one from lifecycles existing should be defined together, and also an empty four-objects should be generated automatically under the lifecycle.

The relationships within and between four domain objects under a lifecycle are complex, since the process assets and the software process itself should be covered by them as shown in Figure 2.

The "Role" object in the lifecycle part is a type of actors responsible for the "Activity" and "Artifact" objects. It has a one-to-many relationship with the others. A "Role" element of one kind may be responsible for multiple "Activity" elements and "Artifact" elements under the support of multiple "Support" elements.

When we add a new "Role" element, in addition to defining some basic properties of the "Role" such as RoleName, Description, etc, the "Activity" elements and "Artifact" elements and the "Support" elements should be assigned to "Role". Moreover, a tailoring guideline such as "IsAllowedTailoring", "TailoringCondition" and so on, should be defined together with "Role" and each element of the "Activity" elements, "Artifact" elements and "Support" elements related to the "Role".

The "Activity" is mainly responsible for the workflows of the software process with tailoring guidelines. That is where the most important meaning of the process tailoring lies and the crux of the process tailoring[7, 14, 1, 6]. For this reason, it has a much more complex relationship than other objects, as listed below.

- Relationship with "Lifecycle". Many "Activity" elements are included in a "Lifecycle" element. They have a many-to-one relationship.
- Relationship with "Role". More than one "Activity" elements are in the charge of one "Role" element. They have a many-to-one relationship too.
- Relationship with "Artifact". In software development process, an activity usually has one or more outcomes, and in order to make this achievement, it may also need a number of incomes as a prerequisite for current activity. The income and outcome talked about here, are input and output "Artifact" shown in the domain model. Meanwhile, an input or output "Artifact" element can serve several "Activity" elements. Therefore, every "Activity" element has a many-to-many (or rather, Zero or more, corresponding to "0...*") relationship with both "'input' Artifact" and "'output' Artifact".

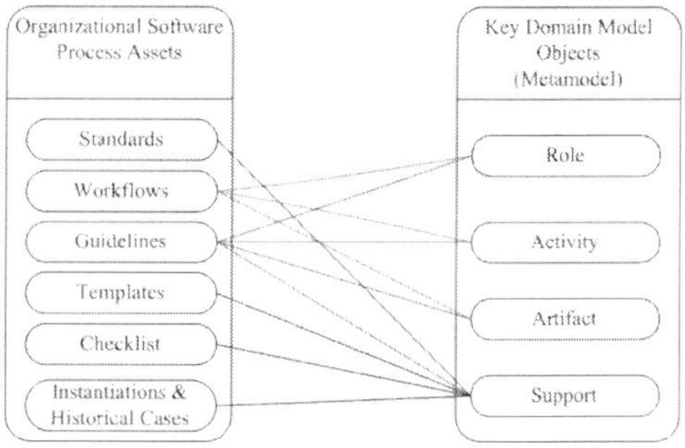

Fig. 2 A Simple Mapping Between Software Process Elements and the Key Domain Objects

- Relationship with "Support". During the process of carrying out an activity, many process assets like tools, templates, checklists and so on, may be needed as a necessity or as an aid. So, an "Activity" element has a one-to-many relationship with "Support" elements.
- Relationship within "Activity". Basically, a software process is composed of a series of orderly activities. To guarantee the orderly development activities and to promote the software development of the organization, two kinds of relationships are defined within "Activity" in the domain model. One is related to "Activity" elements in the sequence, an activity is a previous activity or a successive activity to another specific activity in the software process; the other is related to "Activity" elements in the level, an activity is a higher level activity as a parent activity or a lower level activity as a sub-activity (or child activity) to another specific activity in the software process. It is noteworthy that, a "'Parent' Activity" here corresponds to "Phase" mentioned in some other papers, while a "'Child' Activity" refers to "Task". An "Activity" element may have zero or more previous activities to depend on, and zero or more successive activities to make reliant.Meanwhile, an "Activity" element may have zero or one parent activity, and zero or more children activities.
- Latent relationships. For example, if the activity is a parent activity, then the activity's output artifacts are a union set of its children's.

One thing to make clear, the entrance criteria and exit criteria of each "Activity" element have already been taken into consideration, mainly in form of input/output "Artifact" elements, previous/successive "Activity" elements and "Support" elements. Similar to "Role", when we add a new "Activity" element under a specific lifecycle, besides some basic properties to be defined, there is also a collection of associated elements to consider, for instance,

- to define the tailoring guidelines, such as "IsAllowedTailoring", "TailoringCondition" and "TailoringDescription" and so on;
- to allocate the activity level, identify that this activity is a parent activity or a child activity, if it is a parent activity, specify its children; if it is a child activity, then specify its parent;
- to arrange the activity sequence, specify the previous and successive activities of current activity;
- to designate a "Role" element who is responsible for the activity;
- to assign the "Artifact" elements associated with the activity, including input artifacts and output artifacts;
- to assign the "Support" elements to put in use in the implementation of the activity.

Thus, as an example, a typical "Activity" has a set of properties to meet these needs. Most of the properties we enumerated have already been involved above except four, the "ActivityType(Enum)", "TailoringApproach(string)", "TailoringDescription(string)" and "TailoringTimes(int)", which are also very useful to the process definition, tailoring or improvement for an organization. We will introduce them here as a supplement for the "Activity". An "ActivityType" refers to one of the four categories of process areas according to CMMI Models[2]: Process Management, Project Management, Engineering, and Support. "TailoringApproach" records the tailoring operation, which typically may be DELETE, ADD and so on, to a process element. "TailoringDescription" records the reasons for an element's tailoring. "TailoringApproach" and "TailoringDescription" associate together in the process tailoring to provide a specific tailoring report for the project defined process. "TailoringTimes" counts how many times an activity is used, to lay a foundation for future process improvement.

The "Artifact" is the output of an "Activity", which in return can be used as input of other "Activity". The "Support" mainly provides support for development process. If needed, it can be subdivided into several different types such as tools, templates, checklists, standards, guidelines, instantiation guidance and so on. "Artifact" and "Support" each have a corresponding relationship with other objects just like "Role" and "Activity". For example, if a "Role" element has a one-to-many relationship with "Artifact", in return the "Artifact" has a many-to-one relationship with the "Role" element. Moreover, they can be defined in accordance with the way regarding "Role" and "Activity".

Multiple organization reference pocess models can be developed from the metamodel according to different lifecycle models in organization.

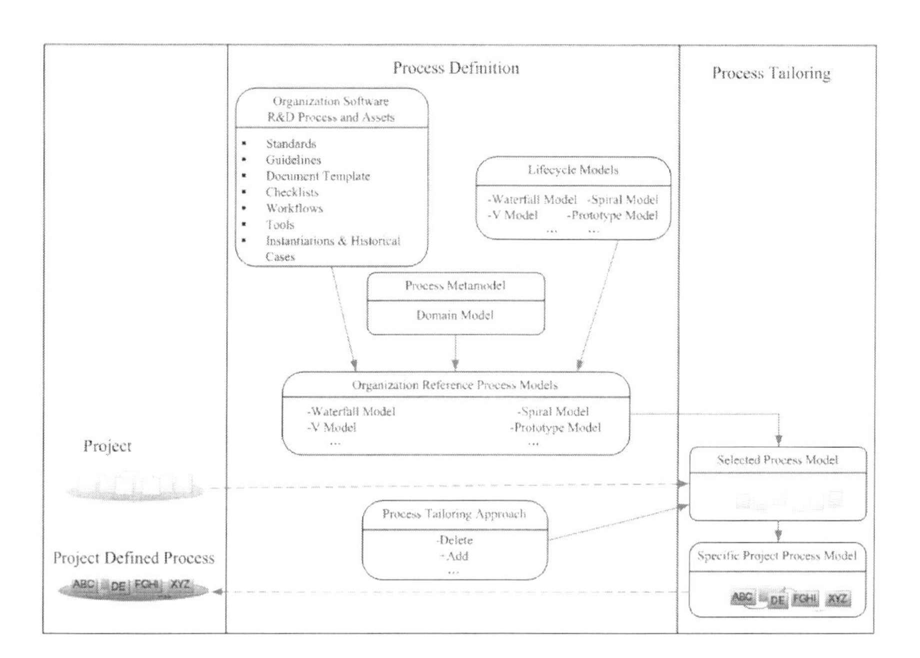

Fig. 3 Steps to arrive at Project Defined Process

Table 1 Tailoring Approaches - an Activity Tailoring Example

	DELETE	ADD*	Else
Project	–	–	–
Role	(1)If the role element responsible for the activity has no other activities responsible for,delete the role element;	(1)Designate the role element responsible for it;	/
Activity	(1) If the activity has a parent activity or has not any child activity, delete the activity only;	(1) Allocate the activity level, if it is a parent activity, specify its children; if it is a child activity, then specify its parent;	/
	(2) If the activity has children, delete the activity together with its children;	(2) Arrange the activity sequence, specify the previous and successive activities;	
	(3) Rearrange the sequences: modify the previous activities' successive activities and the successive activities' precious activities;		
Artifact	(1) Delete the activity's output artifacts;	(1) Assign its input artifacts and output artifacts;	/
	(2) If the activity's children activities or successive activities have input artifacts including the activity's output artifacts, delete the output artifacts from their input artifacts;		
Support	(1) Eliminate the associations with the supports for the activity; (2) Eliminate the associations between supports and those deleted roles, activities, artifacts;	(1) Assign the supports for the implement of the activity;	/

* Some basic tailoring guidelines as mentioned before are defined with the activity itself.

4 Constructing Specific Project Process Model

The metamodel represented as domain model makes a real blueprint of specific organization reference process model for software development process definition and tailoring. Yet, the definition of organization reference process model and process tailoring can not be accomplished in one move. It involves organizational software process assets, lifecycle models, process tailoring approaches, etc. A systematic process definition and tailoring architecture is shown in Figure 3 .

Once multiple organization reference process models are defined and a project is established, the process tailoring could start. The first step of process tailoring is selecting a defined reference process model according to characteristics of the project. To clarify the characteristics of a project, the following aspects should be taken into account[9], Nature, Complexity, Size, Risks, Miscellaneous and so on.

Then the next step is implementing process tailoring on the elements of selected specific reference process model according to the defined tailoring approach, so as to develop a well-cut and optimized project process. Process tailoring in our approach is easy and effective, and many tailoring approaches are supported well. The tailoring approaches are not limited to the most common operations such as DELETE/DROP/REMOVE and ADD as before, and more operations such as REPLACE/MODIFY (update a process element with a new name or a complete new element), CHANGE (change the workflows, for instance, change the sequence of process elements, upgrade/degrade the activity level, etc), MERGE (convert two or more process elements into one, or otherwise, unmerge one into several elements), etc, could be implemented easily too.

Whatever the tailoring approach is, process tailoring should preserve dependency relationships among the elements that compose the software process[11, 4]. An example of activity tailoring with well-formed guidelines and constraint rules is shown in Table 1. Taking "DELETE" operation for example, when an activity is to be deleted, the process elements or relationships related should be either deleted or eliminated. If the role element that is responsible for the activity has no other activities responsible for, then the role elements should be deleted with the activity. If the activity has a parent activity or has not any child activity, then we only need to delete the activity; if the activity has children activities, then we should delete the activity together with its children. In either case, after the activity is deleted, the old activity sequences should be rearranged, that is, we need to modify the previous activities' successive activities and the successive activities' previous activities. The output artifacts of the activity should also be deleted; moreover, if the activity's children activities or successive activities have input artifacts including these output artifacts, they should also be deleted from the input artifacts. Also, we should eliminate the associations between the activity and its supports, and the associations between supports and those deleted roles, activities, artifacts.

Overall speaking, during the implementation of process tailoring to arrive at project defined process, every tailoring operation taken to selected reference process model should guarantee the consistency of software process with elaboration, to avoid leading to process "fragment" or "orphan", since a well-cut or optimized lifecycle process should at least be a continuous and complete activity "chain".

5 Conclusion and Future Work

This paper has presented a validated process metamodel for organizational software process definition and process tailoring, and a tool prototype is also presented to serve to process tailoring by defining reference process models based on the approach.

The process metamodel is also represented as domain model for our process managment system, which is composed of a set of process elements abstracted from organizational software process assets. Through uniting the domain objects as an interconnected web with associations and guidelines, the domain model offers a well-organized reference process model; on the other hand, it also serves as a model template from which all organization reference process models for process tailoring are derived on a basis of existing lifecycle models. The process tailoring starts from selecting an appropriate predefined reference process model from all referenced models available in an organization according to the characteristic of the project. When it is time to implement process tailoring, multiple operation approaches are applied to the selected reference process model to arrive at a well-cut and optimized project defined process.

Our next step is to develop a quantitative approach on the selecting of the defined reference process model. The approach would provide an accurate mapping between the projects and reference process model according to characteristics of the specific project. And furthermore, the new feature could be implemented on current platform.

References

1. Arbaoui, S., Oquendo, F.: Reuse sensitive process models: Are process elements software assets too? In: The 10th International Software Process Workshop (1996)
2. CMMI for Development,Version 1.2. Carnegie Mellon University, Software Engineering Institute (August 2006)
3. Demirors, O., et al.: Tailoring ISO/IEC 12207 for instructional software development. In: The 26th Euromicro Conference (2000)
4. Pereira, E.B., Bastos, R.M., Oliveira, T.C.: A Systematic Approach to Process Tailoring. In: Proceedings of the 2007 International Conference on Systems Engineering and Modeling (2007)
5. Ginsberg, M.P., Quinn, L.H.: Process Tailoring and the Software Capability Maturity Model. Technical Report,SEI (November 1995)
6. Greenwood, R.M., et al.: An Asset View on the Software Process. In: The 10th International Software Process Workshop (1996)
7. Jacobson, I., Booch, G., Rumbaugh, J.: The Unified Software Development Process. Addison-Wesley Longman, Inc. (1999)
8. Rittinghouse, J.W.: Managing Software Deliverables:A Software Development Management Methodology. Digital Press (2004)
9. Motorola: Software Production Process. In: Acmc_Process_Spp_1.1.0, SPP, p. 3. Motorola Software Center, Motorola China (1998)
10. Pedreira, O., Piattini, M., Luaces, M.R., Brisaboa, N.R.: A Systematic Review of Software Process Tailoring. ACM SIGSOFT Software Engineering Notes 32(3) (May 2007)
11. Xu, P., Ramesh, B.: Using Process Tailoring to Manage Software Development Challenges. IEEE Computer Society, Computer.org/ITPro (July/August 2008)
12. Polo, M., et al.: MANTEMA: a Software Maintenance Methodology Based on the ISO/IEC 12207 Standard. In: The Fourth IEEE International Symposium and Forum on Software Engineering Standards (1999)
13. Polo, M., et al.: MANTEMA:a Complete Rigorous Methodology for Supporting Maintenance based on The ISO/IEC 12207 Standard. In: The Third IEEE European Conference on Software Maintenance and Reengineering (1999)
14. Pressman, R.S.: Software Engineering: A Practitioner' s Approach. McGraw-Hill Companies, Inc., New York (2001)

An Extended Role-Based Access Control Model

Zheng Yu

Abstract. By Analysis character of RBAC and problems of practical application, the article presents an extend Role-Based Access Control Model which makes authorizing more flexible and convenient

1 Introduction

Access control plays great role in information system; the strong, high-efficiency access control model makes effective management to all access requests. The Role-Based Access Control Models greatly reduces user right administrative working capacity and complexity, and better meets safe requirement in actual area, it has been developed as one more mature access control strategy. But the scale of information system turns stronger, access control strategy is required more flexible; besides that, when access control strategy is designed, the complexity and variability of users are not into consideration, as information system is developed, it is necessary to consider how to design and realize one access control strategy which can be adopted to complex and variation users and flexible usage.

As traditional access control strategy is lack of flexibility, and it has not enough capacity to deal with the change of users, based on actual requirements, this article designs and realizes access control system which has strong flexibility, adaptability, can be used into various kinds of information system, reduces management of information system on access control.

2 Role-Based Access Control (RBAC) Model

Core thought of RBAC is based on traditional access control strategy, introduces concept of "Role", divides User and Permission from logic, according to different responsibility and post of organization, sets corresponding Role, and then endows relevant access permission to Role, at last distributes Role to User. Through Role, User indirectly accesses permission to access system resources, the Permission is sealed in Role and indirectly contacts with User. The basic thought of RBAC model as showed in Figure 1.

Zheng Yu
Guizhou College of Financial and Economics Guiyang, China

Z. Qian et al. (Eds.): Recent Advances in CSIE 2011, LNEE 125, pp. 539–543.
springerlink.com

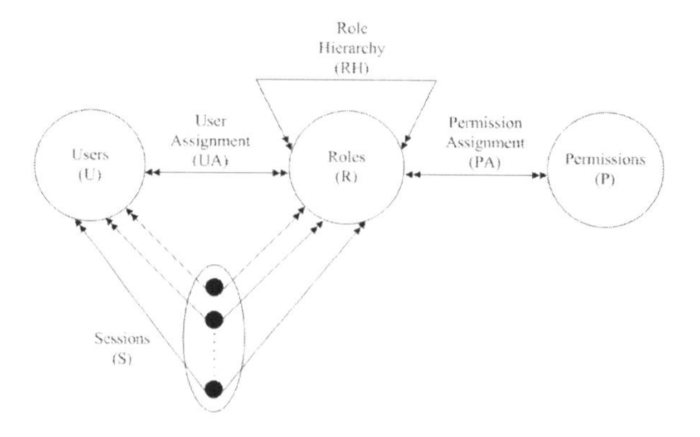

Fig. 1 RBAC Model

Compared with traditional access control, RBAC model can reduce cumbersome degree of authorized working. Such as: User is U, Role is R, and Permission is P. The value of U can be big and changes frequently; this because the Role is set according to function and post within organization, the value of R is limited, and is not influenced by U. The value of P is also limited and not influenced by U. That is, the quantity of Role and Permission is limited, and rarely changes, thus it distributes P kinds of permission to R roles, this need RxP times of authorized working, and not needs frequent change. To each user, it needs to endow one or more roles to it (to multi roles, the permission of this user is the complication of those roles' permission). In the traditional access control strategy which lacks of the Role that is the bridge between User and Permission, it needs UxP times of authorizations, and U is so big; while in RBAC, it needs $R{\cdot}P+U{\cdot}R=R{\cdot}(U+P)$ times, compared with U, R is so small, this can greatly reduce authorized working capacity of User.

In traditional access control, when certain Permission changes, it needs to amend each User who has such Permission, while in RBAC, it just needs to amend the Permission of Role, this can greatly reduce authorized changed working capacity. That is when the staffs are changed or functions of Users are altered within organization, it needs to simply delete original Role of such Role and redistributes new Role.

Above all, RBAC model abandon the method of original access control model which directs User to access Permission, this greatly reduces complexity and working capacity arose from changed Permission of User, and reduces influence on system stability stemmed from the change.

3 An Extended RBAC Model

3.1 Design of Extended RBAC Model

\Author has participated in construction of large-scale information system of certain provincial government. RBAC model can efficiently realize access control in large-scale information system with great amount of Users, this character determines RBAC model as basic model to realize information system access control.

But during specific implementation of such large-scale information system, traditional RBAC model still has defects and shortcomings, in traditional RABC model, the User is used as exclusive authorized method, this needs to reasonably set Role within system, the reasonable installation of Role can simplify authorized working capacity of access control. If the Role is divided so meticulously, the distribution of Role's Permission is so cumbersome, this can not simplify management, and increases probability of errors and wastes resources; if the Role is divided not carefully, this makes many Permissions to each Role, this can not distribute suitable Role to User or make more Permissions to certain User to have difficult to make access control, although it adopts Role-setting method within organization, it is hard to avoid any problems during actual application.

Such as: User A and User B belong to the same department, the work tasks of them are similar, in traditional RBAC model, it needs to set different Roles to User A and User B, and then separately authorizes required Permissions to these two Roles, at last distributes required Roles to User A and User B, this reduces efficiency, and increases probability of error.

Besides that, within organization there are some temporal works, based on original Role, such User needs to has one or several temporal Permissions, because such works are temporal, it can not adjust Permission of such Role, this leads all Users who belong to this Role have one or more Permissions, it is obvious it has no suitable Role within system, if it sets temporal Role to such User, after completing such temporal work, it needs to set Role to this User again, and deletes temporal Role, this can not obtain convenient and high-efficient objective, and increases complexity to working of access control.

Thus, during implementation of system, the traditional RBAC model is extended, based on traditional RBAC model, the fine-tuning mechanism is introduced, through such mechanism; it can fine adjust Permission of User. Designed RBAC model of author is in Figure 2

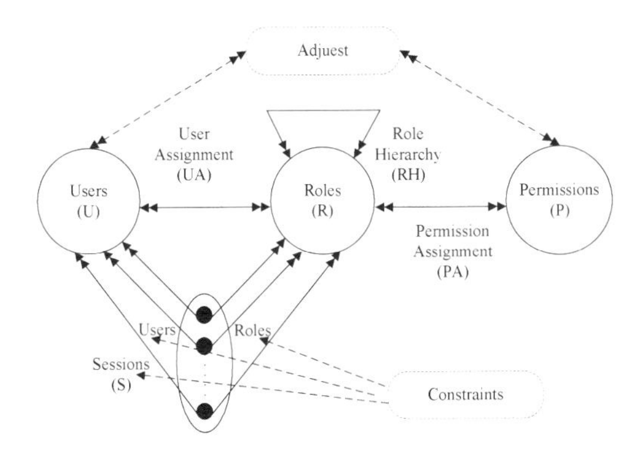

Fig. 2 Extended RBAC Model

Extended RBAC model has advantages of traditional RBAC model and introduces "fine-tuning" mechanism to make implementation of access control more flexible, convenient, it has following advantages in large-scale information system:

(1) Simplifying complexity of setting Role, avoiding meticulous division of setting Role, when the differences of two Users is small, these two Users can be endowed same Role and then they are fine adjusted.
(2) The reply of temporal change during organizing business is more flexible, it can temporally fine adjust the Permission of certain User when it is needed to complete certain temporal working, this may not influence other Users who own the same Role.
(3) The management of access control of temporal User is more convenient, the construction of information system in provincial government, which was participated by author, the business data of department was frequently reviewed by upper department, to such temporal User, it can directly adopt "fine-tuning" method of Permission to authorize such User to avoid complexity of defining Role of such temporal User.
(4) It has all advantages of traditional RBAC model.

3.2 Realization of Extended Model

The extended RBAC model was successfully used in information system of provincial government, during construction of system, the design of main data base of access control is as following:

(1) User Information Sheet. This sheet is used for storing information of Users, they are: User ID, user name, user password, etc.
(2) Role Sheet. This sheet is used for storing information of Role. It has many-to-many relationships with User Information Sheet and Permission Sheet. This information is: Role ID, Role name, Role description.
(3) Permission Sheet. This sheet is only used for storing Permission information. It has many-to-many relationship with User Sheet. The information is: Permission ID, Permission Name, Permission Description, etc.
(4) User-Role Sheet. This sheet is used for storing relationship between User and Role, User Sheet has many-to-many relationship with Role Sheet, this sheet is the relationship sheet of such two entity sheets, main information is: User ID and Role ID, etc.
(5) Role-Permission Sheet. This sheet is used for storing relationship between Role and Permission, Role Sheet has many-to-many relationship with Permission Sheet, main information is: Role ID and Permission ID, etc.
(6) User's Permission Fine-Tuning Sheet. After distribution of Role to User, if the Role's Permission can not completely meet the requirement of User's function and post, it needs to fine adjust its Permission to fit requirements. When the information of such sheet is used to distribute Role to User, it makes use of program to automatically fill in, and then amends it; this reduces difficulties to program to set Role.

When this extended model is used, the steps of distributing the Role and Permission are as following:

The steps of distributing Role are:

(1) Choosing one User, obtaining User's information.
(2) Choosing one or more Roles, obtaining Role's information.
(3) Adding User information and Role information into system.
(4) After adding Role, automatically adding fine-tuning information of User Permission into system for fine adjustment in future.

(1) Choosing one Role, obtaining Role's information
(2) Choosing one or more Permission, obtaining Permission's information.
(3) Adding Role information and Role-Permission information into system.

4 Conclusion

Aiming at shortcomings of traditional RBAC model in practical application, the RBAC model is extended, and the extended model is used in e-government affairs information system of certain provincial government, through practice, the management of User's Permission turns more easy, high-efficient, this greatly reduces difficulty of managing information system access control and expense of system, it is used for reference in designing other large-scale information system access control.

Acknowledgments. This article was subsidized from project Seagull system research and development of Chongqing Information Security Technology Center.

References

Yu, Z.: Role-Based Access Control Modelsin e-Government (2006) (in Chinese)
Zhang, Z.-J., Liu, W.: Research on general secure group communication scheme based on C/S model. Computer Engineering and Des. (2009)
Han, L.-H., Hong, F.: Asiedu Baffour Kojo. Role-based Mandatory Access Control Model. Journal of Detection & Control (August 2009)

Research on Dependability Evaluation and Measurement of Ordnance Safety Critical Software

Ma Sasa, Zhao Yang, Zhou Lei, and Zhao Shouwei

Abstract. It is the impartment method of dependability measurement and evaluation for the ordnance safety critical software to improve the software quality and ensure the whole fighting efficiency. Firstly, some concepts of dependability for the ordnance safety critical software and the evaluation grade of trust software were defined, and some characteristics of its dependability measurement were analyzed. Then the attribute model and measurement model were built. Finally, the evaluation model based on fuzzy comprehensive evaluation was applied in dependability evaluation. The estimation mechanism of dependability integrated indicator was setup to provide foundation for the quality management and dependability control of ordnance safety critical software.

Keywords: Safety critical software, Trust software, Dependability, Measurement, Evaluation.

1 Introduction

The large-scale application of software in the military has greatly improved the equipment's level of information technology and combat effectiveness, but it also broughts new problems. Such as: one gunfire control radar fault leaded to the gun system losing control,because of the clearing of internal data (Availability); because

Ma Sasa
Ordnance Technology Institute, Ordnance Engineering College Shijiazhuang, China

Zhao Yang
Ordnance Engineering College Shijiazhuang, China

Zhou Lei
Mianyang Military Representative Office of Chongqing Mianyang, China

Zhao Shouwei
Hebei Province Examination and Education Institute Shijiazhuang, China

Z. Qian et al. (Eds.): Recent Advances in CSIE 2011, LNEE 125, pp. 545–553.
springerlink.com

of the dead halt of UAV ground control systems, plane crash occurred(Defensive); the messy code of fire-control system softwares and uncertain fault became a unexcludeed risk in joint military exercise(Reliability); system paralysis occurred, because the command system was infected(Survivability). Due to software failure, software-intensive equipment accidents occur frequently. The accidents led to the loss of equipment and the unavailable effecttiveness. And there are shortcomings and loopholes at the process of software quality control in the military.The main reason is: the ordnance safety critical software has oniy a single indicator in quality evaluation system. The correlation and comprehensive of the software quality indexs are overlooked in the reliability test , security test, maintainability test and so on.It leads to that there do not form a multi-dimensional index evaluation system and a comprehensive assessment system.

Considering the operating environment and the special operations of ordnance safety critical software , it is important to analysis the software defects and the reliability of the internal relations,and to study the multi-dimensional index evaluation system, measurement and evaluation the mechanisms and evaluation system.And in order to improve and enhance the quality assurance system of military equipment,it is valuable to establish a evaluation system of multi-dimensional index and to establish the testing procedures , quantitative assessment and prediction model.During the modle established process , real-time embedded software will be used .Because the special features and the unpredictable operating environment always lead to serious consequences.

2 Basic Concept

2.1 *The Safety Critical Systems and Software*

Safety critical systems mean the failure of system functions will cause the significant damage of life , property and the system itself. The systems are widely applied in the aerospace, defense, transportation, nuclear energy, health, and many other key safety areas.

In safety critical systems, the software used in the safety critical control is called safety critical software. In the national military standard[1], safety critical software is defined as softwares whose error could cause a serious risk. Refering to GJB900 --- the grade standards of safety-critical software (Ref.[1]), A and B-class software is defined as the safety critical software. In general, the following software should be defined as safety critical software:

I The priority structure of fault detection and the module of security control , correction logic, processing and response of fault.

II The interrupt handling routine,interrupt priority model and the example program (permiting and prohibiting the interrupt).

III The software that can produce a signal(The signal has a direct impact on the safety-critical functions of hardware's operation or initiating.And the error of such software could cause serious risk to the system function).

IV The software which can generate independent-control signals to the hardware.

V The software whose output is to show the status of safety-critical hardware.

2.2 The Dependability of Software

At present, dependability has not formed a unified definition yet. There are mainly the following statements:

The TCG defines the dependability based on the expected behavior. If it's action is always as expected and towards the desired goal, then the entity is trusted[2].

The ISO / IEC 15408 standard defines dependability as follows: In any condition, components, operation or process involved in the calculation is predictable, and can also resist the virus and the physical interference[3].

According to the 4th reference, dependability means that the services provided by the computer system can be demonstrated that it is reliable. It means that the services provided by the computer system is reliable, and the dependability is demonstrable from a user's perspective.

According to the 5th reference, dependability means that the services provided by the computer system are reliable, available.The information and behavior of it's system is safe. The corresponding dependable-computing-platform is computer hardwares and software entities which can provide dependable computing-services. It also can ensure the reliability and availability of the system, the security of the information and behavior.

Dependable computer systems are generally divided into dependable hardware, dependable software, and dependable network. Dependability is the important factor that influences the combat efficiency of military equipment. It is also a important indicator of the equipment quality. In addition,its basic properties, influencing factors and protection methods include the following content generally, shown in Ref.[1].

3 The Characteristics Analysis of the Ordnance Safety Critical Software Dependability Evaluation

Most ordnance safety-critical softwares is task softwares of hard real-time embedded equipment. The software dependability assessment has the following remarkable characteristics.

3.1 Real-Time Ability

As a task and control software, real-time ability is the most obvious characteristics. Real-time system not only requires the software time is short, but also requires the results of its data processing and data entry time is closely linked together. Therefore, during the testing process of equipment software, we should not only verify the execution logical relation of software sentence, but also compare the process execution time with the given time. And we put forward higher requirements to the choice of test cases. In order to test the software real-time behavior, test case must contain detailed information of the input and output time, and time-related input variables should be defined by stochastic process. The execution results of test cases is always related to its executed order which is must be taken into consideration during the analysis of test results.

3.2 Complexity

As the core of the equipment. the task software emphasis on security, and has a lot error handling and fault-tolerant features in application-level. As a result, software complexity degree has been greatly improved, and the testability has been reduced. Fault-tolerance mechanism is mainly seen in the part of exception event handling, state restoration, and boundary value of the data, accident data processing. This mechanism is rarely executed in general testing process. Many system software defects or hardware failure, operator errors are covered up by a large number of software fault-tolerant mechanisms. As the result, during the testing process,the spread of software defect and effect is difficult to be tracked and located. Therefore, in the process of software testing it is necessary to do fault-tolerant mechanisms' experimental evaluation through fault injection.

3.3 Hardware-Dependent

All real-time embedded software is highly dependent on the hardware. As the result, on the one hand it makes difficult to distinguish between software and hardware error, on the other it brings difficult questions about building of test environment. As the development process of equipment,task softwares is done in the host machine.And its running target machine is not available to software testers. Then it is difficult to detect the software errors caused by the differences of target machine and the host machine's hardware-environment. Therefore, the building of software test platform considers the actual work environment of simulation equipment task software, and also takes advantage of the host's ground comprehensive development and debugging environment. And the test frame construction follows the principle of hardware softening. The building of software test platform makes use of external motivation conditions required by software program completion test. Then it reduces the number of hardware,and uses software program to complete the external motivation.

3.4 *Failure of Unconventional Conditions*

The unconventional conditions (also called Rare Conditions) are generated by the error or unexpected input data and seldom executed fault code. For example: For a small piston engine UAV, the engine spark interference is the main interference. If the interference signal is introduced to UAV computer system, the UAV softwares will produce false data, and cause unexpected failure.When we get the code of equipment task software, we can confirm that the software coding and debugging stages belong to high reliability evaluation work. Therefore, during the validation testing stage, we should make limited time, energy and equipment used for small number of key parts and unconventional events test. Thus the whole software and its work state will be confirmed.

After carrying out traditional conventional software evaluation, we should pay attention to the technical features and dependability demands of high dependability safety-critical software. The software troop users' actual needs and their main focus should be taken into consideration. In the module of this type software's high dependability function, we should add some special test content to the software high dependability and anti-risk nature. Then we can ensure it meets the demands of ordnance safety-critical software dependability evaluation indicator. The special test content includes: dependability test for software defect model, reliability test based on coverage and fault injection, and safty test based on importance sampling.

4 The Dependability Assessment Methods of Ordnance Safety Critical Software

4.1 *The Dependability Atribute Model of Ordnance Safety Critical Software*

Dependability reouires the software provides proper, safe, reliable services according to the user's expectations .Not only it have covered functionality, reliability, easey use, efficiency, maintainability, portability and other software quality characteristics,but also it should include safety, timeliness, survivability and other software features[6][7][8] .Based on Military Software Quality Metrics (GJB 5236-2004), Military Guide to Software Testing (GJB/Z 141-2004),Software Reliability, Security Dsign Gidelines (GJB-Z 102-1997) , Military General Requirements for Software Testing and Evaluation (GJB 2434-95) and other military standars, reliability attributes and evaluation systems are established.It also need refer to ordnance safety critical software's main functions, acceptance criteria, operating environment, security features and other aspects, and refer to quality measurement system.

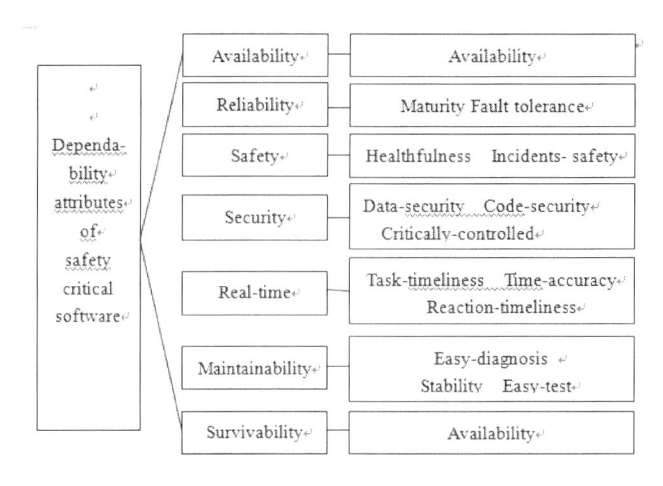

Fig. 1 The main attributes of ordnance safety critical software.

Figure 1 shows that each of these attributes is composed by a number of sub-features,and the attributes constitute thedependability attribute modle of ordnance.

4.2 *The Dependability Measurement Modle of 0rdnance Safety Critical Software*

In cycle process, dependability assessment contains software development phase, software submission and validation phase, software application phase. This article conducts the dependability assessment at the perspective of users.Therefor, the dependability assessment of software submission and validation phase should be concerned mainly.

Furthermore, Military Sftware Ealuation(GJB2434A-2004) says"It is very important to associate the internal software quality attributes with the outside.The developing quality characteristics of software can be estimated according to the final requirment. Unless there is evidence to show that the quality of the internal and the external is relevant, the internal measurment will not be a large number ".

In software submission and validation phase, according to the above principles the dependability attributes and parameter indexs are defined as Ref[8].

4.3 *The Dependability Assessment Modle of Ordnance Safety Critical Software*

As defined above ,AHP (Analytic Hierarchy Process) evaluation system is consisted of target layer, the first grade index layer and the second grad index layer[10].

In the actual environment and at the time of assessing the dependability of a system,the attributes are not treated equally. The importance of each attribute depends on the system application requirements, design and development costs, external environment and other factors[11][12].Therefore, the measurement and evaluation of dependability is actually a comprehensive evaluation process.The formula is showed :

$$\bar{U} = a_1U_1 + a_2U_2 + \cdots + a_nU_i(\sum_{i=1}^{n} a_i = 1, 0 \le a_i \le 1, i = 1,2,3,4,5,6,7) \cdot \quad (1)$$

$\bar{U}$ indicates the dependability of a system, U_i indicates the first grade index parameter of one software dependability attribute(availability, reliability, safety,etc) . $(a_1, a_2, \cdots, a_i)$ indicates the weight of the first grade attribute. Designer or end-user's demand makes the trade-offs according to the application requirements and other factors. For example : for the control software of UAVs, the army is mainly concerned about the safty, real-time and reliability. Therefore, the weight distribution of these attributes will be more than the others. In this way,the system can make credible and quantitative comparison and analysis in different design and configuration.And the different of emphasizing levels and the focus of each attribute will affect the dependability of the system directly. It will be more meaningful when each attribute is a contradiction.

Let $U_i (i = 1,2,\cdots 7)$ contain $K_i = \{k_1, k_2, \cdots, k_i)$ second grade indexs, as follow (u_{ij} is designated as one second grade index of U_i) :

$$U_i = \{u_{i1}, u_{i2}, \cdots, u_{ij}\}, i = 1,2,\cdots 7, j = 1,2,\cdots, k_i \;. \quad (2)$$

Then let the second grade indexs weight as follow:

$$B_i = \{b_{i1}, b_{i2}, \cdots, b_{ik_i}\}, 0 \le b_{ij} \le 1, \sum_{j=1}^{k_i} b_{ij} = 1, i = 1,2,\cdots 7, j = 1,2,\cdots k_i \cdot \quad (3)$$

Because evaluation parameters of each attribute are different, when software dependability is assessed.As the Table 2 saying, most parameter values are ranged of 0-1.And it says that the parameter value closer to 1, then more to meet the requirements of the target.So reliability and safty are necessary to be selected only.Such as:A(t) is the reliability evaluation value,and let S(t) be the safty evaluation value. Therefor the above indicators are subject to the same quantization interval and the the value trend

Then,based on the grading standards of the software reliability, the corresponding evaluation grade is be established as follow: $V = \{v_1, v_2, \cdots, v_m\}, m = 1,2,3,4,5$

According to collected data , information and the experience of expert evaluation are established.This membership tells that the second grad indexs belong to which evaluation grade.Let Ri be the evaluation results of the K_i second grade indexs,as follow:

$$R_i = \begin{bmatrix} r_{i11} & r_{i12} & r_{i13} & r_{i14} & r_{i15} \\ r_{i21} & r_{i22} & r_{i23} & r_{i24} & r_{125} \\ \vdots & \vdots & \vdots & \vdots & \vdots \\ r_{ik_i1} & r_{ik_i2} & r_{ik_i3} & r_{ik_i4} & r_{ik_i5} \end{bmatrix} (\mathrm{i} = 1,2,3,4,5,6,7) \tag{4}$$

As mentioned above, Ri is a single-factor evaluation matrix of the second grad index comprehensive evaluation. r_{ijm} is the membership adout the relationship of u_{ij} and the m evaluation grades(m=1,2,3,4,5).Based on the established set of the second grade weight(B_i),the second grad index comprehensive evaluation is a matrix as follow:

$$H_i = B_i \circ R_i = (b_{i1}, \ b_{i2}, \ \cdots \ b_{ik_i}) \circ \begin{bmatrix} r_{i11} & r_{i12} & r_{i13} & r_{i14} & r_{i15} \\ r_{i21} & r_{i22} & r_{i23} & r_{i24} & r_{125} \\ \vdots & \vdots & \vdots & \vdots & \vdots \\ r_{ik_i1} & r_{ik_i2} & r_{ik_i3} & r_{ik_i4} & r_{ik_i5} \end{bmatrix} = (t_{i1}, t_{i2}, \cdots, t_{i5}) \tag{5}$$

As one merging algorithm, $\circ$ changes as the concrete condition.H is the single factor matrix of second grade index fuzzy evaluation:

$$H = \begin{bmatrix} H_1 \\ H_2 \\ \vdots \\ H_7 \end{bmatrix} = \begin{bmatrix} B_1 \circ R_1 \\ B_2 \circ R_2 \\ \vdots \\ B_7 \circ B_7 \end{bmatrix} = (h_{im})_{7\times 5} \tag{6}$$

D is the fist gradeset of fuzzy comprehensive evaluation:

$$D = A \circ H = (a_1 \quad a_2 \quad \cdots \quad a_7) \circ \begin{bmatrix} h_{11} & h_{12} & \cdots & h_{15} \\ h_{21} & h_{21} & \cdots & h_{25} \\ \vdots & \vdots & \cdots & \vdots \\ h_{71} & h_{72} & \cdots & h_{75} \end{bmatrix} = (d_1, d_2, d_3, d_4, d_5) \tag{7}$$

then, a final evaluation is achieved based on the principle named maximum membership degree.

References

1. GJB/Z 102-97, The Design Criteria of Software Reliability and Security. The National Defense Technology and Industry Committee (1997) (in Chinese)
2. Trusted Computing Group. TCG Specification Architecture Overview [EB/OL] [2005203201], https://www.trustedcomputinggroup.org/groups/TCG_1_0_Architecture_Overview.pdf

3. Pearson, S.: Trusted Computing Platform. In: The Next Security Solution, HP Laboratories, Bristol (2002)
4. Avizienis, A., Laprie, J.C., Randell, B., et al.: Basic Concepts and Taxonomy of Dependable and Secure Computing. IEEE Transaction on Dependable and Secure Computing 1(1), 11–33 (2004)
5. Zhang, H., Luo, J., Jin, G.: Development of Trusted Computing Research. Journal of Wuhan University (Natural Science Edition) 52(5), 513–518 (2006) (in Chinese)
6. GJB 2434-95, The General Requirement of Ordnance Software Test and Evaluation, The National Defense Technology and Industry Committee (1995)
7. GJB/Z 141-2004, Ordnance Software test Directory. PLC General Equipment Department (2004)
8. GJB 5236-2004, Ordnance Software Quality Measurement. PLC General Equipment Department (2004) (in Chinese)
9. Liu, X., Lang, B., Xie, B.: Software Trust- worthiness Classification Specification. TRUSTIE- STC V 2.0 (May 2009) (in Chinese)
10. Zhang, Y., Zhang, L.: Fuzzy Comprehensive Evaluation for Software Trustworthiness. In: The 4th China Management Science AGM- Management Science and Engineering Branch, pp. 26–133 (2009) (in Chinese)
11. Xu, S.: Design and Alaysie of Trusted Computing System. The Publishing House of Qinghua University, Beijing (2006)
12. Yiangyu, D.: Classification of Software Quality Using Tree Modeling with the S-Plus Algorithm. Thesis for Master's Degree, Florida Atlantic University, 77–102 (1999)

The Research of Web Security Model Based on JavaScript Hijacking Attack*

Zhiguang Wang, Chongyang Bi, Wei Wang, and Pingping Dong

Abstract. The era of Web 2.0 has come, which provides users with a large number of interactive features. But network attacks also come thick and fast, which pose a threat to network security. By simply expounding related network security strategy, introducing the principle of JavaScript Hijacking attacks, and proposing preventive measures, this paper has a practice of network security on a certain guiding function.

Keywords: JavaScript Hijacking, Prevention, Attack.

1 Introduction

In recent years, the network has launched a wave of Web 2.0 development boom (Web 2.0 is the business revolution in the computer industry[1]), and many websites have introduced the concept of Web 2.0, providing a large number of interactive features. At the same time, these sites are also likely to suffer from JavaScript Hijacking attacks, which can steal a user's privacy information from the server on the network and result in a great threat. JavaScript Hijacking was first discovered by Jeremiah Grossman, who confirmed the attack in his Google Gmail[2]. As early as 2007, JavaScript Hijacking has been recognized as a high hazard level of attack by CAPCE (Common Attack Pattern Enumeration and Classification)[3].

This paper will briefly discuss network security strategy and introduce the principle of JavaScript Hijacking attack, the similarities and differences with CSRF, as well as the corresponding preventive measures.

Zhiguang Wang
Dept of Computer Science & Technology China University of Petroleum Beijing, China
e-mail: wzg0202@cup.edu.cn

* The paper is supported by the National Natural Science Foundation of China (60803159).

Z. Qian et al. (Eds.): Recent Advances in CSIE 2011, LNEE 125, pp. 555–563.
springerlink.com

2 Web Security

2.1 SOP

SOP (SOP, The Same Origin Policy) is an important security metrics of client-side scripting (especially JavaScript, VbSrcipt, etc.)[4]. It is the first from Netscape Navigator2.0, and its core content as follows: same protocol, same domain name and same port. Its purpose is to prevent a document or script loaded from a number of different sources, in other words, client-side script only can be read or modified with homologous cookie or message.

SOP, to a large extent, ensures the security of Web applications, but it merely prevents the client-side script to access the content from non-homologous websites. Many attacks would still be able to bypass the same origin policy, in other words, client-side script can launch attacks through the non-homologous website. JavaScript Hijacking also takes advantage of this flaw to make SOP do nothing.

2.2 Web Authentication

As we all know, Http is a stateless protocol and different "request / response" operations are independent, which could do nothing with non-connected state of the maintenance and management. In order to get a user-friendly interface in Web 2.0, many websites use cookie for user identity authentication. If someone visits a website using cookie, the site will save a cookie on the client-side after the completion of the authentication. Thus, if the user does not appear to exit the site, or the cookie has not expired, the user would not need to verify identity when visiting the same website next time. The principle of cookie mechanism is very simple--bringing a user-friendly interface. But it also leads to a lot of security implications.

In order to convenience users, many browsers at present are multi-tabbed browsing, for example, Maxthon, Firefox, Opera, Chrome and so on. These browsers are more or less at security risk, for they run a single process and share the cookie between pages in each window or tab. Thus, websites containing malicious code can use cookie for the theft of private information. Recently the so-called "Cookie poison" and "Cookie Thief" could steal cookie, including the content of session cookie, which causes great personal worries on private information. And so is the theory of JavaScript Hijacking. In addition to cookie, browsers will "smartly" and automatically add SSL authentication information to third party, and attach it to other web authentication request, even if they use Secure Sockets (SSL) to encrypt the connection [5].

2.3 P3P

As the existing security problems in cookie, P3P comes into being in the aspect of privacy protection. P3P (Platform for Privacy Preferences Project) is a recommended standard of privacy protection published by World Wide Web Consortium

(W3C), it provides a standard machine-readable format for the privacy policy, as well as an agreement which enables browsers automatically read and process strategies [6]. Now a lot of websites, especially the ones with commercial character, would "transparently" collect users' information for commercial purposes when they visit. Such as online shopping.

We will often see the following tips: Customers Who Bought This Item Also Bought. However, this will reveal users' privacy. Developing P3P standard is to reduce the concerns triggered by violations of the websites which collect users' privacy information. P3P standard's idea is: Website's privacy policy should tell the users the type of the information collected by the website, who will receive the information, how long the information will be retained and what the mode of using the information[7]. Users who visit the website supporting P3P have the right to view the site's privacy reports, and decide whether to accept the cookie continually visit the site.

To a certain extent, P3P has brought the gospel for the web application's privacy protection. However, its flaw is also obvious. For P3P is included in the HTTP header, if the user allow it to access to his privacy, the browser reading the P3P would be tantamount to allowing access of a third party cookie (from other cookie rather than the current visit to the site) . So the site including malicious code can attack vulnerable site or get user's private information through the means such as cross-domain. JavaScript Hijacking can also achieve its aim in such a way.

2.4 JSON

JSON (JavaScript Object Notation) is a lightweight data interchange format[8]. It consists of key-value pairs, or arrays, different values are separated by commas, which simplifies the data access. JSON has the advantage: it is more convenient than the DOM (Document Object Model), and has small data size, fast transmission, without the similar open-closing tags of XML elements. JSON is originally a safe JavaScript subset and does not contain assignment and calling. We use the Eval() function in JavaScript to analyze and parse JSON data, however if JSON contains malicious code, it may bring unexpected security issues. An attacker can also use this function to send malicious JSON data, so that the Eval() function will parse these malicious code. The principle of JavaScript Hijacking is to use the feature of JSON arrays which can be parsed in browsers. But it receives the returned JSON array from the server instead of sending a JSON array.

3 The Principle of Javascript Hijacking Attack

3.1 Javascript Hook

To put it simply, JavaScript Hook is to cover the original function, similar to the concept of high-level programming language. The Object provides JavaScript objects with common functions. FireFox supports __defineSetter__ function in order to set the properties of an object. By using Object.prototype. __defineSetter__ function, JavaScript Hijacking overwrites the original default function. Thus when making use of eval () function to resolve JSON array, the overridden function will automatically be called, which can get the private information of the JSON array.

3.2 JavaScript Hijacking Conditions

From the analysis above, in order to obtain privacy information, JavaScript Hijacking needs certain conditions, approximately as follows:

A malicious site attacks on private information and has specific target.

Browser should support JavaScript. Modern browsers mostly support this script language, and users basically do not forbid the browser to parse JavaScript.

Attacked vulnerable site should return JSON array instead of JSON objects. Because JSON array is considered to be an executable JavaScript script that the browser will execute, while the JSON objects do not. The malicious site also needs to know the format of JSON array.

Vulnerable site only responds to GET requests, that is, if sending POST request, the vulnerable site will not be attacked.

The attack requires a specific browser support.

3.3 Attack Analysis

Figure 1 shows the whole process of JavaScript Hijacking.

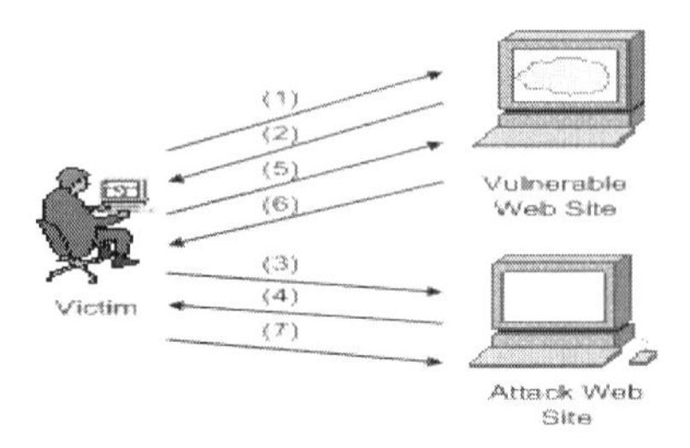

Fig. 1 JavaScript Hijacking attack

The victim visits a vulnerable site （http://www.vulnerable.com/）, and logs in through legal means.

Vulnerable site audits the victim's request, and response the browser with an authentication.

Before exiting the vulnerable site, the victim opens a browser tab, and visits a malicious web site (http://www.attack.com/). The malicious site has settled up malicious code, once the object is setting 'LoginTime' property, it will trigger the execution of the following code:

```
<script type="text/javascript">
Object.prototype.__defineSetter__('LoginTime',
function(obj){
var objString = "";
for (fld in this) {
objString += fld + ": " + this[fld] + ", ";
}
objString += "Time: " + obj;
alert(objString);
});
</script>
<script type="text/javascript"
src="http://www.vulnerable.com
/Default.aspx?oper=details"> </script>
```

The malicious site sends a request to obtain vulnerable site's (http://www.vulnerable.com/) data.

Because the victim does not exit the vulnerable site, the browser using the unexpired authentication information sends a GET request to the vulnerable site (http://www.vulnerable.com/).

The vulnerable site audits the authentication, and returns a JSON array according to the request.

```
<script type="text/javascript">
var object;
var req = new XMLHttpRequest();
req.open("GET", "Default.aspx?oper=details",true);
req.onreadystatechange = function () {
if (req.readyState == 4) {
var txt = req.responseText;
object = eval(" (" + txt + ")");
req = null;
}};
req.send(null);
</script>
```

The browser receives the response from the vulnerable site and sends JSON messages to the malicious site. JSON information is as follows:

```
[{"UserName": "JhonSmith", "Password":"JhonSmith",
"Balance": "1234.56", "LoginTime": "2009.07.22"}]
```

Now, the malicious site obtains the victim's private information on the vulnerable site. Its attack way is very similar with the principle of CSRF, but by different means.

3.4 Conclusion

From the above analysis we can see, if there are loopholes in the server-side code or users' security awareness is not strong, the website is very likely to suffer JavaScript Hijacking attacks. The attack exists not only in the FireFox browser, but also IE. There is no difference between them other than the implemented attack code. Strengthening the prevention of server-side and client-side could avoid such attacks.

4 The Research of Web Security Model

4.1 JavaScript Hijacking and CSRF

CSRF (Cross-Site Request Forgery) is a widespread network loophole, whose principle is also to make use of the deficiencies of web authentication and the malicious code to obtain data from vulnerable site. They both bypass the SOP strategy and simulate the victim's identity to attack the vulnerable site [9]. A site with CSRF vulnerability exists probably JavaScript Hijacking vulnerability, and vice versa. The biggest difference between JavaScript Hijacking and CSRF is: CSRF executes malicious operations by sending a request, such as modifying user accounts, deleting data, etc.; While JavaScript Hijacking is to use the concept of Hook in JavaScript language to steal the victim's privacy information. Although JavaScript Hijacking has more additional constraints, but its attack means is more subtle, more difficult to detect. So it is difficult to estimate its fatalness.

4.2 Authentication Based on Client-Side and Server-Side

Based on the above analysis of JavaScript Hijacking and CSRF, in order to prevent the leakage of user information and data tampering, it is necessary to improve the web authentication. Using the combination authentication of clientside and server-side is a good choice, which can effectively reduce or prevent similar attacks on the probability of the network. The combination authentication of client-side and server-side is shown in Figure 2.

User logins. After the authentication, the server returns a Session and a form (page) with random value (using RN expressed in figure) to the user, and records the user's random value.

User must send the cookie and the form's random value to the server when requesting operation of server data. If the server passes the authentication, then it will response user's request, otherwise the user's request is not legal.

If the user using multi-tabbed browser opens multiple form pages after login, then the server should temporarily maintain more than one form's random values, set a time stamp, and release resources after time-out.

Because the site containing malicious attacks lacks of access to the form of random values, it is very difficult to launch attacks similar to CSRF or JavaScript Hijacking.

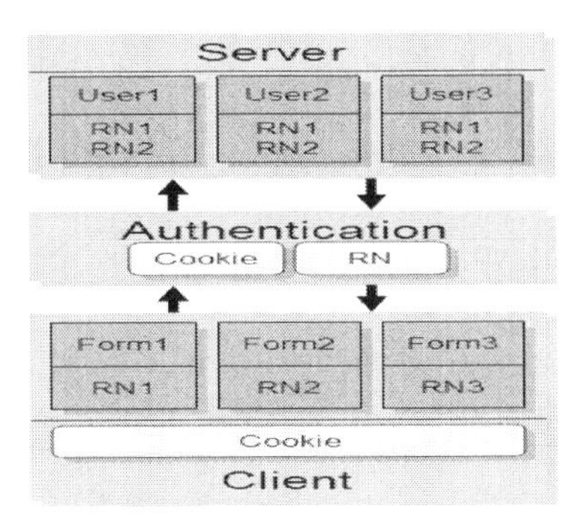

Fig. 2 Combination authentication of client-side and server-side

5 Javascript Hijacking Prevention

5.1 Prevention on the Server-Side

It is necessary to ensure the safety of JSON when using JSON as a data interchange format. One of the common approaches is to use regular expressions to check whether JSON data contain the keyword of malicious code. Use the parseJSON () function provided by the official site (http://www.json.org/) instead of JavaScript's eval () function, which includes a big security risk.

When there is privacy or important data, use POST request instead of GET request. Because <script> uses GET request to obtain data, using POST request can effectively reduce the probability of attack and increase the difficulty of attacks. However, the use of POST can not totally prevent JavaScript Hijacking, under certain conditions a malicious website can also send POST request in the form of the form [10].

Check HTTP-Refferer and make sure that the requests come from the same domain. It can reduce the possibility of attacks although the referrer can be forged.

In achieving the authentication combination of server-side and client-side, we need to notice: "compatible with parallel sessions." If the user visiting a site opens two different forms at the same time, the measures of authentication protection should not affect any other forms' submission. If the site generates a pseudo-random value to cover the previous one when the form is loaded every time, the user can only successfully submit the final opened form, for all the other forms contain the illegal pseudo-random value. We must be careful to ensure that protection measures will not affect tabbed browsing or multiple browser windows browsing the same site.

Add comment symbol before and after the JSON after the server returns a JSON array, such as /*[{" name ":" abc "}]*/, which can be regarded as comment and not implemented by JavaScript. The comment could be removed when needed.

Use XML instead of JSON when the data are extremely important, which is more secure than JSON.

Reduce dynamically generated code and executions. JavaScript's eval () function allows the execution of arbitrary strings as JavaScript code, so it should be used with caution.

Use tools of vulnerability-checking to test possible vulnerable code and reduce the possibility of attacks.

5.2 Prevention on the Client-Side

Use high security and strongly stability browser, update security patches and to the latest version in time.

Exit after visiting the website with privacy (such as Internet Bank). Don't visit other websites before exiting.

Clear history and cookie regularly, or use the browsers with a "privacy browsing" feature, such as Safari, IE8, Chrome, etc.

Visit the legitimate sites instead of unknown ones. It is recommended that using the site authentication function of IE8 or Google toolbar to identify illegal websites.

6 Conclusion and Summary

Network changes with each passing day, more and more websites start to use or are already using Web 2.0 technologies, in order to enhance the interactivity, enrich user experience, and reduce the burden on the server. But the security issues of network attack bring much trouble to developers and users. This article introduces one such attack - JavaScript Hijacking, and describes its principle and prevention strategies in detail, which is a practice of network security on a certain guiding function.

References

1. Web 2.0 Compact Definition: Trying Again, http://radar.oreilly.com/archives/2006/12/web-20-compact.html
2. Chess, B., O'Neil, Y.T., West, J.: JavaScript Hijacking, http://www.fortify.com/servlet/download/public/JavaScript_Hijacking.pdf
3. CAPEC Dictionary (Release 1.3), http://capec.mitre.org/data/dictionary.html#j

4. Justin, S.: Same-Origin Policy Part 1: Why we're stuck with things like XSS and XSRF/CSRF, http://taossa.com/index.php/2007/02/08/same-origin-policy/
5. Chen Zhen, J.: The Ananysis and Solution of CSRF. Fujian Computer, 6–28 (2009)
6. Enabling smarter Privacy Tools for the Web, http://www.w3.org/P3P/
7. Yuan, X.: Privacy Parameters Selection Platform—P3P, http://www.yxl.cn/Info/20060302,212041,5095.html
8. Introducing JSON, http://www.json.org/
9. Understanding JavaScript Hijacking Concept, http://www.cnblogs.com/hyddd/archive/2009/07/02/1515768.htm
10. Submitting Form Automatically When users are not wared, http://blog.roodo.com/rocksaying/archives/2665954.html

Formalization of Risks and Control Activities in Business Process

Yasuhito Arimoto, Shusaku Iida, and Kokichi Futatsugi

Abstract. It has been an important issue in society to evaluate effectiveness of control activities in business processes in order to ensure that control activities are dealing with risks as expected.

In this paper, we propose a method to model business processes with risks and control activities formally in order to verify effectiveness of control activities precisely. Formal descriptions gives us deeper understanding of risks and control activities, and by formal verification, we can evaluate control activities in a precise way. Our approach to formal modeling of business processes with risks and control activities is modeling business processes by focusing on activities related to documents which are handled during business activities.

The main contribution of this work is to show how we can solve problems in risk managements of business domain by application of a technology in Computer Science and Information Engineering.

1 Background

Dealing with risks in business processes by control activities has important roles for achieving a company's goals. Control activities are policies (which establish what should be done) and procedures (the actions of people to carry out policies) help ensure that management directives identified as necessary to address risks are carried out, such as approvals, authorization, reconciliations and segregation of duties [2]. And, it has been more remarkable that evaluation of effectiveness of control

Yasuhito Arimoto · Kokichi Futatsugi
Japan Advanced Intitute of Science and Technology, Asahidai 1-1,
Nomi, Ishikawa, Japan 923-1292
e-mail: {arimotoy,futatsugi}@jaist.ac.jp

Shusaku Iida
Senshu University, Higashimita 2-1-1, Tama-ku, Kawasaki, Kanagawa, Japan 214-8580
e-mail: iida@isc.senshu-u.ac.jp

Z. Qian et al. (Eds.): Recent Advances in CSIE 2011, LNEE 125, pp. 565–575.
springerlink.com

activities should be done more precisely. For example, while improper activities by well-known corporations are revealed one after another, Sarbanes-Oxley Act in the United States [1] and Financial Instruments and Exchange Law in Japan [3] are promulgated for improving reliability of financial reporting, and one of requirements of these laws is that managements of listed companies must issue a report as the result of the evaluation of Internal Control. Since control activities are ones of main elements of Internal Control [2], precise evaluation of effectiveness of control activities is increasingly important.

In order to evaluate effectiveness of control activities in a business process, risks, control activities in the business process must be understood. In general, a flowchart, a risk control matrix, and a business process narrative are created for understanding and evaluating control activities in each business process [11]. These documents are represented in graphical representations and natural languages, and have important roles that they are helpful for stakeholders to understand control activities, risks, business processes, and relations between them.

However, informal descriptions like these documents might include ambiguity and can be evaluated by only humans. Ambiguous descriptions might make people misunderstand what they mean, and evaluation by human might include some errors and it causes a lack of consistency of evaluation. Inconsistent evaluation cannot give us scientific discussions on correctness of evaluation, that is, we cannot formally reason why a set of control activities are necessary or why a set of control activities are not enough in order to avoid a risk.

2 Aim

The aim of this study is to give an approach to precise evaluation of Internal Control by defining what risks and controls activities really mean in the business processes in order to obtain deeper understanding of how risks can be avoided or mitigated by control activities. By this study, we can analyze relations between risks and control activities formally and it realise consistent evaluation.

For that purpose, we apply a formal method technique to formalize and analyze business processes with risks and control activities (BPRC). Applying formal methods to business process modeling is useful because formal models do not leave any scope for ambiguity, and formal models increase the potential for formal analysis [4]. By formal analysis, we can verify some properties of the models. Both formalization and verification are helpful to understand the model that the designers construct. One more point we would like to emphasize is that formal models and formal verification give us potential to have scientific discussions. Informal descriptions do not have this characteristic. For example, a risk control matrix gives information like "risk R is dealt with controls C1, ..., Cn", however it does not give which controls from C1 to Cn are really important and why the risk can be avoided or mitigated by these controls scientifically.

Our Target for Modeling and Verification

In this work we focus on modeling document flows while business activities and control activities performed to documents, and risks about falsification of documents. And, the properties we would like to prove is that a set of control activities are enough for detecting falsified documents because documents play an important role in business processes, that is, all information created during business activities is recorded some documents.

Our work proposes a method to construct formal model of document flows in a specific business process, e.g. sales process, etc., and verify properties about avoiding falsification of documents. In our models, we focus on behaviors related to documents and modeling such behaviors as state transition machines. States are characterized by values of documents, and a behavior is represented as a sequence of events. An event is something which changes a state. Activities like creating a document, chocking a document, etc. are modeled as events. An important feature of our modeling is that our models include irregular event, in this study, forging a document. An irregular event is an event which changes a state of a business process to an undesirable state. By a *desirable state* we mean a state in which no document is forged, and by a *undesirable state* we mean a state in which one or more documents are forged. We verify that states where risks are not tangible be reached because of control activities although irregular events are performed.

Although evaluating control activities related to documents might not be complete way to evaluation of all kinds of control activities, this study can be the foundation for scientific analysis of risks and control activities by using a formal method technique.

Different View Points to Capture Document Flows

For modeling document flows, we need to define what models can include by deciding how we capture the real world to model.

Figure 1 intuitively represents a model of document flows for a whole sales process (WSP) in a UML activity diagram. Behaviours in this model relates to a single order. One of behaviours for a single order is as follows:
Creating order D1 with a master, Sending order D1 to sales, Creating ack D2 in sales, ...
However, in the real world, there is not only one order handled, but many orders are handled concurrently. An exampled of behaviours of this situation is as follows:
Creating order D1 with a master, Creating order D2 with a master, Sending order D1 to client, Creating order D2 with a master, ...

Let us assume that each created document belongs to a *session*. A *session* is a set of behaviours for achieving an instance of a business goal. By a business goal we mean a purpose of business activities in a business. For example, the business goal of the simple sales process is celling a product.

Irregular transitions may occur in not only single session but also multiple sessions. For example, forging an order document is a kind of irregular transition, which appears in single session, and an example of irregular transitions in multiple sessions is handling a document related to a different order. We need to recognize what kinds of risks can be analyzed by out modeling.

In this paper, we focus on analysis of single session document flows. However, the model and the verification technique introduced in this paper is capable of application to formalization and verification of multi-session document flows with some extension.

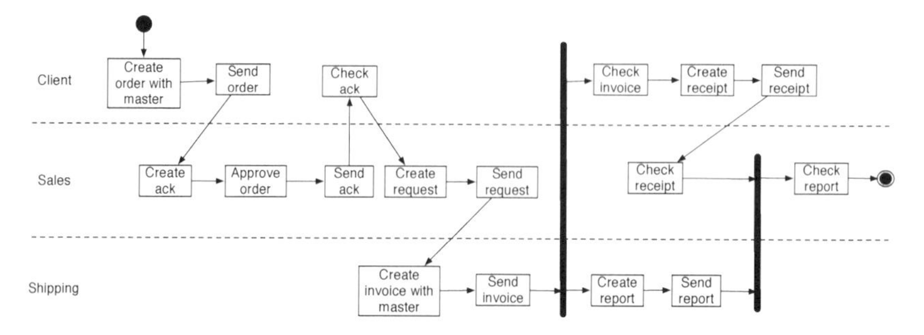

Fig. 1 An example of whole sales process

The structure of this paper is as follows. Section 3 gives introduction to OTS/CafeOBJ Method [6, 10], which is one of formal methods techniques, and section 4 shows how we can describe the formal model of document flows by using our modeling. In section 5, verification of the formal description introduced in section 4 is given. Some related work is introduced in section 6, and section 7 describes conclusion and future work.

3 OTS/CafeOBJ Method

In our work, we have chosen OTS/CafeOBJ Method [6, 10] for formally describing models of document flows and analysis of them. OTS/CafeOBJ Method is for modeling and verification of systems. In this method, systems are modeled as an Observational Transition System (OTS), which is a kind of state transition machines, and the OTS model is described CafeOBJ [6], an algebraic specification language. The reasons why we have chosen OTS/CafeOBJ Method are as follows:

- Since CafeOBJ is an executable language, verification can be done deductively on CafeOBJ system semi-automatically with user interactions.
- Verification with deductions need less computations than model checking.
- In verification with OTS/CafeOBJ Method, interactions with humans can be done intuitively in most of the cases.
- Through Interaction between human and CafeOBJ system, we can have deeper understanding of the problems.

The last reason is the most important for this study because this feature helps us to analyze why control activities are enough or not scientifically.

We introduce the definition of OTS briefly in this subsection. The more precise definition is given in [10].

Definition of OTS We assume that there exists a universal state space denoted Υ and that data types used in OTSs are provided. OTS $\mathcal{S}$ can be defined as $\langle \mathcal{O}, \mathcal{I}, \mathcal{T} \rangle$ such that

- $\mathcal{O}$: A finite set of observers. An observer is a function, which takes a state and values of data types as arguments and returns a value of a data type. States with respect to (wrt) an OTS are characterized by these values.
- $\mathcal{I}$: The set of initial states such that $\mathcal{I} \subseteq \Upsilon$, where Υ is a universal state space.
- $\mathcal{T}$: A finite set of conditional transitions. A transition is a function, which takes a state and values of data types as arguments and returns a state. Each transition has the condition, which is called the effective condition. If the effective condition does not hold, then the transition does not change the state.

Successor States Given an OTS $\mathcal{S}$, two states $\upsilon, \upsilon' \in \Upsilon$, and values of data types $X_1, ..., X_n$, if there exists $t \in \mathcal{T}$ and $t(X_1, ..., X_n, \upsilon) = \upsilon'$, we write $\upsilon \rightsquigarrow_{\mathcal{S}} \upsilon'$ and call υ' a *successor state* of υ wrt $\mathcal{S}$. $\rightsquigarrow^*_{\mathcal{S}}$ is a reflexive transitive closure of $\rightsquigarrow_{\mathcal{S}}$.

Reachable States Reachable states wrt $\mathcal{S}$ are inductively defined: (1) each $\upsilon_0 \in \mathcal{I}$ is reachable wrt $\mathcal{S}$, and (2) for each $\upsilon, \upsilon' \in \Upsilon$ such that $\upsilon \rightsquigarrow_{\mathcal{S}} \upsilon'$, if υ is reachable wrt $\mathcal{S}$, so is υ'. Let $\mathcal{R}_{\mathcal{S}}$ be the set of all reachable states wrt $\mathcal{S}$. From the definition of reachable states wrt $\mathcal{S}$, the following proposition is derived from the definition of reachable states wrt $\mathcal{S}$: given an arbitrary OTS $\mathcal{S}$ and arbitrary two states $\upsilon, \upsilon' \in \Upsilon$ such that $\upsilon \rightsquigarrow^*_{\mathcal{S}} \upsilon'$, if υ is reachable wrt $\mathcal{S}$, then so is υ'.

Verification Procedure Assume that $\mathrm{P} : S \times D_1 \times ... \times D_n \rightarrow$ Bool is the predicate which represents the property which should be proved, where S is a state space wrt an OTS $\mathcal{S}$ and $D_1, ..., D_n$ are data types. An *invariant property* is a property which holds in all reachable states, and proof that shows P is an invariant property wrt OTS $\mathcal{S}$ is constructed by following procedures.

1. For each $\upsilon_0 \in \mathcal{I}_{\mathcal{S}}$ and $d_1 \in D_1, ..., d_n \in D_n$, prove that $\mathrm{P}(\upsilon_0, d_0, ..., d_n) = \text{true}$.
2. For each $\upsilon, \upsilon' \in \mathcal{R}_{\mathcal{S}}$ and $d_1 \in D_1, ..., d_n \in D_n$, such that $\mathcal{R}_{\mathcal{S}}$ is the set of all reachable states wrt $\mathcal{S}$ and $\upsilon \rightsquigarrow_{\mathcal{S}} \upsilon'$, prove that $\mathrm{P}(\upsilon, d_1, ..., d_n)$ implies $\mathrm{P}(\upsilon', d_1, ..., d_n)$ = true.

CafeOBJ system is used for reasoning the result of application of predicates to arguments. Case splitting for covering all possibilities in the model can be done with interactions between users and CafeOBJ system.

4 Modeling Document Flows

In this section, we explain how we can model WSP in OTS as an example.

4.1 Modeling Document Flows

In WSP shown as figure 1, documents which have following types are created: order, a master of order (M(order)), ack, request, invoice, a master of invoice (M(invoice)), receipt, and report.

Events performed in this process is as follows:Creating order with M(order) in client (Create order with master), Sending order from client to sales (Send order), Creating ack from order in sales (Create ack), Approving order in sales (Approve order), Sending ack from sales to client (Send ack), Checking ack with M(order) in client (Check ack), Creating request from order (Create request), Sending request from sales to shipping (Send request), Creating invoice with M(invoice) from request (Create-invoice), Sending invoice from shipping to client (Send invoice), Checking invoice with M(order) in client (Check invoice), Creating receipt from invoice in client (Create receipt), Sending receipt from client to sales (Send receipt), Creating report from M(invoice) in shipping (Create report), Sending report from shipping to sales (Send report), Checking receipt with order which is approved in sales (Check-receipt), Checking report with order which is approved in sales (Check-report).

We assume that sales and shipping are untrusted divisions and the following irregular event can be performed in these divisions.
Forging a document in sales or shipping division (Forge-WSP).

We assume some conditions under which a document can be forged as follows:

- We assume that there are untrusted divisions in the objective document flows and a document is forged when it is in an untrusted division.
- Evidences cannot be forged. If a document who has evidences is forged, the evidences are deleted. This assumption ensures that a forged document with evidence has been forged before getting put the evidence .
- Masters are not forged. A master of a document in this model is the abstraction of documents assumed not to be forged in the real world. For example, it can be electronic documents which are protected by access controls for avoiding outsiders from forging.

4.2 OTS Model of Document Flows for WSP

As we have mentioned before, we focus on models of single session document flows in this paper. In this subsection, we introduce OTS $\mathcal{S}_{\mathrm{WSP}}$ for single session document flows for WSP.

Observations

Since states wrt OTS for document flows are characterised by values of documents, so, observations are defined based on values of documents. Values of documents are defined by a document type, an evidence history, a division name, and information which shows the authenticity of the document. A document type shows what kind of document it is, e.g., order, invoice, etc. On documents, something like signatures or a seals are put to show that documents are approved or checked. We call something put on the document for this purpose evidence and an evidence history is a set of evidence. A division name is the information to show where the document currently exists.

Information which shows the authenticity can be modeled as a flag. If it is not forged, it returns true, otherwise false. This is meta information, i.e., the value of this information can not be seen by objects in the domain. However, from object level, equality of these values between two documents can be checked. A document is consistent with another document if these values are the same. The value itself cannot be observed, but consistency of two document can be observed when event "check" is performed.

OTS $\mathcal{S}_{\mathrm{WSP}}$ for WSP can be defined as follows.

$\mathcal{O}_{\mathcal{S}_{\mathrm{WSP}}}$ = { Evidences : State × DocumentType → EvidenceHistory, Place : State × DocumentType → Division, Legal? : State × DocumentType → Bool, UntrustedSet : State → DivisionSet }

State denotes a state space, and DocumentType, EvidenceHistory, Bool, Division, and DivisionSet data types which represent types of documents, sets of evidences, boolean values, division names, and sets of division names respectively. Attributes of each document are observed by observers Evidences, Place, and Legal?. Evidences(S, T) returns the set of evidences, Place(S, T) returns the place, Legal?(S, T) returns the authenticity value of a document whose type is T in state S respectively. Observation UntrustedSet(S) returns the set of untrusted division.

Initial States

We assume that initial states are desirable states, and in this case study, untrusted divisions are sales and shipping. It can be defined as follows.

$\mathcal{I}_{\mathcal{S}_{\mathrm{WSP}}}$ = { init ∈ State | Evidences(init, T:DocumentType) = empty ∧ Place(init, T) = noDivision ∧ Legal?(init, T) = true ∧ UntrustedSet(init, T) = {sales, shipping} }

Transitions

Events are modeled as transitions. Transitions in WSP are as follows.

$\mathcal{T}_{\mathcal{S}_{\mathrm{WSP}}}$ = { Create-M-order : State → State, Send-order : State → State, Create-ack : State → State, Approve-order : State → State, Send-ack : State → State, Check-ack : State → State, Create-request : State → State, Send-request : State → State, Create-M-invoice : State → State, Send-invoice : State → State, Check-invoice : State → State, Create-receipt : State → State, Send-receipt : State → State, Create-report : State → State, Send-report : State → State, Check-receipt : State → State, Check-report : State → State, Forge-WSP : State × DocumentType → State }

Each event corresponds to an activity in WSP.

By these transitions, return values of observations are changed. In our model, how values are changed by these transitions are described as transition rules. An example of descriptions of transition rules is as follows:

```
ceq Evidences(Check-receipt(S))
    = (if (in?(apv(sales), Evidences(S, order)) and
            (Legal?(S, receipt) = Legal?(S, order))) then
           (ch(order) Evidences(S, receipt))
       else Evidences(S, receipt) fi)
      if c-Check-receipt(S) .
eq Place(Check-receipt(S)) = Place(S) .
eq Legal?(Check-receipt(S)) = Legal?(S, T3) .
eq UntrustedSet(Check-3(S)) = UntrustedSet(S) .
```

S is a variable of State. `receipt` and `order` are document types representing documents whose types are receipt document and order document respectively. `apv(sales)` and `ch(order)` are evidence which is put after approving in sales and after checking with order respectively. Term `in?(apv(sales), Evidences(S, order))` means if order document has a piece of evidence which shows it is approved in sales division. `(Legal?(S, receipt) = Legal?(S, order))` means if the authenticity values of these two document are the same in state S.

Equations started with `ceq` are conditional equations, and this state transition occurs if effective condition, in this case, `c-Check-receipt(S)` holds. The definition of this effective condition is as follows:

```
op c-Check-receipt : SaleAndShip -> Bool
eq c-Check-receipt(S)
   = (Place(S, receipt) = sales) and
     (Place(S, order) = sales) .
```

This equation defines a predicate checking if receipt document and order document are in sales division.

5 An Example of Formal Verification of Document Flows

Verification of a model of document flows for a business is to prove that the control activities are effective to avoid risks or not. By verification, we can ensure if document flows are prescribed right by control activities or not. If a proof returns true, we can say that control activities are effective under some conditions, otherwise, we can say that control activities are not enough for avoiding the risk.

In the rest of this section, we show an example of verification for the formal description introduced in subsection 4.2.

5.1 A Property to Be Proved

One of properties we want to prove in the document flows for WSP is that "report which is in sales and has been checked with order is not forged". This property says report is not forged after all of regular and of control events have been performed in WSP. It does not always hold that all of them can be performed, because control events might stop the process in order to avoid risks. By proving that this property is an invariant property, we can show the effectiveness of control activities in WSP. The property can be described as follows.

```
op inv1 : State -> Bool
eq inv1(S:State)
   = ((Place(S, report) = sales) and
      in?(ch(order), Evidences(S, report)))
     implies (Legal?(S, report) = true) .
```

Place(S, report) = sales represents report is in sales, ch(order) is the evidence which is put on the document after checking a document with order and in?(ch(order), Evidences(S, report)) represents report has been checked with order, and Legal?(S, report) = true represents report is not forged.

In the proof, CafeOBJ system returns false for inv1(s) implies inv1(s'), where report is in sales, order is in sales, receipt is checked by order, report is not checked by order yet, order is approved in sales, both order and report are forged in state s, and s' = Check-report(s). In this case, inv1(s) = true, because report is not checked yet in s. inv1(s') = false, because report is in sales, has been checked by order in s' after application of Check-report, and is forged.

Since case splitting in the proof constructed by interactions between CafeOBJ system and users covers all possible cases which the model can represent, states in some cases are not reachable states. In order to prove inv1 is an invariant property, we need a lemma, which shows state s is not a reachable state wrt OTS $\mathcal{S}_{\mathrm{WSP}}$ for WSP. Lemma discovery is helpful to understand the model of document flows, since it gives users new facts on the model.

5.2 An Example of Lemma Discovery

Through interactive proving, we may need lemmas, and they are helpful for us to obtain new facts about roles of a control activity or of a set of control activities in the model of document flows.

In state s, order is forged and has the evidence, it was forged before the approval because evidences are deleted after forging from conditions of forging a document. From this statement, request and invoice are also forged because order has been forged before the approval. In this situation, any receipts cannot be created because of the control "checking invoice with M(order)". Since M(order) is a master of order and it is created in client, there is no chance to be forged. invoice is forged in this situation, and the check for invoice with M(order) cannot be passed. As long as the check for invoice is not passed, receipt cannot be created in this process. Then we can find one of candidates of lemmas, "if order approved in sales is forged, receipt is not created". In state s, receipt has been checked by order, and it means receipt has been created. Thus, we can prove that s is not a reachable state by proving the lemma is an invariant property.

6 Related Work

There are several work in order to solve problems in Internal Control by using information science technologies such as [7, 9].

[7] explains how the logic behind the obligations and permissions on a business process and contracts can be made explicit in terms of deontic concepts to check compatibility between business processes and business contracts. While this work focuses on risks which are about the violation of contracts, we focus on risks which appear in handling documents in business processes.

[9] introduces a pattern based approach for modeling Internal Control. They show controls patterns used for implementation of control activities on business processes. Their work focuses on how to design control activities in business processes. On the other hand, we focus on formal descriptions and verification of existing business processes.

The idea of modeling business processes by focusing on flows of documents is based on [8]. In their work, business processes are specified in Maude [5] and verification is done by model checking. Modeling business processes in our work is from different point of view compared to their work. While they focus on actors' and attackers' behaviours in business processes, we capture model of DFB domain where control activities are performed. In our work, events are important elements of the models.

7 Conclusion and Future Work

The problem that we set up is that there is a lack of scientific discussions on effectiveness of control activities in business processes. The scheme to solve this problem is as follows: (i) In order to have consistent evaluation and scientific discussions on the result, we need a method to verify effectiveness of control activities in business processes precisely. (ii) Formal verification technique can be useful for a precise verification. (iii) For formal description and verification, we focus on document flows including control activities and risks related to documents.

Through developments of formal descriptions, we can obtain deeper understandings of the model of document flows, which give us ideas about what risks are and what control activities actually are for. While verification, we can formally analyze if control activities are good enough or not. The results of formal analysis reveals relations between risks and control activities precisely and they give us high potential to have scientific discussions on effectiveness of control activities in the models of document flows. That is, we can understand how a set of control activities are supposed to deal with risks in the model of document flows through interactive proving as we show an example of lemma discovery in subsection 5.2.

We introduce formalization and analysis of models of single session document flows in this paper, and there are more risks and control activities in both single session document flows and multi-session document flows. We have not introduce formalization and analysis of multi-session document flows, however, the result of verification of multi-document flows for WSP is different from single session document flows. For future work, we need more case studies in order to construct models of document flows which have more critical risks and control activities and analysis of relations between single session document flows and multi-session document flows.

References

1. Pub. L. 107-204. 116 Stat. 754, Sarbanes Oxley Act (2002)
2. The Committee of Sponsoring Organizations of the Treadway Commision (COSO), Internal Control - Integrated Framework, `http://www.snai.edu/cn/service/library/book/0-Framework-final.pdf`
3. Financial Service Agency, Financial Instruments and Exchange Law, `http://law.e-gov.go.jp/htmldata/S23/S23HO025.html`
4. van der Aalst, W.M.P.: Three Good Reasons for Using a Petri-net-based Workflow Management System. In: Navathe, S., Wakayama, T. (eds.) Proceedings of the International Working Conference on Information and Process Integration in Enterprises (IPIC 1996), Cambridge, Massachusetts (November 1996)
5. Clavel, M., Durán, F., Eker, S., Lincoln, P., Martí-Oliet, N., Meseguer, J., Talcott, C.: Introduction. In: Clavel, M., Durán, F., Eker, S., Lincoln, P., Martí-Oliet, N., Meseguer, J., Talcott, C. (eds.) All About Maude - A High-Performance Logical Framework. LNCS, vol. 4350, pp. 1–28. Springer, Heidelberg (2007)
6. Diaconescu, R., Futatsugi, K.: AMAST Series in Computing, vol. 6. Cafe OBJ Report, World Acientific (1998)
7. Governatori, G., Milosevic, Z., Sadiq, S.: Compliance Checking between Business Processes and Business Contracts. In: Proceedings of the 10th IEEE International Enterprise Destributed Object Computing Conference. IEEE (2006)
8. Iida, S., Denker, G., Talcott, C.: Document Logic: Risk Analysis of Business Processes Through Document Authenticity. In: Proceedings of 2nd International Workshop on Dynamic and Declarative Business Processes. IEEE (2009)
9. Namiri, K., Stojanovic, N.: Pattern-based Design and Validation of Business Process Compliance. In: Meersman, R., Tari, Z. (eds.) OTM 2007, Part I. LNCS, vol. 4803, pp. 59–76. Springer, Heidelberg (2007)
10. Futatsugi, K., Babu, C. S., Ogata, K.: Verifying Design with Proof Scores. In: Meyer, B., Woodcock, J. (eds.) VSTTE 2005. LNCS, vol. 4171, pp. 277–290. Springer, Heidelberg (2008)
11. Sasano, M.: Naibutousei no Nyuumon to Jissen (Introduction and Practice of Internal Control), 2nd edn., Chuuou Keizaisha (2007)

Implementation of Panorama Virtual Browser and Improvements

Ranran Feng, Hongqiang Qian, Batbold Myagmarjav, and Jiahuang Ji

Abstract. As an efficient interactive representation of virtual environment, 2D images show a 3D scene when a **360**° panoramic image is presented. Visitors can virtually walk-around in the scene with a panorama browser with hotspot pop-up enabled. An implementation of panorama browser along with the optimization algorithms involved in the generation of panoramic images is introduced in this article. Also the hotspot pop-up technique is introduced. The browser has been successfully used on Sam Houston State University Virtual Hotel Tour system with timely mouse click response and large interactive scene window.

Keywords: panorama image, cylindrical projection, inverse cylindrical projection. Hotspot pop-up.

1 Introduction

In 1994, Apple Inc. released its first version of QuickTime VR (QuickTime VR). Since then, panoramic images showed a high potential in visual perception and 3D scene representation. Since then cylindrical or sphere panoramic image[1] are widely used. Several notable commercial software systems have been introduced on the market; IBM HotMedia [4] and Tobias Hüllmandel PanoramaStudio [7] are two of them. These systems provide viewer a panoramic view based on cylindrical panoramic images.

Cylindrical panoramic images are developed using cylindrical projection. Such images provides viewer 3D effects. However, viewers may also find that those

Computer Science Department, University of Texas at Dallas,
800 W. Campbell Rd., Richardson, TX 75080 USA
e-mail: rxf090020@utdallas.edu

Computer Science Department, Sam Houston State University,
Huntsville, TX 77340 USA
e-mail: hongqiangqian@gmail.com

Z. Qian et al. (Eds.): Recent Advances in CSIE 2011, LNEE 125, pp. 577–583.
springerlink.com

views are somehow a little bit distorted. The panorama browser introduced in this paper is aimed to solve this problem. The browser uses Inverse Cylindrical Projection algorithm to convert the cylindrical projected panorama images to human's perspectives before the image display. In order to provide a smooth view for customers to look around a scene, bi-directional buffer windows [9] are used; and to further improve the efficiency and performance, optimizations were also developed. Besides, Java Applet was used in the implementation for easy release over internet.

2 Generate Cylindrical Panorama Images

Imagine a photographer is standing at the center of a hotel lobby taking pictures. To get a complete view of the lobby, he needs to take pictures in different directions.

Fig. 1 One kernel at x_S (*dotted kernel*) or two kernels at x_i and x_j (*left and right*) lead to the same summed

The photographer is in a 3D environment; but the pictures he took are all 2D images (Fig. 1); and, it is hardly for him to imagine out the real scene. However, if we wrap this long sheet to a cylinder [3] and let the viewer stand inside the cylinder to look around for a 360°, it will generate him a feeling of standing at the center of the scene and looking around. Then the viewer will consider himself in a 3D environment.

Mapping a 2D image into a cylinder view is called cylindrical projection [3]. The projection is very straight forward as shown in Fig. 2. Let O be the center of the cylinder, f be the radius of the cylinder, image $A_1B_1C_1D_1$ is projected to $A_1'B_1'C_1'D_1'$ on the cylinder. Let W and H be the width and the height of the image $A_1B_1C_1D_1$, let (x, y) be the coordinate values of any given pixel on $A_1B_1C_1D_1$, and (x', y') be the coordinate values of the pixel after projection. Then equation (1) can be used to perform the projection. Fig. 3 shows an example of cylindrical projection.

$$\begin{cases} x = f * \arctan[(x - W/2)/f] + f * \arctan((W/2)/f) \\ y = f * (y - H/2)/\sqrt{(x - W/2)^2 + f^2} + H/2 \end{cases} \tag{1}$$

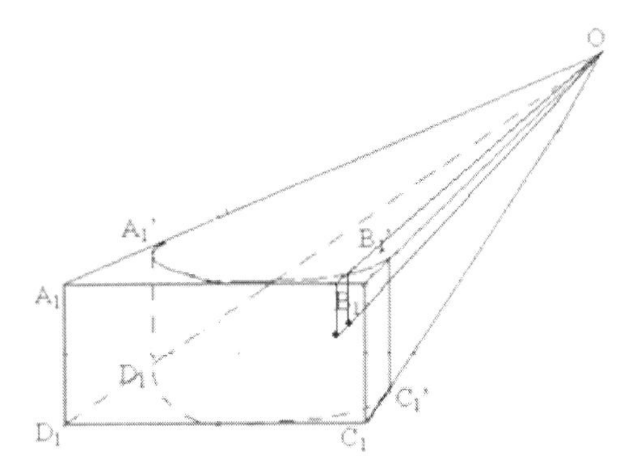

Fig. 2 Cylindrical projection.

In order to generate a 360° panoramic image, more pictures from different perspectives need to be taken. Fig. 4 shows how two adjacent pictures are projected to the cylinder. Image $A_1B_1C_1D_1$ (C_1 and D_1 are not visible) was projected to $A_1'B_1'C_1'D_1'$ (C_1' and D_1' are not visible); while $A_2B_2C_2D_2$ was projected to $A_2'B_2'C_2'D_2'$. There is overlap between these two pictures; but this is necessary not only for the size change of the picture after the projection, but also for the quality of stitching.

Fig. 3 Result of cylindrical projection. (Left) Before projection. (Right) After projection.

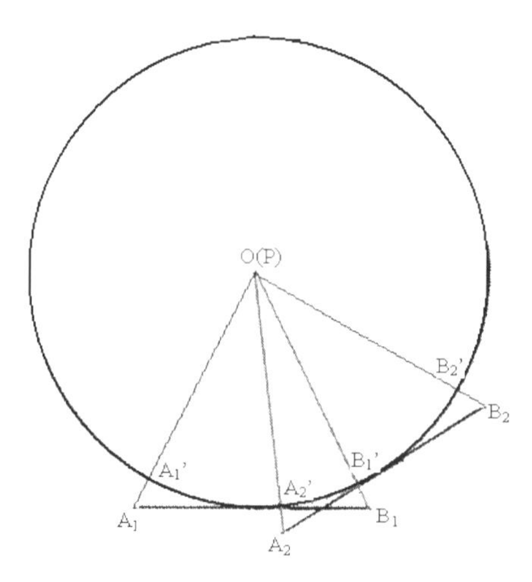

Fig. 4 Two adjacent images A1B1C1D1 and A2B2C2D2 are projected to A1'B1'C1'D1' and A2'B2'C2'D2'. A2A2'B1B1' is the overlap.

After all the individual images were projected, the image stitching is performed to generate a panorama image. Five procedures are taken to complete this task: (1) image adjustment (brightness, contrast, and color adjustment); (2) overlap calculation – As mentioned earlier, adjacent pictures have to have overlaps in order to avoid data loss and properly perform image adjustment; (3) stitching – connect all the projected images together after all the overlaps being removed; (4) interpolation for seamless – using interpolation algorithm to calculate the values on those pixels that had information loss during stitching; (5) cropping – getting rid of blank or useless fields. Fig. 5 shows an example of panorama image generation.

Fig. 5 (a) Stitched image before cropping. (b) After cropping.

3 Inverse Cylindrical Projection

Cylindrical panoramic images provide 3D effect; however, users may not satisfy with what shown on the screen since the view is distorted. This is because users will not consider themselves viewing the scene at the center position of a cylinder. Instead, users position themselves outside of the scene. To resolve this problem, the cylindrical projected panorama image should be converted to human's perspectives before the image display. This can be done through the Inverse Cylindrical Projection. The function of this projection acts like to unfold the image from the cylinder back to a plane. The formula of inverse cylindrical projection is shown in equation (2). Fig. 6 is an example of inverse cylindrical projection.

$$\begin{cases} x = f * \tan[(x' - f * \arctan(W/2/f))/f] + W/2 \\ y = (y' - H/2)/\cos[(x' - f * \arctan(W/2/f))/f] + H/2 \end{cases} \quad (2)$$

Fig. 6 (Left) Image before inverse cylindrical projection. (Right) Image after inverse cylindrical projcetion.

4 Hotspot Pop-Up

As users may have some interesting points while viewing a scene, a hotspot pop-up function is added for this panorama browser. The main idea is that an information window will automatically popped-up while the user moves the mouse to any interesting points.

The hotspot is pre-defined several view area in a scene. These hotspot are assumed to either contains more scene behind (like a door indicates there maybe new space inside) or contains some information that user may would like to know (like the sofa, user may want to know details about it). Based on these hotspot areas, a mouse position detector is first created for showing pop-up window with detailed hotspot information. Once the mouse is moved to the hotspot area, a pop-up window showing either detailed information or further link to another scene is shown; while the user click inside the pop-up window, the viewing window is refreshed and the detailed information and view of the hotspot is shown; on clicking "back" the user can be back to the original panorama view. Fig. 7 shows the procedure to implement this function and a demo of the pop-up information of sofa.

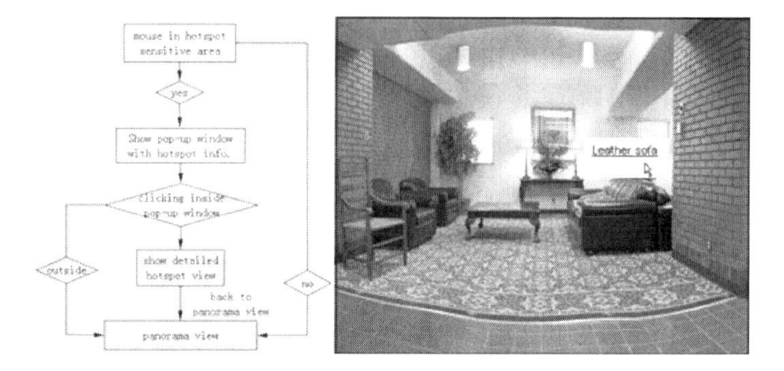

Fig. 7 (Left) Implement procedure. (Right) A demo picture of "sofa" hotspot.

5 Panorama Browser

One of the important issues to design the panorama browser is its real-time feature. Every time when users drag the image left or right, the browser needs to locate the new mouse position and refresh the new part of panorama image on the screen. Two slide-window buffers were used [9] to make the image shown smoothly and timely.

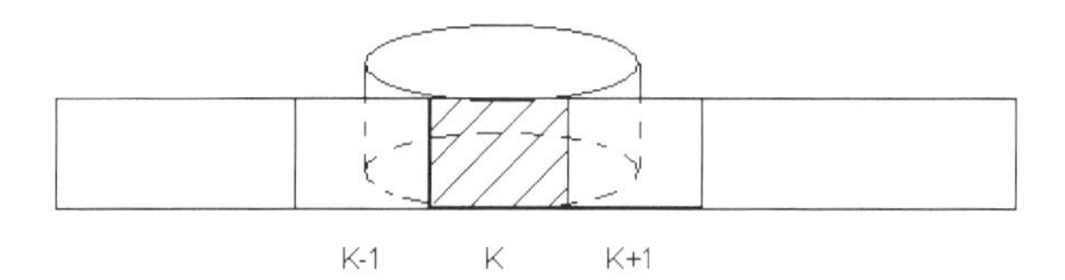

Fig. 8 Two buffers are designed to store pre-calculated image data to generate a smooth dynamic view.

Fig. 8 shows this design. Assume at a given time, image covered by the window K is shown on the screen, then two buffers are used to store the pre-calculated image data covered by the windows *K+1* and *K-1*. Note that the windows will slide dynamically according to the mouse action, so the contents of the two buffers.

It is observed that the calculation of pixel coordinates is symmetric. This fact tells that only one quarter of the pixel position projection is needed, and the coordinates of the pixels in the rest three quarters can be derived accordingly. Thus, further optimization can be done within the quarter of the image. Within the quarter, all the pixels in the same horizontal line have the same ordinate value while all the pixels in the same vertical line have the same abscissa value. Thus for each horizontal (vertical) line, only one calculation is needed to decide their ordinate (abscissa) value. This is a significant improvement because it reduced the time complexity from $(W * H)$ to $(W + H)$. With the optimization, not only the browser responses the mouse clicks "immediately", but also a higher resolution can be adopted to improve the image quality.

6 Conclusions

In this paper, we introduced a panorama browser that provides users 360 degree view of scenes using cylindrical and inverse cylindrical projection algorithm. The browser is easy and quick to build/publish online within Java Applet. The optimization algorithms developed improved both space and time efficiency. This makes the browser having potential in commercial applications.

References

1. Belinda Teoh Soo Phing, M.I.: Panorama - A Better Way To See All Around. ACM SIGCAS Computers and Society, 52–53 (2004)
2. Chen, S.E.: QuickTime VR: an image-based approach to virtual environment navigation. In: International Conference on Computer Graphics and Interactive Techniques, pp. 29–38
3. Cylindrical Projection. Wolfram MathWorld (1995) http://mathworld.wolfram.com/CylindricalProjection.html (retrieved 2008)
4. IBM HotMedia, http://wiki.panotools.org/IBM_Hotmedia
5. Yu, J., Zheng, Y.Z.: Scene tunnels for seamless virtual tour. In: International Multimedia Conference, pp. 448–451. ACM (2004)
6. Nishino, K.: CS 583: Introduction to Computer Vision, http://www.cs.drexel.edu/~kon/introcompvis/lectures
7. Panorama Studio Software, http://www.tshsoft.de/en/index.html
8. QuickTime VR, http://en.wikipedia.org/wiki/QuickTime_VR
9. Szeliski, R., Shum, H.-Y.: Creating full view panoramic image mosaics and environment maps. In: International Conference on Computer Graphics and Interactive Techniques, pp. 251–258. ACM Press/Addison-Wesley Publishing Co. (1997)

Situation Analysis and Policy Research of Software Engineering Standardization of China

Yangyang Zhang, Yuyu Yuan, and Jørgen Bøegh

Abstract. In China, almost all large and medium size software enterprises hope that their software goes international, but this ambition is heavy going. The article compares and describes the different standard system architecture in software engineering. As the core content of this article, it emphasizes to compare the difference between software engineering standardization in China and ISO/IEC standardization. From this analysis it concludes that the main reason, which cause the difference between national self-innovated standardization and international's, are standard system, speed of developing, language and degree of mastering advanced technology.

Keyword: Software Engineering, Standardization, National Standard, International Standard.

1 Introduction

In China, almost all large and medium size software enterprises hope that their software goes international, but this ambition is heavy going.

The lack of software quality has obstructed the development.

On one hand, the software engineering (SE) standards in China are not sufficient. Sometimes we cannot find the appropriate standards when we need to improve the software quality.

On the other hand, because of language problems, we cannot understand the real meaning of International Standards. Therefore International Standards cannot help the companies to improve the software quality.

Yangyang Zhang · Yuyu Yuan · Jørgen Bøegh
Beijing University of Posts and Telecommunications, Beijing, R.P. China

Yangyang Zhang
China Electronics Standardization Institute, Beijing, R.P. China

East Street 1, Andingmen, Dongcheng District, Beijing, China

Z. Qian et al. (Eds.): Recent Advances in CSIE 2011, LNEE 125, pp. 585–593.
springerlink.com

By analyzing the situation of SE standardization, comparing the difference between SE national standards of China with international ones, this article summaries the standardization shortcoming of our country and proposes a policy of SE standardization of China. According to standards the industrial activities of Chinese software industry, we could achieve the goals that strengthen the self-innovation capability of standardization, improve product total quality of Chinese software product to high level and promote Chinese software industry development.

2 The Effect of Software Engineering Standards

As we know, standardization and documentation are two obvious characteristics of software engineering technology.[1]

Standardization of SE is to standards software development, operation, maintenance, through the development, implementation and supervision of the implementation of standards, also to improve software product quality according to the effective management against to software development, production and usage. The effect of software engineering standards embodied in:

- Standardization, which defines the general framework and basic requirements for software engineering activities, ensures the integrity and effectiveness of the activities and improves the transparence, controllability and ordering of management.
- Standardization, which defines the common behavior criterion for software development organizations and person, is helpful to coordinate and harmonize the development activities of software and hardware.
- Standardization, which provides enough proof to test the software development result, is useful to assessing, testing and accepting the software product.
- Standardization, which becomes the choose software engineering method to specific standard, is helpful to overcome the difficult which come from the confusion of understanding of too many methodologies to be there, and to harmonize the software development methods and hardware ones.
- Standardization, which unitized the same or similar attributes of software product, is useful to improve the reuse, multi-operation and protection of software.

3 Standardization Situation of SE

Until now, standards of SE include software product standards and software process standards. Especially developing software process standard is as the important sign that SE is matured. In the recent years, with the internal standardization of SE becoming more active, a variety of standards from enterprise to international level are developed, which guide the worlds software industry into direction of standardization.

Two SE standardization frameworks are popular internationally. One framework is introduced by Software and System Engineering Subcommittee of International Standard Organization (ISO/IEC). It describes the standard request from every basic elements of SE.

Another one is a guide to the SE standard which is introduced by United States Department of Defense. This framework guide puts forward compulsory requirement and proposes to adopt SE standards for SE environment, software life cycle process, SE security and programming languages.

Although the two frameworks are different in scope and level of detail, the same point of view in the two frameworks is taken, namely that standards must be based on the life cycle processes.

Standardization development of SE of China is so late compare with international one. National Standard of China is adopted in relation to the international standard. Standardization system architecture of SE is almost similar with framework of international standardization organization expect the software modeling and interchange standards.

4 Comparison of National and International Standard in SE

4.1 Difference about Adopted Standard

Until now, totally 54 standards have been adopted in China, of which 34 have been published and 20 are under development. All the standards mainly focus on tools and methods, product characteristics, life cycle, process implementation and assessment, process assessment and certification, but the versions are of two different forms:

- The versions of identical adopted standards (same as international standards).

There standards normally are translated directly from respective international standards when they were published, so the publication time lagged about 3-5 years behind international one. The tools and methods was developed in early period by the national standards body were lagged more than 10 year behind international ones.

- The early version identical adoption standards were adopted from canceled international ones.

The main reason that cause the passive situation is that, after 2002, with software engineering technology getting more matured and engineering practices providing a better environment for technology validation, the developing works went to high climax period by the experts of international standardization organization' efforts.

4.2 Difference between Parallel Standards

Currently, there are 3 categories and 5 national standards developed in parallel with international ones.

- Software engineering terminology: GB/T11457 Information technology Software engineering terminology[2]

This standard absorbs not only all terminology from ISO/IEC 24765 Systems and software engineering – Vocabulary[3], ISO/IEC 15504-9 Information technology -- Software process assessment -- Part 9: Vocabulary[4] but also some content of IEEE Std 610.12.1990 Glossary of Software Engineering Terminology[5]. In addition, this standard adds terminologies and definitions of GB/T8566-2001 Systems and software engineering -- Software life cycle processes[6], GB/T 5271.20－1994 Information technology--Vocabulary--Part 20:System development[7], GB/T 18905.1－2002 Information technology – Software product evaluation – Part1: general overview[8], as well UML-Unified Modeling Language and MOF-Meta Object Facility. The scope of the self-defined terminology standard of China is more extensive than the international one, but ISO/IEC 24765 always continues to improve the standard, so the terminology standard of China is far lagged behind ISO in the updated factor.

- Documentation standard

Although international standards keep more close to new engineering practice, the standards only considers the practice requirement that it is less coordination of whole framework.

GB/T 8567-2006 Specification for computer software documentation [9] is based on the provision of GB/T 8566-2001 Systems and software engineering -- Software life cycle processes, it rules the basic requirement to the documentation, content and format of software development process and management process. This standard is applied to all type of software development processes and management processes, among them to the content have more integrity and all documents are more consistent. The development of GB/T9385-1988 Specification for drawing up computer software requirements specification [10] and GB/T9386-1988 Specification for computer software test documentation [11] is effective complement to GB/T 8567-2006. The three standards combine with GB/T16680-1996 Information technology -- Guidelines for the management of software documentation [12] which is adopted standard, and developing adopted project like Systems and software engineering -- Content of systems and software life cycle process information products (Documentation), Software and system engineering -- Guidelines for the design and preparation of user documentation for application software, became the Chinese software engineering standard set of documents. Therefore, the Chinese documentation standard is more consistent and coordinated in whole, and the content also meets the actual demand of current Chinese SE industry.

- GB/T 20918-2007 Systems and software engineering -- Life cycle processes -- Risk management[13]

The content of this standard is based on the standard of software life cycle processes series and IEEE Std 1540-2001[14], DIS ISO/IEC 16085[15]. The start time of developing GB/T 20918-2007 is so early that it keep track with ISO/IEC 16085 from the proposal stage, and form the draft for Approval when ISO/IEC 16085 is in WD stage. There still is a long time from ISO/IEC16085 WD to the last version, so the content might be greatly changed. But the content at this stage is enough to satisfy the need of Chinese software risk management standardization. Considering Systems and software engineering -- System life cycle processes is developed, if only blindly followed the changes in international standards, it is not beneficial to the usage of risk management standard auxiliary. The developing model of GB/T 20918-2007, not only solves the problem that published time of Chinese SE standard is more lagged than international one, but also solves the one that adopted standards are not satisfied with Chinese real condition generally.

4.3 Situation Analysis Where No International Standards Correspond National Standards

So far, 58 ISO standards have no corresponding national standard, the situation includes:

- Standards which do not belong to Chinese SE standard system.

Chinese SE standard system is more similar to the SE standard system of ISO, but open distributed process and interaction standards under tools and methods sub-field does not belong to the scope of Chinese SE standard. Hence, large part of this type of ISO standards does not have corresponding national standards.

- Standards that are still tracked and researched and not considered best opportunity to develop.

For example, SQuaRE series includes International Standards and technical reports for a quality model and measures, as well as on quality requirements and evaluation, which integrate former ISO/IEC9126, ISO/IEC14598 and add content of Data quality and CIF. So far, on one hand ISO/IEC9126 [16], ISO/IEC14598 [17] were identically adopted by China. On the other hand, SQuaRE is being tracked.

Therefore, after Chinese experts track and research against SQuaRE, they all agree that the time to adopted SQuaRE is not mature, and ISO/IEC9126, ISO/IEC14598 are still proper to GB/T16260[18] and GB/T18905[19], if we copy and introduce overall SQuaRE history, it would affect the requirement and evaluation of software quality hugely in China. So for this type of standards, we track and research the developing situation of international, but also combine the self-research technology situation and developed it to relative national standard when time is ripe.

- The related technologies just begin to rise, so the corresponding standardization just work.

This type of standard include IT service standard such as ISO/IEC 20000 IT Service management series[20], ISO/IEC 15940:2006 Information Technology -- Software Engineering Environment Services, ISO/IEC 15504 Information technology -- Process assessment series[21], ISO/IEC26702:2007 Systems engineering -- Application and management of the systems engineering process[22]. For these standards, Chinese experts try to put more efforts to do research and to combine adopted and self-developed method to develop them.

5 The Reason That Conduct the Difference and Countermeasure Policy

From the view of current developing situation of national standard, by introducing international standard, China has initially formed a SE standard system which basically adapt to the development need of Chinese software industry. Without doubt, there exist huge difference between national standard and international. The reasons for the differences are described below:

- Lack of clear concept about SE system

During the usage of standard, the users consider their own condition and find the related specific standard. But this approach could ignore the using environment of standard as well the relevance and complementary among the standards of SE.

- The introduced speed of international standards is slow.

Because of language constraints and other reasons, national standards delay averagely 3-5 years to international ones.

- The introduced international standards are so hard to use and apply

Although the international standards are translated and identified, the standards are so hard to use and apply. To do this translation properly you must understand all details very clearly, and often you will discover that the English text is not clear.

- China is in the state of accepting international standards passively

A few developed countries control the advanced technology in information field, these results in China is always accepts international standards passively.
To solve the above problems, we should take following steps:

- Early intervention, track actively, self-development, publish properly and review on time.

We should insist on the principle that early intervention, track actively, self-development, publish properly and review on time. To promote the healthy development of Chinese information technology, we must actively participate in international standardization activities, track developments in international standards in time, carry out research in important standard properly, combine standardization with development of Chinese information technology, and populate the national standards.

- Refine software engineering standard system

For system construction, we should refine software engineering standard system and confirm the relationship among every standard. According to the developing status of Chinese software industry, self-developed basic standard like software classification standard exists.

- Focus on practical result.

To solve national standards lagging behind internationals', China should increase participation in international hot standard projects and concentrate on the trend in international standard. And then, when we develop national standard with international one at same time, we should absorb and learn the latest international research results, speed up national standard in product quality and try to publish them simultaneously with international standards. On the other hand, we let the research result to be known by discussing, file reply, even proposal for international standards, so that, keep self-developed national standard more close to international standards.

- Strengthen interpretation of the standard

By interpreting and studying the introduced international standards on time, release explanation specification and implantation guide for he description of differences between national and international standard.

6 Conclusions

The main reasons for the difference between national self-innovation standardization and international standards are standard system, speed of developing, language and mastering level of advanced technology. Based on above four causes, this article gives many suggestions to solve the problem. Firstly, the overall policy is early intervention, track actively, self-development, publish properly and review on time, which should be adopted when develop national SE standard. Refined and refresh SE standard system, confirm the relation among all standards and the blank field that should be developed. Focus on practical result and solve the lagging problem that national standards are years behind internationals by all kinds of ways. Finally, strengthen interpretation of the standard and release explanation specification and implantation guide properly.

From technology, SE standards could be refined into many sub-field, basic technology, tools and methods, product characteristics, process implementation and assessment, assessment and certification, life cycle and so on. We do not list the standardization degree and specific situation of these sub-filed. It will be the next step in standardization research how to develop more targeted policies for the sub-field.

Acknowledgment. This work is supported by National Natural Science Foundation of China under Grants No. 90818006 and Foundation of AQSIQ (Administration of Quality Supervision, Inspection and Quarantine) of China under Gants No. 200810627.

References

[1] Wang, B., Feng, H.: Software Engineering Standardization. Information Technology & Standardization, 56–57 (July 2008)
[2] GB/T11457-2006 Information technology Software engineering terminology, SAC (2006)
[3] ISO/IEC 24765: 2009 Systems and software engineering – Vocabulary, ISO/IEC (2009)
[4] ISO/IEC 15504-9: 1998 Information technology – Software process assessment – Part 9: Vocabulary, ISO/IEC (1998)
[5] IEEE Std 610.12.1990 Glossary of Software Engineering Terminology, IEEE (1990)
[6] GB/T8566-2001 Systems and software engineering – Software life cycle processes, SAC (2001)
[7] GB/T 5271.20 1994 Information technology-Vocabulary - Part 20:System development, SAC (1994)
[8] GB/T 18905.1-2002 Information technology – Software product evaluation, SAC (2002)
[9] GB/T 8567-2006 Specification for computer software documentation, SAC (2006)
[10] GB/T9385-1988 Specification for drawing up computer software requirements specification, SAC (1988)
[11] GB/T9386-1988 Specification for computer software test documentation, SAC (1988)
[12] GB/T16680-1996 Information technology – Guidelines for the management of software documentation, SAC (1996)
[13] GB/T 20918-2007 Systems and software engineering – Life cycle processes – Risk management, SAC (2007)
[14] IEEE Std 1540-2001Software Life Cycle Processes—Risk Management, IEEE (2001)
[15] New Proposal for the revision of ISO/IEC 16085 - Software Engineering - Software Life Cycle Processes - Risk Management, ISO/IEC JTC1/SC7 (2003) (unpublished)
[16] ISO/IEC 9126-1: 2001 Software engineering – Product quality – Part 1: Quality model, ISO/IEC (2001)

[17] ISO/IEC14598-1: 1999 Information technology – Software product evaluation – Part 1: General overview, ISO/IEC (1999)
[18] GB/T16260.1-2006 Software engineering – Product quality – Part 1: Quality model, SAC (2006)
[19] GB/T18905.1-2002 Information technology – Software product evaluation – Part 1: General overview, SAC (2002)
[20] ISO/IEC 20000-1: 2005 Information technology – Service management – Part 1: Specification, ISO/IEC (2005)
[21] ISO/IEC 15940: 2006 Information Technology – Software Engineering Environment Services, ISO/IEC (2006)
[22] ISO/IEC26702: 2007 Systems engineering – Application and management of the systems engineering process, ISO/IEC (2007)

Algorithms for Multilevel Analysis of Growth Curves

Xiaodong Wang and Jun Tian*

Abstract. The growth curve analysis is a hotspot in the analysis to understand the characteristics of things change over time. However, the conventional curve fitting method does not apply to the structure data obtained by repeated measurement. Although the outline design matrix computation can be used to treat the structure data in the growth curve analysis, the corresponding SAS programs are very long and hard to understand. In this paper, we present algorithms for multi-level model applied to the analysis of the growth curve. The usage of the algorithms is demonstrated by practical examples.

1 Introduction

The growth curve is very important in repeated measurement data analysis. From growth curve we can continuously observe an object at different time points and then understand its change characteristics and rules. The influence on the object under different conditions changing over time can then be revealed.

For example, the researchers may want to understand the changes of prealbumin for patients with liver cirrhosis after portion of their liver cut. By drawing the prealbumin mean graph of these 8 patients at 6 time points we can see that the relations of prealbumin mean with different time may be expressed by a cubic polynomial curve.

Therefore, by fitting the polynomial curve on relations of pre-albumin and time, we can understand the changing rules of pre-albumin changing with time. However, we cannot fit the polynomial curve by traditional curve fitting method, since the

Xiaodong Wang
Quanzhou Normal University, China
e-mail: wangxiaodong@qztc.edu.cn

Jun Tian
Fujian Medical University, China

* Corresponding author.

Z. Qian et al. (Eds.): Recent Advances in CSIE 2011, LNEE 125, pp. 595–599.
springerlink.com

condition to use the method is the observed data $(t_i, \bar{y}_i), 1 \leq i \leq 6$, at 6 time points must be mutual independent, while in repeated measured data, the observed data are not independent of each other [1]. For repeated measured data, at present to analyze the trends of observations changing with time we have to use plenty of the matrix computation [2]. Although in the SAS (Statistical Analysis System) some special matrix computation modules are available [3], but we need to write complicated procedure [4]. Therefore they are not suitable for the scientists who are not familiar with computer science.

Multilevel model is a data analysis method in social science [5]. It was initially proposed for social science research data in a common multi-layer nested structure of linear multivariate statistical analysis [6]. Since the measured data for each observed object in m points of time can be viewed as one standard unit of level, and each of observed object is of 2 level units, the observation data can be viewed as a two-level nested structure. This means that the multilevel model can be used for growth curve analysis. However, it has not yet reported so far that how the multilevel model can be applied to growth curve analysis, particularly to write a simple program for growth curve analysis in SAS the famous international statistical software. In this paper we introduce with an example application of multi-level model in growth curve analysis and specific steps and the analysis method.

2 The Growth Curve for Single Data Set

Let y be a quantitative observation variable. In the test, the values of y at m time points and n subjects were observed. The test results of y for subject j at time i are denoted by $y_{ij}, j = 1, \cdots, n; i = 1, \cdots, m$.

Suppose in the following repeated measurements, the change rates over time of y values are the same. Then the random intercept model can be expressed by

$$y_{ij} = \beta_{0j} + \beta_1 t_{ij} + \beta_2 t_{ij}^2 + e_{ij} \tag{1}$$

where,

$$\beta_{0j} = \gamma_{00} + u_{0j} \tag{2}$$

Substitute formulae (2) into formulae (1) we get:

$$y_{ij} = \gamma_{00} + \beta_1 t_{ij} + \beta_2 t_{ij}^2 + (e_{ij} + u_{0j}) \tag{3}$$

In the formulae (3): e_{ij} is an estimation error and $e_{ij} \sim N(0, \sigma_e^2)$; u_{0j} is the random effect of intercept; γ_{00}, β_1 and β_2 are fixed parameters.

Suppose in the following repeated measurements, the change rates over time of y values are different. Then the random intercept-coefficient model can be expressed by

$$y_{ij} = \beta_{0j} + \beta_{1j} t_{ij} + \beta_{2j} t_{ij}^2 + e_{ij} \tag{4}$$

where,

$$\begin{cases} \beta_{0j} = \gamma_{00} + u_{0j} \\ \beta_{1j} = \gamma_{10} + u_{1j} \\ \beta_{2j} = \gamma_{20} + u_{2j} \end{cases} \tag{5}$$

Substitute formulae (5) into formulae (4) we get:

$$y_{ij} = \gamma_{00} + \gamma_{10}t_{ij} + \gamma_{20}t_{ij}^2 + (e_{ij} + u_{0j} + u_{1j}t_{ij} + u_{2j}t_{ij}^2) \tag{6}$$

In the formulae (6): e_{ij} is an estimation error and $e_{ij} \sim N(0, \sigma_e^2)$; u_{0j}, u_{1j} and u_{2j} are the random effect of intercept and coefficient and $(u_{0j}, u_{1j}, u_{2j}$ follows the ternary normal distribution $N(0, \Omega)$,

$$\Omega = \begin{bmatrix} \sigma_0^2 & & \\ \sigma_{10} & \sigma_1^2 & \\ \sigma_{20} & \sigma_{21} & \sigma_2^2 \end{bmatrix}$$

is the variance-covariance matrix of u_{0j}, u_{1j} and u_{2j}; γ_{00}, γ_{10} and γ_{20} are fixed effect parameters.

[Example] The changes of prealbumin (mg/L) for 8 patients with liver cirrhosis classified as level 5 according to the international pointrating method after portion of their liver cut are shown. We are supposed to study the changes of prealbumin for the patients with liver cirrhosis after portion of their liver cut. Suppose time=1,2,3,4,5,6 correspond to the data of before operation, 2 days, 5 days, 10 days, 15 days and 20 days after operation respectively.

Usually, in fitting the polynomial curve equation of observation variable y and time t, we may assign t with values in chronological order $t = 1, 2, 3, 4, 5, 6$ as shown. This assignment method can make t_{ij} and t_{ij}^2 become collinear, while this kind of colinearity often have a great influence on the estimation results to random effects and fixed effects parameters. In order to avoid the problem of colinearity, we use the centralized method in time assignment. We set the middle point of the observation interval as $t = 0$. Therefor in the SAS data file the data of before operation, 2 days, 5 days, 10 days, 15 days and 20 days after operation correspond to time=-3,-2,-1,0,1,2 respectively. From the prealbumin mean graph of patients at 6 time points we can see that the relations of prealbumin mean with different time can be expressed by a cubic polynomial curve.

From the results the relations of prealbumin with time can be described by the following polynomial curve equation:

$$\hat{y} = 175.55 + 37.6789t - 6.0044t^3$$

From this cubic polynomial equation we can find the maximum and minimum points as follows. Set $\hat{y}_t' = 37.6789 - 3 \times 6.0044t^2 = 0$ and we get $t^2 = \frac{37.6789}{18.0132}$. The two extreme value points are $t = 1.45$ and $t = -1.45$ respectively.

We can compute by linear interpolation that the time corresponding to $t = -1.45$ is 3.7 days after operation and the time corresponding to $t = 1.45$ is 7 days after operation.

Conclusion: The patients' prealbumin began to decline after the operation and achieved its minimum value at the time of 3.7 days after operation. Then the patients' prealbumin , as time increasing, increased gradually to about 17 days after operation to highest value, and then decreased again.

3 The Growth Curve for Multiple Data Sets

Let y be a quantitative observation variable and the experiment factor *treat* has k levels. The n subjects are divided randomly into k groups with each group of n_l subjects performed with a experiment in one level of *treat*, $l = 1, \cdots, k$. In the test, the values of y at m time points and n subjects were observed. The test results of y for subject j at time i are denoted by $y_{ij}, j = 1, \cdots, n; i = 1, \cdots, m$. For example, if the relation of y and time can be described by a quadratic polynomial equation, then the model will be expressed by

$$y_{ij} = \beta_{0j} + \beta_{1j} t_{ij} + \beta_{2j} t_{ij}^2 + e_{ij} \tag{7}$$

where,

$$\begin{cases} \beta_{0j} = \gamma_{00} + \gamma_{01} treat_j + u_{0j} \\ \beta_{1j} = \gamma_{10} + \gamma_{11} treat_j + u_{1j} \\ \beta_{2j} = \gamma_{20} + \gamma_{21} treat_j + u_{2j} \end{cases} \tag{8}$$

Substitute formulae (8) into formulae (7) we get:

$$\begin{aligned} y_{ij} &= \gamma_{00} + \gamma_{01} treat_j + \gamma_{10} t_{ij} + \gamma_{20} t_{ij}^2 \\ &+ \gamma_{11} treat_j t_{ij} + \gamma_{21} treat_j t_{ij}^2 \\ &+ (e_{ij} + u_{0j} + u_{1j} t_{ij} + u_{2j} t_{ij}^2) \end{aligned} \tag{9}$$

In the formulae (9): e_{ij} is an estimation error and $e_{ij} \sim N(0, \sigma_e^2)$; u_{0j}, u_{1j} and u_{2j} are the random effect of intercept and coefficient and $(u_{0j}, u_{1j}, u_{2j}$ follows the ternary normal distribution $N(0, \Omega)$,

$$\Omega = \begin{bmatrix} \sigma_0^2 & & \\ \sigma_{10} & \sigma_1^2 & \\ \sigma_{20} & \sigma_{21} & \sigma_2^2 \end{bmatrix}$$

is the variance-covariance matrix of u_{0j}, u_{1j} and u_{2j}; $\gamma_{00}, \gamma_{01}, \gamma_{10}, \gamma_{20}, \gamma_{11}$ and γ_{21} are fixed effect parameters.

From the analysis above we know that the regulations of EMG frequency changing over time for group A and group B are different. The relation of frequency and time in EMG is quadratic for group A and linear for group B. From the sign of quadratic term of the equation we know that the frequency has a minimum in EMG for group A. The minimum point can be computed at point $t = 0.9164$ (at about the 5th time point). The linear equation for group B has a negative slope, so the frequency for group B is decreasing with time.

4 Conclusion

We have introduced with example applications of multi-level model in growth curve analysis and specific steps and the analysis method. We have also prepared a clear and simple SAS program for practical application of scientific researchers.

The multi-level model in growth curve analysis is an effective method to treat nested structure data. Using this model in growth curve analysis for repeated measured data can avoid complex and difficult matrix computation in SAS. It is a very useful and practical method for scientific researchers.

Acknowledgment. This work was partially supported by the Natural Science Foundation of Fujian under Grant No.2009J01295 and the Haixi Project of Fujian under Grant No.A099.

References

1. Box, G.: Problem in the analysid of grows and wear curves. Biometrics 6, 362–389 (1950)
2. Davidian, M., Giltinan, D.M.: Nonlinear Models for repeated measurement data. St. Edmundshury Press, Suffolk (1995)
3. Cody, R.P., Smith, J.K.: Applied Statistics and the SAS Programming Language, 5th edn. Prentice Hall, New York (2005)
4. Lindsey, J.K.: Models for repeated measurements. Oxford University Press, Oxford (1993)
5. Park, G., Deshon, R.P.: A multilevel model of minority opinion expression and team decision-making effectiveness. The Journal of Applied Psychology 95, 824–833 (2010)
6. Wenzel, D., Ocana-Riola, R., Maroto-Navarro, G.: A multilevel model for the study of breastfeeding determinants in Brazil. Maternal & Child Nutrition 6, 318–327 (2010)

Effects of Sample Size on Accuracy and Stability of Species Distribution Models: A Comparison of GARP and Maxent

Xinmei Chen and Yuancai Lei*

Abstract. Prediction of species distribution and its changes play an increasingly important role in the fields of ecological protection and application as well as global climate changes. It is impracticable to survey species distribution in large area, especially for rare species; therefore species distribution models are very useful for predicting species distribution. Sampling size has an essential influence on the cost of actual survey and the accuracy of model prediction. Generally, the larger sample size, the higher accuracy of models, but with the more cost of survey. It is necessary to research species distribution models to get the highest accuracy but with the least sample size. Taking 31 different sample sizes (5, 10, 15, 20, 25, 30, 40, 50, 60, 70, 80, 90, 100, 120, 150, 180, 200, 220, 250, 300, 350, 400, 450, 500, 550, 600, 650, 700, 800, 900, 1000 and 1200) of 4 species as an example, this study analyzed the influence of different sample sizes on accuracy and stability of both MaxEnt (Maximum Entropy Species Prediction Model) and GARP(Genetic Algorithm for Rule-set Production). The results showed that both the accuracy and stability of two models increased with the sample size. And the two models had different predictive results to different species. The omission values of GARP were always smaller than that of MaxEnt at the same sample size to all species.

1 Introduction

Species distribution models play an important role in ecology, conservation and management[1]. Close relations exist between species distribution and geographical

Xinmei Chen · Yuancai Lei
Research Institute of Forest Resources Information Techniques,
Chinese Academy of Forestry, Beijing, 100091, P.R. China
e-mail : chxm200299@163.com, yclei@caf.ac.cn

* Corresponding author.

Z. Qian et al. (Eds.): Recent Advances in CSIE 2011, LNEE 125, pp. 601–609.
springerlink.com

environment factors. Taking account of geographical environment factors into species distribution models can help predict species potential distribution. Species distribution models have been widely used in species protection management and monitoring as well as species distribution changes under the influence of climate change. Five models such as Bioclim, CLIMEX, DOMAIN, GARP and MaxEnt are widely used in predicting species distribution [2]. According to Wang et al.[3], among the five models, GARP and Maxent are two best models based on AUC (area under the receiver operating characteristic curve).

A lot of previous studies have applied GARP model to predict species distribution and some researchers have obtain accurate results [4]. This model has been widely applied in China [5]. MaxEnt is a relatively new method of predicting species distribution which can also test the predictive results of the predicting species. Phillips [6] firstly introduces the application of MaxEnt model in predicting species distribution. Then various studies appear following him [7]. The previous researches on GARP and Maxent mainly focus on species distribution and variation in large range and the influence of environment factors [7]. However, little information is available for the effect of sample size of species distribution data on accuracy of GARP and MaxEnt models, especially for large sample sizes of different species.

Due to data of some species is difficult to sample or collection data can not be digitized in time, that species distribution data used for model prediction in deed is highly limited [1]. Many studies have confirmed that the sample size of species distribution data significantly affected accuracy of model simulation results. Small sample sizes face challenges to any statistical analyses and cause decreased predictive potential when compared to models developed with more occurrence data [8,9]. Generally, with sample size increasing, accuracy should also increase; and then extent of increase generally decreases, until the maximum accuracy potential is achieved [11]. Because of limitation of species distribution data, small sample sizes are usually used to study the accuracy of species distribution model, the largest sample sizes of species distribution data are only achieved 100 or 150. Although the trend of simulation accuracy of species distribution with sample size changing can be observed, the sample size is not enough for us to reach the stable value of simulation accuracy [8]. Generally, the larger sample size, the higher the estimate accuracy of species distribution model, but the higher the cost for sample survey. In order to reach the estimate accuracy of models and need much less expense for field survey, it is also essential to simulate and study sample size and model accuracy.

In this paper, four species, *Elaeagnus angustifolia(a), Haloxylon ammodendron(b), Nitraria tangtorum(c) and Tamarix chinensis(d),* and 31 sample sizes (5, 10, 15, 20, 25, 30, 40, 50, 60, 70, 80, 90, 100, 120, 150, 180, 200, 220, 250, 300, 350, 400, 450, 500, 550, 600, 650, 700, 800, 900, 1000 and 1200 records) were used to test accuracy and stability of GARP model and MaxEnt model.

2 Materials and Methods

2.1 Materials

2.1.1 Study Area and Species Occurrence Data

Species occurrence data came from an area in Dengkou County, PR China, along the Huanghe River valley within the arid regions of western Inner-Mongolia. During 2006, the Chinese Academy of Forestry counted all desert shrubs within a 1 km [2] area centered on latitude 40°15′37.8″ and longitude 106°56′28.0″. Six tree species were found, the main survey factors are single tree or the number of clump trees, plant height, ground diameter, crown sizes and spatial relatively coordinates(x, y). Here we focus on Elaeagnus angustifolia(a), Haloxylon ammodendron(b), Nitraria tangtorum(c) and Tamarix chinensis(d). These species showed some degree of rarity and clustering. In which, a is a small arbor tree species, b, c and d are shrub tree species. All of them have important economic and social values for sand-fixing, conserving soil and water and greening.

2.1.2 Environmental Variables

We selected an Advanced Spaceborne Thermal Emission and Reflection Radiometer (ASTER) Digital Elevation Model (DEM) (https://wist.echo.nasa.gov/wist-bin/api/ims.cgi?mode=MAINSRCH&JS=1 free charge Aster 30m G-Dem data) from NASA Jet Propulsion Laboratory which was scanned on 2009 for deriving habitat variables. The ASTER DEM, at 30×30m, offered the smallest pixel size available for the study area. The DEM coordinates were matched to GPS coordinates taken at study site so that the DEM grids corresponded to the sampling units used in the sampling simulation. The topography-based environmental gradients which we derived from the DEM included elevation, slope, aspect, curvature, and distance from roads. Vegetation index extracted from QuickBird, which is the resolution of 0.1 meters standard orthophoto data. That is to say, the QuickBird data, at 0.1×0.1m, offered the smallest pixel size available. These environmental variables have a major impact on species distribution.

Eight variables were selected as environmental variables when models were simulated. Topography data, such as elevation, slope, aspect, curvature and distance to road. Vegetation Index data, as normalized difference vegetation index (NDVI), ratio vegetation index (RVI) and difference vegetation index or environment vegetation index (DVI/EVI).

2.2 Modeling Methods

2.2.1 GARP Model

Desktop GARP is a model based on genetic algorithm built by David Stockwell in the 1990s, which can produce different combined models to judge ecological needs of species by species present distribution data and environment data, and then predict species potential distribution [29-31]. GARP describes environment conditions for the needs of maintaining species populations. It uses present species distribution plots data and environment parameters layer which relate to species survival ability as model input parameter. GARP employs four distinct modeling methods, atomic, logistic regression, bioclimatic envelope, and negated bioclimatic envelope rules, to derive several different rules [11]. GARP uses these rules to iteratively search for non-random correlations between the presence and background absence observations and the environmental predictors, and further forecasts and estimates species potential distribution area [14]. Operation principle of genetic algorithm leads to the random differences in every output, so it needs to generate best-subset. By summarizing these results of best-subset, the species potential distribution in probability distribution can be achieved [9,14]. GARP model operation procedure is: (1) species distribution data and environment data were imported Desktop GARP software; (2) sixty percent species distribution data were selected randomly as training data, and the remained 40% species distribution data as test data. According to Anderson et al. [14] and Zuo et al. [15], we set parameters as follow: 100 runs, 0.01 convergence limit, 1000 maximum iterations, at the same time selecting all of four rules (do not use their combinations), opening best-subset selection parameter options, other parameters are default value of software; (3) at last output format is ASCII raster layer. Run DesktopGarp version 1.1.6, receive 10 simulated results in best-subset, fold the 10 results and receive species potential distribution probability distribution map with 10 ranks. (the GARP model based on DesktopGarp version 1.1.6, the software derived from: http://www.nhm.ku.edu/desktopgarp/Download.html)

2.2.2 Maxent Model

Jaynes have proposed the maximum entropy theory in 1957. Then maximum entropy is used in computer science and statistics field and received practical application, especially widely used in the field of natural language processing and identification [6]. Maxent utilizes a statistical mechanics approach called maximum entropy to make predictions from incomplete information. It is based on actual geographic coordinates of species distribution and the environment variables of species distribution area to run and receive the predictive model, and then applied this model to estimate the possible distribution of the target species in target area [15]. Maxent model operation procedure is: (1) species distribution data and environment data are imported Maxent software; (2) sixty percent species distribution data are selected randomly as training data, and the remained 40% species distribution data as test data, other parameters are default value of software; (3) at last output format is ASCII raster layer (the Maxent model based on MaxEnt software Version 3.3.1, the software derived from http://www.cs.princeton.edu/~schapire/maxent/)

2.2.3 Model Evaluation

Three metrics were used to evaluate the performance of species distribution models. We compared the predictive results of our performance metrics over 10 replications for each models. First, AUC was selected as a method to evaluate the performance of species distribution models. AUC is the area under ROC (receiver operating characteristic curve). ROC curve analysis method was first used to evaluate the ability of radar signal reception [16], and then it was widely used to evaluate performance of medical diagnostic tests. Recently, ROC curve analysis method was more and more widely used in species potential distribution predictive evaluation [2, 9]. AUC provides a threshold-independent measure of model performance; it can compare two or more experiments and select the best method as an evaluation method to judge distribution models' predictive results. The size of AUC value reacts to predictive ability of models. The more AUC value indicates, the better the predictive ability is. According to Swets' [17] criteria, 0.5<AUC<0.7 indicates predictive ability is fair; 0.7<AUC<0.9 indicates predictive ability is good; 0.9<AUC<1 indicates predictive ability is excellent. The AUC can also be used as a measure of the model's overall performance and has values usually ranging from 0.5 (random) to 1.0 (perfect discrimination), but can have values below this range indicating that a model is worse than random [18].

Second, AUC standard deviation was selected as another evaluation criterion. In the course of processing data, there are many times to deal with the species distribution data, which calculated lots of AUC value. The AUC standard deviation was calculated from many AUC value relative to their average. AUC standard deviation is computed as

$$SD = \sqrt{\frac{\sum_{i=1}^{N}(X_i - \overline{X})^2}{N-1}}$$

Where SD is the AUC standard deviation. N is the number of replication times. X_i is the AUC standard deviation in the i th times. $\overline{X}$ is the average of AUC value in N times. If AUC standard deviation is instability, the result data is inaccurate. In the other hand, the lower AUC standard deviation expresses that the AUC value is more stable and the result data is more accurate.

The third evaluation is omission error. Percentage of the training points that are omitted from the prediction, that is, those that are absent predicted but are presence records. Moreover, the smaller omission indicated the better performance. The omission error was calculated under the threshold of 0.01 in this study.

2.2.4 Data Analysis

Four species, *a, b, c* and *d*, were selected. First, we used ArcGIS (geographical information system) to extract topographic variables from DEM data, then we extracted vegetation index variables from QuickBird data, finally the extracted data were translated into the format of .csv for Maxent using. GARP model used excel format data to simulate. In order to match the scale of the species and environment variables, the DEM and QuickBird were resampled to a 10×10m cell size by using a bilinear interpolation in ArcGIS. Then we used ERDAS IMAGINE software to cut, and changed the raster data into ASCII format. Finally, sample sizes of 5, 10, 20, 25, 30, 40, 50, 60, 70, 80, 90, 100, 120, 150, 180, 200, 220, 250, 300, 350, 400, 450, 500, 550, 600, 650, 700, 800, 900, 1000 and 1200 were selected randomly from actual field survey data for every species in select case of SPSS software [19].

3 Results and Discussion

From Figure1 we can see that the predictive accuracy of *a* and *d* were higher than *b* and *c*. And the accuracy of *a* was the best while *c* was the worst. From the Figure1 we also found that predictive performance of Maxent species distribution model of *a* and *d* were much better than *b* and *c*, but predictive performance of GARP species distribution model of b and c were much better than *a* and *d*. In the actual distribution, *a* and *d* belonged to aggregate distribution, and the distribution of *b* and *c* were relatively scattered. The results showed that the bigger the range of species distribution, the worse performance, and vice versa, which was similar to Hernandez et al.[10] and Stockwell et al. [8]. Hernandez et al. [10] find that models for species with broad geographic ranges tend to be less accurate than those with smaller geographic ranges. Stockwell et al. [8] find that maximum predictive accuracy is not independent on range size: widespread species are modeled less accurately.

The accuracy of species predictive models was different and the accuracy within the four species was not compared when sample size was small, because predictive results of Maxent and GARP species distribution models were quite different (Figure1). The AUC changed few with sample size increasing, and changing trends were stable. The results were consistent with those of Wisz et al [11] and Hernandez et al. [10]. Wisz et al. [11] simulate and evaluate 12 models including MaxEnt and GARP for 46 species (from six different regions of the world) at three sample sizes (100, 30 and 10 records). Their results show that the performance of Maxent model is much better and the sensitivity is small to sample size even in the case of small sample size; the performance of GARP model is moderate when sample size is big, while its performance is the best when sample size is small. Also it is found that with sample size decreasing, accuracy of model is decreased and variability is increased. Hernandez et al. [10] test four models including Maxent and GARP for 18 species by six different sample size treatments and three different evaluation measures. The evaluate results show that the performance of Maxent is the best and the performance of GARP is better even in the case of very small sample size.

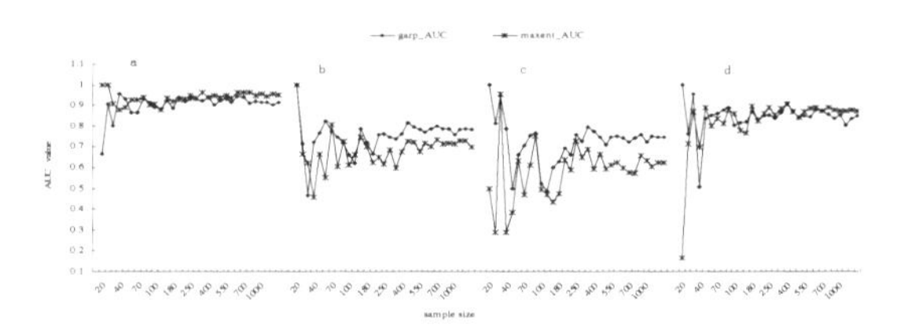

Fig. 1 Comparing the changes of AUC of two species predictive models with the changes of sample size of four species.

With sample size increasing, curves changed more and more gently, and the standard deviations got smaller and smaller. The result showed that the stability of species predictive models got higher and higher (Figure2). Outliers appeared around the AUC standard deviation of each species and different species predictive distribution when sample size was small. And AUC value standard deviation appeared deviation many or few (Figure2). With sample size increasing, AUC value standard deviation was more stable, and more sample size, fewer AUC value standard deviation. AUC value standard deviation of *a* and *d* reached 0.05 earlier than *b* and *c*, furthermore the AUC value standard deviation of *a* and *d* was fewer than *b* and *c* at last. In other words, the predictive accuracy of Maxent and GARP of *a* and *d* performed better than those of *b* and *c*, which was the same as AUC value predictive accuracy, and the phenomenon may be not only relevant to models but also species ecological specialization and the distribution of species themselves.

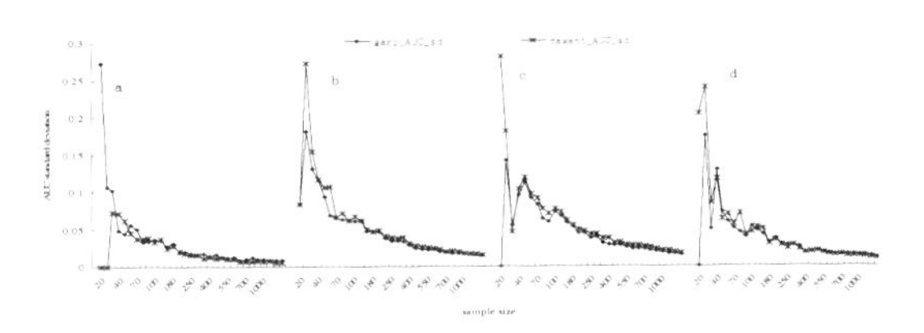

Fig. 2 The changes of AUC Standard Deviation with the sample size increasing of four species.

Figure3 showed changes of omission error with sample size increasing for *a*, *b*, *c* and *d*. The predictive results of GARP and Maxent species distribution models had great discrepancy. First about GARP model, Omission errors on training data and test data had the same variation trend as sample size increasing. Omission errors of *a* and *d* were small and the distribution was aggregated, while those of species *b* and *c* were big and the distribution was dispersed. Predictive performance of *a* and *d* was better than *b* and *c*, which was consistent with AUC evaluation result. Second about Maxent model, Omission errors on training data and test data had different variation trends. Omission errors on the training data decreased with sample size increasing, while that on the test data increased with sample size decreasing. This was consistent with Stockwell et al. [8]. Stockwell et al. [8] show that: the predictive accuracy of species predictive models is clearly related to sample size. Accuracy on the training data decreased with lager sample size, while accuracy on the test data increased with sample size increasing. The accuracy of difference between test data and training data indicated the degree of overfitting, which was greatest with small sample size, and decreased with sample size increasing. Omission error changed few as sample size increasing. Finally we can see that omission errors of *a* and *d* were small, while *b* and *c* were large. In other words, predictive performance of *a* and *d* was better than *b* and *c*, which was consistent with AUC evaluation results.

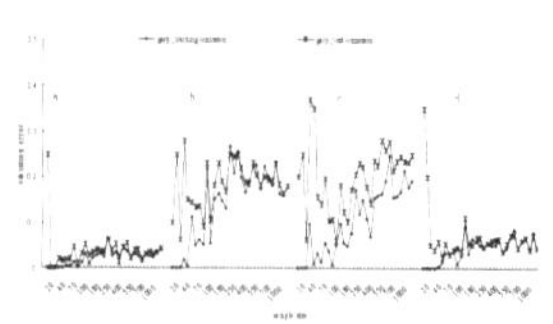
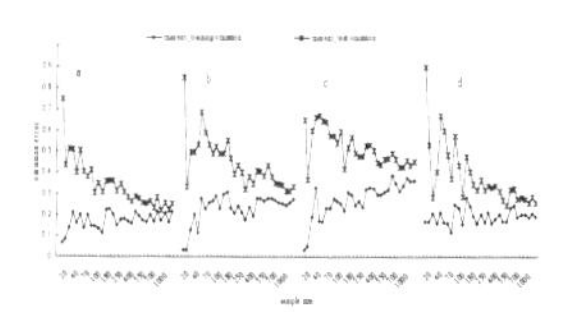

Fig. 3 The changes of omission errors of training data and test data with the sample size increasing for four species. Left is the result of GARP species predictive model, right is the result of Maxent.

4 Conclusions

(1) AUC value changes of GARP and Maxent models are decreased with sample size increasing, and the variation trend is much more stable. The performance of species predictive model is not only relevant to models but also species ecological specialization and the distribution of species themselves.

(2) With sample size increasing, curves change much more gently, and the AUC standard deviations (SD) become small. With sample size increasing, AUC value standard deviations is more stable, and more sample size, less AUC value standard deviations.

(3) The predictive results of GARP and Maxent species distribution models have great discrepancies. Omission errors of GARP model on training data and test data have the same variation trend as sample size increasing. Omission errors of Maxent model on training data and test data have exactly different variation trend; omission errors on the training data decreased with sample size increasing, while omission errors on the test data increased as sample size decreasing. This is consistent with Stockwell et al. [8].

In this paper, we concentrate on the relationship between sample size and accuracy of Maxent and GARP species distribution models. Because the two models are used widely to simulate species distributions and used easily, lots of predictive species distribution models are used now. However, as what Hernandez et al. [10] and Engler et al. [18] say, models are simply an estimate of a species potential distribution. Species distribution model cannot substitute for fieldwork to collect more distributional data but can be a useful tool for data exploration to help identify potential knowledge gaps and provide direction to fieldwork design [10,18].

Acknowledgments. The authors express their appreciation to the Ministry of Science and Technology (2005DIB5J142), National Forestry Bureau (2006-4-34) and National Natural Science Foundation (31170588) for fiscal support.

References

1. Graham, C.H., Ferrier, S., Huettman, F., Moritz, C., Peterson, A.T.: New developments in museum-based informatics and applications in biodiversity analysis. Trends in Ecology and Evolution 19, 497–503 (2004)
2. Elith, J., Graham, C.H., Anderson, P.R., Dudik, M., Ferrier, S., Guisan, A., Hijmans, R.J., Huettmann, F., Leathwick, J.R., Lehmann, A., Li, J., Lohmann, L.G., Loiselle, B.A., Manion, G., Moritz, C., Nakamura, M., Nakazawa, Y., Overton, J.M., Peterson, A.T., Phillips, S.J., Richardson, K., Scachetti-Pereira, R., Schapire, R.E., Soberon, J., Williams, S., Wisz, M.S., Zimmermann, N.E.: Novel methods improve prediction of species' distributions from occurrence data. Ecography 29(2), 129–151 (2006)

3. Wang, Y.S., Xie, B.Y., Wan, F.H., Xiao, Q.M., Dai, L.Y.: Application of ROC curve analysis in eva-luating the performance of alien species' potential distribution models. Biodiversity Science 15(4), 365–372 (2007a)
4. Peterson, A.T.: Predicting species' geographic distributions based on ecological niche modeling. Condor 103(3), 599–605 (2001a)
5. Zhong, G.P., Sheng, W.J., Wan, F.H., Wang, J.: Potential distribution areas of Solanum rostratumin China: A prediction with GARP niche model. Chinese Journal of Ecology 28(1), 162–166 (2009)
6. Phillips, S.J., Dudik, M., Schapire, R.E.: A maximum entropy approach to species distribution modeling. In: Proceedings of the 21st International Conference on Machine Learning, Banff, Canada (2004)
7. Phillips, S.J., Anderson, R.P., Schapire, R.E.: Maximum entropy modeling of species geographic distributions. Ecological Modelling 190, 231–259 (2006)
8. Stockwell, D.R.B., Peterson, A.: Effects of sample size on accuracy of species distribution models. Ecological Modelling 148, 1–13 (2002)
9. McPherson, J.M., Jetz, W., Rogers, D.: The effects of species' range sizes on the accuracy of distribution models: ecological phenomenon or statistical artifact? Journal of Applied Ecology 41, 811–823 (2004)
10. Hirzel, A., Guisan, A.: Which is the optimal sampling strategy for habitat suitability modelling? Ecological Modelling 157, 331–341 (2002)
11. Stockwell, D., Peters, D.: The GARP modeling system: problem and solutions to automated spatial prediction. International Journal of Geographic Information Systems 13(2), 143–158 (1999)
12. Higgins, S.I., Richardson, D.M., Cowling, R.M.: Modeling invasive plant spread: the role of plant environment interactions and model structure. Ecology 77(7), 2043–2054 (1996)
13. Higgins, S.I., Richardson, D.M., Cowling, R.M.: Predicting the landscape —scale distribution of alien plants and their threat to plant diversity. Conservation Biology 13, 303–313 (1999)
14. Stockwell, D.R.B., Noble, I.R.: Induction of sets of rules from animal distribution data: a robust and informative method of data analysis. Mathematics and Computers in Simulation 33, 385–390 (1992)
15. Zuo, W.Y., Lao, N., Geng, Y.Y., Ma, K.P.: Predicting species' potential distribution—SVM compared with GARP. Journal of Plant Ecology (Chinese Version) 31(4), 711–719 (2007)
16. Leshowitz, B.: Comparison of ROC curves from one- and two-interval rating-scale procedures. The Journal of Acoustical Society of America 46, 399–402 (1969)
17. Swets, J.A.: Measuring the accuracy of diagnostic systems. Science 240, 1285–1293 (1988)
18. Engler, R., et al.: An improved approach for predicting the distribution of rare and endangered species from occurrence and pseudo-absence data. J. Appl. Ecol. 41, 263–274 (2004)
19. Yin, H.J., Liu, E.: SPSS for Windows of Social Statistical Package Made Simple. Social Sciences Documentation Publishing House, Beijing (2003)
20. Peterson, A.T., Papes, M., Eaton, M.: Transferability and model evaluation in ecological niche modeling: a comparison of GARP and Maxent. Ecography 30, 550–560 (2007)

Matching Algorithm Based on Characteristic Matrix of R-Contiguous Bit

Jianping Zhao, Hua Li, and Jingshan Liu

Abstract. Computer immunology plays an important role in network security. Matching algorithms based on r-contiguous bits and Hamming distance use the ideas that comparing strings by each bit from start to end in string match, which makes the efficiency of the algorithm low. This paper presents a matching algorithm based on characteristic matrix of r contiguous bit. This method takes bitwise "exclusive-or" operation on random string and "self" string to obtain a result string, and gets a result matrix by doing "and" operation on this result string and each row of a characteristic matrix with special structure. It can be determined quickly whether the random string is matching or not., and the efficiency is greatly improved.

1 Introduction

In the study field of computer immune, there has a very classical negative selection algorithm [1] proposed by American scholar, Forrest, Perelson et al in 1994. This algorithm generates a detector (each detector is a characteristic string) and simulates the negative selection principle in the immune tolerance phase of biological immunology. So the detects are the "receptor" with immune tolerance, which can accurately detect the intrusion. The essence of detecting real-time behavior by detectors is a comparing process of real-time behavior characteristics string and detector string. Matching algorithms based on r-contiguous bits and Hamming distance are two common and classical algorithms in comparing two strings.

2 Introduction of Negative Selection Principle and Negative Selection Algorithm

Negative selection principle is a very important theory in biological immune. Its essence can be summarized as follows: immune cell maturation process must have

Jianping Zhao · Hua Li · Jingshan Liu
School of Computer Science and Technology, Changchun University of Science and Technology, CUST, Changchun, China

Z. Qian et al. (Eds.): Recent Advances in CSIE 2011, LNEE 125, pp. 611–616.
springerlink.com

self-tolerance test. Self- tolerance means that the immune cells from the body do not produce an immune response. All the cells which can recognize autologous antigen immune cells will be eliminated and perish. Only the immune cells which cannot recognize self-antigen will be selected and become ultimately mature, which is called "negative selection". As shown in Fig.1.

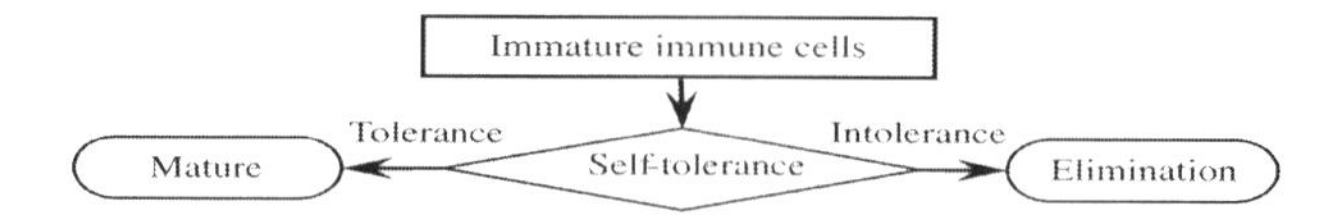

Fig. 1 Negative selection principle diagram

Negative selection algorithm is a core algorithm in computer immune system. It origins from the principle of negative selection [2], and is the key for generating detector database in computer immune system. The mainly processes are as following.

1. Match one by one the characteristic string generated by system randomly with "self" characteristic string.
2. If the matching result can match any characteristic string in "self" set, the random characteristic string is eliminated.
3. If the matching result cannot match any characteristic string in "self" set, the random characteristic string is selected into "Non-self" characteristic string set.

Because the characteristic string in "Non-self" set is different from that in "Self" set, once the actual behavior characteristic string is matched with some string in "Non-self" set, it means that this real-time behavior is exception which can detect invasions.

3 Matching Algorithms Based on Hamming Distance and R-Contiguous Bit

In negative selection algorithm implementation, the string comparison needs another algorithm, and the most commonly used one is mentioned as the title of this section. These two algorithms are introduced in the following.

Hamming distance: the number of positions at which the corresponding bits are different is called the Hamming distance of these two bit strings [3]. The minimal Hamming distance of any two bit strings is called the minimum Hamming distance of the code sets [4]. For example, 10101 and 00110 are different at the first, fourth and fifth bits, then the Hamming distance is 3. The idea of the algorithm is that when the Hamming distance of two strings is less than a given value, it can be called that the two strings match.

R-contiguous bits matching algorithm: If there have at least contiguous r same bits in two bit strings, it can be thought that these two strings are matched [5]. For example: for $r = 4$, bit string 1011010 and 0011101 has 3 contiguous same bitts from the second bit to the fourth bit, so these two bit strings can be considered as matched strings. When the matched bits are less than r, they are unmatched bit strings.

4 Matching Algorithm Based on Characteristic Matrix of R-Contiguous Bits

"XOR" is a very common bit string logical operator. The characteristic of "exclusive or" is that the result can be 0 only when two binary numbers are both 0 or 1. Inspired by this way, when bitwise XOR operation is done to compare two binary strings, the result will have a characteristic, that is, contiguous same bits corresponding two bit strings in the result string will have contiguous 0. For example:

$$M=10111010,\ N=11111101 \tag{1}$$

$$S=M\ XOR\ N = 01000111 \tag{2}$$

For such results, it can be seen that whether two bit strings are matched can be determined by whether the number of contiguous 0 is greater than r. Traditional methods compare detected string and standard string by a loop from the start, and set an integer variable Sum. When the comparison is the same, Sum + = 1, otherwise,

The traditional method is no such expression but from the start with a string loop will detect the bit-string comparison with the standard, and then set an integer variable Sum, when the comparison is equal to Sum + = 1, otherwise Sum = 0. Every time when Sum is assigned, comparison on Sum and r has to be done to know whether Sum is not less than r. If the condition is satisfied, exit the loop; or, continue to loop until reaching the last bit of two strings. Such algorithms will undoubtedly have very low efficiency.

Then are there any easy method can break through such thinking frame to improve the efficiency? Let us see the following matrix:

For given system call and parameter short sequence length L and optimized match threshold r, the following matrix is constructed. For example, L=8, r=3, then:

$$\mathbf{X}_{mxn}=\begin{pmatrix} 1 & 1 & 1 & 0 & 0 & 0 & 0 & 0 \\ 0 & 1 & 1 & 1 & 0 & 0 & 0 & 0 \\ 0 & 0 & 1 & 1 & 1 & 0 & 0 & 0 \\ 0 & 0 & 0 & 1 & 1 & 1 & 0 & 0 \\ 0 & 0 & 0 & 0 & 1 & 1 & 1 & 0 \\ 0 & 0 & 0 & 0 & 0 & 1 & 1 & 1 \end{pmatrix}$$

The number of rows $M = L - r + 1 = 6$, and that of columns $N = L = 8$

Definition: For a given bit string with length L and optimized match threshold value r (L>r), construct matrix X with L-r+1 rows and L columns. The width of X is that the value of the cross line is 1, and the rest are all 0. This matrix is called system characteristic matrix in L, r.

We make bitwise "and" operation on each column of X and S in (2). "And" operation has the characteristic that when two binary numbers are both 1, the result is 1; otherwise, the result is 0. For example:

$$S = 01000111,\ T = 11100000$$
$$S \ \&\ T = 01000000$$

Only when the corresponding bits are both 1, the result is 1; and the rest is 0. Then the final operation obtained is also a matrix:

$$R = S\&X = \begin{pmatrix} 0 & 1 & 0 & 0 & 0 & 0 & 0 & 0 \\ 0 & 1 & 0 & 0 & 0 & 0 & 0 & 0 \\ 0 & 0 & 0 & 0 & 0 & 0 & 0 & 0 \\ 0 & 0 & 0 & 0 & 0 & 1 & 0 & 0 \\ 0 & 0 & 0 & 0 & 0 & 1 & 1 & 0 \\ 0 & 0 & 0 & 0 & 0 & 1 & 1 & 1 \end{pmatrix}$$

It can be seen that only the second row is all 0(counting from the 0 bit, and the following are the same) and other rows are not all 0. It means that string S has a substring with 3 contiguous bits are 0 from the second bit to the fourth bit according to the data construction characteristics of the second bit of matrix X. The result also shows that the contiguous matching bit of string M and N, that is, the optimized match threshold is equal to r. So the two strings are matched. The logic flowchart is illustrated as Fig. 2.

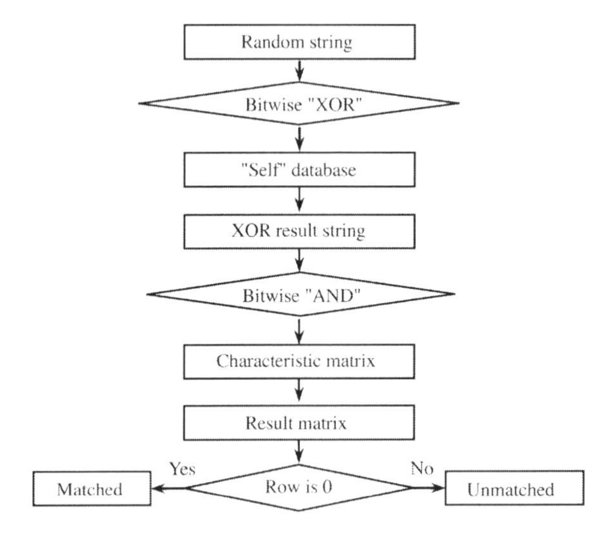

Fig. 2 Logic flowchart of algorithm

5 Time Complexity Analysis and Comparison of Algorithm

Assume that make a real-time monitoring on one process. The result obtains m short sequences of system calls and parameters, n data in database corresponding to the process. The length of short sequence L = 16, optimized match threshold r = 8. For traditional bitwise comparison, the time complexity is O (m x n x k) and m x n x k string comparisons need to be made. The time complexity of the improved algorithm is O (m x n x (k-r+1)) which does not obviously reduce the time complexity intuitively, but the new algorithm is based on "exclusive or" and "and" operation. In computer system, the efficiency of logical operators "exclusive or" and "and" is the same with that of addition and subtraction, which has significant advantages comparing with the efficiency of string comparison operation. Moreover, when the data size increases gradually, the new algorithm has more obvious advantages from the formula of time complexity, so the new algorithm will greatly improve the system efficiency.

6 Experiment and Data Analysis

6.1 Experimental Environment

Hardware environment: CPU Intel core-i3, memory 2G; Software environment: VS2008+SQL2005, programming language C#. Experimental method: Use the same "self" set database to match the same set of random string and repeat the operation several times to compare the two algorithms. The string length L = 48, r = 16. The experimental data are shown in Table 1:

Table 1 Experimental data list

The number of random string	Traditional algorithm (ms)	New algorithm (ms)
50	119	72
100	263	158
150	437	272
200	664	401
250	938	548
300	1272	720

6.2 Experimental Data Analysis and Conclusions

1) From the comparison of two curves in Fig. 3, the new algorithm has obvious efficiency improvement than the traditional method.

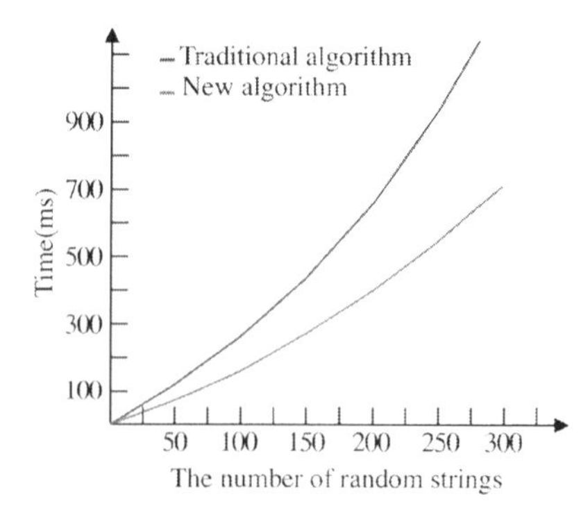

Fig. 3 Curve of experimental data.

2) The new algorithm has no obvious advantage for small scale problem which is consistent with the conclusion analyzed by time complexity O(m x n x k) and O(m x n x (k-r+1)). Only when the scale of problem increases, the advantages of the algorithm will be shown in time complexity

3) The number of random string has linear growth, but time growth rate does not correspond with them. With the increase in the number of random string, the time growth rate becomes greater. These two algorithms are the same in this point. This is caused by the increasing scale of access and addressing in computer system and nothing to do with the algorithm itself.

References

[1] Jiang, E.-L., Zhang, F.-B., et al.: Research of r-Continuous Bit Matching Rule Based on Hamming Matching. Harbin University of Seience and Technology Journal 12, 92–93 (2007) (in Chinese)

[2] Cai, T., Ju, S.-G., Niu, D.-J.: Efficient Negative Selection Algorithm and its Analysis. Joural of Chinese Computer Systems 30, 1172–1174 (2009) (in Chinese)

[3] Zhang, L.: The Influence of Hamming Distance Parameter on Genetic Algorithm Based on Hamming Distance. Journal of BeiHua University 11, 287–288 (2010) (in Chinese)

[4] Zhang, Y., Zhou, X.-C., Shen, H.-B.: Improvement of negative selection algorithm based on hamming distance. Mechanical & Electrical Engineering Magazine 24, 2–4 (2007) (in Chinese)

[5] Zhang, H., Cai, H.-F., et al.: Intrusion detection research based on r-contiguous bits match rule. Computer Engineering and Design 28, 1532–1534 (2007) (in Chinese)

Modeling and Simulation for Electric Field of Electrorotation Microchip with Ring Electrode

Liu Ganghai and Yang Qihua

Abstract. The turning performance of particle in electrorotation microchip experiment is related to its dielectric property. The ring electrodes of the chip are excited by alternating voltage source with a same frequency. The electric field distributed model within the ring electrodes, which is based on the Maxwell equations, is established. The boundary condition is approximated and simplified so that the second-order Laplace equation can be solved analytically to obtain the analytic description of electric field. Then the distribution of the dielectric force is calculated by Matlab when the model is quantized. The case may have some possibility that the alternating signal which exciting the electrorotation microchip will have asymmetric amplitude and phase, and the effect is preliminarily analyzed.

Keywords: Electrorotation, Ring electrode, Laplace equation, Stress distribution, Nonuniform.

1 Introduction

The particle in a nonuniform electric field is dielectric polarized and will be under stress to travel or rotate, which is called dielectrophoresis[1]. The principle of dielectrophoresis has been applied in the technology of fast recognition, characterization, and analyses on micro- or nano-meter scale particles. Compared with the mechanical or optical technique[2], the means of dielectrophoresis can realize noninvasive operation with high flux particles, and the operation mode can be altered by changing parameters like frequency and amplitude of the excitation voltage [3].

When the nonuniform electric field is a rotating field of a certain frequency, it will generate rotary moment upper polarized particle and which results in the rotation of the particle[4]. The attributes such as rotation rate and direction, depend on the rotating electric field and dielectric properties of particle and medium. The

Liu-Ganghai · Yang-Qihua1
China Jiliang University, Hangzhou, China
e-mail: {Liugh,000066}@cjlu.edu.cn

Z. Qian et al. (Eds.): Recent Advances in CSIE 2011, LNEE 125, pp. 617–623.
springerlink.com

rotation rate is also in relation to the mechanical properties of particle liquid medium like viscous force. By detecting the rotation attributes will obtain the indispensable data for analyzing dielectric property or other characteristic of particles. In area of biological research and application, Hölzel[5] used electrorotation experiment to analyze dielectric property of yeast cell, Foster made comparative studies between electrorotation and levitation. Yang analyzed the dielectric parameter of human leucocyte with electrorotation technology[6].

When the condition of particle and medium is definite, the rotation attribute is completely defined by electric field. In this paper, for the ring electrode which is extensively used, the model of its electric field is built, and the analytic expression of this electric field is obtained. Then the stress distribution of dielectric particle in electric field is developed with Matlab.

2 Analyzing and Modeling of Electrorotation Microchip

2.1 Theory Model

The electric dipoles is generated when particle lies in nonuniform electric field, the dipole has tendency to the surrounding electric field. A four-channel alternating signal with same frequency and orthogonal phases will generate rotating electric field. The particle will rotate under rotary moment as shown in Fig.1. The particle will suffer dielectric force $\boldsymbol{F}_{DEP}$ and rotary moment $\boldsymbol{\Gamma}_{\mathrm{ROT}}$.

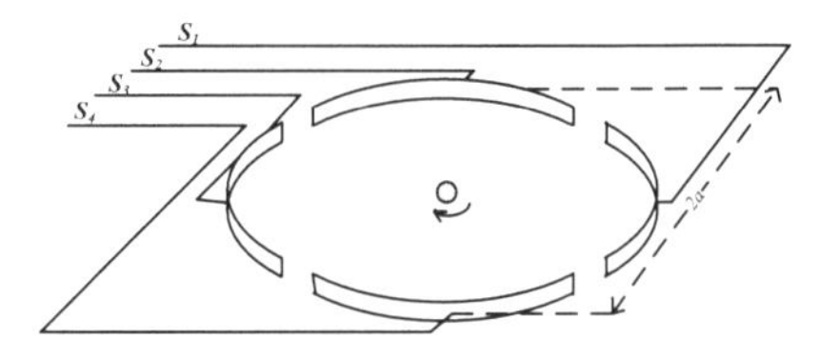

Fig. 1. Rotating electric field generated by alternating signal with orthogonal phases. $S_i=V_i\sin(2\pi ft+(i-1)\pi/2)$. $i=1,2,3,4$.

$$\boldsymbol{F}_{DEP} = 2\pi a_p^3 \varepsilon_m \mathrm{Re}[K_e]\nabla E_{rms}^2, \qquad \boldsymbol{\Gamma}_{ROT} = -4\pi a_p^3 \varepsilon_m \mathrm{Im}[K_e] E_{rms}^2 \cdot \mathbf{r}_0 \tag{1}$$

In the Equation (1), K_e is Clausius-Mossotti factor a is radius, ε is dielectric constant, σ is conductivity, the suffix p and m is related to particle and liquid medium, E_{rms} is root-mean-square value of electric field, $\boldsymbol{r}_0$ is unit vector, f is the frequency of electric field.

Under a steady rotation status, $\boldsymbol{\Gamma}_{ROT}$ will balance with the opposite friction moment $\boldsymbol{\Gamma}_f$. Assume rotation angular velocity as Ω. $\boldsymbol{\Gamma}_f = -8\pi\eta\Omega a_p^3$, and $\boldsymbol{\Gamma}_{ROT} + \boldsymbol{\Gamma}_f = 0$, then rotation angular velocity will be:

$$\Omega = -\frac{\varepsilon_m}{2\eta}\mathrm{Im}[K_e]E_{rms}^2 \tag{2}$$

Equation (2) shows the relationship of particle rotation with dielectric property and electric field.

2.2 Configuration of Electrode and Modeling

The configuration of electrorotation microchip's electrode is shown in left part of Fig.2, and the right part is its structural representation. When proceeding test with electrorotation microchip, if the property of liquid medium and particle has been ascertained, only the voltage applied on the electrode can make changes. So the distribution of electric field is important for the study on particle's electrorotation. Loading alternating voltage S_1~S_4 with the same frequency onto the four electrodes, whose amplitudes and phases can be set respectively, as in Equation (3). V_i and θ_i represent the amplitude and phase of S_i (i =1~4). ω, which is circular frequency of alternating voltage, equals to $2\pi f$.

$$S_i = V_i \sin(\omega t + \theta_i) \quad i = 1, 2, 3, 4 \tag{3}$$

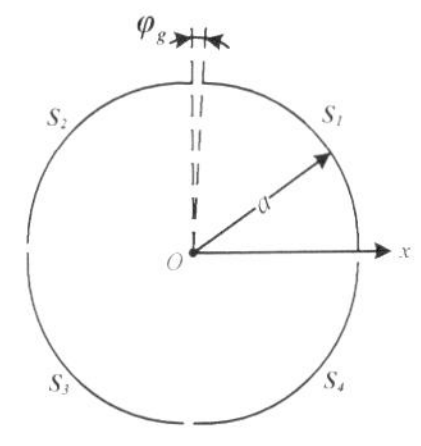

Fig. 2 Configuration of electrodes. The radius of the ring is a. The angle-gap between electrodes is φ_g. The alternating voltage (S_1~S_4) with quadrature phases is applied onto the electrodes.

The region surrounded by electrodes is passive. Based on Maxwell's electromagnetic equations, the equation to describe the electric potential distribution is shown as the equation below, assuming electric potential distribution a symbol of u. The domain being the interior of a circle, the equation can convert into polar form.

$$\Delta u = 0, \text{ that is: } \frac{\partial^2 u}{\partial x^2} + \frac{\partial^2 u}{\partial y^2} = 0 \text{ or } \frac{\partial^2 u}{\partial r^2} + \frac{1}{r}\frac{\partial u}{\partial r} + \frac{1}{r^2}\frac{\partial^2 u}{\partial \varphi^2} = 0 \tag{4}$$

Boundary condition: voltages applied on the four electrodes.

$$u\Big|_{r=a,\ \frac{(i-1)\pi}{2} \le \varphi \le \frac{i\pi}{2} - \varphi_g} = S_i \quad i = 1, 2, 3, 4 \tag{5}$$

E and E^2_{rms} are described in Equation (6).

$$E = -\nabla u, \quad E^2_{rms} = \frac{1}{T}\int_0^T |\boldsymbol{E}(t)|^2\, dt \tag{6}$$

2.3 Solve Laplace Equation

Separating the variables, $u(r,\varphi)=R(r)\cdot\Phi(\varphi)$, and using the natural period condition $\Phi(\varphi+2\pi)=\Phi(\varphi)$, the format of solution comes to be:

$$u(r,\varphi)=(C_0+D_0\ln r)+\sum_{m=1}^{\infty} r^m(A_m\cos m\varphi+B_m\sin m\varphi)+\sum_{m=1}^{\infty} r^{-m}(C_m\cos m\varphi+D_m\sin m\varphi)$$

Obviously, $u(0,0)<\infty$, then the factors of $\ln r$ and r^{-m} will be zero.

$$u(r,\varphi)=C_0+\sum_{m=1}^{\infty} r^m(A_m\cos m\varphi+B_m\sin m\varphi)$$

Then according to the boundary condition: Equation (5), to work out the parameters: C_0, A_m and B_m.

Boundary condition can be plotted as in Fig.3(left). The voltage in region φ_g is unknown. While for φ_g is continues, and if φ_g is small enough, the voltage in φ_g can be considered as linear (see right chart in Fig.3).

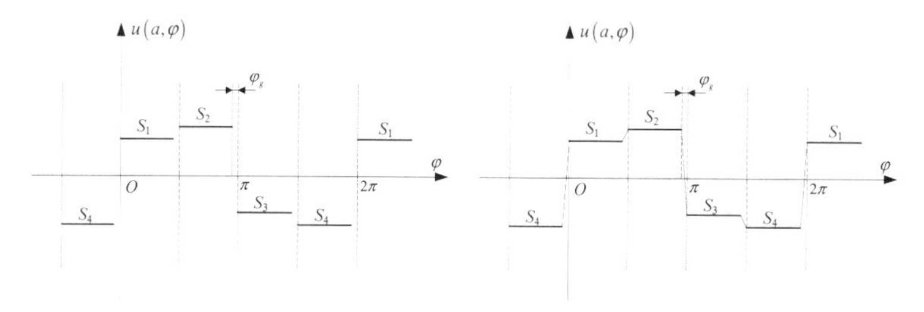

Fig. 3 (Left) Boundary Condition $u(a,\varphi)$. Voltage in φ_g is unknown.(Right) Boundary Condition u(a,φ). Voltage in φg is linear.

Name the right curve in Fig.3 as $f(\varphi)$, then $u(a,\varphi)=f(\varphi)$. $f(\varphi)$ is periodical. According to Fourier expansion:

$$f(\varphi)=C_0+\sum_{m=1}^{\infty} a^m(A_m\cos m\varphi+B_m\sin m\varphi)=C_0+\sum_{m=1}^{\infty}\left(A_m a^m\cos m\varphi+B_m a^m\sin m\varphi\right)$$

$$C_0=\frac{1}{4}\sum_{i=1}^{4}S_i,\quad A_m=\frac{1}{\pi a^m}\int_0^{2\pi} f(\varphi)\cos m\varphi d\varphi=\frac{P_m(t)}{\pi a^m},\quad B_m=\frac{1}{\pi a^m}\int_0^{2\pi} f(\varphi)\sin m\varphi d\varphi=\frac{Q_m(t)}{\pi a^m}$$

$$P_m(t)=\sum_{i=1}^{4}\frac{S_i}{m}\left[\sin m\left(\frac{i\pi}{2}-\varphi_g\right)-\sin\frac{m\cdot(i-1)\pi}{2}\right]+\sum_{i=1}^{4}\frac{S_{i+1}-S_i}{m\varphi_g}\left[\frac{i\pi}{2}\sin\frac{m\cdot i\pi}{2}-\left(\frac{i\pi}{2}-\varphi_g\right)\sin m\left(\frac{i\pi}{2}-\varphi_g\right)\right]$$

$$+\sum_{i=1}^{4}\frac{S_{i+1}-S_i}{m^2\varphi_g}\left[\cos\frac{m\cdot i\pi}{2}-\cos m\left(\frac{i\pi}{2}-\varphi_g\right)\right]+\sum_{i=1}^{4}\frac{1}{m}\left(S_{i+1}-\frac{S_{i+1}-S_i}{\varphi_g}\cdot\frac{i\pi}{2}\right)\left[\sin\frac{m\cdot i\pi}{2}-\sin m\left(\frac{i\pi}{2}-\varphi_g\right)\right]$$

$$Q_m(t)=\frac{1}{m}\sum_{i=1}^{4}S_i\left[-\cos m\left(\frac{i\pi}{2}-\varphi_g\right)+\cos\frac{m\cdot(i-1)\pi}{2}\right]+\sum_{i=1}^{4}\frac{S_{i+1}-S_i}{m\varphi_g}\left[-\frac{i\pi}{2}\cos\frac{m\cdot i\pi}{2}+\left(\frac{i\pi}{2}-\varphi_g\right)\cos m\left(\frac{i\pi}{2}-\varphi_g\right)\right]$$

$$+\sum_{i=1}^{4}\frac{S_{i+1}-S_i}{m^2\varphi_g}\left[\sin\frac{m\cdot i\pi}{2}-\sin m\left(\frac{i\pi}{2}-\varphi_g\right)\right]+\sum_{i=1}^{4}\frac{1}{m}\left(S_{i+1}-\frac{S_{i+1}-S_i}{\varphi_g}\cdot\frac{i\pi}{2}\right)\left[-\cos\frac{m\cdot i\pi}{2}+\cos m\left(\frac{i\pi}{2}-\varphi_g\right)\right]$$

$S_5=S_1$.

Then we obtain the result.

$$u(r,\varphi)=\frac{1}{4}\sum_{i=1}^{4}S_i+\frac{1}{\pi}\sum_{m=1}^{\infty}\left[\left(\frac{r}{a}\right)^m P(m)\cdot\cos m\varphi+\left(\frac{r}{a}\right)^m Q(m)\cdot\sin m\varphi\right]$$

$$\boldsymbol{E} = -\nabla u = -\left(\frac{\partial u}{\partial x}\boldsymbol{a}_x + \frac{\partial u}{\partial y}\boldsymbol{a}_y\right) = -\left(\frac{\partial u}{\partial r}\boldsymbol{a}_r + \frac{1}{r}\frac{\partial u}{\partial \varphi}\boldsymbol{a}_\varphi\right)$$

$$= -\sum_{m=1}^{\infty}\frac{m}{\pi r}\left(\frac{r}{a}\right)^m \left[P_m(t)\cos m\varphi + Q_m(t)\sin m\varphi\right]\cdot \boldsymbol{a}_r - \sum_{m=1}^{\infty}\frac{m}{\pi r}\left(\frac{r}{a}\right)^m \left[Q_m(t)\cos m\varphi - P_m(t)\sin m\varphi\right]\boldsymbol{a}_\varphi$$

$$= E_r \cdot \boldsymbol{a}_r + E_\varphi \cdot \boldsymbol{a}_\varphi$$

$$E_{rms}^2(r,\varphi) = \frac{1}{T}\int_0^T \left|\boldsymbol{E}(r,\varphi,t)\right|^2 dt = \frac{1}{T}\int_0^T \left(E_r^2 + E_\varphi^2\right)dt, \quad T = \frac{1}{f} = \frac{2\pi}{\omega}$$

Till now we have achieve analytic expression of E^2_{rms}, which is used in Equation (1), (2), and can deduce $\boldsymbol{\Gamma}_{ROT}$, $\boldsymbol{F}_{DEP}$ and Ω.

3 Numerical Emulation

First the parameters will be chosen as list in Table 1. Potential line and electric field distribution will be made and plot with the tool Matlab as left part of Fig.4. The right part is obtained in Ansoft Maxwell 2D.

Table 1 Simulation Parameters.

Electrode Radius a	200μm	Angle gap φ_g	5°
Voltage Amplitude v	5V	Original Phase of S_1 θ_1	0°
Original Phase of S_2 θ_2	90°	Original Phase of S_3 θ_3	180°
Original Phase of S_4 θ_4	270°	Simulation Moment t	0

E^2_{rms} is the average value of $|\boldsymbol{E}|^2$ in a complete alternation. Isoline of $\lg(E^2_{rms})$ is shown in Fig.5. Meanwhile, the electric field distribution of multi-point within the region at moments of same interval in a cycle time is plot. It is obviously that E^2_{rms} distribution in central zone is flat enough ($\boldsymbol{F}_{DEP}$ in Equation (1) tends to zero) for the particle rotating smoothly, the rotation rate can be considered to be independent of particle position. While the particle deviates from the center will suffer the force of $\boldsymbol{F}_{DEP}$ and will be traveling rather than rotating.

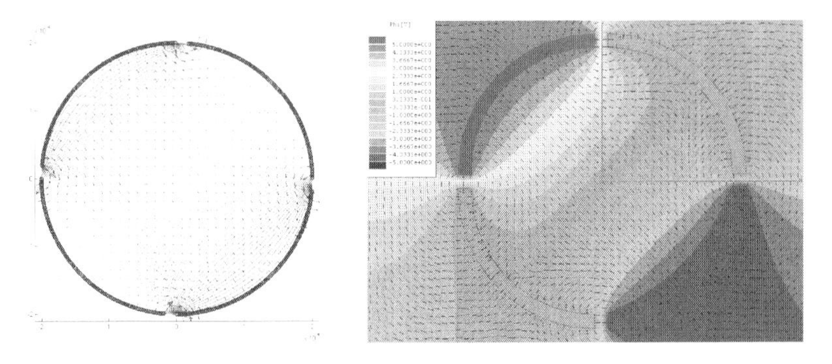

Fig. 4 (Left) Potential-line and Electric-field Distribution, in Matlab.(Right) Potential-line and Electric-field Distribution, in Ansoft Maxwell 2D.

The distribution of electric field within central zone is uniform in one-period time, but is not same as a round presumed. The end point distribution of electric field vector approximately constitutes a parallelogram. While the father departure from central zone in space, the less uniform the electric field distribution in time will be.

In this case, the particles should be put in the central zone. $\lg(E^2_{rms})$ of positioning electric field is shown in Fig.6, which is generated by applying alternating voltages with phase difference of 180°upon adjacent electrodes. The particles will travel towards the central zone under the force $\boldsymbol{F}_{DEP}$ in Equation (1).

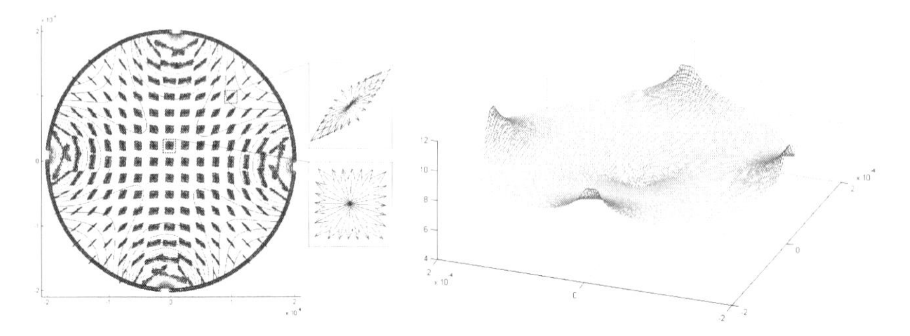

Fig. 5 Isoline of $\lg(E^2_{rms})$ and electric field vectors at moments of same interval in one cycle time

Fig. 6 $\lg(E^2_{rms})$ of Positioning Electric Field.

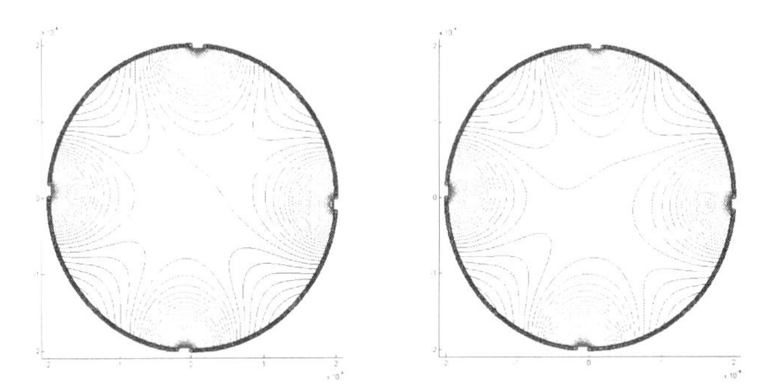

Fig. 7 Isoline of $\lg(E^2_{rms})$ when Amplitude (left) or Phase (right) deviated. Left: v_1=6V, Right: θ_1=15°.The conditions else are accordance with Table 1.

The alternation signal in real test may have a wide band of 1KHz to 100MHz, with a large power output. Cases are that the amplitudes or the phases or both will turn to unbalance among the four channels. Fig.7 shows the distribution of when the amplitude (left) or phase (right) has light deviation. Obviously the flat central zone is excursion consequently. Once the distribution of E^2_{rms} forms a notability gradient so that the effect by dielectric force is neglectable, the particle will travel translationally, and make the electrorotation test fall into involvement.

The cases in Fig.7 are the simplest. Usually the scattering of signal frequency, deviation of amplitude or phase, and wave distortion may come out at the same time, and the effect brought to electrorotation test will need further research and discussion.

4 Conclusions

In this paper, the model of electrorotation microchip with ring electrode structure is established. By solving the second-order Laplace Equation according to its special boundary condition, obtain the analytic expression of electric potential and electric field. Use numerical method to achieve distribution of electric potential and electric field with its root mean square value in space, as well as distribution of electric field in time. The rationality of assumptive boundary condition is demonstrated in the comparison with distribution figures elicited from Ansoft Maxwell 2D, and can spread to the conclusions subsequent.

Rotating electric field vector in space in Fig.5 appears to be a parallelogram distribution. The effect to particle's rotation has not been analyzed, and need further analysis.

References

1. Pohl, H.: Dielectrophoresis. Cambridge University Press, New York (1978)
2. Cen, E.G., Dalton, C.: A combined dielectrophoresis, traveling wave dielectrophoresis and electrorotation microchip for the manipulation and characterization of human malignant cells. Journal of Microbiological Methods 58, 387–401 (2004)
3. Fei, F., Qu, Y.-L.: Simulation of Electric Field for Dielectrophoretic Electrode Arrays. Computer Simulation 25, 314–318 (2008)
4. Arnold, W.M., Zimmermannn, U.: Electrorotation: development of a technique for dielectric measurements on individual cells and particles. Electrostatics 21, 151–191 (1988)
5. Hölzel, R.: Single particle characterization and manipulation by opposite field dielectrophoresis. Electrost. 56, 435–447 (2002)
6. Yang, J., Huang, Y.: Dielectric properties of human leukocyte subpopulations determined by electrorotation as a cell separation criterion. Biophys. 76, 3307–3314 (1999)

Particle Swarm Optimization Algorithm for the Application of Reactive Power Optimization Problem

Dian-Sheng Yang

Abstract. Reactive power system optimization is one of the core thesis of safe and economic operation, reactive power optimization planning is a more complex, multi-target, nonlinear mixed-planning. Its goal is to meet the restrictive conditions under the premise of the system of indicators or more of a target to achieve optimal. Based on the analysis of reactive power distribution network optimization facing difficulties a PSO algorithm is presented, combined with IEEE30 node test system using particle swarm algorithm. The results show that the optimization method is conducive to raising the reactive power distribution network optimization level.

Keywords: PSO algorithm, Reactive Power Optimization, Optimization Design.

1 Introduction

Reactive power optimization planning is a more complex, multi-target, nonlinear mixed-planning. Its object is to realize the optimization of objective function with constraints by means of reactive power compensation devices, switching, load tap of the generator terminal voltage regulation and co-ordination.The key to reactive power optimization focused on the handling of non-linear function, the algorithm convergence and how to solve the optimization problems in three aspects of the problem with discrete variables [1].

Relevant personnel have already provided some methods of Reactive Power Optimization, such as Linear programming, nonlinear programming, mixed integer programming, etc and achieved some effect.

However, as the reactive power optimization is a nonlinear combinatorial optimization which is more complex, the traditional optimization based on gradient can't solve it well in consequence of combinatorial explosion and curse of

DianSheng Yang
Academic affairs Division of EZhou Polytechnic,
436000 HuBei,China
e-mail: YangDianSheng@126.com

Z. Qian et al. (Eds.): Recent Advances in CSIE 2011, LNEE 125, pp. 625–631.
springerlink.com

dimensionality; Having many unique advantages, the Particle Swarm Optimization has been used in this kind of problem and made much progress.

2 Particle Swarm Optimization

The Particle Swarm Optimization is initialized to a group of random solution (random particle),all particles have a fitness decided by a fitness function, every particle has a speed factor to decide its flight direction and distance. And then it find optimal solution by iteration. In each iteration, the particles update themselves by tracking two extremums, one of them is the optimal solution found by the particle itself,it is called individual extreme $P_{Best(i)}$,the other one is the optimal solution found by the whole species currently,it is called global extreme G_{Best} .After achieving the two extremums,the particles update their speed according to the following formula (1), (2), [2-4]:

$$v_i^{k+1} = \omega v_i^k + c_1 rand \times (P_{Best(i)} - x_i) + c_2 rand \times (G_{Best} - x_i) \quad (1)$$

$$x_i^{k+1} = x_i^k + v_i^{k+1} \quad (2)$$

In the formula, ω is an inertia coefficient ; c_1, c_2 are learning factors, $rand$ is a random function whose value is between (0,1).

3 Mathematical Model of Reactive Power Optimization

Based on System Security, the objects of Reactive Power Optimization mainly include: minimize the entire injection or weighted entire injection of reactive power, minimize the bus voltage deviation from the specified value, minimize the overload of the line, maximum the stability margin of the voltage, etc. Based on operating system in economic view, the objects mainly include: minimize the working cost which is also called transmission loss, minimize operating point deviation from the optimal operating point, etc.

In this way, the objective function includes performance index of technologies and economic indicators. Constraints mainly include :Active power flow equations, Reactive power flow equations,upper and lower limits of Control variables and State variables. Mathematically speaking. Reactive Power Optimization is a dynamic, multi-target, multiple constraint, probabilistic nonlinear mixed integer programming.

3.1 Objective Function

This thesis, the minimization of transmission loss is used as the objective function:

$$Min \quad f(x_1, x_2) \quad (3)$$

In the formula, the objective function $Min\ f(x_1,x_2)$ is Active power loss; controlled variable $x_1=[Q_C^T,V_G^T,T_B^T]^T$, $x_1\in R^{(M)}$,M is a sum of constrained optimization variables; Q_C is a vector being composed of Reactive power compensation equipment; V_G is a vector being composed of amplitude of Generator terminal voltage; T_B is a vector being composed of transformation ratio of Adjustable Transformers; state variable x_2 is a vector being composed of all of the voltage amplitude of load node and Generator reactive power, $x_2\in R^{(n)}$, n is the number of System nodes.

3.2 Constraints

There are two types of constraints of Reactive power optimization, one is equality constraints, the other is inequality constraints.

(1)Equality constrain
Power Balance Equation: $g(x_1,x_2)=0$

(2)Inequality constraints
Other constraints to insure the normal operation of system include:

Control variable constraints: bound of Installed capacity of reactive power compensation, bound of Generator terminal voltage magnitude, bound of Adjustable transformer turns ratio

State variable constraints: bound of Reactive power generator, bound of Node voltage.

This thesis uses penalty function to process the out-of-limit of the load voltage amplitude and generator reactive power, then get an augmented objective function and finally build a model as following formula shows.,

$$Min\ F=f(x_1,x_2)+\lambda_1\sum_{i=1}^{Nl}(V_i-V_{i,\lim})+\lambda_2\sum_{i=1}^{Ng}(Q_i-Q_{i,\lim}) \tag{4}$$

In the formula:

$$V_{i,\lim}=\begin{cases}V_{i,\max},V_i\geq V_{i,\max}\\ V_{i,\min},V_i\leq V_{i,\min}\end{cases},\quad Q_{i,\lim}=\begin{cases}Q_{i,\max},Q_i\geq Q_{i,\max}\\ Q_{i,\min},Q_i\leq Q_{i,\min}\end{cases}$$

λ_1,λ_2 are Penalty factors, Nl is the number of Load node, Ng is the number of Generator node

4 Solving of Reactive Power Optimization: Based on Particle Swarm Optimization

Particle Swarm Optimization is a kind of Multi-agent algorithm, it has a Global search capability to solve complicated nonlinear problems and is also simple, universal and robust, for the past few years, PSO algorithm has already been applied in Optimal Power Flow, Economic dispatch and Generator Combination problem.

4.1 Process of Solving of Reactive Power Optimization Using Particle Swarm Optimization

When solving Reactive power optimization by PSO, $x_1 = [Q_C^T, V_G^T, T_B^T]^T$ which is defined in the formula (3) correspond to the position of particles from the Particle Swarm. the boundary of the Feasible region is determined by bound of Q_C, V_G and T_B.

The process of Solving of reactive power optimization. using Particle Swarm Optimization is described as following.

Step 1: Input the system data and initialize the particle swarm. first of all, input Structure of the system,Network data and Control parameters, Feasible region consists of the boundary of Generator node voltage, Capacitor Capacity and Transformer tap. Secondly, determine the dimension of particles M (the Variable number of a group of controlled variable)and generate N particles in M-dimensional Feasible region randomly as Initial Particle Swarm. In this way, the position of each particle is corresponding to the value of a group of controlled variable from the system. At the same time, initialize the flight speed of each particle. The Iterations $k = 0$.

Step 2: Calculate the value of objective function. calculate the power flow of every particle in the swarm respectively and get the active power loss with the value of every group of controlled variable, then identify whether it violate the constraints of node voltage and reactive power of generator and so on. at last the voltage and cross-border value of generator reactive power are substituted in the objective function. as penalty function term.

Step 3: Record two extremums. Compare the corresponding objective function value of all particles. First of all, record current individual extreme $P_{Best(i)}$ and allowable objective function value $F(P_{Best(i)})$; Determine Overall extreme G_{Best} from $P_{Best(i)}$ and record the corresponding objective function value of G_{Best}.

Step 4: Update $k = k + 1$ .Particle update their flight speed according to formula(1),in this way, particle i gets a certain flight speed which tends to $P_{Best(i)}$ and G_{Best} .Then particle update their position in Solution space.

Step 5: Recompute the objective function value of every particle at this time and determine whether to update $P_{Best(i)}$ and G_{Best}.

(1)for particle i, compare the function value of $F(i,k+1)$ and $F(P_{Best(i)})$ after k+1 iterations, if $F(i,k+1) < F(P_{Best(i)})$ then $F(P_{Best(i)}) = F(i,k+1)$, $(i=1,2,\cdots,N)$ and update $F(P_{Best(i)})$ accordingly; or don't update.

After update individual extremes of all particles, if

$\min\{F(P_{Best(i)}),(i=1,2,\cdots,N)\} < F(G_{Best})$ then

$F(G_{Best}) = \min\{F(P_{Best(i)}),(i=1,2,\cdots,N)\}$,and update G_{Best} accordingly; or don't update.

Step 6: Judge whether it is convergent. Stop the iterates when following conditions are satisfied: the best global position don't change for 20 times or Iterations has already come up to maximum $k = Iter_{\max}$ .Or, turn to step four.

Step 7: Output the solution of problems, including the values of generator terminal voltage, the number of compensation capacitors. values of transformer tap and other controlled variables, dates of system node voltage, Reactive power generator and other state variable and the corresponding working cost.

4.2 Parameters Choice in Solving Reactive Power Optimization: by Particle Swarm Optimization

(1)Size of particle group N and choose of Initial Solution group

From the introduction of algorithm principle, we can know that the operation of PSO algorithm deals with a group consists of many individuals at the same time. Size of particle group N is also the number of initial solution, its value decides the diversity of initial solution. and computing time of each iteration decides the whole computing time directly, so N has a strong influence on both of solutions and the whole computation time. In order to make sure of the algorithm's performance, the first thing to do is insure that size of group is big enough. However, excessive size is not good for solving the problem but leads to a big increase in calculates amount, so that the computing time is extended and can nor satisfy the need of projects. So computing time has an important significance in improving the convergence velocity and quality of solution because it can not only make sure of the diversity of initial solution, but also choose the size of group N appropriately. In this thesis, the number of particle is 30,the HrationalityH of choose will be discussed in numerical example.

(2)Inertia coefficient ω and maximum of iterations $I_{\max}$

Inertia coefficient ω is a controlled parameters., it not only controls effect of this flight on next flight, and also reflect the balance between global search and local search. In the early stages of search, big weight coefficient can improve the global search ability; with search tending to global optimal solution, ω dwindles away, it improves local search ability and enhances convergence rate. This thesis assumes that Inertia coefficient ω decreases linearly. It calculates as follows:

$$\omega_i = \omega_{\max} - \frac{\omega_{\max} - \omega_{\min}}{Iter_{\max}} I_k$$

In the formula: $I_{\max}$ is the maximum of iterations, in this thesis, it is set as 100; I_k is iterations at this time; $\omega_{\max}$ and $\omega_{\min}$ are constants. Numerical experiment shows that it can achieve better optimize effective when $\omega_{\max} = 0.9$ $\omega_{\min} = 0.4$ in actual calculation.

maximum of flight speed $v_{\max}$

To achieve better solution, this thesis set maximum of flight speed $v_{\max}$,so the numeric area of speed is $[-v_{\max}, v_{\max}]$ $v_{\max}$ decides the precision of the region between local and global extremism. If $v_{\max}$ is too big, the particle may skim over the optimal solution; if $v_{\max}$ is too small, the particle may can't search sufficiently outside the neighborhood of local optimal solution and be prone to get a local optimal solution. This set $v_{\max}$ as 10% of the numeric area of control variables.

$$v_{\max} = (x_{(1,\max)} - x_{(1,\min)}) / P, \quad P = 10$$

5 Simulation

In order to check the validity of Solving reactive power optimization. using Particle Swarm Optimization, this thesis compute with IEEE 30 node test system. System parameters are all expressed in Per-unit value whose Base Power is 100mw.

Take IEEE 30 node test system for example, node and slip data can be found in the literature[5].This system include 6 generator,4 adjustable transformers and 2 Reactive power compensation points. System parameters are all expressed in Per-unit value whose Base Power is 100mw.bound of P-V node and balance node are set as 0.95 and 1.05.

Form one list values of each variable before and after optimization using primal-dual interior point algorithm and PSO algorithm respectively. Method one is primal-dual interior point algorithm, method two is PSO algorithm. it is observed that with method two, Reactive power generator and Reactive power compensation are both within the limits of constraints,system nodal voltages are improved generally and minimum voltage rise from 0.940 to 1.011.Therefore,the PSO algorithm can yield better economic benefit under the condition of safety operation of system.

Table 1 Initial condition and solutions of controlled variable of IEEE 30 node test system

controlled variable	*upper limits of parameters*	*lower limits of parameter*	*initial value*	*solutions(meth od one)*	*solutions(method two)*
V_{G1}	1.1	0.9	1.0	1.030	1.058
V_{G2}	1.1	0.9	1.0	1.032	1.055
V_{G5}	1.1	0.9	1.0	1.027	1.051
V_{G8}	1.1	0.9	1.0	1.004	1.038
V_{G11}	1.1	0.9	1.0	1.040	1.030
V_{G13}	1.1	0.9	1.0	1.020	1.046
Q_{10}	0	0.05	0	0.037	0.032
Q_{24}	0	0.05	0	0.047	0.033
N_{6-9}	0.9	1.1	0.978	0.992	0.9675
N_{6-10}	0.9	1.1	0.969	0.955	0.950
N_{4-12}	0.9	1.1	0.932	0.978	0.9675
N_{28-7}	0.9	1.1	0.968	0.965	0.9675

References

1. Li, L.Y., Zhou, Q.J., Yang, S.K.: Research on VAR Optimization in Power Systems 1(3), 69–74 (2002)
2. Eherhart, R.C.: A new optimizer using particles swarm theory. In: Proc. Sixth International Symposium on Micro Machine and Human Science, Nagoya, Japan, pp. 39–43 (1995)
3. Shi, Y.H., Eherhatt, R.C.: A modified particle Swarm optimizer. In: IEEE International Conference on Evolutionary Computation, Anchorage, Alaska, May 4-9, pp. 69–73 (1998)
4. Li, A., Qin, Z., Bao, F., He, S.: Particle Swarm Optimization Algorithms. Computer Engineering and Applications 21, 1–3 (2002)
5. Wu, J.S., Hou, Z.J.: Computer Power Flow Calculation Shanghai Jiaotong (1999)

Pulse Wave Detection for Ultrasound Imaging

Xuemin Wang, Wei Wang, Xiaozuo Lu, and Peng Zhou*

Abstract. Pulse Diagnosis is the most important diagnostic methods in Oriental medicine. By finger pressing to the radial artery the doctor can get information in the human body through the fingers sense of the pulse , so they may judge the patient's physiological and pathological situations. Based on the information above, in this study, researchers use the ultrasonic tomography to show the changes in deformation of vessels. By detection the deformation of arterial vessels in a cardiac cycle researchers can calculate patient's pulse wave. This detection method has good consistence with traditional methods.

Keywords: Pulse, ultrasonic pulse diagnosis, edge detection, panoramic ultrasound imaging.

1 Introduction

Pulse diagnosis is an important and unique part of Chinese medicine. Doctors through his figures touch patient's radial artery to feel the Attributes of Position, Rhythm, Shape and Variance of Pulse Diagnosis to understand the information of the patient's potential physiological and pathological conditions. Meanwhile with the look, smell interrogation confirm each other to determine the type and properties of the disease.

Heaps of clinical experience accumulation the Chinese medicine can be used as a basis of diagnosis, but only the doctor's experience and figures feeling cannot but carries too many subjective factors. The same disease pulse, even if an experienced physician may have a different diagnosis in different time. Traditional Chinese medicine pulse diagnosis has to be doubt with the continuous development of

---

Xuemin Wang · Wei Wang · Peng Zhou
School of Precision Instruments and Opto-Electronics, Tianjin University, Tianjin 300072, China

Xiaozuo Lu
Tianjin University of Traditional Chinese Medicine, Tianjin 300193, China

* Corresponding author.

Z. Qian et al. (Eds.): Recent Advances in CSIE 2011, LNEE 125, pp. 633–638.
springerlink.com

modern science and technology. So the much-needed research direction of the traditional Chinese medicine pulse diagnosis is how to combined with engineering and quantify statistics research to make sure the Chinese medicine pulse diagnosis has a higher theory of objectivity. And also keeping the advantages of the overall concept and dialectical of the Chinese pulse diagnosis theory.

In recent years, intravascular ultrasound imaging technology [1] has been increasingly widely applied to many diseases, diagnosis and interventional treatment. The signals through the ultrasound probe of catheter tip to the blood vessels transmitted around. Because of the different component of vascular tissue have different acoustic characteristics, so the resulting echo signal is different. It reflects the difference pixels of gray in the image. Therefore, moving the ultrasound probe in vascular can provide clearly and detailed the lumen, wall (including the intima and media, adventitia) cross-section image sequence in real time. To make doctors estimate vessel wall tissue and vascular components and physical parameters of the geometric parameters of the direct measurement possibility [2].

Real-time Panoramic ultrasound images [3], also known as ultrasound imaging broaden horizons. It obtained a series of two dimensional slices images by moving the probe. Using computer reconstructions [4] make a series of two dimensional slices images mosaic a continuous and time-dependent section images.

In this research, using the ultrasound imaging equipment processed the vessel information. On one hand, it can monitor the real-time information on patients physiological and pathological; on the other hand, to facilitate doctor's diagnosis, it takes the patient information visualization to eliminating the most of the factors of overly dependent on the experience which has been plagued doctors. Finally, it also facilitates the detection of doctors for the noninvasive and less interference on the human body's facts for the ultrasound equipment.

2 Image Acquisition Process

Use of electronic technology Co., Ltd. Tianjin Suoer Full Scale UBM SW-3200 to collected the information of human pulse. Work parameters: transducer frequency 50 MHz; vertical resolution 50 μm; lateral resolution of 50 μm; electronic cursor location, display resolution 0.01mm; scanning range of 10 mm × 6.5 mm; gain range 30 ~ 150dB; signal to noise ratio 100 dB.

When processing we need to put patient's arm in the water. And put the ultrasound probe placed on the top of the radial artery (Fig.1). We can see the picture of the arteries with the mechanical scanning using ultrasonic probe. Keep the ultrasonic probe fixed the position where can see the blood vessels (measured in Fig.3) so we can see the continuous image to the vessels. And displayed the image in the form of two-dimensional to interception the pulse information. This image acquisition equipment can displayed 100 frames image in top with the frequency is 10 frames / s. While the human cardiac cycle is about 0.8 time / s. Which means this system only can take 8 points in one human cardiac cycle. It can't representation of human pulse information very well. So this test takes ECG detection system coupled with the above system. To get different image information in a cardiac cycle, then overlay several cardiac cycles image to obtain the required pulse wave.

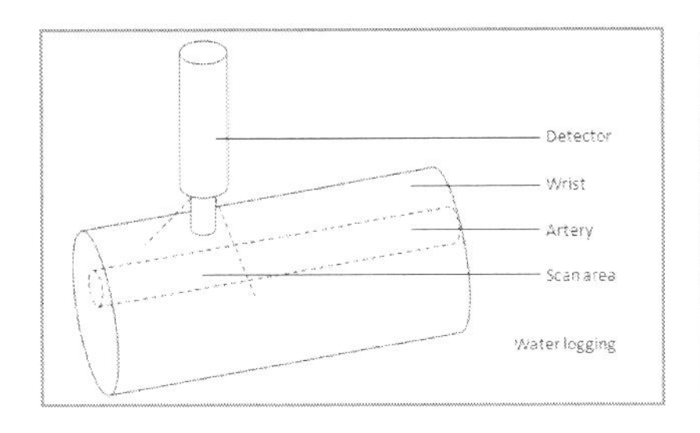

Fig. 1 Image acquisition diagram

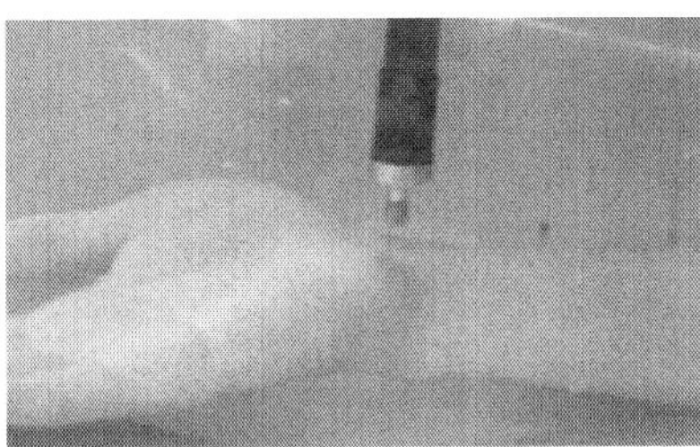

Fig. 2 Measured diagram

3 Image Preprocessing

The image doesn't have high quality and resolution, because the system need collected image in water bath. So the radial artery image has been lost in speckle [5] and texture noise. To obtain complete image information, the testing is doing a series of vascular ultrasound image processing.

We got the three-dimensional color image after the processing in Fig.1. MATLAB can only handle two-dimensional image, so it should first adjust the images to grayscale. And in order to ensure the clarity requirement, it needs to adjust the contrast. Processing: contrast adjustment (color image), gray scale adjustment (black and white image), adjust the contrast (black and white images).

At first time it needs an image denoising. To remove the salt and pepper, spots, texture noise, it takes median filter. Then smoothing the image used mathematical morphology algorithm. The opening and closing are not mutually inverse operation, so they can be cascaded in combination. Finally, use the subtraction algorithm to get the image of vessel section area.

It needs to vessel cross-sectional area changes in continuous data. To obtain the image of changing in pulse wave, it needs deal with separately the group of consecutive image to get the continuous area curve.

Under normal conditions were collected (unpressured) 100 panoramic vascular ultrasound images. From which selected 88 images that more accurate and continuous to processed. Then it can use the ECG detection system through interpolation method to superposition the cardiac cycles [6].

Used the ZD-type pulse measuring instrument (Main technical indicators: MH-IIA single probe pulse sensors: Sensitivity: 0.5mV / Chris (bridge voltage 6V); linear range: 0 ~ 500 Chris (formerly 0 250 Chris); temperature drift: less than 1% (FS) (– 5 °C ~ +40 °C) (original less than 2%).) measured on the same subject (Fig.3). Then compared the two methods.

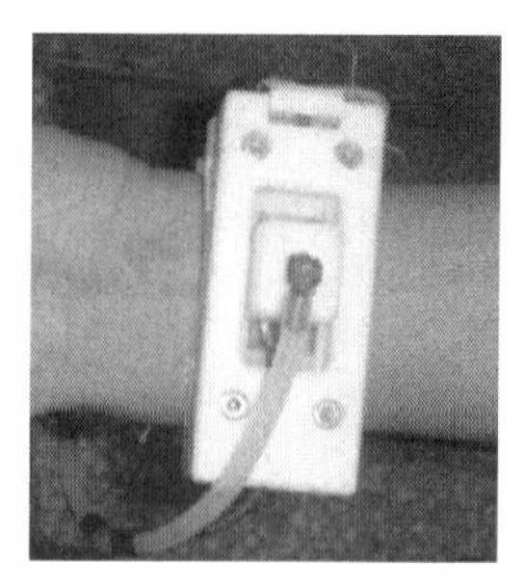

Fig. 3 Test chart of pulse measuring instruments

4 Results

Fig.4 shows the results to adjust the contrast and gray scale. In order to get a clear image, it carried out several tests to adjust the contrast and gray scale. Fig.4 (a) is the source of ultrasound images, which is color image; Fig.4 (b) is the contrast adjustment for the first time, did not deal with colors; Fig.4 (c) is grayscale adjustment, as MATLAB can only handle two-dimensional black and white images, so you need to remove the source image's color components; Fig.4 (d) for the second contrast adjustment. Above is the image of the initial adjustment. Fig.5 shows the results of image denoising, using the median filter remove the salt and pepper noise in Fig.4 (d). Using the median filter get the smooth edge vascular image. Fig.5 (a) is the result for the median filter, which will remove the noise. Fig.5 (b) is the open computing, fill in the small gaps or holes in specific structural elements. Fig.5 (c) is the close computing, filtered out the thrusting smaller than structure element, and cut off the lap to play a separate role slender. Then it has been relatively smooth vessel image. Finally, cut the non-vascular in the image area to get the integrity vessel in Fig.5 (d). Then calculate the area.

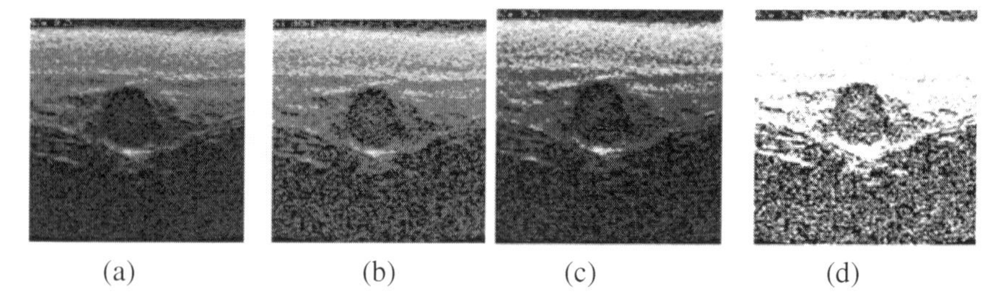

(a) (b) (c) (d)

Fig. 4 Adjust to contrast and gray-scale. (a) A source image; (b) Contrast adjustment; (c) Gray image; (d) contrast adjustment

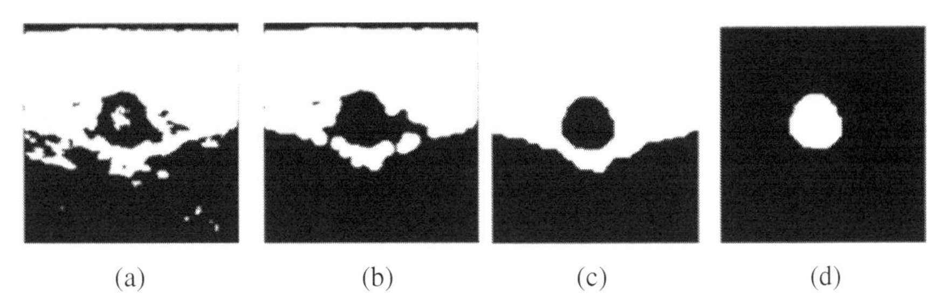

Fig. 5 Image denoising process. (a) Median filter; (b) Open computing; (c) Close computing; (d) Choosing a specific area

After smoothing the pulse wave signal map are shown in Fig.6. To ensure the test data is accurate, it is compared the pulse image that connected with the vessel's area and diameter. Fig.7 shows the diameter pulse image.

As Fig.5 (d) is not a rule of vascular circle, the diameter can only be measured manually. It is increasing the error process lead to a not accurate result. In the same time, through the vascular area to get the information is more accurate information than the diameter (compared with Fig.6 and Fig.7).

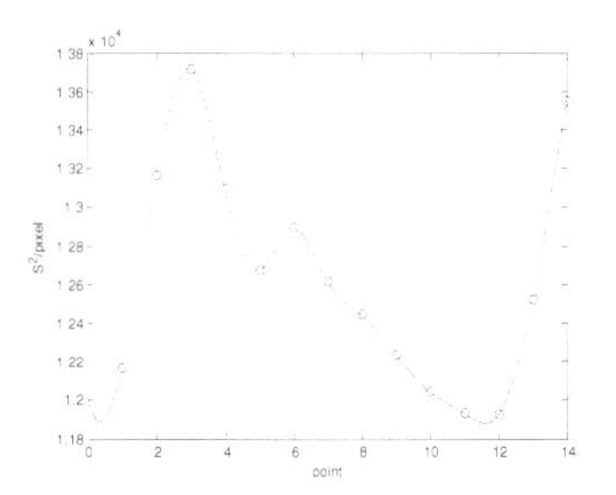

Fig. 6 Area point smoothing pulse

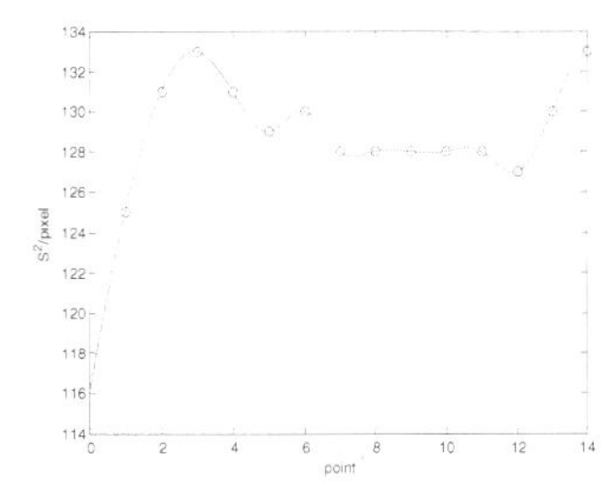

Fig. 7 Diameter smooth pulse point

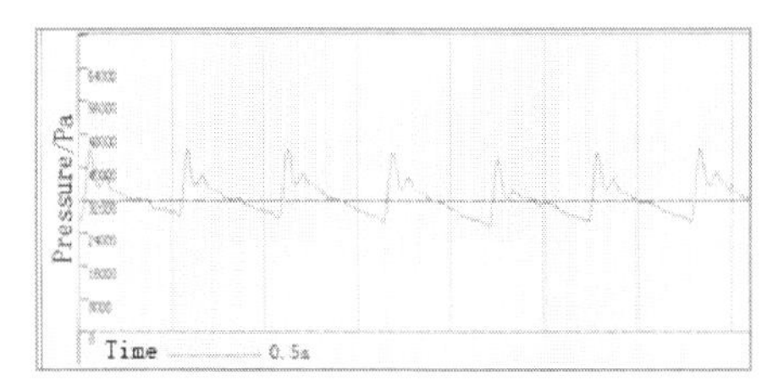

Fig. 8 The example of the measurement map and the results

Table 1 is for 5 normal subjects obtained pulse with ultrasonic sensor (one in F.6) and use the ZD pressure pulse instrument [7] (one in Fig.8). Comparing the results obtained we can see the average error is 8.29%. Because of the non-repetitive in human pulse wave and the two graphs are the results from different software, we can say that the result from the ultrasonic sensor is accurately.

Table 1 Compared with the ultrasonic pulse and stress results

NUM	H2/H1(height of dicrotic notch)		H3/H1(height of dicrotic wave)		H1/T1 (ratio of major peak and cardiac cycle)		error
	Ultrasonic	pressure	ultrasonic	pressure	ultrasonic	pressure	
1	0.50	0.47	0.89	0.83	7.60	7.38	4.46%
2	0.38	0.42	0.72	0.80	6.92	7.01	7.67%
3	0.54	0.60	0.91	0.88	7.35	7.50	5.33%
4	0.42	0.55	0.77	0.65	7.02	6.86	16.33%
5	0.60	0.57	0.80	0.70	8.20	7.80	7.67%

5 Discussion and Conclusion

Through Fig.6 and 8 shows that this method used to obtain the pulse wave is much more clearly and exactly than the traditional instrument.

This proposed acquisition method based on the ultrasonic pulse, it make analyzed the multi-dimensional pulse possible. Using ultrasound pulse sensor to collection the dynamic pulse image, then extracted as single static image. Through the static image processing, to enhance contrast and eliminate noise. Then using the open and close function to order the image to get area of pulse. Processing by the collected images to get the continuous curve of vascular area is laid a good foundation for further research.

References

1. Sonka, M., Zhang, X., Siebes, M., et al.: Segmentation of intravascular ultrasound images: a knowledge-based approach. IEEE Transactions on Medical Imaging 14(4), 719–732 (1995)
2. Li, Y., Yu, D., Lin, J., Xing, G.: Panoramic ultrasound imaging. Information of Medical Equipment 20(10), 39–40 (2005)
3. Bhuiyan, M.I.H., Ahmad, M.O., Swamy, M.N.S.: Spatially adaptive thresholding in wavelet domain for despeckling of ultrasound images. Image Processing, IET 3(3), 147–162 (2009)
4. Aoki, Y., Boivin, A.: Computer reconstruction of images from a microwave hologram. Proceedings of the IEEE 58(5), 821–822 (1970)
5. Yan, X., Li, P., Li, Y., et al.: Image speckle reduction and coherence enhancement in underwater ultrasonic imaging based on wavelet transform. In: Computer Science and Software Engineering, pp. 145–148. IEEE Computer Society, Wuhan (2008)
6. Sava, H., Durand, L.G.: Automatic detection of cardiac cycle based on an adaptive time-frequency analysis of the phonocardiogram. In: Proceedings of Engineering in Medicine and Biology Society, pp. 1316–1319. Pergamon Press, Chicago (1997)
7. Di, J., Chen, S., Wang, X.: Study on the traditional Chinese medicine pulse-taking machine. BME & Clin. Med. 12(6), 503–506 (2008)

The Integration of Security Systems Using WBSC

Sanchai Rattananon and Suparerk Manitpornsut

Abstract. Presently, the aggravated economy and society in many countries can critically lead to various forms of crimes, especially building robbery. Therefore, one of the best ways to prevent the properties' owners from such crime is to install a security system. Basically, the designed building security systems consist of two installation systems, the Monitoring system such as Closed-Circuit Television (CCTV) and the Alarm system. Each system has its own control equipments such as Digital Video Recorder for the CCTV system and Control Panel for the Alarm system, which might constitute high cost and difficulties to control and manage those equipments. In this paper, we introduce new controlled equipment for the both building security systems. It integrates and extends the functions to control both Monitoring and Alarm systems, which is called WBSC (Wireless Building Security Control). In addition, we also propose an incorporation of the wireless network, mobile network, and PSTN (Public Switching Telephone Network) into the use of the building security systems. Finally, the result of this paper presents a functional success of the WBSC equipment that meets the requirements to control the building security systems and to perform the alarming system very efficiently for the properties' owner at anywhere and anytime.

Keywords: WBSC, GSM, Wireless Network, IEEE802.11, Mobile Network.

1 Introduction

The economic crisis all around the world in the present day imposes on not only the financial difficulties but also the social problems to the society. Financial difficulties come to the attention of the government in every country due to its

Sanchai Rattananon · Suparerk Manitpornsut
School of Engineering
University of the Thai Chamber of Commerce
Bangkok, 10400 Thailand
e-mail: {sanchai_rat,suparerk_man}@utcc.ac.th

Z. Qian et al. (Eds.): Recent Advances in CSIE 2011, LNEE 125, pp. 639–645.
springerlink.com

scale, required the macro perspectives to alleviate the financial situation. On the

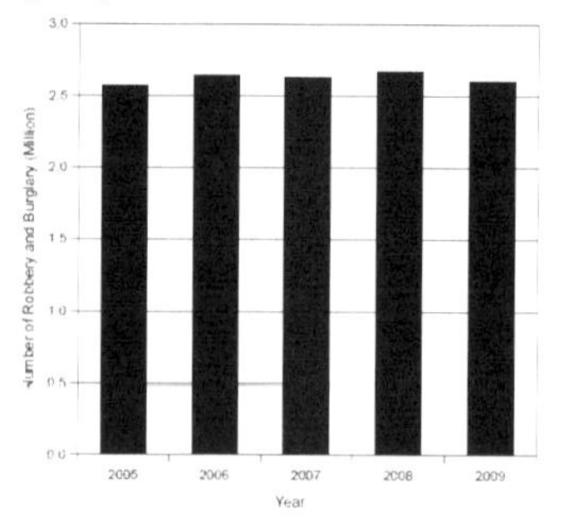

Fig. 1 Number of robbery and burglary in USA during 2005 – 2009

other hand, the social problems need the special attention from everyone in the community to sincerely improve the circumstances. The social problems include conflict, drinking and drugs, physical and mental health, racial, crime, etc.

One of the most critical social problems is the robbery and burglary. In Figure 1, the number of robbery and burglary in the USA during the last five years [1] has constantly occurred more than two and half million times each year. Approximately 83% of such crimes occurred in the building of both residential and non-residential (e.g. store, office, etc) areas.

In order to alleviate the aforementioned crime, a building security system is usually deployed. The building security system can be categorized into two major types, depending on its purposes: Monitoring and Alarming systems.

Conventional building security monitoring system is Closed-Circuit Television (CCTV). CCTV typically consists of camera, video recorder, and monitor. CCTV provides real-time monitoring as well as video playback features. However, it requires security officers in charge for all time to monitor the possible threats and/or identify the problem after someone breaches into the building.

On the contrary, alarming system provides alerts by informing the user via notification so that no need to monitor the building in person. The notification could be in the forms of voices, messages, etc. Alarming system basically consists of sensors (e.g. motion detector, smoke detector, etc.), alarm notification unit (e.g. siren), control unit, and user access unit. However, unlike monitoring system, alarming system generally offers only instantaneous notification without playback feature.

In order to enforce the security in the building, both monitoring and alarming systems are required. However, these systems have their own core technologies and consequently interoperability is hardly possible. This issue introduces high cost of ownership and maintenance drawback.

We, therefore, propose the solution to prevent the robbery and burglary by using the Wireless Building Security Control System (WBSC), which combines the advantages of both monitoring and alarming systems. The proposed system offers 'anywhere and anytime' access, as explained in more details in the next sections.

2 Related Work

In this section, we provide the reviews of available products of monitoring and alarming systems. In addition, we also show a number of proposals for building security service system.

2.1 Monitoring System

Conventional monitoring system is CCTV (Closed-Circuit Television). CCTV consists of camera, video recorder, and monitor. CCTV provides real-time monitoring as well as video playback features.

Camera in CCTV could be as simple as analog camera or as advanced as wireless high-definition (HD) video camera. In the same manner, video recorder ranges from analog video recorder (e.g. VHS) to HD digital video recorder. The connectivity between camera and video recorder could be done through cable, radio connection, or IP network.

2.2 Alarming System

Alarming system provides the alert by informing the user via notification. The notification could be in the forms of voice, message, etc. Alarming system basically consists of sensors (e.g. motion detector, smoke detector, etc.), alarm notification unit (e.g. siren), control unit, and user access unit.

A variety of sensors could be used in the system, depending on the purpose of the alert system, such as motion detector, smoke detector, touch sensor, etc.

Unlike monitoring system, alarming system usually offers instantaneous notification without playback feature.

2.3 Building Security Service System

Jiang *et al* [2] proposed a framework that uses the wireless home alarm system. Therefore, we can make the use of GSM network by adding GSM module [3] which has simple structure and easy to control. Liting *et al* [4] proposed to distribute security system for intelligent building, especially the alarm devices. They provided the mechanisms to develop the detect sensors based on wireless communication technology and embedded system which could be applied the intelligent building [5].

3 WBSC System Architecture

The main objective of WBSC is to support service integration that provides the monitoring capabilities as well as alarming features. It consists of hardware architecture and controller software stack, as explained below.

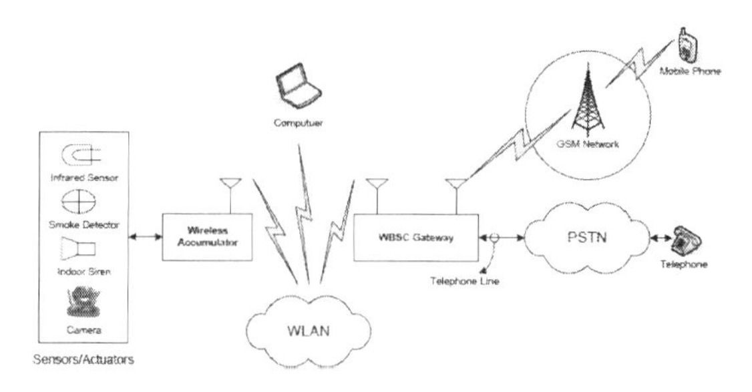

Fig. 2 WBSC System Architecture

As shown in Figure 2, the hardware architecture composes of three main components as follows:

3.1 Sensors/Actuators

In order to be aware of the environment, sensors are used to detect the conditions of their surroundings, e.g. heat, motion, smoke, etc. Actuators are installed so that they can react to the environment upon the preset actions. An example of actuators is a siren, used to produce alert voices.

3.2 Wireless Accumulator (WA)

WA acts as an interface between sensors/actuators and WBSC Gateway (WG), as illustrated in Figure 3. It comprises four subsystems:

1) *WLAN Interface:* IEEE 802.11b WLAN interface module enables WA to communicate to WG via wireless network.
2) *Power Unit:* Rechargeable battery, regulator and power switching circuits are used to supply uninterrupted power to all units.
3) *Processing Unit:* Empowered by PIC Microcontroller from Microchip [6], the processing unit is designed for low power consumption and support a large number of sensors/actuators.
4) *Sensors/Actuators Interface*: This module is used to convert data from sensors to useful information and pass it to the processing unit. Another functionality of this module is to deliver the commands from processing unit to the actuator.

3.3 WBSC Gateway (WG)

WG is an important component that allows authorized users to connect to the system remotely through PSTN, mobile network, or wireless LAN. As depicted in

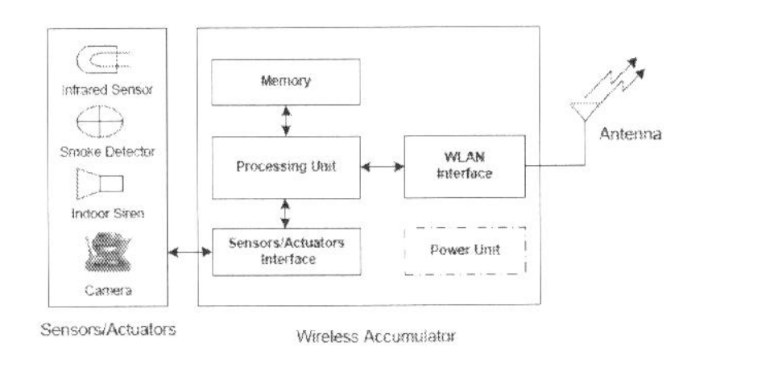

Fig. 2 Architecture of Wireless Accumulator

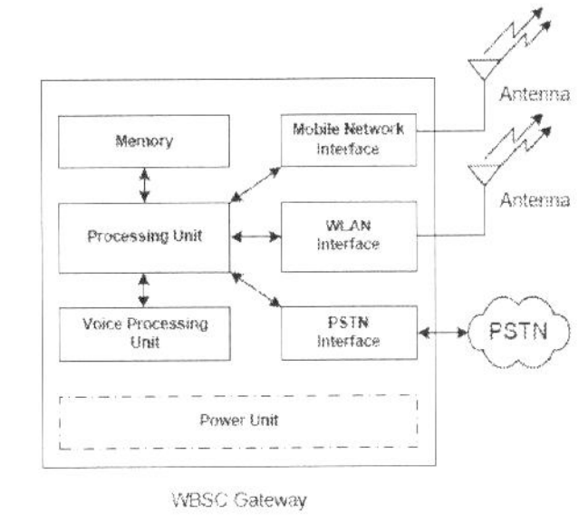

Fig. 3 Architecture of WBSC Gateway

Fig. 5 SMS Alert from Smoke Detector

Fig. 6 Video Monitoring and MMS Alert Experiment Setup

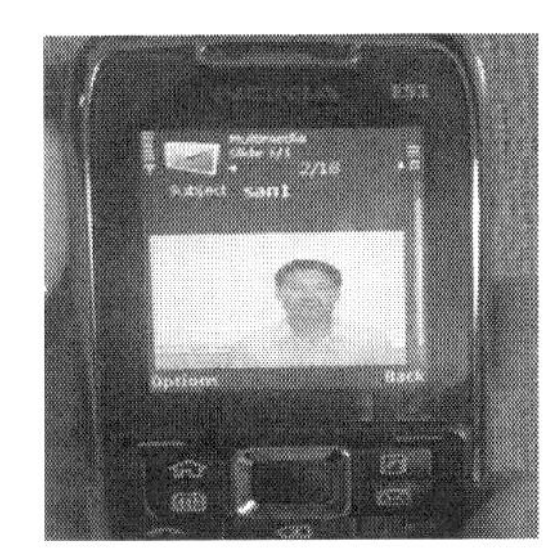

Fig. 7 MMS Alert via Mobile Network

Figure 4, there are six main subsystems but only three of which differ from WA are described here:

1) *Mobile Network Interface:* This module supports GSM network with GPRS, allowing the video monitoring on mobile phones or portable devices through mobile network.
2) *PSTN Interface:* PSTN interface offers a reliable communication channel required in emergency situation, e.g. to inform fire and police stations.
3) *Voice Processing Unit:* This unit is used to process voice message, allowing the system to automatically send pre-recorded voice message to the relevant authorities via PSTN interface.

4 Experimental Results

4.1 Binary Sensor and SMS Alert

This scenario is to test the simplest functionality of the system – alarming system. Binary or on/off sensor is the simplest kind of sensors, which can provide only two output values: on or off. We use smoke detector as a sample of binary sensor and SMS (Short Message Services) as an alert message via mobile network. The result in Figure 5 shows the SMS alert with sensor information.

4.2 Video Monitoring and MMS Alert

In order to illustrate the integration of monitoring and alarming systems, video camera is deployed in WBSC, as shown in Figure 6. During the video recording, WBSC can send the MMS (Multimedia Message Services) alert to the associated mobile phone as shown in Figure 7.

4.3 Binary Sensor and Voice Alert through PSTN Network

Some alert needs an immediate attention, e.g. fire alarm. It is better if this type of alert is sent to the relevant authority when the building owner is not available to monitor the alarm or fail to reset the alarming system within the predefined alert delay. During the test, the recorded voice message is triggered to send to a preset phone number through PSTN network when the alert delay is run out.

5 Discussion and Conclusion

WBSC offers integration of monitoring and alarming system as a single point of services. Wireless accumulator (WA) is a wireless platform that can be interconnected to a variety of sensors and actuators. It decouples the presentation of the sensed data from the WBSC Gateway (WG). This decoupling allows authors to emphasize on the access control on WG. In WBSC architecture, WG support multiple access technologies, including WLAN, PSTN network, and GSM mobile network. Access through WLAN can provide a high quality video monitoring while access via GSM mobile network enables monitoring and alarming to mobile users and access through PSTN network provides alarming function for emergency cases.

The WBSC system itself still has some limitations. The connection between WA and WG is a point- to-point connection. Multicasting connection is required to support multiple WAs associated to the same WG, which we have left it for future work.

References

[1] Federal Bureau of Investigation, Offense Analysis, Table 7 (2005-2009), http://www.fbi.gov/
[2] Lian, J., Xing, D.Y.: A wireless home alarm system based on tele-phone line. Journal of Human University (1), 36–39 (2006)
[3] GSM Module, http://www.etteam.com/

[4] Cao, L., Jingwen, Jiang, W.: Distributed Security System for Intelli-gent Building Based on Wireless Communication Network. In: Proceedings of the 2006 IEEE International Conference on Information Acquisition, IEEE ICIA (August 2006)
[5] Liu, C., Xiao, N.: Design and implementation of home service robot and home intelligent security control. Information of Micro-Computer 22(2), 212–214 (2006)
[6] Microchip, http://www.microchip.com/

Computer Simulation for a Catalytic Reaction on Fractal Surfaces by Monte Carlo Method

He-Bei Gao and Hong Li*

Abstract. In this paper, we adopt a cluster-cluster aggregation model to produce fractal catalyst surface by the computer simulation. And a Monte Carlo simulation was used for studying the CO oxidation on the fractal catalyst surface. The simulation results show that: There is a reactive range for the values of the mole fraction y_{co} of CO in gas phase between two values (y_1,y_2) outside of which the surface is poisoned by O ($y_{co}<y_1$) or CO ($y_{co}>y_2$). With the decrease of active particles on the square lattices (p), the relative coverage rate of reactants increases; while the coverage rate of empty active sites on the catalytic surface decreases; so that the utilization rate of the catalyst decreases and the production rate of CO_2 (R_{co_2}) reduces; and R_{co_2} deceases linearly with the adjacent site number of each active site on the catalyst surface. Decreasing p leads to a shift of reactive windows to smaller value of y_{co} and phase transition at y_2 from fist order on regular lattices to second order on fractal surface.

1 Introduction

The study of chemical reaction models has attracted considerable attention in recent thirty years [1-10]. These models not only exhibit a rich variety of complex behaviors, but also can explain many experimental results with catalysis and could show the way for designing more efficient processes. The ZGB model describes some kinetic aspects of the catalytic oxidation of CO on a crystal surface [2]. The surface is in contact with a CO/O_2 gas mixture which determines the reactive rates

He-Bei Gao
Computer Science Department, Wenzhou Vocational & Technical College, Wenzhou 325035, China

Hong Li
Oujiang College ,Wenzhou University ,Wenzhou 325035, China
e-mail: `lihong@wzu.edu.cn`

* Corresponding author.

Z. Qian et al. (Eds.): Recent Advances in CSIE 2011, LNEE 125, pp. 647–652.
springerlink.com

and the coverage of CO and O adsorbed. CO can adsorb on a single empty active site, O_2 adsorbs and dissociates on two adjacent empty active sites. CO and O adsorbed on neighboring sites immediately react. The model exhibits two kinetic phase transitions, first order one at the upper boundary of the reactive zone and second order one at the lower boundary.

Despite its interesting simulation results, the ZGB model doesn't consider some reaction mechanisms, such as desorption and diffusion of adsorbed species [3-4], finite reaction rates [5-6], multiple crystal planes [7]. In order to overcome the shortcoming, lattice gas models have been developed [8-10] which are more suitable to represent the real physical situation and consider the characteristics of the substrate acting as a catalyst. A homogeneous catalytic surface is assumed for a catalyst formed by a single crystal. In fact, many catalytic surfaces are made up of small clusters of metal particles supported on an inert material, and the catalyst has a non-homogeneous structure in which the catalyst sites are not equivalent. Therefore, the surface often can be described as a fractal [11-19], random fractals such as percolation clusters [11-13], or diffusion-limited aggregates (DLA) [14-15].

Surface heterogeneity can affect the adsorption-process, since all the catalyst sites are uneven, especially in random fractal surface. The interesting thing is how the heterogeneous catalyst influences the reaction, when the reaction is observed to be effective, and whether the clusters of reactants self-organize the fractal structure. On one hand, the scanning tunneling microscope has obtained some knowledge of reactive basic mechanism [16]. On the other hand, in addition to the above-mentioned model, many scholars build and improve models to simulate this domain.

In this paper we study the CO oxidation reaction on fractal catalytic surfaces, where surface irregularity is considerably enhanced. In the present study we neglect diffusion and desorption. We find that the inhomogeneity will have an important effect on the reaction of CO oxidation. We analyze the reactive window and the fractal structures on the surface which are generated as a consequence of the coupling of the reaction and the fractal structure of the catalyst.

2 The Model and Simulation

We use a fractal structure surface on the square lattice $L \times L$ ($L = 400$) to serve as the catalyst surface. The fractal surfaces are formed on a simple square lattice with periodic boundary conditions and they are generated by the standard procedure as following:

At the start of the simulation, the density p of the lattice sites are selected at random and occupied by particles. Clusters (particle or particles) are picked at random (the probability is inversely proportional to the size of the cluster) and diffuse with a probability, which is chosen at random in one of four equally probable directions (±x, ±y). If a cluster contacts other clusters by the nearest neighbor, these clusters are merged to form a single cluster. The process continues until all the particles are formed for only one cluster.

In the catalytic oxidation of CO, the adsorption–reaction processes on these surfaces are considered as follows:

$$CO(g)+v \rightarrow CO(a) \quad \text{with probability } y_{co} \tag{1}$$

$$O_2(g)+2v \rightarrow 2O(a) \quad \text{with probability } 1\text{-}y_{co} \tag{2}$$

$$CO(a)+O(a) \rightarrow CO_2(g)+2v \tag{3}$$

where g and a denoting a species in the gaseous and adsorbed phases respectively, v corresponding to a vacant site on the fractal surface. Let y_{co} be the mole fraction of CO in the gas phase. Then, CO strikes the square lattice $L \times L$ with the probability y_{co} and O_2 with $1\text{-}y_{co}$. A lattice site is chosen at random. If that site is an inactive site or occupied by other CO molecule or O atom, then the trial ends; otherwise, the CO adsorbs and the four nearest neighbors are chosen in random order. If one of the nearest neighbors is occupied by O, CO_2 is formed from these pair lattice sites and is desorbed immediately. If the striking gas molecule is an O_2, two adjacent sites are chosen at random. If either site is inactive or occupied by others, the trial ends; otherwise, the O_2 disassociates and adsorbs on two nearest neighbor sites. If two adjacent sites adsorb CO and O, respectively, Eq. (3) occurs immediately.

The fractal dimension d_f of the final ramified clusters is calculated by the box method [17]. All the values of d_f given below are averaged over at least 20 independent simulation runs.

3 Results and Discussion

We have carried out simulations to form the fractal metal surfaces on the square lattice $L \times L (L=400)$.The relation between the fractal dimension d_f and the density p is shown as in Fig. 1. With the increasing of the density of active particle p, the fractal dimension of the fractal metal surface increase. We can see from Fig. 1, where p <0.3, the fractal dimension d_f increases rapidly with the increase of the active site density p, but, where p >0.3, d_f changes indistinctively (is nearly 2).

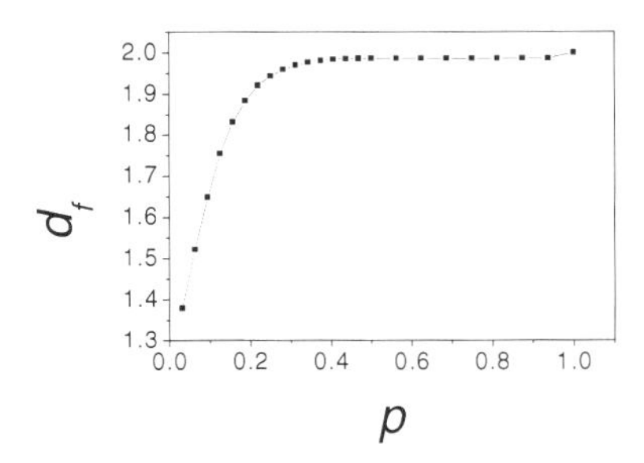

Fig. 1 The dependence of the fractal dimension d_f of the clusters on the density p of the lattice sites occupied by particles.

In the reaction model, p represents the density of the active site. CO and O_2 molecules strike on the square surface in their respective collision probability, which depends on pressure or mole fraction of the reactant, respectively, CO y_{co} and O_2 1- y_{co}. The total number of collisions of CO and O_2 molecules achieves the total number of square lattice sites, which represents a Monte Carlo Step (MCS). The number of adsorbed CO molecules, adsorbed O atoms, empty active sites and CO_2 molecules produced and desorbed are calculated in every MCS. In order to compare these parameters on the catalyst surface with different impurities or defects, the parameters are normalized with the ratio of the total active sites, such as the coverage for CO molecule θ_{co}, O atoms θ_o, empty active site θ_v and the rate of CO_2 production which shows that the utilization of the active sites.

We carry out simulations with the fractal surfaces previously mentioned. For low values of y_{co} ($y_{co} < y_1$), a saturation of the catalyst is filled by atoms O; for high values of y_{co} ($y_{co} > y_2$) by molecules CO. For the range $y_1 < y_{co} < y_2$ there is a reactive window that CO_2 is produced in every MCS and the fractal surface is made up of CO, O and empty active site.

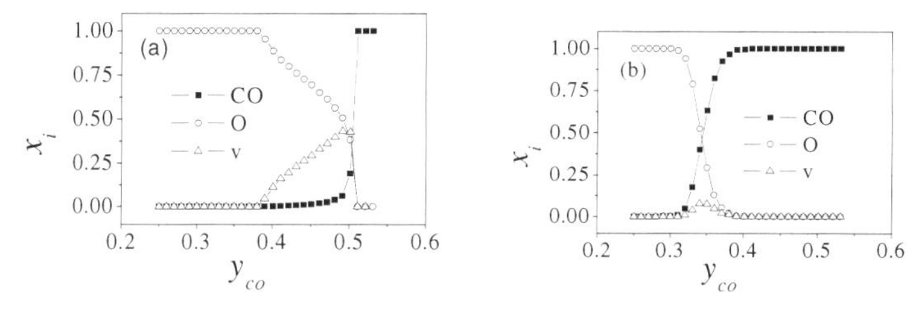

Fig. 2 The coverage of CO, O and vacant active site versus control parameter y_{co}, for the density of the active site is (a) p=0.9375 (b) p =0.5

The dependence y_{co} on the coverage of empty active sites and reactions in the fractal surfaces with different density of active site is shown in Fig. 2. In this model we get some different results compared to Ref [2], with the decrease of the active site density p, the reaction windows moves to the low-end of y_{co}. With the increase of y_{co}, θ_o declines gradually (from 1 to 0), and θ_{co} increases gradually (from 0 to 1), and θ_v first increases and then decreases after it achieves a maximum value. The second order phase transition occurs rather than first-order phase transition.

The dependence of Rco_2 on y_{co} is shown in Fig. 3a for the different density of the active sites in fractal surface. Rco_2 increases linearly with the increase of y_{co} in the low p, which consist with Ref.[2]. But, in the fractal surface, with decreasing active site density p, nonzero regional of Rco_2 gradually moves to the smaller end of y_{co}. At first Rco_2 increases with the increases y_{co} and then decreases after it achieves the maximum value $Rco_{2\,\max}$ ($y_{co}=y_3$) and $Rco_{2\,\max}$ decreases with the increases p. O poisoning occurs for $y_{co}<y_1$, the reaction can be sustained in the time given above for $y_1<y_{co}<y_2$ and CO poisoning occurs for $y_{co}>y_2$. The relationship between the value of y_1, y_2, y_3 and p is shown in Fig. 3b. It shows reaction window moves in the direction of smaller end of the partial pressure of CO gas. The less the density of active site in the surface is, the less the fractal dimension is, so that O_2 is more difficultly adsorbed on the catalyst surface. It is conducive to CO oxidation reaction for the smaller CO pressure, but CO poisoning occurs more easily in the higher CO partial pressure.

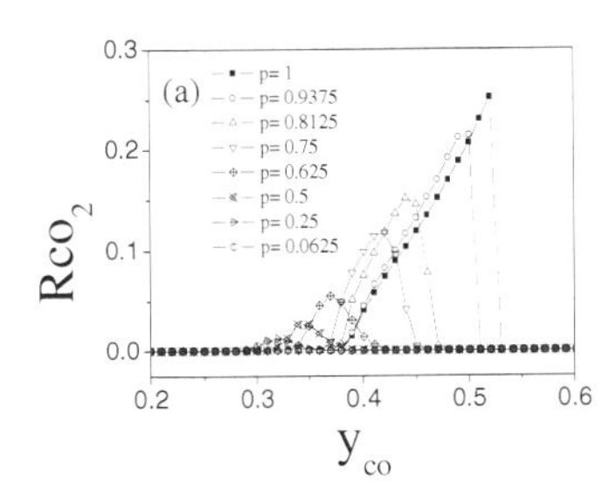

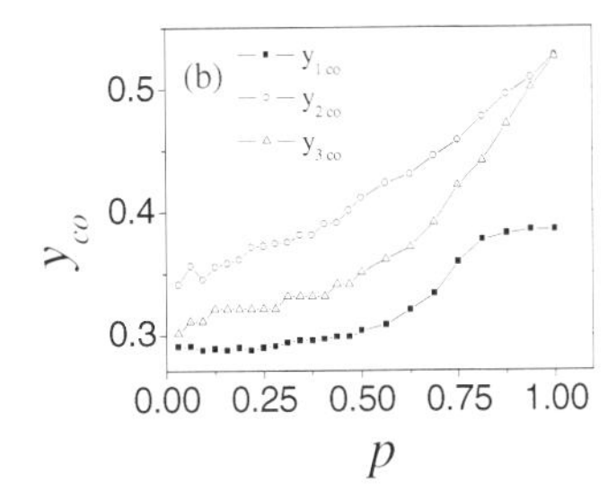

Fig. 3 (a)The production rate of CO_2 versus control parameter y_{co} for different density of the active site (b) The dependence of y_1, y_2, y_3, on the density of the active site p.

The maximum rate of CO_2 desorbed from the catalyst surface in each MCS when the system reaches the stable state depends on the density of active sites, as illustrated in Fig. 4a. With the increase of p, $Rco_{2\,\max}$ also increases. For p <0.5, $Rco_{2\,\max}$ increases very slowly with the increasing density of active sites, however, for p >0.5, $Rco_{2\,\max}$ increases rapidly.

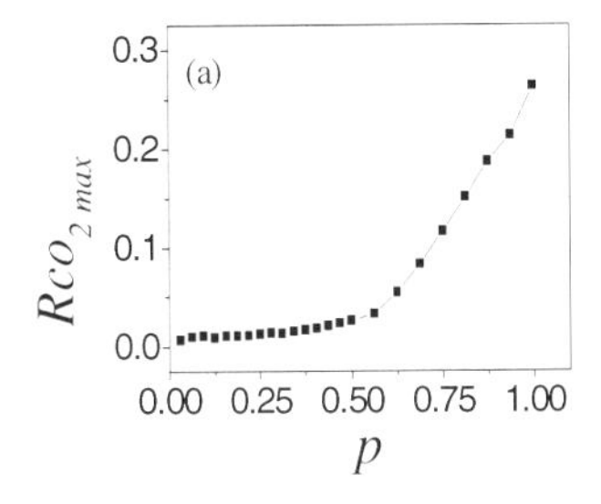

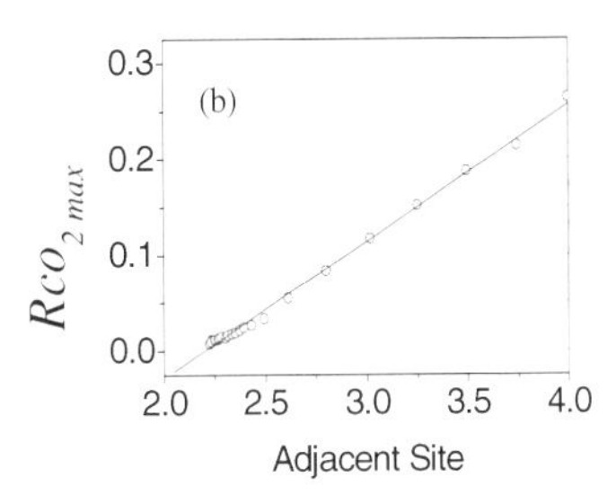

Fig. 4 (a) The dependence of the maximum production rate of CO_2 Rco_2 max (at y_{co}=y_3) on the density of active site p. (b) Rco_2 max on the average adjacent degree of the active site

We found a linear relationship between the maximum production rate of CO_2 and the number of adjacent active sites, as shown in Fig. 4b. The adjacent degree presents the average number of the nearest neighbor site of each site on the fractal surface. In fractal surface, the greater the adjacent degree is benefited to O_2 adsorption and dissociation, and the smaller one is not conducive to the adsorption and dissociation of O_2. This means an increase of the adsorption of CO occurs with the decrease of the adjacent degree, so that the coverage of CO increases on the fractal surface. In this case, the clusters of CO easily aggregate on the fractal surface (local CO poisoning) and the contact of CO and O reduces so as to reduce the CO_2 production rate. Therefore the reactive window moves in the direction of the smaller end of the partial pressure of CO (appears displaced toward the lower values of y_{co}).

4 Conclusion

The simulation results in fractal surface show that, excluding desorption of the reactants and the interaction between the reactant on the catalyst surface, there are O poisoning phase ($y_{co}<y_1$), the reaction phase ($y_1<y_{co}<y_2$) and CO poisoning phase ($y_{co}>y_2$) on the catalytic reaction. Only if $y_1<y_{co}<y_2$, the catalyst surface is adsorbed a certain percentage of CO and O coexisted where the reaction sustained.

The coverage of the reactants on the fractal catalyst surface is greater than the one on the regular surface, so that the coverage rate of active site reduces and the production rate of CO_2 reduces. We also found that the formation of CO_2 depend on the adjacent points of the active site. O_2 is difficultly adsorbed on the fractal surface with the small adjacent degree, which is conducive to CO oxidation reaction in low pressure of CO and is easy to CO poisoning in high pressure of CO.

The clusters of CO and O are self-organized on the catalytic surface, where the reaction only occurs in the junction between the clusters of different reactant. In the fractal surface, the coverage of the active sites reduces and the junction of different reactants becomes the bottleneck of the reaction.

Acknowledgments. This work was supported by the Foundation of Wenzhou Vocational and Technical College, China (WZYJG0939 and WZY2010034,).

References

1. Zhadanov, V.P., Kasemo, B.: Surf. Sci. Rep. 20, 111 (1994)
2. Ziff, R.M., Gulari, E., Barshad, Y.: Phys. Rev. Lett. 56, 2553 (1986)
3. D'Ajello, P.C.T., Hauser, P.R., Figueiredo, W.: J. Chem. Phys. 118, 6003 (2003)
4. Mai, J., von Niessen, W., Blumen, A.: J. Chem. Phys. 93, 3685 (1990)
5. Tammaro, M., Evans, J.W.: Phys. Phys. Rev. E 57, 5087–5094 (1998)
6. Leitner, D.M., Wolynes, P.G.: Chemical Physics 329, 163–167 (2006)
7. Huang, W., Zhai, R., Bao, X.: Langmuir 17, 3629–3634 (2001)
8. Vicsek, T.: Fractal Growth Phenomena. World Scientific, Singapore (1992)
9. Dickman, R.: Phys. Rev. A 34, 4246 (1986)
10. Luque, J.J.: Phys. Rev. A 42, 3319 (1990)
11. Casties, A., Mai, J., von Niessen, W.: J. Chem. Phys. 99, 3082 (1993)
12. Hovi, J.P., Vaari, J., Kaukonen, H.P., Nieminen, R.M.: Comput. Mater. Sci. 1, 33 (1992)
13. Cortés, J., Valencia, E.: Physica A 309, 26 (2002)
14. Mai, J., Casties, A., von Niessen, W.: Chem. Phys. Lett. 211, 197 (1993)
15. Gomez, A., Luque, J.J., Cordoba, A.: Chaos, Solitons and Fractals 24, 151 (2005)
16. Hahn, J.R., Ho, W.: Phys. Rev. Lett. 87, 166102 (2001)
17. Bunde, A., Havlin, S.: Fractals in Science. Springer, Berlin (1994)

A New Interpolation Criterion for Computation of Two-Dimensional Manifolds

Hengyi Sun, Yangyu Fan, Jing Zhang, Huimin Li, and Meng Jia

Abstract. In this paper, a multi-threshold criterion was proposed for computation of two-dimensional manifolds. By taking the minimum threshold as the reference standard, thresholds are adapted according to the corresponding growth rate in different directions. With the study of distance changes between adjacent orbits, prior knowledge can be got and used to guide the current interpolation to prepare data for the next loop. Minimum threshold reflects details of the manifold structure. To meet the geometric scale of current loop in processing of the computation of manifold, the size of minimum threshold is required to be proportional to the size of the loop. Ratio is recorded as the control factor. Due to the introduction of control factor, the changes of thresholds can adapt to the changes of manifold better, and the structure of manifold can be constructed in different geometric scales. Lorenz system and Duffing system are taken as examples to demonstrate the effectiveness of the proposed approach.

1 Introduction

With the development of science and technology people are paying more and more attention in computation of autonomous system and non-autonomous system manifolds. From the tendency of manifolds stability can be gained easily and this feature could be used to optimize control parameters[1,2]. Therefore the computation has become an effective way to understand system performance. At present there have been some algorithms to compute manifold. Such as Normalization method[3], Geodesics method[4], PDE method[5], Segmentation method[6] etc. In these methods flow and loop as two main styles build the frame of manifold. Advantage of flow to show manifold is that no error exists caused by interpolation and point adjustment. And disadvantage is that in some direction flows are very centralized so to interpolate a new flow become very hard. In constant loop need fewer points and could use relationship between adjacent points. It is facility to

Hengyi Sun · Yangyu Fan
School of Electronic Information of Northwestern Polytechnical University
e-mail: sunnyfly@mail.nwpu.edu.cn, fan_yangyu@sina.com

Z. Qian et al. (Eds.): Recent Advances in CSIE 2011, LNEE 125, pp. 653–658.
springerlink.com

control the growth of manifold in different directions. As time evolves loop resizes. How to choose the key points to build the loop that is what we need to study. Common criterions and interpolation methods[7] which fit two-dimensional or three-dimensional curves very well are helpless to reflect the relationship among loops. In order to improve accuracy of key frame we adopt multi-threshold criterion to learn priori knowledge a moment before to conduct operation.

This paper includes five sections: in next section we explain straightforward why density of flows is uneven. In section 3 and section 4 algorithm of new interpolation criterion will be elaborated and two examples are taken as experiments to check the new criterion. In last section conclusion would be made.

2 Fundamental Analyses

Take a three-dimensional hyperbolic saddle autonomous system as example:

$$\dot{x} = f(x) \tag{1}$$

Where $x \in R^3$, $f : R^3 \rightarrow R^3$. Assume that $\lambda_i \ (i = 1,2,3)$ represents eigenvalues of equ.(1) while $v_i \ (i = 1,2,3)$ as eigenvectors which will be considered as linear independence. For convenience of description we suppose that at the origin of coordinates of three-dimensional autonomous system two-dimensional unstable manifold exists and corresponding eigenvalues are real. Choose a point in basic region $P = [p_1, p_2, p_3]^T$ and $P = \alpha v_1 + \beta v_2 \ (\alpha \cdot \beta \neq 0)$.With relevant knowledge of differential equation flow of equ.(1) can be solved out as follows.

$$x(t) = C\left[\alpha e^{\lambda_1 (t-t_0)}, \beta e^{\lambda_2 (t-t_0)}, 0\right]^T \tag{2}$$

To meet the constraints of linear conditions, the discussion above is confined in a neighborhood of the equilibrium. From equ.(2) we know that due to the impact of different eigenvalues density of flows in different directions may be different in the processing of computation of manifolds.

3 Algorithm

By above, typically, since distribution of flows in stable and unstable manifolds is uneven it raises difficulty to compute manifold and build its structure. Improper distribution of the point density on manifold can bring about deformation of manifold. In order to solve this problem a new criterion in this section is proposed and two-dimensional unstable manifold will be taken to illustrate the algorithm in detail.

Step 1. Calculate base manifold domain and choose N points $x_{t_0} = \{x_{t_0}^1, x_{t_0}^2, \cdots, x_{t_0}^N\}$. Arc length of adjacent points can be ciphered out $D_{t_0} = \{D_{t_0}^1, D_{t_0}^2, \cdots D_{t_0}^N\}$.

Step 2. $x^j_{t_{-1}} = f^{-1}(x^j_{t_0})$, arc lengths of adjacent points are recorded in $D_{t_{-1}}$. Let $d_{t_{-1}}$ stand for the minimum of $D_{t_{-1}}$, R represents the rate of changes of arc length $R^j = D^j_{t_{-1}} / D^j_{t_0}$. S_{t_i} sums D_{t_i}, that is $S_{t_i} = \sum_j D^j_{t_i}$. λ is set as control factor, so at t_0 minimal arc length: $d_{t_0} = d_{t_{-1}} \cdot (\lambda \cdot \mathrm{S}_{t_0}/\mathrm{S}_{t_{-1}} + 1 - \lambda)$. r is the minimum in R. ρ_{t_0} as the threshold of arc length: $\rho_{t_0} = d_{t_0} \cdot R / r$. Interpolate point if $D^j_{t_0} > \rho^j_{t_0}$ or delete point if $D^{j-1}_{t_0} + D^j_{t_0} < (\rho^{j-1}_{t_0} + \rho^j_{t_0})/2$. Obtain initial data x_{t_0} after adjustment.

Step 3. x_{t_i} as point set for time t_i. $x^j_{t_{i-1}} = f(x^j_{t_i})$, arc lengths of adjacent points of $x^j_{t_{i-1}}$ are $D_{t_{i+1}}$. Detailed procedure imitates step2.

4 Simulation

Generally for a given system its density of flows may be uneven and growth in different direction may be unbalanced. In order to demonstrate the role of interpolation criterion proposed in this paper radial normalization method is adopted to calculate. Specific procedures are as follows:

Assume equilibrium as $x_0 = (ox, oy, oz)$; basis vectors of two-dimensional unstable subspace $\vec{v}_1$, $\vec{v}_2$; coordinate of the unknown point $P = [x(t_i), y(t_i), z(t_i)]$; tangent vector at P is $\vec{t} = [f_1, f_2, f_3]$. Let $\vec{n} = \vec{v}_1 \times \vec{v}_2$ and $\vec{r} = P - x_0$ with the presupposed parameters the expression of normalization factor can be deduced as follows:

$$\mu = \frac{C_0 |\vec{n} \times \vec{r}|^2}{\left| |\vec{n} \times \vec{r}|^2 \vec{t} - \vec{t} \bullet (\vec{n} \times \vec{r})(\vec{n} \times \vec{r}) \right|} \tag{3}$$

4.1 Example 1

The well-known Lorenz system is defined as[3,4]

$$\begin{cases} \dot{x} = \sigma(y - x) \\ \dot{y} = \alpha x - y - xz \\ \dot{z} = xy - \beta z \end{cases} \tag{4}$$

Where the standard parameters $\sigma = 10$, $\alpha = 28$, $\beta = 8/3$. The origin is always an equilibrium, and it has a stable manifold for which the corresponding eigenvalues are $\lambda_1^s = -8/3$ and $\lambda_2^s \approx -22.8$, and eigenvectors are $v_1 = [-0.6148, 0.7887, 0]^T$, $v_2 = [0,0,1]^T$.

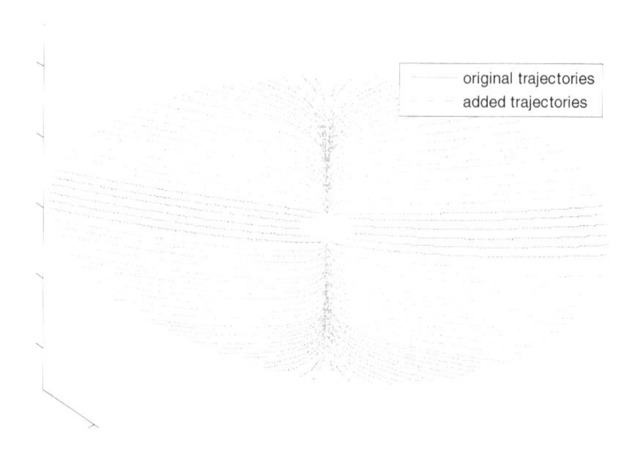

Fig. 1 Solid lines evolved from initial data; dotted lines interpolated

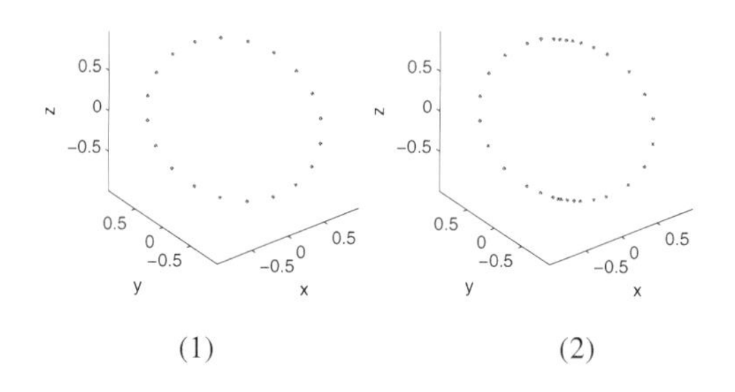

Fig. 2 (1) Initial points without adjustment (2) initial data after adjustment

This two-dimensional stable manifold has a complicated structure. We take two-dimensional stable manifold as example to show the new criterion. According to the values of eigenvalues and eigenvectors in the processing of computation it has low increment speed and large extended range along v_2 while in v_1 it grows fast and extend narrowly. Figure 1 is the result by discrete track method to illustrate necessity of interpolation in the process of manifold computation. Where solid lines are tracks evolving from initial data on base manifold and dotted lines are interpolated tracks. Intuitively we can learn the dilatation in different directions. Figure 2(1) shows the initial isometric data on base manifold; figure 2(2) gives out the initial data after adjustment. It is the aim of adjustment that let points produced by initial data become uniform distribution on next loop. Figure 3 reveals a whole manifold from 0 to -80 with step -5.

Fig. 3 Lorenz manifold

4.2 Example 2

Duffing system

$$\begin{cases} \frac{dx}{dt} = y \\ \frac{dy}{dt} = x - x^3 + \varepsilon \sin(t) \\ \frac{dz}{dt} = z + \varepsilon \sin(t) \end{cases} \tag{5}$$

Its DHT that can be solved by disturbing theory[8,9]. At point $x_{DHT}(0)$ three dimensional Duffing system could be linearized and its eigenvalues and eigenvectors are represented in $\lambda_1 = -1$, $\lambda_2 = 1$, $\lambda_3 = 1$, $v_1 = [-0.7071, 0.7071, 0]^T$, $v_2 = [0.7071, 0.7071, 0]^T$, $v_3 = [0, 0, 1]^T$. The two-dimensional unstable manifold we take as example is constructed tangent to eigenspace which spans by v_2 and v_3. Parameters of figure 4 are $\varepsilon = 0.01$, $t \in [0, 4]$, $\lambda = 0.8$, $step = 1$. It reveals the structure of Duffing system manifold among time range $t \in [0, 4]$.

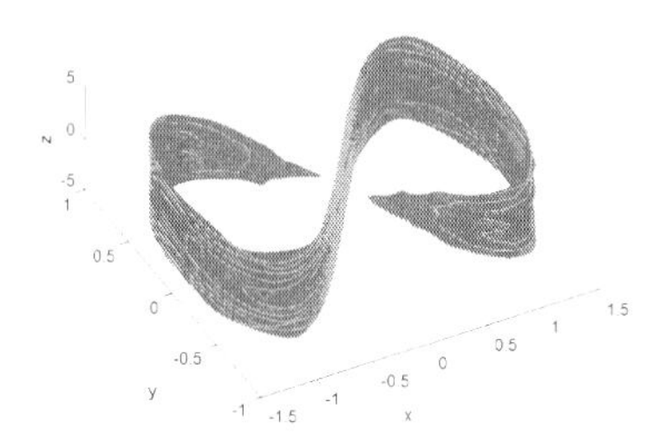

Fig. 4 Duffing manifold

5 Conclusion

In allusion to the anisotropic appearance of manifold computation a new criterion which supports multi-threshold and control factor was proposed and checked by lorenz system and Duffing system. Results demonstrate that new interpolation criterion not only is able to grasp the overall size of manifold but also could build the details according to the size of it.

Acknowledgment. This work is supported by the National Nature Science Foundation of China (Grant Nos. 60872159).

References

1. Serban, R., Koon, W.S.: Proceeding of IFAC Workshop, A 1
2. Chiang, H.D., Hisch, M.W., Wu, F.F.: IEEE Trans. Automat. Contr., doi: 10.1109/9.357
3. Johnson, M.E., Jolly, M.S., Kevrekidis, I.G.: Numer. Algorithms, doi: 10.1023/A:1019104828180
4. Krauskopf, B., Osinga, H.M.: Bifurc. Chaos, doi: 10.1142/S0218127498000310
5. Guckenheimer, J., Vladimirsky, A.: Appl. Dyn. Sys., doi: 10.1137/030600179
6. Dellnitz, M., Hobmann, A.: Num. Math., doi: 10.1007/s002110050240
7. Mancho, A.M., Small, D., Wiggins, S., Ide, K.: Physica D, doi: 10.1016/S0167-2789(03)00152-0
8. Ide, K., Small, D., Wiggins, S.: Nonlinear Proc. Geoph., doi:10.5194/npg-9-237-2002
9. Jiménez, J.A., Mancho, A.M.: Chaos, doi:10.1063/1.3056050

Computer Simulation for Viscous Fingering Occurred in a Hele-Shaw Cell

Jun Luo and Jianhua Zhang

Abstract. The viscous fingering that occurred in a Hele-Shaw cell was simulated using a modified DLA model with sticking probability on off-lattices. The present computer algorithm can also simulate the morphological evolution of viscous fingering from skeletal patterns to fleshy patterns. Both the shapes and the fractal dimension of fractal clusters can be obtained and they are in good agreement with experiments. The present work indicates that the effective fractal dimension of viscous fingering decreases and the corresponding morphologies of finger growth vary from a fleshy pattern with great fractal dimension to a skeletal pattern with low fractal dimension for decreasing model parameter A or increasing model parameter B.

1 Introduction

When a less-viscosity fluid drives a more-viscous one, viscous fingering occurred[1]. Viscous fingering has affection on some fields such as electro deposition, colloidal aggregation, and so on. Especially the viscous fingering has a dramatic impact on enhanced oil recovery (EOR) and reduces displacement efficiency[2]. The general results about the viscous fingering come from the devices called Hell-Shaw cell[3], which is constructed by two parallel plates and the interface of two immiscible viscous fluids in the narrow gap. Viscous fingering phenomenon is caused by hydrodynamic instability, hence it is difficult to characterize the viscous fingering quantitatively by traditional theory and scaling methods though there were numerous publications dealing with the problem of viscous fingering[4].

Damme *et al.*[5] showed that viscous fingering patterns could be described by a fractal relationship between area swept and radius of gyration (maximum finger length). Fractal geometry can be used to study the phenomena of viscous fingering

Jun Luo · Jianhua Zhang
Science College, Xi'an Shiyou University, Xi'an, China
e-mail: {luojun,jhzhang}@xsyu.edu.cn

Z. Qian et al. (Eds.): Recent Advances in CSIE 2011, LNEE 125, pp. 659–664.
springerlink.com

under certain circumstances, especially in the shape evolution of viscous fingering and the calculation of fractal dimension by diffusion-limited aggregation (DLA). Fractal geometry provides a new method to study the dynamic behavior of viscous fingering in fluids.

The present paper used a modified DLA model with sticking probability to generate various clusters for viscous fingering and to simulate the experimental result of viscous fingering in a Hele-Shaw cell, and the fractal dimension was calculated simultaneously. The present simulation is an effective method to show the morphological evaluations of a viscous fingering pattern by controlling the model parameters.

2 Viscous Fingering and Computer Simulation

In a horizontal Hele-Shaw cell, the movement of fluids, which cause the viscous fingering, satisfied the Darcy's law[4]. This cell has a behavior which is described by the very same set of equations as those for two-dimensional flow in a porous medium. Both the driving and displaced fluids were incompressible and the fluid pressure P will satisfied the Laplace's equation:

$$\nabla^2 P = 0 \tag{1}$$

At time t and position $\vec{r}$, the velocity of interface for driving fluid is

$$V_n = -b\vec{n} \cdot \nabla P(\vec{r}, t) \tag{2}$$

Where b is the distance between the two parallel plates in Hele-Shaw cell, $\vec{n}$ is the normal direction of the fluid interface. In addition, in the infinite position, the pressure will be

$$P_\infty = const. \tag{3}$$

And the pressure of the phase interface P_η will satisfied the following boundary condition

$$P_\eta = d_0 \kappa - \beta V_n^\eta \tag{4}$$

Where, d_0 is length of capillary tube and proportion to surface tension, κ is the local curvature in the interface, both d_0 and β are related to surface tension and the interface curvature. β and η are constants. Eqs.(1)~(4) are the nonlinear equations for the problem of fluid driving process.

The shapes of viscous fingering obtained from the Hele-Shaw cells are similar to the fractal aggregation generated by DLA model in a two-dimension Euclidean space. In a traditional DLA model suggested by Witten and Sander[6], a seed particle is put at the origin of an on-lattice or an off-lattice. Then a new particle is released at a random site at a distance far from the seed. This second particle walks stochastically and becomes part of the growing cluster after it reaches the seed site. The procedure is repeated until a cluster of sufficiently large size is formed. The density U of particle number satisfied Laplace equation and boundary condition:

$$\nabla^2 U(\vec{r},t)=0 \tag{5}$$

$$V_n=-a\vec{n}\cdot\nabla U(\vec{r},t) \tag{6}$$

$$U_\infty=c. \tag{7}$$

Where, r is radial distance, t is time, and n means the vertical direction to the interface of fluid. V_n is the normal velocity of the interface. Both a and c are constants, and Eq.(7) means the value of the density far from the interface is a constant. Although the non-linear differential equation (5) is difficulty to be solved, an aggregation with fractal shape can be obtained from a DLA model easily. The fractal dimension D of a DLA cluster can be obtained from the density-density or pair correlation function[6]

$$c(r)=\sum\rho(r)\rho(r+r)\propto r^{D-d} \tag{8}$$

Where, d is Euclidean dimension, $\rho(r)$ denotes the density function of a cluster. When a particle locates position r, $\rho(r)=1$; otherwise $\rho(r)=0$.

In order to save computer time, we present use a modified DLA model[7] on off-lattices to simulate the morphological evolution of finger growth process. Different from the traditional DLA model, a seed was put on the center of an off-lattice. Then a particle at (x, y) is launched at a random site in an annular region with radius between R_{max} and $R_{max}+R_0$. Here R_{max} is the maximum radius of the growing cluster and R_0 is an adjustable parameter that denotes the diffusion radius of particles. The walks of a particle are limited within a circle of radius R_0 until it touches the seed (at beginning) or the aggregate and becomes a part of the cluster. The center of the circle, which is named as "diffusion circle", is determined as following formulae

$$x_0=\left(1-R_0\big/\sqrt{x^2+y^2}\right)\cdot x \tag{9}$$

$$y_0=\left(1-R_0\big/\sqrt{x^2+y^2}\right)\cdot y \tag{10}$$

Thus, some branches of the cluster were kept within the diffusion circle so that these branches might be touched by a walking particle. During the simulation, a particle jumps stochastically at step s within the circle centered on the position (x_0, y_0), and the next position (x', y') will be

$$x'=x+s\cdot\cos\theta$$
$$y'=y+s\cdot\sin\theta$$

Here, θ is a random number at region $[0, 2\pi]$.

The dynamic Eqs.(1)~(3) for viscous fingering are similar to the DLA simulation equations (5)~(7), hence, it is possible to use the DLA model for the computer simulation of viscous fingering.

However, a DLA model is related to the flow of two immiscible fluids with a vanishing interfacial tension. In fact, the surface-tension effects are generally non-negligible in two-fluid-flow phenomena, as illustrated in Eq.(4). In order to simulate experimental situations of viscous fingering with varying conditions, the surface-tension effect must be included in the boundary of the driving fluid and the displaced fluid. A simple method to resolve the problem is to introduce a sticking probability[8].

$$p_s(\kappa) = A\kappa + B \tag{11}$$

This equation responds Eq.(4). Here, both A and B are constants. κ is the local surface curvature of a cluster. If a walker reaches the adjacency of cluster, it will be incorporated into the growing cluster at probability p_s determined by equation (11) and move away at probability $1-p_s$.

During our simulation, R_0, A and B can change for generating different aggregates. Fig.1 illustrated simulation results for R_0=10. The cluster will be more open and dendritic with increasing value of B when A is fixed, as shown in the Fig.1a and Fig.1b, thus less fractal dimension is obtained. Same characters were observed with decreasing value of A when B is fixed, as shown in Fig.1c and Fig.1d. In our simulation for Fig.1, the number N of particles was set to be 20000.

Increasing A or decreasing B means a stronger effect of the surface curvature and the randomness of the growth interface becomes less significant, so a well-developed regular cluster can be obtained.

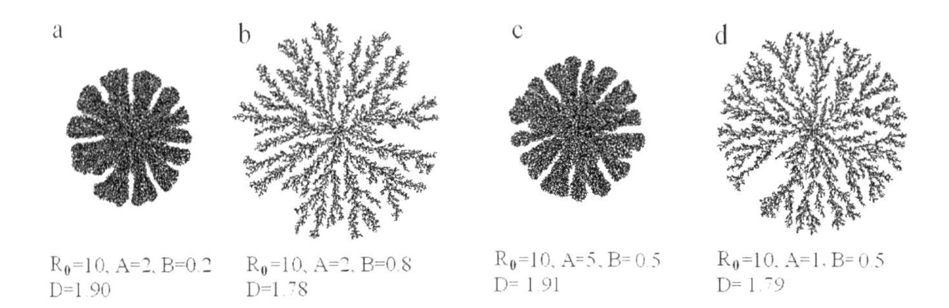

Fig. 1 Various shapes and dimension of cluster for R_0=10

Another adjustable parameter in our simulation is the diffusion radius R_0. The size of the diffusion circle (or the value of radius R_0) also controls the overall shape and the fractal dimension of the aggregates. The small value of R_0 means that a particle incorporates into the cluster at the vicinity of the position where it is launched. The limited diffusion path of the particle leads to an aggregate with a random form and high fractal dimension. The large R_0 permits a walker jumping in a wide region so that the tips of most advanced branches capture the incoming diffusing walker most effectively. Thus, small fluctuations due to the random walks are enhanced and a more open structure with low fractal dimension is formed. Therefore, the change in the value of the parameter R_0 leads the cluster from a dendritic structure to an open pattern. Fig.2 shown the results of the same A and B as Fig.1(R_0=10) but for less R_0 (here R_0=5 in Fig.2).

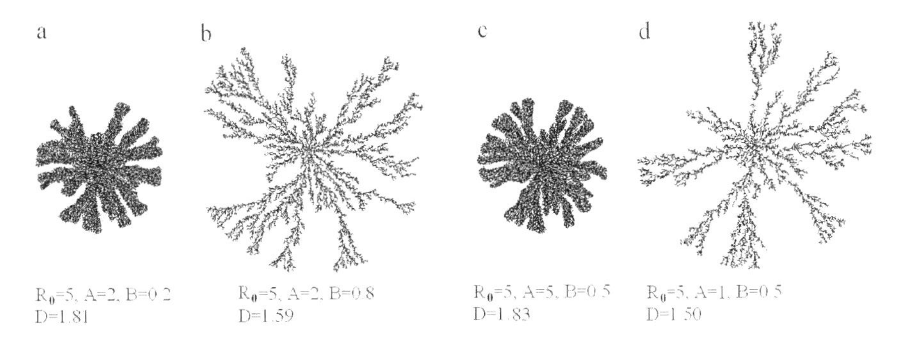

Fig. 2 Various shapes and dimension of cluster for R_0=5

3 Computer Simulation for an Experimental Viscous Fingering

Since it is possible to control the growing procedures during simulation by the parameters R_0, A and B, a series of aggregates with different fractal dimensions and various morphologies can be obtained by altering the values of R_0, A and B in order to match experimental results. Figure 3 gave the simulation results for an experimental viscous fingering while a low viscosity fluid displacing a high viscosity fluid immiscibly obtained by Fanchi and R.L. Christiansen[9] using a Hele-Shaw cell. Fig.3a is the experimental result, and Fig.3b is the present simulation results, they are in good agreement.

In our present computer simulation, we set A=4, B=0.4 and R_0=10. The number of particles is N=30000 and fractal dimension D=1.89, which was calculated from Eq.(8).

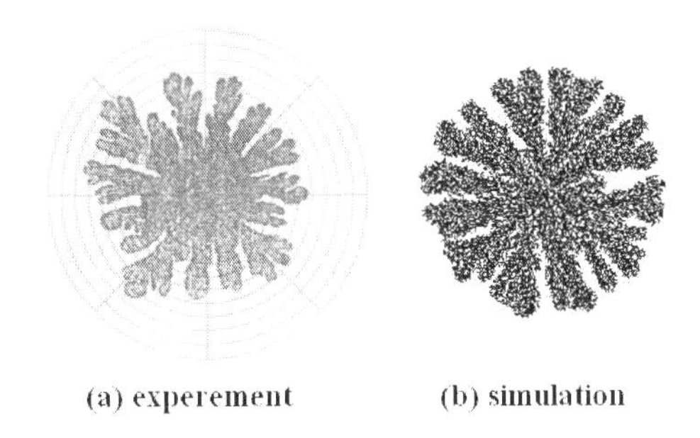

(a) experement (b) simulation

Fig. 3 (a) Experimental result[9] and (b) the present simulation for viscous fingering

4 Conclusions

In the present modified DLA model with sticking probability, by setting values of R_0, A and B, the overall shape of a cluster can be arbitrarily changed from a ramified configuration with a low fractal dimension to a regular structure with a high fractal dimension. The fractal dimension of a cluster decreases and the corresponding morphologies of finger growth vary from a fleshy pattern with great fractal dimension to a skeletal pattern with low fractal dimension for decreasing A or increasing B.

By the present computer simulation, the growing process of aggregations can be controlled efficiently in order to simulate the evolution of experimental viscous fingering. It is convenient to generate different fractal clusters by controlling the parameters in the present algorithm to simulate the morphological evolution of viscous fingering from skeletal patterns to fleshy patterns for various experimental conditions.

The fractal dimension of viscous fingering is difficult to determine in experiments, but it can be obtained easily during the simulation procedures. The computer simulation provides an effective way to describe the behaviors of fluid dynamics for viscous fingering.

References

1. Bensimon, D., Kadanoff, L.P., Liang, S., Shraiman, B.I., Tang, C.: Viscous flows in two dimensions. Rev. Mod. Phys. 58, 977–999 (1986)
2. Peter, E.J., Cavalero, S.R.: The fractal nature of viscous fingering in porous media. In: The 65th Ann. Tech. Conf. & Ex. of Soc. Pet. Eng., SPE, September 23-26, vol. 20491, pp. 225–227. Held in New Orleans, LA, U.S (1990)
3. Hele-Shaw, H.S.: The flow of water. Nature 58, 34–36 (1898)
4. Saffman, P.G.: Viscous fingering in Hele-Shaw cells. J. Fluid Mech. 173, 73–94 (1986)
5. Damme, H., Obrecht, F., Levitz, P., Gatineau, L., Laroche, C.: Fractal viscous fingering in clay slurries. Natural 320, 731–733 (1986)
6. Witten, T.A., Sander, L.M.: Diffusion-limited aggregation: a kinetic critical phenomenon. Phys. Rev. Lett. 47, 1400–1403 (1981)
7. Zhang, J.H., Liu, Z.H.: Study of the relationship between fractal dimension and viscosity ratio for viscous fingering with a modified DLA model. J. Pet. Sci. & Eng. 21, 123–128 (1998)
8. Vicsek, T.: Pattern formation in diffusion-limited-aggregation. Phys. Rev. Lett. 53, 2281–2284 (1984)
9. Fanchi, J.R., Christiansen, R.L.: Applicability of fractals to the description of viscous fingering. In: The 64th Ann. Tech. Conf. & Ex. of Soc. Pet. Eng., SPE, October 8-11, vol. 19782, pp. 105–120. Held in San Antonio, TX, U.S (1989)

Numerical Simulation of 3D Free Overfall Flows

Jyh-Haw Tang and Ming-Kuan Sun

Abstract. 3D Free overfall flows by the two-phase least-squares finite element method (LSFEM) have been presented. The dynamic and kinematic boundary conditions of free surface are described in an Eulerian coordinate system. The governing 3D Navier-Stokes equations in association with the color function are solved by element-by-element scheme. In this simulation, the volume of fluid (VOF) method and continuous stress force models are applied for the determination of the interface between liquid and gas. The free surface is determined by the distribution of the color function. The formation of the dam-break flow is carefully examined for model verification by the experiment done in our hydraulic laboratory and the quantitative comparison is shown in good agreement. For the free overfall flows, the regression of the free surface distribution is compared with previous studies. Finally, the 3D free overfall on a expanded channel is presented for demonstration that the LSFEM can effectively simulate the 3D flow fields.

Keywords: free overfall, dam break, VOF, numerical simulation, least-squares finite element method.

1 Introduction

The simulation of two-phase flows problems is one of the most active topics in computational hydraulics, and computational fluid mechanics. It covers a wide range of two-phase flows problems in engineering applications, including wave flows around a ship or structure and free surface flows in open channel. Many problems in fluid flow applications require the consideration of incompressible immiscible fluids and, consequently, the treatment of the free surface between these two phases. Making use of numerical simulations and analyses to some suitable simplified models, scientists can get a lot of significant information for various complicated two-phase flows phenomena. In recent years, computational tools are being

Jyh-Haw Tang · Ming-Kuan Sun
Chung-Yuan Christian University Dept. of Bioenvironmental Engineering No. 200, Chung-Bei Rd. Chung-Li, Taiwan, R.O.C.
e-mail: jyhhaw@yahoo.com

Z. Qian et al. (Eds.): Recent Advances in CSIE 2011, LNEE 125, pp. 665–671.
springerlink.com

developed for the study of such phenomena, especially, many wonderful numerical simulated results for 2D or 3D discontinuous problems are continually reported.

For researches of two-phase flows problems, Harlow and Welch [1] developed the popular Marker and Cell method (MAC) which uses massless marker particles to explicitly represent the flow domain and thus its free surface. Daly [2] extended this method to two-phase flows. Various modifications of the original MAC scheme were developed by Daly [3], Amsden and Harlow [4] over the years and have further improved this approach. The volume of fluid (VOF) method from Hirt and Nichols [5] modifies the MAC method by replacing the discrete marker particles with a continuous field variable – the color function or the level set function. This function assigns a unique constant (color) to each fluid. At fluid interfaces this color function has a sharp gradient. The continuum surface force (CSF) model interprets surface tension as a continuous, three-dimensional body force across an interface, rather than as a boundary condition on the interface (Brackbill et al [6]). The computer implementation of the CSF model is therefore relatively simple compared with other approaches. Recently, a numerical method for the simulation of 3D incompressible two-phase flows is presented by Marchandise and Remacle [7]. The proposed algorithm combines an implicit pressure stabilized finite element method for the solution of incompressible two-phase flow problems with a level set method implemented with a quadrature-free discontinuous Galerkin method. By doing so, we can compute the discontinuous integrals without neither introducing an interface thickness nor reinitializing the level set.

In this study, the free surface is determined by the distribution of the color function. The numerical calculation procedures based on the least squares finite element method (LSFEM) are employed to study the dam-break flow and the free overfall problems. The formation of the model is carefully examined; and the quantitative comparisons of the 3D numerical simulations with experimental measurements and previous studies are verified in good agreement. It is shown in this study that the LSFEM can effectively simulate the 3D flow fields.

2 Mathematical Model

In this study, with the assumptions that the fluid is incompressible both air and water and governed by the unified three-dimensional Navier-Stokes equations. The governing equations are rearranged into the velocity-pressure-vorticity form in rectangular coordinates and expressed as follows:

Continuity equation:

$$\frac{\partial u_x}{\partial x} + \frac{\partial u_y}{\partial y} + \frac{\partial u_z}{\partial y} = 0 \tag{1}$$

Momentum equation:

$$\begin{aligned} &\rho\frac{\partial u_x}{\partial t} + \rho u_x\frac{\partial u_y}{\partial x} + \rho u_y\frac{\partial u_z}{\partial y} + \rho u_z\frac{\partial u_1}{\partial z} + \frac{\partial p}{\partial x} + \mu\left(\frac{\partial \omega_z}{\partial y} - \frac{\partial \omega_y}{\partial z}\right) \\ &-2\frac{\partial \mu}{\partial x}\frac{\partial u_x}{\partial x} - \frac{\partial \mu}{\partial y}\left(\frac{\partial u_y}{\partial x} + \frac{\partial u_x}{\partial y}\right) - \frac{\partial \mu}{\partial z}\left(\frac{\partial u_z}{\partial x} + \frac{\partial u_x}{\partial z}\right) = \rho f_x \end{aligned} \tag{2}$$

$$\rho\frac{\partial u_y}{\partial t}+\rho u_x\frac{\partial u_y}{\partial x}+\rho u_y\frac{\partial u_y}{\partial y}+\rho u_z\frac{\partial u_y}{\partial z}+\frac{\partial p}{\partial y}+\mu\left(\frac{\partial \omega_x}{\partial z}-\frac{\partial \omega_z}{\partial x}\right)$$
$$-2\frac{\partial \mu}{\partial y}\frac{\partial u_y}{\partial y}-\frac{\partial \mu}{\partial x}\left(\frac{\partial u_x}{\partial y}+\frac{\partial u_y}{\partial x}\right)-\frac{\partial \mu}{\partial z}\left(\frac{\partial u_z}{\partial y}+\frac{\partial u_y}{\partial z}\right)=\rho f_y \tag{3}$$

$$\rho\frac{\partial u_z}{\partial t}+\rho u_x\frac{\partial u_z}{\partial x}+\rho u_y\frac{\partial u_z}{\partial y}+\rho u_z\frac{\partial u_z}{\partial z}+\frac{\partial p}{\partial z}+\mu\left(\frac{\partial \omega_y}{\partial x}+\frac{\partial \omega_x}{\partial y}\right)$$
$$-2\frac{\partial \mu}{\partial z}\frac{\partial u_z}{\partial z}-\frac{\partial \mu}{\partial x}\left(\frac{\partial u_x}{\partial z}+\frac{\partial u_z}{\partial x}\right)-\frac{\partial \mu}{\partial y}\left(\frac{\partial u_z}{\partial y}+\frac{\partial u_y}{\partial z}\right)=\rho f_z \tag{4}$$

Vorticity equation:

$$\frac{\partial \omega_x}{\partial x}+\frac{\partial \omega_y}{\partial y}+\frac{\partial \omega_z}{\partial z}=0 \tag{5}$$

where p is the pressure, u_x, u_y, u_z are the velocities, ω_x, ω_y, ω_z are the vorticities in the x, y and z direction respectively; ρ is the density, μ is the dynamic viscosity, and f_x, f_y, f_z is the body forces. The surface tension effect is interpreted as a continuous body force spread across the transition region [8]: According to the volume of fluid (VOF) method, we need another governing equation for identifying the value of different phase of fluid, which is expressed as follows:

$$\frac{\partial C}{\partial t}+(\vec{u}\cdot\nabla)C=0 \tag{6}$$

where C is the colour function, usually range from value 0 for air to the value of 1 for water.

In the slow flow problem such as the liquid drop or the bubble motion, the surface tension effect is important. However, the surface tension can be neglected in the simulation of dam-break and free overfall flows for its rapid motion character. The difficulty in imposing the interface condition caused by surface tension was alleviated by applying the CSF model [6]. The basic idea of the CSF model is to regard the interface between two fluids as a transition region with a finite thickness, instead of a zero-thickness membrane. By using the CSF model, the interface condition no longer needs to be explicitly imposed, as it is already implied in the momentum equations [9].

$$f=\frac{\sigma}{[C]}k\nabla C \tag{7}$$

In the above formulation, $[C]$ denotes the jump of C across the interface, σ is the surface tension coefficient, the curvature k is calculated from: $k=-(\nabla\cdot\vec{n})$ where $\vec{n}$ is the unit normal to the surface.

Fluid properties such as the viscosity and the density are assumed to be the linear distribution as C.

The equation (1)-(5) used in LSFEM is expressed as follows:

$$A_x \frac{\partial W}{\partial X} + A_y \frac{\partial W}{\partial Y} + A_z \frac{\partial W}{\partial Z} + A_0 W = F \qquad (8)$$

where the expression of each terms can be found in the reference [8,9]. The governing equations can be discretized with a fully implicit scheme in time and be linearized for the nonlinear convection terms. The time dependent problem must be solved by the iteration procedure for each time step to obtain the converged solution. The general procedure for solving the equation (8) by employing the element-by-element method is referred to [8,9].

3 Results and Discussion

The numerical results of the 3D free surface flow problems will be discussed in this section. The numerical model is calibrated by the dam break problem first to demonstrate the ability for the simulation of 3D flow. Then, the simulation of free overfall flow problems will be discussed.

3.1 Model Verification by Dam Break Problem

We computed the well-known collapsing water column benchmark problem and it is solved as a two-fluid problem involving water and air. To validate our numerical simulation we also compared our simulation with experimental data obtained from the measurement of the hydraulic laboratory in Chung-Yuan Christian University, see Fig. 1. In our numerical simulation, we rebuild the experimental setup in the computer. The computational domain is 1.5m × 0.3m × 0.3m and with a uniform grid size,0.05. Initially, water occupies a 0.25m × 0.3m × 0.3m column at the lower left corner. Zero surface tension and slippery walls are assumed. As boundary conditions we used outflow boundaries for the wall on the right side, on all other walls we used classical slip boundary conditions. On the top and right side zero pressure is imposed. The nondimensionalized gravitational acceleration is taken to be unity, the dimensionless time-step is $\Delta t = 0.01$. The dynamic viscosity is set at $1.8\times10^{-5}\, kg/m\cdot s$ and $1\times10^{-3}\, kg/m\cdot s$ for water and air. The densities for water and air are 1 and 0.001.

The snapshot of experimental measurement for the free surface distribution are shown in Fig. 1. This is also clear from the graphs depicted in Fig 2, we can observe the time evolution of the collapsing column, which is in good agreement with the experimental data. The water column height and water front position with respect to time are depicted in Fig. 3 (a) - (b). From the graph, we can observe that the simulated results from our numerical scheme are in very good agreement with the experiment.

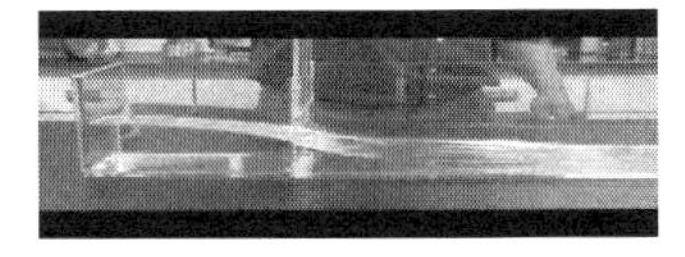

Fig. 1 The experimental set-up for the measurement of the free surface in a dam-break flow.

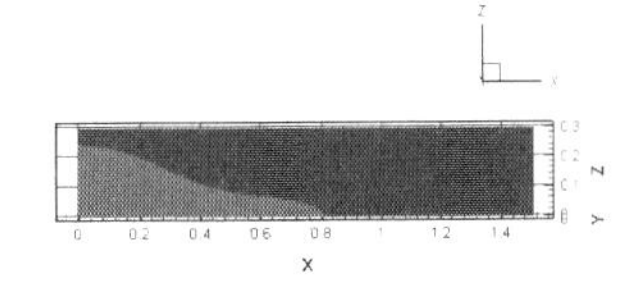

Fig. 2 Transient free surfaces of the collapsing water column problem.

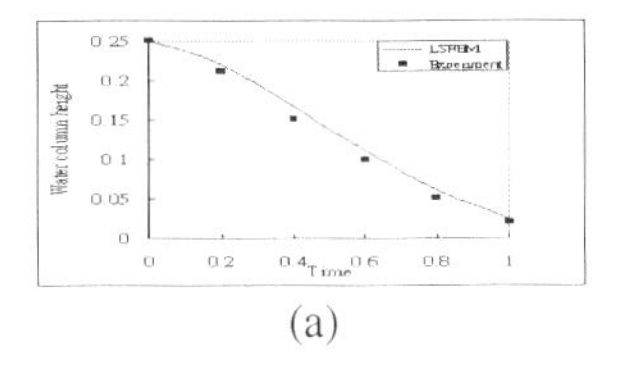

(a)

(b)

Fig. 3 Numerical simulation results of (a) water column height and (b) surge front positions compared with experimental results

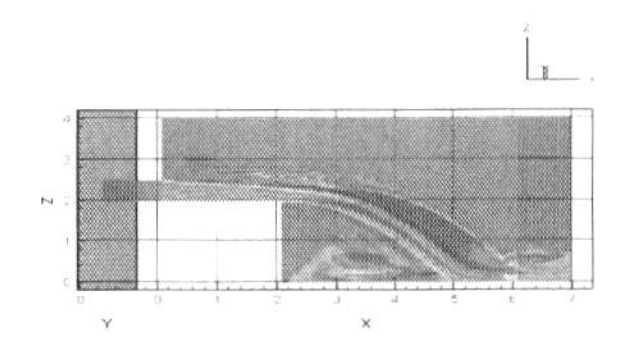

Fig. 4 Computational result of 2D free overfall.

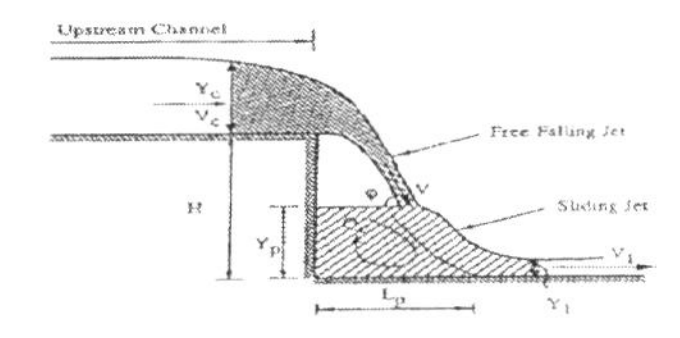

Fig. 5 A typical profile of free falling jet[13]

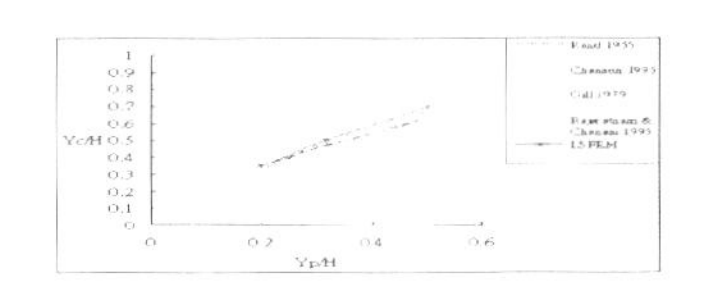

Fig. 6 Dimensionless critical depth compared with still water height.

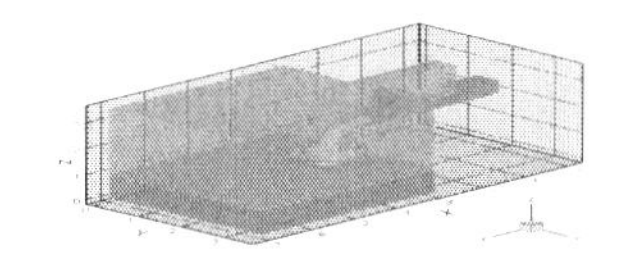

Fig. 7 The snapshot of water flowing down at T = 5.

3.2 2D Free Overfal

A typical 2D free overfall phenomenon is studied by our 3D numerical code to find the profile of the overfall water jet, the simulated result of a certain snapshot is presented in Fig. 4. For comparison, in Fig. 5, there shows a typical figure for some definitions of terminology. In Fig. 6, the dimensionless critical depth compared with still water height from many different experimental results are in good agreement with our results from LSFEM.

3.3 3D Free Overfall

To show the ability for the simulation of 3D free surface flows, we consider the free overfall flow in water flume moving into a sudden expanded vertical drop pool. The computational domain with dimensions (length × width × depth) of the water flume and vertical drop pool are 3 m×0.5 m×0.5 m and 4 m×3.5 m×3.5 m, respectively. The elevation of water flume is 2 m higher than the drop pool. Initial depth in the pool is 0.8 m. The domain is discretized into 50750 elements and 55416 nodes for uniform grid △X=△Y=△Z=0.1. The nondimensionalized gravitational acceleration is taken to be unity, inlet velocity is 1 m/s and the dimensionless time-step is $\Delta t = 0.1$,the dynamic viscosity is set at $1.8\times10^{-5}\ kg/m\cdot s$ and $1\times10^{-3}\ kg/m\cdot s$ for water and air. The densities for water and air are 1 and 0.001. Surface tension is assumed to be zero in this flow problem. For boundary conditions, we used closed boundary at the outflow section, and slippery walls are assumed for the boundary conditions, zero pressure boundaries for the wall on the top.

Some numerical simulation results from LSFEM for the free surface flows distribution for different time evolutions are shown in Fig. 7 to Fig. 9. The 3D free overfall phenomena for water flow hit the drop pool can be clearly found through the figures. It is clearly that the flow distribution show 3D phenomena and will be useful for future study for the engineering applications.

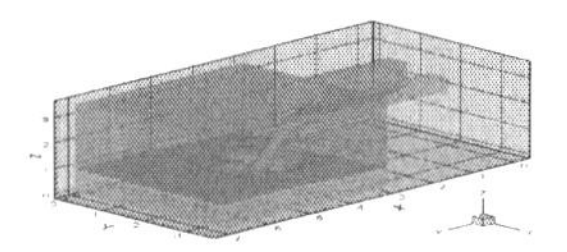

Fig. 8 The snapshot of water flowing down at T = 6.

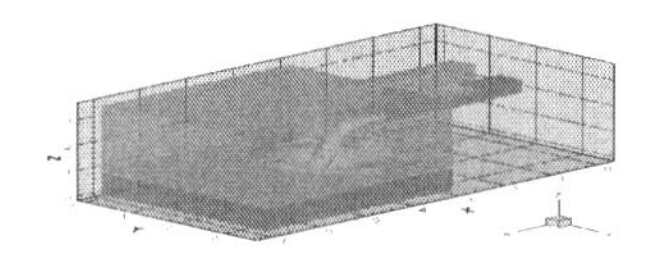

Fig. 9 The snapshot of water flowing down at T = 7.

4 Conclusions

In this paper, it is believed that this is the first time the LSFEM be applied to investigate the 3D free overfall flow phenomena in an open channel. The incompressible Navier-Stokes equation is solved for two-phase flow problems inthree dimensions. The free surface between the two fluid phases is tracked with a color function approach. Here, the interface conditions are implicitly incorporated into the momentum equations by the CSF method. The results of our numerical simulations clearly show the anticipated convergence behavior and good agreement with the experimental data. This study can present some dynamic results such as the forces and flow fields for the practical application in the design of a hydraulic structure downstream of the free overfall.

Acknownledgment. The authors would like to express appreciation of the support of the National Science Council for grant NSC-98-2221-E-033-040.

References

[1] Harlow, F.H., Welch, J.E.: Numerical calculation of time-dependent viscous incompressible flow of fluid with a free surface. Phys. Fluids 8, 2182–2189 (1965)
[2] Daly, B.J.: Numerical study of two fluid Rayleigh-Taylor instability. Phys. Fluids 10, 297–307 (1967)
[3] Dal, B.J.: A technique for including surface tension effects in hydrodynamic calculation. Comp. Phys. 4, 97–117 (1969)
[4] Amsdem, A.A., Harlow, F.H.: SMAC method. A Numerical Technique For calculating Interal Fluid Flows. In: Los Alamos Scientific Laboratory, University of California, Los Alamos (1970)
[5] Hirt, C.W., Nichols, B.D.: Volume of Fluid (VOF) Method for the Dynamics of Free Boundaries. Comp. Phys., 39–201 (1981)
[6] Brackbill, J.U., Kothe, D.B., Zemach, C.: A continuum method for modelings surface tension. Comput. Phys. 100, 335–354 (1992)
[7] Marchandise, E., Remacle, J.F.: A stabilized finite element method using a continuous level set approach for solving two phase incompressible flows. Comput. Phys. 219, 780–800 (2006)
[8] Jiang, B.N.: The Least-Squares Finite Element Method, Theory and Applications in Computational Fluid Dynamics and Electromagnetics. Springer, Heidelberg (1998)
[9] Tang, J.H., Sun, M.K.: Application of the least-squares finite element method on the simulations of 3D incompressible free surface flows. In: WCCM/APCOM 2010, IOP Conf. Materials Science and Engineering (2010), doi:10.1088/1757-899x /10/1/012025
[10] Rand, W.: Flow geometry at straight drop spillways. Hydraulic Engineering, ASCE 81, 1–13 (1955)
[11] Gill, M.A.: Hydraulics of Rectangular Vertical Drop Structures. IAHR 17(4), 289–302 (1979)
[12] Chanson, H.: Hydraulic design of stepped cascades, channels, weirs and spillways. Pergamum, Oxford (1995)
[13] Rajaratnam, N., Chamani, M.R.: Energy loss at drops. Journal of Hydraulic Research, IAHR 33(3), 373–384 (1995)

Simulation Research on Effect of Coach Top-Window Opening on Internal and External Flow

Xingjun Hu, Fengtao Ren, Peng Guo, and Yang An

Abstract. For three limit conditions of the top-window mode of the coach: All three top-windows are closed, all windward opened and all leeward opened, the numerical simulation is done, and the results of all the three different conditions that we get show the preliminary data of the internal and external flow for the coach, and the results are analyzed. Finally the effect of different ways of top-windows opening on the internal and external flow of the coach is discussed.

Keywords: Automobile aerodynamics, CFD, Numerical simulation, Coach, Internal and external flow.

1 Introduction

Because of the van modeling characteristics, the coach has a more serious problem of aerodynamic drag than cars. Due to the top-window opening, the internal and external flow field are coupled. Therefore the research of coach flow field characteristics and aerodynamic characteristics become more difficult. Scholars domestic and overseas have done lots of research both on external flow field and drag reduction accessories. However, there are seldom scholars that have done research on the coupling of internal and external flow field when the top-windows are opened. The significance of the research is to make clear the difference between the coupling of the internal and external flow field and the external flow field. Then we can obtain the detailed data of the flow field when the top-window are opened, and provide reference for evaluating thermal comfort. In this paper, we will do the numerical simulation to explore the influence of the flow field when the top-windows are opened with FLUENT, and try to simulate the true flow field condition.

Xingjun Hu · Fengtao Ren · Peng Guo · Yang An
State Key Laboratory of Automobile Dynamic Simulation,
Jilin University, Changchun 130022
e-mail: hxj@jlu.edu.cn, {tao-ss,ppnfs}@163.com, anpremier@yahoo.com

Z. Qian et al. (Eds.): Recent Advances in CSIE 2011, LNEE 125, pp. 673–679.
springerlink.com

2 Control Equation

Normally, the highest speed of the coach is less than 400km/h, namely less than one third of the velocity of sound. Therefore the flow on the surface is steady incomepress-ible flow, and the physics parameters of the air medium are constant. Considering about the separation phenomenon caused by the complicated shape of the coach, we take this flow as the turbulent flow. And the control equations are:

Continuum equation:

$$\nabla \bullet V = 0 \tag{1}$$

Momentum equation:

$$\frac{\partial V}{\partial t} + (V \bullet \nabla)V = -\frac{\nabla p}{\rho} + \frac{\mu}{\rho}\nabla^2 V \tag{2}$$

In the equations, V is the velocity; p is the pressure, $\rho$is the density of the air, and equals to 1.225kg/ m^3, μ is the air dynamic viscosity equals to $1.81\text{x}10^{-5}$kg/m·s.

3 Geometry and Grid Condition

A coach usually has three top-windows. There are three modes when a top-window opens, the windward mode, the leeward mode, the parallel mode. There are totally 28 modes in total. In this paper, we will choose 3 special modes, scheme A, the windward mode, scheme B, the leeward mode, scheme C the parallel mode.

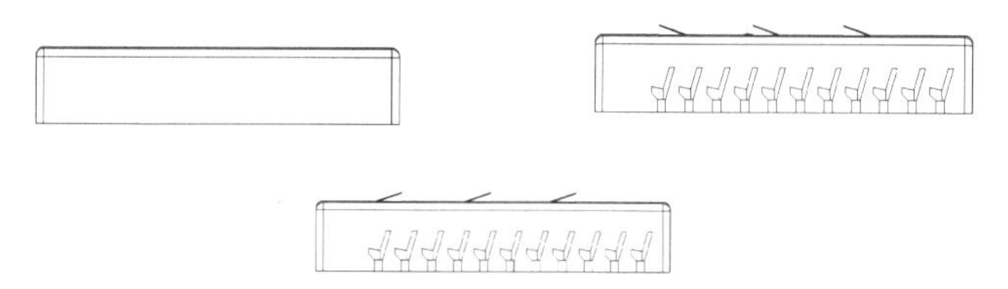

Fig. 1 The different geometry scheme

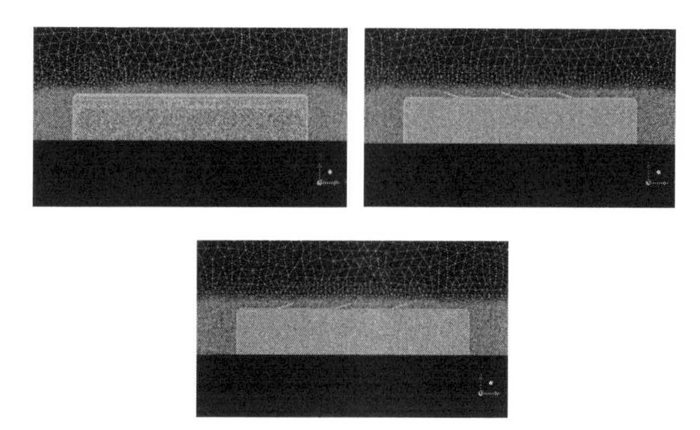

Fig. 2 Grid meshing conditions

We built the model of scheme A, scheme B and scheme C in CATIA. In order to analysis, we simplified the model in some extent. This article focuses on the aerodynamic characteristics of the model influenced by the opening of the top-windows, so we only consider the front, the roof, the back of the coach, and ignore the bottom and the wheels. Scheme A ignores the occupant space and the internal structure. Scheme B and C take the internal structure of the coach into consideration. The length of the coach is 12616mm, and the width is 1210mm, the height is 2357mm. As the structure of the coach is symmetrical, so only half of the model is simulated, and the calculation area is set in this way: the length in front of the coach is two times of the model length and the width of the calculation area is 3.5 times of the width of the coach, and the height is 5 times of the height of the coach. After the establishment of three-dimensional model, the solid model data is exported in the MODEL form.

In ANSYS ICEM CFD, we import the model in MODEL file form, and separate the space in tetrahedral grids, and the boundary layer is prism which is got through the extrusion of the tetrahedral.

4 Turbulence Model and Boundary Condition

This article uses the $\kappa-\omega$ as the turbulence model ,and chooses the second order upwind difference scheme as the spatial dispersion method, and employs the SIMPLEC pressure correction method to do the interaction. The flow field inlet boundary condition is velocity inlet, and the speed is 20m/s, outlet is pressure outlet. The top and both sides of the flow field are all set to symmetry plane, and the bottom is the moving ground with the relative speed of 20m/s, as for the surface of the main body we choose the wall boundary condition.

5 Analysis of the Results

5.1 The Pressure Distribution Result without Consideration of the Internal Flow

Figure 3 is longitudinal symmetrical plane pressure distribution without the conside-ration of the internal flow. The air is hampered in the front of the coach. As the velo-city of the airflow falls off, the pressure rises up, an obvious positive pressure area forms in front of the coach. The airflow separates in the front of the roof and in the rear of the roof. As the velocity of the airflow rises up, two negative pressure area forms in two places .There is a positive pressure area a certain distance away from the rear of the coach. The pressure on the top of the coach is almost equally distributed.

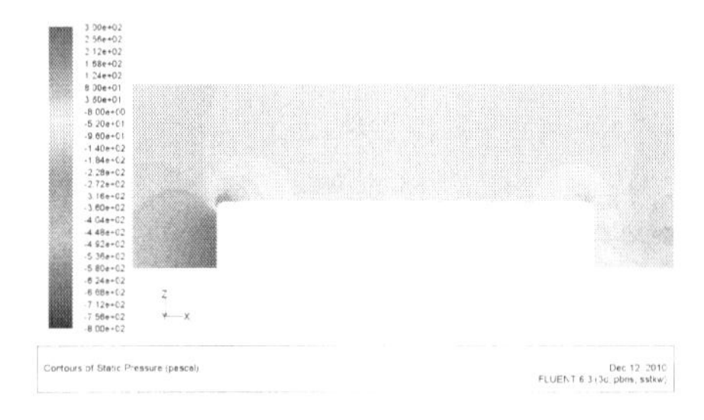

Fig. 3 The longitudinal symmetrical plane pressure distribution without the consideration of the internal flow.

5.2 *Pressure Distribution Result Considering about the Internal Flow*

Figure 4 is the longitudinal symmetrical plane pressure distribution for the windward and leeward mode. Compared with Figure 3, the pressure distribution of windward mode in front of the coach is almost the same. As the top-windows open, the airflow is hampered on the windward side of the top-windows, the velocity of the air flow falls off, a positive pressure area forms on the windward side of the top-windows; the air flow separates at the top-windows, the velocity of the airflow rises up, and the pressure falls off; And, on the leeward side of the first top-window, the negative pressure is connected with the negative pressure in front of the coach and forms a larger negative pressure area. While the negative pressure area of the rear of the coach becomes lager and extends to the negative pressure area on rear of the coach roof, the positive pressure becomes smaller. So the drag coefficient becomes larger.

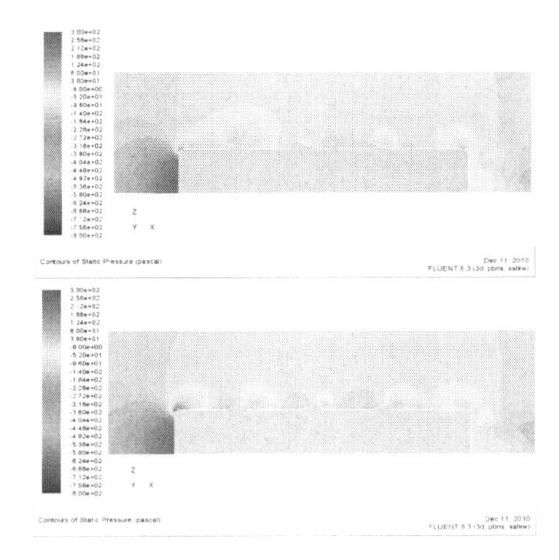

Fig. 4 The longitudinal symmetrical plane pressure distribution for the windward and leeward mode

Due to the top-window opening, the pressure in the van is almost the same as the windward side of top-window, and the pressure inside the van is larger than the outside.

Compared with Figure 3, the pressure distribution of leeward mode in front of the coach is almost the same. As the top-windows open, the positive pressure area becomes small, the two negative pressure area becomes small; and positive pressure and negative pressure one by one pattern occurs. The reason for the occurrence of this pattern is described as follows: when the air flows to the top-window, because of the hamper of the windward side of the top-window, the velocity of the airflow falls off, and the pressure rises up; as the air flow separates at the top-windows, so the velocity of the air flow at the leeward side of the top-window rises up, but the pressure falls off. The positive pressure area becomes small, and the drag coefficient becomes larger.

5.3 *The Velocity Distribution Result without Consideration on the Internal Flow*

Figure 5 shows the longitudinal symmetrical plane velocity distribution when the top-windows are closed, Obviously we can see from the figure that the air flow separates at rear of the coach, and form a speed divergence area.

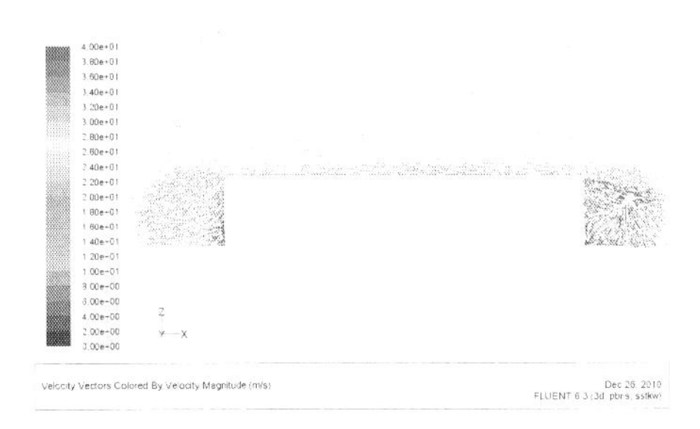

Fig. 5 The longitudinal symmetrical plane velocity distribution when the top-window is closed

5.4 *The Velocity Distribution Result with Consideration on the Internal Flow*

Figure 6 shows the longitudinal symmetrical plane velocity distribution in windward and leeward mode. Compared with figure 5, the velocity of windward mode divergence point appears earlier. Due to the top-windows opening, the airflow separates at the first top-window, and the airflow in the van mainly move in vertical direction, either forward or backward.

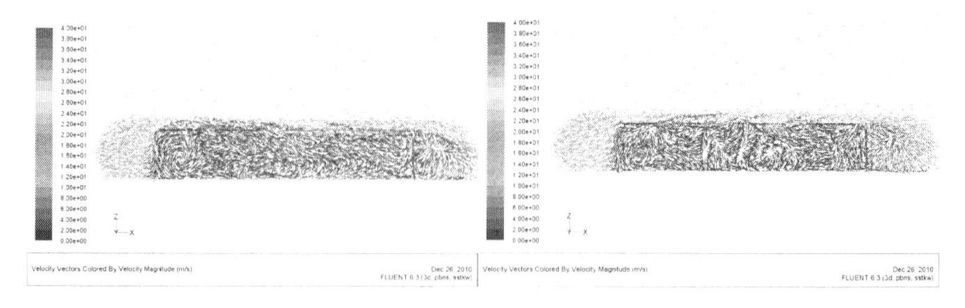

Fig. 6 The longitudinal symmetrical plane velocity distribution in windward and leeward mode

Compared with figure 5, the divergence point of leeward mode appears earlier. And compared with the windward mode, the airflow separates behind the first top-window. The air flow in the van is mainly outward divergence movement with certain points as the center.

The drag coefficients for A, B and C show in the table below.

Table 1 The drag coefficients for A, B and C

Scheme	Drag coefficient
A	0.288
B	0.363
C	0.315

6 Conclusions

When the top-window is closed, there is large positive pressure area in the front of the coach, and two negative pressure areas exist in the front and rear of the coach roof respectively. The quality of the flow field of the other places of the roof of the coach is relatively good. And also there is a positive pressure area not far away from the rear of the coach.

Due to the opening of the top-window, the original flow field of the coach roof is destroyed, and this kind of destruction amplifies the negative pressure area, and reduces the positive pressure area of the coach, which makes the drag coefficient bigger, However the change of the flow field of the coach roof is very small.

The pressure in the coach is bigger while the top-windows are windward opened than they are leeward opened.

The drag coefficient is bigger while the top-window are windward opened than they are leeward opened.

When the top-window are windward opened, large amount of air flows into the coach which make the internal pressure relatively high, and the internal flow loss increase, and so the drag coefficient becomes large.

When the top-window are leeward opened, the air flow doesn't separate until it reaches the rear of the top-window which makes the pressure in the rear of the coach low and so as the pressure in the coach.

Acknowledgement. Thanks for the supports from National Science Foundation of China (50805062), Development Programs in Science and Technology of Jilin Province (20080114, 20096005) and Foundation of State Key Laboratory of Automobile Dynamics Simulation, Jilin University.

References

1. Chen, Z., Chu, X., Liu, X.: The application of CFD simulation in the research of analysising the external flow of the coach. Agricultural Equipment and Vehicles 4, 4 (2010)
2. Ma, F.: Automobile Aerodynamics. China machine press, Beijing (1993)
3. Wang, F.: Computational Fluid Dynamics. Tsinghua university press, Beijing (2004)
4. Jiang, L., Gu, Z.: The application of CFD in the research of Automobile aerodynamics. Hunan university Journal 8 (1997)
5. Guilmineau, E.: Computational study of flow around a simplified car body. Journal of Wind Engineering and Industrial Aerodynamics 96, 1207–1217 (2008)
6. Kim, M.-H.: Numerical Study on the Wake low and Rear-Spoiler Effect of a Commercial Bus Body. SAE Technical Paper Series 2003-01-1253

Ventilation Control and Risk Assessment for Fire Accident of Qing-Cao-Sha Water Tunnel

Baoliang Zhang, Jue Ding, Yi Liu, Qingtao Wang, and PeifenWeng

Abstract. Fire accident in the ultra-long tunnel could bring significant losses to human lives and economic.In this paper, by using computational fluid dynamics method and the theory of ventilation, physical and mathematical models were established, and a numerical study had been conducted to analyze characteristics of smoke flow near the shield tunneling machine in case of fire.Furthermore, temperature distribution of tunnel field in the middle cross-section of construction was discussed, then corresponding ventilation plans to control the heat around the shield and smoke spread were also extensively proposed, which providea theoretical foundation to ensure security, and economy of constructionin long-distance tunnel.

1 Introduction

In order to solve the long-term water supply requirements of Shanghai and improve the urban water supply quality, raw water construction project of Yangtze River has been built in Shanghai which starts at Pudong, crossing the Yangtze River and ending at the Changxingisland. By using shielding method, two tunnels of 5.84m diameter (inner) and 7.2km length were built.As for the characteristics of long-distance and middle cross-section,when fire accident appears during tunnel construction, it is easy for cause poisoning of people, burn and explosion accidents for relatively small cross-section and simple ventilation system existed. Thus, fire ventilation control, includingthe effective ventilation time causesresearches concerns.

Plenty of tunnel fire cases show that smoke and toxic gases are the main reason for causing people death[1-2].When fire accident happens, smokecan spread to certain areas, affected by fire size and construction ventilation. If the ventilation design and layout arereasonable,and correct, effective ventilation solutions can be adopted at the first moment, it is possible to control fire extension and decrease the scope of fire to a minimum. What's more, much more time could be saved for staff escaping.Therefore, optimization and control of ventilation, as well as

Baoliang Zhang · Juc Ding · Yi Liu · Qingtao Wang · PeifenWeng
Shanghai Institute of Applied Mathematics and Mechanics
Shanghai University Shanghai, China

Z. Qian et al. (Eds.): Recent Advances in CSIE 2011, LNEE 125, pp. 681–686.
springerlink.com

emergency ventilation is distinctly important for reducing casualties and property losses for the case of fire.

So, on base of the Reynolds-averaged Navier-Stokes equations differentiated by the finite volume method, the 3-D turbulent flow of tunnel and temperature of smoke under the control of ventilation section were simulated numerically with the RNG k-εturbulent model and by the two-layermodel for the near-wall region, then it was solved by SIMPLE scheme. The numerical simulation should reveal the features of smoke flow and heat diffusion.

(a) Flexible duct (b)Rigid ducton the shield rack

Fig. 1 Maps of Duct

2 Theoretical Models and Numerical Methods

1. Physical model

The fire size has a great influence on the setting of emergency prevention facilities of the tunnel and choosing ventilation rescue programs. Different types of fuels have different time constants of fire growth. The speed of heat release rate follows the laws of T^2, which is shownin the table 1. And T^2 heat release curvescan be seen in figure 2. Moreover, the process of growth curve of heat release rate to smooth is the important stage for staffs evacuating.In our study, the magnitude of the heat power is 8MW, and the heat release rate of the time takes the constant as0.04444.Fire occurred in the vicinity of the tunnel shield was chosen by considering the most dangerous position, and parts of the jet fans wereadded toexhaust smoke.

Table 1 Reference and selection values of heat release rate

Traffic type	Rate of heat release (MW)	a (MW/min^2)	t_{max} (min)	Corresponding to Figure 2 curve
Sedan car	4-10	0.036	10	1
Bus	30	0.036	10	3
2-3 cars	8-12	—	5	2
Many buses	> 30	—	5	4

Small-scale fire could produce a lot of smoke,and the release rate of smoke and amount of reactive products changeduring combustion process.Becausethe process is complex and difficult to describevariations,smoke release rate was postulatedto be steady, and CO mass percentage of smoke was 2%.

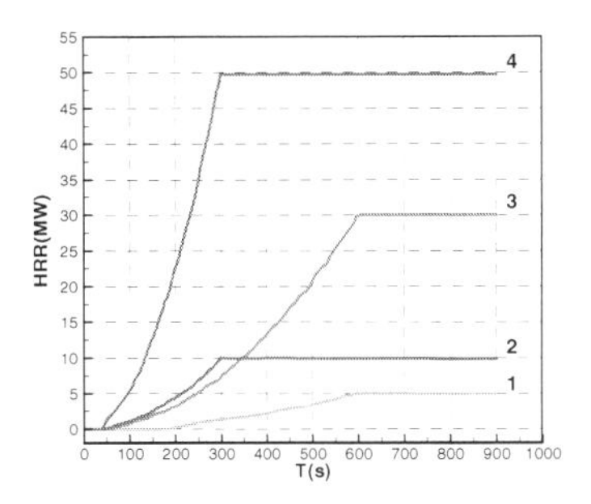

Fig. 2 The Curves of Heat Release RateMathematical model

Three-dimensional mathematical model of air flow in tunnel was established,according to the basic conservation law.Governing equations [3]are as followed.

$$\frac{\partial(\rho\phi)}{\partial t}+\nabla(\rho\phi U)=\nabla(Tgrad\phi)+S \quad (1)$$

Where ϕ is the general variable, which denotes solution variables, such as three velocity components (u,v,w) in x, y, z directions, the turbulent kinetic energy (k), the turbulence dissipation rate (ε) and enthalpy (H) separately. U is velocity vector, and ρ is the density. $\Gamma$is the general diffusion coefficient. S is the source term of gas.

In this paper,RNG $k - \varepsilon$ model[4] is adopted for depicting complex turbulent characteristics. Furthermore, the computed fieldhad been divided into wall flow area and turbulent core area, and two-layer wall model was adopted for near wall region. By finite volume method,above governing equations were solved by SIMPLE algorithm.

3 Numerical Results and Discussion

Smokediffusion characteristics and temperature distributions in the shield region under fire accident with different ventilation plans were numerically investigated for fire-8MW. The streamline distance of computation field is 90m including shield rack. The coordinate origin is set at the middle of surface for rack rear, and fire source is far 40.25m from coordinate origin. In order to increase the ventilation effect, the exhaust equipments are set in the middle of shield (seeing position 2)and near the shield rack(seeing position 1), which were shown as figure 2. The computed results show that the first exhaust way (position 2) has better ventilation effect. Table 2 gives CO average concentrations and temperatures at t = 15min under the energy release 8MW, smoke emissions 0.05kg / (m^3s).

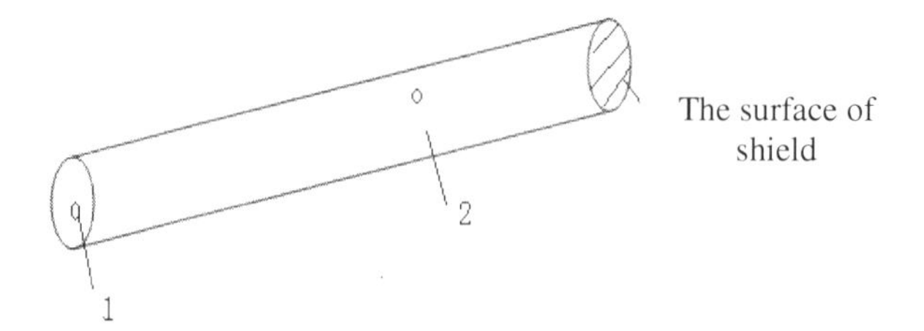

Fig. 3 Computation field

Table 2 Average concentrations and temperatures (t = 15min)

Ventilation quantity (m^3/min)	Exhaust pressure (Pa)	Shield rack (g/m^3)			Downstream of fire ventilation (g/m^3)		
		CO	*smoke*	*temperature (K)*	*CO*	*smoke*	*temperature (K)*
237.78	200	3.51e-2	1.76	397.01	3.63e-2	1.82	396.32
380.00	200	2.98e-2	1.49	386.89	3.89e-2	1.94	397.82
837.02	200	2.05e-2	1.03	333.72	2.95e-2	1.48	349.82
1302.32	200	1.26e-2	0.63	320.01	2.19e-2	1.09	341.05

The shield rackrefers to entire flow for the shield rack. Downstream of fire ventilation is the flow region from the fire source location to the shield rack rear.

Characteristics of smoke diffusion under the same rate of gas release and different ventilation quantity are simulated numerically in this paper. Figure 4 shows distribution of smoke mass concentration near the shield rack when the smoke release rate is 0.05kg / (m3s), ventilation quantities in pipes which is located near the middle of the flow field are237.78 m3/min、380.00 m3/min、837.20 m3/min, 1302.32 m3/min separately, and exhaust pressuresare same (200Pa).

The computed results show that for a smaller ventilation of 237.78 m^3/min, 380.00 m^3/min, the smoke concentrations gradually spread near the source of fire with the development of time. But there is still certain concentration of smoke in the upstream of fire ventilation. For a larger ventilation quantity of 837.20 m^3/min, 1302.32 m^3/min, the backflow of smoke has been blocked. Therefore, there is almost no smoke in the upstream of the fire ventilation, which shows that the larger vertical wind of tunnel cancontrol smoke to side of the fire sourcemore easily. However, when the vertical wind is too large, it not only could bring investment increased in equipment, but also enhances the degree of turbulent flow movement which would cause the smoke layer down to the road earlier and the tunnel section is full of smoke in advance. So, ventilation quantity of 837.20 m^3/min is optimum condition.

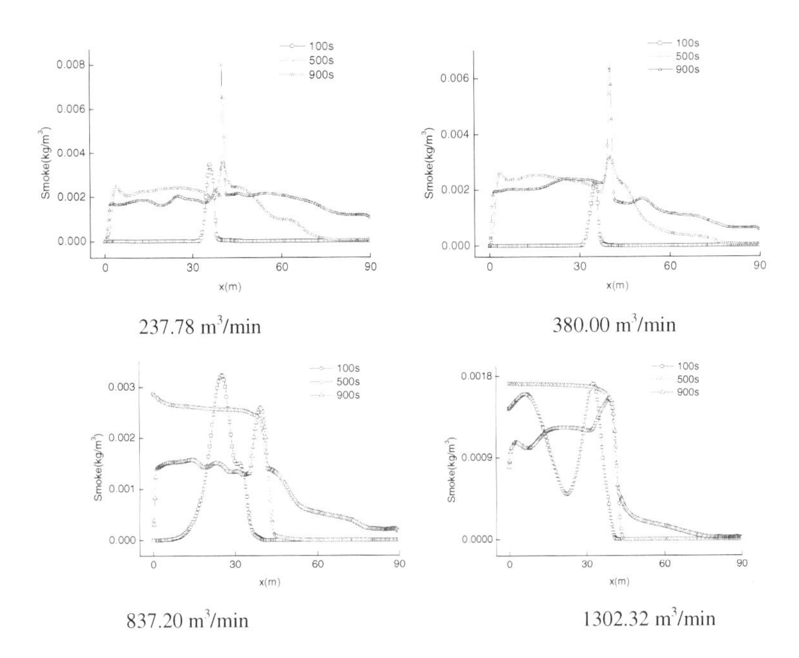

Fig. 4 Mass concentration profiles of smoke along X axisof the flow field (y=0、z=0)

Fig.5 shows temperature distributions near the shield with different ventilation quantities onthe 1.8mof height (from the bottom of tunnel) and smoke release rate of 0.05kg/(m^3s). Commonly, 1.8m is characteristic standing height of people. When the temperature is below80°C, it's possible for staff survival with moving in the smoke layer. On the contrary,when temperature is above80°C, human would be potentially dangerous or even death. So,80°C is used to be the critical temperature of dangerous smoke for the characteristic height of people eye. In this paper, for the smoke release rate 0.05kg/(m^3s), different ventilation quantities of pipes were discussed for 237.78m^3/min, 380.00m^3/min, 837.20m^3/min, 1302.32m^3/min respectively. The exhaust pipe locates in the middle of the flow field, and the exhaust pressure is 200Pa.

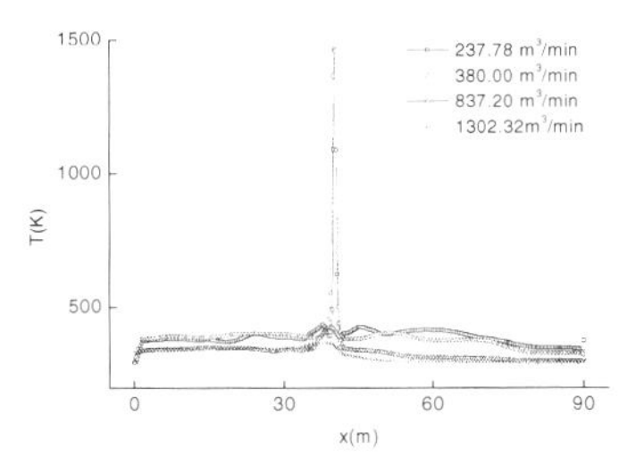

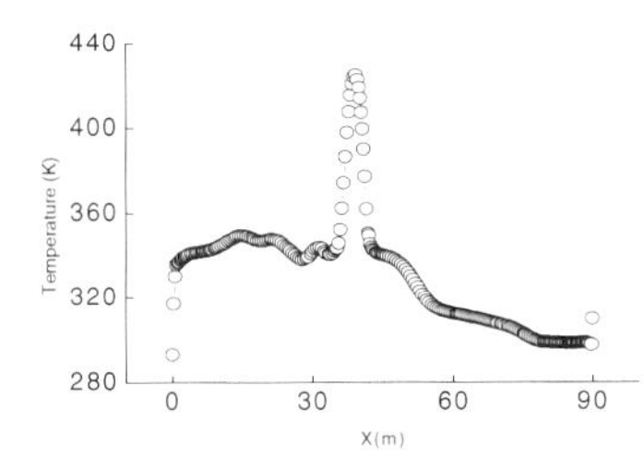

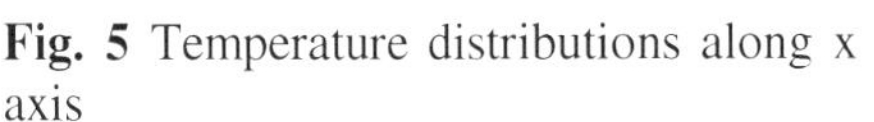

Fig. 5 Temperature distributions along x axis

Fig. 6 Temperature distributions(837.20 m^3/min)

It is shown that the flow temperature of fire source is very high,with closing to 1500K. Moreover, the average temperature of flow field downstream are 396.32、397.82、349.82、341.05K under ventilation conditions mentioned above. With the increasing of ventilation mass, the flow temperature is gradually decreased. But, if ventilation quantity increased largely, temperature gradient is yet slowly decreased. So, ventilation of 837.20 m^3/min could control the temperature of flow field preferably (seeing figure 5、6).The computed results show that temperature near the firesource reaches about 430K. People should pay attention to it, and try their best to move toward the opposite direction of smoke spread.

4 Conclusion

In this paper, on base of the Reynolds-averaged Navier-Stokes equations differentiated by the finite volume method, the 3-D turbulent flow of tunnel and temperature of smokefor fire accident under the control of ventilation were simulated numerically. The numerical simulation reveals the features of smoke flow and heat diffusion.Moreover, the effects of different ventilation plans were discussed, which provide a reference for determining the ventilation control techniques for long-distance and middle-section tunnel during fire accident, and assessment of fire resistance for tunnel structureand arrangement of fire fighting facilities. Computed resultsshow that the diffusioncharacteristics for the harmfulgas of smoke are very similar to thetemperature. Moreover, for gasrelease rate of 0.05kg / (m3s), the bigger ventilation quantities of ventilationtubes837.20 m^3/min,the backflowsofsmoke-have been restrained, which are better to control the temperature of the flow field.The smoke temperature of people standing position reaches about 430K. Consequently, ventilation quantity of 837.20 m^3/minisoptimization for fire accident.

References

[1] Carvel, R.O., Beard, A.N., Jowitt, P.W.: Fire spread between vehicles in tunnels: effects of tunnel size, longitudinal ventilation and vehicle spacing. Fire Technology 4(41), 271–304 (2005)

[2] Hwang, C.C., Edwards, J.C.: The critical ventilation velocity in tunnel fires—a computer simulation. Fire Safety 40, 213–244 (2005)

[3] Tao, W.Q.: Numerical Heat Transfer, pp. 186–237. Xian Jiao Tong University Press, Xi'an (2001)

[4] Yakhot, V., Orszag, S.A.: Renormalization group analysis of turbulence 1, Basic theory. Journal of Scientific Computing 1(1), 39–51 (1986)

A New Method for Reducing Metal Artifacts of Flat Workpiece in Cone-Beam CT

Feng Zhang, Li-zhong Lu, Qing-liang Li, Bin Yan, and Lei Li

Abstract. Metal artifacts of flat workpiece often exist in the horizontal direction in cone-beam Computed Tomography (CT) caused by beam-hardening, photon starvation, etc. In this paper, a novel method for reducing metal artifacts of flat workpiece is presented. In the proposed algorithm, firstly the 3-D image is reconstructed after normal scanning. Secondly, the flat workpiece which was rotated by ninety degrees is scanned again and the new 3-D reconstructed image is obtained. Thirdly, Scale-Invariant Feature Transform (SIFT) algorithm is adopted to match the slice image of the first reconstructed image with the slice image of the second one exactly. Finally, image fusion technique is applied to produce a fused image with less metal artifacts. The experiments have shown that the proposed method improves the reconstructed image quality greatly, and provides more detailed information compared to the current Metal Artifacts Reduction (MAR) algorithms.

1 Introduction

Non-destructive testing (NDT) is widely used in industry inspection, manufacturing, etc. Cone-beam CT with cone X-ray source and X-ray flat detector which has been one of the most effective ways for NDT. However, MAR remains a major challenge in cone-beam CT. Metal artifacts closely depends on the structure of objects, the spectrum of X-ray source and so on, the flat workpiece such as Printed Circuit Board (PCB) whose length or width to thickness ratio is more than 10, which is a special problem in the field of CT.

Metal artifacts of flat workpiece are severe in horizontal direction, this feature is critical. But what caused these artifacts? The beam-hardening effect is the major source. In practice, the X-ray beam contains continuous energy spectrum which

Feng Zhang
National Digital Switching System Engineering & Technology Research Center,
Zhengzhou, China
e-mail: langzicjx@gmail.com

Z. Qian et al. (Eds.): Recent Advances in CSIE 2011, LNEE 125, pp. 687–692.
springerlink.com

does not comply the hypothesis of reconstruction theory. Especially, when the front plane of flat workpiece is perpendicular to that of the X-ray flat detector, much fewer X photons can reach the flat detector, producing corrupted projection data. Besides, The flat workpiece contains many high-density objects, scatter can bring scatter artifacts. As a result, images reconstructed by use of FDK algorithm are polluted seriously, particularly near the metal surfaces. Therefore, it's a valuable work that reducing metal artifact of flat workpiece.

This paper is organized as follows: MAR methods are briefly reviewed in section 2. In section 3, our method will be introduced in detail. The reminder of this paper is composed by experiment results and discussion in section 4 and the conclusion in section 5.

2 Related Work

Many algorithms have been proposed to reduce metal artifacts in the literature. To date, these methods can be classified into two categories: the projection-interpolation method and iterative methods.

The projection-interpolation method consists of four steps, which can be described as follows: Firstly, image (3D) is reconstructed by use of the traditional Filtered Back-Projection (FBP) method such as FDK algorithm. Secondly, metal objects in reconstruction image are segmented using image segmentation algorithms such as k-means cluster, etc. Thirdly, the image without metal objects is projected forward, then fitting missing data in the shadows of metal objects in projection using interpolation methods. Fourthly, image is reconstructed again using FBP algorithm, and the metal objects would be insert into it for compensation. Many researchers have done much work on this family of MAR methods. For example, Kalender [1] investigated removing metal artifacts basing on linear interpolation. Zhao et al.[2] use wavelet to interpolate the miss-ing data rather than linear interpolation. Because Kalender used threshold to segment metal objects that maybe coarse, Wei et al.[3] smooth image before image segmentation and then a second-order polynomial interpolation was adopted to interpolate metal area of projection. However, these methods are efficient when the number of metal objects are few and the complexity of the structures is low.

Another family of MAR algorithms is Iterative-type method. Compared to projection interpolation methods, iterative methods such as Algebraic reconstruction technique (ART)[4], simultaneous ART (SART)[5], etc, can often provide better reconstruction image with less number of projections. Mathematically, image reconstruction based on iterative method is an optimization problem. MAR based on iterative methods has been discussed in detail by Wang et al.[6]. Whereas, the computation complexity of this family of MAR algorithms is high, and it's time consuming, besides it requires a large amount of computer resources, so it's not practical. To combine the strengths of these metal artifacts correction methods, some hybrid approaches have been introduced [7][8].

The purpose of this paper is to develop an effective approach to improve image quality of flat workpiece in cone-beam CT which named double scan algorithm.

3 Double Scan Algorithm

3.1 The Overview of Double Scan Algorithm

Unlike the metal artifacts of other objects, those of flat workpiece locate in the horizontal direction of image, which is very key for depressing it. Hence, double scan algorithm is presented to reduce metal artifacts of flat workpiece based on the analysis above.

Our method consists of four steps. Firstly, scanning the flat workpiece and the image I_0 is reconstructed by use of FDK algorithm. Secondly, scanning the flat workpiece after rotating ninety degrees and image I_1 is reconstructed using FDK method. Thirdly, the interested S_0 of I_0 and S_1 of I_1 are registered using SIFT algorithm. At last, the two images matched exactly are fused, consequently the final image S is obtained. The flowchart of this method is shown in Fig. 1.

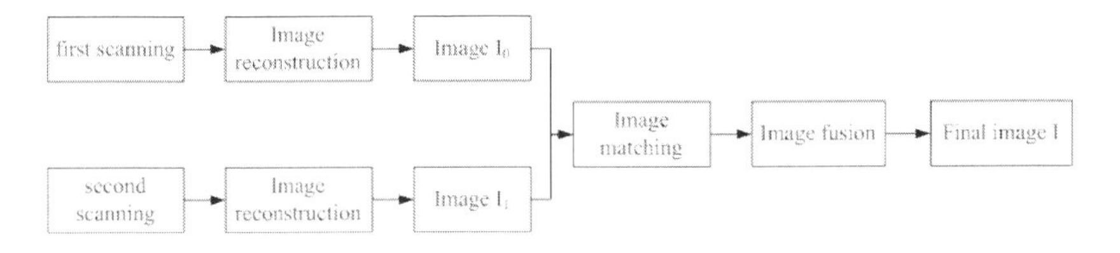

Fig. 1 Flowchart of the double scan algorithm which contains double scan and reconstruction, image matching using SIFT algorithm and image fusion.

3.2 Double Scan and Reconstruction

In this paper, Circular trajectory scanning mode is adopted. The flat workpiece is scanned in cone-beam CT system, as shown in Fig. 2, and the image I_0 and image I_1 are reconstructed respectively by use of FDK algorithm.

3.3 Image Matching

Because the position of flat workpiece in the first scanning is different from that in the second scanning, we can not match I_0 with I_1 directly. Some image matching algorithm must be applied to match I_0 to I_1.

Image matching is an important computer vision problem in order to get a measure of their similarity. SIFT algorithm which proposed by David G.Lowe [9] is a

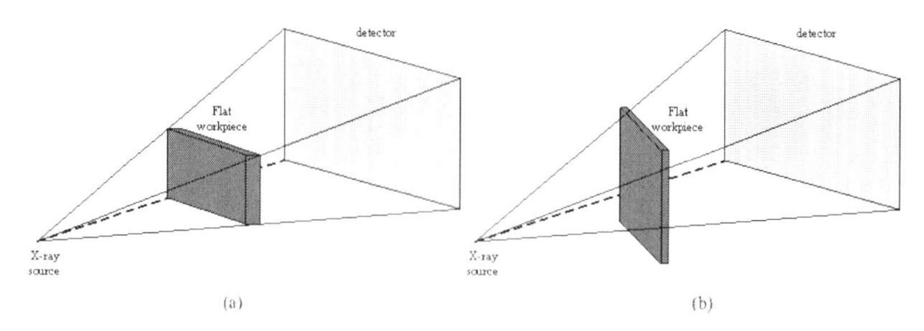

Fig. 2 (a) The first scan.(b) The second scan. The flat workpiece in (a) and (b) is from an angle of ninety degree.

fast and efficient way to compute the scale invariant feature transform. SIFT provides robust matching across affine distortion, addition of noise, etc. In this paper, SIFT algorithm is adopted for image matching.

We extract features of I_0 and I_1 using SIFT algorithm respectively, and then find candidate matching features based on Euclidean distance of these two feature vectors. After eliminating outliers, we will get the transform matrix M, then $S_1^{'}$ is obtained through $S_1^{'} = MS_1$.

3.4 *Image Fusion*

Image fusion technique combines the information of the same scene of multiple images, the goal of that is to generate a composite image which can provide more detailed information.

I_0 and $I_1^{'}$ contain different information, I_0 is polluted in horizontal direction and I_1 is polluted in vertical direction. Therefore, it's important to fuse them and generate a new image. Depending on the stage at which fusion takes place, image fusion can be done at different levels of information representation, namely, pixel level, feature level and decision level [10]. At pixel level fusion, all input images must be registered exactly. Generally speaking, image fusion at pixel level means at the lowest processing level, but always it can provide good fused image.

Our image fusion algorithm is implemented as follows,

$$I(i,j) = \begin{cases} I_0(i,j) & if\ I_0(i,j) \geq I_1(i,j) \\ I_1(i,j) & if\ I_1(i,j) < I_0(i,j) \end{cases} \tag{1}$$

The advantages of this method includes low computation complexity and good image quality.

4 Results and Discussion

This paper uses a cone-beam CT system which was made by our laboratory. The X-ray source (Hawkeye 130, Thales Corporation, France) has a micro-focus continuously adjustable from 10μm to 20μm. The maximum voltage of the X-ray tube is 130KV, with maximum output power 30W. The X-ray flat panel detector (4343F, Thales Corporation, France) with a column CsI scintillator plate has a $430mm \times 430mm$ photodiode area with $148\mu m$ pixel size. The source to object distance (SOD) is $477mm$ and source to detector distance (SDD) is $1265mm$. 360 frames of projections are grabbed with interval of 1 degree for circular trajectory, and FDK algorithm is adopted for three dimension reconstruction, the size of volume is $512 \times 512 \times 512$.

The flat workpiece in our experiment is shown in Fig. 3. After the first scan, I_0 is reconstructed. The 256th slice image of I_0 i.e. S_0 is displayed in Fig. 4(a). After the second scan, image I_1 is reconstructed. The 256th slice image of I_1 i.e. S_1 after rotating ninety degrees is shown in Fig. 4(b).

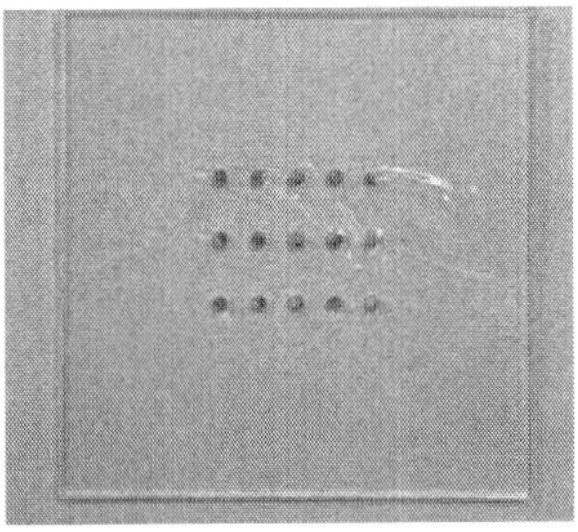

Fig. 3 The flat workpiece with 5×3 metal balls machined from iron is used in our experiment.

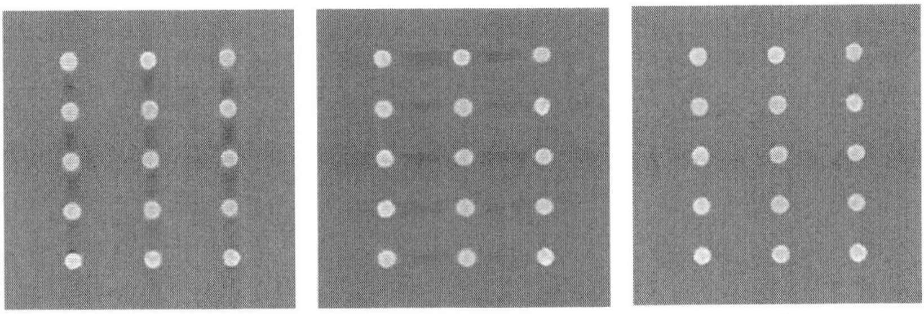

(a) (b) (c)

Fig. 4 (a) S_0. There are much black streak artifacts in horizontal direction, which degrades quality of image. (b) S_1. There are much black streak artifacts in vertical direction, which pollute reconstructed image. (c) S.

SIFT method is applied to match S_0 and S_1, and then the image fusion approach which was proposed in this paper is used to fuse them. Finally, the final image S is shown in Fig. 4(c).

5 Conclusions

In this paper, we propose a novel method which named the double scan method to depress metal artifacts of flat workpiece in cone-beam CT. This approach is a new way to reduce metal artifacts of flat workpiece avoiding interpolation in projection-interpolation algorithms and complicated computation in iterative methods.

The results of experiment demonstrate that this method can modify image quality of flat workpiece greatly. Further more, our method is simple, and hence it is practical.

References

1. Kalender, A., Hebel, R., Ebersberger, J.: Reduction of CT artifacts caused by metallic implants. Radiology 2, 576–577 (1987)
2. Zhao, S., Douglas, D., et al.: X-ray CT metal artifacts reduction using wavelets: An application for imaging total hip prostheses. IEEE Transactions on Medical Imaging 19, 1238–1247 (2000)
3. Wei, J., Chen, L., Sandison, G., Liang, L.: X-ray CT high density artifact suppression in the presence of bones. Physics in Medicine and Biology 49, 5407–5418 (2004)
4. Gordon, R., Bender, R., Herman, G.: Algebraic reconstruction techniques (ART) for the three-dimensional electron microscopy and X-ray photography. Theor. Biol. 29, 471–481 (1970)
5. Andersen, A., Kak, A.: Simultaneous algebraic reconstruction technique (SART): A new implementation of the ART algorithm. Ultrason. Imag. 6, 81–94 (1984)
6. Wang, G., Snyder, D.L., O'Sullivan, J.A., Vannier, M.W.: Iterative Deblurring for CT Metal Artifact Reduction. IEEE Transactions on Medical Imaging 15, 657–664 (1996)
7. Xia, D., Roeske, J., et al.: A hybrid approach to reducing computed tomography metal artifacts in intracavitary brachytherapy. Brachytherapy 4, 18–23 (2005)
8. Lemmens, C., Faul, D., Nuyts, J.: Suppression of Metal Artifacts in CT Using a Reconstruction Procedure That Combines MAP and Projection Completion. IEEE Transactions on Medical Imaging 28, 250–260 (2008)
9. Lowe, D.: Distinctive Image Features from Scale Invariant Keypoints. International Journal of Computer Vision 60, 91–110 (2004)
10. Pohl, C., Genderen, J.: Multisensor image fusion in remote sensing: concepts, methods and applications. International Journal of Remote Sensing 5, 823–854 (1998)

Configure Scheme of Mixed Computer Architecture for FMM Algorithm*

Min Cao and Zhen Cao

Abstract. Along with the scale expansion of high performance computing, accelerators are increasingly viewed as computer coprocessors that can provide significant computational performance at low price. Thus, research of mixed computer architecture is becoming popular. This paper presents a mixed configurable computer architecture which can run fast multipole method (FMM) algorithm of N-Body problem well. Each sub-procedure of FMM algorithm is implemented and tested on GPU, FPGA and CELL. FMM is optimized on the proposed configure scheme through decomposing its task flow. The probable solution for different task flow is also put forward. The conclusion is significant to the research on the mixed computer architecture of high performance computing.

1 Introduction

In the so-called N-Body problem, the system is described by a set of N particles, and the dynamics of the system is the result of the interactions that occur for every pair of particles. To calculate all such interactions, the total number of operations required normally scales as N^2 [1]. One class of methods provides efficient computation of the interactions by means of a hierarchical or multilevel approach. Fast multipole method (FMM) achieving O(N) complexity is the typical algorithm of this class.

With the development of technology, the problem scale of N- Body is becoming larger and larger, thus challenges the computing speed of high performance computer. Recently, more and more accelerators, such as graphical processing units (GPU), Field Programmable Gate Array (FPGA), and Cell Broadband Engine (Cell B.E), become increasingly usable for high-performance numerical

Min Cao · Zhen Cao
School of Computer Engineering & Science, Shanghai University
email: mcao@shu.edu.cn, cz00000@126.com

* This research is supported by Nation 863 Program under No. 2009AA012201-CFA2009SHDX01, Shanghai Leading Academic Discipline Project under No. J50103.

Z. Qian et al. (Eds.): Recent Advances in CSIE 2011, LNEE 125, pp. 693–698.
springerlink.com

computations [2, 3]. The methods of the N-Body problem, which is the typical high performance numerical computation, are implemented on accelerators. The implemented methods mainly include the SPH [2], PM [4] and BHA [5]. The research on accelerating FMM algorithm focuses on GPU [6, 7].

This paper concerns with the implementation of FMM algorithm on accelerators. Based on the decomposition of the algorithm, all the sub procedures (or the computational steps) are classified as three sorts with different characteristics and tested on GPU, FPGA, and Cell BE, respectively. According to the experimental result, a mixed configurable computer architecture which can run FMM algorithm well is presented. In addition, the probable task flow decomposition run on the special mixed architecture is analyzed and proposed. Thus, FMM algorithm is optimized further.

2 Algorithm Analysis and Test of FMM on Accelerators

The acceleration of FMM algorithm is researched deep in this section.

2.1 Algorithm Theory and Decomposition

The FMM works by approximating the influence of a cluster of particles by a single collective representation, under the assumptions that the influence of particles becomes weaker as the evaluation point is further away, and that the approximations are used to evaluate far distance interactions. To accomplish this, the FMM hierarchically decomposes the computational domain and then it represents the influence of sets of particles by a single approximate value [1]. Thus, the following terminology with respect to the mathematical tools used to agglomerate the influence of clusters of particle is introduced.

Definition 1: Multipole Expansion (ME): a series (e.g, Taylor series) expansion truncated after p terms which represents the influence of a cluster of particles, and is valid at distances large with respect to the cluster radius.

Definition 2: Local Expansion (LE): a truncated series expansion, valid only inside a sub-domain, which is used to efficiently evaluate a group of MEs.

In order to calculate the interactions among particles by using ME and LE, FMM provides a spatial decomposition scheme to represent the near-field and far-field domains, and each sub-space is called box in the algorithm. That is, for a complete set of particles, FMM finds the clusters that will be used in conjunction with the MEs to approximate the far field, and the sub-domains where the LEs are going to be used to efficiently evaluate groups of MEs. This spatial decomposition is accomplished by a hierarchical subdivision of space associated to a tree structure (quadtree structure in two dimensions, or an octree structure in three dimensions) to represent each subdivision [1]. Thus, we divide the FMM into 11 steps as shown in figure 1. (1)Construct the tree structure, that is, finish the spatial decomposition. (2)Find a particle's neighbors, which mean the near-field, for calculating the near distance interactions. (3)Search a particle's far-field domain. (4)Initialize the tree, that is, locate each particle into related sub-space according to its initial coordinate, velocity, and so on. (5) P2M, transformation of particles into MEs (particle-to-multipole). (6) M2M, translation of MEs (multipole-to-multipole). (7) M2L, transformation of an ME into an LE (multipole-to-local). (8) L2L, translation of an LE (local-to-local). (9) L2P, evaluation of LEs at particle locations (local-to particle). (10)PP, directly compute the interaction between all the particles in the near domain of the box. (11)SUM, get the sum of the near and far distance interactions.

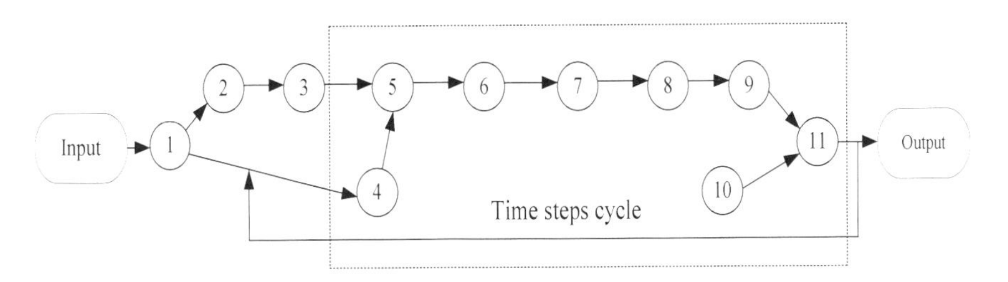

Fig. 1 Flow chart of steps of the FMM

2.2 Experimental Environment of Selected Accelerators

We test all the 11 steps of FMM on the accelerators include GPU, FPGA and Cell B.E. The experimental environment is as follows.

(1) Test of GPU: GPU is NVIDIA GeForce 9800 GTX+(512MB); CPU is Intel(R) Core(TM)2 Quad Q8200 @2.33GHz, 2GB RAM; Windows XP SP2; Developing platform includes CUDA 2 and Microsoft Visual Studio .NET 2008.
(2) Test of FPGA: CPU is Intel Core 2 Quad Q8200 2333 MHz (7x 333), Intel Eaglelake G45, 2GB RAM, 64 bit bandwidth, 500GB hardware disk; Windows XP SP2; Developing platform includes CETOS 5.4, gcc 4.1.2, petsc-3.0.0-p8 and cppunit-1.12.1; FPGA is Xilinx Virtex-5 ML507, time step sets to 10ns; Matlab7.0 is used as calculation verification.
(3) Test Cell B.E: host is CECHP12, 160GB PS3 with system of version 3.10; coprocessor is PowerXCell 8i 3.2GHz, 1PPE, 8SPE (6 SPE is usable), 256MB XDR RAM, Gigabit Ethernet; Yellow Dog Linux 6.1 ppc-new-ps3; Developing platform includes IBM SDK for Multi-core Acceleration v3.1, with ALF programming model.

2.3 Test Result of Classified Sub Procedures

The 11 steps of FMM are classified as three groups according to their function and computing characteristics.

The first group includes M2M, M2L and L2L, whose calculation is related to the traversal of the tree structure, and whose calculation is based on the box. Calculating the LE of a box from ME, namely, M2L, spends 0.040ms on GPU, and 0.016ms on FPGA. The performance of FPGA is 2.5 of GPU. Calculating the ME of a box from ME of its children, namely, M2M, spends 0.415ms on GPU, 0.27ms on FPGA, and 0.513ms on Cell B.E, respectively. The performance of FPGA is 1.537 and 1.9 of GPU and Cell B.E respectively, and the performance of GPU is 1.2 of Cell B.E. Calculating the LE of a box from LE of its ancestor, viz., L2L, spends 0.0102ms on GPU, and 0.004ms on FPGA. The performance of FPGA is 2.55 of GPU. By all appearances, the performance of GPU and Cell B.E for calculating the first group steps is the same class, and GPU goes beyond Cell B.E. Therefore, only M2L is tested on Cell B.E.

The second group includes P2M, L2P and PP, whose calculation is independent of the traversal of the tree structure, and whose calculation is based on particle, which is called blob in the algorithm. P2M (for a blob) spends 143ns on GPU, and 136ns on FPGA. The performance of FPGA is 1.05 of GPU. L2P spends 193ns on GPU, and 272ns on FPGA. The performance of FPGA is 0.7 of GPU. As for PP, its calculation relates to the number of blobs in one box. Assume that there are 8 and 32 blobs in one box, respectively, and suppose the blobs are distributed evenly in the box, then, PP spends 1805ns and 11572ns on GPU, 576ns and 2304ns on FPGA. The performance of FPGA is about 3.13~5.02 of GPU. Because GPU beats Cell B.E in numerical computation, this group of steps is not tested on the latter.

The third group includes the other steps, which needs more logic rather than numerical computation. Thus, it is not suitable to run this group of steps on FPGA. Locating a blob, that is, judging which box a blob is in, spends 20.09ns on GPU, and 168.6ns on CPU. The performance of GPU is 8.39 of CPU. Finding far-field and near-field of the blobs in one box spends 135.8ns on GPU, and 1120ns on CPU. The performance of GPU is 8.25 of CPU. Initialization of tree structure, includes data arrangement to which GPU does not fit, spends 29.09ns on Cell B.E. for one blob, and the performance of Cell B.E. is 17.62 of CPU. As a whole, Cell B.E. gains 1.8~6.4 performance of CPU, which varies with the number of level of the tree structure.

3 Optimization of FMM Based on Mixed Computer Architecture

Aimed at the mixed configurable computer architecture, a configuration scheme run FMM algorithm well and optimization of the FMM is presented.

3.1 Conclusion Based on Test Result

The above experimental result is not precise enough because of poor performance of the FPGA we selected, programming model without using multi-core, and average of running time for calculating one box or blob, it is still worth guiding the analysis of the characteristics of the sub procedures, namely, steps.

The experimental result shows that running M2L, M2M and L2L on FPGA is obviously faster than running them on GPU, even if the FPGA is poor performance. Running P2M, L2P and PP on FPGA achieves 1.05, 0.7 and 3.13~5.02 of GPU, respectively, while L2P occupies a thimbleful of computing time that can be ignored. GPU and Cell B.E. get hold of the same level of performance while running all the mentioned steps, though GPU is a little faster.

Because FPGA is good at numerical computation instead of logic computation, the initialization of the tree structure does not adapt to be run on FPGA. As for the other steps in the third group, though the performance of GPU is 8.39, 8.35 of CPU, there are still two choices, namely, GPU or multi-core CPU. Compare with 0.27ms/box for M2L run on GPU, the 1120ns/box of far-field calculation can be ignored. Judging the box number of a blob spends 168.6ns on CPU and 20.09ns on GPU. However, multi-core CPU is able to accelerate the calculation. Besides, it is only 2% of the PP calculation which spends 2304ns/blob when there are more than 32 blobs in a box. Therefore, it is unnecessary to run it on GPU. Though it is faster to pack up the initial data on Cell B.E than on CPU, it is still able to abrogate the necessity of using Cell B.E. by packing up data asynchronously.

3.2 Discussion on Scheme of FMM Algorithm

According to the analysis and decomposition of the FMM, two task flows can be considered.

The first task flow is made up of P2M, M2M, M2L, L2L and L2P. The second task flow includes PP only. Both are run on the mixed processor composed by CPU and FPGA, where CPU provides data to PFGA and receive the computing result from FPGA. The difference between the two task flows is that they need different proportion of CPU and FPGA. The operating load of data transfers from CPU to FPGA in PP is very heavy. The solution includes debasing proportion of CPU and FPGA or reducing the needful data transmission quantity of FPGA. As for the first task flow, running P2M on FPGA needs an awful lot of data transmission and lesser computing time. M2L is just the reverse. We improve on code, and make it possible to run P2M and M2L on different FPGA of the mixed processor at the same time when necessary. The proposed solution makes the mixed processor composed of one CPU and more than one FPGAs work better for the FMM.

4 Conclusions

This paper presents a configure scheme of mixed computer architecture fits to FMM algorithm based on the analysis of the algorithm and the test of the decomposed steps on CPU, GPU, FPGA and Cell B.E. The probable task flow decomposition run on the special mixed architecture is analyzed further. The future work includes distributing the computational accelerators to different task flows dynamically and reasonably by considering more factors.

References

1. Cruz, F.A., Barba, L.A.: Characterization of the accuracy of the fast multipole method in particle simulations. Journal of Numerical Methods in Engineer (79), 1577–1604 (2009)
2. Spurzem, R., Berczik, P., Marcus, G., et al.: Accelerating astrophysical particle simulations with programmable hardware (FPGA and GPU). Computer Science - Research and Development (23), 231–239 (2009)
3. Che, S., Li, J., Sheaffer, J.W., et al.: Accelerating Compute -Intensive Applications with GPUs and FPGAs. In: Proc. of the IEEE Symposium on Application Specific Processors, SASP (June 2008)
4. Aubert, D., Amini, M., David, R.: A Particle-Mesh Integrator for Galactic Dynamics Powered by GPGPUs. In: Allen, G., Nabrzyski, J., Seidel, E., van Albada, G.D., Dongarra, J., Sloot, P.M.A. (eds.) ICCS 2009, Part I. LNCS, vol. 5544, pp. 874–883. Springer, Heidelberg (2009)
5. Hamada, T., Nitadori, K., Benkrid, K., et al.: A novel multiple-walk parallel algorithm for the Barnes–Hut treecode on GPUs – towards cost effective, high performance N-body simulation. Computer Science - Research and Development (24), 21–31 (2009)
6. Gumerov, N.A., Duraiswami, R.: Fast Multipole Methods on Graphics Processors. Journal of Computational Physics (227), 8290–8313 (2008)
7. Xu, K., Ding, D.Z., Fan, Z.H., et al.: Multilevel Fast Multipole Algorithm Enhanced by GPU parallel Technique for Electromagentic Scattering Problems. Microwave and Optical Technology Letters 52(3), 502–507 (2010)
8. Stokes, M.L.: A Brief Look at FPGAs, GPUs and Cell Processors. ITEA Journal (7), 9–11 (2007)

Radiative Properties Modeling for Complex Objects Using OpenGL

Yu Ma, Shikui Dong, and Heping Tan

Abstract. Based on the Z-Depth buffer and occlusion query functions of OpenGL hardware graphics accelerator, a new approach is presented, by which the valid radiative facets of complex objects can be found out exactly and quickly, through rendering the object twice. Combining the new approach with the Light of Sight method which is used for radiation transfer calculation, the radiative properties calculation model is constructed. Then the radiative properties of typical complex objects are obtained with the new model.

1 Introduction

OpenGL has been used in the calculation of radar scattering section as its stronger raphics processing ability [1, 2]. Like with the radar, as an important judgement basis to determine the targets, infrared characteristics of the targets have very important research value. Thermal radiation transmission mainly research heat transfer process caused by electromagnetic waves, and is one of three typical heat transfer process. With the recent development of the numerical heat transfer, radiation transfer has been not only an important energy transport carrier, but also an important signal propagation process, and has been widely used in aerospace related fields, such as remote sensing, guidance and others.

Based on OpenGL's graphics hardware function and with wysiwyg means, Ref.3 and Ref. 4 calculated the target surface radiation including emission by itself and incidence reflection. But few objects are in a vacuum environment, the absorption and emission characteristics of medium surrounding the target have obvious influences on the target spectral radiation characteristics. Therefore, target radiation characteristics research must be combined with radiation translation.

Based on the radiative transfer equation, many numerical methods for radiative transfer calculation have been developed systematically such as the Monte Carlo

Yu Ma · Shikui Dong · Heping Tan
No. 92, West Da-Zhi Street, School of Energy Science and Engineering,
Harbin Institute of Technology, Harbin, China
e-mail: {mayu,dongsk,tanheping}@hit.edu.cn

Z. Qian et al. (Eds.): Recent Advances in CSIE 2011, LNEE 125, pp. 699–704.
springerlink.com

method, Discrete Ordinates method, Finite Volume method, and Finite Element method [5, 6]. However, due to the complexity of radiative transfer equation, most of the radiative transfer numerical methods are more time-consuming. And as with the simple basic principles and high-speed calculation, The Light of Sight (LOS) method is especially suitable for radiative transfer calculations without scattering. One of the key conditions for LOS calculation is to obtain the effective emitting surfaces of the targets, which is very difficult and time-consuming for complex shape objects. OpenGL's powerful graphics processing ability and LOS method are combined in this file to model radiative transfer properties of complex targets.

2 Target Emission and Radiative Transfer Calculations

Simulation of target radiation characteristics mainly includes two parts: the first, calculation of target's surface emission and the reflection of incident radiation from the background environment; the second, calculation of radiative transfer between the target and the probe position.

According to Planck's law, spectral emissive power distribution from target's surface can be writen as

$$E_{i,\lambda,emi} = \varepsilon_\lambda \frac{c_1 \lambda^{-5}}{\exp\left[c_2 / (\lambda T_i)\right] - 1} \tag{1}$$

where T_i is tempecture, ε_λ is surface's spectral emissivity, λ is spectral wavelength, and c_1 and c_2 are the first and second radiation constants.

According to Lambert's law [5], spectral radiation intensity for Isotropic diffuse surface can be evaluated as

$$I_{i,\lambda,emi} = E_{i,\lambda,emi} / \pi = \frac{\varepsilon_\lambda}{\pi} \frac{c_1 \lambda^{-5}}{\exp\left[c_2 / (\lambda T_i)\right] - 1} \tag{2}$$

Integrate Eq. (1) in the whole spectral range, according to Stefan-Boltzmann Law, the total emissive power can be written as

$$E_{i,emi} = \varepsilon \, \sigma T_i^4 \tag{3}$$

where ε_λ is the surface average emissivity in the whole spectral range, and σ is the Stefan-Boltzmann constant.

Target surface reflection is affected by many factors, including the angle of incidence and the characters of outside light, and the property of the surface. Details about the BRDF please refer to the Ref. 7.

Radiative transfer calculation is an important role for modeling target radiation property. As absorption, emission characteristics of gas medium around the target, there would be some influences on the target radiation property. So an accurate calculation of radiative transfer process on the target radiation simulation is essential.

On the basis of obtaining the medium spectral radiation properties, the model of infrared radiation transmission in the flow field outside the target has been established in this article.

Radiative transfer process in the flow field is showed in Fig. 1, where dotted lines indicate the flow field grids. By the radiative transfer theory, in the direction $\vec{\Omega}$, s position within the media to consider the absorption, emission, scattering, the equation of radiative transfer in participating media can be described as:

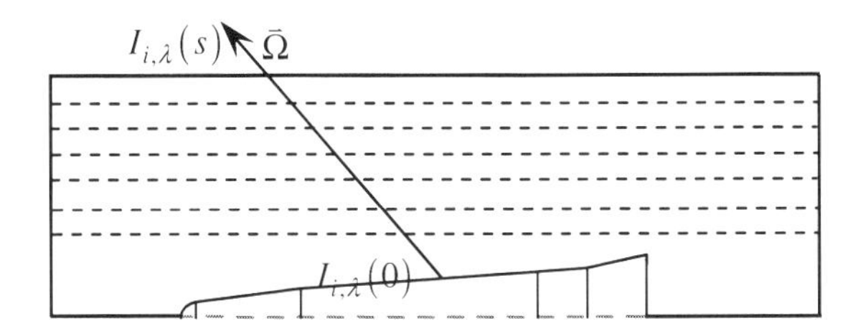

Fig. 1 Radiation transfer in the field

$$\frac{dI_{\lambda}(s)}{ds} = -\kappa_{a\lambda}I_{\lambda}(s) - \kappa_{s\lambda}I_{\lambda}(s) + \kappa_{a\lambda}I_{b\lambda}(s) + \frac{\kappa_{s\lambda}}{4\pi}\int_{4\pi} I_{\lambda}(s,\vec{\Omega}')\Phi_{\lambda}(\vec{\Omega}',\vec{\Omega})d\Omega' \tag{4}$$

where $I_{\lambda}(s)$ is spectral intensity of radiation, $I_{b\lambda}(s)$ is blackbody emission spectral intensity, $\kappa_{a\lambda}$ and $\kappa_{s\lambda}$ are spectral absorption coefficient and scattering coefficient, $I_{\lambda}(s,\vec{\Omega}')$ is the spectral intensity of radiation projecting from $\vec{\Omega}'$ direction, and $\Phi_{\lambda}(\vec{\Omega}',\vec{\Omega})$ is scattering phase function.

In Eq. (4), the first term of the right-hand side is attenuation due to absorption, the second term is attenuation due to scattering, the third term is augmentation due to emission, and the last term is augmentation due to scatter from other directions. Without considering the contribution of other directions, and combining the attenuation term caused by absorption and scattering, eq. (4) can be simplified as:

$$\frac{dI_{\lambda}(s)}{ds} = -\kappa_{e\lambda}I_{\lambda}(s) + \kappa_{a\lambda}I_{b\lambda}(s) \tag{5}$$

where $\kappa_{e\lambda} = \kappa_{a\lambda} + \kappa_{s\lambda}$ is attenuation coefficient.

Then we can lead to LOS methods with differential forms

$$I_{\lambda}(s_0 + \mathrm{d}s) = I_{\lambda}(s_0) + \mathrm{d}I_{\lambda}(s) \tag{6}$$

in which

$$dI_\lambda(s) = -\kappa_{e\lambda} I_\lambda(s_0) ds + \kappa_{a\lambda} I_{b\lambda}(s_0) ds \tag{7}$$

It can be seen from the eq. (6) and eq. (7) that the LOS method is very simple and efficient, which starts the calculation from the target surface, and calculating of radiative transfer layer by layer through eq. (6).

However, the effective emitting surfaces of the target to be detected, which are in the detection range and not blocked by other surfaces, must be accurately determine. This is particularly important for the target with complex surface shape, and directly affects the simulation accuracy and computational efficiency.

3 OpenGL-Based Effective Emitting Surface Calculation

OpenGL [8] is a software interface to graphics hardware, which is an open common share graphics standard of three-dimensional based on SGI's GL three-dimensional graphics library. In this paper, determining effective emitting surfaces of complex geometry and modeling radiative transfer are achieved through OpenGL's graphics computing power

At present, solid targets are generally modeling with a variety of CAD software package, by which usually a large number of small surface elements are used to describe the solid targets, especially for more complex structure of target surface. Surface elements only in the detection range and not obscured by other surface elements are effective emitting surface elements, and can be determined through OpenGL's depth test and query functions effectively.

Depth (z) values are encoded during the viewport transformation, and stored in the depth buffer. For each pixel on the screen, depth buffer in charge of recording the distance between observating point and object occupy the pixel. If the depth te st was passed through for the object, the depth values of source fragment would replace pre-existing values in the depth buffer. Depth buffer is generally used for eliminating hidden surface. If there was a new candidate color fragment in a pixel, it would be drawn only when the corresponding object closer than the original one. Therefore, only the objects that are not obscured by other objects will be preserved during rendering.

The depth test results can be called out by OpenGL's occlusion query function, which can provide the number of pixels of the queried object shown on the screen. The object is visible if only its pixel number is greater than zero.

In this paper, OpenGL's occlusion query function is applied to accurately obtain the effective emissive surface elements. We use rendering with occlusion queries twice to achieve effective detection surface elements. Starting with rendering the target by each small element set of vertices one by one, then began a second rendering, rendering each small surface element, according to the first rendering of the state, we can query informed that the surface element is valid or not. That is the basis for calculation emission and transmission of radiation.

4 Simulation of Complex Target Radiation

A complex shape of aircraft was built with CAD software, and shown in Fig. 2. The aircraft was structured with a ball cone, a cylinder and four trapezoidal tails, and its surface was combined with a large number of small surface elements. The polar angle θ and azimuthal angle φ were shown in Fig. 3.

Fig. 2 The aircraft with complex shape

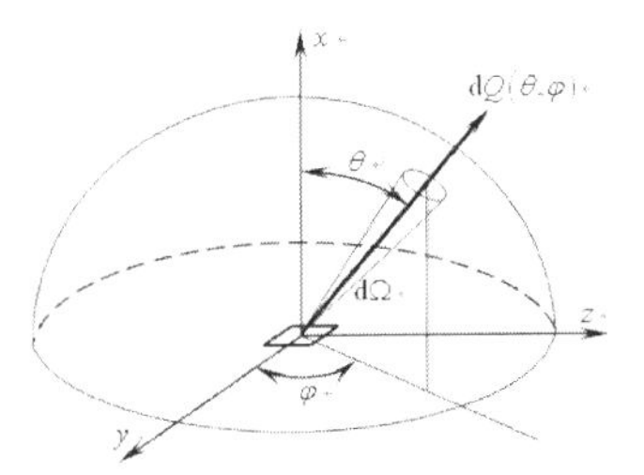

Fig. 3 Sketch of polar angle and azimuthal angle

Obviously, different detecting directions lead to different effective emitting surface elements. OpenGL's occlusion query function can efficiently and accurately solve this problem. Assuming the wall of the aircraft is gray body with emissivity 0.7 and even temperature 1000K. The total radiation energy of the aircraft detected from different directions is calculated in this paper, and the normalized result is shown in Fig. 4.

It can be seen from the simulation results that the total radiation energy of aircraft with complex shape is sensitive to detection direction, which determines the effective emission elements corresponding to the total radiant energy of the target.

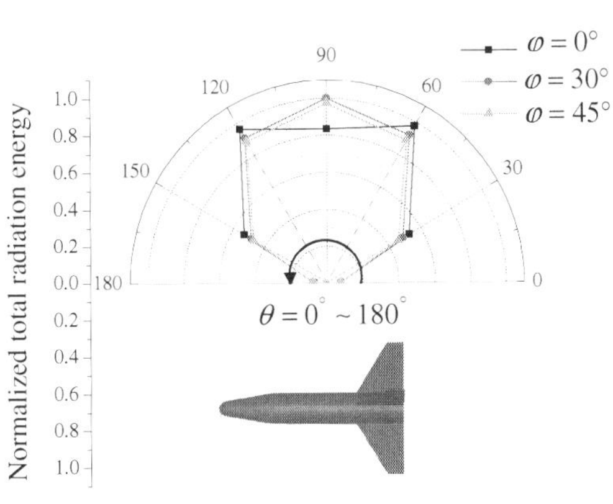

Fig. 4 Normalized total radiation energy of the aircraft in different directions

5 Conclusions

Based on the OpenGL graphics hardware features, valid radiative facets of complex objects are found out through rendering the target twice, with using OpenGL's depth buffer and occlusion query functions; on this basis, combining with LOS method, the aircraft radiation calculation model is built, and the radiative properties of complex objects are calculated efficiently and accurately.

Acknowledgements. This research is supported by the National Natural Science Foundation of China (Grant. No. 50906016). The authors are indebted to them for their financial support.

References

1. Rius, J.M., Ferrando, M., Jofre, L.: GRECO: Graphical Electromagnetic Computing for RCS Prediction in Real Time. IEEE Antenn. Propag. M. 35(2), 7–17 (1993)
2. Lu, Z.Y., Su, D.L., Qi, W.Q., Zeng, G.Q.: RCS Calculation for Complex Objects Using GRECO. In: 7th International Symposium on ISAPE 2006, Guilin (2006)
3. Fang, N., Zhang, X.L., Gao, J., Qi, Z.Y.: Infrared Radiant Computation by OpenGL. J. Beijing Univ. Aero. Astro. 35(12), 1473–1476 (2009)
4. Huang, Q., Zhang, T., Lv, Y.H., Cui, W.N.: Research of Infrared Target Images Simulation Based on Visual C++ and OpenGL. Infrared Technol. 32(2), 101–104 (2010)
5. Modest, M.F.: Radiative Heat Transfer. McGraw-Hill Inc., New York (1993)
6. Tan, H.P., Xia, X.L., Liu, L.H., Ruan, L.M.: Numerical Calculation of Infrared Radiative Properties and Transfer. Harbin Institute of Technology Press, Harbin (2006)
7. Yu, Q.Z.: Theory of Radiative Heat Transfer. Harbin Institute of Technology Press, Harbin (2000)
8. Tom, M., David, B.: Advanced Graphics Programming Using OpenGL. Morgan Kaufmann Publishers, San Francisco (2005)

Bounds on Pair-Connected Reliability of Networks with Edges Failure

Hu Zhao, Wen Lu, and Haixing Zhao

Abstract. For undirected simple networks G with perfectly reliable vertices and unreliable edges, we consider the Pair-Connected Reliability of G for which edges fail independently of each other with a constant probability p. The Pair-Connected Reliability of G denoted by $PC(G\ ,\ p)$, is defined to be the expected number of pairs of vertices that keep connected in network G. Using the definition of probability Random Events that basing on the Path Sets and Cut Sets between arbitrary pair connected nodes, we give a approximate calculating method to get the upper and lower bound on Pair-Connected Reliability with edges failure and find the concise bounds expression of Complete-Networks K_n.

Keywords: Networks, Edges Failure, Pair-Connected Reliability, Upper bound, Lower bound.

1 Introduction

Lots of networks such as distributed Multi-Processor System & Computer Communication Networks may become inefficient even damaged in entire network system because of failure on computing nodes and/or communication links. Analyzing the reliability of existing networks and designing a more reliable network would have been more valuable in theory and practice. We represent a network with a undirected simple probabilistic graph G with vertex set V(G) and edge set E(G), the Reliability of networks $R(G,\phi,\psi)$is defined as the probability that graph G keeps connected. The accurate computation of reliability $R(G,\phi,\psi)$ has been proven as the NP-hard problems[1].

Amin etc.[2]. invented another network reliability measure named Pair-Connected Reliability which concerned the expected number of pairs of connected vertices. The definition is:

Hu Zhao · Wen Lu
Department of Computer, Qinghai Normal University Xining China
e-mail: {zhaohu361,luwen897}@163.com

Z. Qian et al. (Eds.): Recent Advances in CSIE 2011, LNEE 125, pp. 705–710.
springerlink.com

Definition 1. [2] For $S \subseteq E(G)$ with $|s|=k$, let $p^k q^{m-k}$ denote the probability that G is in state S. Further, let $PC(S)$ denote the number of pairs of vertices that are connected in state S, then the pair-connected reliability of G is defined to be:

$$PC(G,p) = \sum_{S \subseteq E(G)} PC(S)P(S), \tag{1}$$

The meanings of definition above is that the expected number of pairs of vertices keeping connected in network G whose vertices are fail-safe but each edge is down independently with probability $q(q=1-p)$. The reliability measure based on the expected number has more advantages then other measures, in particularly it can describe network's reliability exactly. Such as, for two networks $G(n,m)$ and $H(n,m)$, G is a connected network and H is non-connected, the Pair-Connected Reliability may describe the special situation that network H is more reliable then G, but the other reliability measures can not effectively discriminate this. Few of researches has been made in Pair-Connected Reliability with edges failure recently, Siegrist[6] gave the efficient calculation method on polynomial coefficient of Pair-Connected Reliability of networks with edges failure, and proved the non-existence of uniform optimal networks $G(n,m)$, when $n \le m \le C_n^2 - 2$.

In this paper, we investigate the upper and lower bound on Pair-Connected Reliability of networks with edges failure by Using the definition of probability Random Events basing on the Path Sets and Cut Sets between arbitrary pair connected nodes and find the concise bounds expression of Complete-Networks K_n.

2 Relationship between Two-Terminal Reliability and Pair-Connected Reliability

The essential meanings of Pair-Connected Reliability $PC(G,p)$ is the expected number of pairs of vertices when network G is in state $S(S \subseteq E(G))$, we regard the number of pairs of vertices which keep connecting as the value of disperse random variables. On the other hand, if we define the meaning of disperse random variable with arbitrary pairs of vertices (u,v) in network G, and the probability that keep connecting between these two vertices is just the Two-Terminal Reliability $R_{u,v}(G,p)$, then we have the following definition:

Defination 2.[5] Two-terminal reliability $R_{u,v}(G,p)$ is the probability that two specified distinct vertices (u,v) remain connected in network $G(m,n)$, where $(u,v)\in V(G)$ and edges in G are failure independently with probability $q(q=1-p)$.

It can construct another new probability distributing function of disperse random variables when we count over all of the pairs of vertices in network G. Let number 1 denote the disperse random variable that a pair of specified vertices remain connected, and $R_{u,v}(G,p)$ denote the probability of disperse random variable, then we have

$$PC(G,p) = \sum_{(u,v)\in V^2(G)} R_{u,v}(G,p), \tag{2}$$

where $(u,v)\in V^2(G)$ denotes the traversing on every pair of vertices in network .

3 Upper and Lower Bound on Pair-Connected Reliability

Calculation of Pair-Connected Reliability is very difficult because it has been shown by Provan [4] that precise calculation of $R_{u,v}(G,p)$ is NP-hard.

We can transform the problem of the bounds on Pair-Connected Reliability into the bounds on Two-Terminal Reliability through equition (2). When we traverse every pairs of vertices in network G and get probability $P_i^{'}(i=1,2,\cdots C_n^2)$ of every pairs of vertices remain connected, it would be equal to define a new random event that *a certain pair of vertices is connected*. If we define a disperse random variable X denote this random event, the probability that the variable X has value of 1 is $P_i^{'}$, so the expected number of pairs of vertices remain connected is $\sum_{i=1}^{C_n^2} X_i \bullet P_i^{'}$. Then we have the theorem followings.

Theorem 1. *The Upper and Lower Bounds on Pair-Connected Reliability of networks G(n,m) with edge failure is given by*

$$\sum_{(u,v)\in V^2(G)} \prod_{i=1}^{r}(1-q^{|C_i|}) \le PC(G,p) \le \binom{n}{2} - \sum_{(u,v)\in V^2(G)} (\prod_{i=1}^{k}(1-p^{|P_i|})),$$

where $|C_j|$ denote the number of edges in jth cut set between a specified pair of vertices $(u,v)\in V^2(G)$ and $|P_i|$ is the length of the ith path in path set between a pair of specified vertices $(u,v)\in V^2(G)$.

Proof: As the consideration above, Let $P_1,P_2,\ldots,P_k$ denote every u-v path in network G respectively, define probability event E_i="every edges in path P_i are all reliable", so an arbitrary occurrence of event E_i may result in connection between vertice u and v. It's obviously easy to get that

$$R_{u,v}(G) = \Pr(E_1 \cup E_2 \cup \cdots E_k) = \sum_{i=1}^{t}\Pr(E_i) - \sum_{i\langle j}\Pr(E_iE_j) + \cdots + (-1)^{k+1}\Pr(E_1E_2\cdots E_k) \quad (3)$$

But the probability in equition (3) is hard to calculating and it is impossible to confirm that all of the paths between u and v have no conjunct edges or vertices. But we can make a hypothesis that the probability event $E_i(i=1,2,\ldots k)$ are all independently each other.

Through this hypothesis, we can transform the hard calculating probability $\Pr(E_1 \cup E_2 \cup \ldots E_k)$ into the probability of opposite event intersection $1-\Pr(\overline{E_1}\,\overline{E_2}\cdots\overline{E_k})$. on the assumption above and the basic property of paths between two vertices, the probability events $\overline{E_i}$ are all independently. According to the properties of probability multiplication, the probability

$1-\Pr(\overline{E_1}\,\overline{E_2}\cdots\overline{E_k})$ will become lower than $1-\prod_{i=1}^{k}\Pr(\overline{E_i})$. So we have

$$R_{uv}(G) = \Pr(E_1 \cup E_2 \cup \cdots E_k) = 1-\Pr(\overline{E_1}\,\overline{E_2}\cdots\overline{E_k}) \le 1-\prod_{i=1}^{k}\Pr(\overline{E_i}) = 1-\prod_{i=1}^{k}(1-\Pr(E_i)) \quad (4)$$

Let $PC(G,p)_{UP}$ denote the Upper Bounds of Pair-Connected Reliability, we have

$$PC(G,p)_{UP}=\binom{n}{2}-\sum_{(u,v)\in V^2(G)}(\prod_{i=1}^{k}(1-p^{|P_i|})), \tag{5}$$

where $n=|V(G)|$. k is the number of paths between a specified pair of vertices $\{u,v\}$.

Similarly, we denote u-v cut sets in network G by $C_1,C_2,\ldots,C_r$ and probability events "the edges in cut set C_i are all failure" by F_i, so an arbitrary occurrence of event F_i may result in non connection between u and v. then

$$R_{u,v}(G)=1-\Pr(F_1\cup F_2\cup\cdots F_r), \tag{6}$$

If we make a hypothesis that probability event $F_i(i=1,2\ldots r)$ are all independently each other and the probability $\Pr(\overline{F_1}\,\overline{F_2}\cdots\overline{F_r})$ will become greater then $\prod_{j=1}^{r}\Pr(\overline{F_j})$. So we have

$$R_{u,v}(G)=1-\Pr(F_1\cup F_2\cup\cdots F_r)=\Pr(\overline{F_1}\,\overline{F_2}\cdots\overline{F_r})\geq\prod_{j=1}^{r}\Pr(\overline{F_j})=\prod_{j=1}^{r}1-\Pr(F_j), \tag{7}$$

Let $PC(G,p)_{DN}$ denote the Lower Bounds of Pair-Connected Reliability, we have

$$PC(G,p)_{DN}=\sum_{(u,v)\in V^2(G)}(\prod_{j=1}^{r}(1-q^{|C_j|})), \tag{8}$$

where $|C_j|$ denotes the number of edges in jth cut set between a specified pair of vertices $(u,v)\in V^2(G)$. This complete the proof.

According to the equation (5) and (8), we have the corollary1 obviously below.

Corollary 1. *The deviation between Upper and Lower Bounds on Pair-Connected Reliability of networks $G(n,m)$ with edge failure is*

$$\Delta PC(G,p)=\binom{n}{2}-\sum_{(u,v)\in V^2(G)}(\prod_{i=1}^{k}(1-p^{|P_i|})+\prod_{i=1}^{r}(1-q^{|C_i|}))$$

In reality, most Topological structure of networks were similar to complete graph K_n. By the basic attributes of complete graph K_n: There are many paths with different length and just one cut sets with n-1 edges between an arbitrary pair of vertices, the length of various paths between an arbitrary pair of vertices is $i(i=1,2,\ldots n\text{-}1)$ and the number of paths with length i is P_{n-2}^{i}. So we have the followings.

Corollary 2. *The Upper and Lower Bounds on Pair-Connected Reliability of completed networks K_n with edge failure is:*

$$\binom{n}{2}\bullet(1-q^{n-1})\leq PC(G,p)\leq\binom{n}{2}\bullet(1-\prod_{i=1}^{n-1}(1-p^{i})^{P_{n-2}^{i-1}})$$

where $n=|V(K_n)|$ and $P_{n-2}^{i-1}=(n-1)!/(n-i-1)!$.

4 Validation

Now we can get Lower and Upper bounds on a specified network shown in Figure 1 by Theorem 1. Replace corresponding items of Theorem 1 with the length of connected paths and the number of edges in cut sets between all of the pairs of vertices, The curves of $PC(G,p)$, $PC(G,p)_{DN}$ and $PC(G,p)_{UP}$ was shown in Fig. 2 – Fig. 3.

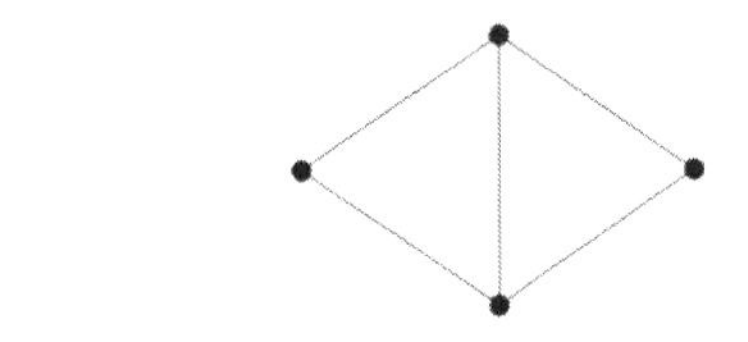

Fig. 1 A specified network with 4 vertex and 5edges.This sample graph has been often cited by other authors.

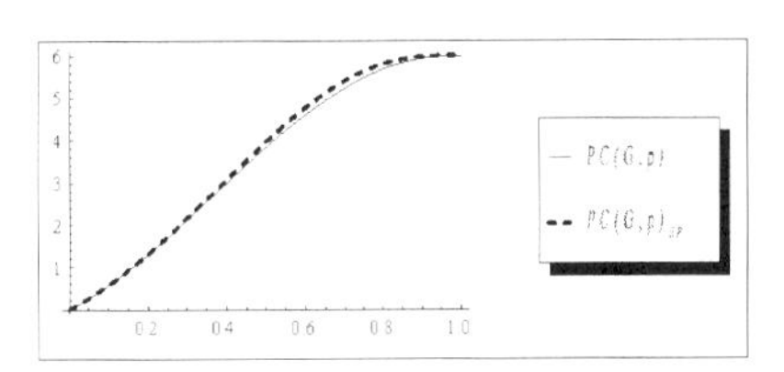

Fig. 2 Curve of polynomial $PC(G,p)$ and $PC(G,p)_{UP}$.The upper bound polynomial is more closer to the exact value of reliability polynomial than lower bound polynomial. Upper bound is tending to amalgamation when probability $p \to 0$ or $p \to 1$.

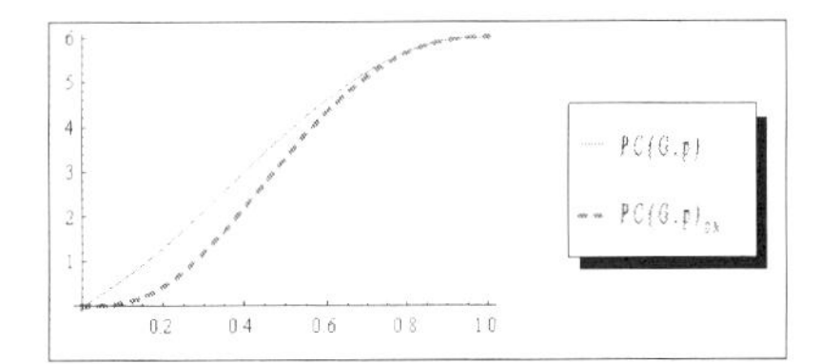

Fig. 3 Curve of polynomial $PC(G,p)$ and $PC(G,p)_{DN}$. The difference between lower bound polynomial and real reliability polynomial is larger than that on upper bound. Lower bound is tending to amalgamation when probability $p \to 0$ or $p \to 1$ too. .

5 Discussion

The upper bound polynomial is more closer to reliability polynomial than lower bound polynomial. The most difference between Upper and Lower bound is less then 1, that means the deviation of expected number of pairs would not more than one pair.

To calculate bounds on Pair-Connected Reliability by Theorem 1, it need to traverse every pairs of vertices in network G and to adopt calculation on Paths searching and Cut Sets searching, so the computation complexity is much higher when the number of edges and vertices is to a certain degree of huge. For a undirected simple network with n vertices and m edges, the time complexity of path searching one time is up to $O(n^2)$ using Dijkstra algorithm[3]. The space complexity can be reduced too by method of storing temporary data in hard disk and releasing the RAM memory in time. A modified Dijkstra algorithm was given in article[7], the time complexity and space complexity can be up to $O\ (n(logn+m))$ and $O(4n+m)$ respectively.

References

1. Amin, A.T., Siegrist, K.T., et al.: On uniformly optimal reliable graphs for pair-connected reliability with vertex failures. Networks 23, 791–794 (1993)
2. Amin, A.T., Siegrist, K.T., et al.: The expected number of pairs of connected nodes pair-connected reliability. Mathl. Comput. Modelling 17, 1–11 (1993)
3. Bondy, J.A.: Graph theory with applications. Elsevier Science press (1982)
4. Provan, J.S.: The complexity of reliability computation in planar and acyclic graphs. University of North Carolina Technical Report 2, 279–290 (1983)
5. Shier, D.R.: Network Reliability and algebraic structures. Clarendon, Oxford (1991)
6. Siegrist, K.T., Slater, P.J.: On the nonexistence of uniformly optimal graphs for Pair-connected reliability. Networks 21, 369–371 (1991)
7. Wang, Z.: Analysis and Improvement of Dijkstra Algorithm. Journal of Hubei University of Education 8, 12–14 (2008)

Rapid Calculation Preprocessing Model of N-Body Problem

Shaoping Chen and Shesheng Zhang

Abstract. In the paper, rapid calculation preprocessing model of particle's interacting N-body problem is discussed. Coordinate of particle is approximately replayed by grid center. According to the similarity of grid's topological structure, concept and algorithm of basic solvable range and basic insolvable range of radial distribution function are put forward. In advance, distance function and class partition function between cell centers is found, radial distribution function is computed by translation transformation. In this paper, examples indicate that approximation algorithm is of computing time short and error small, so the approximation algorithm has stronger practicability.

1 Introduction

Literature [1] focuses on a kind of particle simulation data analysis plays an extremely important role in the two-body correlation function query – spatial distance histogram (SDH). Query directly the original two-dimensional algorithm SDH time complexity is $O(N^2)$. As the particle simulation system usually consists of a large number of individual particles, this complexity is unacceptable. This situation has seriously hindered the large-scale data processing and simulation applications, thus affecting the particle-based simulation of scientific discovery process. This raises the urgent need to address technology issues, namely fast SDH algorithm design problems.

SDH is described in a series of important physical quantities of the main building blocks, is known as the radial distribution function (RDF) of a continuous statistical distribution function of the direct estimates. Thus, RDF can be regarded as continuous for SDH. RDF for the calculation of thermodynamic characteristics of physical systems is essential. Some important physical quantities such as total pressure, energy cannot be calculated without RDF. In practice, in order to

Shaoping Chen · Shesheng Zhang
Wuhan University of Technology, 122Luoshi Road, 430070 Wuhan, China
e-mail: chensp@whut.edu.cn, sheshengz@yahoo.com

Z. Qian et al. (Eds.): Recent Advances in CSIE 2011, LNEE 125, pp. 711–718.
springerlink.com

simplify the calculation, RDF is usually showed as a non-continuous form--SDH. There is no doubt that according to the distribution of particle analysis and calculation of expression construct RDF makes sense. Literature [1] studied the three-dimensional SDH approximation algorithm for a given error condition, according to the tree structure and density of layer structure, layer by layer to determine the solvable and understand the region. The theoretical analysis and numerical calculations show that the calculated amount $O(N^{5/3})$.

Based on the work of literature [1], we apply the error theory to analyze the tree structure and density-layer structure, to find the length of the grid of density-layer, to establish basic solvable range and basic insolvable range for the origin of RDF, with a simple translation transformation method, quickly find the RDF on the whole region, while the calculation volume is $O(N)$.

2 Preprocessing Method for Single Particle

Radial distribution function (RDF) [2] is defined as $g(r)=\frac{F_h(r)}{dV\rho^*}$, where $F_h(r)$ is the number of particles in the shell for the radius r, thickness h, dV is their corresponding volume, ρ^* is the average particle density. As a discrete distribution, taking distance $r=nh-h/2$, $n=1,2,\ldots,N_n$, N_n is the maximum value of the distance classification. Let the particle coordinates are X_i, $n=1,2,\ldots,N$, $F_h(r)$ can be expressed as

$$F_h(r)=\sum_{j=1}^{N}\sum_{i=1}^{N}\delta_h(r^*-r) \tag{2.1}$$

where $r^*=|X_i-X_j|$ is the distance between X_i and X_j, the function $\delta_h=1$ for $|r^*-r|\le h/2$, the function $\delta_h=0$ for $|r^*-r|>h/2$. As the distance $r=nh-h/2$, physical meaning of the above formula is of the distance classification.

Let H be a step. $\text{int}[a]$ is integral function of a. the particle coordinate $X=(x,y,z)$ is approximately replaced by $Y=(iH,jH,kH)$, where $i=\text{int}[x/H]$, $j=\text{int}[y/H]$, $k=\text{int}[z/H]$. Can prove that the distance from X to Y is less than $0.5\sqrt{3}H$. From the foregoing, we can use integer point (i,j,k) approximation Expressed space point (x,y,z). And obviously the number of integer points is less than the number of space points, and easy to construct approximation algorithms. Let basic classification function $Bf(i,j,k)=[r/h]$, where $r=H\sqrt{m}$, $m=i^2+j^2+k^2$ is the distance from Y to the origin point. $Bf(i,j,k)$ function call is quite simple, as long as the given integer point (i,j,k), $Bf(i,j,k)$ outputs value of the point classification. Optional two points $A(x,y,z)$, $B(x',y',z')$, expressed their integer points $A(i,j,k)$, $B(i',j',k')$, the distance between A and B is classified as t $Bf(i-i',j-j',k-k')$. Thus, according to the basic classification function, we can quickly find the function $F_h(r)$.

Example: Suppose particles obey uniform distribution in the unit cube, total number of particles is N, step $h = 1/10$, $h = 5H$, when the total number of particles $N = 10^4$, the exact solution of (2.1) used CPU time is 2.201 seconds, while the time of the approximate solution was 0.8243 seconds. When the total number of particles $N = 10^5$, the exact solution of (2.1) used CPU time is 220.3 seconds, while the time of the approximate solution was 82.42 seconds. Differed by about 2.7 times. Or approximation method is 2.7 times faster than the precise method. Approximation error is 0.08.

3 Basic Solvable Range and Unsolvable Range

Literature [4] studied the N-body problem a number of numerical simulation methods that use tree structure code not only reduces the storage overhead, and more conducive to fast computing and parallel partitioning. Literature [1] used SDH to construct density map approximate calculation method, and used tree structure to compute solvable range and unsolvable range. According to the tree structure mesh similarity, this paper will studied basic solvable range and basic unsolvable range, through the translation transform, and will quickly computed solvable and unsolvable range.

We know that the radial distribution function is the distance to be classified with step h, define nth class range of the basis

$$\Omega_{dn} = \{X \mid nh - h < r \le nr; r = |X|\}, n = 1,2,\dots,N_n \tag{3.1}$$

It is the shell range with radius from $(nh-h)$ to (nh), its volume is $4\pi[n(n-1)+1/3]h^3$. Let us take a cube of space, it contains all the particles. The cube is divided into a grid of small cubes with edge length H. Small cube is called as cell. When a cell is belong to nth class range of the basis, said basic solvable cell. When a cell is partly belong to nth class range of the basis, said basic unsolvable cell. Easy to get cell diagonal length $\sqrt{3}H$. The maximum distance from cell-central of basic unsolvable cell to nth class range of the basis boundary is $0.5\sqrt{3}H$.

Define the cell located at the origin as O cell. Let O cell center locate at the origin and r be the distance from the point X to the origin O. Define range as

Ω_1 : $0 < r < h - 0.5\sqrt{3}H$ $\qquad$ Ω_n : $nh - h + 0.5\sqrt{3}H < r < nh - 0.5\sqrt{3}H$

Ω_n is called the nth class solvable basic range, it is belong to basic nth class range Ω_{dn}. Ω_n is the shell with radius from ($nh - h + 0.5\sqrt{3}H$) to ($nh - 0.5\sqrt{3}H$)and thickness $(h - \sqrt{3}H)$, Its volume is $4\pi H^3[(n\tau + 0.5\sqrt{3})(n\tau - 0.5\sqrt{3} - \tau) + \tau^2/3]\tau$, where $\tau = h/H$ .Sum of all class basic solvable range is collectively referred to as basic solvable range and it is defined as $\Omega_s = \cup\Omega_n$. The nth class basic unsolvable range Ω_{on} is defined as Ω_{on} : $nh - 0.5\sqrt{3}H \le r \le nh + 0.5\sqrt{3}H$, basic unsolvable range is defined as $\Omega_b = \cup\Omega_{on}$, where Ω_{on} is the nth class basic unsolvable range and is the shell with radius from $\left(nh-0.5\sqrt{3}H\right)$ to $\left(nh+0.5\sqrt{3}H\right)$ and thickness $\left(\sqrt{3}H\right)$, which is consist of basic nth class range and basic (n+1)th class range, its volume is $4\pi H^3\left(n^2\tau^2+0.5\right)\sqrt{3}$. Summing to n from 1 to N_n,we get basic unsolvable range volume V_b : $V_b=4\pi H^3\left[N_n(N_n+1)(2N_n+1)\tau^2/6+0.5N_n\right]\sqrt{3}$,

V_b 'half divided by sphere with a radius $R=N_nh$ is $Er_2=0.866/\tau+o(1/R)$.

This paper will use above formula to do error estimation formula. When a given error δ, the cell edge length H and the ratio of the classification length h of the radial distribution function is $\tau=H/h$, taken as $\tau=0.866/\delta$.

Viewing from section plane through the origin O, solvable and unsolvable set is a set of rings set a ring. Ring in two solvable set is ring of unsolvable set, and ring in two unsolvable set is ring of solvable set. When $h<\sqrt{3}H$, we take Ω_n is empty set, then Ω_b is the total calculated range. When H =0, Ω_b is empty set, then Ω_y is the total calculated range. For other conditions, let $\tau=h/H$, Range of basic solvable and unsolvable can be respectively expressed as

$$\Omega_y=\cup\Omega_n \text{ and } \Omega_b=\cup\Omega_{on} \tag{3.2}$$

where Ω_n : $n\tau-\tau+0.5\sqrt{3}<r/H<n\tau-0.5\sqrt{3}$, Ω_{on} : $n\tau-0.5\sqrt{3}<r/H<n\tau+0.5\sqrt{3}$, $n=1,2,\ldots,N_n$.

where N_n is the maximum layer number. As r/H is not less than 0, then the lower limit of Ω_1 is 0.As the coordinate origin is center of O cell, then the other cell midpoint coordinates can be expressed as (iH, jH, kH),and the distance from (iH, jH, kH) to the origin is

$$r=H\sqrt{m},\quad m=i^2+j^2+k^2$$

As a result, set consisted of all cell center point located in nth class basic solvable range Ω_n is called basic nth class solvable cell center point set:

$$\Omega_{P_n}=\left\{(i,j,k)\middle|(iH,jH,kH)\in\Omega_n, i,j,k,\text{are integers}\right\},\quad n=1,2,\ldots,N_n$$

All basic nth class solvable cell center point set is called basic solvable cell center point set:

$$\Omega_P=\cup\Omega_{P_n} \tag{3.3}$$

Similarly, set consisted of all cell center point located in nth class basic unsolvable range is called basic nth class unsolvable cell center point set:

$$\Omega_{U_n}=\left\{(i,j,k)\middle|(iH,jH,kH)\in\Omega_{on}, i,j,k\in\mathrm{N}\right\},\quad n=1,2,\ldots,N_n$$

All basic nth class unsolvable cell center point set Ω_{on} is called basic unsolvable cell center point set :

$$\Omega_U=\cup\Omega_{U_n} \tag{3.4}$$

Easy to see, basic solvable cell center point set Ω_P included in basic solvable range Ω_y, basic unsolvable cell center point set Ω_U included in basic unsolvable range Ω_b. Arbitrary point in basic nth class solvable cell center point set (i, j, k) corresponds to the spatial point $X = (iH, jH, kH)$, and the small cube with the point as cell center and side length H included in basic nth class range. This is because the distance from arbitrary point in cell to cell center is less than $0.5\sqrt{3}H$, and the boundary distance from spatial point $X = (iH, jH, kH)$ to the boundary basic nth class range is not less than $0.5\sqrt{3}H$. Similarly, basic unsolvable cell center point set Ω_U included in basic unsolvable range Ω_b. Arbitrary point in basic nth class unsolvable cell center point set (i, j, k) corresponds to the spatial point $X = (iH, jH, kH)$, and the small cube with the point as cell center and side length H may be across basic nth class range and basic (n+1)th class range. This is because the distance from existing point in cell to cell center may equal $0.5\sqrt{3}H$,and the boundary distance from spatial point $X = (iH, jH, kH)$ to the boundary basic nth class range is not greater than $0.5\sqrt{3}H$ .let the distance from arbitrary point corresponding to the spatial point in basic nth class unsolvable cell center set to the origin be $H\sqrt{m}$, $m = i^2 + j^2 + k^2$,and the dimensionless distance from the point to the boundary point of basic nth class boundary range is defined as

$$r' = \left(nh - H\sqrt{m}\right)/\left(0.5\sqrt{3}H\right) = 2\left(n\tau - \sqrt{m}\right)/\sqrt{3} \tag{3.5}$$

Easy to calculate the value r' from -1 to 1. When cell center point (iH, jH, kH) is in basic nth class range, r' values from -1 to 0; When the point (iH, jH, kH) is in basic (n+1)th class range, r' values from 0 to 1. Let $r'' = 0.5 + 0.5r'$, then r'' values from 0 to 1. When r'=-1, r''=0; when r'=0, r''=0.5; when r'=1, r''=1. According to the value r' (or r''), we approximately find volume of cell in basic nth class range $H^3 r''$, and volume of cell in basic (n+1)th class range $H^3(1 - r'')$. When the number of particles is N_{ijk}, we approximately compute the number of particles of cell $N_{ijk}\ r''$ in basic nth class range, and the number of particles of cell N_{ijk} (1- r'') in basic (n+1)th class range. When cell center point is belong to basic nth class solvable cell center point set, taking r''=1, we can find volume of cell in basic nth class range H^3. Therefore, Cellular location, distance classification and scale of particles can be denoted by vector (i, j, k, n, r'', u), where (i, j, k) indicates cell location, n indicates distance classification of particles, r'' is the ratio, and u is a discriminate function of solvable set. When u=1, X is belong to solvable set; when u=-1, X is not belong to solvable set. n, r'', u is decided by i, j, k (given later method that u is replaced by r'').

From the above discussion, we know the total number of particles in basic nth class range is the total number of particles corresponding to cell in basic nth class solvable cell center point set, plus weighted sum of the number of particles corresponding to cell basic nth and (n-1)th class unsolvable cell center point set. It can be described by mathematical formulas:

$$\sum N_{ijk}\left[(i,j,k)\in\Omega_{P_n}\right]+\sum(1-r'')N_{ijk}\left[(i,j,k)\in\Omega_{U_{n-1}}\right]+\sum r''N_{ijk}\left[(i,j,k)\in\Omega_{U_n}\right]$$

multiplied by the number of particles N_{000} of cell, basic nth class distance total is

$$f_n = N_{000}\left\{\sum N_{ijk}\left[(i,j,k)\in\Omega_{P_n}\right]+\sum(1-r'')N_{ijk}\left[(i,j,k)\in\Omega_{U_{n-1}}\right]+\sum r''N_{ijk}\left[(i,j,k)\in\Omega_{U_n}\right]\right\} \quad (3.6)$$

4 Radial Classification of Non-O Cell

When cell center is not the origin, we can move on cell center to the origin, then compute basic nth class distance total by the above formula. Let numbering s cell center point be (i',j',k'), for arbitrary point (i'',j'',k'') of the calculation region, we take translation transformation $(i,j,k)=(i'',j'',k'')-(i',j',k')$,then use the above method to calculate basic nth class distance total f_{n_s}. Summing for s, we obtain nth class radial function value:

$$F_h(r)=\sum_s f_{n_s} \quad (4.1)$$

Here we study the computational complexity. For particles N, its coordinate is $X=(x,y,z)$, and multiplication operation number of computing particle number in cell (i,j,k) is 3 N. The rest of the calculation is related to number of cell NH. In the basic classification, computing $i^2+j^2+k^2$ uses 3 NH multiplications, use the database to compute integer square root $\sqrt{m}$, when do calculation, pre-read into an array. According to value m, directly search $\sqrt{m}$, search number is NH. Multiplication number of computing nth class distance number is less than NH. When cell center is not the origin, because of using base solution, through coordinate translation transformation to compute distance classification, reduce the amount of calculation, and the total number of multiplications is less than NH^2 .When $NH^2<N$ or $NH^2=O(N)$, the total amount of calculation is $O(N)$, i.e., the amount of calculation is direct proportion to N .When the number of particles N is large, we take the total number of cell is less than N. Actually, this method makes sense for Large N.

Error analysis: As this paper constructs distribution proportion function, it is easy to know at least half the particles in the unsolvable range calculations are correct. So when calculation accuracy (1- δ) is given, the paper uses formula τ =0.866/ δ to calculate the ratio τ.

5 Results

As numerical example, in cube with side length 1, randomly choose particles N obeying uniform distribution, and calculate radial distribution function of particles N .Respectively use the exact method and approximate method to calculate radial distribution function. Accurate method is to directly calculate the distance between arbitrary two points, then according to classification step of radial distribution function, calculate distance classification, and finally calculate the radial distribution function. The approximate method in this paper is preprocessing method of above discussion. The preprocessing method finally considers basic solvable range and basic unsolvable range, according to the ratio of a given step τ, and integer coordinate (i,j,k), calculate m, $m'=\sqrt{m}$, and radial classification $n=[m'/\tau]+1$, and calculate r' or r'', and solvable set discriminate function u. Then use coordinate translation transformation to find radial distribution function. When cell center is (i,j,k), it may be in nth class solvable range, or (n-1)th class unsolvable range, or nth class unsolvable range. According to definition of r', when $r'\le 1$, particles are in nth class unsolvable range; when $1\le r'<\tau/0.866-1$, particles are in nth class solvable range; when $\tau/0.866-1\le r'<\tau/0.866$, particles are in (n-1)th class unsolvable range. When computing functions $n(i,j,k,\tau)$ and $r''(i,j,k,\tau)$, $i,j,k=1,2,3,\ldots,1/H+1$; stored as an array, called directly in the calculation. Then calculate the number of particles in each cell N_{ijk}, and use translation transformation to find radial distribution function. Table 1 shows calculated CPU time of change relationship with the ratio τ. For particles $N=2\times10^6$ and $h=0.1$, Table 1 shows the two approximation methods. Approximation method 2 uses preprocessing methods of this paper and approximation method 1 directly calculates distance between cell and cell, and determines solvable range of cell. Table 1 shows computing time of approximation method 1 is as about four times as one of approximation method 2, and calculating result of the two approximate methods is the same. Approximation method 2 takes into account the preprocessing time, where preprocessing time is only 2% of total computing time. But time of precise method is greater 200 times than one of the approximate method 2. We also calculated error of method of this paper and exact method. When τ=4, error is 0.0824; when τ=8, error is 0.0315.

Table 1 Calculated CPU time of change relationship with the ratio τ

Ratio τ	4	5	6	7	8
Exact method	312.7	312.7	312.7	312.7	312.7
Approximation method 1	0.826	3.028	9.401	23.74	55.42
Approximation method 2	0.201	0.704	2.041	5.396	12.35

6 Conclusion

That use spatial grid to approximately compute radial distribution function can significantly reduce the computation time, and the calculation error is in the controllable range. Based on the spatial grid cube topology similarity, this paper given preprocessing calculation method, discussed the basic solvable range and basic unsolvable range, given the classification value and pre-calculation value of distribution function, then used translational transform to calculate all the cell distance classification.

References

1. Tu, Y.C., Chen, S.P., Pandit, S.: Computing Distance Histograms Efficiently in Scientific Databases. In: Procedings of ICDE, Shanghai, China, pp. 796–807 (March 2009)
2. Bamdad, M., Alavi, S., Najafi, B., Keshavarzi, E.: A new expression for radial distribution function and infinite shear modulus of Lennard Jones fluids. Chem. Phys. 325, 554–562 (2008)
3. Chen, N.Y., Xu, C., Li, T.H., Jiang, N.X.: Comptuterized simulation of molten LiFKCl solution by Monte Carlo method-radial distribution function and thermodynamic properties. Science in China, Ser.B (10), 21–26 (1987) (in Chinese)
4. Wang, W., Feng, Y.D., Chi, X.B.: Application of tree structure in N-body problem. Application Research of Computers 25(1), 42–44 (2008) (in Chinese)

Stability of a Continuous Type of Neural Networks with Recent-History Distributed Delays

Yuejin Zhou, Juan Zhang, and Yunjia Wang

Abstract. In this paper a class of neural networks with recent-history distributed delays and impulses are studied. Sufficient conditions are obtained for the existence and globally exponential stability of a unique equilibrium of the impulsive system by using Banach's fixed point theorem, matrix and associated theory. The results of this paper generalize and improve the previous known results of non-impulsive systems.

Keywords: a class neural networks, distributed delays and impulses, Banach's fixed point theorem, global exponential stability, unique equilibrium.

1 Introduction

Consider a class of neural networks with recent-history distributed delays and impulses as follows.

$$\begin{cases} \dot{x}_i(t) = -c_i x_i(t) + \sum_{j=1}^{n} a_{ij} g_j(x_j(t)) + \sum_{j=1}^{n} \int_0^{\tau} b_{ij}(s) g_j(x_j(t-s)) ds + d_i \\ \Delta x_i(t_k) = I_k(x_i(t_k)) \end{cases} \tag{1}$$

with the initial values $x(\theta) = \varphi(\theta)$,$\theta \in [-\tau, 0]$.

Yuejin Zhou . Yunjia Wang
School of Environment Science and Spatial Informatics,
China University of Mining and Technology, 221008 Xu Zhou, China

Yuejin Zhou
School of Information and Electronics Engineering,
Jiangsu Institute of Architectural Technology, 221008 Xu Zhou, China

Juan Zhang
Audio-visual Library Information Center, Jiangsu Provincial Xuzhou Pharmaceutical Vocational College, 221008 Xu Zhou, China
e-mail: yuejinzh@163.com

Z. Qian et al. (Eds.): Recent Advances in CSIE 2011, LNEE 125, pp. 719–725.
springerlink.com

Where n is the number of neurons in the indicated neural networks, $x(t)=(x_1(t),x_2(t),...x_n(t))^T$ is the state vector of the network at time t, $g(x(t))=(g_1(x_1(t)),...g_n(x_n(t)))^T$ is the output vector of the network at time t, $C=diag(c_1,...c_n)$ is a diagonal matrix with $c_i>0$ for i=1,…,n, $A=(a_{ij})_{n\times n}$ is the feedback matrix, τ denotes the maximum possible transmission delay from one neuron to another, $B(s)=(b_{ij}(s))_{n\times n}(0\le s\le\tau)$ is the delayed feedback matrix, d_i is a constant and denotes external inputs, Φ is a continuous function that maps [- τ ,0] into R^n . $\Delta x_i(t_k)=x_i(t_k^+)-x_i(t_k^-)$ is the impulses at moment t_k and $t_1<t_2<...$ is a strictly increasing sequences such that $\lim_{k\to+\infty}t_k=+\infty$. Furthermore, we assume that the following three hypotheses hold.

(H1) the signal transmission functions g_i (i=1,2,…,n) is Lipschitz continuous on R with Lipschitz constant L_j ,that is, $|g_j(u)-g_j(v)|\le|u-v|$ for all u, $v\in$ R.

(H2) for i=1,2,…,n, $\int_0^\tau|b_{ij}(s)|ds$ is existent, and there exist nonnegative constant b_{ij}^+ such that

$$\int_0^\tau|b_{ij}(s)|ds\le b_{ij}^+ \tag{2}$$

(H3) for j=1,2,…,n, $\int_0^\tau|x_j(t-s)-y_j(t-s)|ds$ is existent, and there exist nonnegative constant q_j such that

$$\int_0^\tau|x_j(t-s)-y_j(t-s)|ds\le q_j|x_j(t)-y_j(t)| \tag{3}$$

Hopfield neural networks, cellular neural networks, and bi-directional associative memory networks with recent-history distributed delays and impulses have been applied in pattern recognition[1~4]. We summarize these situations and obtain more generalized conclusions.

In this paper, by using the Banach's fixed point theorem, matrix and associated theory, we will give some new sufficient conditions for the existence and globally exponential stability of a unique equilibrium point for the impulsive system (1). The results of this paper are new and more generalized. Moreover, an example is also provided to illustrate the effectiveness of the new results.

we assume that $(x_1(t),x_2(t)...,x_n(t))^T=(x_1(t-0),x_2(t-0)...,x_n(t-0))^T$.

An equilibrium point of (1) is a constant vector $(x_1^*,x_2^*,...x_n^*)^T\in R^n$ which satisfies the system

$$C_ix_i^*=\sum_{j=1}^n a_{ij}g_j(x_j^*(t))+\sum_{j=1}^n\int_0^\tau b_{ij}(s)g_j(x_j^*(t-s))ds+d_i \tag{4}$$

When the impulsive jumps $I_k(\cdot)$ as assume to satisfy $I_k(x_i^*)=0$, $i=1,2,...,n,k\in Z^+$, where Z^+ denotes the set of all positive integers. We use following norm of R^n .

Definition 1. Let $X^* = \left(x_1^*, x_2^*, ... x_n^*\right)^T$ be an equilibrium point of system (1). If there exist constants $\alpha > 0$ and $M > 1$ such that for every solution $X(t) = \left(x_1(t), x_2(t), ... x_n(t)\right)^T$ of system (1) with any initial value $\varphi = \left(\varphi_1(t), \varphi_2(t), ..., \varphi_n(t)\right)^T \in C\left([-\tau, 0]; R^n\right)$, $\left|x_i(t) - x_i^*\right| \le M \left\|\varphi - x^*\right\| e^{-at}, \forall t > 0$ Where i=1,2,…,n. Then X^* is said to be globally exponentially stable point.

Definition 2. (*Berman and Plemmons [2], Lasalle [9]*). A real $l \times l$ matrix $K = (k_{ij})$ is said to be an M-matrix if $k_{ij} \le 0; i, j = 1, 2, ..., l; i \ne j$, and $K^{-1} \ge 0$.

Lemma 1.1. (*Berman and Plemmons [2], Lasalle [9]*). Let $K = (k_{ij})_{l \times l}$ with $k_{ij} \le 0; i, j = 1, 2, ... l; i \ne j$ .Then the following statements are equivalent.

K is a M-matrix.

There exists a vector $\eta = (\eta_1, \eta_2, ... \eta_l) > (0, 0, ... 0)$ such that $\eta K > 0$.

There exists a vector: $\zeta = \left(\zeta_1, \zeta_2, ..., \zeta_n\right)^T > (0, 0, ..., 0)^T$ such that $K\zeta > 0$.

Lemma 1.2. (*Berman and Plemmons [2], Lasalle [9]*).Let $A \ge 0$ be a $l \times l$ matrix and $\rho(A) < 1$, then $\left(E_l - A\right)^{-1} \ge 0$,where $\rho(A)$ denotes the spectral radius of A.

2 Existence and Uniqueness of the Equilibrium Point

Theorem 2.1. Let conditions (H1), (H2) and (H3) hold and $\rho\left(C^{-1}(A + BQ)L\right) < 1$,where $C = diag\left(c_1, c_2, ..., c_n\right)$, $A = \left(\left|a_{ij}\right|\right)_{n \times n}$, $B = \left(\left|b_{ij}^+\right|\right)_{n \times n}$, $Q = diag\left(q_1, q_2, ..., q_n\right)$, $L = diag\left(L_1, L_2, ..., L_n\right)$

Then there exists a unique solution of the system (1), which is a unique equilibrium point of this system.

Proof. It follows from (1) that

$$x_i^* = \frac{1}{c_i}\left[\sum_{j=1}^{n} a_{ij} g_j\left(x_j^*(t)\right) + \sum_{j=1}^{n} \int_0^{\tau} b_{ij}(s) g_j\left(x_j^*(t-s)\right) ds + d_i\right] (i = 1, 2, \cdots, n) \quad (5)$$

We therefore need to show that (5) has a unique solution. Now consider a mapping $\Phi : R^n \to R^n$ defined by

$$\Phi(x_1,x_2,...,x_n)=(\Phi_1(x_1,x_2,...,x_n),...,\Phi_n(x_1,x_2,...,x_n))^T$$
$$=\left(\frac{1}{c_1}\begin{bmatrix}\sum_{j=1}^{n}a_{1j}g_j(x_j)\\+\sum_{j=1}^{n}\int_0^{\tau}b_{1j}g_j(x_j(t-s))ds+d_1\end{bmatrix},...,\times\frac{1}{c_n}\begin{bmatrix}\sum_{j=1}^{n}a_{nj}g_j(x_j)\\+\sum_{j=1}^{n}\int_0^{\tau}b_{nj}g_j(x_j(t-s))ds+d_n\end{bmatrix}\right)^T \tag{6}$$

Then, for any $x=(x_1,x_2,...,x_n)^T\in R^n$, $y=(y_1,y_2,...,y_n)^T\in R^n$,
One has

$$\left(\left|(\Phi(x)-\Phi(y))\right|_1,...,\left|(\Phi(x)-\Phi(y))\right|_n\right)^T\le F\left(\left|(x-y)_1\right|,...,\left|(x-y)_n\right|\right)^T \tag{7}$$

Let m be a positive integer, then it follows from (7) that

$$\left(\left|(\Phi^m(x)-\Phi^m(y))_1\right|,...,\left|(\Phi^m(x)-\Phi^m(y))_n\right|\right)^T\le F^m\left(\left|(x_1-y_1)\right|,...\left|(x_n-y_n)\right|\right)^T \tag{9}$$

Since $\rho(F)<1$, we see that $\lim_{m\to+\infty}F^m=0$.

Which implies that there exist a positive integer N and a positive constant $r<1$ such that $F^N=\left(c_i^{-1}\left(|a_{ij}|+|b_{ij}^+|q_j\right)L_j\right)^N=(h_{ij})_{n\times n}$ and

$$\sum_{j=1}^{n}h_{ij}\le r, i=1,2,...,\text{n}. \tag{10}$$

In view of (9) and (10), For all i =1,…,n, we have

$$\left|(\Phi^N(x)-\Phi^N(y))_i\right|\le\sum_{j=1}^{n}h_{ij}\left|x_j-y_j\right|\le\sum_{j=1}^{n}h_{ij}\max_{1\le j\le n}\left|x_j-y_j\right|=r\|x-y\|_\infty.$$

It follows that

$$\left\|\Phi^N(x)-\Phi^N(y)\right\|_\infty=\max_{1\le i\le n}\left|(\Phi^N(x)-\Phi^N(y))_i\right|\le r\|x-y\|_\infty \tag{11}$$

By r<1, this implies that mapping: $\Phi^N:R^n\to R^n$ is a contraction mapping. By the fixed point theorem of Banach space, Φ has a unique fixed point x^* in R^n, such that $\Phi:x^*=x^*$.Therefore, system (1) has a unique solution.

3 Global Exponential Stability

Definition 3.1. The equilibrium state $X^*=(x_1^*,x_2^*,...,x_n^*)^T$ of system (1) is said to be globally exponentially stable if there exist constants $\lambda>0$ and $M>0$ such that $\left|x_i(t)-x_i^*\right| \le Me^{-\lambda t}\left\|\varphi_i - x_1^*\right\|, t>0$, Where, $\left\|\varphi_i - x_i^*\right\| = \max_{1\le i\le n}\sup_{s\in[-\tau,0]}\left|\varphi_i(s)-x_i^*\right|$.

Theorem 3.2. Assume that the assumptions in Theorem 2.1 hold. Furthermore, suppose that the impulsive operators $I_k\left(x_i\left(t_k\right)\right)$ satisfy

$$\begin{cases} I_k\left(x_i\left(t_k\right)\right) = -\gamma_{ik}\left(x_i\left(t_k\right)-x_i^*\right), i=1,2,...,\mathrm{n}\ ,\ k\in\{1,2,...\} \\ 0\le\gamma_{ik}\le 2, i=1,2,...,n; k\in Z^+ \end{cases} \tag{12}$$

Then the unique equilibrium point x^* of (1) is globally exponentially stable.

Proof. Let $X(t)=\left(x_1(t),x_2(t),...,x_n(t)\right)^T$ be an arbitrary solution of system (1) with any initial value $x(\theta)=\varphi(\theta)$ and $\varphi=\left(\varphi_1(t),\varphi_2(t),...,\varphi_n(t)\right)^T\in C\left([-\tau,0];R^n\right)$.

Set $y_i(t)=x_i(t)-x_i^*, i=1,2,...,\mathrm{n}$, Then, from (1), we get

$$\begin{cases} \dfrac{dy_i(t)}{dt} = -c_i y_i(t)+\sum_{j=1}^n a_{ij}f_j\left(y_j(t)\right)+\sum_{j=1}^n\int_0^\tau b_{ij}(s)f_j\left(y_j(t-s)\right)ds+d_i, t\ne t_k \\ \Delta y_i\left(t_k\right) = -\gamma_{ik}y_i\left(t_k\right), t=t_k, k\in Z^+, i=1,2,...,\mathrm{n} \end{cases} \tag{13}$$

Also $\left|y_i\left(t_k+0\right)\right| = \left|1-\gamma_{ik}\left|y_i\left(t_k\right)\right|\right| \le \left|y_i\left(t_k\right)\right|$, where $f_j\left(y_j(t)\right)=g_j\left(x_j(t)\right)-g_j\left(x_j^*\right)$. Due to (H2), we know that $0\le\left|f_i\left(y_i(t)\right)\right|\le L_i\left|y_i\right|$. The initial condition of (13) is $\Psi(s)=\varphi(s)-x^*$.

Since $\rho(F)<1$, it follows from Lemma 1.2 that E_n-F is an M-matrix. In view of Lemma 1.1, there exists a constant vector, $\bar{\zeta}=\left(\bar{\xi}_1,\bar{\xi}_2,...,\bar{\xi}_n\right)^T>(0,0,...,0)^T$

Such that, $\left(E_n-F\right)\bar{\zeta}=\left(E_n-C^{-1}(A+BQ)L\right)\bar{\zeta}>(0,0,...,0)^T$

Let $F_i\left(\lambda_i\right)=\left(c_i-\lambda_i\right)\bar{\zeta}_i-\sum_{j=1}^n\bar{\zeta}_j\left(\left|a_{ij}\right|+\left|b_{ij}^+\right|q_je^{\lambda_i\tau}\right)L_j$, for $\lambda_i\in[0,\infty)$. Since $F_i(0)>0$, $F_i(\infty)<0$, we can choose a constant $0<\lambda<1$ such that

$$\left(\lambda-c_i\right)\bar{\zeta}_i+\sum_{j=1}^n\bar{\zeta}_j\left(\left|a_{ij}\right|+\left|b_{ij}^+\right|q_je^{\lambda\tau}\right)L_j<0 \tag{14}$$

It follows from that (14) that there exists a constant $l>0$ such that $\zeta_i=l\bar{\zeta}_i>\sup_{-\tau\le s\le 0}\left|\Psi_i(s)\right|$ and

$$(\lambda - c_i)\zeta_i + \sum_{j=1}^{n} \zeta_j \left(|a_{ij}| + |b_{ij}^+| q_j e^{\lambda\tau}\right) L_j = \left[(\lambda - c_i)\bar{\zeta}_i + \sum_{j=1}^{n} \bar{\zeta}_j \left(|a_{ij}| + |b_{ij}^+| q_j e^{\lambda\tau}\right) L_j \right] l < 0 \quad (15)$$

We define a Lyapunov function $V = (V_1, V_2, ..., V_n)^T$ by $V_i = e^{\lambda t} |y_i(t)|, i = 1, 2, ..., \text{n}$. In view of (13), we obtain

$$\frac{d^+ V_i(t)}{dt} = (\lambda - c_i) V_i(t) + \sum_{j=1}^{n} \left(|a_{ij}| + |b_{ij}^+| q_j\right) L_j V_j(t) \quad (16)$$

For $t \geq 0, t \neq t_k$ .we claim that

$$V_i(t) = e^{\lambda t} |y_i(t)| < \zeta_i, t \geq 0, t \neq t_k, i = 1, 2, ..., \text{n} \quad (17)$$

Suppose not. There must exist some $i \in \{1, 2, ..., \text{n}\}$ and $t_i > 0$ such that $V_i(t_i) - \zeta_i = 0$. And

$$V_j(t) < \zeta_j, \forall t \in [-\tau, t_i), j = 1, 2, ..., \text{n} \quad (18)$$

This implies that $V_i(t_i) - \zeta_i = 0$,

$$V_j(t) - \zeta_j < 0, \forall t \in [-\tau, t_i), j = 1, 2, ..., \text{n} \quad (19)$$

It follows from (16) and (19) that

$$0 \leq D^+ (V_i(t_i) - \zeta_i) = D^+ (V_i(t_i)) \leq (\lambda - c_i)\zeta_i + \sum_{j=1}^{n} \zeta_j \left[|a_{ij}| + |b_{ij}^+| q_j \right] L_j, t \geq 0, t \neq t_k$$

This is a contradiction to (15). Hence, (17) holds. It follows that

$$|y_i(t)| < \max_{1 \leq i \leq n} \{\zeta_i\} e^{-\lambda t}, \quad t \geq 0, t \neq t_k, k = 1, 2, ..., i = 1, 2, ..., \text{n} \quad (20)$$

$$V_i(t_k + 0) = e^{\lambda t} |y_i(t_k + 0)| \leq e^{\lambda t} |y_i(t_k)| = V_i(t_k) \quad k \in Z^+ \quad (21)$$

Letting $\|\Psi\| = \|\varphi - x^*\| > 0$, it follows from (20) and (21) that we may choose $M_\varphi > 1$ such that $\max_{1 \leq i \leq n} \{\zeta_i\} \leq M_\varphi \|\varphi - x^*\|, t \geq 0$. In view of (20) and (21), $|x_i(t) - x_i^*| = |y_i(t)| \leq \max_{1 \leq i \leq n} \{\zeta_i\} e^{-\lambda t} \leq M_\varphi \|\varphi - x^*\| e^{-\lambda t}, t \geq 0, i = 1, 2, ..., \text{n}$. This completes the proof.

4 Example

Consider the following system:

$$\begin{cases} x_1'(t) = -x_1(t) + \frac{2}{3} g_1(f_1(t)) + \int_0^1 2s g_1(f_1(t-s))ds - 1, \\ x_2'(t) = -6x_2(t) + g_2(f_2(t)) - \int_0^1 \frac{1}{2} s g_1(f_1(t-s))ds - 1, t > 0, t \neq t_k, \\ \Delta x_1(t_k) = -\gamma_{1k}(x_1(t_k - 1)), \Delta x_2(t_k) = -\gamma_{2k}(x_2(t_k - 1)), k = 1, 2, ..., \end{cases} \tag{22}$$

Where $g_1(x) = g_2(x) = |x|$, and $t_1 < t_2 < \cdots$ is a strictly increasing sequences such that $\lim_{k \to +\infty} t_k = +\infty$, $\gamma_k = 1 + \frac{1}{2}\sin(1+k)$, $\gamma_{2k} = 1 + \frac{2}{3}\cos(2k)$.

Take $L_1 = L_2 = 1$, $q_1 = q_2 = 1$, by easy computation, (22) satisfy the conditions (H1), (H2) and (H3), and $\rho(C^{-1}(A+BQ)L) = \frac{1}{6} < 1$. Therefore, from Theorem 3.1, (22) has exactly one equilibrium point (2), which is globally exponentially stable.

5 Conclusions

In this paper, a class of neural networks with recent-history distributed delays and impulses has been studied. Some sufficient conditions are obtained for the existence and globally exponential stability of a unique equilibrium of the impulsive system. When the impulsive jumps are absent, the results reduce to those of the non-impulsive systems. Our results generalize and improve the previous known results.

References

1. Yang, X.F., Liao, X.F.: Global exponential stability of a class of neural networks with recent-history distributed delays. Chaos, Solitons and Fractals 25, 441–447 (2005)
2. Liu, B.W., Huang, L.H.: Global exponential stability of BAM neural networks with recent-history distributed delays and impulses. Neuro Computing 69, 2090–2096 (2096)
3. Xia, Y.H., Cao, J.D.: Global exponential stability of delayed cellular neural networks with impulses. Neurocomputing, 2495–2501
4. Sun, J.T., Zhang, Y., Wu, Q.D.: Less conservative conditions for asymptotic stability of impulsive control systems. IEEE Trans. Automat. Control 48, 829–831 (2003)
5. Lasalle, J.P.: The Stability of Dynamical System. SIAM, Philadephia (1976)

The Numerical Fitting and Program Realization for Ordinary Differential Equation with Parameters to Be Determined

Changlong Yu, Jufang Wang, and Xianglin Wei

Abstract. Based on the theoretical basis of the numerical solution of ordinary differential equation(s), the principle of least squares and the basic idea of unconstrained optimization, the paper proposes a method for the problem of numerical fitting for ordinary differential equation(s) with parameters to be determined. In according to mixed programming with Delphi and Matlab, we achieve the visual calculation of the numerical fitting for ordinary differential equation(s) with parameters to be determined. Calculation is easily to be realized and operated, but also suitable for the numerical fitting problem of parameters to be determined in non-linear differential equation(s) which contains more than one parameters to be determined.

1 Introduction

In modern science and technology areas, the movement of many things can be expressed by ordinary differential equation(s). Ordinary differential equation is the basic mathematical theory and method to study the motion, evolution and variation of things in natura science and social science, it has increasing wide application. At present, lots of principles and rules in physics, chemistry, biology, engineering, aerospace, medical science, economy and finance can be expressed by proper ordinary differential equation(s), see [1,2]. However, ordinary differential equation(s) which satisfies some rules often has parameters $\lambda_1, \lambda_2, \cdots, \lambda_k$, people hope to gain m sets of data $(t_i, y_{ij})(i = 1, 2, \cdots, m; j = 1, 2, \cdots, n)$ by experimental observation to approximately determine the values of $\lambda_1, \cdots, \lambda_k$ in ordinary differential equation(s). This problem can be described as follow.

Suppose that we can get m sets of data

$$(t_i, y_{ij}), (i = 1, 2, \cdots, m; j = 1, 2, \cdots, n) \tag{1}$$

Changlong Yu · Jufang Wang
Heibei University of Science nd Technology, 70 Yuhua East Rd, Shijiazhuang, Hebei, (050018)PRC
e-mail: {changlongyu, wangjufang}@126.com

Z. Qian et al. (Eds.): Recent Advances in CSIE 2011, LNEE 125, pp. 727–732.
springerlink.com

where t_i is the value of independent variable t, y_{ij} is the experimental observed value of function $y_j(t)$ at $t = t_i$. In fact, the real functional relation

$$y_j = y_j(t), (j = 1, 2, \cdots, n) \tag{2}$$

is unknown, we just only know the differential equation(s) of μ order for the function(s) of (2) with $k(k \geq 1)$ parameters $\lambda_1, \lambda_2, \cdots, \lambda_k$ to be determined,

$$\frac{d^\mu \overrightarrow{Y}}{dt^\mu} = \overrightarrow{f}(\lambda_1, \lambda_2, \cdots, \lambda_k; t, \overrightarrow{Y}, \frac{d\overrightarrow{Y}}{dt}, \cdots, \frac{d^{\mu-1}\overrightarrow{Y}}{dt^{\mu-1}}) \tag{3}$$

or

$$\overrightarrow{f}(\lambda_1, \lambda_2, \cdots, \lambda_k; t, \overrightarrow{Y}, \frac{d\overrightarrow{Y}}{dt}, \cdots, \frac{d^{\mu-1}\overrightarrow{Y}}{dt^{\mu-1}}, \frac{d^\mu \overrightarrow{Y}}{dt^\mu}) = 0 \tag{4}$$

where $\overrightarrow{Y} = [y_1, y_2, \cdots, y_n]^T, \overrightarrow{f} = [f_1, f_2, \cdots, f_n]^T, k \leq m, n \geq 1, \mu \geq 1, k, n$ and μ are positive integer.

According to known observed data of (1), we should determine the values of unknown parameters $\lambda_1, \lambda_2, \cdots, \lambda_k$ in (3) (or (4)), which are denoted by $\lambda_1^*, \lambda_2^*, \cdots, \lambda_k^*$. This problem is called numerical fitting problem of ordinary differential equation(s) with parameters to be determined.

In[3], Xinyuan Wu proposed the numerical fitting method for ordinary differential equation(s) with parameters to be determined, but it only considered $\overrightarrow{f}$ about $\lambda_1, \lambda_2, \cdots, \lambda_k$ is linear, so it has certain limitations. For general, we give a new numerical fitting method for ordinary differential equation(s) in our paper.

2 The Basic Idea of Solution

The common and direct method is least squares. The least squares is to find best function for data by minimizing the error sum of squares, it is a common method to deal with experimental data. We use least squares to make the curve of $y_j = y_j(t)$ which is the solution of the differential equation(s) of (3) approximate to discrete points $(t_i, y_{ij})(i = 1, 2, \cdots, m)$ for every $j(j = 1, 2, \cdots, n)$. Before curve fitting by least squares, we should firstly determine the form of fitted curve. In fact, the form of the curve of $y_j = y_j(t)$ has been indirectly given by the differential equation(s) of (3), it's not necessary to determine the form. Obviously, y_j has parameters $\lambda_1, \lambda_2, \cdots, \lambda_k$ and is denoted by $y_j = y_j(\lambda_1, \lambda_2, \cdots, \lambda_k; t)(j = 1, 2, \cdots, n)$.

We often solve some special type equation to solve ordinary differential equation(s), but the analytic solution of many problems is difficult to be solved, so for general $y_j = y_j(\lambda_1, \lambda_2, \cdots, \lambda_k; t)$ can not be solved. Sometimes, for a large number of calculation, the analytic solution we get is not feasible, so it is not suitable to calculate ordinary differential equation(s) by solving analytic solution. Therefore, we can use the numerical algorithm of ordinary differential equation(s) to get the numerical solution $y_j^*(t_i)$ of $y_j(t)$ at $t = t_i$, see [4,5]. Apparently, y_j^* has parameters $\lambda_1, \lambda_2, \cdots, \lambda_k$ and is denoted by $y_j^* = y_j^*(\lambda_1, \lambda_2, \cdots, \lambda_k; t)(j = 1, 2, \cdots, n)$. We can substitute $y_j^* = y_j^*(\lambda_1, \lambda_2, \cdots, \lambda_k; t)$ for $y_j = y_j(\lambda_1, \lambda_2, \cdots, \lambda_k; t)$ to approximately solve, and change this problem to the problem of solving the minimum point of multivariate function

$$I(\lambda_1, \lambda_2, \cdots, \lambda_k) := \sum_{j=1}^{n} \sum_{i=1}^{m} [y_j^*(\lambda_1, \lambda_2, \cdots, \lambda_k; t_i) - y_{ij}]^2, \tag{5}$$

then it can be summed up as unconstrained optimization problem, see [6-8].

3 Algorithm and Program Realization

3.1 Algorithm

In this section, we give the algorithm for the problem above.

The process of algorithm is as follow:

(1) Give the ordinary differential equation(s) with parameters for special problem;

(2) Input experimental data $(t_i, y_{ij}), (i = 1, 2, \cdots, m; j = 1, 2, \cdots, n)$;

(3) Compute $h_i = t_i - t_{i-1}$ and the numerical solution $y_j^*(\lambda_1, \lambda_2, \cdots, \lambda_k; t_i)$ of function $y_j(t)$ at $t = t_i$ through Euler formula

$$y_{(i+1),j} = y_{ij} + h_i f(x_i, y_{i1}, y_{i2}, \cdots, y_{in},);$$

(4) Based on the principle of least squares, we construct multivariate function

$$I(\lambda_1, \lambda_2, \cdots, \lambda_k) = \sum_{j=1}^{n} \sum_{i=1}^{m} [y_j^*(\lambda_1, \lambda_2, \cdots, \lambda_k; t_i) - y_{ij}]^2;$$

(5) Based on Matlab, we make the solution of the minimum point $(\lambda_1^*, \lambda_2^*, \cdots, \lambda_k^*)$ of multivariate function $I(\lambda_1, \lambda_2, \cdots, \lambda_k)$ by the method of unconstrained optimization;

(6) We make the fitting curves of corresponding functions by Matlab and comparatively analyze with original data;

(7) Realize the visualization of algorithm by the program language of Delphi.

3.2 Program Realization

The program is with the developing software of Delphi as platform, the mathematical calculation software of Matlab as tools, and optimization above as basic idea to design program and realize the processing of visualization.

The main interface is as follow:

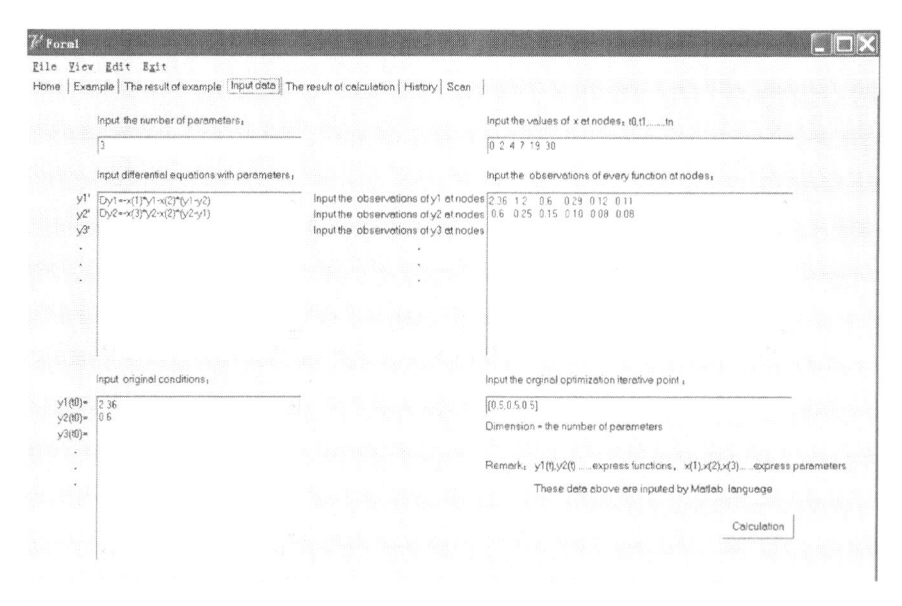

Fig. 1 Interface

The program is showed by the form of software. This software not only realize the visualization of calculation, but also give the graph of corresponding fitting curve and realize the function of recording many sets of data.

4 Numerical Example

In a chemical reaction, the change of reactants y_1 and y_2 are satisfied differential equations

$$\begin{cases} \dfrac{dy_1}{dt} = -\lambda_1 y_1 - \lambda_2(y_1 - y_2) \\ \\ \dfrac{dy_2}{dt} = -\lambda_3 y_2 - \lambda_2(y_2 - y_1) \end{cases}$$

where $\lambda_1, \lambda_2, \lambda_3$ are parameters to be determined. The experimental data are as below:

We use designed program to solve the numerical fitting problem of ordinary differential equation(s) with parameters to be determined. First, we should input some data in the interface of "Input data":

Table 1 The experimental observed values of t

t	0	2	4	7	19	30
y_1	2.36	0.86	0.51	0.29	0.12	0.11
y_2	0	0.44	0.37	0.28	0.08	0.08

(1) Input the number of parameters: 3;
(2) Input the differential equations with parameters:

$$Dy1 = -x(1) * y1 - x(2) * (y1 - y2),$$
$$Dy2 = -x(3) * y1 - x(2) * (y2 - y1);$$

(3) Input the values of independent variable t at known nodes:
0 2 4 7 19 30;
(4) Input the observed values of every functions at known nodes:
2.36 0.86 0.51 0.29 0.12 0.11
0 0.44 0.37 0.28 0.08 0.08;
(5) Input the initial iterative points of unconstrained optimization: [0.5, 0.5, 0.5].

After inputting data above, please click "calculation", the interface of "calculation result" for program can be showed as follow:

The values of parameters are $x(1) = 0.3077$, $x(2) = 0.2019$ and $x(3) = 0.0869$.

According to the values of parameters, we give the graphs of functions $y_1(t)$ and $y_2(t)$ and point out the discrete points of experimental observations.

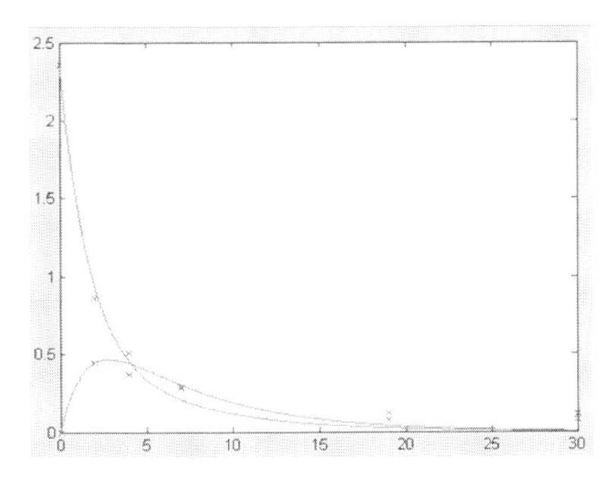

Fig. 2 Fitted curve

It can be showed from graph that the approximating level of discrete experimental observed points and the curves of functions $y_1(t)$ and $y_2(t)$ is much better, so the numerical fitting method which we put forward for ordinary differential equation(s) with parameters to be determined is effect and the program designed is feasible. In this program, we can solve the numerical fitting problem of nonlinear differential equation(s) with many parameters to be determined. The process of calculation is simple, practical and easy to be operated, even one who doesn't know Delphi and Matlab can use the software to solve the numerical fitting problem of ordinary differential equation(s) with parameters to be determined.

Remark: In center algorithm, what we use to compute the numerical values at corresponding nodes of ordinary differential equation(s) is famous EULER algorithm. At the beginning, our idea is that we can use RUNGE-KUTTA method of four order, its approximating effect is much better to deal with this problem. However, the number of terms we get from program for objective function is exponential growth, it makes running speed much slower and this is not our will. If we can solve problem above, I think the result will be better.

Acknowledgements. This paper is supported by National Natural Science Foun- dation of China (10901045)(10926054) and the Natural Science Foundation of Hebei Province (A2009000664) and the Foundation of Hebei University of Science and Techno- logy (XL200757) (XL201047) .

References

1. Walter, M.: Ordinary Differential Equation. Springer, Beijing (2003)
2. Yicang, Z., Zhen, J., Qin, J.: Ordinary Differential Equation and Application—-Method, Theory, Modelling and Computor. Sceince Press, Beijing (2003)
3. Xinyuan, W.: The Numerical Fitting Problem of Ordinary Differential Equation(s) with Parameter to be Determined. Journal of Nanjing University (3), 557–562 (1985)
4. Gerald, C.F., Wheatley, P.O.: Numerical Analysis, 7th edn. Beijing Higher Education Press (2006)
5. Yanjun, L., Ke, Z.: The Theory and Application of System Identification. National Defense Industry Press, Beijing (2004)
6. Qingyang, L., Nengchao, W., Dayi, Y.: Numerical Analysis, 4th edn. Huazhong University of Science and TechnologyPress, Wuhan (2006)
7. Yanli, W.: The Five Mixed Programming Methods of Delphi and Matlab. Journal of Heze University (2), 100–103 (2006)
8. Gang, T., Fei, R.: The Method of Realizing Matlab Based on Delphi. Jouranl of Shenyang Institute of Aeronautical Engineering (1), 38–39 (2002)

Dynamics Analysis of Blast in the Concrete by SPH Method

Guannan Wu, Baoliang Zhang, Jue Ding, Qingtao Wang, and Pu Song

Abstract. Smoothed particle hydrodynamics (SPH) method can deal with a larger deformation problem under the Lagrange framework, especially for simulating extend of cracks, so it shows its advantage gradually in investigating effect of blast. For the blast in the concrete, detonation wave model and dynamics model of concrete were established, and the SPH method was adopted to research the damage effect of concrete by TNT blast, which widens SPH method for study of blasting in a confined medium, and provide an important way to evaluate the damage effect of blast in the concrete.

1 Introduction

Numerical simulation by computer has gradually become a significant way to resolve modern engineering and scientific analysis problems. Recently, numerical method based on meshing system, like the finite difference method and the finite element method, they have been used widely in many fields of computational fluid mechanics (CFD) and computational solid mechanics (CSM), which get a great of achievement. But there still have shortcomings in some fields. For example, it is difficult to dispose many complicated problems, which involve free surface, deformed boundary, movement interface (in FDM) and maximum deformation problem (in FEM). And, a great deal of time and energy are needed in producing or reproducing high quality meshes. In recent years, the meshfree technology of new-type numerical method has caused people's attention, which is better than

Guannan Wu · Baoliang Zhang · Jue Ding · Qingtao Wang Shanghai
Institute of Applied Mathematics and Mechanics, Shanghai University,
Shanghai, China

Pu Song
China North Industries Group Corporation 204 Research InstituteXi'an, China

Z. Qian et al. (Eds.): Recent Advances in CSIE 2011, LNEE 125, pp. 733–738.
springerlink.com

traditional FDM and FEM based on meshing [1] in many applications. Therefore, the dynamics analysis of blast in the concrete is conducted by the smoothed particle hydrodynamics (SPH) method in this paper.

2 Theoretical Models and SPH Method

The process of internal-blast in the concrete is as followed. When explosive is detonated, detonation gas is produced under high temperature and high pressure. Then, detonation production expands immediately, and pushes surrounding air to form blast wave, which acts on the concrete medium to become stress wave. With development of waves, the movement, deformation and damage effects of medium happen. Moreover, blasting effect in the concrete is much more complicated than in the air and water. But so far, the researches on blast theory and damage effect in the concrete are imperfect [2-4]. In this paper, energy release model of TNT blast and dynamics model of concrete are proposed, and the numerical research on propagation of detonation wave and the process of damage in the confined media is conducted by SPH method.

1. Energy Release Model of TNT Detonation

For the basic equations of explosion mechanics, a state-equation describes a relation between pressure, temperature and volume of detonation production under the conditions of high temperature and high pressure. The afterburning effect model based on JWL equation of aluminized explosive is established.

Pressure in JWL state equation of detonation is defined as a function of specific volume and initial specific volume [5]:

$$p = A\left(1-\frac{\omega}{R_1 V}\right)e^{-R_1 V} + B\left(1-\frac{\omega}{R_2 V}\right)e^{-R_2 V} + \frac{\omega E}{V} \tag{1}$$

Where: ω, A, B, R_1, R_2 are input parameters .

Reactive Rate Model-Miller Extension Model [6]

A reactive rate model is needed for depicting numerically the development and propagation of detonation wave. Based on Lee-Tarver model, the researcher of Miller [6] proposes a simplified model, which considers only burning item for energy release. So, the speed of computation is greatly advanced.

Miller energy release model of detonation is as followed:

$$\frac{d\lambda}{dt} = a(1-\lambda)^m p^n \tag{2}$$

In this model, the burning process is controlled by reactive rate λ and pressure p. In addition, a is energy releasing constant. m is energy releasing index; n is pressure index, which are related to grain sizes of aluminum particles and specific surfaces

Table 1 Constitutive Model of RHT

RHT	Description	Gruneisen state equation is adopted, which can describe limit-surface more detailed. The partial surface in the stress space is not circular shape of HJC model, but is a triple folding symmetry shape. Three limit-surfaces involve: yield limit-surface, maximum limit-surface and residual strength limit-surface. Moreover, the model considers the evolvement of compression damage.
	Advantages	RHT model can predict remaining velocity more accurately, with considering integrally strain hardening, failure-surface, softening, damage of compression and effect of strain rate effect. Moreover, it can reveal concrete characteristics of initial yield strength, peak of failure intensity and residual intensity after peak. Furthermore, The angle θ is taken into consideration of limit-surface effect for yield-surface, which is consistent with the mechanical properties of concrete multi-axis.
	Disadvantages	Consideration of tensile damage is not very well.

2. Dynamic Model of Concrete

Concrete is a kind of mixed materials combined by cement, sand, stone and water. Mechanics characteristics of concrete is inhomogeneous, nonlinear and brittleness of damage. Under dynamic load of impulsion, constitutive relationship of concrete is very complicated. Currently, majority constitutive models of concrete are based on damage mechanics, which are phenomenological models from experiences, and they are lack of rigorous and completed analysis. Consequently, the RHT constitutive model of concrete material is adopted in this paper [7] .(Tab. 1)

3. Numerical Method

Smoothed particle hydrodynamics is one type of meshfree particle methods. The main ideal of this method is as following. Integral equation and partial differential equation are solved by a series of particles distributed arbitrarily, which gets accurate and steady values. Moreover, connection of mesh between particles is not needed.

SPH form for Euler equation is followed with artificial viscosity term:

$$\frac{D\rho_i}{Dt} = \sum_{j=1}^{N} m_j (v_i - v_j) \cdot \nabla_i W_{ij} \tag{3}$$

$$\frac{Dv_i}{Dt} = -\sum_{j=1}^{N} m_j \left(\frac{p_i}{\rho_i^2} + \frac{p_j}{\rho_j^2} + \Pi_{ij}\right) \cdot \nabla_i W_{ij} \tag{4}$$

$$\frac{De_i}{Dt}=-\frac{1}{2}\sum_{j=1}^{N}m_j(\frac{p_i}{\rho_i^2}+\frac{p_j}{\rho_j^2}+\Pi_{ij})(v_i-v_j)\cdot\nabla_i W_{ij} \tag{5}$$

$$\frac{Dr_i}{Dt}=v_i \tag{6}$$

$$p_i=p(\rho_i,u_i) \tag{7}$$

Where $W_{ij}=W(x_i-x_j,h)$ is smoothing function.

ρ, v, p, e is density, vector of particle velocity, pressure and internal energy of per mass respectively. Π_{ij} represents artificial viscosity term.

3 Analysis on Effect of Internal-Blast Damage in the Concrete

As a reference to a blast test (Fig 1), the explosive charge (TNT) is placed in the concrete target (the cylinder of radius 0.5m, and 0.5m height), then the charge is buried with sand. When TNT is detonated in the concrete, cavity or pit in the concrete target is formed. (Fig. 2) Where, R is radius of blast funnel, and H_1 is depth of blast funnel. Moreover, H_2 and H_0 are buried depths of charge and concrete height.

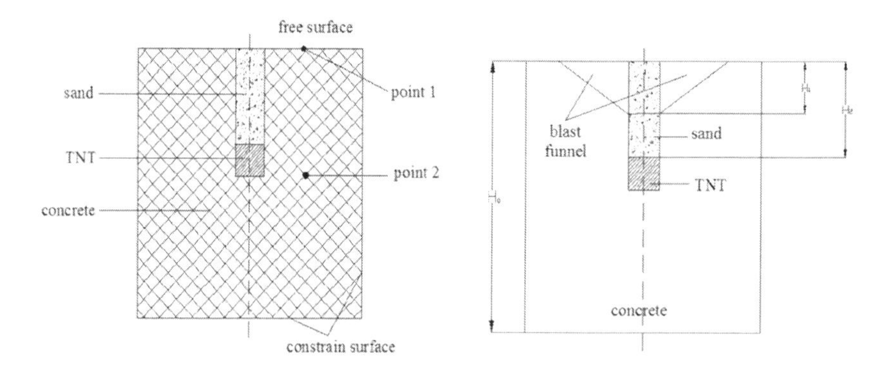

Fig. 1 Schematic diagram of blast in the concrete

Fig. 2 Schematic diagram of blast funnel

By SPH method, the particle size of SPH 2.5mm and 5mm are adopted separately. Computed radius of blast funnels (R, seeing Fig. 2) is consistent with measured data, which suggests particle-size 5mm of SPH can meet the engineering practical requirements. In order to save computing resource, particle-size 5mm of SPH is adopted.

Below, the different buried depths of charges TNT (H_2) blasted in the concretes are simulated numerically, which involves 1/8 and 2/8 of concrete height (H_0) separately. (from top to bottom) Fig. 3 shows the profiles of pressures and motion velocities for the position at the free surface (point 1), which is far from the axis of concrete about 90mm along horizontal direction.

Fig. 4 shows the profiles of pressures and velocities for the position at the half section of concrete height (point 2), which is far from the axis of concrete about 90mm along horizontal direction.

From Fig. 3 and Fig. 4, it shows the pressures and absolute velocities of point 2 are larger than point1, which is mainly due to point 1 affected by atmospheric environment.

Fig. 5 gives the damage distributions of concrete blasted by TNT, which buried depth is 2/8 of concrete height (H_0) and 1/4 structure is shown clearly at 8.05ms. From the computed results reveal blast funnel appearing, which dimension is 118.34mm of R-radius and 198.44mm of H_1-depth.

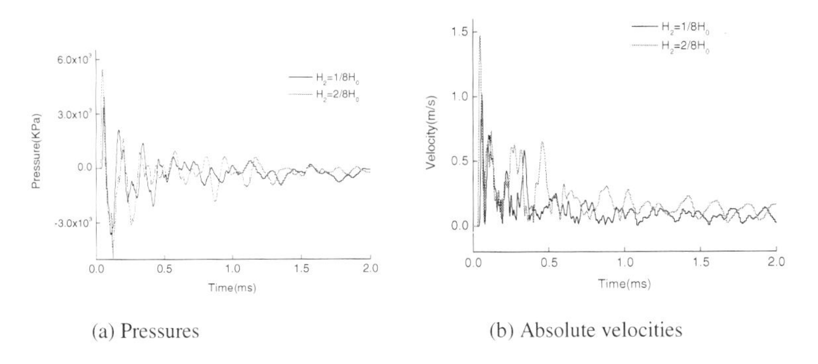

(a) Pressures (b) Absolute velocities

Fig. 3 Pressures and velocities development with time. (position (90, 0, 250), unit: mm)

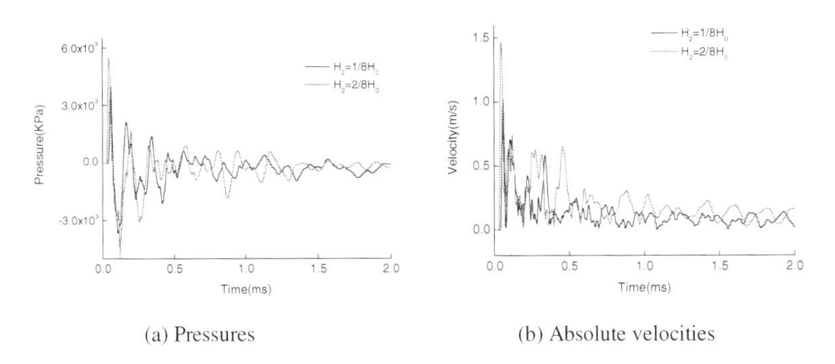

(a) Pressures (b) Absolute velocities

Fig. 4 Pressures and velocities development with time. (position (90, 0, 0) unit: mm)

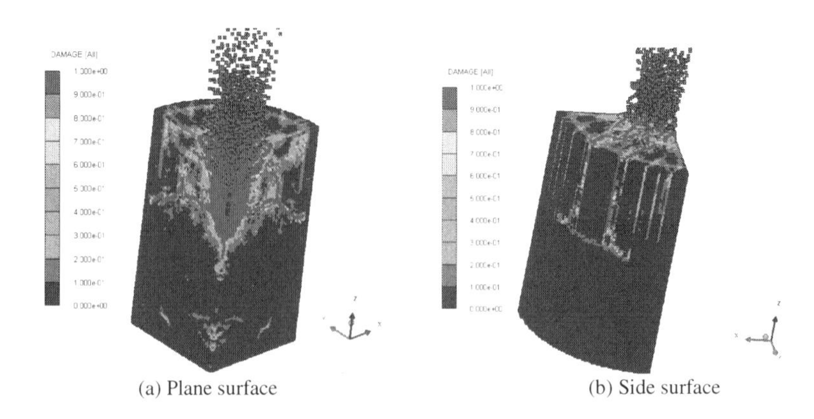

(a) Plane surface (b) Side surface

Fig. 5 Blast-internal-damage distributions of concrete by TNT

4 Conclusions

By the dynamic load of blast and impact, the mechanical characteristics of concrete material have essential difference with one by quasi-static load. In this paper, energy release model of TNT detonation with reactive rate, and dynamic model of concrete are established. The numerical simulation on dynamics internal-blast of the concrete is conducted by smoothed particle hydrodynamics method. Moreover, internal-blast damage effects are discussed deeply, which provides theory references for carrying out assessment of TNT blast damage for internal-blast of concrete.

References

[1] Liu, G.R., Liu, M.B.: A Meshfree particle method. In: Xu, H., Yang, G., Qiang, H.F. (eds.) Smoothed Particles Hydrodynamics, pp. 50–100. Human University Press, Hu Nan (2005)

[2] Brara, A., Camborde, F., Klepaczko, J.R.: Experimental and numerical study of concrete at high strain rates in tension. In: Mechanics of Materials, pp. 33–45 (2001)

[3] Bischoff, P.H., Perry, S.H.: Compression Behavior of Concrete at High Strain-rates. Material and Structure 24(144), 425–450 (1991)

[4] Lu, Y., Xu, K.: Modeling of Dynamic Behavior of Concrete Materials under Blast Loading. International Journal of Solids and Structures 41, 131–143 (2004)

[5] Long, C.L., Xin, P., Feng, C.G.: Detonation of Aluminized Explosive. National Defense Industry Press, Beijing (2004)

[6] Miller, J.: A Reactive Flow Model with Coupled Reaction Kinetics for Detonation and Combustion in Non-ideak Explosives. In: Mat. Res. Soc. Symp. Proc., vol. 418 (1996)

[7] Hansson, H., Skoglund, P.: Simulation of Concrete Penetration in 2D and 3D with the RHT Material Model. Swedish Defense Research Agency, FOI-R-0720-SE (2002)

Free Vibration Analysis of Ring-Stiffened Cylindrical Shells Based on Transfer Matrix Method

Guanmo Xie

Abstract. Transfer matrix method is applied to investigate the natural frequencies and natural modes of ring-stiffened cylindrical shells in this paper. Field transfer matrix of the shell segments is obtained using the Flugge shell theory, point transfer matrix of the rings is derived considering their in-plane vibrations and coupled vibrations between bending and torsional deformations simultaneously. The precise integration technique, instead of the Runge-Kutta-Gill method used in previous papers, is employed to calculate the elements of the field transfer matrix. Numerical results reveal that transfer matrix method using the Flugge shell theory is very accurate, and influence of the rings on the shells' frequencies is complicated and depends upon the ring eccentricities.

Keywords: transfer matrix method, free vibration, ring-stiffened cylindrical shells, Flugge shell theory, precise integration technique.

1 Introduction

A great amount of papers were reported on the free vibration of circular cylindrical shells. Leissa [1] classified and reviewed the extensive literature dealing with the vibration of circular cylindrical shells up to 1972. But for ring-stiffened cylindrical shells, their free vibration characteristics are difficult to be investigated analytically.

Transfer matrix method is very suitable to treat the vibrations of structures whose stiffness and mass components are distributed successively along a line. It owns many advantages such as requiring no displacement functions, being easy to

Guanmo Xie
Department of Engineering Structures and Mechanics,
Wuhan University of Technology, Wuhan 430070, P.R. China
e-mail: {Guanmo.Xie,LNCS}@Springer.com

Z. Qian et al. (Eds.): Recent Advances in CSIE 2011, LNEE 125, pp. 739–745.
springerlink.com

deal with complicated boundary conditions, concentrated masses, supports and springs. Irie, Yamada and Tanaka [2] used it to analyze the free vibration of a circular cylindrical double-shell interconnected by several springs. Huang and Xiang [3] calculated the free vibration of circular cylindrical tanks with variable wall-thickness by this method.

In the present paper, transfer matrix method is applied to analyze the free vibration of ring-stiffened cylindrical shells, in which the Flugge shell theory is used to obtain the field transfer matrix. To obtain the ring's point transfer matrix, its in-plane vibrations and coupled vibrations between bending and torsional deformations are considered simultaneously. The precise integration technique [4], more accurate and convenient than ordinary Runge-Kutta-Gill method [2], is used to calculate the field transfer matrix. Numerical results show that transfer matrix method using the Flugge shell theory can obtain the shells' frequencies accurately, and effect of the rings on the shells' frequencies is complex and depends on the ring eccentricities.

2 Theoretical Formulas

2.1 *Field Transfer Matrix and Point Transfer Matrix*

Consider a circular cylindrical shell with the axial length l, the radius R and the wall thickness h, mass density ρ. The cylindrical coordinates (x, θ, z) is chosen. For harmonic free vibrations, the distributed loadings caused by inertial forces are

$$p_x = -\rho h\frac{\partial^2 u}{\partial t^2} = \rho h\omega^2 u, \quad p_\theta = -\rho h\frac{\partial^2 v}{\partial t^2} = \rho h\omega^2 v, p_r = -\rho h\frac{\partial^2 w}{\partial t^2} = \rho h\omega^2 w \tag{1}$$

Using the Flugge shell theory [5], the shell's equilibrium equations are

$$\frac{\partial N_x}{\partial x} + \frac{1}{R}\frac{\partial N_{\theta x}}{\partial \theta} + \rho h\omega^2 u = 0 \tag{2}$$

$$\frac{1}{R}\frac{\partial N_\theta}{\partial \theta} + \frac{\partial N_{x\theta}}{\partial x} - \frac{Q_\theta}{R} + \rho h\omega^2 v = 0 \tag{3}$$

$$\frac{N_\theta}{R} + \frac{\partial Q_x}{\partial x} + \frac{1}{R}\frac{\partial Q_\theta}{\partial \theta} - \rho h\omega^2 w = 0 \tag{4}$$

The shearing forces, Kelvin-Kirchhoff membrane forces are the following

$$Q_\theta = \frac{1}{R}\frac{\partial M_\theta}{\partial \theta} + \frac{\partial M_{x\theta}}{\partial x}, \quad Q_x = \frac{\partial M_x}{\partial x} + \frac{1}{R}\frac{\partial M_{\theta x}}{\partial \theta}, \quad V_x = N_{x\theta} - \frac{M_{x\theta}}{R}, \quad S_x = Q_x + \frac{1}{R}\frac{\partial M_{x\theta}}{\partial \theta}$$

The components of the membrane force and moment are

$$N_x = D\left(\frac{\partial u}{\partial x} + \frac{\mu}{R}\left(\frac{\partial v}{\partial \theta} + w\right)\right) - \frac{K}{R}\frac{\partial \psi}{\partial x}, N_\theta = D\left(\frac{1}{R}\left(\frac{\partial v}{\partial \theta} + w\right) + \mu\frac{\partial u}{\partial x}\right) + \frac{K}{R^3}\left(w + \frac{\partial^2 w}{\partial \theta^2}\right)$$

$$N_{x\theta} = \frac{1-\mu}{2}D\left(\frac{1}{R}\frac{\partial u}{\partial \theta} + \frac{\partial v}{\partial x}\right) + \frac{K}{R^2}\frac{1-\mu}{2}\left(\frac{\partial v}{\partial x} - \frac{\partial \psi}{\partial \theta}\right), N_{\theta x} = \frac{1-\mu}{2}D\left(\frac{1}{R}\frac{\partial u}{\partial \theta} + \frac{\partial v}{\partial x}\right) + \frac{K}{R^2}\frac{1-\mu}{2}\left(\frac{1}{R}\frac{\partial u}{\partial \theta} + \frac{\partial \psi}{\partial \theta}\right)$$

$$M_x = K\left(\frac{\partial\psi}{\partial x} + \frac{\mu}{R^2}\frac{\partial^2 w}{\partial\theta^2} - \frac{1}{R}\frac{\partial u}{\partial x} - \frac{\mu}{R^2}\frac{\partial v}{\partial\theta}\right), M_\theta = K\left(\frac{1}{R^2}w + \frac{1}{R^2}\frac{\partial^2 w}{\partial\theta^2} + \mu\frac{\partial\psi}{\partial x}\right)$$

$$M_{x\theta} = \frac{1-\mu}{R}K\left(\frac{\partial\psi}{\partial\theta} - \frac{\partial v}{\partial x}\right), M_{\theta x} = \frac{1-\mu}{R}K\left(\frac{\partial\psi}{\partial\theta} + \frac{1}{2R}\frac{\partial u}{\partial\theta} - \frac{1}{2}\frac{\partial v}{\partial x}\right)$$

Where K, D are the shell's bending, membrane stiffness respectively.

The relation between the radial displacement w and the rotation angle ψ is

$$\psi = \frac{\partial w}{\partial x}$$

Now, there are totally 16 unknowns, in the meantime there are just 16 equations to eliminate 8 unknowns $N_\theta, N_{x\theta}, N_{\theta x}, M_\theta, M_{x\theta}, M_{\theta x}, Q_x, Q_\theta$ and keep the state vector elements $u, v, w, \psi, N_x, M_x, V_x, S_x$.

For the purpose of simplicity and convenience, it is necessary to introduce the following dimensionless quantities

$$(u, w, \psi) = h(\bar{u}, \bar{w}, \bar{\psi}/R)\cos n\theta, v = h\bar{v}\sin n\theta$$

$$(N_x, N_\theta, Q_x, V_x) = (K/R^2)(\bar{N}_x, \bar{N}_\theta, \bar{Q}_x, \bar{V}_x)\cos n\theta$$

$$(N_{x\theta}, N_{\theta x}, Q_\theta, S_x) = (K/R^2)(\bar{N}_{x\theta}, \bar{N}_{\theta x}, \bar{Q}_\theta, \bar{S}_x)\sin n\theta$$

$$(M_x, M_\theta) = (K/R)(\bar{M}_x, \bar{M}_\theta)\cos n\theta, (M_{x\theta}, M_{\theta x}) = (K/R)(\bar{M}_{x\theta}, \bar{M}_{\theta x})\sin n\theta$$

$$\xi = x/R, \bar{l} = l/R, \bar{h} = h/R, \lambda^2 = \rho(1-\mu^2)R^2\omega^2/E$$

Extensive simplification aiming to keep only the state vector elements leads to

$$\frac{d}{d\xi}\{Z(\xi)\} = [A]\{Z(\xi)\} \tag{5}$$

Where $\{Z(\xi)\}$ is the state vector. Non-zero elements of the matrix $[A]$ are given by

$$A_{12} = -A_{78} = -\mu n, A_{13} = A_{68} = \frac{\mu(n^2\bar{h}^2 - 12)}{12 - \bar{h}^2}, A_{15} = A_{18} = A_{48} = \frac{\bar{h}}{12 - \bar{h}^2}, A_{21} = -A_{87} = \frac{4n}{\bar{h}^2 + 4}$$

$$A_{24} = -A_{57} = -\frac{n\bar{h}^2}{\bar{h}^2 + 4}, A_{27} = \frac{2\bar{h}}{3(\bar{h}^2 + 4)(1-\mu)}, A_{34} = A_{56} = 1, A_{43} = A_{65} = \frac{12\mu(n^2 - 1)}{12 - \bar{h}^2}$$

$$A_{45} = \frac{12}{\bar{h}(12 - \bar{h}^2)}, A_{51} = -A_{54} = \frac{n^2\bar{h}(1-\mu)(\bar{h}^2 + 16)}{2(\bar{h}^2 + 4)}, A_{62} = -A_{73} = -\frac{12(1-\mu^2)n}{\bar{h}}$$

$$A_{63} = \frac{12}{\bar{h}}\lambda^2 - \frac{12(1-\mu^2)}{\bar{h}} - \left(1 - n^2\right)^2\bar{h} + \frac{12\mu^2\left(1 - n^2\right)^2\bar{h}}{12 - \bar{h}^2}, A_{72} = \frac{12(1-\mu^2)n^2}{\bar{h}} - \frac{12\lambda^2}{\bar{h}}$$

$$A_{81} = \frac{(1-\mu)n^2\bar{h}}{2}\left(1 + \frac{12}{\bar{h}^2 + 4}\right) - \frac{12}{\bar{h}}\lambda^2, A_{84} = \frac{(1-\mu)n^2\bar{h}}{2}\left(1 + \frac{12}{\bar{h}^2 + 4}\right)$$

The general solution to (5) is as follows

$$\{Z(\xi)\} = e^{[A]\xi}\{c\} \tag{6}$$

In which $\{c\}$ is a non-zero constant vector.

For the left end of the shell segment, $\xi = 0$, from (6) one can obtain

$$\{c\} = \{Z(0)\} = \{Z(\xi)\}_L \tag{7}$$

For the right end of the circular cylindrical shell segment, $\xi = l/R = \bar{l}$, from (6) the following relation can be got

$$\{Z(\xi)\}_R = \{Z(\bar{l})\} = e^{\bar{l}[A]}\{c\} = e^{\bar{l}[A]}\{Z(\xi)\}_L \tag{8}$$

Therefore the field transfer matrix of the cylindrical shell segment, which communicates the state vectors at two ends, is

$$[C_F] = e^{\bar{l}[A]} \tag{9}$$

Since the matrix $[C_F]$ is significant to transfer matrix method, the precise integration technique [5], rather than ordinary Runge-Kutta-Gill method [2,3], is utilized to calculate the matrix $[C_F]$ accurately.

At the shell cross-section where a circular ring is located, the internal forces at the left side differ from that at the right side, whereas the shell displacements are the same, therefore point transfer matrix should be introduced.

At first, four differential equations governing the ring's in-plane vibrations and its coupled vibrations between lateral bending and torsional deformations can be written respectively. Then the continuity relations between the displacements and forces of the ring and that of the shell can be established. By means of the ring equations and the continuity conditions [6], in addition to the circumferential modal expansion, the point transfer matrix of the ring can be derived. For the limit of this paper's length, detailed course of the extensive simplifications is neglected, only the result is given as follows

$$\begin{Bmatrix} \bar{u} \\ \bar{v} \\ \bar{w} \\ \bar{\psi} \\ \bar{M}_x \\ \bar{S}_x \\ \bar{V}_x \\ \bar{N}_x \end{Bmatrix}_R = \begin{bmatrix} 1 & 0 & 0 & 0 & 0 & 0 & 0 & 0 \\ 0 & 1 & 0 & 0 & 0 & 0 & 0 & 0 \\ 0 & 0 & 1 & 0 & 0 & 0 & 0 & 0 \\ 0 & 0 & 0 & 1 & 0 & 0 & 0 & 0 \\ t_3 & 0 & 0 & t_4 & 1 & 0 & 0 & 0 \\ 0 & t_7 & t_8 & 0 & 0 & 1 & 0 & 0 \\ 0 & t_5 & t_6 & 0 & 0 & 0 & 1 & 0 \\ t_1 & 0 & 0 & t_2 & 0 & 0 & 0 & 1 \end{bmatrix} \begin{Bmatrix} \bar{u} \\ \bar{v} \\ \bar{w} \\ \bar{\psi} \\ \bar{M}_x \\ \bar{S}_x \\ \bar{V}_x \\ \bar{N}_x \end{Bmatrix}_L = [C_P]_{8\times 8} \begin{Bmatrix} \bar{u} \\ \bar{v} \\ \bar{w} \\ \bar{\psi} \\ \bar{M}_x \\ \bar{S}_x \\ \bar{V}_x \\ \bar{N}_x \end{Bmatrix}_L$$

In which

$$t_1 = \frac{hn^2R^2}{KR_1^4}(EI_2n^2 + GJ) - \frac{12}{h^2}A\lambda^2, t_2 = \frac{hn^2R^2}{KR_1^4}(EI_2(R_1 - e_1n^2) - GJ(e_1 - R_1)) + \frac{12}{Rh^2}Ae_1\lambda^2$$

$$t_3 = \frac{hn^2R}{KR_1^3}(EI_2(1 + \frac{e_1}{R_1}n^2) + GJ(1 + \frac{e_1}{R_1})) - \frac{12}{Rh^2}Ae_1\lambda^2$$

$$t_4 = \frac{h}{KR_1^3}(EI_2(R_1 - e_1n^2)(1 + \frac{e_1}{R_1}n^2) - GJn^2(1 + \frac{e_1}{R_1})(e_1 - R_1)) - \frac{12}{R^2h^2}I_p\lambda^2(1 - Ae_1^2)$$

$$t_5 = \frac{Ehn^2R}{KR_1}(A + \frac{I_1}{R_1^2}) - \frac{12}{Rh^2}R_1A\lambda^2, t_8 = -\frac{EhR^2}{KR_1^2}(\frac{I_1n^4}{R_1^2}(1 + \frac{e_1}{R}) + A(1 + \frac{e_1}{R}n^2)) + \frac{12}{h^2}A\lambda^2$$

$$t_6 = \frac{EhnR^2}{KR_1^2}(A(1 + \frac{e_1}{R}n^2) + \frac{n^2I_1}{R_1^2}(1 + \frac{e_1}{R})) - \frac{12n}{Rh^2}Ae_1\lambda^2, t_7 = \frac{EhnR}{KR_1}(A + \frac{I_1n^2}{R_1^2})$$

In which, R_1 is the radius of the ring's neutral axis, $R_1=R+e_1$, e_1 is the ring eccentricity which is the distance from the ring's neutral axis to the shell's middle surface, positive for exterior rings and negative for interior rings; I_1, I_2 are the ring's moments of inertia with respect to longitudinal symmetry axis and radial symmetry axis, respectively; I_p is the polar moment of inertia of the ring; J is the torsional constant of the ring; A is the cross-sectional area of the ring; G is the ring's shear modulus.

2.2 *Frequency Equations of Ring-Stiffened Cylindrical Shells*

For the i-1 th to the i th shell segment, there is the field transfer matrix $[C_{iF}]$ to connect the state vector of the left end and that of the right end; For the i th cross section where concentrated mass or stiffener exists, or concentrated force or moment is applied, or change of radius or thickness occurs, there is the point transfer matrix $[C_{iP}]$ connecting the state vectors of the two sides of this cross section.

Then there is the following relation to connect the state vector of the left end of the shell and that of the right end of the shell

$$\{Z_n^R\}=[C_{nF}][C_{n-1P}][C_{n-1F}]\cdots[C_{1P}][C_{1F}]\{Z_0\}=[C]\{Z_0\} \tag{10}$$

In which $[C]$ is called the total transfer matrix of the shell, depending upon ω.

For simply-supported ring-stiffened cylindrical shells, since the elements v, w, N_x, M_x at both ends are all zero, the following matrix equation can be obtained

$$\begin{bmatrix} C_{21} & C_{24} & C_{26} & C_{27} \\ C_{31} & C_{34} & C_{36} & C_{37} \\ C_{51} & C_{54} & C_{56} & C_{57} \\ C_{81} & C_{84} & C_{86} & C_{87} \end{bmatrix} \begin{Bmatrix} u \\ \psi \\ S_x \\ V_x \end{Bmatrix}_L = [C_1] \begin{Bmatrix} u \\ \psi \\ S_x \\ V_x \end{Bmatrix}_L = \begin{Bmatrix} 0 \\ 0 \\ 0 \\ 0 \end{Bmatrix} \tag{11}$$

Because there should be non-trivial solution to the homogeneous linear equations (11), therefore the following frequency equation can be got from (11)

$$Det[C_1]=0 \tag{12}$$

3 Numerical Discussion and Conclusion

Consider a simply-supported ring-stiffened cylindrical shell with parameters as

$$E=2.1\times10^{11}Pa, \rho=7850kg/m^3, \mu=0.3, L=1m, R=0.5m, t=0.005m$$

The cross-section of the ring is a rectangle, with height h_r=0.025m and width b_r=0.002m. The rings are uniformly distributed, and the distance between two adjacent rings is 0.02m. For the purpose of comparison, three cases are taken into consideration: exterior rings, outside the shell's middle surface, is called Case 1; interior rings, inside the shell's middle surface, is called Case 2; rings with no eccentricities, that is, the axes of the rings are coincident with the shell's middle surface, is called Case 3.

As it is known, analytical solutions to the natural frequencies and modes of a ring-stiffened cylindrical shell are difficult to be found. To evaluate the accuracy of transfer matrix method and the influence of the rings, the natural frequencies of this ring-stiffened cylindrical shell using transfer matrix method are compared with those obtained by the finite element method. Table 1 shows the frequency comparison for these three cases.

Table 1 Natural frequency and mode comparison for three cases

Frequency order	Frequency of Case 1; wavenumber n; relative error	Frequency of Case 2; wavenumber n; relative error	Frequency of Case 3; wavenumber n; relative error
1	212.12; 5; 1.42%;	225.70; 5; -2.02%	218.34; 5; -0.37%
2	225.47; 4; 0.56%;	231.33; 4; -0.92	228.16; 4; -0.24%
3	254.36; 6; 1.74%;	275.79; 6; -2.44%	264.09; 6; -0.35%
4	320.62; 3; -0.10%;	321.23; 3; -0.02%	320.89; 3; -0.08%
5	327.50; 7; 1.179%;	357.05; 7; -2.59%	340.77; 7; -0.29%
6	418.95; 8; 1.713%;	431.76; 6; -0.27%	425.94; 6; -0.16%

From the above table, it can be seen that the transfer matrix method to solve the natural frequencies and modes of ring-stiffened cylindrical shells is very accurate, using the Flugge shell theory and the precise integration technique. The natural modes (indicated by wavenumber n) are completely in agreement with those by using the finite element method. The maximum relative error between them is only 2.59% for interior rings. This accuracy can not be achieved by other analytical methods, as long as displacement functions are necessary to be assumed. Similar to the results by other analytical methods, for higher order frequencies the errors will gradually increase.

Comparing these three cases, it is evident that influence of the rings on the natural frequencies of the ring-stiffened cylindrical shells is complicated and it is uneasy to summarize a definite and unified law. Existence of the rings increases the structural stiffness, on the other hand it also contributes to the mass of the structure. The ring eccentricities have considerable effect on the natural frequencies. As it is expected, the relative errors are very small for zero eccentricities of Case 3.

References

1. Leissa, A.W.: Vibration of Shells, NASA SP-288, 185 (1973)
2. Irie, T., Yamada, G., Tanaka, T.: Free vibration of circular cylindrical double-shell system interconnected by several springs. J. Sound and Vib. 95(2), 249–259 (1984)

3. Huang, Y., Xiang, Y.: Transfer matrix method on dynamics analysis of circular cylindrical tanks with variable wall-thickness. J. Vib. Engrg. 2(4), 23–32 (1989) (in Chinese)
4. Zhong, W.: On precise time-integration method for structural dynamics. J. of Dalian University of Technology 34(4), 131–136 (1994) (in Chinese)
5. Flugge, W.: Stresses in Shells. Springer, Berlin (1996)
6. Xie, G., Li, J., Luo, B., Luo, D.: Sound radiation from fluid-loaded infinite cylindrical shells with rings, bulkheads and longitudinal stiffeners. J. Ship Mech. 8(2), 101–108 (2004) (in Chinese)

Implementation of ActiveX Control of Three-Dimensional Model Based on OpenGL

Xiuli Gong and Youqing Guan

Abstract. Three-dimensional model is an important carrier between designers. This article researches the method of developing an ActiveX control to view three-dimensional model based on OpenGL with visual C++6.0 environment. After analysing and storing different classes of the VRML files generated by Pro/E, the API of OpenGL in VC++ is called, rendering the three-dimensional model dynamic. This control can be easily used in ActiveX-supported browser, and provides a good solution for distributed browsing system.

1 Introduction

With the increasing popularity of network technology and distributed systems, application of computer technology is transferring from traditional single-user mode to multi-user mode. CSCW (Computer Supported Cooperative Work) is a new product development model proposed in recent years. It is an effective way to solve complex problems, and this co-design platform can improve the effect of communication between designers.

Three-dimensional model is an important carrier between designers and its development tool is basically based on commercial software such as Pro / E. Electronic engineers normally do not have the permission to install such a three-dimensional software (three-dimensional software license files are usually very expensive), which will lead to the following results: mechanical engineers will not be able to keep up communication with the electronic engineers to improve their designs, and like wise electronic engineers are not aware of the progress of

Xiuli Gong · Youqing Guan
Institute of Information Network Technology,
College of the Internet of Things,
Nanjing University of Posts and Telecommunications,
Nanjing 210003, China
e-mail: ashan5000@163.com, guanyouq@njupt.edu.cn

Z. Qian et al. (Eds.): Recent Advances in CSIE 2011, LNEE 125, pp. 747–753.
springerlink.com

mechanical designs. It is very unfavorable for the improvement of work efficiency and product design quality. In order to shorten the development time and reduce unnecessary communication problems, the development for common three-dimensional software without relying on specialized browser is very necessary. In this area, Zhen-yu LI researches the method of developing an ActiveX control to view 3DS files based on OpenGL [1]. Yu RAN researches a number of key technologies about Web3D visualization based on MAPGIS-TDE platform by adopting Microsoft ActiveX control [2]. This article provides an ActiveX plug-ins realization based on OpenGL, which entails the reconstruction and rendering of Pro / E generated files. Commonly used methods of developing ActiveX control are MFC and ATL. ATL provides a small and concise framework, but is lacking in supporting the classes and tools; while MFC provides a very large and robust class library, and most problems in development can be solved by using MFC. Therefore, MFC is employed in ActiveX control development.

2 Key Technologies

Commercial Three-dimensional software has a separate output format, but since internal details are not open, it requires specialized software to open the three-dimensional model. In addition to the independent three-dimensional model format, there are also many common formats, such as STEP, Parasolid, VRML, etc. These formats have their own unique advantages [3]. STEP is a more common format for the CAD software and its compatibility is better. Parasolid is developed by Unigraphics Solutions Inc, mainly used for Unigraphics and Solid Edge, receives support from other three-dimensional modeling software. VRML which is the abbreviation for Virtual Reality Modeling Language is a standard format to represent interactive, three-dimensional virtual world on a computer display. It has been designed particularly to realize 3D interaction in Web browser window, so it is represented in a text form, in a manner similar to HTML [4]. It takes little space and it is suitable for network transmission.

Two methods for visualization of three-dimensional model are OpenGL and Java3D. These two technologies have their own advantages and disadvantages. OpenGL the abbreviation for Open Graphics Library, which is one of the most widely supported 3D technologies, defines a set of interfaces used to implement three-dimensional applications. It has great advantages in interactive three-dimensional modeling and external device management [5]. The Java3D API is a set of classes for writing three-dimensional graphics applications and 3D applets. It gives developers high level construction for creating and manipulating 3D geometry and for constructing the structures used in render that geometry .Compared with OpenGL, the flexibility and run-time of Java3D programming affected to some extent.

Based on the above comparison, VRML is selected as a common output format for three-dimensional model. OpenGL is selected to reconstruct and render the three-dimensional model in VC++.

3 ActiveX Control Implementation

ActiveX control implementation is roughly divided into the following four steps: OpenGL rendering context initialization, OpenGL rendering, Interactive interface design, and three-dimensional model properties design.

3.1 OpenGL Rendering Context Initialization

The ways of graphic drawing are different for OpenGL and Windows. With Windows GDI (Graphics Device Interface) drawing is employed and must specify which device context is being used; while with OpenGL, when its function is called, a rendering context should be specified. The first step to generate OpenGL, rendering context is to define window pixel format. Graphic storage in memory is determined by the pixel format, and the parameters including color depth, buffering model and painting interface are also determined by the pixel format. OpenGL rendering context initialization is as follows:

1. Create a new project named "GL" by MFC (Microsoft Foundation Class) ActiveX Control Wizard.
2. Add a message response function to "CGLCtrl" class to achieve OpenGL rendering environment initialization.
3. Define the function "WindowPixelFormat(HDC hDC) " to set the window pixel format.
4. Define the function "InitGL()" to set OpenGL rendering way.

Several major OpenGL API functions are set as follows:

```
glShadeModel(GL_SMOOTH)        // Set Shade Model
glClearDepth(1.0f)             //Clear Depth Buffer
glEnable(GL_DEPTH_TEST)        //Enable Depth Test
glLightModelf(GL_LIGHT_MODEL_TWO_SIDE, 1.0)    //Set Light Model
glEnable(GL_LIGHTING)
glEnable(GL_LIGHT0)            //Enable Light
```

3.2 Implementation of OpenGL Rendering

"onDraw" function in "CGLCtrl" class is to achieve three-dimensional model redraw when ActiveX is refreshed. This redraw function calls the custom function "GLRender()".

Before the implementation process is introduced, it is necessary to give a brief explanation of the concept "node" in VRML. Node is the basic unit in VRML file. Each node has a child node with field or field value to describe it. The role of the node is to describe the spatial form and its properties. Shape node is used to define of the geometry, shape, size, color, texture, and appearance of three-dimensional objects in VRML.

1. Call the function glViewport() of OpenGL to set viewport; call the function glMatrixMode() to select the desired matrix and initialize; call the function gluPerspective() to define projection way.

2. Call the custom function DrawCoordinates() to draw axis. This function calls the glBegin(GL_LINES), glEnd() of OpenGL API to implement axis drawing.

3. Call the function glPushMatrix(), glPopMatrix() of OpenGL API to operate matrix stack.

4. Call the function DrawScene() to realize three-dimensional model reconstruction and rendering.

DrawScene() function achieves parser of the VRML file and three-dimensional model reconstruction and rendering in ActiveX. See Figure 1.

Define a "CVrmlParse" class that contains the following main functions:

```
int ReadVrmlFile(CString filename)   //Read VRML file
int ReadLine()                       // Read one line of VRML file
int ReadWord()                       // Read one string of VRML file
int OffsetToString(char *string)     // Offset to the specified string of VRML
```

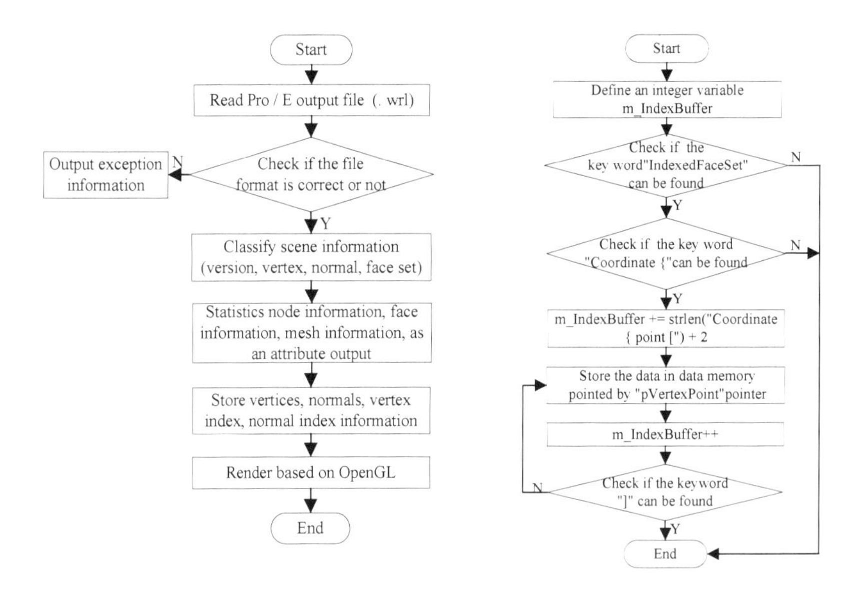

Fig. 1 DrawScene() Implementation Process

Fig. 2 Vertex Coordinates Analysis

1. Call the function ReadVrmlFile() to read the VRML file output by Pro/E, and check the format of the VRML file.

2. Check Version Information. In each VRML file, header file is necessary and is located in the first line of VRML file. Syntax is: # VRML V2.0utf8. Design custom function "CheckVersion ()" to check if the keywords "VRML V2.0" exist in source file.

3. Information Statistics. Design the function CountMesh() to count the mesh information; design the function CountFace() to count the face information; design the function CountVertex() to count the vertex information.

4. Syntax analysis and classification storage. Take vertex data storage for example. The analysis process is shown in Figure 2, as described below:

Set integer variable "m_IndexBuffer" which represents the order of the string and is initialized to zero. Call the function "OffsetToString()" to check whether the key word "IndexedFaceSet" can be found. If it can be found, then face sets are there. Otherwise, exit the storing operation. Call the function "OffsetToString()" to check whether the key word "CoordIndex {" can be found. If it can be found, then the coordinates of the vertex are there. Otherwise, exit the storing operation. Call the function "ReadWord ()" three times to read vertex information, store the date in turn in the data memory pointed by "pVertexPoint" pointer. Call the function "ReadWord ()" to read one string and check if the string is "]". If it is, vertex array reading is completed. Then exit the storing cycle. If not, continue to read and store the vertex coordinates.

The method of storing the vertex array index, normals, normal vector index is similar to the storage of the vertex data. The vertex array index is stored in data memory pointed by "pCoordIndex" pointer.

5. Render based on OpenGL. Use API functions of OpenGL in VC++ for scene redraw. The process goes like this:

Use vertex arrays: Call the function "glEnableClientState(GL_VERTEX_ARRAY)" to activate vertex arrays. This function indicates which arrays need to be used.

Specify the data in vertex array: Call the function "glVertexPointer(3, GL_FLOAT, 0, pVertexPoint)" to specify the data of the vertex arrays. This function specifies which coordinate data need to be accessed. "pVertexPoint" is the memory address of the first coordinate from the first vertex in vertex arrays. "GL_FLOAT" represents the data type of each coordinate. "3" indicates the number of coordinates in each vertex. "0" indicates that the vertexes are closed related in the array.

Rendering: Call the function "glDrawElements(GL_TRIANGLES, 3*m_NbFace, GL_UNSIGNED_INT, pCoordIndex)" to render the data. This function use "3 * m_NbFace" (m_NbFace represents the number of faces) elements to define a set of graphical sequence. These elements are stored in the array pointed by "pCoordIndex" pointer. "GL_UNSIGNED_INT" indicates the data type of array. "GL_TRIANGLES" specifies what type of graphics is created.

3.3 Interactive Interface Design

Add custom method "RotateX()", "RotateY()", "RotateZ()"in "_DGL" class, and provide the implementations in "CGLCtr" class.

Design the function "RotateX()" to implement the three-dimensional model of free rotation around the X-axis.

Design the function "RotateY()" to implement the three-dimensional model of free rotation around the Y-axis.

Design the function "RotateZ()" to implement the three-dimensional model of free rotation around the Z-axis.

3.4 Three-Dimensional Model Properties

Set properties dialog box in "Dialog" class; add the member function "SetEdirValue()" in "CProperty" class . See Figure 3:

void CProperty::SetEdirValue(int *NbMesh, int *NbFace, int *NbVertex)

This function outputs the number of vertexes, the number of faces and the number of the meshes of the model.

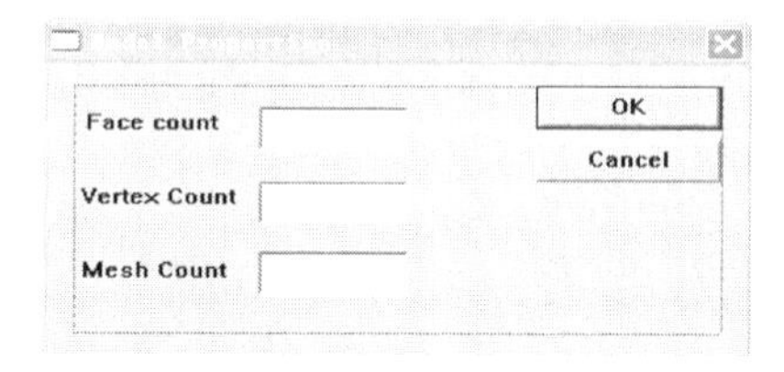

Fig. 3 Three-dimensional Model Properties

4 Results

When we have created three-dimensional model in Pro /E, saved the output file as VRML format, opened the file in ActiveX, we will get the following results as shown in Figure 4 and Figure 5.

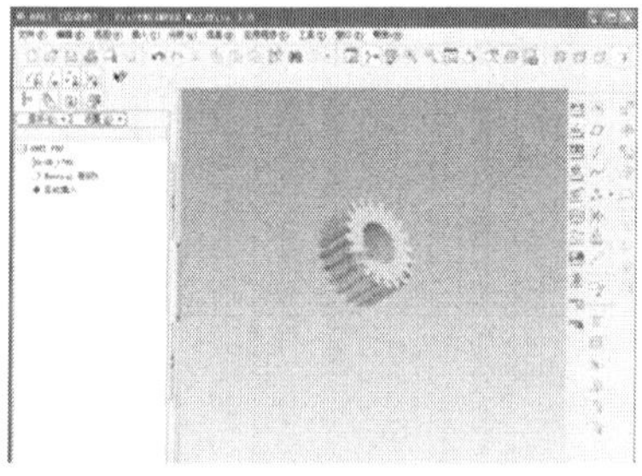

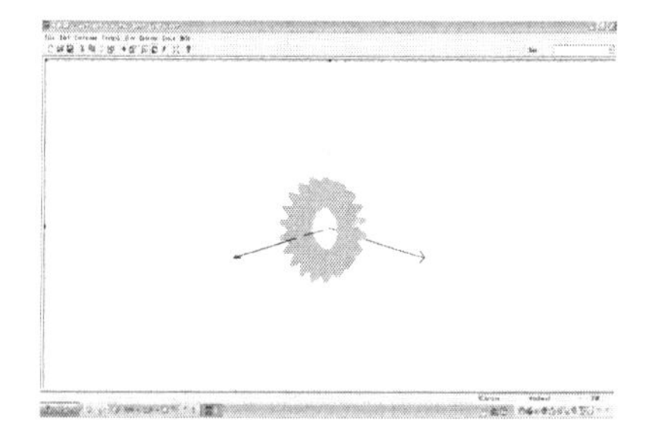

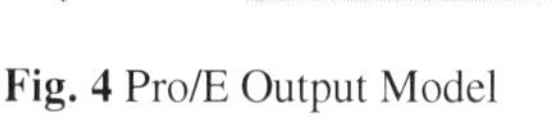

Fig. 4 Pro/E Output Model

Fig. 5 ActiveX Output Model

From the results we can see that ActiveX control achieves the browsing function to browse three-dimensional model generated by Pro / E.

5 Conclusion

Information sharing of three-dimensional model is one of the main means of communication between designers. In this paper, one ActiveX control is developed based on Visual C++ 6.0 platform. This software implements the reading, displaying, rendering, free rotation around axis and attribute query functions of three-dimensional model generated by Pro / E. This control can be easily used in ActiveX supported browser, and it provides a good solution for distributed browsing system. At present, in addition to Pro / E, there are other three-dimensional modeling software such as AutoCAD and UG. Therefore, the follow-up work is to consider increasing support for a variety of output files of mainstream three-dimensional modeling software so as to enhance its versatility.

References

1. Li, Z., Zhuang, C.H.: Programming an ActiveX control viewing 3DS Model in Three-dimension Configuration software Based on OpenGL and ActiveX. In: The 8th Industrial Instrumentation and Automation Conference, vol. 2, pp. 490–495 (2007) (in Chinese with English abstract)
2. Ran, Y., Zheng, K., Liu, X.G.: Research and Implementation of Three-Dimensional Visualization Based on Internet. In: Geoinformatics International Conference, vol. 18, pp. 1–4 (2010)
3. Wikipedia, Comparison of CAD editors for architecture, engineering and construction (November 2010), `http://en.wikipedia.org/wiki/Comparison_of_CAD_editors_for_architecture_engineering_and_construction_AEC`
4. Foit, K.: Introduction to use virtual reality visualizations in the exploitation and virtual testing of machines. Journal of Achievement in Materials and Manufacturing Engineering 25, 57–60 (2007)
5. de Oliveira, J.C.: Issues in Large Scale Collaborative Virtual Environments, Ottawa, Ontario, Canada (2001)

Meshless Method Based on Wavelet Function

Dengfeng Wu, Kun He, and Xin Ye

Abstract. The wavelet theory is introduced into the meshless method, the design of the form of wavelet function is derived based on wavelet function meshless method formula. Daubechies wavelet function is used in the calculation of wavelet algorithm, which verifies the reasonableness. The error of the example computed result and the exact solution is small, which has confirmed that the meshless methods based on wavelet function is validate.

Keywords: wavelet function, meshless methods, multi-resolution analysis.

1 Introduction

Recently, there are more than ten kinds of the meshless methods, in which methods the trial function and the equivalent form of differential equations in the meshless methods are different. Nayroles[1] introduces moving least square into the Galerkin method in 1992. According to function integral transformation Liu W K[2] proposes reproducing kernel particle method based on Galerkin method in 1995. In 1996, Duarte[3] decomposes the function using moving least square method, and establishes discrete format using Galerkin method, then proposes hp-clouds method. Liu[4] G R proposes point interpolation method in 2001. Zhang Xiong[5,6] applies compact support function to collocation method, and establishes the corresponding meshless method; then proposes least square

---

Dengfeng Wu · Kun He
Department of Computer Science and Technology,
Changchun Normal University Chang Chun, China,
e-mail: wdf_@163.com

Xin Ye
2 College of Mathematics, Jilin University, Chang Chun, China
e-mail: Evan673@sina.com

Z. Qian et al. (Eds.): Recent Advances in CSIE 2011, LNEE 125, pp. 755–760.
springerlink.com

collocation method and weighted lease-squares method based on lease-squares method.

Wavelet theory developed rapidly in recent years, in many fields such as signal processing, pattern recognition, function approximation, and other fields with a wide range of applications, and some have achieved good results in finite element. Daubechies, who is a famous mathematician, from the compactly supported orthogonal wavelet filter satisfied conditions, puts forward compactly supported wavelets orthogonal design method, so that Daubechies wavelet with compact support and orthogonal becomes a wavelet of the most widely used one. But Daubechies wavelet has not significant mathematical expression, the scaling function and wavelet function normally use numerical methods to a few tables and curves gives way.

In this paper, the wavelet theory is introduced into the meshless method, the design of the form of wavelet function is derived based on wavelet function meshless method formula. The displacement and the error are calculated and the result is reasonable.

2 The Field Function Using Meshless Method Based on Wavelet

If there are N groups of the nodes in solution region Ω, then the form of the $U(x)=(u_x, v_x)$ is

$$U(x)=\sum_{L=1}^{N}\sum_{k\in Z}\alpha_k^L\phi_k(r_L(X)) \tag{1}$$

Where, $X_L=(x_L, y_L)$ is an interpolation nodes, $r_L(X)=\left\|X-X_L\right\|/R_{def}$, R_{def} is influence radius of the interpolation nodes, α_k^L is the coefficient of $\phi_k(r_L(X))$, $\phi_k(r_L(X))$ is the scaling function.

$$\phi(x)=\sum_{k=-\infty}^{\infty}p_k\phi(2x-k)$$

If using the Daubechies wavelet, the scaling function is (2).

$$\phi(x)=\sum_{k=0}^{2N-1}p_k\phi(2x-k) \tag{2}$$

3 Governing Equations Using Meshless Method Based on Wavelet

Equilibrium equations of elasticity

$$\begin{cases} \sigma_{ij,j} + \overline{f}_i = 0 & in\Omega \\ \sigma_{ij} n_j - \overline{t}_i = 0 & on\Gamma_t \\ u_i - \overline{u}_i = 0 & on\Gamma_u \end{cases} \tag{3}$$

Where, Ω is solution region, Γ_t is force boundary, Γ_u is displacement boundary.

Considering the scaling space of the function, if there are n groups of the nodes in the whole domain, and there are 2N nodes in each group. The center point in the group is interpolation node $X_L = (x_L, y_L)$, and the relative radius $r_L(X) = \|X - X_L\| / R_{def}$ between this point and the others is different, then $U(x) = (u_x, v_x)$ is

$$U(x) = \sum_{L=1}^{N} \sum_{k \in Z} \alpha_k^L \phi_k(r_L(X)) \tag{4}$$

In the solution region, the strain of any point is

$$\varepsilon(X) = (\varepsilon_x, \varepsilon_y, r_{xy})^T = B(X)a \tag{5}$$

$$B_L(X) = \begin{bmatrix} \frac{\partial \phi_1(r_{L,1})}{\partial x} & 0 & \cdots & \frac{\partial \phi_{2N}(r_{L,2N})}{\partial x} & 0 \\ 0 & \frac{\partial \phi_1(r_{L,1})}{\partial y} & \cdots & 0 & \frac{\partial \phi_{2N}(r_{L,2N})}{\partial y} \\ \frac{\partial \phi_1(r_{L,1})}{\partial y} & \frac{\partial \phi_1(r_{L,1})}{\partial x} & \cdots & \frac{\partial \phi_{2N}(r_{L,2N})}{\partial y} & \frac{\partial \phi_{2N}(r_{L,2N})}{\partial x} \end{bmatrix} \tag{6}$$

The displacement of any point is

$$\sigma(X) = (\sigma_x, \sigma_y, \tau_{xy})^T = D\varepsilon(X) = DB(X)a$$

Where, D is the elastic constant matrixes,

$$D = \frac{E}{1-\mu^2} \begin{bmatrix} 1 & \mu & 0 \\ \mu & 1 & 0 \\ 0 & 0 & \frac{1-\mu}{2} \end{bmatrix} \tag{7}$$

The governing equation is

$$K\alpha = F \tag{8}$$

The stiffness matrix of the nodes is

$$K = \begin{bmatrix} K_{11} & K_{12} & \cdots & K_{1n} \\ \vdots & \vdots & \cdots & \vdots \\ K_{n1} & K_{n2} & \cdots & K_{nn} \end{bmatrix} K_{ij} = \int_{\Omega} B_i^T D B_j d\Omega \tag{9}$$

4 Numerical Examples

As an example using cantilever beam, the length L=8m, the height and the width are all D=1m, E=20Mpa, $\mu = 0.25$, p=500Pa(fig.1).

The scheme of the arrangement points as follows (fig.2)

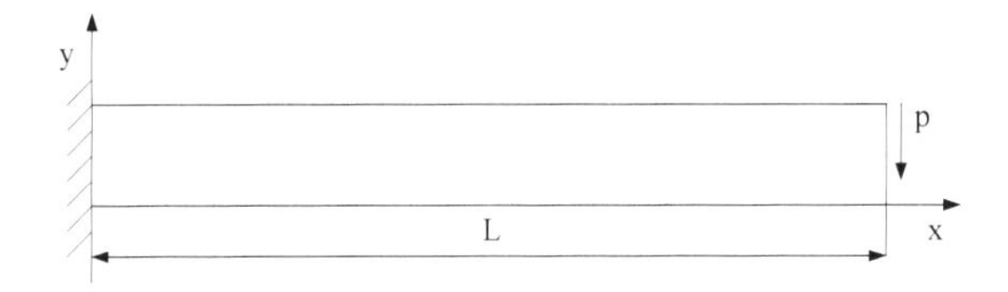

Fig. 1 Cantilever beam subject to end load

Fig. 2 The scheme of the arrangement points

The displacements of the Y-direction and the X-direction are figured out, and the figure is formed (fig 3, 4).

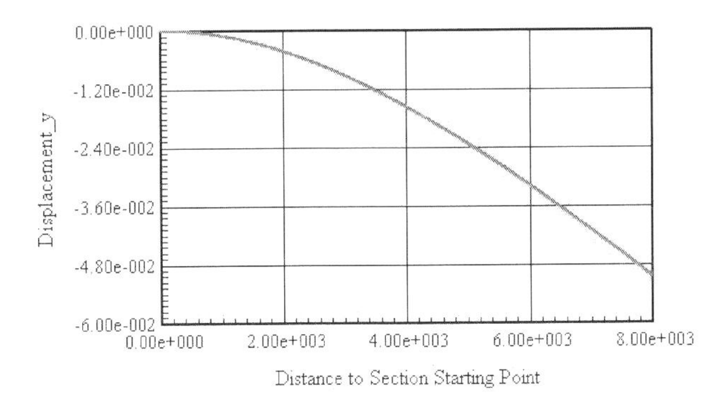

Fig. 3 Y-direction of Cantilever beam

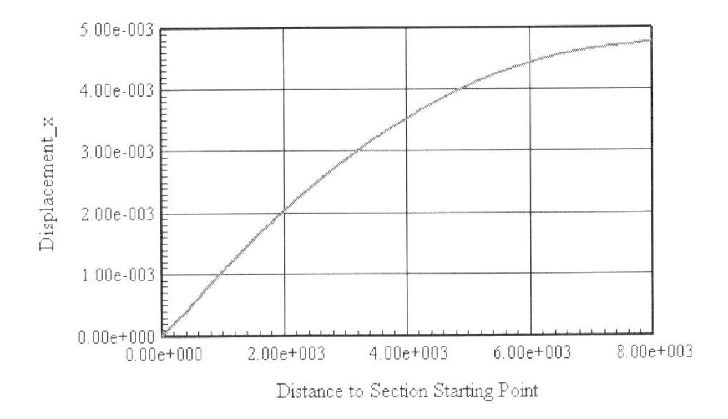

Fig. 4 X-direction of Cantilever beam

Defining the calculation formula of the error norm

$$L = \frac{\sqrt{\sum_{i=1}^{N}\left[u(x_i) - u^h(x_i)\right]^2}}{\sqrt{\sum_{i=1}^{N}\left[u(x_i)\right]^2}} \tag{10}$$

Where, $u(x_i)$ are exact results, $u^h(x_i)$ are wavelet meshless method results

The error norm of x-direction is 1.08 and the error norm of y-direction is 1.19, which shows that meshless method based on wavelet is effective.

5 Conclusion

Because wavelet functions own the characters of compact support and orthogonally, which can overcome the redundancy of other field function in the calculation, and can reduce calculation or increase accuracy. Therefore, to construct the field function of meshless method using wavelet basis function is a new idea in theory, and is an improvement in algorithm. The example proves the feasibility and effectiveness of this method.

Acknowledgments. The corresponding author is Ye Xin. This work is supported by Research Fund for Educational Commission of Jilin Province of China.

References

1. Nayroles, B., Touzot, G., Villon, P.: Generalizing the finite element method: diffuse approximation and diffuse elements. Comput. Mech. 10, 307–318 (1992)
2. Liu, W.K., Jun, S., Zhang, Y.F.: Reproducing Kernel Particle Methods. Int. J. Numer. Meth. Fluids. 20, 1081–1106 (1995)
3. Duarte, C.A., Oden, J.T.: Hp clouds: a h-p meshless method. Numerical Methods for Partial Differential Equations 12, 673–705 (1996)
4. Liu, G.R., Gu, Y.T.: A point interpolation method for two-dimensional solids. Int. J. Num. Meth. Engng. 50, 937–951 (2001)
5. Zhang, X., Liu, X.H., Song, K.Z.: Least-square collocation meshless method. Int. J. Num. Meth. Engng. 51(9), 1089–1100 (2001)
6. Zhang, X., Hu, W., Pan, X.F.: Meshless Weighted Least-Square Method. Chinese Journal of Theoretical and Applied Mechanics 35(4), 425–431 (2003)

The Research and Development on the Mould Virtual Assembly System

Dongfeng Xu, Yan Chen*, and Yun Liang

Abstract. This paper presents a mould virtual assembly system based on the 3DS MAX and Inventor. The mould virtual assembly system can simulate the mutual relationships between all the accessories. Firstly, we build the 3D model based on the Inventor; then we employ the 3DS MAX to make the animation course; lastly, we use VB to integrate all the 3D dynamic sources to develop a display system. Our display system about the mould virtual assembly system can be displayed automatically or step by step. This system is very valuable for teaching the course of building moulds.

Keywords: Stamping Mould, 3D Mould, Inventor, 3DS MAX.

1 Introduction

The course about "Technology of stamping forming and mould design" for enginery in college is very practical course. The traditional teaching method giving the course by stationary instances and graphs is very abstract and absent of realist. Without perceptual knowledge, the complicated technics as deforming, stretching, warping, are difficult for college students to deeply understand their contents, especially for the structures and working courses of some familiar Stamping mould [1].

There are some successful examples to develop the software of the media instruction based on the 3D model in our country, such as the course of the

---

Dongfeng Xu · Yun Liang
College of Informatics, South China Agricultural University,
Guangzhou, P.R. China

Yan Chen
College of Engineering, South China Agricultural University,
Guangzhou, P.R. China

* Corresponding author.

Z. Qian et al. (Eds.): Recent Advances in CSIE 2011, LNEE 125, pp. 761–767.
springerlink.com

technics of manufacturing machines, the course of 3D modeling and Animation Simulation, the course of dynamic simulate moulds.

In order to enrich teaching sources and improve the teaching quality, we develop the 3D dynamic mould and display system. With the help of this system, students can intuitively, deeply comprehend the principle of operation, the combination of structure and assemblage. By simulating the course of processing Stamping mould, teachers can help the students to understand and master the knowledge [2-5].

2 An Overview of the System Development

There are three steps to develop the mould virtual assembly system. Firstly, we construct the model of constructing mould components based on the 3D geometrical modeling software Inventor. Secondly, we import the 3DS MAX to make the animation and render. Finally, we use the VB as the developing language to integrate all the dynamic sources and achieve the display system. The overview of developing this system is described in Fig.1.

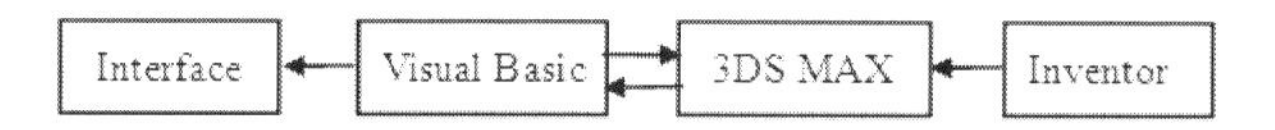

Fig. 1 The overview of developing our system

3 Construct the Dynamic 3D Mould of Stamping Mould

3.1 File Modeling and Assembling the Mould Components

(1) Preparing the Modeling

Stamping mould is a special equipment of the technics. It structure and classes varies based on the content of the mould. Stamping mould can be divided into three mould, including blanking die, blending die and drawing die and son on. In the course of the constructing the mould, all the mould do not have the specified mould components.

Each establishment of 3D model was based on their assembly drawing even schematic drawing and the size of each part was self-identified. The assembly could not be carried out if some of associated sizes weren't determined enough accuracy, which brought certain difficulties for creating assembly model. Therefore, to guarantee accurate assembly, the shape size and position size of each component must be determined before creating the model. Moreover, if there were any problems in the assembly process, the size of components model should be revised repeatedly, so that it can accurately match other parts.

(2) Modeling and Assembly

Inventor, Pro/E, Solid Works, UG were common 3D design software in mechanical design. The models build in Pro/E, UG had high accuracy but the process of modeling was complex or troublesome relatively. While the models build in Inventor, Solid Works had low accuracy but the process of modeling was simple relatively. Moreover, it was easy to operate and the effect of rendering was better. As large number of parts and heavy workload, Inventor software was used to build solid model.

Assembly mainly reflected the mutual constraint relations of components and the correct relations of assembly was the premise for creating animation. When assembling, the benchmark must be determined first, that is, to choose the surface with assembly request or position request for benchmark. Then, according to the coordination or the request of position, the installation location of each component can be determined, which assembled into 3D model. Establishment of the assembly can be divided into two steps: ① Creating components, transferring model of parts or components and adding constraints; ② Editing components, such as arraying, mirroring, digging holes in the parts and so on. After assembly, each part can be given the definition of color and texture.

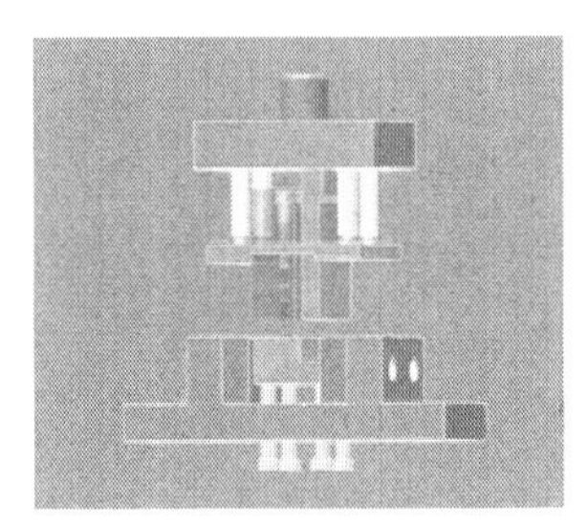

Fig. 2 The assembly effect produced by Inventor

3.2 Producing Animations

The 3DS MAX is the most popular software during modeling the products. Its advantage is that it can create the complicated model and render the result naturally. For the design of some mechanical parts, it is very difficult to build the relation of the relative Relations of Assembly Constraints. As a result, we firstly build the model based on the Inventor software, and then import the 3DS MAX to develop animation process of producing moulds.

Before importing the moulds, we first change the original data of the mould into the data that the 3DS MAX can accepted based on the Inventor. Here, we use the obj format. The obj files are the universal data. After importing the moulds, the relationship between different parts are built and not be changed in the future.

There are two basic methods to produce animations:

(1) Animation based on the key frames. This is the basic and popular method to produce the animation. Its principle is very simple and mainly related to the rule of recording the change of the angles, locations and ratios.

(2) Animation based on constraint. This animation is produced by controlling objects and their parameters to control the locations, rotations and the scaling. The constraint of an object is related to a target object. The target object is constrained on the original objects. Because the animation relate to warping or deforming the mechanical part, we build all the moulds under the software of 3DS MAX [6].

3.3 System Output

The output of our mould system is based on the rendering tools. Using the rendering tool 3DS MAX, we can rotate, move, scaling up/down the objects in producing the animations. We change the file format into avi to display. Fig.3 is the rendering course.

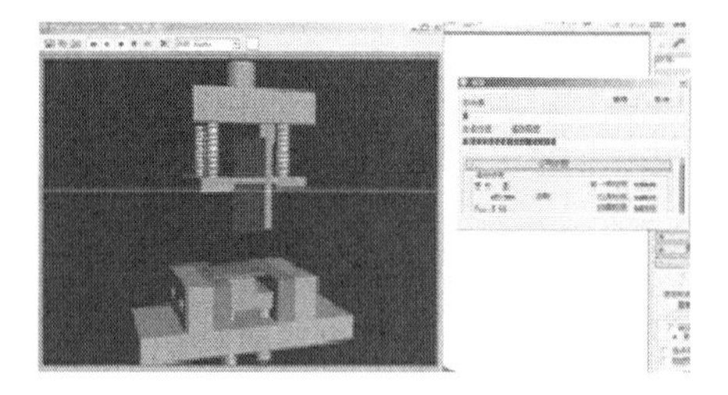

Fig. 3 The Output by using the rendering tool

4 Develop the System

4.1 Functions and Flow of the System

The functions of the display system are equal to assemble the functions of the 3D dynamic model produced by the methods in the last section. We can automatically display the system or display it step by step. If we display the system step by step, we add the explanation by words on each step. Every step is explained to be forward or backward to describe the working state. This system is very easy to add new moulds.

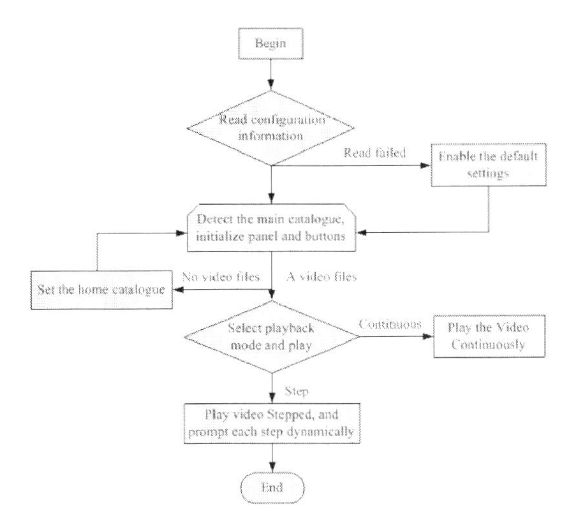

Fig. 4 An overview of our method

After beginning the system, it can read the configuring information automatically and produce the dynamic display panels. If to read the configuring information is failure, the system starts the default configuration. If we successfully start the system, we can select the model to display the video files and sequent display video files until stop the video files. The flow of the display system is described on Fig.4.

4.2 Run the System

After opening the main interface of the display system, users can click the button of "select the main directory". Then, we select a video file to form a new panel. The number of buttons is equal to the number of the video files. We also can select the video files by the toolbar, such as Fig.5.

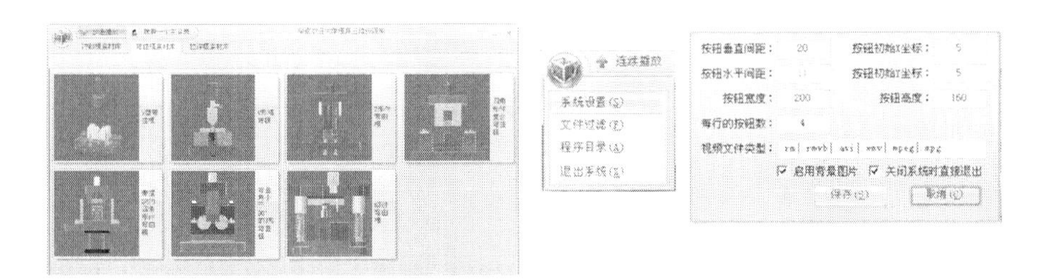

Fig. 5 Form the panel based on the video files

Fig. 6 To set the parameters of our system

Click the button on the upper left corner and select the "setting the system", we can set the parameters of the system, such as Fig.6. Click the button on the panel, we can directly display the video files. There are two methods to display the video files, including displaying video file step by step or consequently. Fig.7 shows the method to display the animation step by step.

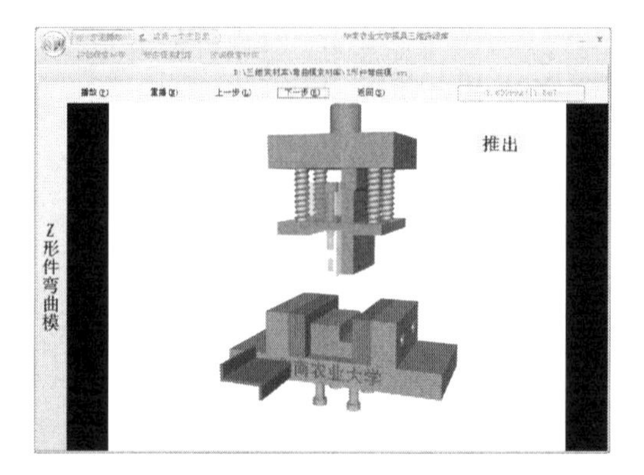

Fig. 7 The display step by step with illustrative text

Our system is developed under the platform of Visual Studio, i.e., VB. As the developing tool is universal, we can promote the system to other media software. Many media file format, such as rm, wmv, mpg, avi, mpeg can be used.

4 Conclusion

Our method has three advantages to construct the Mould Virtual Assembly System. First, the course of forming an animation becomes easily by combining the Inventor and 3DS MAX. Second, the course to display the work of the mould becomes very natural. Third, the system built based on the VB programming tool becomes flexible and diversity.

To display the abstract teaching content by the 3D simulation system, we can improve the teaching method and promote the reformation of teaching method. Following the good teaching method, we can help students overcome the difficulties and produce good teaching effect. The Mould Virtual Assembly System developed by us has the guiding means in Multimedia teaching.

Acknowledgment. This paper was supported by the Spark Program of China (No. 2010GA780049).

References

1. Liu, X.F.: Development and realization of CA I courseware for typical mould structures. Journal of Guangxi University of Technology 16(3), 77–79 (2005)
2. Wang, X.Q., Xiao, S.D.: Three dimensions CAI of the punch tool. Machinery 31(S1), 105–106, 108 (2004)

3. Zhao, T.T.: The dynamic imitation of blanking dies assembling using integration of UG and 3DSMAX. Journal of Shandong University of Technology (Sci. & Tech.) 19(3), 46–48 (2005)
4. Chen, M.: Application of 3DSMAX to Mechanical Manufacturing Engineering Three dimensional Animation Emulation. Journal of Shanghai University of Engineering Science 18(6), 148–151 (2004)
5. Hua, Y.S., Chen, M.: Method of Three Dimensional Modeling and Animation. Journal of Shanghai University of Engineering Science 18(3), 199–202 (2004)
6. Wang, Y.J.: Application of Multimedia Creation of Casting Mould Teaching Based on Software 3DSMAX. Rock Drilling Machines Pneumatic Tools (2), 55–58 (2010)

Breast Measurement of EIT with a Planar Electrode Array

Wang Yan and Sha Hong

Abstract An EIT method with planar electrode array has been proposed and a hand-held planar electrode system has developed. Based on a 3-dimentional finite element model, a modified weighted backprojection algorithm, which is back-projected along equipotential surfaces of the 3D electric filed, is used to reconstruct slice images of 3D conductivity distribution under electrode array. The electrical field distribution has been calculated and several slice images beneath the electrode plane have been reconstructed. The result showed that superficial object can be distinguished well, but image of deeper object needs to be improved sequentially in the future.

1 Introduction

According to the electrical characteristic of tissue or organs of human body, electrical impedance tomography (EIT) extracts impedance and its change information relative to physiology and pathology situation of human body, provides the functional image result reflecting tissue or organs state [1,2]. In the early stage of breast cancer, although the tumour entity has not formed, because of pathological changes of the tumour tissue, the corresponding biology variations have happened, the electrical characteristic of breast tissue in the area has already changed, even the change is much evident. Use of EIT to detect the breast cancer is entirely feasible [3]. EIT has outstanding advantages of noninvasive and convenient, it is favorable for use of EIT that to discovery breast cancer in its early stage and carry out general screening of breast tumour for women [4].

Wang Yan · Sha Hong
Institute of Biomedical Engineering,
Chinese Academy of Medical Sciences & Peking Union Medical College, Tianjin, China
e-mail: sarahyky@163.com, sha_hong_2000@yahoo.com

Z. Qian et al. (Eds.): Recent Advances in CSIE 2011, LNEE 125, pp. 769–774.
springerlink.com

2 Planar Electrode Array

Nowadays most of EIT measurement is based on the annulus electrode [5]. For EIT measurement with annulus electrode, only one image located in electrode plane can be reconstruct in one measurement, the information obtained is limited. To Get the information of more planes, 3D imaging of EIT using more annulus electrodes with different location are needed. This is much difficult for breast measurement because the volume of breast is too small compare with the head, chest and abdomen. Human breast has semisphere form and limited size or volume. To construct a high variability electrode system to suit the breasts with different spheres and different size, there are too many technique problems to overcome [6]. In breast measurement, by the means of raising electrode number and using more electrode planes to improve the distinguishability of EIT reconstructed image, to realized 3D EIT measurement is much difficult.

Based on analysis and study of measurement technique and the development trend of EIT [7, 8], according to the character of beast measurement, an approach of 3D-EIT with a planar electrode array has put forward in this paper. This is a new method combined EIT technology with T-scan electrode structure. The method may overcome some limitations of EIT method using annulus electrode, effectively raise the measurement sensitivity and the image resolution of EIT breast detection. The method draws the flexibility of electrode structure from T-scan [9] technique and adopts EIT method which is different from T-scan and is more accurate, effective than T-scan [10]. 3D-EIT developed in the paper is going to realize multilayer impedance distribution measurement for breast detection by 3-dimension EIT technology. The method is expected to detect breast cancer in early stage and realize the large-scale general investigation of breast tumour. Early discovery, early diagnosis, early treatment of the breast cancer is possible.

Planar electrode array developed in the paper is shown as Fig.1. 211 electrodes of the array are perpendicular to the plane. The electrodes of 19 rows in the array arrange as the form of equilateral triangles. Each electrode has 6 same electrodes around with equal angles and equidistance. The distance between each two electrodes is 10 mm [11].

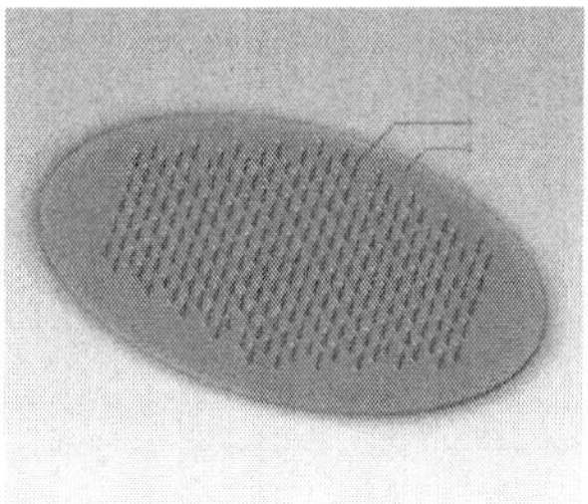

Fig. 1 Planar electrode array

3 Imaging Algorithm

All the electrodes in planar electrode array can serve as the excitation electrode or measurement electrode. The far-end electrode or reference electrode is placed on back of the measured patient. The excitation current is applied via one electrode in planar electrode array and the far-end electrode, the voltages between each electrode in the array and the far-end electrode are measured in turn except the excitation electrode. Then change the excitation current from the first electrode to the second electrode, then the third electrode, the forth electrode and so on of the array, the same voltage measurement procedures above are executed. The measurement procedure is finished when all electrodes in the array have been used as excitation electrode.

The 3-D subdivision of finite-element method (FEM) for the measurement area consists of 6 plane layers and 5 unit layers. The plane FEM subdivision of 6 plane layers adopts equilateral triangle method, Fig.2. The nodes of the equilateral triangle with the same number in 6 plane layers are connected to form 5 unit layers between 6 plane layers, numbered 1, 2, 3, 4, 5 from top unit to bottom unit, Fig.3.

Set the excitation current is applied via one electrode in planar electrode array and the far-end electrode, the equipotential area close to the electrode array plane can be considered as a half sphere surface, Fig.4. Based on this consideration, the simplified equipotential lines back projection (LBP) algorithm can be applied [12]. For a dot S(X, Y, Z) in the imaging area, the distance r, between the dot and the electrode plane, is given. The distance r is also the radius of the equipotential half sphere containing the dot. The intersecting line of the equipotential half sphere and the electrode plane is a roundness in the electrode plane(x, y), the centre is the excitation electrode, the radius is r.

The conductivity of the imaging dot can be calculated from the following equation.

$$S(x,y,z) = 1 + W_1(z)\sum_i \left(\int_{L(x,y,z,i)} W_2(l)dl\right)^{-1} \int_{L(x,y,z,i)} \frac{W_2(l)\left(E_r(l)-E_m(l)\right)}{E_r(l)dl} \quad (1)$$

Here i is the serial number of the excitation electrode, L(x, y, z, i) is the intersecting line of the equipotential half sphere and the electrode plane, a roundness with radius r. Em, Er are the electric field distributions of the uniform medium and the disturbance field respectively.

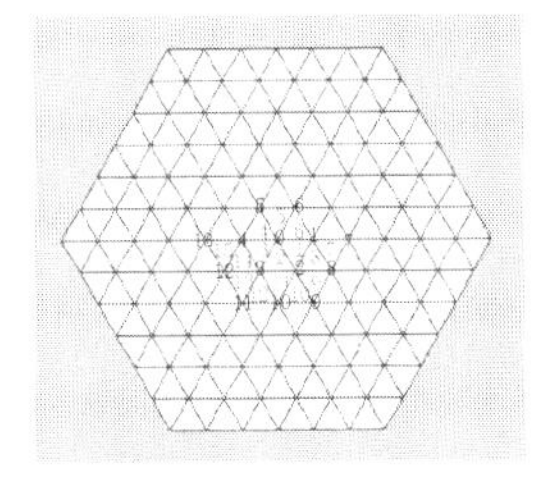

Fig. 2 The plane subdivision of FEM

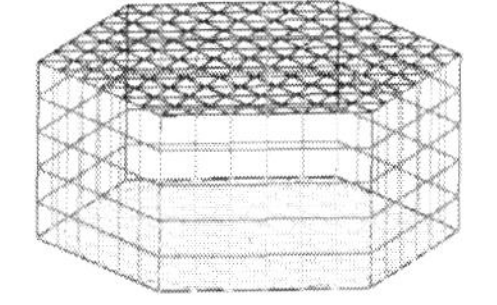

Fig. 3 The 3-D subdivision of FEM

4 Simulation Experiments and Results

According to the 3D subdivision of FEM above, the simulation study of 3D electric field distributions using a planar electrode array is carried out. The simulation results of the electric field distribution are shown in Fig.5.

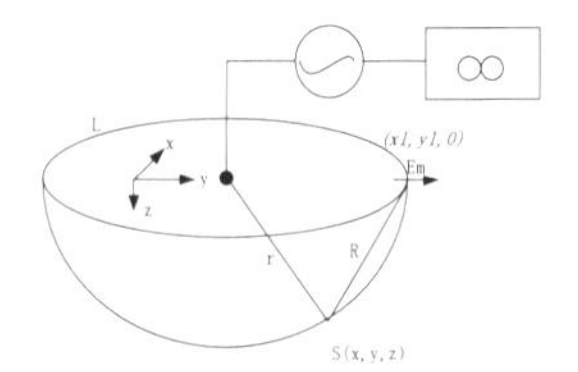

Fig. 4 The algorithm principle of 3-D EIT with planar electrode array

Considering the inhomogeneous of electric field distribution, in order to improve effect of image reconstruction, 2 weight coefficients, W_1 and W_2 are set. Coefficient $W_1(z)$ used to correct of measurement sensitivity with detection depth.

$$W_1 = \frac{z+\alpha}{z+\beta}, \alpha \ll \beta \tag{2}$$

Coefficient W_2 is to make the dots located at the intersecting line L have bigger contribution to the reconstructed image when the distance from the dots to the imaging dot is shorter.

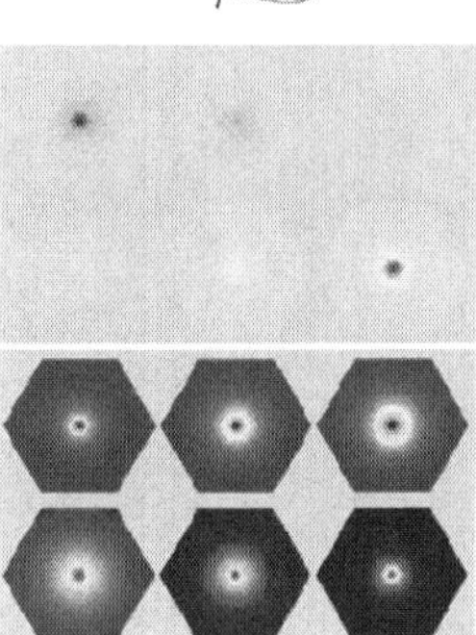

Fig. 5 Simulation results of 3D electric field distributions with the planar electrode array. Top: simulation pattern, mid: electric field distribution from top layers to bottom, bottom: the electric field distribution, the images normalization is according to different layers.

$$W_2 = \frac{1}{R^4} = \frac{1}{(x-x_i)^2 + (y-y_i)^2 + z^2} \tag{3}$$

Here R is distance between L and the dot being calculated. $(x_i, y_i, 0)$ is the coordinate of calculating dot in L.

Excitation current is applied via two nodes located at the top plane and bottom plane respectively, the simulation results of 3D EIT using a planar electrode array are shown in Fig.6. The conductivity of imaging target is 255, and the conductivity around the target is 1.

3 Discussion

The simulation results in Fig.5 and Fig.6 show that based on the half sphere LBP algorithm, the EIT method using planar electrode array can distinguish impedance change in an area under the electrode plane.

Although the result indicate that imaging effect of superficial object is better, the reconstructed images of deep object is not yet good enough, sometimes the image is even diffusion, it should be feasible that the method used for beast measurement. In fact, for the beast measurement, the far-end reference electrode is in the dorsum of measured subject, excitation current is applied via electrode of planar electrode array and the reference electrode. Compared with the distance between the breast and the far-end reference electrode, breast area is only a superficial layer, the assumption of half sphere equipotential line can be fulfilled. It is reasonable that the 3D EIT method with planar electrode array, developed in the paper, can gain good imaging effect for breast detection.

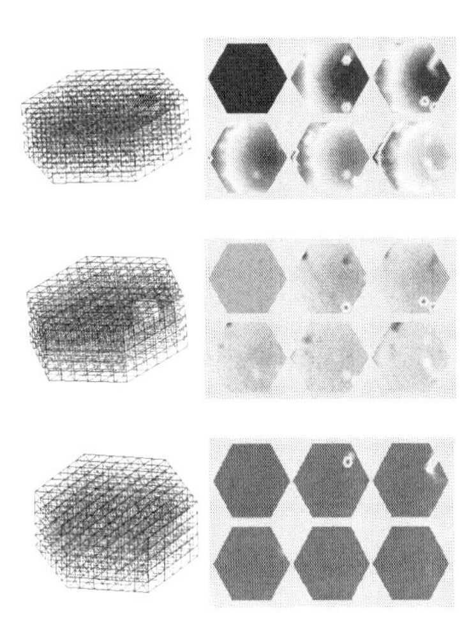

Fig. 6 Reconstructed results of 3D EIT with planar electrode array

In simulation study of the paper, seeking of L and choosing of weight coefficient W1 and W2 have a great influence on the imaging result. The 3D algorithm developed in the paper need sequentially improved and developed. The work in future is to pay attention to raise the imaging effect of the object in deeper area and carry out some practical application study.

Acknowledgment. This work was supported by the National Science and Technology Support Program of China, No. 2006BA103A14, and the key Projects of Science and Technology Support Plan of Tianjin, No. 10ZCGYSF00100.

References

1. Brown, B.H.: Medical impedance tomography and process impedance tomography: a brief review. Meas. Sci. Technol. 12, 991–996 (2001)
2. Holder, D.S.: Electrical impedance tomography: methods, history and applications. IOP Publishing Ltd., London (2005)
3. Chaoshi, R.: Bioelectricity impedance measurement technology. China Medical Device Information 10, 25 (2004)
4. Newell, J.C., Saulnier, G.J., Isaacson, D., Kao, T.-J., Xia, H., Ross, A.S., Kim, B.S., Moore, R.H., Kopans, D.B.: Regional Admittivity Spectra with Mammograms in Breast Biopsy Patients. World Congress on Medical Physics and Biomedical Engineering Seoul Korea, 3780–3783 (2006)
5. Wang, H., Wang, C., Yin, W.: Optimum design of the structure of the electrode for a medical EIT system. Meas. Sci. Technol. 12, 1020–1023 (2001)
6. Wang, W., Tang, M., Yang, J.: Electrical impedance mammography a new imaging modality for breast cancer screening. In: Proceedings of the 9th Annual Conference of LSSCB Leicester, UK, pp. 58–59 (December 2000)
7. Kao, T.-J., Isaacson, D., Newell, J.C., Saulnier, G.J.: A 3D reconstruction algorithm for EIT using a handheld probe for breast cancer detection. Physiol. Meas. 27, 1–11 (2006)
8. Hartinger, A.E., Gagnon, H., Guardo, R.: A method for modelling and optimizing an electrical impedance tomography system. Physiol. Meas. 27, 51–64 (2006)
9. Kerner, T.E.: Electrical impedance tomography for breast imaging. Phd thesis, Dartmouth College (2001)
10. Assenheimer, M., Laver-Moskovitz, O., Malonek, D., Manor, D., Nahaliel, U., Nitzan, R., Saad, A.: The T-scan technology: electrical impedance as a diagnostic tool for breast cancer detection. Physiol. Meas. 22, 1–8 (2001)
11. Sha, H., Wang, Y., Zhao, S., Ren, C., Wei, J., Wang, L.: A electrode array probe used in electrical impedance mammography. Utility Model Patent of China 200620025265.8
12. Cherepenin, V., Karpov, A., Korjenevsky, A., Kornienko, V., Mazaletskaya, A., Mazourov, D., Meister, D.: A 3D electrical impedance tomography (EIT) system for breast cancer detection. Physio. Meas. 22, 9–18 (2001)

Finite Difference Time Domain Method Based on GPU for Solving Quickly Maxwell's Equations

Zhen Shao, Shuangzi Sun, and Hongxing Cai

Abstract. Based on Graphics Processing Units (GPUs) as main computational core, the Finite Difference Time Domain(FDTD) is presented for solving quickly Maxwell's equations for electromagnetic. Firstly, the FDTD algorithm is developed by analyzing the direct time-domain solution like FDTD to Maxwell's curl equations. Then, it is analyzed and compared with CPUs that how GPUs can be used to greatly speedup FDTD simulations, so enormous computation problem in FDTD simulations is resolved. At last, leveraging GPU processing power for FDTD update calculations, researchers can simulate much longer pulse lengths and larger models than was possible in the past, and computationally expensive simulations are completed in reasonable time. It is proved that FDTD simulations based on GPUs is accurate and high efficiency compared with CPUs.

1 Introduction

The Finite Difference Time Domain (FDTD) method is one of important ways for solving Maxwell's equations of electromagnetism, and has long been used widespreadly[1]. FDTD can solve arbitrary geometries models comprised of dielectric materials. Additionally, we can easily handle sinusoidal, transient and pulsed sources through FDTD, which is a useful tool for many problems in electromagnetics. However, simulations of large model spaces or long nonsinusoidal waveforms can require a tremendous amount of floating point calculations and executing times of several months or more even on large HPC systems. Even for small models with approximately 10^9 cells, a simulation of a millisecond duration waveform could

Zhen Shao
School of Computer Science and Technology, Changchun University of Science and Technology, Changchun, China
e-mail: custsz@126.com

Shuangzi Sun · Hongxing cai
Changchun University of Science and Technology, Changchun, China
e-mail: ciomsz@126.com, sshz_2000@yahoo.com.cn

Z. Qian et al. (Eds.): Recent Advances in CSIE 2011, LNEE 125, pp. 775–780.
springerlink.com

easily take over a month even on 1,000 processors on a traditional high-performance computing (HPC) cluster[2]. It is obvious that current FDTD codes running on existing HPC clusters are not viable for this class of problems, prompting a need to investigate methods of increasing computational speed for these simulations.

2 Graphics Processing Units(GPUs)

GPUs with low-cost and high-performance are the early form of graphics processors, which are also parallel data flow processor highly optimizing vector calculation. GPUs will no longer need a large number of complex control logic for Explicit data parallelism. Data flow through high-speed memory interface no longer requires a lot of caches. This can empty out a lot of space to storage computational units. For a few years now, GPUs have been used for general purpose computation. Just a few years, GPUs show the overwhelming potential for large-scale computation such as FDTD.

In terms of raw floating point computation capability, the performance of GPUs is more powerful than CPUs. Parallel processing of data only requires computation streams, no longer need to share memory space. Branching logic shows higher performance if hardware is low-cost. Since streamed data not maintained in cache, the arithmetic how to computing each instruction with much computation is particularly important. Due to these restrictions, many traditional algorithms cannot be directly ported to run on streaming architectures, but for algorithms that are data parallel and are arithmetically intense, great speed up can be achieved[3].

Based on existing GPUs products, a new FDTD simulation program has been developed and achieved speedup of many times than CPUs, solving Maxwell's equations and simulating large model space.

3 Finite Difference Time Domain(FDTD)

FDTD is used to solve Maxwell's equations for arbitrary model spaces and can actually help us to solve the relatively difficult or impossible models with analytical methods [4]. FDTD Maxwell's curl equations for solving the direct time domain as follows:

$$\Delta \times \vec{E} = -\mu \frac{\partial \vec{H}}{\partial t} \tag{1}$$

$$\Delta \times \vec{H} = \varepsilon \frac{\partial \vec{E}}{\partial t} + \bar{J} \tag{2}$$

In the FDTD scheme, Maxwell's curl equations are first scalarized into their x, y, and z field components. Then, centered finite difference expressions are used to approximate the spatial and time derivatives [5].

This method was first proposed in the original FDTD paper by Kane Yee. In particular, he introduced the next important FDTD concept known as the Yee Space Grid, as illustrated in Fig.1.

Seen from Fig.1, each component of a magnetic field H urrounded by the four electric field components E and vice versa [6]. The E and H field are staggered to one another with respect to time by one half of the time step.

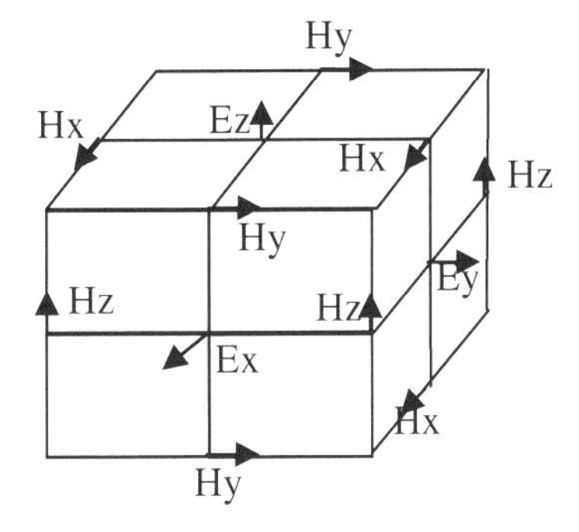

Fig. 1 The Yee three-dimensional spaces

4 The Algorithm for Finite Difference Time Domain

The FDTD algorithm solves Maxwell's equations by first performing the E field update equations for each voxel at time step n, and then performing the H field update equations for each voxel at time step n+1/2 [7]. The model's spatial resolution determines time resolution of the simulation, and the waveform and duration of the source being modeled determine the number of time steps [8].

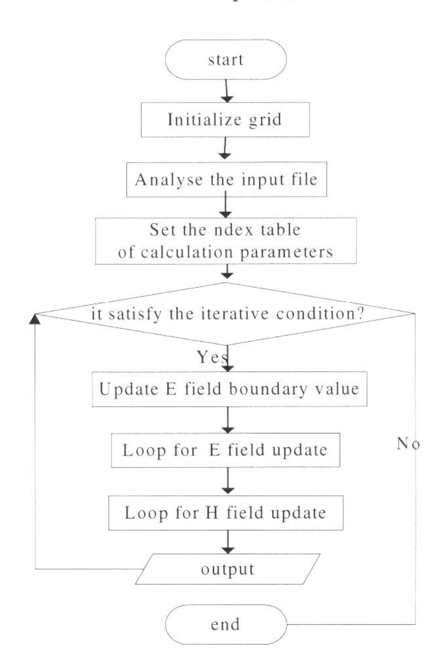

Fig. 2 FDTD top algorithm

The FDTD model space is stored in a three dimensional grid. Each cell in the grid is assigned a material type with corresponding dielectric properties. The x, y, and z components are stored in corresponding cell for both the E and H field initialized to 0. After initializing the model space, the basic FDTD algorithm used is shown in Fig. 2.

This algorithm takes up most of the running time in loops of E and H field updates. E field update and H field update are independent and can be executed in parallel.

5 The Implementation and Data Analysis on GPUs

FDTD update calculations are both of data parallelism and high-intensity computing, so that the FDTD algorithm is very suitable to run on GPUs.

For every grid of the model space satisfying the conditions of data parallelism, the same E and H update calculations are executed. Every calculations for each update require at least 18 floating-point operations per second(FLOPS),thus enough to ignore memory latency problem on most GPUs.

The initial GPU FDTD code was implemented in C and Cg for GeForce 7 series based GPUs, which have distinct vertex and fragment processors[9]. The vector E and H stored in the memory were as two-dimensional textures of 32-bit floating-point red-green-blue (RGB) in which color channels the X, Y, and Z field components were stored.

Using Linux, C, OpenGL, Cg, and commodity GeForce 7 series GPUs, we developed a new FDTD code based on GPUs for solving quickly Maxwell's equations. The graphics hardware was accessed through standard OpenGL. Then the FDTD model space was transferred by OpenGL textures to the GPU memory and host readable via frame buffer objects which were exposed by the OpenGL2.0 application programming interface (API). GPU fragment processors were utilized for the FDTD update computations via Cg fragment programs.

In the simulation test, stability limit of 0.9,a 0-20 GHz point source was placed at the centre of the mesh as illustrated in Fig.3. The tests were run on an Intel P4 2.4GHz CPU with 512MB RAM and NVIDIA's GeForce 7800Gs GPU with 256MB video memory.

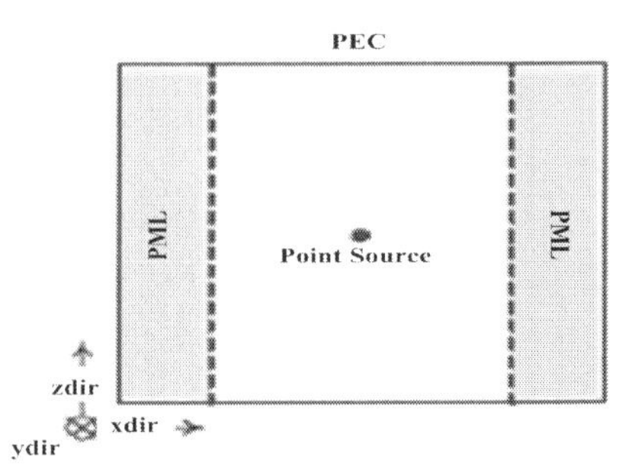

Fig. 3 Simulation region

Fig.4 shows the accuracy of GPU and CPU simulations as a function of time-step. On average the GPU simulator accumulated error 12.3943 times faster than the CPU simulator.

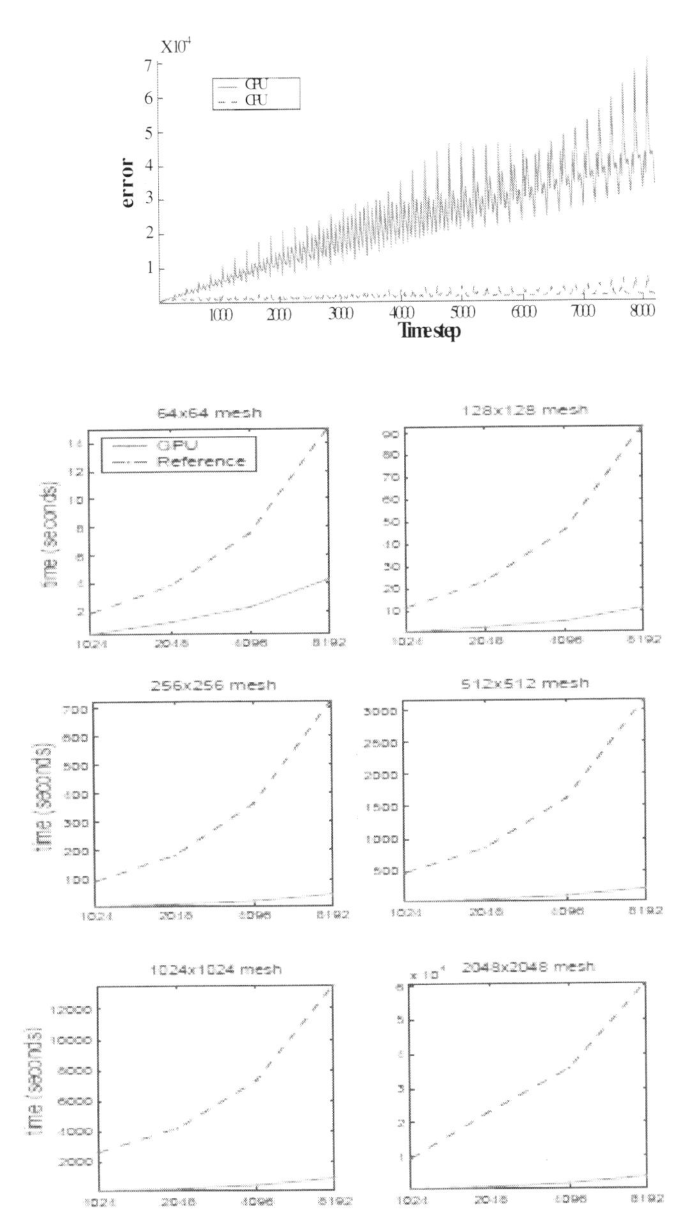

Fig. 4 Accuracy

Fig. 5 Simulation results in different meshes

Fig.5 shows the simulation time in seconds of GPU and CPU simulators for different meshes as a function of time-step. The GPU simulator approaches an average sustained 17.5733 times speed-up as simulation size increases.

6 Conclusion

The use of GPUs shows great promise for high performance computing applications like FDTD that have high arithmetic intensity and limited or no data dependencies in computation streams. Generally if the model space was large enough to keep the fragment processors busy, the GPUs version was always several times faster, even for the slowest seven series GPU tested, the GeForce Go 7400. For models that are not big enough to keep the GPUs fragment processors busy, the GPUs would run slower than on the CPU. Therefore, using the GPUs achieves the result of a faster execution time, if the FDTD model is large enough. Another interesting observation is that with a small number of iterations, the performance penalty can be clearly seen in initializing the streams on GPUs, but along with the increasing number of iterations, this initialization cost is amortized. Since typical FDTD simulations require tens of thousands of iterations to produce results, our GPU implementation produces significant speedup over the CPU implementation in almost all cases except very small model sizes.

In this paper, we present the innovative solution to solve quickly equations represented by Maxwell's equations with complicated boundary conditions, arithmetically intension and no data dependencies, which will facilitate the development of such solving equation technology.

References

1. Taflove, A., Hagness, S.C.: The Finite-Difference Time-Domain Method. In: Computational Electrodynamics, Artech House Publishers, Boston (2000)
2. Harris, M.: GPU: General-Purpose Computation on Graphics Hardware. In: Introduction to CUDA, pp. 301–312. ACM, San Diego (2007)
3. Pharr, M. (ed.): GPU Gems 2. Addison-Wesley, Upper Saddle River (2005)
4. Durbano, J.P., Ortiz, F.E., Humphrey, J.R., Curt, P.F., Prather, D.W.: FPGA-Based Acceleration of the 3D Finite-Difference Time-Domain Method. In: FCCM, pp. 156–163 (2004)
5. Jackson, D.J.: Classical Electrodynamics. John Wiley & Sons, Inc., New York (1999)
6. Yee, K.: Numerical Solution of Initial Boundary Value Problems Involving Maxwell's Equations in Isotropic Media. IEEE Transactions on Antennas and Propagation AP-14(3), 802–807 (1966)
7. Kunz, K.S., Luebbers, R.J.: The Finite Difference Time Domain Method for Electromagnetics. CRC Press, Boca Raton (1993)
8. Chen, P., Wu, C.: Numerical Simulation of Electromagnetic Waves Propagation in Stratified Stratum. Computer Engineering 36(15), 262–264 (2010) (in Chinese)
9. Göddeke, D.: GPGPU–Basic Math Tutorial (November 2005), `http://www.mathematik.uni-dortmund.de/~goeddeke/`

TERPRED: A Dynamic Structural Data Analysis Tool

Karl Walker, Carole L. Cramer, Steven F. Jennings, and Xiuzhen Huang

Abstract. Computational protein structure prediction mainly involves the main-chain prediction and the side-chain confirmation determination. In this research, we developed a new structural bioinformatics tool, TERPRED for generating dynamic protein side-chain rotamer libraries. Compared with current various rotamer sampling methods, our work is unique in that it provides a method to generate a rotamer library dynamically based on small sequence fragments of a target protein. The Rotamer Generator provides a means for existing side-chain sampling methods using static pre-existing rotamer libraries, to sample from dynamic target-dependent libraries. Also, existing side-chain packing algorithms that require large rotamer libraries for optimal performance, could possibly utilize smaller, target-relevant libraries for improved speed.

1 Introduction

Proteins, such as enzymes, carrier proteins, receptors and antibodies, play important roles in the cell. There are four levels of protein structures: primary structure (protein sequence), secondary structure, tertiary structure and quaternary structure. Protein tertiary structure is essential for its correct function in the cell. Protein tertiary structure prediction is a very important research area. Its applications include drug design in medicine and the design of enzymes in biotechnology. To make fast and accurate predictions of protein tertiary structures remains very challenging. Computational protein structure prediction mainly involves the main-chain prediction and the side-chain confirmation determination.

Karl Walker · Steven F. Jennings
University of Arkansas at Little Rock, Little Rock, AR 72204 USA
e-mail: {kawalker1,sfjennings}@ualr.edu

Carole L. Cramer · Xiuzhen Huang
Arkansas State University, State University, AR 72467 USA
e-mail: {ccramer,xhuang}@astate.edu

Z. Qian et al. (Eds.): Recent Advances in CSIE 2011, LNEE 125, pp. 781–786.
springerlink.com

Finding the lowest-energy, side-chain conformation is considered as a separate problem: the protein side-chain packing problem. The major difficulty with predicting side-chain conformations is that there are an extremely large number of possible conformations for even small residues. In order to reduce the potential search space, the number of allowed conformations for each type of side-chains is restricted to a limited number of configurations called rotamers. The typical approach is to generate a collection of these rotamers for each type of residue, called a rotamer library, from structural bioinformatics or other statistical analyses of side-chain conformations in experimentally determined protein structures (Dunbrack, 2002). These libraries generally contain information about the conformation, its frequency, and the variance about the mean dihedral angle. Rotamer libraries can be backbone independent, secondary structure dependent, or backbone dependent.

In this paper, we present a new structural data analysis tool, called TERPRED, which implements a hybrid idea to protein structure prediction. The approach we describe applies the idea of template-based modeling in that it uses predetermined protein structures as a parameter. However, it differs in that it does not require direct alignment of the target sequence onto template sequences or structures. It also applies the idea of ab initio methods in that it uses physiochemical properties of amino acids to drive the prediction towards the lowest free energy. It differs in that it does not require computationally expensive simulations to predict structure. We present an algorithm that generates a tertiary prediction search space from the amino acid sequence of a target protein. Compared to various modern rotamer sampling methods (Dunbrack, 2002; Xiang and Honig, 2001; Peterson, Dutton, and Wand, 2004), our work is unique in that it provides a method to generate a rotamer library dynamically based on small sequence fragments of a target protein.

2 Dynamic Structural Data Analysis

Our new structural data analysis tool called TERPRED is composed of two components: the Database and the Rotamer Generator, which are discussed in the following two subsections. TERPRED can perform well even with very low sequence identity ($< 10\%$). The algorithm accepts various customizable parameters including amino acid sequences, a set of predetermined structures (global search space), secondary structure prediction, motifs, and domains.

2.1 The TERPRED Database

The purpose of the construction of the TERPRED platform was to enable detailed analysis of structural data in order to determine the factors that define the structure of a protein. In order to do this, a database was constructed to hold tertiary structure data in a relational format so as to facilitate the querying of structures based upon various combinations of attributes. Over thirty database tables were constructed with interconnectivity at the forefront of the design. Database tables were designed to hold sequence data, Cartesian coordinate data, polar coordinate data, secondary structure information, binding sites information, amino acid properties, and metadata about a protein such as its type, function, organism, etc. There are also tables to hold data about the encoding and translation of a protein such as mRNA transcript, genetic code, and codon-usage tables. These tables are provided in order to facilitate queries about the co-translational effect on protein structure.

Data to be loaded into the tables were downloaded from the Protein Data Bank, NCBI, and other sources. In order to facilitate loading data into the TERPRED database, several Perl scripts were designed to read, extract, and store data into the appropriate tables. The data from a single PDB file were stored across several database tables.

Predetermined structures in the Protein Data Bank were downloaded to the TERPRED server. The file organization structure used on the PDB server was retained after the data was migrated. Initial testing of the TERPRED system was first done with approximately twenty randomly-selected proteins, and benchmark testing was later done using the Lindahl dataset (Lindahl and Elofsson, 2000), which is a set of 976 non-redundant protein structures widely used to test prediction algorithms. The PDB record IDs corresponding to each of the records in the Lindahl dataset were extracted and each of the structure files were loaded directly from the original PDB files. This was done so that all of the metadata or descriptive information could be extracted in addition to the structural information.

Other information needed for the TERPRED system was derived from data stored within PDB records. Secondary structure assignments were made using the algorithm developed by Wolfgang Kabsch and Chris Sander called DSSP (Define Secondary Structure of Proteins). The standard method for assigning secondary structures to the amino acids of a protein is the DSSP algorithm using the atomic-resolution coordinates of the protein. By means of an electrostatic definition, the hydrogen bonds of the protein structure are first identified. A hydrogen bond is identified by the DSSP algorithm if the energy E is less than -0.5 kcal/mol. The torsion (or dihedral) angles of bonds were also derived from the atomic coordinates within the PDB files.

After the data were successfully downloaded, derived, and loaded into the system, queries could be run against it. The following are several example queries, "What proteins have a specific pattern within their sequences?" or "What are the various sequence segments that fold into a right-handed alpha-helix?" The answers of the queries are limited by the quality and quantity of the data within the database. As a result, TERPRED was designed to hold a large number of predetermined structures while providing the capability of various structures to be grouped into datasets that allow users to query a subset of the total available records. Users may utilize all of the records loaded into the system, a public dataset such as the Lindahl dataset, or create their own custom dataset (collection of PDB identifiers, model numbers, and chain identifiers). This allows users to generalize the search or limit the scope of the search to a set of non-redundant structures or to a set of specific structures like globular proteins, enzymes, receptors, etc. based upon information known (or hypothesized) about a target protein. There are various preloaded datasets within the TERPRED Rotamer Generator.

2.2 The TERPRED Rotamer Generator

The concept of finding the local amino acid influence has been integrated into an algorithm that generates a tertiary prediction search space from an amino acid sequence. The TERPRED Rotamer Generator is primarily designed to address the search space representation aspect of the protein side-chain packing problem. However, since torsion angles for both side-chain and main-chain bonds were computed from the atomic coordinates in PDB files and integrated into the TERPRED Database, it is flexible enough to use in virtually any project that involves structural analysis: protein main-chain prediction, protein design, homology modeling, and the protein docking problem. As the free energy landscape limits the potential folding hyperspace in vivo to those energetically reachable, TERPRED seeks to limit the prediction search space in silico to those configurations observed in proteins with similar sequence, structural elements, or function. To accomplish this goal, TERPRED provides several parameters that may be customized based upon known properties of a target protein. These parameters facilitate the dynamic search and retrieval of structural data relevant to a specific target. The customizable parameters allow a tertiary predictor to select a subset of model structures and extract conformational data based on the influence of local amino acids, secondary structure predictions, and key words (motifs) or phrases (domains) found within an amino acid sequence. TERPRED provides several parameters whose use can be customized to search the database of known structures using more information than just the amino acid sequence of the target protein.

The window of influence that neighboring amino acids have on the conformational space can be set to an odd value between 3 and 7. This type of influence may be considered as general influence because it is not based upon a specific pattern of amino acids known to be associated with a motif or domain. Within TERPRED, the influence of neighboring amino acids is captured using a sliding window approach. In this approach, a window of the selected size is scanned across the amino acid sequence of the target beginning at the first residue and incrementing by one until the last window of that size is reached. In our example, a window size of 3 is chosen to capture the influence of the amino acids on either side (1) (Figure 1a). Each pattern extracted from the window scan is searched for in the database (Figure 1b). After the patterns are found, the associated dihedral angles are extracted and assigned to the appropriate residue of the target sequence (Figure 1c). With a window size of three, each residue (except for the first and last two) will appear in three positions: as the first, second, and last residue of a window. The angle conformations of a particular residue are stored selecting all the conformations or only those where the residue is in the center of the pattern (position 2 in Figure 1c), for example.

Torsion-angle conformations may also be selected for a specific pattern corresponding to a motif or domain. Rotamers selected from such specific patterns are not grouped or filtered along with the rotamers selected from non-specific patterns. They are considered separately by the algorithm and are assumed to be more likely candidates than non-specific rotamers. Selecting rotamers based upon motifs and domains first requires that these elements be found in the target sequence. For this

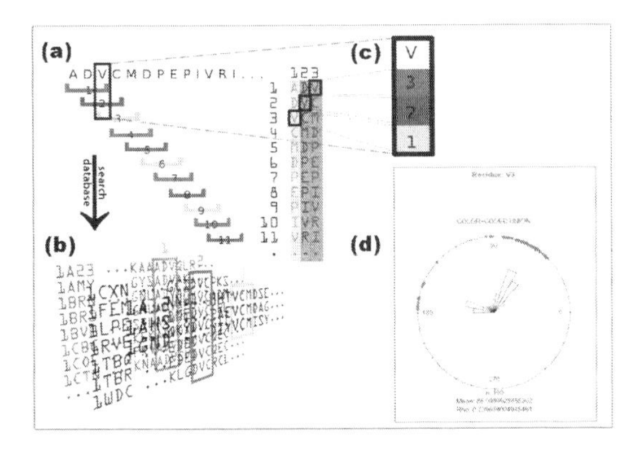

Fig. 1 (a) The amino acid sequence is scanned in windows of size 3 and (b) the database is searched for each triplet of amino acids. (c) The angles found for each amino acid are assigned to the corresponding residue of the target amino acid sequence. (d) The angles are color coded according to the position they are in and graphed.

purpose, TERPRED integrates the analysis with third-party tools designed to locate such patterns within an amino acid sequence. To search for motifs within a sequence, TERPRED uses the PROSITE scanning tool, ScanProsite (de Castro et al., 2006). Domains can be located using tools like SCOP (Murzin et al., 1995) or CATH (Orengo et al., 1997). These tools use databases of known sequence patterns or tertiary structure elements to search for motifs or domains, respectively. The motifs or domains matched within a target sequence are then submitted to TERPRED for the extraction of rotamers based upon their associated patterns.

3 Discussion and Further Work

We have developed a web-based tool capable of selecting rotamers, not only by amino acid and secondary structure but also by factoring in the influence of neighboring amino acids. We have presented this new structural data analysis tool and have shown ways in which it may be used to analyze the protein structures maintained in the Protein Data Bank. The initial testing of this tool has shown promising success in generating rotamer libraries containing values very close to the actual values of the test structures. The Rotamer Generator provides a means for existing side-chain sampling methods that use pre-existing, static rotamer libraries to sample from dynamic, target-dependent libraries. The filtration step of the Rotamer Generator is of great importance in reducing the computational load on protein side-chain packing algorithms. The secondary structure and mutual influence methods have shown to be quite effective at reducing the number of possibilities while retaining accurate values. A combination of filters generally applicable to any target protein is yet to be determined.

We would further work on the third component of TERPRED, the Structure Modeler. The goal of the Structure Modeler is to find the lowest-energy conformation, given the rotamers of the Rotamer Generator as parameters. However, existing side-chain packing algorithms that require large rotamer libraries for optimal performance could possibly utilize smaller, target-relevant libraries produced by the Rotamer Generator for improved speed. Furthermore, the TERPRED platform provides researchers with the tools to closely analyze the relationship between sequence and structure and also allows researchers to search for the protein folding code.

Acknowledgements. This publication was made possible partly by NSF Experimental Program to Stimulate Competitive Research (EPSCoR) Arkansas Center for Plant-Powered Production (P3) seed grant Fund# 224050 and NIH Grant# P20 RR-16460 from the IDeA Networks of Biomedical Research Excellence (INBRE) Program of the National Center for Research Resources.

References

1. Chou, P.Y., Fasman, G.D.: Prediction of the secondary structure of proteins from their amino acid sequence. Adv. Enzymol. Relat. Areas Molecular Biology 47, 45–148 (1978)
2. De Castro, E., Sigrist, C.J.A., Gattiker, A., Bulliard, V., Petra, S., Langendijk-Genevaux, P.S., Gasteiger, E., Bairoch, A., Hulo, N.: ScanProsite: detection of PROSITE signature matches and ProRule-associated functional and structural residues in proteins. Nucleic Acids Research 34, 362–365 (2006)
3. Dunbrack, R.L.: Rotamer libraries in the 21st Century. Current Opinion Structural Biology 12(4), 431–440 (2002), doi:10.1016/S0959-440X(02)00344-5
4. Lindahl, E., Elofsson, A.: Identification of related proteins on family, superfamily and fold level. J. Mol. Biol. 295(3), 613–625 (2000)
5. Murzin, A.G., Brenner, S.E., Hubbard, T., Chothia, C.: SCOP: a structural classification of proteins database for the investigation of sequences and structures. Journal of Moecular Bioogyl. 247, 536–540 (1995)
6. Orengo, C.A., Michie, A.D., Jones, D.T., Swindells, M.B., Thornton, J.M.: CATH: A Hierarchic Classification of Protein Domain Structures. Structure 5, 1093–1108 (1997) ISSN: 0969-2126
7. Peterson, R.W., Dutton, P.L., Wand, A.J.: Improved side-chain prediction accuracy using an ab initio potential energy function and a very large rotamer library. Protein Sci. 13, 735–751 (2004)
8. Xiang, Z., Honig, B.: Extending the accuracy limits of prediction for side-chain conformations. Journal of Molecular Biology 311, 421–430 (2001)

An ECG Signal Processing System Based on MATLAB and MIT-BIH

Tao Lin and Shuang Tian

Abstract. Massachusetts Institute of Technology created MIT-BIH Arrhythmia Database in 1980. In this database, the ECG signal is stored and transported in MIT-BIT format, which has been an important universal format standard. The database contains abundant typical cases with detailed annotations, thus enjoys international impact. To save storages space, MIT-BIH database adopt a specific-defined format, storing each ECG signal data in a sealed file respectively, therefore unable to conduct large scale based on content. In order to query the MIT-BIH ECG data and images, this paper mainly concerns this problem. We propose to use MATLAB to read MIT-BIH ECG signals, then store them in SQL database, again, apply the MATLAB software to access the information stored in the SQL . providing a novel and convenient method for other software to access, use and revise the MIT-BIT ECG signal. The proposed method can ensure the convenient communication and also facilitate the task of data mining for the ECG signals.

1 Introduction

Electrocardiography (ECG) is a kind of transthoracic data recording the electrical activity of the heart, which is detected by electrodes attached on the skin. ECG is an objective measurement for the activation, transportation and recovery of heart activities.[1] Since ECG demonstrates the activity of heart, it is a very important reference to diagnose cardiovascular diseases.

MIT-BIH Arrhythmia Database, recently created by Massachusetts Institute of Technology, is widely used to study Arrhythmia. Due to the representative cases and abundant annotations, MIT-BIH database has become the benchmark database

Tao Lin · Shuang Tian
Hebei University of Technology, Tianjin
e-mail: lintao@scse.hebut.edu.cn, tianshuang2009@gmail.com

Z. Qian et al. (Eds.): Recent Advances in CSIE 2011, LNEE 125, pp. 787–792.
springerlink.com

in the research community.[2] Currently, how to make use of the ECG signals is a very critical issue for the diagnosis and medical information. The aim of this paper is to read and store MIT-BIH format ECG signals to facilitate the further information processing and transformation.

2 Read and Store the Header File of MIT-BIH ECG Signals

Each MIT-BIH format signals are comprised of several files instead of a single file, including a header file (extension format: .hea), a data file (extension format: .dat) and several annotation files (.atr).[3, 4] The header file and the data file are two basic files to store the signal data. We will discuss these two files in detail in the following. In this chapter, we will discuss how to read data from these two files using MATLAB.

There has been much research work on how to build ECG database. In the paper, Establishment of 12 - lead Synchronous ECG Database Based on Virtual Instrumentone novel method is proposed.[5] Some researchers have focused on building ECG databases using MIT-BIH database. Qiong Wang and Zhengguo Zhang translate the MIT-BIH format to a new format of ECG data which is consistent with HL7-aECG standard, which can be used for retrieve. However, their work ignores how to display ECG data.[6]

2.1 Header File

Each header file in MIT-BIH is composed of one or multiple row of ASCII characters, the maximum number of characters are 255. The first row contains four fields for some basic information as overall description for an EEG data. The following rows are the presentation of leads information and the annotations for the patient.[7]

The following example is 100.hea:

100 2 360 650000

Four fields starting from left side are file name, lead number, sample frequency and the number of sampling points.

100. dat 212 200 11 1024 995-22131 0 MLII

From left to right are file name, storage format, gain, AD resolution, ADC zero value, the first sample value, verification number, block size and the number of lead.;

100.dat 212 200 11 1024 1011 20052 0 V5

The format is the same as the above row.

69 M 1085 1629 x1
AldometInderal

Rows starting with "#" are comments for some basic information of the patient such as age, gender and previous prescription.

2.2 Reading and Saving Header Files

In the previous chapter, we have analyzed the data structure of header files. In this chapter we will discuss how to save the header file into SQL.

In database, table is the main storage form for various kind of information. First we define a table named HEADFILE. Then we can import the data into this table. The implementation of this step is easy. We use fget1 function here. This function is used to process pure text file usually, how-ever, it is also appropriate for the header file of MIT-BIH. Then we store the data into a matrix.Then we correspond each element with a single field in the database for storing.

nextcombining matrix values with the Field of SQLis suitable for storage.

Each field, from left to right, are:'filename', 'num', 'freq', 'nsamp', 'dformat', 'gain', 'bitres', 'zerovalue', 'firstvalue', 'chec', 'ofsize', 'lead', 'age' and 'sex', representing file name, the numbers of lead, sampling frequency, number of sampling points, file format, gain, resolution, ADC zero value, the first value of the lead, verification number, block size, lead number, age and gender respectively. After importing the header files of the whole MIT-BIH dataset into the database, it is easier and faster to store, update and manipulate data.

After importing data into "HEADFILE" table SQL database using MATLAB, we have following table as shown in fig 1.

filename	leadnum	freq	nsamp	dformat	gain
100	2	360	650000	212	200
100	2	360	650000	212	200
101	2	360	650000	212	200
101	2	360	650000	212	200
102	2	360	650000	212	200
102	2	360	650000	212	200
103	2	360	650000	212	200
103	2	360	650000	212	200
104	2	360	650000	212	200
104	2	360	650000	212	200

Fig. 1 Part of the screenshot of table HEADFILE in SQL

3 Reading and Saving the MIT-BIH EEG File

Introduction: data file is also called waveform file. In MIT-BIH dataset, there are 8 types including Format8, Format16, Format80, Format212, Format310 and so on. In this paper, we mainly discuss Format212 data format. Acutely, in each header file of MIT-BIH EEG file, there are detail descriptions for this data. Please refer to last chapter.

3.1 Data File

MIT-BIH ECG database is a kind of arrhythmia database"212" format stores two datasets each three bytes.

Take 100.dat as example. According to '212' format, read 3 bytes first"E3 33 F3"then the two values are: 0x3E3 and 0x3F3, after translating to decimal, 995 and 1011. These two values are the first sample point of the two signals, multiplying his corresponding gain; the amplitudes are 4.975mv and 5.055m. The remaining data are saved in the same format.[8]

3.2 Saving and Reading Data File

We can read sample points using unit8 format and store into matrix M with SAMPLE rows and 3 columns. Each column is one byte. Each row use 3 bytes to represent two numbers m1 and m2. Each number has 12bits This is so called 212 format. The lower 8 digit of m1 are saved in M(:,1) while the lower 8 digit of m2 saving in M(:,2), the higher 4 digits of m1 saved in M(:,1) and the higher 4 digits of m2 saved in M(:,2).

We can extract decimal dual channel signal data using a serial of procedure like displacement and AND. Then they can be transformed into EEG signal value which can be used in practical. According to the values of ECG Signals we can calculate true value, also we can use these EEG data to calculate the real EEG data and draw the waveform, as shown in figure 2.

The image can be viewed as the ensemble of many coordinates. The horizontal axis is time and the vertical axis is the voltage value of EEG signal. In this way we can transform the image file into coordinates of points. First we define some time points saving in matrix T. Then we save the corresponding image file into matrix M. After importing matrix M and time matrix T into database, the EEG image is saved into the SQL. Figure 3 shows the table called DATAFILE after imported into SQL.

As we know, SQL don't require users to use any specific storing methods for data. The system will determine the fastest method for a particular type of data. it has excellent transportability universal property ,and it's easy to carry out relational data manipulation with the help of data inquiry function. So we can use C, C++ and C# to access to the EEG data. What's more, we can access to the database in

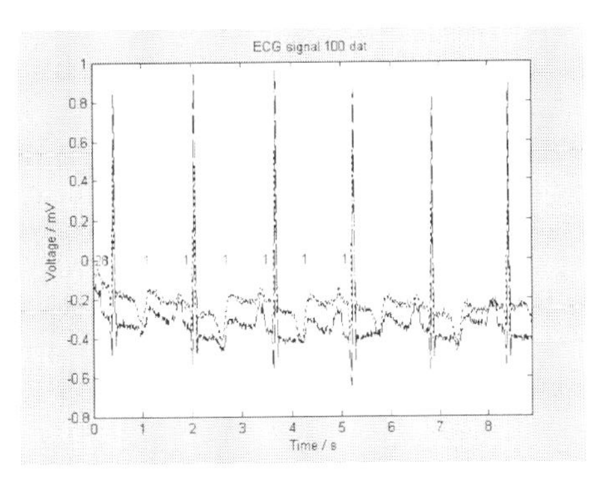

Fig. 2 The image display of the data file in MATLAB

Time	R100	B100	R101	B101
0	-0. 144999995827675	-6. 49999976158142E-02	-0. 344999998807907	-0. 159999996423721
2. 77777784503996E-03	-0. 144999995827675	-6. 49999976158142E-02	-0. 344999998807907	-0. 159999996423721
5. 55555569007993E-03	-0. 144999995827675	-6. 49999976158142E-02	-0. 344999998807907	-0. 159999996423721
8. 33333376795053E-03	-0. 144999995827675	-6. 49999976158142E-02	-0. 344999998807907	-0. 159999996423721
1. 11111113801599E-02	-0. 144999995827675	-6. 49999976158142E-02	-0. 344999998807907	-0. 159999996423721
1. 38888889923692E-02	-0. 144999995827675	-6. 49999976158142E-02	-0. 344999998807907	-0. 159999996423721
1. 66666675359011E-02	-0. 144999995827675	-6. 49999976158142E-02	-0. 344999998807907	-0. 159999996423721
1. 94444451481104E-02	-0. 144999995827675	-6. 49999976158142E-02	-0. 344999998807907	-0. 159999996423721
2. 22222227603197E-02	-0. 119999997317791	-7. 99999982118607E-02	-0. 330000013113022	-0. 150000005960464
0. 025000000372529	-0. 135000005364418	-7. 99999982118607E-02	-0. 319999992847443	-0. 144999995827675
2. 77777779847384E-02	-0. 144999995827675	-8. 50000008940697E-02	-0. 319999992847443	-0. 170000001788139
3. 05555555969477E-02	-0. 150000005960464	-8. 50000008940697E-02	-0. 324999988079071	-0. 180000007152557
3. 33333350718021E-02	-0. 159999996423721	-7. 50000029802322E-02	-0. 330000013113022	-0. 170000001788139
3. 61111126840115E-02	-0. 155000001192093	-7. 00000002980232E-02	-0. 324999988079071	-0. 150000005960464
3. 88888902962208E-02	-0. 159999996423721	-7. 00000002980232E-02	-0. 314999997615814	-0. 159999996423721
4. 16666679084301E-02	-0. 174999997019768	-6. 49999976158142E-02	-0. 305000007152557	-0. 170000001788139
4. 44444455206394E-02	-0. 180000007152557	-5. 49999970197688E-02	-0. 300000011920929	-0. 185000002384186

Fig. 3 the coordinates of the data file after importing into SQL

different OS. So we can access to the database as indirect access, which make the access easier. And the data files are easier to access. Based on this, we can conduct further research for EEG signal, which is meaningful for medical development.

3.3 *Read EEG Data from SQL Using MATLAB*

We can use different tools to read the EEG information we have stored in the database for other processing. Then we will use MATLAB as an example, to read the EEG data from SQL.

According to M1, M2 and T from database, we get figure 4 and figure 5. Figure 4 is the waveform of MLlead while Figure 5 is the waveform of V5 lead.

From the experiment results, we can still do various processing after importing data into SQL. Although we have only used MATLAB to process data in this paper, we can choose proper tools to access to SQL according to the practical requirements and situations.

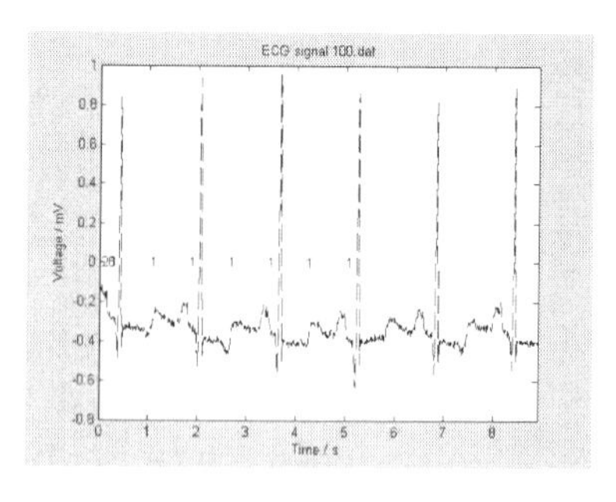

Fig. 4 The ECG of lead ML2

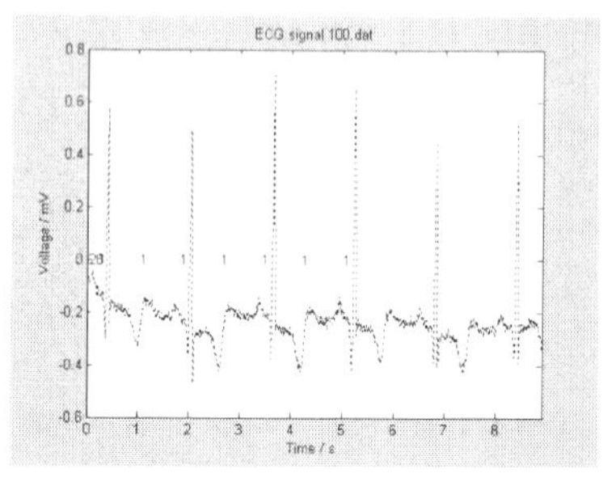

Fig. 5 The ECG of lead V5

4 Conclusions

Utilizing Matlab to read MIT-BIH Arrhythmia dataset indirectly, we have reduced the work for accessing to the data, which make it easier to process data. EEG researchers can process MIT-BIH data flexibly or conduct further development.

References

1. `http://baike.baidu.com/view/20611.html`
2. Geddes, K.O., Czapor, S.R., Labahn, G.: Algorithms for Computer Algebra. Kluwer, Boston (1992)
3. Moody, C.B., Mark, R.G.: The Impact of the MIT-BIH Arrhythmia Database. IEEE Eng. Med. Biol.Mag. 120(3), 40–60 (2001)
4. Moody, G.B.: WFDB Applications Guide, 10th edn., Havard-MIT Division of Health Sciences and Technology (May 5, 2010)
5. Liu, S.-M., Pan, Y., Li, Q., Meng, Y., Sun, W.-H., Zhang, Y.-H., Li, X., Qi, Y., Zhu, X.-L.: Establishment of 12 - lead Synchronous ECG Database Based on Virtual Instrument. Journal of Biomedical Engineering Research 27, 400–404 (2008)
6. Wang, Q., Zhang, Z.-G.: Installation of Electrocardiogram Database with HL7. Chinese Journal of Biomedical Engineering 27(3) (June 2008)
7. Xinlin: The Design and Implementation of 12-Lead ECG Database Generating System. ShanDong Ocean University of China (2007)
8. Song, X.-G., Deng, Q.-K.: On the Format of MIT-BIH Arrhythmia Database. Chinese Journal of Medical Physics 21(4) (July 2004)

Cone-Beam Computed Tomography Image Reconstruction Based on GPU

Hongwei Xu, Fucang Jia, Wenyan Chen, and Xiaodong Zhang

Abstract. As so long, three-dimensional cone-beam computed tomography(CBCT) image reconstruction is a hot issue in medical imaging field. Often the computation operation of CBCT reconstruction is huge and the reconstruction time is long. Now with the development of computer technology, especially the rapid development of Graphics Processing Unit (GPU) based general-purpose computing technology enables fast CBCT reconstruction possible. In this paper, a CBCT reconstruction algorithm-ordered subset expectation maximization (OSEM) based on GPU is developed. Experiment on the Shepp-logan phantom model showed that a good speedup is received.

Keywords: beam reconstruction, OSEM, GPU-based acceleration.

1 Introduction

x-ray computed tomography imaging technology has been used not only in diagnostic radiology medical field, but also as an important tool in modern industrical nondestructive detection and exploration domain. it use x-ray beams to scan the object and present the internal object density distribution through 2-dimensional or 3-dimensional image.

There are two categories of reconstruction algorithm: the analysis and the iteration method. The former, such as algorithm FDK(Feldkamp, Davis and Kress)[1], the iteration algorithm, such as ART(algebraic reconstruction techniques)[2], SART (simultaneous algebraic reconstruction technique) [3], and also the algorithm based on statistics, MLEM(maximum likelihood expectation maximization)[4], OSEM(ordered subsets expectation maximization)[5]. when using analytic algorithm, the reconstruction computation is small, but it need a complete projection data(often the projection angle $>180^0$) .and using iteration

Hongwei Xu · Fucang Jia · Wenyan Chen · Xiaodong Zhang
Center for Human Computer Interaction
Shenzhen Institutes of Advanced Technology
Shenzhen, China
e-mail: {hw.xu,Fc.jia}@siat.ac.cn

Z. Qian et al. (Eds.): Recent Advances in CSIE 2011, LNEE 125, pp. 793–798.
springerlink.com

algorithm, the computation is big, it does not request a complete data, and the reconstruction quality is better when projection data contain noise.

The objective function of MLEM is a likelihood function, which is the joint probability density function of Poisson random variables. the reconstruction image is the solution maximize of this likelihood function. and 1994,Hudson [5]provide the OSEM algorithm, it combined the ordered subset and MLEM algorithm, in theoretically, the convergence speedup reaches n times when compared with MLEM containing n subsets.

To accelerate iterative reconstruction speed, many domestic and foreign scholars concern the GPU technology. K.Mueller[6] fulfill ART and SART algorithm using texture technology. In 2007, D.Riabkov accelerate the reverse projection part of FDK algorithm using GPU-CPU hardware. In 2006, domestic scholar Liang Liang complete FDK on the GeForxe FX5700 display card. Zhisheng Dai(Tsinghua University)[7] realized T-FDK on the GeForce 6800T.

In this paper ,we provide the OSEM algorithm based on GPU, in section 2, we give the brief process of OSEM, in section 3, we discuss the GPU-based OSEM algorithm, in section 4, we give some experiment based on our method.

2 OSEM Algorithm

Cone beam imaging system can be described as a linear system. This system can be re-written in matrix form as:

$$W * v = p$$

where $v = [v_1 \quad v_2 \quad \cdots \quad v_n]^T$, and this vector forms a $n_1 * n_2 * n_3$ reconstruction object, $p = [p_1 \quad p_2 \quad \cdots \quad p_r]^T$ is the projection vector, W is the coefficient matrix, The element w_{ij} represents the weight of the contribution of the j-th pixel v_j to the i-th projection p_i. in this paper, the contribution is the segment length of the projection ray within the pixel intersected. detail of choosing W reference[4], subset selection reference[5].In OSEM algorithm, the projection data are grouped in different sets. And go through the subset in a specified order. so p is updated after every subset is considered.

We consider T_1, T_2, ... T_k as the subsets selected. and $\tilde{v}^m$ as the result iteration m times. $\tilde{v}^0$ is the initial non-negative estimate vector.

1) iter=0, $v^1 = \tilde{v}^m \geq 0$, m=m+1
2) for subset $T_i, i = 1 \ldots k$
   a) projection. For the projection ray t in subset T_i ,and denoted μ_t^i

$$\mu_t^i = \sum_{j=1}^{n} W_{tj} v_j^i \qquad t \in T_i \tag{1}$$

b) back projection . update voxel

$$v_j^{i+1} = v_j^i \sum_{t \in T_i} \frac{p_t W_{tj}}{\mu_t^i} / \sum_{t \in T_i} W_{tj} \quad (2)$$

3) The result above as the initial value of the next iteration

$$\tilde{v}^m = v_j^{k+1}$$

3 GPU-Based OSEM Algorithm

In section 2, we give out the process of OSEM algorithm, in the iterative process, there are two step occupying much time. In the projection step, Commonly used projector operator are often ray-driven and pixel-driven . In the formula(1), the projection image is calculated using different projection rays. In this paper ,we use the ray-based method, so we can take the advantage of CUDA parallel computing power. By the formula (1):

$$\mu_t^i = \sum_{j=1}^{n} W_{tj} v_j^i \quad t \in T_i$$

W_{tj} is the ray distance reached projection t through pixel j, and we replace this equation as:

$$\mu_t^i = \sum_{ip} step * v_{ip} \quad ip^{i+1} = ip^i + step * ray$$

ray is the direction vector form ray-source to the projection t, and ip^0 is the first point intersected with the reconstruction object, $step$ is the step length along direction ray . v_{ip} is the pixel value of ip , in actual calculation, we store the pixel value in three-dimensional CUDA array, and bind the array with texture, and using the texture memory to fast access the floating-point coordinate ip . In this paper , $step$ is equal to 1/3 the actual size of pixel. If the projection image size is m*m, then we get m*m threads.

Also in the process of backprojection , from formula(2), we need calculate $\sum_{t \in T_i} W_{tj}$, the sum length of all the ray in subset T_i through pixel j. as the reconstruction object is formed by $n_1 * n_2 * n_3$ voxel, and we will get $n_1 * n_2 * n_3$ threads, In each thread, we calculate the sum ray length through voxel j. and we use the overlapping non-uniform ball to represent voxel. See in figure 1.

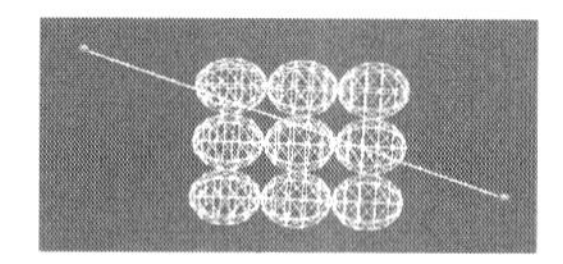

Fig. 1 Voxel modal

We calculate the radiation length through voxel j using the method [8]. And we also need to know which ray pass though pixel j. we do as follow, we first calculate the maximum projection bounding box on the projection image of voxel ball j, traverse all the ray passing through voxel j reaching in bounding box.

4 Experiment

Experiment was carried out by notebook. The hardware configuration of notebook is: CPU: Intel Core2 Duo T6600(2.2GHz), memory :2G, display card : nVidia GeForce GT 240M.,the graphic card has 96 stream processor and its memory is 512mb. The operating system is Windows XP Professional. Development tool: Visual Studio2008, we write CUDA program using c language.

We use shepp-logan model[9] for our test, the projection image resolution is 256*256, the reconstruction object resolution is 128^3, the projection angle is 360°, and there are a total 100slides evenly spaced 3.6°.see figure 2, a, b, c, d, e respectively is the 1st,25st,50st,75st,100st slide. so it is the image taken at projection angle 0°, 90°, 180°,270°,360°.

Figure 3, the left are the three orthogonal direction projection image of the original three-dimensional object, and from left to right, followed by the reconstruction slide in three direction. And we can see that , with the increase number of iteration, the ball edge of the model increase clearly.

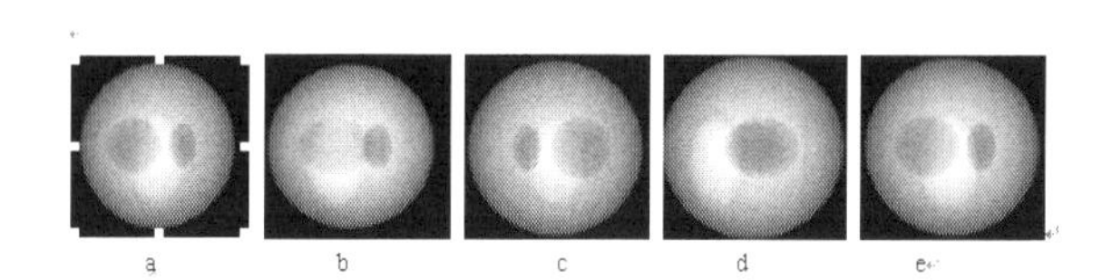

Fig. 2 Image projection resolution is 256*256,and a,b,c,d,e respectively is the 1st, 25st,50st,75st,100st slide

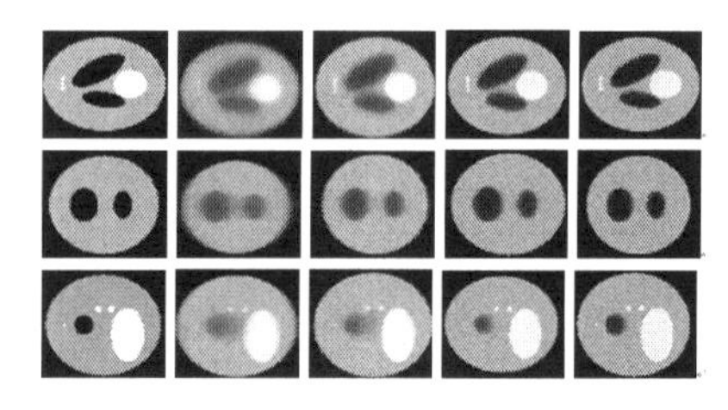

Fig. 3 The left original object, followed by the reconstruction slide in three direction ofter 1,2,3,4 iteration

Also, we fulfill the OSEM CPU-based algorithm. The follow table is the reconstruction time based on CPU and GPU. Though which we find, GPU-based method can achieve speedup>50 times, and ensure the reconstruction quality.

Table 1 Reconstruction time comparison

method	Reconstruction size	Image resolution	Iterative3 times (ms)	Iterative 6times (ms)
CPU	128^3	256*256	851890	1738250
GPU	128^3	256*256	14272.289	27392.568

To describe the reconstruction between the original object and the reconstruction object, we calculate the correlation coefficient and average difference to describe this property.

AD (average difference):

$$AD = \sum_j (v_j - \tilde{v}_j)^2 / \sum_j v_j$$

CC (correlation coefficient)

$$CC = (v - \overline{v}) \cdot (\tilde{v} - \overline{v}) / (\|v - \overline{v}\| \cdot \|\tilde{v} - \overline{v}\|)$$

Where $\overline{v}$ is the average voxel, $\tilde{v} = \{\tilde{v}_j\}$ is the reconstruction object. $v = \{v_j\}$ is the original model.

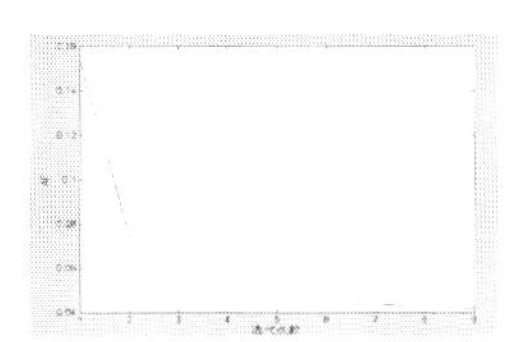

Fig. 4 AD

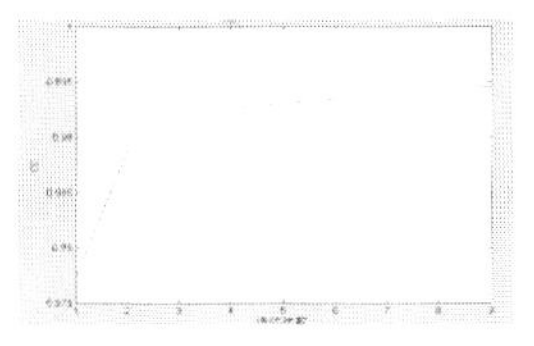

Fig. 5 CC

From table 1, figure 4, figure 5, the reconstruction object can achieve optimum after several iterations, the average pixel error decrease with the increase number of iterations, also the correlation coefficient. from figure 5, after 3 iteration ,the correlation coefficient reaches 0.99.

Similarly, we give some other examples of reconstruction, which the projection angle are less than180^0. And we see that, using the OSEM CUDA-based method, we also give a clear reconstruction result. All the experiment data are provided by Beijing Ji ShuiTan hospital. See figure 6.

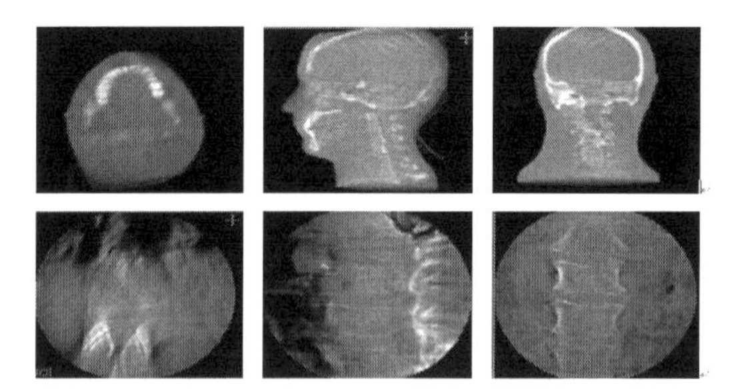

Fig. 6 Projection angle is 120^0, there are 100 projection slices. Up the reconstruction of head, down the spine

5 Conclusion

In this paper, we propose the GPU-based OSEM algorithm for cone beam reconstruction. And in our experiment ,we achieve a speedup>50 times, also when deal with the limit angle projection data(projection angle less than 180^0) , we still get a clear result. In the projection model, we choose the ray-driven method ,and in back projection process, we choose the overlapping non-uniform ball to represent object voxel. also we could model more accurate imaging geometry and physical modal in the projector and back projector, and this is what we will do next. In future, we want to achieve real-time reconstruction of higher resolution (512*512*512).in this situation, memory access speed will be a bottleneck. We may use a more high-end graphics cards or optimization algorithm to deal with it.

References

1. Feldkamp, L.A., Davis, L.C., Kress, J.W.: Practical cone beam algorithm. Journal Of Optical Society America A 1, 612–619 (1984)
2. Gordon, R., Bender, R., Herman, G.T.: Algebraic reconstruction techniques (ART) for three-dimensional electron microscopy and X-ray photography. Journal of Theoretical Biology 29, 471–481 (1970)
3. Andersen, A.H., Kak, A.C.: Simultaneous algebraic reconstruction technique (SART). Ultrasound Imaging 6, 81–94 (1984)
4. Sheep, L.A., Vardi, Y.: Maximum likelihood reconstruction for emission tomography. IEEE Transactions on Medical Imaging 2, 113–122 (1982)
5. Hudson, H.M., Larkin, R.S.: Accelerated Image reconstruction using ordered Subsets of Projection Data. IEEE Transactions on Medical Imaging 13(4), 601–609 (1994)
6. Mueller, K.: Fast and accurate three-dimensional reconstruction from cone-beam projection data using algebraic methods (1998)
7. Dai, Z., Chen, Z., Xing, Y.: Accelerated 3-D T-FDK reconstruction algorithm using the commordity graphics hardware. Journal of TsingHua University 46(9), 1589–1592 (2006)
8. Ray-BoxIntersection, `http://www.siggraph.org/education/materials/HyperGraph/raytrace/rtinter0.htm`
9. 3D Shepp-Logan phantom, `http://tomography.o-x-t.com`

Analysis and Design on Environmental Risk Zoning Decision Support System Based on UML

Weifang Shi and Weihua Zeng

Abstract. Regional environmental risk zoning is an important measure of environmental risk management but a complex project. It will become easier and more convenient to make environmental risk zoning with the help of the decision support system (DSS). In this work, the Unified Modeling Language (UML) is used to analyze and design the DSS of environmental risk zoning, using use case diagrams to analyze the system requirements, package diagrams to design the framework of the system and activity diagrams to analyze the sub-systems. Finally, the DSS was built. The processes of the analysis and design of the DSS show that the UML is a convenient implement to analyze and design the DSS.

1 Introduction

With the fast developing of the industry, the chemical releasing has Actual or potential threat of adverse effects on creatures and environment. The circumstance man living nowadays is surrounded with the environmental risk. Regional environmental risk zoning is an important measure of environmental risk management and has positive significance for preventing and reducing the risk. However, regional environmental risk zoning is a complex project and involves many aspects, which demands of advanced technology to help.

A decision support system (DSS) is a computer-based system that aids the process of decision-making [1]. In a more precise way, it is defined by Turban [2] as an interactive, flexible, and adaptable computer-based information system, especially developed for supporting the solution of a non-structured management problem for improved decision making. It utilizes data, provides an easy-to-use interface, and allows for the decision-maker's own insights.

Obviously, it will become easier and more convenient to make environmental risk zoning with the help of the DSS, but it is very difficult to analyze system requirements completely. The Unified Modeling Language (UML) is a visual modeling

Weifang Shi · Weihua Zeng
State Key Laboratory of Water Environment Simulation, School of Environment, Beijing Normal University, Beijing, 100875, China

Z. Qian et al. (Eds.): Recent Advances in CSIE 2011, LNEE 125, pp. 799–804.
springerlink.com

language that enables system builders to create blueprints that capture their visions in a standard, easy-to-understand way, and provides a mechanism to effectively share and communicate these visions with others [3]. UML now has been adopted as a standard by the Object Management Group (OMG), and familiarity with it seems certain to become a core skill for software engineers [4]. UML consists of a collection of the best engineering practices that have been proven successful in the modeling of large and complex systems, and supports a set of graphical and textual modeling tools that aim to offer a common understandable language for developers and customers.

The main focus of this work lies in the analysis of the system requirements and the design of environmental risk zoning DSS through using UML model, and then the system was built.

2 Environmental Risk Zoning and Decision Support System

The goal of environmental risk zoning work is dividing the specific region into several subareas according to the data and information of the region. First, it is need to address the sources of risk, making explicit the risk type of the each source, instance of fire, explosion, toxic gas releasing or noxious liquid leaking and so on.

Then the influence scope and intensity should be estimated. In this step, gas or liquid diffusion model is demanded. To accurately calculate is much complex, since the diffusion model is different in different condition. So it is necessary to resort to a DSS to generate and select proper model to estimate the intensity and influence scope.

The possible and potential loss by the risk should also be estimated, which need to consider the injury of the persons and the damage of the ecological environment.

Finally, calculating the risk value and classifying into several classes or clustering the characteristic of the region, the zoning result is mapped.

For a person, the knowledge is limited. A DSS is viewed as an anthropocentric, adaptive and evolving information system which is meant to implement the functions of a Human Support System (HSS) that would be necessary to help the decision-maker to overcome his/her limits and constraints he/she may encounter when trying to solve complex and complicated decision problems that count [5].

In this work, introducing the environmental risk zoning technique and method to the DSS and building environmental risk DSS, The characteristic of this DSS different from other DSS is that the database must contain the spatial relation data, and the models have the ability to calculate and process the spatial data. To achieve the function of the zoning proper, the knowledge database also is necessary to store the basic knowledge of the environmental risk, rules of zoning and the experience of the experts.

3 System Analysis and Design

3.1 System Requirements Analysis and Framework Design Analysis

The requirements of the environmental risk zoning DSS are crucial and should be supported by an appropriate method. A design method is essential not only to ensure the DSS quality, but also to facilitate its frequent evolutions imposed by the environment or the decision makers' changing requirements.

UML provide users with a ready-to-use expressive visual modeling language, provide extensibility and specialization mechanisms to extend the core concepts, and to be independent of particular programming languages and development processes, and useful for understanding problems and communicating with everyone involved in a project, promoting better understanding of requirements, cleaner designs, and more maintainable systems.

In UML, diagrams such as use case diagrams, class diagrams, object diagrams, sequence diagrams, state chart diagrams and activity diagrams are supported. In this work, system requirements are modeled with use case diagrams of UML as Fig. 1a.

In the use case diagrams, the users are represented through the actors, and the use cases represent the series of operation come from the users in the system. In this system, the main requirements is for decision makers making environmental risk zoning decision, the decision makers are the main users of the system. In Fig. 1a, the decision maker actor is generalization. In fact, we can divide the decision makers into several classes, and give different permission to different class of the decision makers.

For a system, the system administrators are advanced users who coordinate and control the operation of the whole system, having the highest permission.

For the DSS this paper research on, the data and information of the environmental risk sources are important in order to zoning the region. So the actor of risk owner in this system could bring benefit to making decision, keeping the data and information of the risk sources latest.

A framework is a partially complete software (sub-) system that is intended to be instantiated. It defines the architecture for a family of (sub-) systems and provides the basic building blocks to create them. It also defines the places where adaptations for specific functionality should be made [6].

The architecture of the DSS is built with the package, which contains different modules and/or services, each of which wraps a part of the application logic. The package diagrams of the UML are used to describe the architecture of the DSS as Fig. 1b. The package of the user interface process the information interaction with the user, including input and output, and is as an interface for packages of application. The packages of the data service provide the interface between the database and the application, and could reserve the data object permanently, while the package of the application provide services of classes and component for other packages.

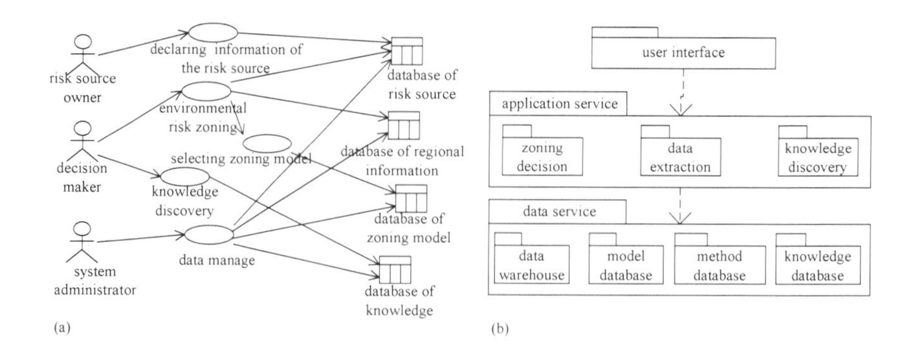

Fig. 1 (a) The use case diagrams of the DSS. (b) The package diagrams of the DSS.

3.2 Sub-system Activity Analysis

The ultimate goal of the DSS is mapping the subareas in the specific region, So sub-system of the zoning is the core of the system, when the decision makers make environmental zoning, selecting zoning model, obtaining and reading the risk source data and regional data, running the zoning model and getting the zoning result, then judging the reasonable and proper of the result, if the result is proper, the zoning sub-system is terminating with mapping the subareas in the region, otherwise going back to selecting the zoning model and running the zoning sub-system again. The proper of the result usually is judged by the knowledge and experience of the decision makers and the knowledge database. The activity of the zoning sub-system is shown as Fig. 2a.

Indicated by Fig. 2a, when the zoning sub-system runs to get the environmental risk zoning result, the zoning model must be selected first. So the sub-system of the model database is essential of the system, it should insure that there are suitable and reasonable zoning models to select, otherwise zoning result coming out from the zoning sub-system is not proper even error. Thus selecting and generating zoning models is the key to the model database sub-system.

The processes of selecting and generating zoning models are generalized as follow: first, raising the question, then analyzing the question, finding whether there is suitable model according to the question, if suitable model existing in the database of zoning mode, selecting and building the model, otherwise, decomposing the question into several questions to select model, if still no suitable model according to the sub-question, continuing to decomposing the question till there are suitable models to the sub-questions, then selecting and combining the models to generating model chains. The activity of the model database sub-system is shown as Fig. 2b.

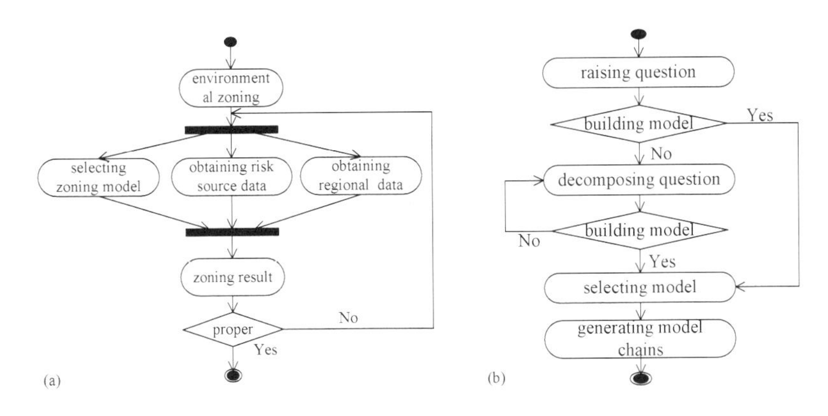

Fig. 2 (a) The activity diagrams of the zoning sub-system. (b) The activity diagrams of the model database sub-system.

Although DSS could provide support to solve semi-structured and unstructured problem efficiently, there are still many problems which could not be solved or completely solved, and some problem solution solved completely by model is not reach the demand of the decision makers. So it is necessary to build knowledge in the DSS to storage various rules, causality, experience of the decision makers, moreover, it should have function of knowledge reasoning in order to achieve knowledge discovery in the knowledge database sub-system.

In the knowledge data base of the environmental risk zoning DSS, there are knowledge of the environmental risk, basic knowledge of zoning, rules of zoning and experience of the environmental risk experts stored. The expression of the knowledge is composed of Meta-Knowledge, then Storing the Meta-Knowledge and reasoning strategy in the knowledge database.

4 System Integration and Application

After analyzing and designing the system with UML, generating the program code framework according to the model, then modifying and complementing it, the implementation model was built. In this DSS, spatial data calculation and management process is a part service of the system, inducing the GIS component as MapObjects or ArcGISODE. Using Visual C# as integrated development environment, taking the technique of ActiveX data object (ADO), data access object (DAO), data environment (DE) to integrate the database, through dynamic link library (DLL) method to achieve data transfer and data presentation with implication model programs, and forming unified interface, the system was built finally.

5 Discussion and Conclusion

In the UML, there are various types of diagrams which are convenient to analyze and design the software system according to each phase of the system development. In this work, use case diagrams are used to analyze the requirements of the system, the package diagrams to show the architecture of the system, the subsystems of the system and correlation and communication mechanism between the sub-systems, and the activity diagrams to illustrate the work flow of the subsystems and the model database sub-system in emphasis.

In order to get the reasonable and suitable model to a question, combining the models and generating the model chains is the usual method, which used to be an efficient way to resolve complex decision problems in DSS.

Analysis and design the DSS with UML is concise and efficient. UML give a way of thinking about problems using models that depict real-world ideas in a visual manner, with graphical notations, it easy to understand the requirement, conceptual, architecture of the system.

For the inherent visualization of the UML, making the system analysis, system design and implementation combining organically, the specific functionality of the DSS is adaptation by extension implementation when the framework of the system is designed based on UML.

Acknowledgment. This study is funded by The National 863 program (2007AA06A404).

References

[1] Finlay, P.N.: Introducing Decision Support Systems, p. 274. Blackwell, Cambridge (1994)
[2] Turban, E.: Decision Support and Expert Systems: Management Support Systems, 4th edn., p. 7, 36, 887. Prentice-Hall, Englewood Cliffs (1995)
[3] Schmuller, J.: Sams teach yourself UML in 24 hours, p. 7. Sams Publishing, Indianapolis (2004)
[4] Koo, S.R., Son, H.S., Seong, P.H.: A method of formal requirement analysis for NPP I&C systems based on UML modeling with software cost reduction. The Journal of Systems and Software 67, 213–224 (2003)
[5] Filip, F.G.: Decision support and control for large-scale complex systems. Annual Reviews in Control 32, 61–70 (2008)
[6] Buschmann, F.: Pattern-oriented Software Architecture: A System of Patterns, p. 457. Wiley, New York (1996)

Evaluation on Pollution Grade of Seawater Based on the D-S Evidence Theory

Hengzhen Zhang, Xiaofeng Wang, Cuiju Luan, and ShiShuang Jin

Abstract. At present, with the rapid increasing of oil spills at sea, it will be helpful to effectively deal with the oil pollution by establishing a scientific evaluation mechanism for the oil spill damage. In the paper, we apply the D-S Evidence Theory to evaluate the degree of seawater pollution in oil spill, then resolve the evaluation problem due to various attributes and uncertainty of data which result in the difficulty to analyze the indicators comprehensively and quantitatively, and make the final decision. In addition, according to the average support of each proposition, we allocate the probability of the D-S Evidence conflict which not only has reasonable physical meaning, but also effectively compensate the disadvantage of D-S Evidence combination formula to improve the reliability and rationality of the combining result. At last, we use the case results to verify the effectiveness and feasibility of our proposed method.

Keywords: Oil spill, Seawater grade, Dempster-Shafer theory.

1 Introduction

It makes the sixty percents of world oil consumption to be transported by sea because of the world oil distribution, imbalance of consumption and relatively cheaper shipping. In addition, coupled with the vast ocean area and the abundant undersea petroleum resources, the oil under the sea is exploited actively in the countries all over the world so far, which leads to more frequent oil spills incident and causes serious ecological destruction and huge economic losses.

The influence of oil spill on marine environment is complicated and indefinite, which increases the difficulty of evaluation of seawater pollution to some extent.

Hengzhen Zhang · Xiaofeng Wang · Cuiju Luan · Shishuang Jin
College of Information Engineering, Shanghai Maritime University Shanghai, China
e-mail: {hzzhang,xfwang,cjluan,ssjin}@shmtu.edu.cn

Z. Qian et al. (Eds.): Recent Advances in CSIE 2011, LNEE 125, pp. 805–810.
springerlink.com

Timely, comprehensive and accurate evaluation of seawater pollution degree plays an important role in emergency response and improvement. Therefore, it is necessary to design and build a scientific and effective evaluation system according to the degree of seawater damage from oil spill.

2 Related Works

The natural resource damage evaluation is the primary damage evaluation of marine eco-environment in abroad. Evaluation technologies are mainly divided into experience formula method and numerical model method . In experience formula method, there are mainly the Florida evaluating formula (1992) and Washington evaluating model (1993) . And among them, the oil spill damage evaluation technologies which are widespread used in judicial claim are mainly natural resources damage assessment (NRDA) [1] and habitat equivalent analysis (HEA) [2].

In the paper, on one hand, considering the features of the oil spill seawater quality evaluation: data uncertainty (i.e., the uncertainty and randomness of some field data collection in a short time after oil spill accident happening), multiple attribute data fusion (i.e., comprehensive consideration of the multiple indexes in the evaluation the degree of seawater damage from oil spill); On the other hand, considering the advantages of evidence theory in the uncertainty of the data (i.e., the uncertainty arising from the randomness in the process of data collection), fuzziness and multiple attribute data fusion, we use the evidence theory to evaluate the degree of seawater damage from oil spill, and the experimental results verify that the method which we proposed is feasible and effective.

3 D-S Evidence Theory

The D-S evidence theory [3] is based on a nonempty set. Let Θ a finite nonempty set of mutually exclusive alternatives, and be called the frame of discernment. This frame of discernment contains every possible hypothesis.

Definition 1: If function $m : 2^{\Theta} \longrightarrow [0,l]$ meets the following conditions:

$$\begin{cases} m(\varphi) = 0 \\ \sum\limits_{A \subset \Theta} m(A) = 1 \end{cases} \tag{1}$$

The $m(A)$ is defined as the Basic Probability Assignment (BPA) of A. The $m(A)$ denotes the belief level of the evidence committed to the proposition A.

Definition 2: Function $Bel : 2^{\Theta} \longrightarrow [0,1]$, for any $A \subseteq \Theta$, $Bel(A) = \sum\limits_{B \subseteq A} m(B)$. The Bel is defined as the belief function for Θ.

The Bel is also called lower limit function. $Bel(A)$ adds the $BPAs$ of all subsets of A.

Definition 3: A is a subset in the frame of discernment, if the basic probability assignment $m(A) > 0$, the subset A is called focal element. The union of all the focal elements is called core.

Definition 4: Function $Pl: 2^{\Theta} \longrightarrow [0,1]$, for any $A \in \Theta$, $Pl(A) = \sum_{A \cap B \neq \phi} m(B)$. The Pl is defined as the plausibility function for Θ.

The Pl is also called the upper limit function. $Pl(A)$ measures the extent to which we fail to disbelieve the hypothesis of A.

The combining results [4] of n basic probability assignment values $m_l, m_2, \ldots, m_n$ are as follows: $m = m_1 \bigoplus m_2 \bigoplus \ldots \bigoplus_n$ that is

$$\begin{cases} m(\varphi) = 0 \\ m(A) = \frac{1}{k} \sum_{\bigcap A_i = A, 1 \leq i \leq n} \prod m_i(A_i) \end{cases} \tag{2}$$

where $k = \sum_{\bigcap A_i \neq \phi, 1 \leq i \leq n} m_i(A_i)$

4 Evaluation Model on Pollution Grade of Seawater

4.1 *The Evaluation Index System Analysis*

According to the dissolved oxygen, chemical oxygen demand (cod), heavy metal content (copper, mercury, lead, cadmium, etc), the oil, the inorganic nitrogen, suspended solids, the sea water was divided into four categories in China's national sea water quality standards [5], as shown in figure 1 below. It identified the categories of polluted sea water only according to one indicator when evaluating the sea water in previous studies. There is certain one-sidedness in the authors opinion. And it is neither scientific nor comprehensive. In the paper, according to the characteristics of the oil spill pollution, considering the various pollution indexes, it has established a set of sea water category evaluation system. And the experimental results have verified the proposed method can reflect the degree of seawater pollution more scientifically, more comprehensively and more systematically. Our method includes the following steps:

First, determine the evaluation factor set, and then establish evaluation system, as follows:

Take the indicator state set as the recognition framework set Θ;

$X = \{X_1, X_2, X_3, X_4, X_5\}$:According to the Chinese national standard, the sea water is divided into four categories water, and the fifth category is the index value which exceeds the other four indicators;

$A = \{a_1, a_2, a_3, a_4, a_5, a_6\}$: Representing the dissolved oxygen, cod, oil content, heavy metal content, inorganic nitrogen and suspended solids;

$A_6 = \{a_{61}, a_{62}, a_{63}, a_{64}\}$: Representative first index heavy metal and contains copper, mercury, lead and cadmium four secondary evaluation index;

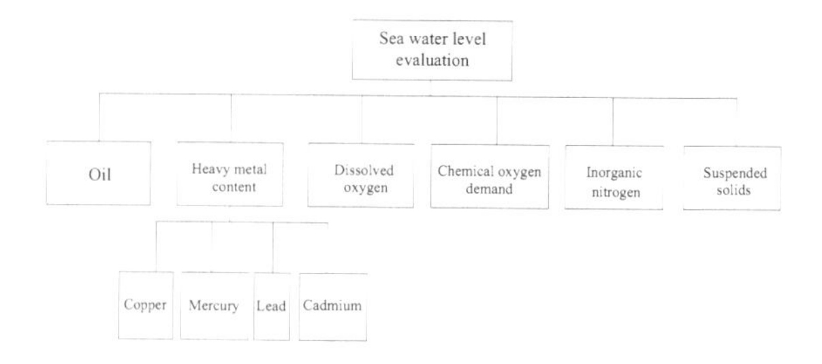

Fig. 1 Water level evaluation system.

Basic belief assignment m_{ij}: On behalf of the sea water sample of the corresponding index,which can constitute to a basic belief assignment by statistical analysis, and $m_{ij} \leq 1$.

Detailed steps in building assessment system are shown as below:

Step 1: Establish a basic belief assignment list according to the sea water samples of each index. Firstly, classify the data of sample set of one index according to the national standard of sea water; then count the sample number of each category; then figure out the percentage of each kind of sample over the sample set, which is m_{ij} and indicates basic belief assignment of the index in a particular sea category.

Step 2: Use formula (2) combination rule to fuse the basic belief assignment of level 2 index and then concludes that the level 1 index credibility distribution list. However, due to a great conflict between the evidence, the result K value is very large and the conclusion is unreasonable, namely the paradox. And that may occur on the D-S evidence theory when evidence synthesis. In addition, evidence theory cannot deal with the extreme case when the evidence is consistent. So it introduces the method of conflict resolution in this paper. And it selects the Bicheng Li method [6] by analyzing the theories and doing experiments. In this method, the evidences conflicting probability is weighted distributed according to each average supported degree, such as formula (3) shows:

$$\begin{cases} m(\varphi) = 0 \\ m(A) = \sum\limits_{A_i \cap B_j \cap C_l \cap \ldots = A} m_1(A_i) \cdot m_2(B_j) \cdot m_3(C_l) \cdot \ldots + k \cdot q(A) \end{cases} \tag{3}$$

where

$$\begin{cases} \forall A \neq \phi \\ k = \sum\limits_{A_i \cap B_j \cap C_l \cap \ldots = \phi} m_1(A_i) \cdot m_2(B_j) \cdot m_3(C_l) \cdot \ldots \\ q(A) = \frac{1}{n} \sum\limits_{1 \leq i \leq n} m_i(A) \end{cases} \tag{4}$$

Use the improved Dempster combination rule (formula (3)) to fuse the level 2 index and then concludes the level 1 index belief assignment list.

Step 3: Use formula (3) to fuse the first index to draw water Category basic belief assignment table.

4.2 Application Examples and Analysis

An oil spill accident happened in Bohai Sea in "the tasman sea" tankers on November 23, 2002. The North Sea test center examined and investigated the related Sea area from November 26, 2002 to December 1. And it got the oil, heavy metal content, dissolved oxygen, chemical oxygen demand (cod), inorganic nitrogen and suspended solids data from 33 monitoring sites. And then it brought in the basic belief assignment list of water classification by statistical analysis, as shown in table 1.

Table 1 Basic belief assignment list.

	First level index	Second level	Basic belief assignment m_{ij}					
			X_1	X_2	X_3	X_4	X_5	Θ
Water level evaluation	Dissolved oxygen	/	1	0	0	0	0	0
	Chemical oxygen demand (cod)	/	0.424	0.273	0.303	0	0	0
	Oil	/	0	0	1	0	0	0
	Inorganic nitrogen	/	0	0.368	0.632	0	0	0
	Suspended solids	/	0	0.167	0.833	0	0	0
	Heavy metal	Copper	1	0	0	0	0	0
		Lead	0.333	0.583	0.084	0	0	0
		Cadmium	0.75	0.25	0	0	0	0
		Mercury	0.08	0.23	0.54	0.15	0	0

Table 2 Level 2 Index Synthetic Credibility After Allocation Schedule.

First level index	Basic belief assignment m_{ij}					
	X_1	X_2	X_3	X_4	X_5	Θ
Dissolved oxygen	1	0	0	0	0	0
Chemical oxygen demand (cod)	0.424	0.273	0.303	0	0	0
Oil	0	0	1	0	0	0
Inorganic nitrogen	0	0.368	0.632	0	0	0
Suspended solids	0	0.167	0.833	0	0	0
Heavy metal	0.549926	0.26044	0.152883	0.036751	0	0

Table 3 The Credibility List Of The Sea Level Distribution.

X_1	X_2	X_3	X_4	X_5	Θ
0.328988	0.178098	0.486814	0.06125	0	0

Then, we use formula (3) combination rule to fuse the second level index of heavy metal to get the basic belief assignment list only containing the first level index,as shown in table 2.

Finally, we'll use formula (3) combination rule to fuse the first index to get the belief assignment list of oil spill sea class, such as shown in table 3.

According to the results of synthesis, the credibility of the third kind in this sea water is 0.4868. After oil spill of "the tasman sea" tankers accident, the state oceanic administration classified this sea water to the third types since the oil content in this sea water is 2.2 times higher than not polluted before. Experimental data are basically consistent with conclusions of the State Oceanic Administration. And the multiple attributes fusion method adopted in this paper makes the result more scientific.

5 Conclusions

As a kind of uncertainty reasoning methods, evidence theory can fuse multi-dimension data, which can make the result of one proposition more distinct for decision-making. In the paper, this method is applied in oil spill on water pollution degree evaluation, and solves the problem that it is difficult to comprehensively and quantitatively analyze each index to get the final decision because of multiple attribute and uncertainty data. The distribution of the evidence conflict probability according to the extent of support on average every proposition is reasonably in physical meaning, and can effectively solve the shortage of evidence synthesis formula, improve the reliability and rationality of the synthesis. And example results verify the feasibility and effectiveness of the method in the paper.

Acknowledgements. This paper by Shanghai scientific research plan project fund (08240510800), Shanghai maritime university scientific research fund (20100094) support.

References

1. NOAA. Natural Damage Assessment Guidance Document: Sealing Compensator Restoration Actions (oil Pollution Act of 1990) Damage Assessment and Restoration Program, National oceanic and Atmospheric Administration. Silver Spring, MD (1997)
2. NOAA. Habitat Equivalency Analysis: An overview. NOAA Damage Assessment and Restoration Program, National oceanic and Atmospheric Administration, Silver Spring, MD (2000)
3. Xiong, H.J., Chen, D.J.: Intelligent Information Processing. National Defense Industry Press (2006) (in Chinese)
4. Wang, W.J., Ye, S.W.: Principle and application of artificial intelligence. Posts & Telecom Press, Beijing (2004) (in Chinese)
5. Chinese National Standards: Sea water quality standard, GB3097-1997 (in Chinese)
6. Li, B.C., Wang, B., Wei, J., Qian, C.B., Huang, Y.Q.: An efficient combination rule of evidence theory. Journal of Data Acquisition & Processing 17(1), 33–36 (2002) (in Chinese)

Study on Simulation of Sandstorm Based on OpenGL

Wei-wei Gan and Xi-tang Tan

Abstract. In order to realistically simulate sandstorm, which is one kind of natural scene, by computer, the shape and motion state of sand when the sandstorm occurs are studied and analyzed in this topic. The sandstorm simulation system is based on VC++ 6.0 Build environment with OpenGL graphics library as its application program interface. It is adopted particle system to simulate the motion of sand particle and fog function in OpenGL to simulate dim environment. The simulation results show that the system is able to realistically simulate the scene of sandstorm and achieve the balances between real time and authenticity.

1 Introduction

Sandstorm is a severe weather phenomena that a large mass of ground objects such as sand and dust are blown into the air by strong wind so that the air is especially cloudy and the visibility is less than 1000 meters. The sandstorm mainly occurred in spring and early summer season. Once it happens, the sandstorm put overwhelmingly forward like tideway and rolled up enough sand to inundate farms, attack villages, decrease temperature, pollute the atmosphere, blow away the surface soil, hurt animals and damage facilities. Meanwhile, it will speed up the desertification and impact cross-border ecosystems and bio-environment through the atmosphere. Therefore, the simulation of sandstorm plays an important role in disaster prevention and scientific education.

2 OpenGL and Particle System

2.1 OpenGL

OpenGL is a kind of software interface between graphics hardware and users which is strictly designed according to the principles of computer graphics and in conformity with the principles of optics and vision. It is suitable for visual simulation systems.

Wei-wei Gan · Xi-tang Tan
Department of Electrical Engineering ,Tongji University, Shanghai, China

Z. Qian et al. (Eds.): Recent Advances in CSIE 2011, LNEE 125, pp. 811–816.
springerlink.com

In OpenGL, it is allowed to use graph method to represent visual object, present the three-dimensional features of the objects by illumination, and provides a method to use image data directly, or define the image data as texture in order to strengthen simulation effect. In the thesis, the mapping method is used to improve the computing speed.

2.2 Particle System

Particle system is considered the most successful graphics generation algorithm to simulate irregular fuzzy objects by far. It fully reflects the dynamic and randomness on fuzzy objects and simulates the dynamic three-dimensional complex scenes. Particle system makes it possible to simulate complex objects by using simple units. It provides a powerful technique for simulating natural phenomena such as fire, water, rain, snow, grass, trees and so on.

A particle system consists of a large number of simple units called particles. Each particle contains a set of attributes which depend on its specific application, such as location, speed, color and life cycle. Initial values of particle are generated by random process. Particle source located somewhere in space produces all the particles.

Particle system changes continuously. Each particle contains its own life cycle and experiences three stages from production to evolution and extinction. During every moment in the life cycle, it must complete the following four steps:

- Produce new particles from particle source.
- Update particle attributes.
- Delete dead particles.
- Draw the particles.

In general, the geometry of particle is so simple that it can be represented by a pixel or a small polygon.

3 Realization of Sandstorm

3.1 Overall Design

The sandstorms can be divided into weak sandstorms, sandstorms of moderate intensity, strong dust storms and exceptionally strong sandstorm by its visibility.

The visibility of weak sandstorm is from 500m to 1000m when the sandstorm happens. Its particles are as small as yellow fog without significant movement, thus, it can be realized by fog function in OpenGL.

The wind of force 8 caused by the strong sandstorm will roll up the stone and sand. As there are more and more sand and dust into the air, the visibility is from 50m to 200m and it cannot been seen a period of time. The thick dust can be achieved by fog function and flying stone by particle system. Fog function superimposed on particle system can achieve the simulation of strong sandstorm phenomenon.

3.2 Realization of Fog Function

Fog function is an graphics effects in OpenGL in order to provide the effect of atomization. Selecting fog function can generate the hazy effect in the implementation of the sandstorm.

Firstly, set global variables needed in the fog function:

```
glFogi(GL_FOG_MODE, fogMode[fogfilter]);   // Fog mode
glFogfv(GL_FOG_COLOR, fogColor);           // Set Fog Color
glFogf(GL_FOG_DENSITY, 0.2f);      // How Dense Will the Fog Be
```

Enables fog effect:

```
glEnable(GL_FOG);
```

3.3 Realization of Sandstorm Particle Function

The data structure of particles includes the life of particles, velocity, direction of motion, coordinate position and particle types, etc. The particle structure is:

```
struct particle
{
        bool active;            // Is active
        float fade;             // Fade speed
        float life;             //Particle Life
        float x,y,z;            // X, Y, Z Position
        float xi,yi,zi;         //X, Y, Z Direction
        float xg,yg,zg;         // X, Y, Z Gravity
};
```

Generally, particle properties include spatial location attributes, appearance attributes, sports attributes, survival attributes, etc.

3.3.1 Spatial Location Attributes

Spatial location attributes include the initial spatial location coordinate and its direction.

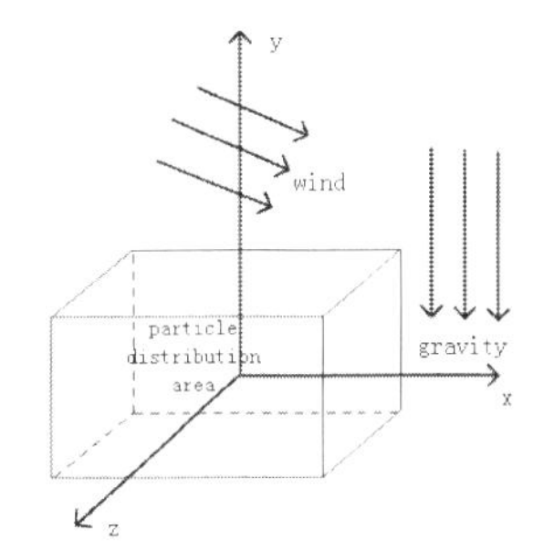

Fig. 1 Distribution of dust particles and the force diagram. Fig. 1 shows that in OpenGL, the world coordinate used to describe the scene is right-handed Descartes coordinate system. X axis is from left to right, Y from bottom to upside, Z from screen inside to the outside. Observer's viewpoint is in the positive direction of z axis and just in front of the particle distribution region, which is represented by a cuboid in front of the observer.

Given that particle[loop] is the spatial location attribute of the NO. loop particle, the current position of the particle is the sum of last position and current speed, namely:

```
particle[loop].x+=particle[loop].xi;
particle[loop].y+=particle[loop].yi;
particle[loop].z+=particle[loop].zi;
```

In the above equation, particle[loop].xi, particle[loop].yi, particle[loop].zi are current particle velocities respectively.

3.3.2 Appearance Attributes

The appearance attributes of particle includes color, size, shape and so on. In order to reduce the computation, we use the texture mapping method and set the color of sand particle (0.5, 0.5, 0.0), and transparency 0.0.

The glaux library functions in OpenGL are used to load a bitmap and convert it into a texture.

When all the necessary textures are created in texture[], we can just bind the texture which needed. A different texture should be re-bound:

```
glBindTexture(GL_TEXTURE_2D, texture[i]);// bind the texture which needed
```

3.3.3 Sport Attribute

Sports attribute is the speed of dust particles. The initial outbreak state of the sands is flying from the ground, so there are speeds in the positive direction of y axis and random directions of x, z axis:

```
particle[loop].xi=float((rand()%3)-0.1f);
particle[loop].yi=float((rand()%3)-0.0f);
particle[loop].zi=float((rand()%3)-0.1f);
```

In the above equation, particle[loop].xi, particle[loop].yi, particle[loop].zi are the initial speeds in x, y, z directions.

Sand particles are in constant motion in the air. According to physics, we know that the sand particles are acted upon by its own gravity, buoyancy, air resistance, external forces (such as wind force), the interaction force between particles and so on. In this thesis, the gravity and wind force are the two chief considerations.

In the simple gravity field, the settling velocity of sand is:

$$u = \sqrt{\frac{2v(\rho_1 - \rho_2)g_{\sim}}{\mu s \rho_2}} \tag{1}$$

The settling velocity of sand is affected by equation (1). Therefore, y-axis acceleration is:

```
particle[loop].yg=float(rand()%10)*(-0.01f);
```

Assumed that the direction of wind in the current scene is northwest with random deflection of 45(Fig. 2). X-axis and z-axis accelerations are as follows:

```
particle[loop].xg=float(rand()%100)*0.001f;
particle[loop].zg=atan(float((rand()%45)*(PI/180)))*particle[loop].xg;
```

Fig.2 Diagram of wind direction. Fig. 2 shows that the level wind force of particles mainly impact x-axis and z-axis correlation parameters, and the gravity impact y-axis correlation parameters.

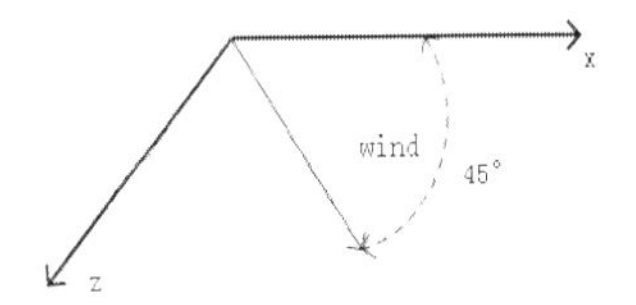

3.3.4 Survival Attributes

The survival attributes are the conditions whether the particles survive or not. The constraint conditions of particles` survival are mainly life cycle and fade rate. In the whole process of scenario simulation, sands as well as its spatial location are in constant motion and change. Particles life are ended when the values of life decrease to 0.When the particle's life cycle expires, or its spatial position is beyond the scope of its distribution area, the calculation and mapping of the particle stop.

```
bool active;    //Active
float life;     //Particle Life
float fade;     // Fade Speed
```

3.4 Sandstorm Simulation

This thesis simulates weak sandstorm and strong sandstorm.The visibility of weak sandstorm is from 500 meters to 1000 meters. The sand dust is so thin that sunlight can go through it. Therefore, the color appears red and yellow and its RGB value sets (242, 155, 23), density 0.2f, which is shown in Fig 3.

The visibility of strong sandstorm is from 50 meters to 200 meters. The color of sand dust appears yellow and it can be rolled up large sand or stone. Thus, its RGB value sets (190, 185, 39), density 0.35f, particle number 6000, and glaux library functions in OpenGL are used to load a bitmap, shown in Fig 4.

Fig. 3 Simulation effect figure of weak sandstorm

Fig. 4 Simulation effect figure of strong sandstorm

4 Conclusions

In this subject, the phenomenon of sandstorm is simulated by the combination of particle system and fog function. The mapping method is used to accelerate the speed of the computation and achieve real-time requirements.

It is only considered the situation of raised sandstorm in the thesis, however, the process of raising sandstorm will be taken into consideration in the future.

References

[1] Shreiner, D., Woo, M.: OpenGL Programming Guide. China Machine Press (2009)
[2] Lv, F.Z.: C++ Programming. China Tsinghua University Press (2007)
[3] Peace Dove Studios. Advanced Programming and Development of Visualization System in OpenGL. China Waterpub Press (2006)
[4] Huang, L.: Modeling and Rendering the Authenticity of Ray Scattring Effect. Hunan University (2007)
[5] Jiang, X.: Mathematical Model and Analysis of Duststorms Spread. Environmental Study and Monitoring 17(3), 32–33 (2004)
[6] Guo, Y., Li, Y.T., Li, J.: Numerical Simulation of Dust Transport by Wind. Journal of Arid Land Resources and Environment 18(4), 34–38 (2004)
[7] Langer, M.S., Zhang, L., Klein, A.W.: A Spectral Panicle Hybrid Method for Rendering Falling Snow. In: Eumgraphics Symposiurn on Rendering (2004)

3D Morphology Algorithm Implementation and Application in High Quality Artificial Digital Core Modeling

Guoping Luo and Linzhu Wang

Abstract. This text introduces image morphological method into digital core modeling. Five key morphological functions: Voronoi tessellation, dilation, erosion, opening and closing are interpreted to simulate the forming of sandstone. Morphology methods, cooperating with periodic boundary idea, produce a high-quality artificial sandstone core which resembles a real core not only in the respect of procedures, but also the amazing visual effects and the reservoir analytic parameters similarity. Sequential morph closing operations are applied to simulate the microscopic remaining oil distribution pattern. Future reservoir Lattice-Boltzmann[1] is promising with a firm backup of such perfect digital core.

Keywords: morphological image processing, 3D digital core, rock microstructure.

1 Intro

Digital core is the computer image of a physical rock or core. Digital core technology is rational subsequence of the shift from macroscopic research to microscopic means in geology and petroleum engineering [2]. Such microscopic-scaled geology study is also inevitably boosted by the modern image technology and computer technology. Advanced digital core modeling technology makes it possible that some important physical parameters of a rock be recognized under lenses, on the computer screen, rather than via expensive physical experiments in current and past laboratories. Some microscopic geological process also can be reproduced virtually on computer. Especially, digital core acts as the indispensable media of Darcy flow modeling, which is the fundamental theory of modern petroleum engineering.

Guoping Luo
Faculty of Resources, China University of Geosciences, Wuhan, China

Guoping Luo
Key Laboratory of Tectonics and Petroleum Resources, Ministry of Education, Wuhan, China

Linzhu Wang
\School of E&M,China University of Geosciences,Wuhan, China

Z. Qian et al. (Eds.): Recent Advances in CSIE 2011, LNEE 125, pp. 817–822.
springerlink.com

There lies a long way before we can do that, but building high-quality digital core is the first step for all these research work. Amount of modeling technologies for digital core modeling come up in recent years, such as annealing algorithm [3], procedure modeling [4]. Although these technologies produce 3D or 2D digital core, which used in further research, they show too much difference from real rock core, visually or physically. This text introduced image morphology algorithm as a main technology into digital core modeling. This technology can produce a 3D digital core ,which resemble the physical rock core in visual effect .Even more important, the operations taken in the procedure of modeling corresponds to the geological procedure in the nature. Finally, the statistical parameters of the virtual cores are much similar to those from true rock.

The image operations involved in this article come as Voronoi decomposition and four basic morphology operations: dilation, erosion, opening and closing. Those image morphology operations correspond to digenetic processes of clastic rock. This study uses MATLAB as programming platform and a extension toolbox IPT (image processing toolbox) of Rafael C.Gonzalez, etc. Before exploring the details of the five algorithms, this text would like to present the routine to build a 3D digital core, evolving the main algorithms.

2 Building and Analysis of 3D Digital Core

The processes of 3D digital core modeling are devised simulate the realistic geological forming processes of a sandstone. In geology, a matrix rock experienced weathering, sediment transportation, deposition, lithification and diagenesis, and then sandstone was formed. As so far, not all such processes are replicated in this text, but the following describing really managed to show some important stages on computer, to some extent, even if not fully demonstrated the main thought of a new microscopic geological modeling technology.

Firstly, a 3-dimensional binary digital image scaled as 128 by 128 by 128 pixels, initialized with zero everywhere, was built to represent a matrix cubic as 1.28 by 1.28 by 1.28 cm, generally acceptably taking the metric as one pixel for 100μm. Finer or larger-scaled digital core request more memory and computational time, usually beyond the capacity of a common PC. This text resorts to duplicating such easy model to meet such high requirement, rather than further device. This solution will be explained in detail in following text.

Secondly, randomize certain number of points (N) in the cubic (V), with label ONE. These disperse points are the centers of coming grains. The number N can be estimated from the granularity of objective rock.

Thirdly, decompose the cubic with the disperse points by Voronoi algorithm and label each grain pixels with an integer n (1…N), where n is the index of the grain center. This operation resembles the first stage of realistic weathering, the rock crashed and there are angular gravels occupying the space of the original whole rock.

Next, extract a subset P(i) ,of the cubic, with all the pixels labeled as i. This subset is a pure grain. Rounding the grain with a designated sphericity ,by the morphological algorithm Opening. The value of sphericity is computed according the shape maturity of the realistic rock. Here, sphericity takes Wadell's definition[5]. The formula:

$$\Psi = \frac{\pi^{1/3}(6V_p)^{2/3}}{A_p} \tag{1}$$

compute the radius for each grain(Where:Vp, grain volume; Ap, grain surface area.), which define the opening structural element. This text uses simple spherical structural element, isotropic or anisotropic, depending on the objective realistic rock. When the rock is obviously directional, the radiuses in 3 axis directions can take different values and the structural element would rotate to certain direction. After this step, the grains become round with certain sphericity, the rock looks porous.

The porosity of the rock is not defined by the sphericity alone. Usually, the rock has a smaller porosity relative to the realistic core with the same sphericity. To increase the pore space of digital core, especially the fissure space, it is feasible to erode the grains morphologically. After all, by using the morphology algorithm, it seems possible to produce all kind of virtual core as wished.

Finally, by now, if the digital core we got has too little similarity to the realistic one, we can use other operations, e.g. morphology algorithm to improve the effect. Dilation also can apply here to decrease pore volume and make the grains cemented together to some extent.

Now, a 3D digital core was completed on computer. The stereoscopic picture (fig. 1), produced by MATLAB, shows the outline of the digital core.

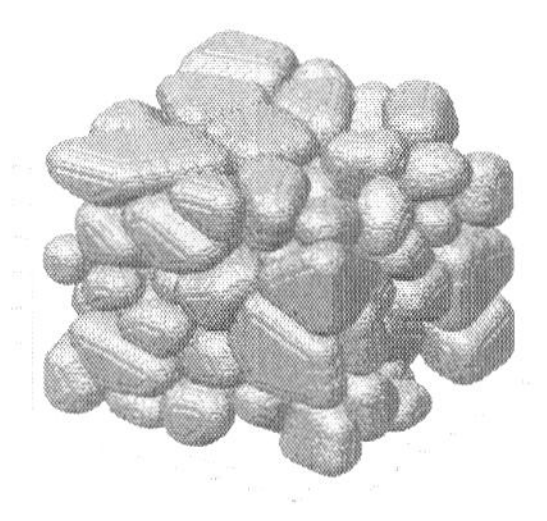

Fig. 1 Digital core 3D outline

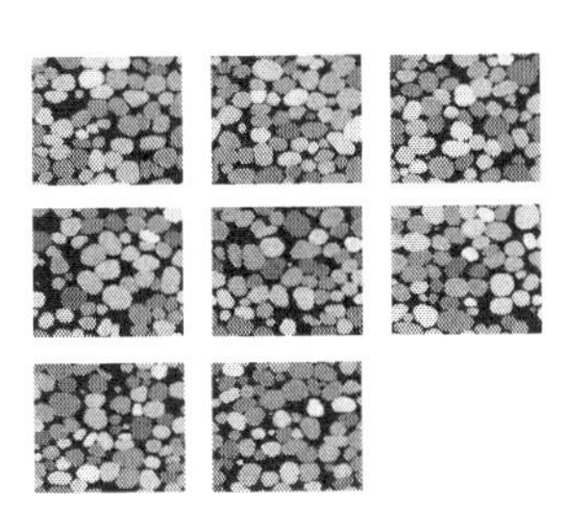

Fig. 2 Digital core sections on Z=1,16,32,48,64, 80,96,112.

Several sections (fig. 2) are presented to show the inner structure of it. Owing to similarity of above operation to the realistic geological process, the virtual core resembles well the core photo, visually at least.

Some promising operations corresponding to realistic geological processes maybe redundant to improve the effect further, they are not implemented fully yet. The implementation details of the available ones as so far may give some clues for future work.

This text also analyzed some parameters of the core model. The Closing operation was applied to compute the porosity size distribution. The result image can represent the cement stage of increasing intense. In the realistic geological diagenesis, cement is prone to occur in smaller hole space as showing in the figure (3).

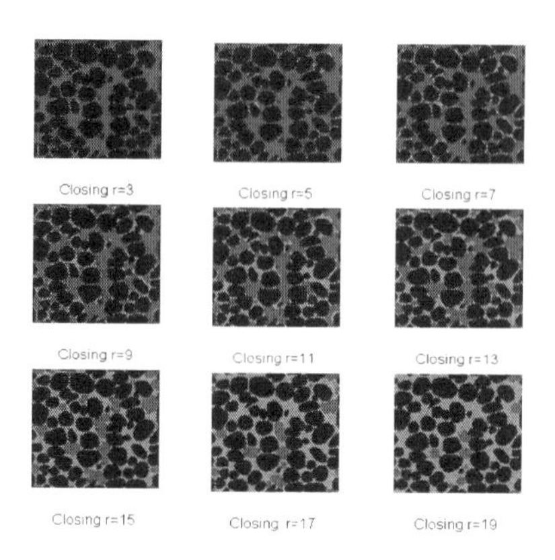

Fig. 3 Section at Z=49 of the digital core after closing with different radius

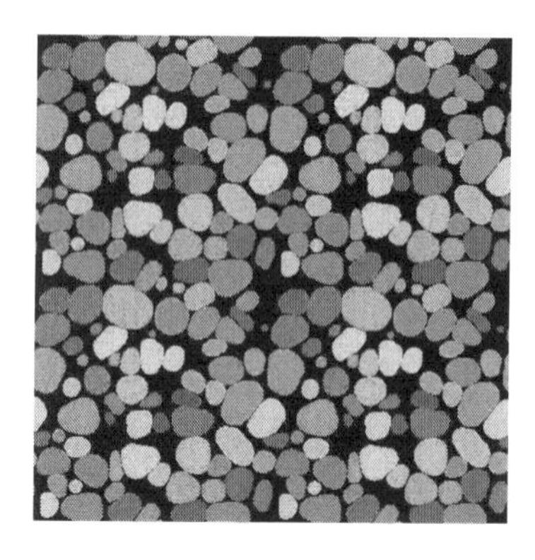

Fig. 4 Seamless mosaic of digital core (2 by 2) on the plane xy

3 Algorithm Implementation

The basic conceptions of Voronoi tessellation and computational morphology operations are not new; they were innovated and have been applied for decades in several domains. The implementation in different computer languages can be collected publicly. Especially, MATLAB basic toolboxes and the IPT provided by Rafael C.Gonzalez gave respectively the implicit algorithm in the language M. Therefore, all the basic information about them is omitted for concise. The readers concerned can refer to documents[6,7,8,9].

Nevertheless, the algorithms of Voronoi tessellation and morphology operations implemented in MATLAB and IPT are revised to fit the digital core modeling in this text.

Voronoi tessellation algorithm in MATLAB was modified to generate 3D digital core without boundary effect. To overcome the boundary effect, the modified Voronoi algorithm circulates across opposite side when a decomposition cell meets certain boundary. To understand fully the circulative boundary Voronoi algorithm, we can imagine that a straight line, a entity of one-dimensional space, bent to a circle, so the line has no ends any more. When we bent a two-dimension plane in the same way, we get a global space without boundary edges. Although we cannot find physical corresponding object for such bending space when it comes to a high dimensional space, e.g. 3D or higher, it is possible for us to imagine it as an abstract one and use such abstract consideration in later algorithm implementation. Using this bending technique, Vononoi decomposition therefore performs in an abstract bended, circular 3D space. Since there are no boundaries anymore, the boundary effect was eliminated absolutely. The output digital core (fig 3) shows consistent visual effect and structural parameters in both the inner area and the boundary area. What is more, since the boundaries are circulative, we can mosaic one or more simple cores in all directions to output a seamless core as big as you want in following Darcy flow simulation or geological procession modeling, without unaffordable cost of computer memory and clock time.

The improved 3D Voronoi algorithm helps to build original virtual rock similar to a crashed physical rock without further weathering processes, with all breccias grains in situ. The next morphology algorithms act as such weathering processes to modify the grains to make them ones that are more natural, comparable to those found in physical rock. The main contents of this text really come up as the inspiration of the similarity of the geological process name and the computer morphology algorithm, e.g. erosion, dilation.

The 3D image morphology algorithms in this text derive from the 2D ones from the IPT toolbox of Rafael C.Gonzalez, etc. In IPT, The morphology algorithms just deal with 2D binary images or even grayed and colored ones. However, 3D algorithms remain unavailable in the box toolbox. It is one of the most important and difficult jobs of this text to extent the 2D algorithms to 3D domain. The following piece of M codes gives the implementation of 3D Dilation.

```
function b = dilate3(a,se)
%erode3    General    sliding-neighborhood
operations.
nhood=size(se);b = false(size(a)+nhood-1);
a=logical(a); siza = size(a);
% Find out what output type to make.
rows = 0:(nhood(1)-1);
cols = 0:(nhood(2)-1);
dpths= 0:(nhood(3)-1);
for i=1:siza(1),
   for j=1:siza(2),
      for k=1:siza(3),
     if a(i,j,k)
           b(i+rows,j+cols,k+dpths)=
           b(i+rows,j+cols,k+dpths) | se;
        end
      end
   end
end
nhood=(nhood+1)/2;
b=b(nhood(1):end-nhood(1)+1,nhood(2):end-
nhood(2)+1
,nhood(3):end-nhood(3)+1);
End
```

The M code for dilation is demonstrated below(fig. 5),it is much like the erosion except in the inner loop.

Erosion and dilation are the most elementary operations of al morphology algorithm. Other operation such as opening, closing, hit or missing can be derived by the two basic ones.Their M code go much like the code of Dilation above.

4 Conclusions

Mathematic morphology algorithm is a feasible technology to model 3D digital core. The result models built in this way bear overwhelming visual effect. The building procedure is more controllable than other pure stochastic ones since the operations resemble well the realistic geological process. Some analyses tell the some physics parameters are very close to those of the realistic rock. Besides the building of digital core, morphology algorithm also can act well in parameter analysis.

Some problems remain unsolved. First, 2D morphology can be accelerated dramatically by structuring element decomposition; then is there a general decomposition algorithm adaptable to all kind of structuring elements? Second, this technique seems suitable to just sandstone modeling what will be for other type of rock? Last, not the all, some realistic geological processes remain not replicated on computer yet. All these problems will be the orientation for later research.

References

1. Shan, X., Chen, H.: Lattice Boltzmann model for simulating flows with multiple phases and components. Phys. Rev. E 47, 1815–1819 (1993)
2. Li, H., Pan, C., Miller, C.: Pore-scale investigation of viscous coupling effects for two-phase flow in porous media. Phys. Rev. E 72, 26705 (2005)
3. Yao, J., Zhao, X.-C., Tao, J.: Analysis methods for reservoir rock's microstructure. Journal of China University of Petroleum (edition of natural science) 31(1) (2007)
4. Bo, Q.-W., Dong, C.-Y., Zhang, Q., Li, Z., Zhao, D.-W.: Visual simulation of porous structure in packed gravels. Petroleum Exploration and Development 8(30) (2003)
5. Wadell, Hakon: Volume, Shape and Roundness of Quartz Particles. Journal of Geology 43, 250–280 (1935)
6. Serra, J.: Image Analysis and Mathematical Morphology (1982), ISBN: 0126372403
7. Soille, P.: Morphological Image Analysis; Principles and Applications, 2nd edn (1999/2003), ISBN: 3540-65671-5
8. Mathematical Morphology and its Application to Signal Processing. In: Serra, J., Salembier, P. (eds.) 1st International Symposium on Mathematical Morphology, ISMM 1993 (1993), ISBN: 84-7653-271-7
9. Gonzalez, R.C.: Digital image procession using MATLAB (2004), ISBN: 0130085197

Computation of Time-Dependent AIT Responses for Various Reservoirs Using Dynamic Invasion Model

Jianhua Liu, Zhenhua Liu, and Jianhua Zhang

Abstract. When a formation was opened, the drill mud was used and invasion occurred during the process of well drilling in the petroleum prospection. Since the dynamic invasion process changes with time during drilling while mud filtrate was used, the logging response is time dependent. Dynamic invasion responses of Array Induction Tools (AIT) were modeled and programmed to study the effects of mud filtrate invading into a reservoir dynamically on AIT and sight into the behaviors of responses during dynamic invasion for the new array induction tools,. The characters of AIT responses were discussed for the cases of mud filtrate invading into an oil reservoir, a water zone, and a reservoir containing both water and hydrocarbon, respectively. The present computation model provides the foundational behaviors for reservoir evaluation.

1 Introduction

Formation water saturation is an important parameter for reservoir evaluation during petroleum exploration, because it determines the hydrocarbon saturation. An oil reservoir, water zone, or oil/water formation can be evaluated from water saturation. But it is not directly measurable in situ. Estimation of its value depends on the measurement of other parameters such as formation resistivity, porosity,

Jianhua Liu
Key Laboratory of Read Construction Technique and Equipment of the Ministry of Education, School of Construction Machinery, Chang'an University, Xi'an, China
e-mail: liujianhua1007@163.com

Zhenhua Liu
School of Mechanical Engineering, Xi'an Shiyou University, Xi'an, China
e-mail: jhzhang@xsyu.edu.cn

Jianhua Zhang
Science College, Xi'an Shiyou University, Xi'an, China
e-mail: jhzhang@xsyu.edu.cn

Z. Qian et al. (Eds.): Recent Advances in CSIE 2011, LNEE 125, pp. 823–828.
springerlink.com

etc. Resistivity logging is an effective tool used to determine the formation resistivity, and then to estimate fluid saturation and to evaluate a reservoir.

During well drilling, mud filtrate was used. It invades into permeable and porous formations, so that the resistivity logging data are significantly influenced by the invasion of drilling-mud filtrate into the reservoir adjacent to the borehole. A realistic invasion process is related to time. At the beginning of drilling a formation, the rate of invasion is fast. With the lapse of time, mud cake is built at the wall of borehole; then the rate of invasion becomes slowly. The dynamic invasion process and its effect on resistivity logging tools have been investigated in computer simulation and field application[1-4]. These works modified conventional step-invasion model[5] that presumes the resistivity varies sharply at the boundary of invaded zone and uninvaded formation.

Induction log is an efficient method to measure the formation. The combination of various arrays can detect the formation from shallow to deep distance in a reservoir. In the resent two decades, AIT (Array Induction Tool)[6], has been developed and used in fields wildly[7]. However, the measurement is influenced by invasion effect still. In a dynamic invasion process, the logging data of an AIT measurement are time dependent. Various logging times will lead to different results of measurement.

A computer model of dynamic invasion based on fluid flowing theory was solved numerically in the present work, and then the dynamic variation of AIT responses with invasion time was simulated. The characters of time-dependent AIT logging data were discussed for oil-bearing reservoirs, water zone, and oil/water formation, respectively. The time-dependent invasion profiles and related resistivity responses can provide more information for log analysts to evaluate a reservoir.

2 Computer Model of Dynamic AIT Responses

The time-dependent resistivity-logging model first bases on dynamic invasion formalism. We consider a thick bed without shoulder effects. The mud filtrate invades into a formation radially and displaces the native fluids in porous volumes. The displacement between filtrate and hydrocarbon is immiscible. During drilling, mud filtrate pours into the formation under the pressure differential between borehole and formation after a bed was reached. The invasion process is also related to formation permeability k, porosity φ, original oil and water saturation S_o and S_w, native-water salinity C_w and mud-filtrate salinity C_{mf}, and so on. And the pressures of oil, water, and capillary in formation P_o、 P_w and P_C also affect on the invasion process. At the beginning of bit penetration, mud cake is built up within a short time stage and mud-cake permeability controls the invading into formation. With the extension of invading geometry area and the building of mud cake, the invasion rate decreases with time. The shape of invasion profiles will change at various time stages. They can be calculated from fluid flow equations[1,4] according to following steps.

First, at invasion time t and radial distance from the wellbore r, the original oil and water saturation S_o and S_w, the formation water salinity C_w can be obtained by solve flowing equations.

Second, the petrophysics theory is used to generate the formation resistivity, $R_f(r,t)$, as functions of time t and radial coordinate r :

$$R_f(r,t) = \frac{abR_w(r,t)}{S_w^n(r,t)\varphi^m} \quad (1)$$

Where, $R_w(r,t)$ is the formation water resistivity, both a and b are constants, m is the cementation factor, and n is the saturation exponent.

The last step is to calculate the conductivity response of AIT:

$$\sigma_a = \mathrm{Re}\sum_{j=1}^{n}\sum_{k=1}^{m} w_{kj}\sigma_a^{(j)}(z - z_{kj}) \quad (2)$$

Where w is weight coefficience of the array, $\sigma_a^{(j)}$ is the measured log from the j-th channel, and n is the total number of measure channels.

These steps were repeated numerically in a PC computer for various times, and then the time-dependent AIT responses were yielded.

3 Simulation an AIT Logging Curve in an Artificial Formation

The present computer model provides time-dependent radial invasion profile of formation resistivity R_f, and AIT responses, as shown in Fig.1. The necessary input dada are: initial formation-water salinity C_w=120,000 mg/L, initial water saturation S_w = 0.2, the pressure difference between borehole and original reservoir ΔP=5MPa, C_{mf}=10000mg/L, φ=0.12, k=10μm^2, T=100°C, m=2, n=2, and a=b=1 in Archie's equation.

The four curves in Fig.1a illustrated the radial profile of formation resistivity at various time stages (5, 12, 28, and 60 days respectively) after the bed was reached. A low-resistivity annulus in the front of invaded zone forms because high-salinity formation water is displaced by low-salinity filtrate. With time lapse, resistivity invasion profile moves into the formation with a reduced rate owing to the extending of invasion geometry and the building of mud cake at the wall of borehole.

The variation of resistivity profiles with time makes a notable impact on AIT responses. It is easy to understand that the resistivity logging measurement departs from the actual formation parameter and it is not a constant, but variable data depending on the time, as shown in Fig.1b.

An AIT tool provides five depths of investigation, 10, 20, 30, 60, and 90in. (25, 51, 76, 152, and 229 cm) , respectively. We use AIT10, AIT20, AIT30, AIT60, and AIT90 to denote the five responses of AIT arrays. For an oil-bearing reservoir with less water saturation S_w, during the whole invasion period, AIT90>AIT60>AIT30 >AIT20>AIT10, as shown in Fig.1b. Since an oil zone contains little movable water, the high-resistivity hydrocarbon is displaced by low-resistivity filtrate; thus the resistivity in the zone near the borehole is low. It affects the shallow detective arrays, hence they have low responses.

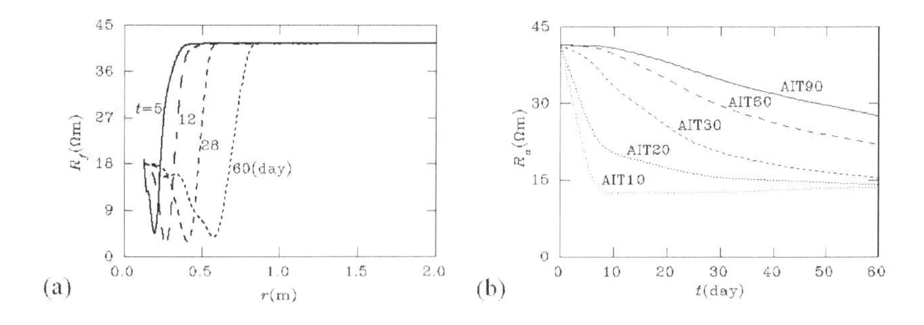

Fig. 1 (a) Formation-resistivity invasion profiles and (b) time-dependent AIT responses for an oil zone

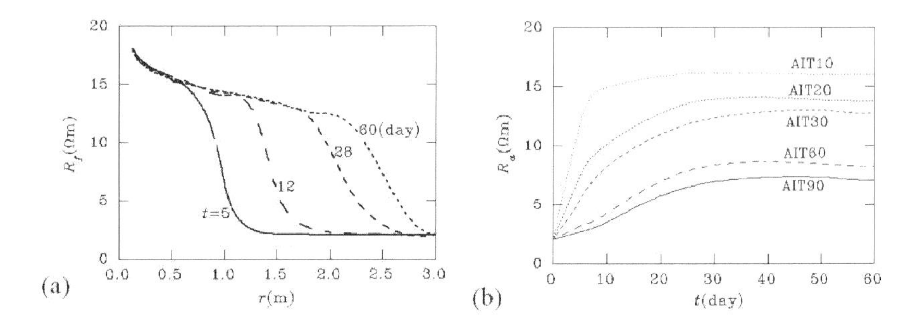

Fig. 2 (a) Formation-resistivity invasion profiles and (b) time-dependent AIT responses for a water zone

A water-bearing layer which has high formation-water saturation, such as S_w=0.9 in our present calculation, the low-resistivity formation water is displaced by high-resistivity filtrate, creating a high-resistivity region adjacent to the wellbore, shown as Fig.2a. Thus, the data of array with shallow investigation depth is greater than the one with deep investigation depth. Hence, the present computer simulation gave the result AIT10>AIT20>AIT30>AIT60>AIT90, as shown in Fig.2b.

The left figure in Fig.3 illustrated an artificial formation with true formation resistivity R_t (real line) and invaded zone resistivity R_{xo} (dash line). The right chart in Fig.3 showed the present simulation results for AIT responses (R_a) as the function of depth at a fixed time.

An oil reservoir with large formation resistivity located the depth from 3m to 5m. In this region, the array responses of deep detective depth are greater than those of shallow detective depth, as shown in the right chart in Fig.3.

A water zone with low formation resistivity located the depth from 5m to 7m. In this region, the array responses of deep detective depth are less than those of shallow detective depth according to the analysis of Fig.2.

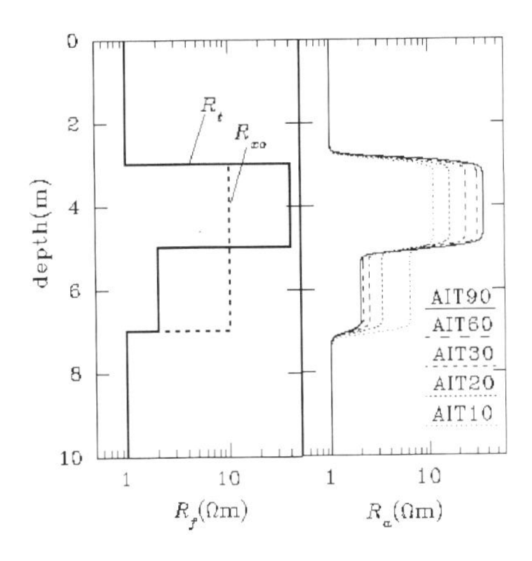

Fig. 3 The computer simulation of AIT logging curves for an artificial formation model.

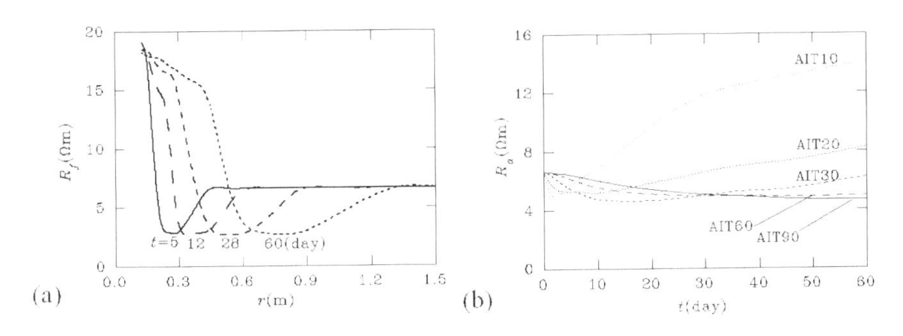

Fig. 4 (a) Formation-resistivity invasion profiles and (b) time-dependent AIT responses for an oil-water zone

A much complicated situation occurs in the case that a reservoir contains both hydrocarbon and water equivalently. For this case, the filtrate near the borehole has high water saturation and low salinity; thus a high-resistivity peak formed. In addition, in the front of invasion zone, low-salinity filtrate mixes with high-salinity native water. This mixture lead to a low-resistivity annulus, as shown in Fig.4a, here S_w=0.5 in calculation. At the beginning of bit penetrate, this low-resistivity annulus is close to borehole and influences on shallow detective array significantly. Therefore, the response of shallow detective array is greater, AIT10>AIT20>AIT30>AIT60> AIT90 (within t<3 day) were recorded, as shown in Fig.4b. With the filtrate propagates into the depth of formation, this low-resistivity annulus migrates beyond the shallow detective array and reaches the area of deep detective array. At last (t>30day in Fig4b), AIT90>AIT60>AIT30>AIT20>AIT10 occurs.

4 Conclusions

The dynamical invasion process of drilling mud filtrate into a reservoir results in the responses of resistivity logging tools are time dependent. The dynamic invasion process can not be solved by conventional step-invasion model. The present computer model suggested an efficient algorithm to calculate the dynamic invasion.

Since the five arrays of AIT measurement have various detective depths, their responses show some characters at certain invasion time for different reservoirs. While fresh water mud was used in an oil-bearing reservoir, the deep detective array responses were usually greater than these of shallow detective arrays. Whereas the recorded shallow detective array measurement were greater than these of the deep detective arrays measurements for a water zone. And more complicated behaviors were obtained for a reservoir containing both water and hydrocarbon simultaneously.

The characters obtained from the present computer model of time-dependent AIT responses provide a reasonable and efficient method for log analysts to simulation the dynamic invasion and evaluate reservoirs[8]. The present principle can be extended to other instruments of resistivity logs as well.

Acknowledgments. The authors would like to thank the Foundation of the Special Fund for Basic Scientific Research of Central Colleges, Chang'an University (CHD2010JC092), the Special Fund for Basic Research Support Program of Chang'an University and Key Laboratory of Read Construction Technique and Equipment of the Ministry of Education, Chang'an University for the support which enabled to perform this research.

References

1. Tobola, D.P., Holditch, S.A.: Determination of reservoir permeability from repeated induction logging. SPE Formation Evaluation, 20–27 (1991)
2. Liu, Z.H., Oyang, J., Zhang, J.H.: Dynamic dual-laterolog responses: model and field applications in the Bohai gulf of China. J. Pet. Sci. Eng., 1-2, 1–11 (1999)
3. Yao, C.Y., Holditch, S.A.: Reservoir permeability estimation from time-lapse log data. SPE Formation Evaluation, 69–74 (1996)
4. Zhang, J.H., Hu, Q., Liu, Z.H.: Estimation of true formation resistivity and water saturation with a time-lapse induction logging method. The Log Analyst 2, 38–148 (1999)
5. Head, E., Allen, D., Colsdon, L.: Quantitative invasion description. In: SPWLA 33rd Annual Logging Symposium, pp. B1–B21 (1992)
6. Barger, T.D., Rosthal, R.A.: Using a multiarray induction tool to achieve high-resolution logs with minimum environmental effects., the 66th Annual Techniqucal Conference and Exihibition of the Society of Petroleum Engineers, Dalas, U.S. Paper SPE 22725 (1991)
7. Martin, L.S., Buchman, J., Bittar, M., et al.: Application Of New Asymmetrical Array Induction Tool In Hostile Environments. In: 48th Annual Logging Symposium, Austin, Texas (2007)
8. Leroux, V., Dahlin, T.: Time-lapse resistivity investigations for imaging saltwater transport. Environ. Geol. 49, 347–358 (2006)

Petroleum Reserve Quality Evaluation System on Offshore Fault-Block Oil and Gas Field

Chenghua Ou, Nan Sun, and Han Xiao

Abstract. To adapt to the rapidly growing offshore international petroleum business, China's oil & gas companies have to change the artificial and qualitative means to carry out petroleum reserve quality evaluation on offshore fault block oil and gas field. For this reason, Petroleum reserve quality evaluation system on offshore fault block oil and gas field has been developed specially. On the basis of the bi-level fuzzy comprehensive evaluation method, and the flow process design mentality, this system was built-in the completely petroleum reserve quality evaluation factor system on offshore fault block oil and gas field, as well as first-level and the second-level factor weight. And user only need to fill up the system factor table with the related second-level factor picked up from the collected data of study area, then to accomplish one project evaluation easily by following the flow process. Through the practical application of specific cases, the system, with simple, rapid and flexible features, has been revised and improved, and become powerful tool for petroleum reserve quality evaluation on offshore fault block oil and gas field.

Keywords Petroleum Reserve, Quality Evaluation, System, Offshore, Fault Blocks.

1 Introduction

Asia-Pacific region is rich in offshore oil and gas resources [1], fault-block oil and gas field is one of the main types [2]. Developing offshore fault-block petroleum reserves quality evaluation system have to be important work for the oil and gas business in Asia-Pacific region.

However, due to the relatively short development history by Chinese petroleum company, the research work about petroleum reserve quality evaluation on offshore fault-block oil and gas field is weak, and a set of prepared evaluation method and software has not been reported till now. To this end, fully to digest and absorb fault-block petroleum reserves quality evaluation method from domestic and foreign,

Chenghua Ou · Nan Sun · Han Xiao
Department of Petroleum Engineering, Southwest Petroleum University, Chengdu, China

Z. Qian et al. (Eds.): Recent Advances in CSIE 2011, LNEE 125, pp. 829–833.
springerlink.com

especially Chinese onshore, to consider the specific characteristics of offshore fault-block oil and gas field in Asia-Pacific region, the petroleum reserve quality evaluation system was developed with the greatly improved efficiency and reliability for offshore fault-block oil and gas reserves evaluation.

2 System Components

2.1 Method and Evaluation Processes

To consider the specific characteristics of offshore fault-block oil and gas field in Asia-Pacific region, the 8 first level factors set and the 43 second level factors set for offshore fault-block gas field were developed, the 9 first level factors set and the 45 second level factors set for oil field were also built up, and the 5 level evaluation Index system was finally created

The evaluation index system is characterized as follows: i evaluation factors are as multi-factor set with two levels; ii there are quantitative evaluation index with the certain classification boundaries, also qualitative evaluation index with the uncertain classification boundaries , iii evaluation results are multi-objective for both the first level evaluating and the second level evaluating. Therefore, petroleum reserve quality evaluation on offshore fault-block oil and gas field is typical of a fuzzy multi-objective, multi-factor comprehensive question.

At present, the multi-objective, multi-factor comprehensive evaluation method includes fuzzy comprehensive evaluation, the conventional multi-index comprehensive evaluation, multivariate statistic, and so on[3-5]. Taking into account the feature of petroleum reserve quality of offshore fault-block oil and gas field, the fuzzy comprehensive evaluation[6-8] was chosen as the main evaluation method, and the technical processes was shown as Figure 1.

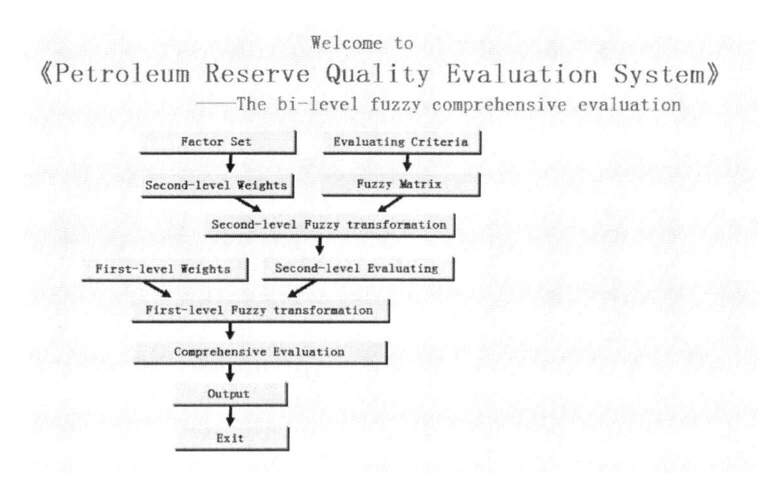

Fig. 1 The system main interface and evaluation process

2.2 Modules

The system consists of oil subsystems and gas subsystems, and each subsystem includes 5 modules shown as Figure 2.

Input Data Module	Weight calculation module	Fuzzy matrix module	Fuzzy transformation module	Comprehensive evaluation module
		Gas Field	Oil Field	
Reserves of oil and gas fields offshore block quality rating system				

Fig. 2 Evaluation system modules

3 System features

3.1 Friendly Interface and Simple Operation

For ease of use, the evaluation system was design in the way of flow process (Figure 1). and user can easily fulfill one evaluation project by hitting a mouse following the process.

3.2 Complete Index, Fast and Flexible Evaluation

There were 225 evaluation indexes from 45 second level factors embedded into oil subsystems, and 215 evaluation indexes from 43 second level factors embedded into gas subsystems. Petroleum reserve quality of offshore fault-block oil or gas field was completely characterized by these indexes from macro and micro, dynamic and static point of view.

At the same time, the system also provides the weights of 8 first level factors and 43 second level factors from offshore fault-block gas field, and 9 first level factors and 45 second level factors from oil field, by analytical hierarchy process[9].

The main job that users need to do is to analyze, extract and establish the factor set from an oil or gas field example. And the other jobs will be finished by the evaluation system automatically. So the system saves user's much time and effort of work by improving efficiency.

4 Application

Vitality and applicability of the system was enhanced by continuous development and improvement on the evaluation index system and the system functions, to rely on application of a large number of evluation projects about petroleum reserve quality of different offshore fault-block oil and gas field. Due to space limitations, only one petroleum reserve quality evaluation project from an Australia offshore fault-block oil field was given as an example.

The structure of the oil field, composed by a series of tilted fault-block area and with 350m depth below sea level, is a rectangular form to spread from westward to eastward. The evaluation factor set (Table 1) was established according to information provided by interest.

The evaluating results, with third class petroleum reserve quality, were shown as Table 2. It is recommended to fully analyze the development and acquisition cost of this project, and then to decide whether to cooperate or buy this gas field (The gas field has now been abandoned).

Table 1 The evaluation factor set from one Australian offshore fault-block oil field

First level factor	Second level factor	Index value
Fault block features	Block size(km^2)	17
	Block oil ratio(f)	0.7
	Block form (f)	0.8
	Fracture complexity within block (f)	0.7
	Nature and elements of the fault(f)	0.7
Reserves features	Oil saturation (%)	55
	Irreducible water saturation (%)	39.6
	Gas-oil ratio(m^3/m^3)	147
	Reserves abundance ($10^8 m^3/km^2$)	400
	Reserves size ($10^8 m^3$)	0.476
	Burial depth (m)	3105
	Reserves concentration (f)	0.487
Reservoir energy	Pressure coefficient (f)	0.96
	Maximum deliverability ($10^4 m^3/d$)	32
Reservoir flow properties	Average permeability ($10^{-3} \mu m^2$)	79.1
	Reservoir heterogeneity (f)	2.52
Rock pore properties	Porosity (%)	13.3
	Pore- throat sorting coefficient (f)	0.89
Distribution of oil sandbody	Effective thickness (m)	78.2
	Average layer thickness (m)	11.18
	Effective coefficient of sandbody (f)	0.835
	Sandbody facies (f)	5
Grain properties	Particle size (mm)	0.24
	Particle sorting (f)	1.5
Crude oil properties	Viscosity (mPa·s)	0.25
	Sulfur (%)	0.001

Table 2 Evaluation results of Australian offshore fault-block oil field

Fuzzy Comprehensive Evaluation Matrix					Evaluation Results
1	2	3	4	5	
0.197	0.251	0.261	0.177	0.114	3

5 Conclusions

i Petroleum reserve quality evaluation on offshore fault-block oil and gas field is typical of a fuzzy multi-objective, multi-factor comprehensive question, and the fuzzy comprehensive evaluation is the method to suit to solve the question.

ii The Petroleum reserve quality evaluation system, with the features of friendly interface and simple operation, and complete index, fast and flexible evaluation, consists of oil subsystems and gas subsystems. And each subsystem includes 5 modules such as input data module, weight calculation module, fuzzy matrix module, fuzzy transformation module and comprehensive evaluation module.

iii The system gradually became one powerful tool for the petroleum reserve quality evaluation on offshore fault-block oil and gas field, by continuous development and improvement on the evaluation index system and the system functions, to rely on a large number of practical evluation example.

Acknowledgments. Author gratitudes the financial assistance by the Chinese National Science and Technology project (2008ZX05030-005-04), and Doctoral Fund (20095121120003) of Chinese Education Ministry.

References

[1] BP companies, BP statistical review of world energy (June 2010), http://www.bp.com/statisticalreview

[2] Yang, F.Z., Xue, L.Q.: Basin classification and oil-gas distribution in south Asia-Pacific area. China Petroleum Exploration (5), 65–70 (2006) (in Chinese)

[3] Mirrazavi, S.K., Jones, D.F., Tamiz, M.: Multi Gen: an integrated multiple-objective solution system. Decision Support Systems 36(2), 177–187 (2003)

[4] Dominy, S.C., Noppe, M.A., Annels, A.E.: Errors and Uncertainty in Mineral Resource and Ore Reserve Estimation: The Importance of Getting it Right. Exploration Mining Geology 11(1-4), 77–98 (2002)

[5] Li, L.J., Shen, L.T.: An improved multilevel fuzzy comprehensive evaluation algorithm for security performance Original Research Article. The Journal of China Universities of Posts and Telecommunications 13(4), 48–53 (2006) (in Chinese)

[6] Běhounek, L., Cintula, P.: From fuzzy logic to fuzzy mathematics: A methodological manifesto. Fuzzy Sets and Systems 157(5), 642–646 (2006)

[7] Pedrycz, W.: Fuzzy modelling: Fundamentals, construction and evaluation Original Research Article. Fuzzy Sets and Systems 41(1), 1–15 (1991)

[8] Laanaya, H., Martin, A., Aboutajdine, D., Khenchaf, A.: Support vector regression of membership functions and belief functions – Application for pattern recognition. Information Fusion 11(4), 338–350 (2010)

[9] Vaidya, O.S., Kumar, S.: Analytic hierarchy process: An overview of applications. European Journal of Operational Research 169(1), 1–29 (2006)

The Application of BPR in Seismic Data Processing and Interpretation Management

Jianyuan Fu, Jianku Sun, Huan Huang*, Jinbiao Zhang, Hongde Yang, Kexin Deng, and Zixin Luo

Abstract. Seismic data processing and interpretation is the core business of geophysical research, relating to project management, processing, interpretation, and computer departments. The project quality control system of seismic data processing and interpretation needs to establish a unified information management platform, with the rapid development of geophysical study. The idea of BPR(Business Process Reengineering) guides re-carding of seismic data processing and interpretation business chain, designing of deployable two-way information flow control model of management. And process quality control management achieves real-time, visual, traceable, fine and continuous optimization scientific management mode, then improves the quality and efficiency of geophysical project management. It lays a good foundation for better and faster to complete more oil & gas projects.

Keywords: BPR, business process, seismic processing and interpretation, information management, automation, optimization.

1 Introduction

The world market is in oil and natural gas exploration boom. New technology, new ways, new equipment on geophysical exploration have been emerging, The world

---

Jianyuan Fu · Jianku Sun · Jinbiao Zhang · Hongde Yang · Kexin Deng
Geophysical Prospecting Company, CNPC Chuanqing Drilling Engineering Co., Ltd., Chengdu 610213, China

Huan Huang
State Key Laboratory of Geohazard Prevention and Geoenvironment Protection (Chengdu University of Technology), Chengdu 610059, China
e-mail: huan@cdut.edu.cn

Zixin Luo
Sichuan Academy of Social Sciences, Cheng Du, 610071

* Corresponding author.

Z. Qian et al. (Eds.): Recent Advances in CSIE 2011, LNEE 125, pp. 835–841.
springerlink.com

petroleum exploration intense market competition. Main geophysical exploration companies, through their core proprietary technology have high-end market, enhance research on new technology of geophysical exploration and improve their comprehensive technology service ability. Since 1972 since the application of the reflected wave geophysical exploration technology in oil exploration industry , the development of geophysical exploration has experienced the points of light ,The Times digital era and the digital age simulation, the future is about to enter optical fiber age .[1]The motility of Oil prospecting technology's development, along with the end of "cheap oil age ",the comprehensive development of geophysical exploration technology in the field of prospecting is vital important in the field of petroleum exploration, which has become the main method of lower production cost and make great profit for oil companies.

During the rapid development of oil and gas exploration, technology innovation and management innovation are both badly needed, to improve comprehensive competitiveness and give full play to the advantages of fast and efficient information intelligence. The business Process Reengineering BPR(Business Process Reengineering) to introduce the idea to restore seismic data processing and interpretation services chain, to achieve automation and to achieve automation and quality control process management to optimize the scientific management of change, to improve the efficiency of project management ,to lay good foundation for more and bigger projects being completed better and faster.

2 The Theory of BPR

First the theory of BPR is presented in a paper called Reengineering Work: Don't Automate but Obliterate [2]115-116, by Professor Michael Hammer from Massachusetts institute of Technology (MIT), , in 1993, Professor Michael Hammer and James Champy define the BPR in a book called "Reengineering The Corporation" like this: "BPR refers to ,to care and satisfy the needs of customs (the recipients and users including users both within and outside the enterprise),in the process of the core business, rethinking the existing business processes fundamental and redesign them radically is necessary. The use of advanced manufacturing technology and information technology and modern management tools to maximize the functional integration of technical and management functions of the integration and breaking the traditional Function Oriented Organization structure and establishing Process Oriented Organization structure [3]95 to make cost, quality, service and efficiency got significant improvements to allow companies to adapt to the customers, competition and change, which is characterized by a modern business environment."Since then, BPR was considered as a new management ideas, just like wave swept across the United States and other industrialized countries.BPR has even been called "restore the competitiveness of the united States the only way".[2]116

In general, the enterprises to implement business process reengineering is guided by the following principles.

1) to achieve the change from functional management to process management.
2) to focus on the best idea of the overall process.
3) to optimize the management level
4)to give full play to each person in the role of the entire business process
5) integrate businсss processes
6)use IT technology to coordinate the relationship between dispersion and concentration.[2]116

The rapid development and the popularization of information technology require enterprises to adapt to the new model that is necessary for new information society. The BPR is put forward under the information age. Information technology is the driving force of BPR and the tool to help enterprises to reform. We should give full play to the role of information technology-driven, to help companies plan business process better, but also take full account of the changes the BPR concept will bring to companies to form a new attitude to make use of information technology so that we can make information and management ideas combine to create perfect business operation [4]45

3 The Application of BPR in the Earthquake Project Management

In traditional geophysical project, the lack of unified information platform in seismic data processing make managers can't make timely and effective progress of the project operation and management for dynamic tracking and real-time monitoring. The use of traditional communication ways such as: telephone and people themselves to exchange information has affected the visibility and timeliness in the project management. And there are some non-logical processes and repetitive content to fill in original system, which created numerous unnecessary repeatable work and the risk of copied data for the staff. It not only can't guarantee the consistency, traceability and timeliness of paper-record data, resulting in the bottlenecks of quality control.

In traditional management model, project management process is not so fine and workflow can't be completely in accordance with industry standards, making it difficult to achieve standardization and refinement of project management. During the business, we found non-logical business issues exist in traditional model for example: during the process of dealing with results archive and interpretation archive phase the actual archiving process logic is not very clear, which make it difficult to achieve information technology platform to support the geophysical aspects of the process of quality control has not been fully implemented.

3.1 The Basic Requirement of Project Management System

In practice, BPR is not necessary to break all existing completely to rebuilt but stressed the need to break though thinking. we should not take what is only right and proper for granted, because anything is changing. Only in this way can we hope to get rid of businesses that exist in certain long-standing disadvantages and establish a vibrant competitive.[2]118

In order to make information management of geophysical research project management processes possible, we can establish seismic data processing and interpretation of the integrated management system by the idea of BPR (Business Process Management, BPM) system. On the base of database, we can get the relevant personnel in relevant departments and work together to form collaborative work environment, to make project quality be under control and leadership can monitor and track the process all the way. [5]The basic requirement are the followings:

1) The seismic processing and interpretation of process management and quality control follow the principle of "process of sub-sub-levels of management time". Archived files should do electronic signatures at all levels ,which has the function of verifying and it can be supervised by all parties.

2) To all projects, especially for key projects, extended real-time project management and monitoring.

3) Collaborative communication platform and integrated business process management system can communicate with each other and remind us of messages at any time.

3.2 The Practice of BPR

As the work and departments related to geophysical research projects are numerous, to fulfill multi-level quality control and monitor project progress effectively , we should conduct a comprehensive study according to working process and examine business process over and over again. According to seismic data processing and interpretation process management and quality control processes, it's necessary to redesign business process management to make the platform based on seismic data processing and interpretation work.

3.3 The Effect of BPR

Seismic data processing and interpretation project progress and quality control operation can be automatic management. Management system provides intelligent electronic forms to take the place of paper forms. Database provides storage space for a large number of old data and the network provides fast and efficient method to convey information. The inherent computing power of computer quicken the process of management.

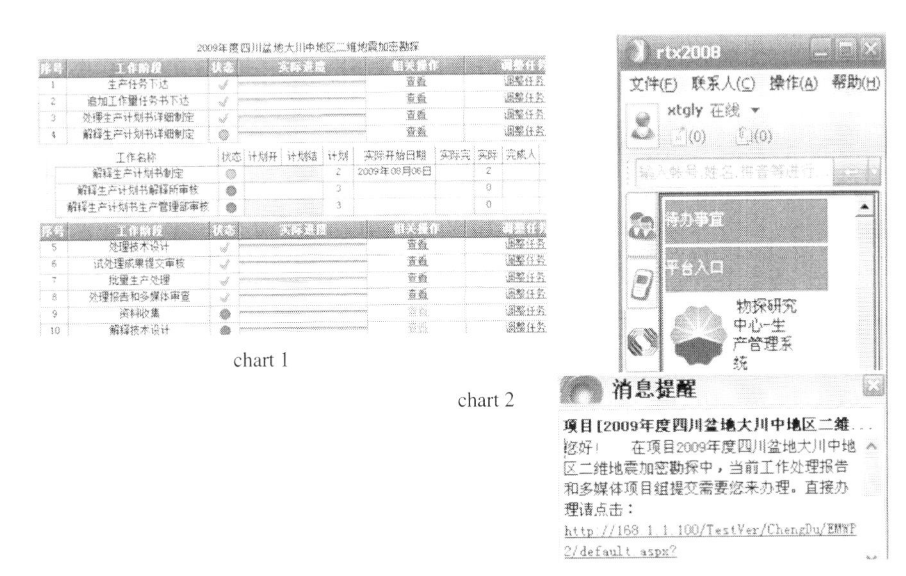

chart 1

chart 2

1) Providing visual, traceable, seismic data processing and interpretation projects to monitor in real time and three tracking and monitoring ways , namely, list gantt chart and flowchart mode. Shown in Chart 1 , in the chart ,for example, in the chart as to during the run of monitoring, there are three state for each three node, did not start (red dot), underway(green dot), closed(green check mark). Each stage has a corresponding progress bar to show, green represents normal progress and pink represents exceed the time limit, which is easy and intuitive. Seen from the chart 1,we can know itemprocessing part has been completed and explaination plan has began. Each node has six indicators to show state of the node, namely: start date, end date, days, actual start date, actual end date, actual number of days.

Get much dynamic command of seismic exploration project accurately, integrally and timely and do real-time statistical analysis to provide technical support for management decisions.

2) The record of quality reviewing and seismic data processing and interpretation of process management is finished automatically without human intervention, which can record suggestions or comments, processing time and automatic electronic signatures in the process of handling the audit.

The automation and standardization of seismic exploration project management provide technical support and system services for business process management and quality control .

3) The combination between processing and interpretation of seismic data management system and collaboration platform RTX(Real Time Exchange), as in chart 2, it can exchange information with customers through to-do things enter interface and real-time news alert, which improved the timeliness significantly.

Highly integrated information management system improve efficiency so that employees can devote into more valuable scientific research and analytical work.

4 The Conditions of BPR Implementation

According to the application of BPR in practice ,we can confer BPR implementation requires the following four conditions.

4.1 BPR Project Implementation Needs the Support of the Leadership

In the implementation of BPR , the staff need a process to change work mode. As the work of various departments goal is to achieve a trade or make a department can be more, faster, better to fulfill its mandate, as a result, one may focus on improving local interests instead of overall interests. Now more and more detailed division of labor and steps also more tasks have made business operation and management process complicated, resulting in affecting the efficiency of management.[3]94 Clear support from leadership of optimizing business processes through BPR project and approving of execution can facilitate the completion of the management changes.

4.2 Departments Need to Implement the BPR Project in a Team

In the process of BPR project implementation , departments' staff will take some time to update the concept. so it's necessary to make people in different department and different positions work together to adapt to the new process management model, in order to realize the core business process integration and optimization.

4.3 Implementation of the BPR Project Requires Interaction with Information Technology

Information technology provides a powerful tool for the application of BPR. BPR also provide strong guidance and support for the information solutions in enterprise. BPR can't success without information technology and it's hard for companies to achieve the desired goals without the guidance and coordination of BPR. Therefore, in the process of in impelled information, leadership should consider what change information technology will bring in and how BPR guide information planning and even implementation of the plan. Enterprises should take full account of the role of information technology and make it a powerful driving force to help companies to advance the process of BPR. [4]50

4.4 Implementation of the BPR Project Needs Continuing Innovation

With internal and external changes, enterprises need constantly change and improvement so that they can promote the quality and efficiency of core business. [6]

5 Conclusion and Understanding

1) The implement of BPR needs effective integration and optimization to relatively business processes.

2) To make workflow easy and automatic and achieve interoperability within geophysical project management system and to improve business operating efficiency and the overall benefits, it's vital to find out the key processes that affect efficiency.

3) In the highly competitive information age, turning geophysical exploration project management into information and creating more effective, which is the final result of geophysical technology, information technology and management techniques to merge together.

References

1. Wang, Y., Feng, L.Y., Niu, Y.: Trend analysis of world petroleum geophysical prospecting market and technical development. Oil Geophysical Prospecting 43(6), 736–741 (2008)
2. Wang, D.Y., Wan, Y.: The practice and inspiration of BPR project on A company. Industrial Technology & Economy 26(6), 115–118 (2007)
3. Hua, H.M.: The real meaning of Reengineering and BPR. China Soft Science 11, 94–95 (1996)
4. Mei, S.Z., Feng, J.Z.: BPR and Information Technology. Systems Engineering-Theory & Practice 2, 45–50 (2003)
5. Duncan, W.R.: A guide to the project management body of knowledge. Project Management Institute, USA (1996)
6. Wang, Y.R.: Five-Step method of Research & Development innovation. Harvard Business Review, http://www.cemmok.com/case/html/yunyingguanli/chanpinguanli/200909/24-4109.html

Author Index